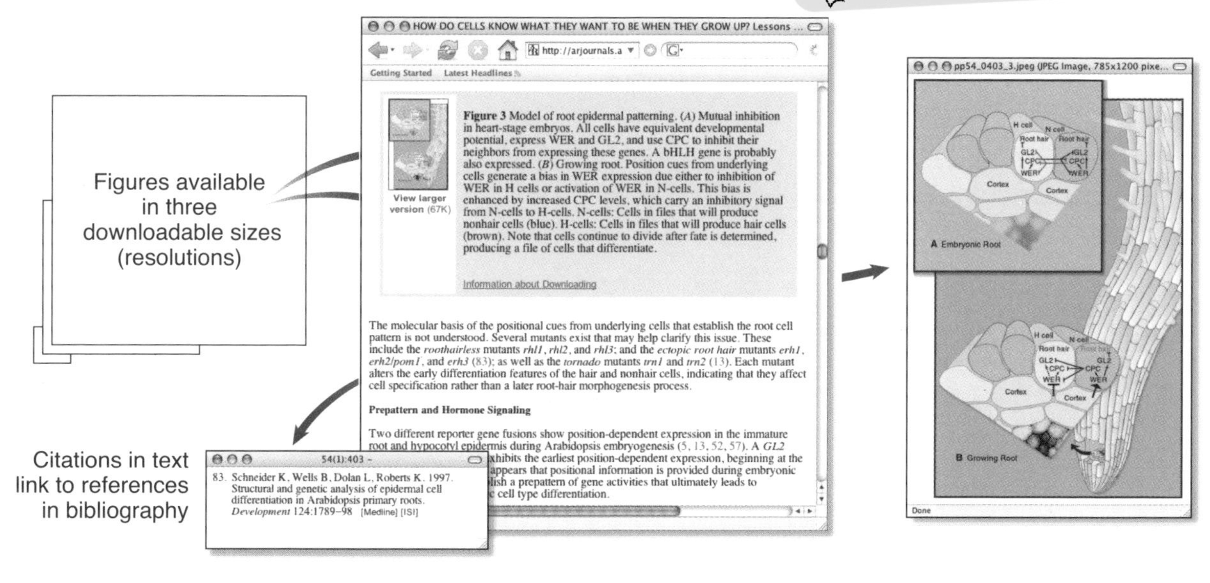
Figures available in three downloadable sizes (resolutions)
Citations in text link to references in bibliography
HOW DO CELLS KNOW WHAT THEY WANT TO BE WHEN THEY GROW UP? Lessons ...
Figure 3 Model of root epidermal patterning.
View larger version (67K)
Information about Downloading
Prepattern and Hormone Signaling
83. Schneider K, Wells B, Dolan L, Roberts K. 1997. Structural and genetic analysis of epidermal cell differentiation in Arabidopsis primary roots. Development 124:1789–98 [Medline] [ISI]
A Embryonic Root
B Growing Root

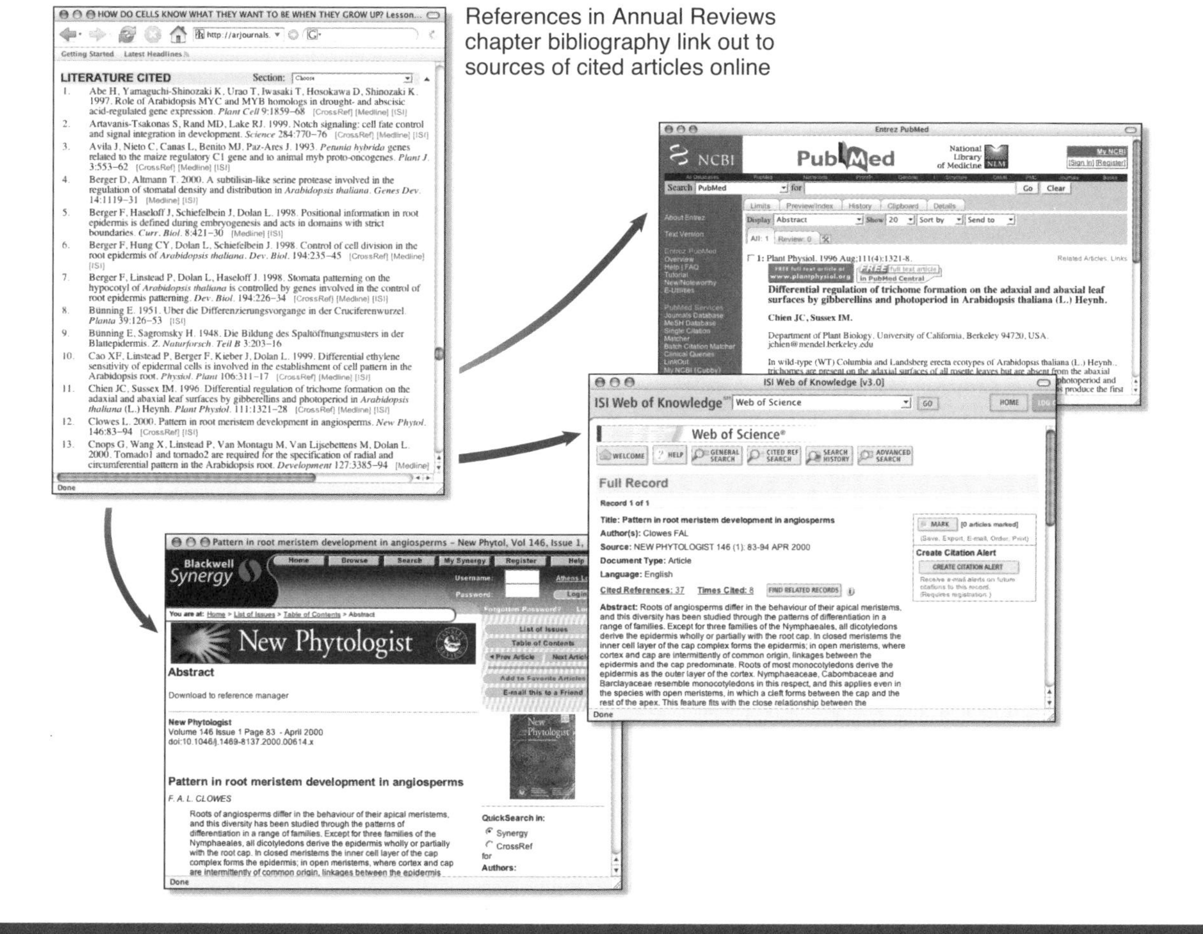
References in Annual Reviews chapter bibliography link out to sources of cited articles online
LITERATURE CITED
Entrez PubMed
Differential regulation of trichome formation on the adaxial and abaxial leaf surfaces by gibberellins and photoperiod in Arabidopsis thaliana (L.) Heynh.
ISI Web of Knowledge
Web of Science
Full Record
Title: Pattern in root meristem development in angiosperms
Blackwell Synergy
New Phytologist
Abstract
Pattern in root meristem development in angiosperms
F. A. L. CLOWES

AR Annual Review of Plant Biology

Production Editor: Ellen Terry
Bibliographic Quality Control: Mary A. Glass
Electronic Content Coordinator: Suzanne K. Moses
Subject Indexer: Kyra Kitts

Annual Review of Plant Biology

Volume 56, 2005

Sabeeha Merchant, *Editor*
University of California, Los Angeles

Winslow R. Briggs, *Associate Editor*
Carnegie Institution of Washington, Stanford, California

Vicki L. Chandler, *Associate Editor*
University of Arizona

www.annualreviews.org • science@annualreviews.org • 650-493-4400

Annual Reviews
4139 El Camino Way • P.O. Box 10139 • Palo Alto, California 94303-0139

Annual Reviews
Palo Alto, California, USA

International Standard Serial Number: 1543-5008
International Standard Book Number: 0-8243-0656-2
Library of Congress Catalog Card Number: 50-13143

♾ The paper used in this publication meets the minimum requirements of American National Standards for Information Sciences—Permanence of Paper for Printed Library Materials, ANSI Z39.48-1992.

TYPESET BY TECHBOOKS, FAIRFAX, VA
PRINTED AND BOUND BY QUEBECOR WORLD KINGSPORT, KINGSPORT, TN

Annual Review of Plant Biology

Volume 56, 2005

Contents

INDEXES

ERRATA

An online log of corrections to *Annual Review of Plant Biology* chapters may be found at http://plant.annualreviews.org/

Related Articles

From the ***Annual Review of Biochemistry***, Volume 74 (2005)

The Mammalian Unfolded Protein Response
Martin Schröder and Randal J. Kaufman

The Structural Biology of Type II Fatty Acid Biosynthesis
Stephen W. White, Jie Zheng, Yong-Mei Zhang, and Charles O. Rock

From the ***Annual Review of Biophysics and Biomolecular Structure,***
Volume 33 (2004)

Enzyme-Mediated DNA Looping
Stephen E. Halford, Abigail J. Welsh, and Mark D. Szczelkun

Conformational Spread: The Propagation of Allosteric States in Large Multiprotein Complexes
Dennis Bray and Thomas Duke

A Function-Based Framework for Understanding Biological Systems
Jeffrey D. Thomas, Taesik Lee, and Nam P. Suh

Structure, Molecular Mechanisms, and Evolutionary Relationships in DNA Topoisomerases
Kevin D. Corbett and James M. Berger

Three-Dimensional Electron Microscopy at Molecular Resolution
Sriram Subramaniam and Jacqueline L.S. Milne

Taking X-Ray Diffraction to the Limit: Macromolecular Structures from Femtosecond X-Ray Pulses and Diffraction Microscopy of Cells with Synchrotron Radiation
Jianwei Miao, Henry N. Chapman, Janos Kirz, David Sayre, and Keith O. Hodgson

Molecules of the Bacterial Cytoskeleton
Jan Löwe, Fusinita van den Ent, and Linda A. Amos

The Use of In Vitro Peptide-Library Screens in the Analysis of Phosphoserine/Threonine-Binding Domain Structure and Function
Michael B. Yaffe and Stephen J. Smerdon

Rotation of F_1-ATPase: How an ATP-Driven Molecular Machine May Work
Kazuhiko Kinosita, Jr., Kengo Adachi, and Hiroyasu Itoh

Model Systems, Lipid Rafts, and Cell Membranes
Kai Simons and Winchil L.C. Vaz

Mass Spectral Analysis in Proteomics
John R. Yates, III

Information Content and Complexity in the High-Order Organization of DNA
Abraham Minsky

The Role of Water in Protein-DNA Recognition
B. Jayaram and Tarun Jain

Force as a Useful Variable in Reactions: Unfolding RNA
Ignacio Tinoco, Jr.

The Thermodynamics of DNA Structural Motifs
John SantaLucia, Jr. and Donald Hicks

From the ***Annual Review of Cell and Developmental Biology,*** Volume 20 (2004)

Bi-Directional Protein Transport Between the ER and Golgi
Marcus C.S. Lee, Elizabeth A. Miller, Jonathan Goldberg, Lelio Orci, and Randy Schekman

The Role of *KNOX* Genes in Plant Development
Sarah Hake, Harley M.S. Smith, Hans Holtan, Enrico Magnani, Giovanni Mele, and Julio Ramirez

New Developments in Mitochondrial Assembly
Carla M. Koehler

Membrane Trafficking in Plants
Gerd Jürgens

The Genetics and Molecular Biology of the Synaptonemal Complex
Scott L. Page and R. Scott Hawley

Mechanisms of Polarized Growth and Organelle Segregation in Yeast
David Pruyne, Aster Legesse-Miller, Lina Gao, Yuqing Dong, and Anthony Bretscher

Cellular Length Control Systems
Wallace F. Marshall

Membrane Domains
Sushmita Mukherjee and Frederick R. Maxfield

G Protein Control of Microtubule Assembly
Yixian Zheng

From the ***Annual Review of Genetics,*** Volume 38 (2004)

Mobile Group II Introns
Alan M. Lambowitz and Steven Zimmerly

The Genetics of Maize Evolution
John Doebley

Light Signal Transduction in Higher Plants
Meng Chen, Joanne Chory, and Christian Fankhauser

Chlamydomonas reinhardtii in the Landscape of Pigments
Arthur R. Grossman, Martin Lohr, and Chung Soon Im

Closing Mitosis: The Functions of the Cdc14 Phosphatase and Its Regulation
Frank Stegmeier and Angelika Amon

Recombination Proteins in Yeast
Berit Olsen Krogh and Lorraine S. Symington

The Function of Nuclear Architecture: A Genetic Approach
Angela Taddei, Florence Hediger, Frank R. Neumann, and Susan M. Gasser

Genetic Models in Pathogenesis
Elizabeth Pradel and Jonathan J. Ewbank

Epigenetic Regulation of Cellular Memory by the Polycomb and Trithorax Group Proteins
Leonie Ringrose and Renato Paro

Mitochondria of Protists
Michael W. Gray, B. Franz Lang, and Gertraud Burger

Mechanisms of Pattern Formation in Plant Embryogenesis
Viola Willemsen and Ben Scheres

Duplication and Divergence: The Evolution of New Genes and Old Ideas
John S. Taylor and Jeroen Raes

Proteolysis as a Regulatory Mechanism
Michael Ehrmann and Tim Clausen

Principles of MAP Kinase Signaling Specificity in *Saccharomyces cerevisiae*
Monica A. Schwartz and Hiten D. Madhani

Comparative Genomic Structure of Prokaryotes
Stephen D. Bentley and Julian Parkhill

Species Specificity in Pollen-Pistil Interactions
Robert Swanson, Anna F. Edlund, and Daphne Preuss

From the ***Annual Review of Phytopathology,*** Volume 43 (2005)

Being at the Right Place at the Right Time for the Right Reasons: Plant Pathology
Robert K. Webster

Biology of Plant Rhabdoviruses
Andrew O. Jackson, Ralf Dietzgen, Michael M. Goodin, Min Deng, and Jennifer N. Bragg

Signal Crosstalk and Induced Resistance: Straddling the Line Between Cost and Benefit
Richard M. Bostock

Exploiting Chinks in the Plant's Armor: Evolution and Emergence of Geminiviruses
Maria R. Rojas, Charles Hagen, W.J. Lucas, and Robert L. Gilbertson

Lipids, Lipases, and Lipid-Modifying Enzymes in Plant Disease Resistance
Jyoti Shah

Plant Disease: A Threat to Global Food Security
Richard Strange and Peter R. Scott

Plant Pathology and RNAi: A Brief History
John A. Lindbo and William G. Doughtery

RNA Silencing in Productive Virus Infections
Robin MacDiarmid

Resistance of Cotton Toward *Xanthomonas campestris* pv. *malvacearum*
E. Delannoy, B.R. Lyon, P. Marmey, A. Jalloul, J.F. Daniel, J.L. Montillet, M. Essenberg, and M. Nicole

Molecular Interactions Between Tomato and the Leaf Mold Pathogen *Cladosporium fulvum*
Susana Rivas and Colwyn Thomas

Viroids and Viroid-Host Interactions
Ricardo Flores, Carmen Hernández, Angel-Emilio Martínez, José-Antonio Daròs, and Francesco Di Serio

Genetics of Plant Virus Resistance
Byoung-Cheorl Kang, Inhwa Yeam, and Molly M. Jahn

Erratum

Annual Review of Plant Biology, Volume 54

CRYPTOCHROME STRUCTURE AND SIGNAL TRANSDUCTION
Chentao Lin and Dror Shalitin

Please note that the following is the legend for Table 1, p. 477:

TABLE 1 Amino acid homology of cryptochrome molecules from the plant kingdom and E. coli PHR.

The designated molecules were aligned and divided to the N terminus, PHR, amino acids: At CRY1, 1-494; At CRY2, 1-489; Tomato 1, 1-487; Tomato 2, 1-488; Rice 1a, 1-504; Rice 1b, 1-502; Fern 4, 1-488; Moss 1a, 1-490, and the C-terminal extension (except for E. coli, which does not have the extension). The homology of PHR and C-terminal extension of every two molecules were calculated for the identical/similar amino acids. The PHR homology is shown in the lower left side of the table and the C-terminal extension homology is shown in the upper right side.

Annual Reviews is a nonprofit scientific publisher established to promote the advancement of the sciences. Beginning in 1932 with the *Annual Review of Biochemistry*, the Company has pursued as its principal function the publication of high-quality, reasonably priced *Annual Review* volumes. The volumes are organized by Editors and Editorial Committees who invite qualified authors to contribute critical articles reviewing significant developments within each major discipline. The Editor-in-Chief invites those interested in serving as future Editorial Committee members to communicate directly with him. Annual Reviews is administered by a Board of Directors, whose members serve without compensation.

Annual Reviews of

Anthropology
Astronomy and Astrophysics
Biochemistry
Biomedical Engineering
Biophysics and Biomolecular Structure
Cell and Developmental Biology
Clinical Psychology
Earth and Planetary Sciences
Ecology, Evolution, and Systematics
Entomology
Environment and Resources
Fluid Mechanics
Genetics
Genomics and Human Genetics
Immunology
Law and Social Science
Materials Research
Medicine
Microbiology
Neuroscience
Nuclear and Particle Science
Nutrition
Pharmacology and Toxicology
Physical Chemistry
Physiology
Phytopathology
Plant Biology
Political Science
Psychology
Public Health
Sociology

SPECIAL PUBLICATIONS
Excitement and Fascination of Science, Vols. 1, 2, 3, and 4

P. Starlinger

Fifty Good Years

Peter Starlinger

Institut für Genetik der Universität zu Köln, 50923 Köln, Germany;
email: peterstarlinger4@compuserve.de

Annu. Rev. Plant Biol.
2005. 56:1–13

doi: 10.1146/
annurev.arplant.56.032604.144236

First published online as a
Review in Advance on
December 6, 2004

1543-5008/05/
0602-0001$20.00

Contents

EARLY YEARS IN POSTWAR GERMANY

Compared to our parents' generation, my generation has been a lucky one. Our parents lived through really hard times: two World Wars, the big inflation of 1922/23 that wiped out any savings—if there were any, the Great Depression, the Nazi rule with its crimes (and with its temptations), and for many, long years spent as prisoners of war[1].

True, we also lived during the war, but we were children, and afterwards things gradually became better, and soon much better. After the turmoil of the last months of the war, I found myself in northern Germany. I finished high school, enrolled in the university as a medical student, and, by a stroke of luck, entered the department of biochemistry, where I spent most of my time during the next two years under the guidance of the late Hans Netter, a very amiable and knowledgeable man with a strong bent toward physical chemistry. He was an enthusiastic teacher, very interested in the application of physical chemistry in biology, about which he wrote books well known in Germany. The little I have learned of thermodynamics and kinetics I owe to him. However, after two years, I felt that I should move on to something different. I started looking around for other institutes in Germany to finish my medical studies and to earn an M.D. and decided that Butenandt's institute in Tübingen was my first choice.

Adolf Butenandt was head not only of the Max Planck Institute for Biochemistry, which had many research groups but was also professor of physiological chemistry in the medical faculty. This was a different world. Nucleic acids, which in Kiel had been but a footnote, were here already the topic of the day. Butenandt, best known for his work on steroid hormones, which earned him a Nobel Prize[2], and later as president of the Max Planck Society, had also initiated research on other topics. Collaborating with Alfred Kühn, the director at a neighboring Max Planck Institute, Butenandt became interested in gene action and found that kynurenin was an intermediate in the formation of the red eye color of *Drosophila* at about the same time that Ephrussi and Beadle were engaged in such work in France (1). Butenandt also initiated a virus group after he heard Stanley talking on the crystallization of tobacco mosaic virus, a finding that at that time excited many discussions on the nature of life.

When I entered the department in 1952 and became a student in this virus group, nucleic acid was a household word. Interesting and vivid lectures were given on this topic, e.g., by Hans Friedrich-Freksa (who became my thesis supervisor), by Gerhard Schramm (who together with Alfred Gierer and Heinz Schuster later showed that the RNA of tobacco mosaic virus was infectious by itself and that deaminating its cytosine caused mutations), and by Wolfhard Weidel (who had been a coauthor of the 1940 paper on biochemical genetics in *Drosophila* with Butenandt and Becker). Avery's experiment was well known, as was Chargaff's discovery that the base composition of DNA is different in different organisms, but identical in different tissues of the same organism, which was considered in keeping with a genetic role for DNA. The A = T and G = C rule was also known, but was not interpreted any more in Tübingen than by the author himself. That

[1]These were the lives of the average German family like the one from which I came. The fate of those who were persecuted by the Nazi regime for race or political reasons and of whom a vast majority perished in the Holocaust was incomparably harder.

[2]He was not allowed to accept this prize because Hitler had decreed that no German could accept a Nobel Prize after a well-known journalist, Carl von Ossietzky, tortured to death in a concentration camp, had been awarded the Nobel Peace Prize. Butenandt obeyed and stayed in Germany as head of the then Kaiser Wilhelm Institute for Biochemistry. Recently, it has been asked whether a scientist in such a high position could have avoided knowledge of the unethical and even criminal research going on in Nazi Germany and even in other institutes of the Kaiser Wilhelm Society. Such investigations may be necessary to understand those dark times. However, not having been around at that time, I do not know how to judge. I do not know what I would have done, in those days, hearing about such research. What I do know is that it is a great privilege to grow up in a society that offers many—and conflicting—views and actions to choose from. This was another privilege of my generation in Germany.

had to await Watson and Crick, a year later, but before that, Hershey and Chase's experiment on DNA being the genetic constituent of bacteriophage was much hailed. The double helix was even more admired, not as a strange surprise, but rather as a wonderful confirmation of the DNA's role in self-replication.

The first reading I was assigned was not an old textbook, but the 1951 volume of the Cold Spring Harbor symposia, in which I discovered a very interesting but, at that time, certainly not well understood (by me) article by Barbara McClintock (9).

THE COLOGNE INSTITUTE FOR GENETICS

I spent four years in Tübingen, after which I moved to a medical research laboratory in an effort to begin a medical career (which turned out not to suit me). By a new stroke of luck I met Max Delbrück, who invited me to give a seminar in Cologne. Here, a small chair of microbiology was just being established at the botany department, and its new head, Carsten Bresch, offered me a position. This offer could not be declined. Soon I learned that much more was to happen. Josef Straub, at that time director of the Botany Institute, had the ambition to improve the Cologne biology. He had tried to persuade Delbrück to take the chair of this new microbiology group, but that was no real offer for a full professor at Caltech. However, Delbrück agreed to teach one of his famous phage courses in Cologne[3], where he and Straub learned to like each other. Delbrück not only had suggested Bresch for the chair, but, together with Straub, conceived the plan for a big institute of genetics that not only should bring the still new science of molecular biology to Cologne, but which also should be organized more like a department at an American research university than as a traditional German institute.

The times were favorable for such an endeavor. Germany had recovered from the destruction of the war and was now able to invest in less immediate goals. Still, it needed the insight of people to initiate new developments at the university. The combination of the scientific reputation of Delbrück and the skill and amiability of Straub as a negotiator carried the day with the ministry in Düsseldorf (responsible for the University of Cologne). A main point of the deal was that Delbrück was to be the director of the newly founded institute. The small print said that Delbrück's directorship would last only for two years, the maximum time for an extended sabbatical from Caltech, but all people concerned chose to overlook this.

Five small research groups were established and, besides Delbrück and Bresch, the radiobiologist Walter Harm (working on DNA repair) and the biochemist Hans Zachau (soon to be famous for the sequencing of tRNAs) were hired, and I also was given the task of establishing such a group. Later Ulf Henning (who had done distinguished work in Yanofsky's lab at Stanford) joined the institute.

These were exciting years, and they passed only too quickly. After two years, in 1963, Delbrück left, as announced, but unfortunately all the other group leaders also got very tempting offers either in Germany or in the United States, and they decided to leave also. For this reason, the people in my small research group found ourselves in the hapless position of being somewhat stranded, waiting for the departure of our colleagues. Most probably, I thought I would also have to leave because I did not have tenure, and in Germany you cannot get tenure unless you receive a comparable offer from another university. This might have led to the closure of the institute that had inspired so many hopes. I was lucky, however. I got an offer from the Max Planck Society and, consequently, was offered a position at Cologne.

This alone hardly mitigated the aforementioned scenario, but the faculty eventually

[3]Delbrück was German and came from a well-known family of scholars in Germany. His brother Justus was involved in the resistance against Hitler, was imprisoned, and died shortly after the war. Max Delbrück, not liking and not being liked by the Nazi regime and not seeing much of a career there, emigrated to America in 1937 and became an U.S. citizen. He kept contacts in Germany, however, and after the war visited the country several times to encourage young scientists.

succeeded, over the course of the following years, in attracting very capable new professors to the department. Among these were the virologist Walter Doerfler, one of the early investigators of DNA methylation, Benno Müller-Hill[4], who had just, together with Walter Gilbert, isolated the lac repressor, and Klaus Rajewsky, who, within a few years, established the immunology group so well known.

WORK ON TRANSPOSABLE ELEMENTS

In all these years I had not begun to earn the credentials to write a prefatory chapter for the *Annual Review of Plant Biology*. Apart from my thesis work with tobacco mosaic virus, I had always worked with bacteria and bacteriophages, the genetics of which were so interesting in those days, and which were so suitable as laboratory organisms.

In 1959, I met Werner Arber and Jean Weigle during a postdoc year at Caltech. Arber and Weigle had managed to move the *galactose* (*gal*)-operon of *Eschericia coli* to a bacteriophage and I thought that infecting a galactose-negative strain with such a phage might allow the study of gene activity in novel ways. Then I, later joined by a small group in Cologne, worked with the *gal*-operon.

In 1963, Heinz Saedler joined the laboratory and began to experiment with mutants that abolish the activity of the whole operon rather than of single ones of its genes, analogous to the o°-mutations described by Jacob and Monod in the *lac*-operon. They mapped to the proximal end of the operon but also abolished the activity of the distal genes, which was called a polar effect. These mutants proved unusual. They reverted readily to the wild type and, hence, they were not deletions. However, they did not react to mutagens and therefore could not be the usual base substitutions. What else could they be? Higher organisms show chromosome rearrangements, but these were not well known in bacteria at the time. Was it conceivable that the mutants were chromosome aberrations, like insertions, inversions, or duplications (12)?

This could not be investigated very directly in those days, but Heinz Saedler, soon joined by Elke Jordan, who came as a postdoc from Herman Kalckar's lab, found a way to distinguish inversions from insertions or duplications. Inversions would not alter the length of the phage DNA carrying the *gal* operon, whereas the latter two would. By measuring this length with an indirect method, inversions could be excluded: There was extra DNA (7)!

The same result was obtained at the same time by an American scientist, James A. Shapiro (14). Such coincidences are none too rare in science, showing that the time is ripe for a certain finding. Independent confirmation enhances the confidence in a new result.

Would it also be possible to distinguish duplications from insertions? Georg Michaelis, another graduate student, did this by exhaustingly hybridizing RNA transcribed randomly from the mutant-carrying phage to the wild-type DNA and showing that some RNA remained that only bound to the DNA carrying the mutation in the *gal* operon. Hence, the extra DNA was not a duplication of preexisting sequence in the wild-type DNA. There were even indications that not all of these insertions were different DNAs. If that were true, there might be a bias for certain sequences to be translocated to the *gal*-operon (10).

This was eventually shown by a third graduate student, Heinz-Josef Hirsch, who applied heteroduplex mapping to this question after learning the technique during a visit to Waclav Szybalski, the expert of this technique. In our sample of mutants as well in a sample of bacteriophage λ mutants provided by Ph. Brachet, there were only two distinct sequences that hybridized among themselves but not between the groups. We suggested the designation IS elements (insertion sequence elements) for these entities (6).

[4]He also became well known for his historical work on the euthanasia crimes in Nazi Germany, on which a book appeared in an English translation (11).

It was only then that we realized the relation of these elements to McClintock's transposable elements, in spite of the fact that I had known McClintock's work since my student days, and in spite of a series of seminars that we had held on this topic in the institute in Cologne. Sometimes we are blind (17)!

Nonetheless, our research on the IS elements went on both in our lab in Cologne and in that of Heinz Saedler's, who did some breakthrough experiments in Norman Davidson's lab at Caltech showing that IS elements are normal constituents of bacterial chromosomes and building blocks of plasmids. Upon his return to Germany he became a professor at the University of Freiburg.

While this was keeping us busy and happy, the temptation grew to also get a hand at McClintock's elements. Sure, there were many problems in the mid 1970s, but the new gene technology with its cloning methods had had its first successes, and we reasoned that starting now our work would be slow enough for improvements in these techniques to come along. This proved true, and it was also true that in these years I had the chance to meet Barbara McClintock[5], who helped us with advice and with mutants, as did Peter Peterson from Iowa, who was (and is) also an expert on maize transposable elements.

The speculation that the techniques of gene technology would improve rapidly was correct. Although they did not lend themselves directly to the isolation of a transposable element with completely unknown properties, the techniques to isolate genes with known protein products improved. We chose to isolate the Sh-gene of maize, which encodes sucrose synthase, by looking for cDNA clones binding an mRNA from endosperm that could then be translated in vitro into a protein reacting with an antiserum to sucrose synthase (5). We then looked for novel DNA within the isolated gene in a mutant caused by the insertion of transposable element Ds. This worked eventually (2) and opened the way to study both the gene (18) and the transposable element (3).

What was much more difficult to master for an M.D. and bacterial geneticist like myself was the work with live plants growing in the field or in the greenhouse. Again I was lucky. I met Francesco Salamini in Bergamo, Italy, who not only let us grow our maize plants there (important if your labs are in Cologne, where the climate does not guarantee that maize will grow to maturity), but who taught me and my students (particularly Hans-Peter Döring) to do our crosses right.

Not only were we helped by Salamini in Bergamo, but big changes were ahead in Cologne. The two directors of the big Max Planck Institute for Breeding Research were retiring. Should somebody succeed them or should the institute close? This question is always formally discussed when the director of a Max Planck Institute retires. If the institute were to continue, who should the new director(s) be?

I served a term on the senate of the Max Planck Society at that time and was assigned to the Green Committee, which had to make recommendations on the above questions. I vividly remember the discussions that covered many topics and persons, until, largely influenced by Georg Melchers, then the doyen of the Max Planck botany research, a daring decision was finally made. The first of the new directors would be Jeff Schell, soon followed by Heinz Saedler, by Klaus Hahlbrock, and also by Francesco Salamini. For me, the years to come saw a close association with the Max Planck Institute, an association that proved highly rewarding, both personally and scientifically.

As mentioned, we eventually succeeded in isolating maize transposable elements, as did Heinz Saedler (15, 16) and Nina Fedoroff (4)

[5] Barbara was an antithesis to the modern research professor who runs a big lab. She did everything by herself, not only because she lacked the means to employ many people, but also as a matter of personal choice. She was admirable. After she had been awarded the long-deserved Nobel Prize in 1983, it was often written that she might serve as a role model for scientists. In a certain sense this is true. Still, I think that biology is best served by a multitude of different characters, and that working with graduate students and thus running a bigger lab also has its merits.

in the United States, and working with these elements kept my lab and those of several associates, particularly Reinhard Kunze [who later started an independent group (8)], busy until my retirement[6].

TEACHING GENETICS AND ADDRESSING THE PUBLIC

A university professor not only has to do research but, of course, also has to teach. We gave lectures to our students and also organized courses for people coming from other places, pursuing the tradition of Delbrück's phage course in Cold Spring Harbor. Here we met many outstanding people, later to become leaders in the field of molecular biology in Germany, and thus these courses were, at least for us, very rewarding. We also organized a yearly "Spring Meeting" on the recent progress in the field, which became quite popular, especially among students and postdocs, because we did not charge fees. However, it is often said that we also have an obligation to serve the public and that we have to convey this attitude to our students. How could we do this?

One possibility was to talk about the consequences of our own field on the society. I tried to do this by preparing lectures on plant breeding and on the influence of this on the nutrition of the growing world population. I also happened to be a professor not only of genetics but also of radiobiology, and in this area I gave many lectures on the field in the narrow sense as well as, together with my colleague Hubert Kneser, on the problems, both positive and negative, of using atomic energy.

These activities were not restricted to the university. I made it a rule for myself to accept all invitations to talk about these topics to committees concerned with these questions, to public groups, or to the media—an exercise that was time-consuming, and increasingly so in the wake of the Asilomar conference in 1975, after which the debate on gene technology began in earnest. I not only waited for invitations to speak, but, together with a group of colleagues, tried to access high school teachers and to organize seminars, courses, and eventually labs for them.

All this belongs in the realm of genetics, as seen in a wider sense. Is there a limit to the topics on which a professor is entitled to speak in public? This question was brought before me, when I did not confine my lectures on radiation damage to DNA, but also included considerations of the economic and ecologic consequences of nuclear reactors.

These questions took on an evermore urgent appeal when I joined colleagues to speak publicly about the dangers of atomic weapons and the arms race, much like the Union of Concerned Scientists in the United States or the Scientists Against Nuclear Arms in the United Kingdom. Our aim was to raise awareness among otherwise busy scientists about the problems of an ever-accelerating arms race and the ensuing destabilization among scientists.

For a while, we were quite successful. We organized a national scientists' congress on these matters in 1983, followed by an international congress in 1986, which saw the massive participation of scientists from Europe, the United States, and the Soviet Union, and at each occasion the audience was in excess of several thousand. Throughout these years, the problems of the arms race were, for a time, really a matter of discussion among scientists, including many who would otherwise refrain from involving themselves in politics. Understandably, these activities declined after the end of the Cold War.

Was all this of any avail? In terms of immediate results, the answer is certainly no. I still think, however, that such activities are more than justified. Take, for example, Germany.

[6]Space does not allow me to describe this work in detail, and a short enumeration would be boring. It has been reviewed (13). However, I want to say that work with many graduate students and postdoctoral fellows both from Germany and from abroad was very rewarding for me both personally and scientifically and I want to express my gratitude to them for their very good work and for many hours of discussion, of which I learned much and of which, I hope, they also learned something.

Nowadays, the general population is inclined toward the peaceful resolution of conflicts and is certainly against war (so much so that this is sometimes criticized in other countries). The situation was very different, however, before the First World War, when the youth of several European countries, and certainly of Germany, longed for a war that would give them the opportunity to put their heroism on display. We do not know how these changes in attitude come about, but certainly not without many people voicing them in public.

Of course we were questioned, and we also questioned ourselves, as to our legitimacy to speak on such topics really distant from our professional expertise. Those politically opposed to us denied us this legitimacy outright.

It is true that, in our role as scientists, we are very strongly asked to restrict our opinions to our area of professional expertise. For scientific progress, a highly collective enterprise, each of us is supposed to add her or his contribution, however small, with utmost care. Who, however, is then entitled to synthesize these bits and pieces? In science, this is done by respected scientists, often in the form of reviews, sometimes textbooks, and by reaching a consensus with the community, which then either accepts or rejects this synthesis. Trained in this tradition, we leave the political decisions to others. Without much asking we believe that a highly technical subject like the destabilization of the strategic situation by novel arms is treated by authors in the media in the same masterly way that physics is treated by Richard Feynman in his *Lectures on Physics* or molecular biology by Jim Watson in his *Molecular Biology of the Gene*.

Ultimately, the community that accepts or rejects their conclusions is the citizenry. This is a group, to which we, the scientists, also belong. And at the very least it is incumbent on those of us who possess an established position like a tenured professorship to cut out from our busy schedule the time to think about these matters. This is a privilege not shared by all other citizens. I think we should use this privilege and join those who scrutinize what is offered to us. If we do not do it, this is left to the media. The journalists certainly do not ask themselves, and also are not asked by the public, whether they have the competence to speak out on the most difficult political questions including the arms race. Without believing that our credits are any better than theirs, we should insist that, in some instances, they are not worse.

SCIENCE IN MODERN SOCIETY

Everybody will agree that science (and recently also biology) is shaping our material world to a great extent. Is it also influencing the way in which our society looks at this world? Here I am not so sure.

When we read about the sciences in the public media or follow the pronouncements of politicians (and here I speak mainly about Germany), the discussion is usually about the question of whether the science departments in the universities have close enough ties to the economy, whether they churn out a sufficient number of patents, and whether they train their students in a way most suitable for their future employment in industry. There is no question that all of these are important goals. Are they sufficient, though? Isn't it also the task of the natural sciences, as of all other branches of the universities, to participate in the intellectual debates of our times, in shaping the way in which we look at the world at large?

The natural sciences, as opposed to the humanities, are not the most vociferous participants in these debates. Other fields, from sociology to psychology (not to mention the arts or religion), are much more prone to explain to all of us how to look at the world. Often the spokesmen of the humanities are even telling us what our science is all about.

Surely we have learned a lot from the cultural sciences. We are all aware now that the sciences do not (only) stem from the pursuit of our curiosity about the outer and inner world. We know that other human characteristics, like ambition and greed, also come into play. They also tell us that the way we speak about what we do is most influential in shaping our understanding of the world and that, consequently, no deep understanding of the world is possible unless philosophers and linguists have found the right metaphors that guide us in our explorations.

Important as this is, we must not forget that not everything is the result of a way of speaking, and that our understanding of the world is not (only) dictated by discourse and majority opinion, but also by the properties of the world outside. There is a world outside of us, and although we cannot see and understand it, but through our senses and through our brains, we must never forget that the world outside is shaping our interaction with and our understanding of it, and that we are not free to speak and think as we like. To remind our colleagues in the humanities of this is the task of all natural scientists.

A particularly important question to mention in this context is ethics. Very often we hear and read that scientists are the "doers," often rather blind and single-minded, and that the professionals of philosophy, especially of ethics, have to restrain the most exuberant actions of the former. I do not believe that this claim (which is certainly not that of the best minds of that science, but a simplified view often heard in the media and in the political debate) is justified. If ethics comprise the set of rules required for a good and just life, and if the lives of all of us are, among other parameters, also shaped by our environment, which constantly changes under the influence of the discoveries of scientists, the latter must not be excluded from considering these questions.

This is a particular point to discuss. Although not exclusively concerned with the erection of barriers to human activities, ethics has a strong inclination toward doing so. The Ten Commandments are a famous example: The "You Shall Not's" carry much more weight than the few active admonitions of the Third and the Fourth Commandments.

Should this also be so in the modern world? Science has certainly provided us with many improvements in our material lives. Without them we would be worse off, and some of them are indispensable to the future existence of mankind. I need only mention the impact of science on agriculture, from the introduction of artificial fertilizer in Justus von Liebig's times in the nineteenth century to the "green revolution" of the twentieth. It is possible that future generations will take a similar view of scientific achievements of the twenty-first century. If this is so, it becomes the overwhelming duty of ethics not only to inhibit wrongdoing, but also to show where the omission of doing something becomes reproachable.

If this is so, we must not only deliberate how to restrict harmful actions, but also the costs incurred, should we fail to do what is becoming possible for us. I am not saying that scientists are the only, or even the best, people to decide this. However, I equally strongly believe that they must be involved in this debate. To do so, we must first work toward a climate in the general debate where the division is not between the "doers" and the "ethicists," but between the proponents of different courses of action.

Why is this not yet so? In my opinion, too many scientists refrain from such debates. Nobody forbids that they participate. However, the very nature of science, where it is not sufficient to think well, but where each thought must be checked against nature by means of painstaking experiments, occupies so much of our daily activities and often of our nightly thoughts that not enough time is left for anything else. Still, it would be good if many of us set aside sufficient time for these considerations because shaping the opinion of an educated public is greatly important, as the debate about genetically modified organisms in agriculture has so amply demonstrated.

However, there may be a deeper reason for the often-heard belief that scientists should do their work and otherwise remain silent. The distrust of the "doers" may be deeply ingrained in the human mind. After all, it was the yielding to the temptation to eat from the tree of knowledge that led to the expulsion of Adam and Eve from paradise. In ancient Greek mythology, the blacksmith of the Gods, Hephaistos, lived deep under the earth with the Cyclopes, as did his Germanic counterpart, Alberich, who took shelter with his dwarfs beneath the Rhine river. It seems that people, as witnessed in their sagas, harbor a deep distrust of change wrought by skilled hands and investigative minds.

This, however, should not discourage us. Despite these prejudices, science has advanced over the centuries and has bestowed on us many gifts, from modern agriculture, so vital in the fight against starvation, to antibiotics, curbing to a large extent the onslaught of infectious diseases. This shows that even ingrained prejudices can be and have been overcome successfully, and I think it is our task as scientists not to shrink back from this and to recognize the responsibility to participate in the great intellectual debates of our time.

LOOKING BACK...

Looking back at these 50 years of molecular biology, one cannot but be impressed. We now know much about the ways in which DNA is transcribed, RNA is processed, mRNA is translated, and all of these processes are regulated. We also know much about the necessary enzymes, etc., used in these processes and the proteins that are all encoded by their proper genes.

However, genetics is older and initially was not a science about the biochemistry of macromolecular biosyntheses. In its beginnings, it was the description of traits (later named phenotypes) due to the presence of factors (later named genes) that in either of two allelic states caused different phenotypes. The first years after the rediscovery of Mendel's work confirmed this concept in an impressive way.

Soon, however, there were difficulties. Did one gene make one phenotype or several? Was a phenotype caused by one gene or by many? Were the phenotypes caused solely by the genes, or did the environment play a role?

The attempt to answer these questions led to many auxiliary concepts without leading to a satisfactory unified picture. There was polygeny and pleioptropy, there were suppression and epistasis, there was penetrance and there was quantitative genetics as opposed to single gene inheritance. Did we then understand how the gene "makes" its phenotype? Do we know this today?

Let us consider some of the phenotypes for which genes are reported nowadays not only in the lay press, but also in the scientific literature. There is talk about the gene "for" male homosexuality, "for" female breast cancer, and "for" the obesity that haunts both sexes. It is clear by now that the gene is not a direct cause of these phenotypes. It is linked to them via several steps by what philosophers call a "chain of causes." However, the farther we move down this chain of causes, the more we are forced to acknowledge that other causes come into play, too. It is like the pedigree of Abraham in Genesis. There are nine generations between Shem and Abraham. Could we say that Shem was the progenitor of Abraham? He certainly was one, but there must have been 511 others, of whom we hear little. Still, the character of Abraham must have been influenced by them as well as by the known progenitor Shem.

The language in which we describe the relation between a gene and a distant phenotype is deceptive. It prompts us to assume that the first is the cause of the latter rather than a contributing factor[7]. All the discussions about a difference between single gene inheritance and quantitative genetics, all the disappointments if one group discovers a gene "for" schizophrenia linked to one chromosome and another group has similar findings for another chromosome, and all the quarrels about "Nature versus Nurture" become meaningless if we acknowledge

[7]The absurdity of this becomes apparent if we try to use such language in everyday life. Imagine a man who is in trouble with his beloved girl, and he knows that he has to write to her and apologize. He cannot do so, however, because his pencil is broken, and to his dismay he discovers that his pencil sharpener does not work. The letter is not written and the girl takes her phone and rings another person to console herself. If we do not know this story completely, if we only see the broken pencil sharpener and the reaction of the girl, and if we are trained in the tradition of genetic nomenclature, we might call the broken gadget the suppressor of true love. Ridiculous as this sounds, that is exactly what geneticists do every day!

the role of the gene as one actor among many for the distant phenotype, and if the only direct phenotype is the RNA, the sequence of which is encoded in the gene.

It might be better not to talk about cause and effect, but rather about interactions (like the physicists do), mostly between two players. The gene interacts with the transcription apparatus, and the result is an RNA that may or may not be altered either by a mutation in the DNA sequence or by an alteration of the transcriptase, causing faulty transcription.

This will sound to many as sheer word playing, philosophy at best (and scientists are often not fond of philosophy and use the word in a rather derogatory sense). However, if we take this suggestion seriously, many of the concepts elaborated by geneticists in the first half of the last century cease to be puzzling. No longer is it necessary to discuss at great length the proposition that a gene causes a certain result not with certainty but only with probability. The gene does its work and acts as an information-carrying template in transcription. But when we look at a trait, it may be removed from the gene that we investigate by several steps (splicing, translation, protein action, cellular events), and moreover, many genes may participate in all of these steps. Therefore, it need not disturb us when the final outcome, depending on so many other interactions, is different in different individuals, because the probability of differences between members of a group even of limited size increases strongly with the number of genes playing a role. The description of a particular gene having a certain effect with only a limited probability is then understandable.

There is no longer a reason to distinguish between a normal gene and a quantitative trait locus (which accounts for only part of the variance of this trait within a population). Both of them produce their RNA, and it is only the role of this RNA and of its subsequent products that differ among different genes. If, in a complicated chain of reactions, a particular RNA and the protein made from it have only a small modifying effect on the eventual outcome, the gene is said to have a minor quantitative effect. At the level of the gene, however, a deletion of this gene is not different from any other.

The acknowledgment of such a distinction between a direct interaction of two players from the eventual, unpredictable outcome of this should not only influence our thinking about genetics, but can reach even further. Nowadays, it has become popular to hold people, and in particular scientists responsible for distant consequences. Otto Hahn is said to be responsible for the atom bomb, and once I heard, in a parliamentary committee, the seriously posed question of whether the federal government could guarantee that genetically manipulated crops, even if benign by present standards, would not cause harmful evolutionary alterations a few hundred years hence.

In my opinion, we should refrain from such ideas because they reach beyond our capabilities. The decision to once and for all refrain from scientific activities that could be harmful for mankind in an unforeseeable and unknown future must lead to complete inactivity, and this would be counterproductive. On the other hand, we should watch the outcome of our work carefully. Should a dangerous development become discernible, we, as the scientists, should be the first to spot it and to alert the public about it (as has often been the case in the second half of the last century, e.g., in the Asilomar conference in 1975, where scientists themselves first discussed the possibility of dangers of the newly emerging gene technology!).

...AND LOOKING AHEAD

This is certainly more difficult than evaluating the events of the past. Should one still try a little step in this direction? I will do so and hopefully not become either too lengthy or too speculative.

We hear much about the overwhelming role of the genes in biology these days. Genes are said to carry the blueprint of the organism, to possess all of the information to explain life. Sequencing the genome and finding out about all of the genes in it should, as we sometimes read, reveal the way in which organisms develop and function. Is this a reasonable expectation? I have my doubts.

The study of the genomic sequences has already produced many surprises, among them the finding that the number of genes is smaller than expected, and is, in human beings, not much larger than in much simpler organisms, like the nematode *Caenorhabditis elegans* or the crucifer *Arabidopsis thaliana*. Similarly interesting is the observation that a large number of the genes, including many that encode proteins of most important and basic functions, are very old, dating back to a time when eucaryotes had not yet split from either eu- or archaebacteria. Still, it cannot be denied that recent mammals, including man, are very different from bacteria. Do we really expect genes to make us mammalian or even human?

Another puzzle: While the concept of pleiotropy, the involvement of a gene in seemingly unrelated functions, is old, the involvement of the same gene products in a plethora of functions has by now become overwhelming. The advent of the microarray techniques has shown that the addition or deletion of a single gene involved in a developmental process can have an influence on the expression of hundreds, if not thousands, of other genes.

If a single process, e.g., eye formation, needs a very large fraction of all genes, and if the number of genes is smaller than previously thought, what do we have to think about for the relation of the genes to the phenotypes?

I suggest that already at the level of cells, and more so at the level of whole organs or of organisms, new forms of complex organization with new, emergent properties will be found. In these complex organizations determining certain levels of life the genes will be building blocks for regulatory modules.

Is such a suggestion amounting to the claim that life is not based on the functions of the genes? Certainly not! I illustrate my view by discussing the relation of nucleotides to genes. Is it conceivable to have a gene without the constituent nucleotides? No! Is it conceivable to alter an important biological trait by exchanging a single one of these nucleotides in a certain sequence? Obviously yes! Is it then reasonable to say that a certain trait, say sickle cell anemia, is caused by the T now present instead of an A at a particular position in the mRNA for the ß-chain of globin? This would hardly be claimed by a geneticist. Are we surprised that, given the absolute importance of the nucleotides for the structure of the genes, there are only four of them? Not at all. With these four nucleotides it is possible to build an unlimited number of sequences, of which many contain the information for a biologically important RNA, often a mRNA. The information is the new, emerging property, making use of the relations of many nucleotides to each other, that distinguishes a gene from an ensemble of nucleotides.

By this way of thinking, the nucleotides do not lose their importance, but this importance lies, in the example mentioned above, in their being part of a larger ensemble of them, namely the gene. The role of the nucleotide in this instance could not be guessed from any of the properties of the molecule looked at in isolation. Its role as determining a codon within a gene is a new, emerging property.

I think it is possible that one day, perhaps in the not too distant future, scientists will describe networks of proteins, many of them regulatory, others with enzymatic, transport, or structural functions, all being expressed at particular places and at specific times. They will unravel the function of cells, and later of whole organs. The genes necessary for the formation of these proteins will be numerous, and they will show up again and again in the different networks to be found and analyzed. Their role in a particular module, with its emerging properties, will not be predictable from the most thorough analysis of its sequence.

If these ideas have any merit, there must be an important difference between the genes as building blocks of higher-order networks and the nucleotides as building blocks of genes. There are only four nucleotides and none of them have alleles compatible with life and, thus, do not allow Darwinian evolution. It will also be impossible to find a gene lacking one of these. Even if there were such a gene in a particular organism, we would not find a mutant removing one nucleotide from the makeup of genes, because the many other genes could not be made and life would immediately collapse.

With genes, the situation is different. There are many more of them, and many of them will be involved in many biological functions, and therefore will be indispensable. Even the relatively simple production of an enzyme depends not only on the gene encoding the information for its sequence, but also on the genes for the whole transcription-translation machinery, and others not enumerated here. Most of them will belong to the indispensable class, the mutations of which will be early lethals and will, for this reason, never show up in a genetic analysis of this particular enzyme. The gene, however, which carries the sequence information for the enzyme, will undergo discernible mutations, as will some genes involved in the regulation of this particular synthesis.

Looking this way, the genes discovered by genetic analysis of (nonlethal!) mutants will not be the most important ones for a particular function, but rather the ones most easily discovered (and by that property possibly of only limited general importance for the cell, though of great value to breeders!).

The constitution of the networks will be unraveled by, e.g., microarray techniques, which are able to show the proteins and the genes involved without mutating or deleting them.

In genetics, the new concept of information is the key to understanding. The chemical and base-pairing properties of nucleotides play only a minor role when genetic information is discussed. In a similar manner, the discussion of the properties of genes may be delegated to the background when higher-order networks are investigated. These will probably be discussed in a novel language yet to be developed. The properties of the genes, will, when necessary, be looked up in data bases, which by then will hopefully be very comprehensive.

Although the study of genetics proper will certainly not end, the main concepts to be gathered from this science may already be at hand, and new vistas will open the view in a new world of biology, which may be similarly adventurous for a new generation of biologists, as the molecular biology of the gene was in the second half of the twentieth century for my generation. I can only express my hope that they will have the same opportunities and the same exhilaration using them that my generation was privileged enough to enjoy.

LITERATURE CITED

1. Butenandt A, Weidel W, Becker E. 1940. Kynurenin als Augenpigmentbildung auslösendes Agens bei Insekten. *Naturwissenschaften* 28:63–64
2. Döring HP, Starlinger P. 1984. Barbara McClintocks controlling-elements: now at the DNA level. *Cell* 39:263–59
3. Döring HP, Geiser M, Starlinger P.1981. Transposable element Ds at the shrunken locus in Zea mays. *Mol. Gen. Genet.* 184:377–80
4. Fedoroff NV, Wessler S, Shure S. 1983. Isolation of the transposable maize controlling elements *Ac* and *Ds*. *Cell* 35:235–42
5. Geiser M, Döring HP, Wöstemeyer J, Behrens U, Tillman E, Starlinger P. 1980. A cDNA clone from Tea mays endosperm sucrose synthase mRNA. *Nucleic Acids Res.* 8:6175–88
6. Hirsch HJ, Starlinger P, Brachet P. 1972. Two kinds of insertions in bacterial genes. *Mol. Gen. Genet.* 119:191–206
7. Jordan E, Saedler H, Starlinger P. 1968. O^0 and strong polar mutations in the *gal* Operon are insertions. *Mol. Gen. Genet.* 102:353–63
8. Kunze R, Saedler H, Loennig W-E. 1997. Plant transposable elements. *Adv. Bot. Res.* 27:331–470

9. McClintock B. 1951. Chromosome organization and genic expression. Cold Spring Harbor Symp. *Quanti. Biol.* 16:13–47
10. Michaelis G, Saedler H, Venkov P, Starlinger P. 1969. Two insertions in the Galactose Operon having different sizes but homologous DNA sequence. *Mol. Gen. Genet.* 104:371–77
11. Müller-Hill B. 1998. *Murderous Science: Elimination by Scientific Selection of Jews, Gypsies and Others in Germany, 1933–1945*. Cold Spring Harbor, NY: Cold Spring Harbor Lab.
12. Saedler H, Starlinger P. 1967. Mutations in the Galactose Operon in *E.coli*: I. Genetic characterization. *Mol. Gen. Genet.* 100:178–89
13. Saedler H, Starlinger P. 1992. Twenty-five years of transposable element research in Köln. In *The Dynamic Genome*, ed. N Fedoroff, D Botstein. pp. 243–63. Cold Spring Harbor, NY: Cold Spring Harbor Lab.
14. Shapiro JA. 1969. Mutations caused by the insertion of genetic material into the Galactose Operon of *Escherichia coli*. *J. Mol. Biol.* 40:93–105
15. Schwarz-Sommer Zs, Gierl A, Klösgen RB, Wienand U, Peterson PA, Saedler H. 1984. The *Spm* (*En*) transposable element controls the excision of a 2-kb DNA insert at the *wx* m-8 allele of *Zea mays*. *EMBO J.* 2:1021–28
16. Shepherd NS, Schwarz-Sommer Zs, Wienand U, Sommer H, Deumling B, et al. 1982. Cloning of a genomic fragment carrying the insertion element *Cin1* of Zea mays. *Mol. Gen. Genet.* 188:266–71
17. Starlinger P, Saedler H. 1972. Insertion mutations in microorganisms. *Biochimie* 54:177–85
18. Werr W, Frommer WB, Maas M, Starlinger P. 1985. Structure of the sucrose synthase gene on chromosome 9 of Zea mays L. *EMBO J.* 4:1373–80

Phytoremediation

Elizabeth Pilon-Smits

Biology Department, Colorado State University, Fort Collins, Colorado 80523;
email: epsmits@lamar.colostate.edu

Annu. Rev. Plant Biol.
2005. 56:15–39

doi: 10.1146/
annurev.arplant.56.032604.144214

First published online as a
Review in Advance on
January 11, 2005

1543-5008/05/0602-
0015$20.00

Key Words

pollution, decontamination, metals, organics, bioremediation

Abstract

Phytoremediation, the use of plants and their associated microbes for environmental cleanup, has gained acceptance in the past 10 years as a cost-effective, noninvasive alternative or complementary technology for engineering-based remediation methods. Plants can be used for pollutant stabilization, extraction, degradation, or volatilization. These different phytoremediation technologies are reviewed here, including their applicability for various organic and inorganic pollutants, and most suitable plant species. To further enhance the efficiency of phytoremediation, there is a need for better knowledge of the processes that affect pollutant availability, rhizosphere processes, pollutant uptake, translocation, chelation, degradation, and volatilization. For each of these processes I review what is known so far for inorganic and organic pollutants, the remaining gaps in our knowledge, and the practical implications for designing phytoremediation strategies. Transgenic approaches to enhance these processes are also reviewed and discussed.

Contents

INTRODUCTION

Phytoremediation: Advantages, Limitations, Present Status

Phytoremediation: the use of plants and their associated microbes for environmental cleanup

TCE: trichloroethylene

TNT: trinitrotoluene

PAH: polycyclic aromatic hydrocarbon

MTBE: methyl tertiary butyl ether

Phytoremediation is the use of plants and their associated microbes for environmental cleanup (99, 107, 108). This technology makes use of the naturally occurring processes by which plants and their microbial rhizosphere flora degrade and sequester organic and inorganic pollutants. Phytoremediation is an efficient cleanup technology for a variety of organic and inorganic pollutants. Organic pollutants in the environment are mostly man made and xenobiotic to organisms. Many of them are toxic, some carcinogenic. Organic pollutants are released into the environment via spills (fuel, solvents), military activities (explosives, chemical weapons), agriculture (pesticides, herbicides), industry (chemical, petrochemical), wood treatment, etc. Depending on their properties, organics may be degraded in the root zone of plants or taken up, followed by degradation, sequestration, or volatilization. Organic pollutants that have been successfully phytoremediated include organic solvents such as TCE (the most common pollutant of groundwater) (90, 111), herbicides such as atrazine (22), explosives such as TNT (61), petroleum hydrocarbons such as oil, gasoline, benzene, toluene, and PAHs (4, 93, 110), the fuel additive MTBE (26, 59, 128), and polychlorinated biphenyls (PCBs) (53).

Inorganic pollutants occur as natural elements in the earth's crust or atmosphere, and human activities such as mining, industry, traffic, agriculture, and military activities promote their release into the environment, leading to toxicity (91). Inorganics cannot be degraded, but they can be phytoremediated via stabilization or sequestration in harvestable plant tissues. Inorganic pollutants that can be phytoremediated include plant macronutrients such as nitrate and phosphate (60), plant trace elements such as Cr, Cu, Fe, Mn, Mo, and Zn (76), nonessential elements such as Cd, Co, F, Hg, Se, Pb, V, and W (15, 60), and radioactive isotopes such as ^{238}U, ^{137}Cs, and ^{90}Sr (34, 35, 87).

Phytoremediation can be used for solid, liquid, and gaseous substrates. Polluted soils and sediments have been phytoremediated at military sites (TNT, metals, organics), agricultural fields (herbicides, pesticides, metals, selenium), industrial sites (organics, metals, arsenic), mine tailings (metals), and wood treatment sites (PAHs) (8, 41, 93, 101, 129). Polluted waters that can be phytoremediated include sewage and municipal wastewater (nutrients, metals), agricultural runoff/drainage water (fertilizer nutrients, metals, arsenic, selenium, boron, organic pesticides, and herbicides), industrial wastewater (metals, selenium), coal pile runoff (metals), landfill leachate, mine drainage (metals), and groundwater plumes (organics, metals) (38, 42, 52, 60, 74, 101). Plants can also be used to filter air, both outdoors and indoors, from, e.g., NO_x, SO_2, ozone, CO_2, nerve gases, dust or soot particles, or halogenated volatile hydrocarbons (64, 86).

Phytoremediation has gained popularity with government agencies and industry in the past 10 years. This popularity is based in part on the relatively low cost of phytoremediation, combined with the limited funds available for environmental cleanup. The costs associated with environmental remediation are staggering. Currently, \$6–8 billion per year is spent for environmental cleanup in the United States, and \$25–50 billion per year worldwide (47, 122). Because biological processes are ultimately solar-driven, phytoremediation is on average tenfold cheaper than engineering-based remediation methods such as soil excavation, soil washing or burning, or pump-and-treat systems (47). The fact that phytoremediation is usually carried out in situ contributes to its cost-effectiveness and may reduce exposure of the polluted substrate to humans, wildlife, and the environment. Phytoremediation also enjoys popularity with the general public as a "green clean" alternative to chemical plants and bulldozers. Thus, government agencies like to include phytoremediation in their cleanup strategies to stretch available funds, corporations (e.g., electric power, oil, chemical industry) like to advertise their involvement with this environment-friendly technology, and environmental consultancy companies increasingly include phytoremediation in their package of offered technologies.

The U.S. phytoremediation market now comprises ~\$100–150 million per year, or 0.5% of the total remediation market (D. Glass, personal communication). For comparison, bioremediation (use of bacteria for environmental cleanup) comprises about 2% (47). Commercial phytoremediation involves about 80% organic and 20% inorganic pollutants (D. Glass, personal communication). The U.S. phytoremediation market has grown—two- to threefold in the past 5 years, from \$30–49 million in 1999 (47). In Europe there is no significant commercial use of phytoremediation, but this may develop in the near future because interest and funding for phytoremediation research are increasing rapidly, and many polluted sites in new European Union countries (Eastern Europe) await remediation. Phytoremediation may also become a technology of choice for remediation projects in developing countries because it is cost-efficient and easy to implement.

Phytoremediation has advantages but also limitations. The plants that mediate the cleanup have to be where the pollutant is and have to be able to act on it. Therefore, the soil properties, toxicity level, and climate should allow plant growth. If soils are toxic, they may be made more amenable to plant growth by adding amendments, as described below. Phytoremediation is also limited by root depth because the plants have to be able to reach the pollutant. Root depth is typically 50 cm for herbaceous species or 3 m for trees, although certain phreatophytes that tap into groundwater have been reported to reach depths of 15 m or more, especially in arid climates (88). The limitations of root depth may be circumvented by deep planting of trees in boreholes (up to 12 m) or pumping up polluted groundwater for plant irrigation. Depending on the biological processes involved, phytoremediation may also be slower than the more established remediation methods like excavation, incineration, or pump-and-treat systems. Flow-through phytoremediation systems and plant degradation of pollutants work fairly fast (days or months), but soil cleanup via plant accumulation often takes years, limiting applicability. Phytoremediation may also be limited by the bioavailability of the pollutants. If only a fraction of the pollutant is bioavailable, but the regulatory cleanup standards require that all of the pollutant is removed, phytoremediation is not applicable by itself (43). Pollutant bioavailability may be enhanced to some extent by adding soil amendments, as described below.

Nonbiological remediation technologies and bio/phytoremediation are not mutually exclusive. Because pollutant distribution and concentration are heterogeneous for many sites, the most efficient and cost-effective remediation solution may be a combination of different technologies, such as excavation of the most

Rhizofiltration: use of plants in hydroponic setup for filtering polluted water

Phytoextraction: use of plants to clean up pollutants via accumulation in harvestable tissues

contaminated spots followed by polishing the site with the use of plants. Such an integrated remediation effort requires a multidisciplinary team of knowledgeable scientists.

This review aims to give a broad overview of the state of the science of phytoremediation, with references to other publications that give more in-depth information. After an introduction to the various phytoremediation technologies, the plant processes involved in uptake, translocation, sequestration, and degradation of organic and inorganic pollutants are reviewed in the context of phytoremediation. Finally, new developments including genetic engineering are discussed with respect to their prospects for phytoremediation.

Phytoremediation Technologies and Their Uses

Plants and their rhizosphere organisms can be used for phytoremediation in different ways (see **Figure 1**). They can be used as filters in constructed wetlands (60) or in a hydroponic setup (100); the latter is called rhizofiltration. Trees can be used as a hydraulic barrier to create an upward water flow in the root zone, preventing contamination to leach down, or to prevent a contaminated groundwater plume from spreading horizontally (90). The term phytostabilization denotes the use of plants to stabilize pollutants in soil (13), either simply by preventing erosion, leaching, or runoff, or by converting pollutants to less bioavailable forms

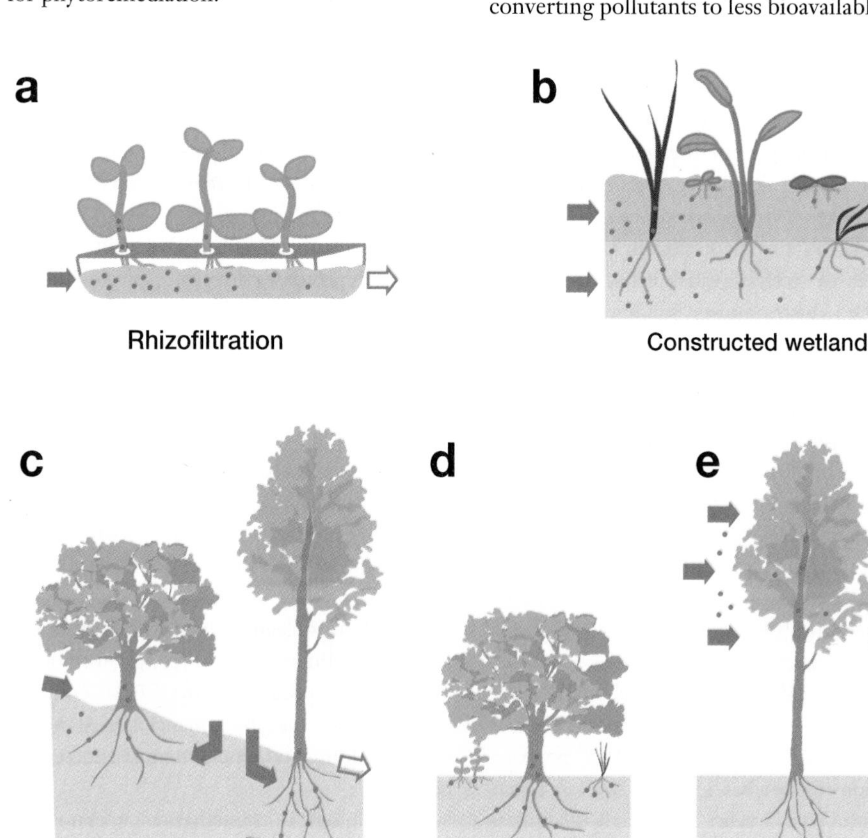

Figure 1
Phytoremediation technologies used for remediating polluted water, soil, or air. The red circles represent the pollutant.

(e.g., via precipitation in the rhizosphere). Plants can also be used to extract pollutants and accumulate them in their tissues, followed by harvesting of the (above ground) plant material. This technology is called phytoextraction (15). The plant material can subsequently be used for nonfood purposes (e.g., wood, cardboard) or ashed, followed by disposal in a landfill or, in the case of valuable metals, recycling of the accumulated element. The latter is termed phytomining (23).

Plants can facilitate biodegradation of organic pollutants by microbes in their rhizosphere (see **Figure 2**). This is called phytostimulation or rhizodegradation (82). Plants can also degrade organic pollutants directly via their own enzymatic activities, a process called phytodegradation (82). After uptake in plant tissue, certain pollutants can leave the plant in volatile form; this is called phytovolatilization (118). These various phytoremediation technologies are not mutually exclusive; for instance, in a constructed wetland, accumulation, stabilization and volatilization can occur simultaneously (52). Because the processes involved in phytoremediation occur naturally, vegetated polluted sites have a tendency to clean themselves up without human interference. This so-called natural attenuation is the simplest form of phytoremediation and involves only monitoring.

The different phytoremediation technologies described above are suitable for different classes of pollutants. Constructed wetlands have been used for a wide range of inorganics including metals, Se, perchlorate, cyanide, nitrate, and phosphate (52, 60, 92), as well as certain organics such as explosives and herbicides (60, 63, 83, 110). Rhizofiltration in an indoor, contained setup is relatively expensive to implement and therefore most useful for relatively small volumes of wastewater containing hazardous inorganics such as radionuclides (35, 87). The principle of phytostabilization is used, e.g., when vegetative caps are planted on sites containing organic or inorganic pollutants, or when trees are used as hydraulic barriers to prevent leaching or runoff of organic or inorganic contaminants. Trees can also be used in so-called buffer strips to intercept horizontal migration of polluted ground water plumes and redirect water flow upward (82). Natural attenuation is suitable for remote areas with little human use and relatively low levels of contamination. Phytoextraction is mainly used for metals and other toxic inorganics (Se, As, radionuclides) (9, 15). Phytostimulation is used for hydrophobic organics that cannot be taken up by plants but that can be degraded by microbes. Examples are PCBs, PAHs, and other petroleum hydrocarbons (62, 93). Phytodegradation works well for organics that are mobile in plants such as herbicides, TNT, MTBE, and TCE (21, 128). Phytovolatilization can be used for VOCs (128) such as TCE and MTBE, and for a few inorganics that can exist in volatile form, i.e., Se and Hg (52, 105).

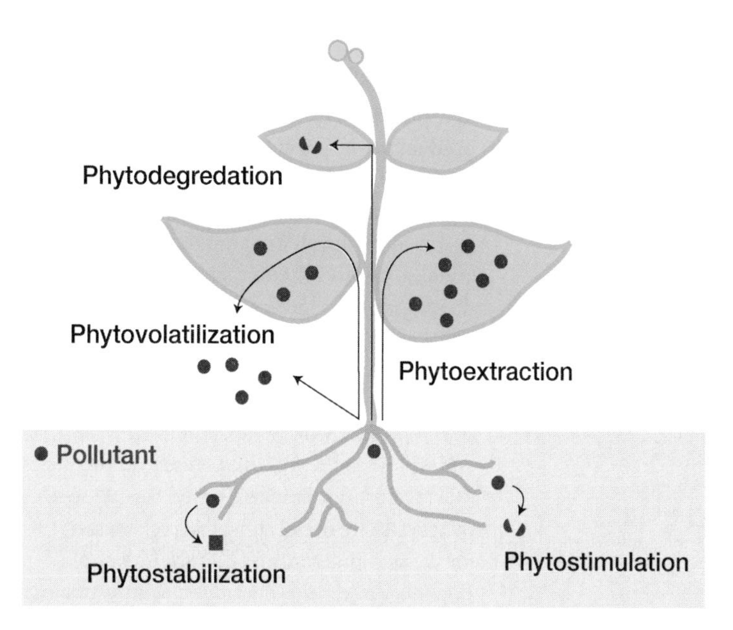

Figure 2

Possible fates of pollutants during phytoremediation: the pollutant (represented by *red circles*) can be stabilized or degraded in the rhizosphere, sequestered or degraded inside the plant tissue, or volatilized.

Rhizodegradation/ phytostimulation: degradation of pollutants in the rhizosphere due to microbial activity

Phytodegradation: breakdown of pollutants by plant enzymes, usually inside tissues

Phytovolatilization: release of pollutants by plants in volatile form

VOC: volatile organic compound

PCB: polychlorinated biphenyl

Different phytotechnologies make use of different plant properties and typically different plant species are used for each. Favorable plant properties for phytoremediation in general are to be fast growing, high biomass, competitive, hardy, and tolerant to pollution. In addition, high levels of plant uptake, translocation, and accumulation in harvestable tissues are important properties for phytoextraction of inorganics. Favorable plant properties for phytodegradation are large, dense root systems and high levels of degrading enzymes. A large root surface area also favors phytostimulation, as it promotes microbial growth; furthermore, production of specific exudate compounds may further promote rhizodegradation via specific plant-microbe interactions (93).

In constructed wetlands for phytoremediation, a variety of emergent, submerged, and floating aquatic species are used. Popular genera/species are cattail (*Typha* sp.), parrot feather (*Myriophyllum* sp.), *Elodea* sp., *Azolla* sp., duckweed (*Lemna* sp.), water hyacinth (*Eichhornia crassipes*), and *Spartina* sp. Poplar (*Populus* sp.) and willow (*Salix* sp.) can be used on the edges of wetlands. For brackish water, certain species of *Spartina* are useful, as well as pickleweed (*Salicornia* sp.) and saltgrass (*Distichlis spicata*) (74). For inorganics, the floating species water hyacinth, Azolla, and duckweed are popular because they are good metal accumulators and can be harvested easily; cattail and poplar are also used because they are tolerant, grow fast, and attain a high biomass. Aquatic plants that work well for organics remediation include parrot feather and Elodea (83) because they have high levels of organic-degrading enzymes. Rhizofiltration involves aeration and therefore is not limited to aquatic species; it often makes use of terrestrial species with large roots and good capacity to accumulate inorganics, such as sunflower (*Helianthus annuus*) or Indian mustard (*Brassica juncea*) (35).

In a vegetative cap for phytostabilization, a combination of trees and grasses may be used. Fast-transpiring trees such as poplar maintain an upward flow to prevent downward leaching, while grasses prevent wind erosion and lateral runoff with their dense root systems. Grasses tend to not accumulate inorganic pollutants in their shoots as much as dicot species (12), minimizing exposure of wildlife to toxic elements. Poplar trees are very efficient at intercepting horizontal groundwater plumes and redirecting water flow upward because they are deep rooted and transpire at very high rates, creating a powerful upward flow (27, 82).

Popular species for phytoextraction are Indian mustard and sunflower because of their fast growth, high biomass, and high tolerance and accumulation of metals and other inorganics (15, 107). A special category of plants are the so-called hyperaccumulators: plant species that accumulate one or more inorganic elements to levels 100-fold higher than other species grown under the same conditions (19). Hyperaccumulators have been reported for As, Co, Cu, Mn, Ni, Pb, Se, and Zn (7, 11, 77). These elements are typically hyperaccumulated up to 0.1–1% of dry weight even from low external concentrations. Despite these properties hyperaccumulators are not very popular for phytoremediation because they are often slow growing and attain low biomass. So far only one hyperaccumulator species, the Ni hyperaccumulator *Alyssum bertolonii*, has been used for phytoremediation in the field (23, 73). The recently discovered As hyperaccumulating fern *Pteris vittata* may also show promise for phytoextraction of As (77).

For phytostimulation of microbial degraders in the root zone, grasses such as fescue (*Festuca* sp.), ryegrass (*Lolium* sp.), *Panicum* sp., and prairie grasses (e.g., *Buchloe dactyloides*, *Bouteloua* sp.) are popular because they have very dense and relatively deep root systems and thus a large root surface area (4). Mulberry trees also enjoy popularity for use in phytostimulation because of their reported ability to produce phenolic compounds that stimulate expression of microbial genes involved in PCB and PAH degradation (44, 72, 93). For phytodegradation of TCE and atrazine, poplar has been the most popular and efficient species so far, owing to its high transpiration rate and capacity to degrade and/or volatilize these pollutants (22, 110).

Poplar is also the most-used species for phytovolatilization of VOCs because of its high transpiration rate, which facilitates the movement of these compounds through the plant into the atmosphere. For volatilization of inorganics, only Se has been investigated in detail. In general, plant species that take up and volatilize sulfur compounds also accumulate and volatilize Se well because S and Se are chemically similar and their metabolism occurs via the same pathways (2). Members of the

Brassica genus are particularly good volatilizers of Se (117). Among the aquatic species tested, rice, rabbitfoot grass, Azolla, and pickleweed were the best Se volatilizers (52, 74, 97, 133).

Finally, when choosing plant species for a certain site, it is advisable to include species that grow locally on or near the site. These species are competitive under the local conditions and, if they are growing on the site, can tolerate the pollutant.

BIOLOGICAL PROCESSES AFFECTING PHYTOREMEDIATION

Phytoremediation effectively removes pollutants, but in many cases the underlying biological mechanisms remain largely unknown. To increase the efficiency of phytoremediation technologies, it is important that we learn more about the biological processes involved. These include plant-microbe interactions and other rhizosphere processes, plant uptake, translocation mechanisms, tolerance mechanisms (compartmentation, degradation), and plant chelators involved in storage and transport. Other processes that need more study are movement of pollutants through ecosystems via the soil-water-plant system to higher trophic levels. In the following sections we follow the path of pollutants toward, into, and within the plant during phytoremediation. For each step I discuss what is known and not known about factors influencing remediation, potential limiting steps for organic and inorganic pollutants, and the practical implications for phytoremediation. Also, I discuss transgenic approaches that have been or may be used to enhance phytoremediation efficiency at each step.

Pollutant Bioavailability

For plants and their associated microbes to remediate pollutants, they must be in contact with them and able to act on them. Therefore, the bioavailability of a pollutant is important for its remediation. Pollutant bioavailability depends on the chemical properties of the pollutant, soil properties, environmental conditions, and biological activity. Soils with small particle size (clay) hold more water than sandy soils, and have more binding sites for ions, especially cations (CEC) (116). The concentration of organic matter (humus) in the soil is also positively correlated with CEC, as well as with the capacity to bind hydrophobic organic pollutants. This is because humus mainly consists of dead plant material, and plant cell walls have negatively charged groups that bind cations, as well as lignin that binds hydrophobic compounds (21).

Two important chemical properties of a pollutant that affect its movement in soils are hydrophobicity and volatility. Hydrophobicity is usually expressed as the octanol:water partition coefficient, or log K_{ow} (121). A high log K_{ow} corresponds with high hydrophobicity. Extremely hydrophobic molecules such as PCBs, PAHs, and other hydrocarbons (log $K_{ow} > 3$) are tightly bound to soil organic matter and do not dissolve in the soil pore water. This lack of bioavailability limits their ability to be phytoremediated, leading to their classification as recalcitrant pollutants. Nonaqueous liquids may sink down to the ground water and, depending on whether they are more or less dense than water, end up below the aquifer (DNAPLs) or on top of the aquifer (LNAPLs). Organics with moderate to high water solubility (log $K_{ow} < 3$) will be able to migrate in the soil pore water to an extent that is inversely correlated with their log K_{ow}.

Pollutant volatility, expressed as Henry's law constant (H_i), is a measure of a compound's tendency to partition to air relative to water (26). Pollutants with $H_i > 10^{-4}$ tend to move in the air spaces between soil particles, whereas pollutants with $H_i < 10^{-6}$ move predominantly in water. If H_i is between 10^{-4} and 10^{-6}, compounds are mobile in both air and water. Both water-mobile and air-mobile organic contaminants can diffuse passively through plants. While the fate of water-mobile organics is phytodegradation or sequestration, volatile organics can be rapidly volatilized by plants without chemical modification (18).

CEC: cation exchange capacity

Log K_{ow}: the octanol: water distribution coefficient, a measure for pollutant hydrophobicity

DNAPL: dense nonaqueous phase liquid

LNAPL: light nonaqueous phase liquid

Inorganics are usually present as charged cations or anions, and thus are hydrophilic. The bioavailability of cations is inversely correlated with soil CEC. At lower soil pH, the bioavailability of cations generally increases due to replacement of cations on soil CEC sites by H^+ ions (116). The bioavailability of ions is also affected by the redox conditions. Most terrestrial soils have oxidizing conditions, and elements that can exist in different oxidation states will be in their most oxidized form [e.g., as selenate, arsenate, Cr(VI), Fe^{3+}]. In aquatic habitats more reducing conditions exist, which favor more reduced elemental forms [e.g., selenite, arsenite, Cr(III), Fe^{2+}]. The oxidation state of an element may affect its bioavailability (e.g., its solubility), its ability to be taken up by plants, as well as its toxicity. Other physical conditions that affect pollutant migration and bioavailability are temperature and moisture. Higher temperatures accelerate physical, chemical, and biological processes in general. Precipitation will stimulate general plant growth, and higher soil moisture will increase migration of water-soluble pollutants. The bioavailability of pollutants may also be altered by biological activities, as described in the next section. In polluted soils the more bioavailable (fraction of) pollutants tend to decrease in concentration over time due to physical, chemical, and biological processes, leaving the less or nonbioavailable (fraction of) pollutants. Consequently, pollutants in aged polluted soils tend to be less bioavailable and more recalcitrant than pollutants in soil that is newly contaminated, making aged soils more difficult to phytoremediate (93).

EDTA: ethylene diamine tetra acetic acid

Understanding the processes affecting pollutant bioavailabilty can help optimize phytoremediation efficiency. Amendments may be added to soil that make metal cations more bioavailable for plant uptake. For instance, adding the natural organic acids citrate or malate will lower the pH and chelate metals such as Cd, Pb, and U from soil particles, usually making them more available for plant uptake. The synthetic metal chelator EDTA is also extremely efficient at releasing metals from soil. This principle is used in chelate-assisted phytoextraction where EDTA is added to soil shortly before plant harvesting, greatly increasing plant metal uptake (108). Before chelate-assisted phytoextraction is used in the field, it is important to do a risk assessment study to determine possible effects of the chelator on metal leaching. In other situations it may be desirable to decrease metal bioavailability if metals are present at phytotoxic levels or in phytostabilization. In such cases lime may be mixed in with the soil to increase the pH or organic matter to bind metals (12, 20). Adding organic matter also decreases the bioavailability of hydrophobic organics, whereas adding surfactants (soap) may increase their bioavailability. For organics that can exist in more or less protonated forms with different charges, manipulation of soil pH can also affect their solubility and ability to move into plants. Finally, water supply may be optimized to facilitate pollutant migration while preventing leaching or runoff.

Rhizosphere Processes and Remediation

Rhizosphere remediation occurs completely without plant uptake of the pollutant in the area around the root. The rhizosphere extends approximately 1 mm around the root and is under the influence of the plant. Plants release a variety of photosynthesis-derived organic compounds in the rhizosphere that can serve as carbon sources for heterotrophic fungi and bacteria (16). As much as 20% of carbon fixed by a plant may be released from its roots (93). As a result, microbial densities are 1–4 orders of magnitude higher in rhizosphere soil than in bulk soil, the so-called general rhizosphere effect (108). In turn, rhizosphere microbes can promote plant health by stimulating root growth (some microorganisms produce plant growth regulators), enhancing water and mineral uptake, and inhibiting growth of other, NO pathogenic soil microbes (65).

In rhizosphere remediation it is often difficult to distinguish to what extent effects are due to the plant or to the rhizosphere microbes. Laboratory studies with sterile plants and microbial isolates can be used to address this question. Rhizosphere remediation may be a passive process. Pollutants can be phytostabilized simply via erosion prevention and hydraulic control as described above. There is also passive adsorption of organic pollutants and inorganic cations to the plant surface. Adsorption of lipophilic organics to lignin groups in the cell walls is called lignification (82). Rhizosphere remediation may also be the result of active processes mediated by plants and/or microbes. These processes may affect pollutant bioavailability, uptake, or degradation.

Pollutant bioavailability may be affected by various plant and/or microbial activities. Some bacteria are known to release biosurfactants (e.g., rhamnolipids) that make hydrophobic pollutants more water soluble (126). Plant exudates or lysates may also contain lipophilic compounds that increase pollutant water solubility or promote biosurfactant-producing microbial populations (113). Furthermore, plant- and microbe-derived enzymes can affect the solubility and thus the bioavailability of organic pollutants via modification of side groups (131).

Bioavailability of metals may be enhanced by metal chelators that are released by plants and bacteria. Chelators such as siderophores, organic acids, and phenolics can release metal cations from soil particles. This usually makes the metals more available for plant uptake (116) although in some cases it can prevent uptake (28). Furthermore, plants extrude H^+ via ATPases, which replace cations at soil CEC sites, making metal cations more bioavailable (116). Some plant roots release oxygen, such as aquatic plants that have aerenchyma (air channels in the stem that allow oxygen to diffuse to the root); this can lead to the oxidation of metals to insoluble forms (e.g., FeO_3) that precipitate on the root surface (60). Conversely, enzymes on the root surface may reduce inorganic pollutants, which may affect their bioavailability and toxicity (e.g., CrVI to CrIII) (76).

Organic pollutants may be degraded in the rhizosphere by root-released plant enzymes or via phytostimulation of microbial degradation. Examples of organics that are degraded in the rhizosphere by microbial activity include PAHs, PCBs, and petroleum hydrocarbons (62, 93). Plants can stimulate these microbial degradation processes. First, plant carbon compounds released into the rhizosphere facilitate a higher microbial density—the general rhizosphere effect. Second, secondary plant compounds released from roots may specifically induce microbial genes involved in degradation of the organic compound, or act as a cometabolite to facilitate microbial degradation (44, 72, 93). Better knowledge of these plant-microbe interactions is needed to more efficiently design phytoremediation strategies or engineer more efficient plant-microbe consortia.

Rhizosphere processes that favor phytoremedation may be optimized by the choice of plant species, e.g., plants with large and dense root systems for phytostimulation, or aquatic plants for metal precipitation. If a certain exudate compound is identified to enhance phytoremediation (e.g., a chelator or a secondary metabolite that stimulates microbial degradation) plants can be selected or genetically engineered to produce large amounts of this compound. In one such study, overexpression of citrate synthase in plants conferred enhanced aluminum tolerance, probably via enhanced citrate release into the rhizosphere, which prevented Al uptake due to complexation (28). In another approach to stimulate rhizosphere remediation, certain agronomic treatments may be employed that favor the production of general and specific exudate compounds, such as clipping or fertilization (72). Inorganic fertilizer is preferred over organic fertilizer (manure) for use in phytostimulation because the latter provides an easy-to-digest carbon source that microbes may prefer to use instead of the organic pollutant.

If the microbial consortia responsible for the remediation process are known, it may be possible to increase the abundance of these species by the choice of vegetation. An alternative approach is to grow these microbial isolates in large amounts and add them to the soil, a process called bioaugmentation. Introducing nonnative microbes to sites is considered ineffective because they tend to be outcompeted by the established microbial populations. In another approach to optimize rhizosphere remediation, the watering regime may be regulated to provide an optimal soil moisture for plant and microbial growth. If redox reactions are involved in the remediation process, periodic flooding and draining of constructed wetlands may be effective to alternate reducing and oxidizing conditions (62).

Plant Uptake

Root concentration factor (RCF): the ratio of pollutant concentration in root relative to external solution, used as a measure for plant uptake

Uptake of pollutants by plant roots is different for organics and inorganics. Organic pollutants are usually manmade, and xenobiotic to the plant. As a consequence, there are no transporters for these compounds in plant membranes. Organic pollutants therefore tend to move into and within plant tissues driven by simple diffusion, dependent on their chemical properties. An important property of the organic pollutant for plant uptake is its hydrophobicity (17, 121). Organics with a log K_{ow} between 0.5 and 3 are hydrophobic enough to move through the lipid bilayer of membranes, and still water soluble enough to travel into the cell fluids. If organics are too hydrophilic (log $K_{ow} < 0.5$) they cannot pass membranes and never get into the plant; if they are too hydrophobic (log $K_{ow} > 3$) they get stuck in membranes and cell walls in the periphery of the plant and cannot enter the cell fluids. Because the movement of organics into and through plants is a physical rather than biological process, it is fairly predictable across plant species and lends itself well to modeling (26). The tendency of organic pollutants to move into plant roots from an external solution is expressed as the root concentration factor (RCF = equilibrium concentration in roots/equilibrium concentratrion in external solution).

In contrast, inorganics are taken up by biological processes via membrane transporter proteins. These transporters occur naturally because inorganic pollutants are either nutrients themselves (e.g., nitrate, phosphate, copper, manganese, zinc) or are chemically similar to nutrients and are taken up inadvertently (e.g., arsenate is taken up by phosphate transporters, selenate by sulfate transporters) (1, 112). Inorganics usually exist as ions and cannot pass membranes without the aid of membrane transporter proteins. Because uptake of inorganics depends on a discrete number of membrane proteins, their uptake is saturable, following Michaelis Menten kinetics (80). For most elements multiple transporters exist in plants. The model plant *Arabidopsis thaliana*, for instance, has 150 different cation transporters (6), and 14 transporters for sulfate alone (56). Individual transporter proteins have unique properties with respect to transport rate, substrate affinity, and substrate specificity (low affinity transporters tend to be more promiscuous) (80). These properties may be subject to regulation by metabolite levels or regulatory proteins (e.g., kinases). Furthermore, the abundance of each transporter varies with tissue-type and environmental conditions, which may be regulated at the transcription level or via endocytosis. As a consequence, uptake and movement of inorganics in plants are complex species- and conditions-dependent processes, and difficult to capture in a model.

When inorganic pollutants accumulate in tissues they often cause toxicity, both directly by damaging cell structure (e.g., by causing oxidative stress due to their redox activity) and indirectly via replacement of other essential nutrients (116). Organics tend to be less toxic to plants, partly because they are not accumulated as readily and because they tend to be less reactive. Thus, when soils are polluted with a mixture of organics and metals the inorganics are most likely to limit plant growth and phytoremediation. Phytoremediation of mixed pollutants (organics and inorganics) is an understudied area, but very relevant because many sites contain mixed pollution.

The presence of rhizosphere microbes can affect plant uptake of inorganics. For instance, mycorrhizal fungi can both enhance uptake of essential metals when metal levels are low and decrease plant metal uptake when metals are present at phytotoxic levels (46, 104). Also, rhizosphere bacteria can enhance plant uptake of mercury and selenium (29). The mechanisms of these plant-microbe interactions are still largely unclear; microbe-mediated enhanced plant uptake may be due to a stimulatory effect on root growth, microbial production of metabolites that affect plant gene expression of transporter proteins, or microbial effects on bioavailability of the element (30).

Depending on the phytoremediation strategy, pollutant uptake into the plant may be desirable (e.g., for phytoextraction) or not (e.g., for phytostabilization). For either application, plant species with the desired properties may be selected. Screening studies under uniform conditions are a useful strategy to compare uptake characteristics of different species for different pollutants. Agronomic practices may also be employed to maximize pollutant uptake. Plant species may be selected for suitable rooting depth and root morphology (88). Furthermore, plant roots can be guided to grow into the polluted zone via deep planting in a casing, forcing the roots to grow downward into the polluted soil and to tap into polluted water rather than rainwater (88). Supplemental water (via irrigation) and oxygen (via air tube to roots) may also facilitate pollutant uptake, and soil nutrient levels may be optimized by fertilization. Not only will nutrients promote plant growth and thus uptake of the pollutant, they may also affect plant uptake of pollutants via ion competition at the soil and plant level. For instance, supplying phosphate will release arsenate from soils, making it more bioavailable; on the other hand, phosphate will compete with arsenate for uptake by plants because both are taken up by phosphate transporters (1).

It may also be possible to manipulate plant accumulation by genetic engineering. A transgenic approach that may be used to alter uptake of inorganic pollutants is overexpression or knockdown of membrane transporter proteins. This approach was used successfully to enhance accumulation of Ca, Cd, Mn, Pb, and Zn (5, 58, 123). The specificity of membrane transporters for different inorganics may also be manipulated via protein engineering (102). Furthermore, altering plant production of chelator molecules can affect plant metal accumulation (39, 49, 54, 134, 135). Hyperaccumulator species offer potentially interesting genetic material to be transferred to high-biomass species. Constitutive expression of a Zn transporter in the root cell membrane is one of the underlying mechanisms of the natural Zn hyperaccumulator *Thlaspi caerulescens* (94). Research is ongoing to isolate genes involved in metal hyperaccumulation and hypertolerance.

Chelation and Compartmentation in Roots

As mentioned above, plants can release compounds from their roots that affect pollutant solubility and uptake by the plant. Inside plant tissues such chelator compounds also play a role in tolerance, sequestration, and transport of inorganics and organics (103). Phytosiderophores are chelators that facilitate uptake of Fe and perhaps other metals in grasses; they are biosynthesized from nicotianamine, which is composed of three methionines coupled via nonpeptide bonds (57). Nicotianamine also chelates metals and may facilitate their transport (115, 127). Organic acids (e.g., citrate, malate, histidine) not only can facilitate uptake of metals into roots but also play a role in transport, sequestration, and tolerance of metals (70, 107, 127). Metals can also be bound by the thiol-rich peptides GSH and PCs, or by the Cys-rich MTs (24). Chelated metals in roots may be stored in the vacuole or exported to the shoot via the xylem. As described in more detail below, organics may be conjugated and stored or degraded enzymatically. An overview of these processes is depicted in **Figure 3**.

GSH: glutathione

PC: phytochelatin

MT: metallothionein protein

Chelation in roots can affect phytoremediation efficiency as it may facilitate root sequestration, translocation, and/or tolerance. Root sequestration may be desirable for phytostabilization (less exposure to wildlife) whereas export to xylem is desirable for phytoextraction. If chelation is desirable, it may be enhanced by selection or engineering of plants with higher levels of the chelator in question. Root

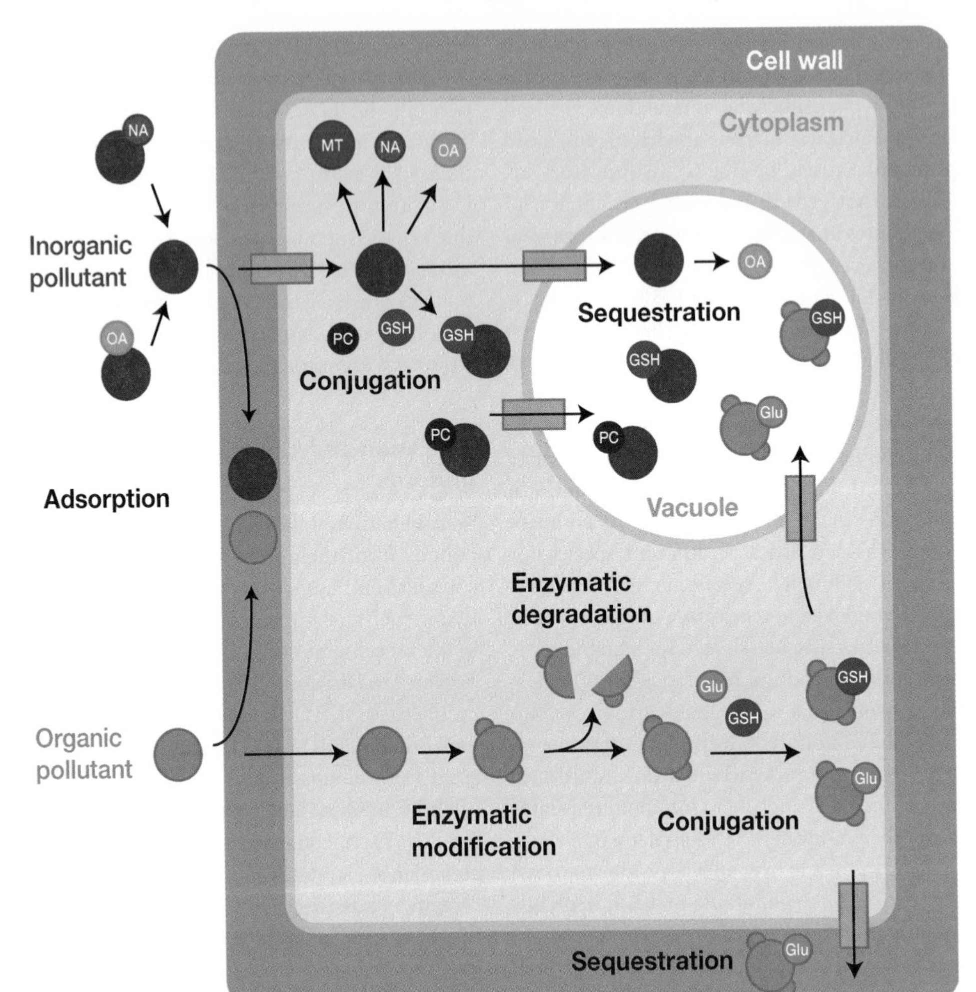

Figure 3

Tolerance mechanisms for inorganic and organic pollutants in plant cells. Detoxification generally involves conjugation followed by active sequestration in the vacuole and apoplast, where the pollutant can do the least harm. Chelators shown are GSH: glutathione, Glu: glucose, MT: metallothioneins, NA: nicotianamine, OA: organic acids, PC: phytochelatins. Active transporters are shown as boxes with arrows.

sequestration and export to xylem might be manipulated by overexpression or knockdown of the respective membrane transporters involved. Unfortunately, little is known about these tissue-specific transporters of inorganics. The completion of the sequencing of the *Arabidopsis* and rice genomes should accelerate the analysis of transporter gene families.

Transpiration stream concentration factor (TSCF): the ratio of pollutant concentration in xylem fluid relative to external solution, used as a measure for plant translocation

Translocation

Translocation from root to shoot first requires a membrane transport step from root symplast into xylem apoplast. The impermeable suberin layer in the cell wall of the root endodermis (Casparian strip) prevents solutes from flowing straight from the soil solution or root apoplast into the root xylem (116). Organic pollutants pass the membrane between root symplast and xylem apoplast via simple diffusion. The TSCF is the ratio of the concentration of a compound in the xylem fluid relative to the external solution, and is a measure of uptake into the plant shoot. Entry of organic pollutants into the xylem depends on similar passive movement over membranes as their uptake into the plant. Thus, the TSCF for organics shows a similar correlation with hydrophobicity as RCF: Compounds with a log K_{ow} between 0.5 and 3 are most easily transported to the xylem and translocated to the shoot (121).

Inorganics require membrane transporter proteins to be exported from the root endodermis into the root xylem. Some inorganics are chelated during xylem transport by organic acids (histidine, malate, citrate), nicotianamine, or thiol-rich peptides (67, 95, 115, 127). For most inorganics it is still unclear via which transporter proteins they are exported to the root xylem and to which—if any—chelators they are bound during transport. Better knowledge of the transporters and chelators involved in translocation of inorganics would facilitate the development of transgenics with more efficient phytoextraction capacity.

Bulk flow in the xylem from root to shoot is driven by transpiration from the shoot, which creates a negative pressure in the xylem that pulls up water and solutes (116). Plant transpiration depends on plant properties and environmental conditions. Plant species differ in transpiration rate, due to metabolic differences (e.g., C3/C4/CAM photosynthetic pathway) and anatomical differences (e.g., surface to volume ratio, stomatal density, rooting depth) (116). Species such as poplar are phreatophytes, or water spenders; they have long roots that tap into the ground water (27). Mature poplar trees can transpire 200–1000 liters of water per day (38, 132). In addition to plant species composition, vegetation height and density affect transpiration, as well as environmental conditions: Transpiration is generally maximal at high temperature, moderate wind, low relative air humidity, and high light (116). Consequently, phytoremediation mechanisms that rely on translocation and volatilization are most effective in climates with low relative humidity and high evapotranspiration.

Chelation and Compartmentation in Leaves

Import into leaf cells from leaf xylem involves another membrane transport step. Inorganics are taken up by specific membrane transporter proteins. Organics enter the leaf symplast from the shoot xylem by simple diffusion, the rate of which depends on the chemical properties of the pollutant, as discussed above. Once inside the leaf symplast, the pollutant may be compartmentized in certain tissues or cellular locations. In general, toxic pollutants are sequestered in places where they can do the least harm to essential cellular processes. At the cellular level, pollutants are generally accumulated in the vacuole or cell wall (21, 24). At the tissue level they may be accumulated in the epidermis and trichomes (50, 69).

When pollutants are sequestered in tissues, they are often bound by chelators or form conjugates (see **Figure 3**). Toxic inorganics are usually metals. Chelators that are involved in metal sequestration include the tripeptide GSH (γ-glu-cys-gly) and its oligomers, the PCs. XAS has shown that inorganics that were complexed by PCs in vivo include Cd and As (95); there may be others since PC synthesis is induced by various other metals (24). After chelation by GSH or PCs, an ABC-type transporter actively transports the metal-chelate complex to the vacuole, where it is further complexed by sulfide (24, 75). Organic acids such as malate and citrate are also likely metal (e.g., Zn) chelators in vacuoles, as judged from XAS (70). Ferritin is an iron chelator in chloroplasts (120). Additional metal-chelating proteins exist (e.g., MTs) that may play a role in sequestration and tolerance (e.g., of Cu) and/or in homeostasis of essential metals (48). There is still much to be discovered about the roles of these different chelators in transport and detoxification of inorganic pollutants.

XAS: X-ray absorption spectroscopy

Conjugation to GSH also plays a role in sequestration and tolerance of organic pollutants (78). A large family of GSTs with different substrate specificities mediate conjugation of organics to GSH in the cytosol (55, 68, 89). The glutathione S-conjugates are actively transported to the vacuole or the apoplast by ATP-dependent membrane pumps (79, 81, 109, 130). An alternative conjugation-sequestration mechanism for organics in plants involves coupling a glucose or a malonyl-group to the organic compound, followed by transport of the conjugate to the vacuole or the apoplast (25). These conjugation steps are mediated by

GST: GSH-S-transferases

a family of glucosyltransferases and malonyl-transferases, and the transport steps by ATP-dependent pumps (21).

To be conjugated, the organic compound may need chemical modification to create suitable side groups for conjugation. These modification reactions can be oxidative or reductive. For example, cytochrome P450 monooxygenases catalyze an oxidative transformation, incorporating an O atom from oxygen into an organic molecule such as atrazine to create a hydroxyl side group (25). Nitroreductases are an example of enzymes that mediate a reductive transformation, converting a nitro group of, e.g., TNT to an amino group (83). Other enzymes that mediate modifications of organic pollutants include dioxygenases, peroxidases, peroxygenases, and carboxylesterases (21). Thus, accumulation of organic pollutants typically comprises three phases: chemical modification, conjugation, and sequestration (**Figure 3**). This sequence of events has been summarized as the "green liver model" because of its similarity to mammalian detoxification mechanisms (21, 109). Some natural functions of the enzymes and transporters involved are to biosynthesize and transport natural plant compounds such as flavonoids, alkaloids, and plant hormones, and to defend against biotic stresses (78, 98).

Uptake and accumulation in leaves without toxic effects are desirable properties for phytoextraction. To maximize these processes, plants may be selected or engineered that have higher levels of transporters involved in uptake of an inorganic pollutant from the xylem into the leaf symplast. Better knowledge of the transporters involved in the process would be helpful because this is still a largely unexplored area. Similarly, plants with high transporter activities from cytosol to vacuole can be more efficient at storing toxic inorganics (58, 114, 123). Sequestration and tolerance may also be enhanced by selection or engineering of plants with higher production of leaf chelators or conjugates. This can be mediated by higher levels of enzymes that produce these conjugates, e.g., enzymes synthesizing GSH, PCs, glucose, organic acids, or chelator proteins (49, 54, 134, 135). In addition, enzymes that couple the chelator or conjugant to the pollutant (GSH transferases, glucosyltransferases) may be overexpressed (40) or enzymes that modify organics to make them amenable to conjugation (32, 33, 51).

In all cases where potentially toxic pollutants are accumulated in plant tissues, phytoremediation in the field should include a risk assessment study because the plant material may pose a threat to wildlife. The degree of toxicity will depend on leaf concentration but also on the form of the pollutant that is accumulated. During accumulation the toxicity of the pollutant may change. To test the potential toxicity of the plant material, a laboratory digestibility study may be done using model organisms or in vitro simulations of animal digestion systems. In the field, exposure to wildlife may be minimized by, e.g., fencing, netting, noise, and scarecrows.

Degradation

Only organic pollutants can be phytoremediated via degradation. Inorganic elements are undegradable and can only be stabilized or moved and stored. In phytodegradation plant enzymes act on organic pollutants and catabolize them, either mineralizing them completely to inorganic compounds (e.g., carbon dioxide, water and Cl_2), or degrading them partially to a stable intermediate that is stored in the plant (82). This enzymatic degradation of organics can happen in both root and shoot tissue. Degradation within plant tissues is generally attributed to the plant, but may in some cases involve endophytic microorganisms (10).

Phytodegradation involves some of the same classes of enzymes responsible for accumulation in tissues. The modifying enzymes that create side groups on organics that increase solubility and enable conjugation also play a role in the initial steps of phytodegradation. Thus, enzyme classes involved in phytodegradation include dehalogenases, mono- and dioxygenases, peroxidases, peroxygenases, carboxylesterases, laccases, nitrilases, phosphatases, and nitroreductases (131). Also, if pollutants are only partially degraded and the degradation products stored in plants, these are often conjugated and sequestered by the same mechanisms described above, involving GSH-S-transferases, malonyl- and glucosyltransferases, and ATP-dependent conjugate-transport pumps (21). These degradation products of pollutants that accumulate in vacuoles or apoplast of plant tissues are called bound residues (21). Atrazine and TNT are examples of organic pollutants that are partially degraded followed by storage of the degradation products as bound residues (14, 22). For TCE, different results were obtained in different studies: Overall, TCE appears to be in part volatilized by the plant, part is stored as bound residue, and part may be completely degraded (111). Phytoremediation of TCE is a much-studied process, and the remaining uncertainty about its fate illustrates that still much remains to be learned about the metabolic fate of organics in plants. Better knowledge in this respect would be beneficial not only for further improvement of phytoremediation efficiency, but also for better estimating the potential risks involved.

Phytodegradation of organic pollutants may be optimized by selecting or engineering plant species with higher activities of the enzymes thought to be involved and rate-limiting. There are some examples of promising transgenic approaches. The expression in plants of bacterial enzymes involved in reductive transformation of TNT (tetranitrate reductase or nitroreductase) resulted in enhanced plant tolerance and degradation of TNT (45, 51). Also, the constitutive expression of a mammalian cytochrome P450 in tobacco resulted in an up to 640-fold higher ability to metabolize TCE (33).

Volatilization

Phytovolatilization is the release of pollutants from the plant to the atmosphere as a gas. Inorganic Se can be volatilized by plants and microorganisms. Volatilization of Se involves assimilation of inorganic Se into the organic selenoaminoacids selenocysteine (SeCys) and selenomethionine (SeMet). The latter can be methylated to form dimethylselenide (DMSe), which is volatile (119). Volatilization of the inorganics As and Hg has been demonstrated for microorganisms, but these elements do not appear to be volatilized to significant levels by (nontransgenic) plants (105).

Many VOCs can be volatilized passively by plants. Volatile pollutants with a Henry's law constant $H_i > 10^{-6}$ that are mobile in both air and water can move readily from the soil via the transpiration stream into the atmosphere (18). In this way, plants act like a wick for VOCs to facilitate their diffusion from soil. Examples of organic pollutants that can be volatilized by plants are the chlorinated solvent TCE and the fuel additive MTBE (26, 90).

Because volatilization completely removes the pollutant from the site as a gas, without need for plant harvesting and disposal, this is an attractive technology. In the case of Se, the volatile form was also reported to be 2–3 orders of magnitude less toxic than the inorganic Se forms (119). Volatilization may be promoted in several ways. Although volatilization of VOCs is passive, the process may be maximized by using phreatophyte species with high transpiration rates and by promoting transpiration (preventing stomatal closure through sufficient irrigation). For Se, enzymes of the S assimilation pathway mediate Se volatilization, and overexpression of one of these, cystathionine-γ-synthase promotes Se volatilization (124). In another approach, the enzyme SeCys methyltransferase from a Se hyperaccumulator species was expressed in a nonaccumulator, also significantly enhancing Se volatilization (71). Volatilization of mercury by plants was achieved by introducing a bacterial mercury reductase (MerA). The resulting plants volatilized elemental mercury and were significantly more Hg-tolerant (105).

If a toxic volatile pollutant is emitted by plants during phytoremediation, the fate of the gas in the atmosphere should be determined as part of risk assessment. Such a study was done for volatile Se and Hg, and the pollutant was reportedly dispersed and diluted to such an extent that it did not pose a threat (74, 85).

NEW DEVELOPMENTS IN PHYTOREMEDIATION

In the past 10 years phytoremediation has gained acceptance as a technology and has been acknowledged as an area of research. There has already been a substantial increase in our knowledge of the mechanisms that underlie the uptake, transport, and detoxification of pollutants by plants and their associated microbes. Still, large gaps in our knowledge await further research, as indicated above. Phytoremediation efficiency is still limited by a lack of knowledge of many basic plant processes and plant-microbe interactions. There is also a need for more phytoremediation field studies to demonstrate the effectiveness of the technology and increase its acceptance.

Continued phytoremediation research should benefit from a (more) multidisciplinary approach, involving teams with expertise at all organization levels, to study the remediation of pollutants from the molecule to the ecosystem. Phytoremediation research at universities is generally carried out by scientists with expertise at a certain organizational level (e.g., plant molecular biology, plant biochemistry, plant physiology, ecology, or microbiology) and of a certain subset of pollutants (e.g., heavy metals, herbicides, TNT, or PAHs). Because research on phytoremediation of organics and inorganics requires different expertise they are carried out in different research communities, with more engineers studying organics and more biologists studying inorganics. These researchers do not interact optimally, in part because of a lack of phytoremediation conferences and scientific journals that cover inorganics and organics equally. Because 64% of polluted sites contain mixtures of organics and inorganics (36), phytoremediation would benefit from more collaborative studies by teams of researchers from different backgrounds, to combine expertise in phytoremediation of both types of pollution and at multiple organization levels.

Despite the remaining gaps in our knowledge, research has yielded much useful knowledge for phytoremediation, as described above. This has also resulted in practical phytoremediation resources, such as online databases of plant species that may be useful for cleanup of different types of pollutants (84) (PHYTOPET lists species particularly useful for cleanup of petroleum hydrocarbons and PHYTOREM lists plants that are recommended for metals and metalloids). The U.S. Environmental Protection Agency also maintains a phytoremediation Web site (http://www.clu-in.org) with a wealth of information for researchers and the general public (e.g., citizen's guides, phytoremediation resource guide) (37, 38).

Future field phytoremediation projects should benefit from (more) collaboration between research groups and industry so that they can be designed to address hypotheses and gain scientific knowledge in addition to meeting cleanup standards. Future field phytoremediation projects will also benefit from coordinated experimental design across projects so that results can be better compared.

An interesting development in phytoremediation is its integration with landscape architecture. Remediation of urban sites (parks, nature areas) may be combined with an attractive design so that the area may be used by the public during and after the remediation process while minimizing risk (66). Other sites that are phytoremediated may be turned into wildlife sanctuaries, like the Rocky Mountain Arsenal in Denver, once one of the most polluted sites in the United States (http://www.pmrma.army.mil/).

Another new development in phytoremediation is the use of transgenic plants. Knowledge gained from plant molecular studies in the past 10 years has led to the development of some promising transgenics that show higher tolerance, accumulation, and/or degradation capacity for various pollutants, as described above. So far, these transgenics have mainly been tested in laboratory studies using artificially contaminated medium rather than soils from the field, let alone field studies. However, this is starting to change. One field phytoremediation study using transgenic Indian mustard plants that overexpress enzymes involved in sulfate/

selenate reduction and in accumulation of GSH was just completed (96, 134, 135). Three types of transgenic Indian mustard plants that overexpress enzymes involved in sulfate/selenate reduction and in accumulation of GSH showed enhanced Se accumulation in the field when grown on soil polluted with Se, B, and other salts (G. Banuelos, N. Terry, D. LeDuc, E. Pilon-Smits & C. Mackey, unpublished results). Earlier, these same transgenics showed enhanced capacity to accumulate Se and heavy metals (Cd, Zn) from polluted soil from the field in greenhouse experiments (12, 125). Another field experiment testing Hg volatilizing (MerA) poplar trees is presently underway (D. Glass, personal communication).

In the coming years, mining of the genomic sequences from *Arabidopsis thaliana* and rice and availability of new genomic technologies should lead to the identification of novel genes important for pollutant remediation, including regulatory networks (e.g., transcription factors) and tissue-specific transporters. The expression of these genes may then be manipulated in high-biomass species for use in phytoremediation. Other new developments in plant genetic engineering are tailored transgenics that overexpress different enzymes in different plant parts (e.g., root-specific expression of one gene and shoot-specific expression of another) or that express a transgene only under certain environmental conditions (31). Also, genetic engineering of the chloroplast genome offers a novel way to obtain high expression without the risk of spreading the transgene via pollen (106). In another totally new approach, it was shown to be possible to genetically manipulate an endophytic microorganism, leading to enhanced toluene degradation (10).

As transgenics are being tested in the field and the associated risks assessed, their use may become more accepted and less regulated, as has been the case for transgenic crops. Also, as more information becomes available about the movement of pollutants in ecosystems and the associated risks, the rules for cleanup targets may be adjusted depending on future use of the site, bioavailability of the pollutant, and form of the pollutant. Because phytoremediation only remediates the bioavailable fraction of the pollution, stringent cleanup targets limit the applicability of this technology. If targets can be adjusted to focus on the bioavailable (i.e., toxic) fraction of the pollutant, phytoremediation could become more widely applicable. This would reduce cleanup costs and enable the cleanup of more sites with the limited funds available.

SUMMARY POINTS

1. Plants and their associated microbes can remediate pollutants via stabilization, degradation in the rhizosphere, degradation in the plant, accumulation in harvestable tissues, or volatilization.
2. Phytoremediation offers a cost-effective and environment-friendly alternative or complementary technology for conventional remediation methods such as soil incineration or excavation and pump-and-treat systems.
3. Although phytoremediation works effectively for a wide range of organic and inorganic pollutants, the underlying biological processes are still largely unknown in many cases. Some important processes that require further study are plant-microbe interactions, plant degradation mechanisms for organics, and plant transport and chelation mechanisms for inorganics.
4. New knowledge and plant material obtained from research is being implemented for phytoremediation in the field. The first field tests with transgenic plants are showing

promising results. As more results demonstrating the effectiveness of phytoremediation become available its use may continue to grow, reducing cleanup costs and enabling the cleanup of more sites with the limited funds available.

ACKNOWLEDGMENTS

The author's research is supported by National Science Foundation Grant MCB9982432 and U.S. Department of Agriculture NRI grant #2003-35318-13758.

LITERATURE CITED

1. Abedin MJ, Feldmann J, Meharg AA. 2002. Uptake kinetics of arsenic species in rice plants. *Plant Physiol.* 128:1120–28
2. Anderson JW. 1993. Selenium interactions in sulfur metabolism. In *Sulfur Nutrition and Assimilation in Higher Plants—Regulatory, Agricultural and Environmental Aspects*, ed. LJ De Kok, pp. 49–60. The Hague, The Netherlands: SPB Academic
3. Anderson TA, Guthrie EA, Walton BT. 1993. Bioremediation. *Environ. Sci. Technol.* 27: 2630–36
4. Aprill W, Sims RC. 1990. Evaluation of the use of prairie grasses for stimulating polycyclic aromatic hydrocarbon treatment in soil. *Chemosphere* 20:253–65
5. Arazi T, Sunkar R, Kaplan B, Fromm H. 1999. A tobacco plasma membrane calmodulin-binding transporter confers Ni^{2+} tolerance and Pb^{2+} hypersensitivity in transgenic plants. *Plant J.* 20:171–82
6. Axelsen KB, Palmgren MG. 2001. Inventory of the superfamily of P-type ion pumps in *Arabidopsis*. *Plant Physiol.* 126:696–706
7. Baker AJM, McGrath SP, Reeves RD, Smith JAC. 2000. Metal hyperaccumulator plants: A review of the ecology and physiology of a biological resource for phytoremediation of metal-polluted soils. In *Phytoremediation of Contaminated Soil and Water*, ed. N Terry, G Bañuelos, pp. 85–108. Boca Raton: Lewis
8. Bañuelos GS. 2000. Factors influencing field phytoremediation of selenium-laden soils. In *Phytoremediation of Contaminated Soil and Water*, ed. N Terry, G Bañuelos, pp. 41–61. Boca Raton: Lewis
9. Bañuelos GS, Meek DW. 1990. Accumulation of selenium in plants grown on selenium-treated soil. *J. Environ. Qual.* 19:772–77
10. **Barac T, Taghavi S, Borremans B, Provoost A, Oeyen L, et al. 2004. Engineered endophytic bacteria improve phytoremediation of water-soluble, volatile, organic pollutants. *Nat. Biotechnol.* 22:583–88**
11. Beath OA, Gilbert CS, Eppson HF. 1939. The use of indicator plants in locating seleniferous areas in Western United States: I. General. *Amer. J. Bot.* 26:257–69
12. Bennett LE, Burkhead JL, Hale KL, Terry N, Pilon M, Pilon-Smits EAH. 2003. Analysis of transgenic Indian mustard plants for phytoremediation of metal-contaminated mine tailings. *J. Environ. Qual.* 32:432–40
13. Berti WR, Cunningham SD. 2000. Phytostabilization of metals. In *Phytoremediation of Toxic Metals. Using Plants to Clean up the Environment*, ed. I Raskin, BD Ensley, pp. 71–88. New York: Wiley

***Burkholderia cepacia*, a bacterial endophyte of yellow lupine, was transformed with a plasmid from a related strain containing genes that mediate toluene degradation. After infection of lupine with the modified strain, the resulting plants were more tolerant to toluene and volatilized less of it through the leaves. This is the first example of genetic modification of an endophyte for phytoremediation.**

14. Bhadra R, Wayment DG, Hughes JB, Shanks JV. 1999. Confirmation of conjugation processes during TNT metabolism by axenic plant roots. *Environ. Sci. Technol.* 33:446–52
15. Blaylock MJ, Huang JW. 2000. Phytoextraction of metals. In *Phytoremediation of Toxic Metals. Using Plants to Clean up the Environment*, ed. I Raskin, BD Ensley, pp. 53–70. New York: Wiley
16. Bowen GC, Rovira AD. 1991. The rhizosphere—the hidden half of the hidden half. In *Plant Roots—The Hidden Half*, eds. Y Waisel, A Eshel, U Kaffkafi, pp. 641–69. New York: Marcel Dekker
17. Briggs GG, Bromilow RH, Evans AA. 1982. Relationships between lipophilicity and root uptake and translocation of non-ionized chemicals by barley. *Pestic. Sci.* 13:405–504
18. Bromilow RH, Chamberlain K. 1995. Principles governing uptake and transport of chemicals. In *Plant Contamination: Modeling and Simulation of Organic Chemical Processes*, ed. S Trapp, JC McFarlane, pp. 37–68. Boca Raton: Lewis
19. Brooks RR. 1998. Plants that hyperaccumulate heavy metals. Wallingford: CAB Intl. 381 pp.
20. Brown SL, Henry CL, Chaney R, Compton H, DeVolder PM. 2003. Using municipal biosolids in combination with other residuals to restore metal-contaminated mining areas. *Plant Soil* 249:203–15
21. Burken JG. 2003. Uptake and metabolism of organic compounds: green-liver model. In *Phytoremediation: Transformation and Control of Contaminants*, ed. SC McCutcheon, JL Schnoor, pp. 59–84. New York: Wiley
22. Burken JG, Schnoor JL. 1997. Uptake and metabolism of atrazine by poplar trees. *Environ. Sci. Technol.* 31:1399–406
23. Chaney RL, Li YM, Brown SL, Homer FA, Malik M, et al. 2000. Improving metal hyperaccumulator wild plants to develop commercial phytoextraction systems: approaches and progress. In *Phytoremediation of Contaminated Soil and Water*, ed. N Terry, G Bañuelos, pp. 129–58. Boca Raton: Lewis
24. Cobbett CS, Goldsbrough PB. 2000. Mechanisms of metal resistance: phytochelatins and metallothioneins. In *Phytoremediation of Toxic Metals. Using Plants to Clean up the Environment*, ed. I Raskin, BD Ensley, pp. 247–71. New York: Wiley
25. Coleman JOD, Blake-Kalff MMA, Davies TGE. 1997. Detoxification of xenobiotics by plants: chemical modification and vacuolar compartmentation. *Trends Plant Sci.* 2:144–51
26. Davis LC, Erickson LE, Narayanan N, Zhang Q. 2003. Modeling and design of phytoremediation. In *Phytoremediation: Transformation and Control of Contaminants*, ed. SC McCutcheon, JL Schnoor, pp. 663–94. New York: Wiley
27. Dawson TE, Ehleringer JR. 1991. Streamside trees do not use stream water. *Nature* 350:335–37
28. **De la Fuente JM, Ramírez-Rodríguez V, Cabrera-Ponce JL, Herrera-Estrella L. 1997. Aluminum tolerance in transgenic plants by alteration of citrate synthesis. *Science* 276:1566–68**
29. De Souza MP, Huang CPA, Chee N, Terry N. 1999. Rhizosphere bacteria enhance the accumulation of selenium and mercury in wetland plants. *Planta* 209:259–63
30. **De Souza MP, Chu D, Zhao M, Zayed AM, Ruzin SE, et al. 1999. Rhizosphere bacteria enhance selenium accumulation and volatilization by Indian mustard. *Plant Physiol.* 119:565–73**
31. **Dhankher OP, Li Y, Rosen BP, Shi J, Salt D, et al. 2002. Engineering tolerance and hyperaccumulation of arsenic in plants by combining arsenate reductase and gamma-glutamylcysteine synthetase expression. *Nat. Biotechnol.* 20:1140–45**

This is one of the first examples of manipulation of plant rhizosphere processes.

This was one of the first investigations of plant-microbe interactions in the context of phytoremediation.

This is an example of a tailored transgenic that overexpresses more than one enzyme and expression is targeted to specific tissues to give maximal effect.

32. Didierjean L, Gondet L, Perkins R, Mei S, Lau C, Schaller H, et al. 2002. Engineering herbicide metabolism in tobacco and Arabidopsis with CYP76B1, a cytochrome P450 enzyme from Jerusalem artichoke. *Plant Physiol.* 130:179–89
33. **Doty SL, Shang TQ, Wilson AM, Tangen J, Westergreen AD, et al. 2000. Enhanced metabolism of halogenated hydrocarbons in transgenic plants containing mammalian cytochrome P450 2E1. *Proc. Natl. Acad. Sci. USA* 97:6287–91**

This study was one of the first to show the potential of genetic engineering plants for organics phytoremediation.

34. Dushenkov S. 2003. Trends in phytoremediation of radionuclides. *Plant Soil* 249:167–75
35. Dushenkov S, Kapulnik Y. 2000. Phytofiltration of metals. In *Phytoremediation of Toxic Metals. Using Plants to Clean up the Environment*, ed. I Raskin, BD Ensley, pp. 89–106. New York: Wiley
36. Ensley BD. 2000. Rationale for use of phytoremediation. In *Phytoremediation of Toxic Metals. Using Plants to Clean up the Environment*, ed. I Raskin, BD Ensley, pp. 3–12. New York: Wiley
37. EPA publ. 542-F-98–011. 1998. A citizen's guide to phytoremediation.
38. EPA publ. 542-B-99–003. 1999. Phytoremediation resource guide.
39. Evans KM, Gatehouse JA, Lindsay WP, Shi J, Tommey AM, Robinson NJ. 1992. Expression of the pea metallothionein-like gene $PsMT_A$ in *Escherichia coli* and *Arabidopsis thaliana* and analysis of trace metal ion accumulation: implications for gene $PsMT_A$ function. *Plant Mol. Biol.* 20:1019–28
40. Ezaki B, Gardner RC, Ezaki Y, Matsumoto H. 2000. Expression of aluminum-induced genes in transgenic *Arabidopsis* plants can ameliorate aluminum stress and/or oxidative stress. *Plant Physiol.* 122:657–65
41. Ferro AM, Rock SA, Kennedy J, Herrick JJ, Turner DL. 1999. Phytoremediation of soils contaminated with wood preservatives: greenhouse and field evaluations. *Int. J. Phytoremed.* 1:289–306
42. Ferro A, Chard J, Kjelgren R, Chard B, Turner D, Montague T. 2001. Groundwater capture using hybrid poplar trees: evaluation of a system in Ogden, Utah. *Int. J. Phytoremed.* 3:87–104
43. Flechas FW, Latady M. 2003. Regulatory evaluation and acceptance issues for phytotechnology projects. *Adv. Biochem. Engin./Biotechnol.* 78:172–85
44. Fletcher JS, Hegde RS. 1995. Release of phenols by perennial plant roots and their potential importance in bioremediation. *Chemosphere* 31:3009–16
45. French CE, Rosser SJ, Davies GJ, Nicklin S, Bruce NC. 1999. Biodegradation of explosives by transgenic plants expressing pentaerythritol tetranitrate reductase. *Nat. Biotechnol.* 17:491–94
46. Frey B, Zierold K, Brunner I. 2000. Extracellular complexation of Cd in the Hartig net and cytosolic Zn sequestration in the fungal mantle of *Picea abies—Hebeloma crustuliniforme* ectomycorrhizas. *Plant Cell Environ.* 23:1257–65
47. Glass DJ. 1999. *U.S. and International Markets for Phytoremediation, 1999–2000*. Needham, MA: D. Glass Assoc.
48. Goldsbrough P. 2000. Metal tolerance in plants: the role of phytochelatins and metallothioneins. In *Phytoremediation of Contaminated Soil and Water*, ed. N Terry, G Bañuelos, pp. 221–34. Boca Raton: Lewis
49. Goto F, Yoshihara T, Shigemoto N, Toki S, Takaiwa F. 1999. Iron fortification of rice seed by the soybean ferritin gene. *Nature Biotechnol.* 17:282–86
50. Hale KL, McGrath S, Lombi E, Stack S, Terry N, et al. 2001. Molybdenum sequestration in *Brassica*: a role for anthocyanins? *Plant Physiol.* 126:1391–402
51. Hannink N, Rosser SJ, French CE, Basran A, Murray JA, et al. 2001. Phytodetoxification of TNT by transgenic plants expressing a bacterial nitroreductase. *Nat Biotechnol.* 19:1168–72

52. Hansen D, Duda PJ, Zayed A, Terry N. 1998. Selenium removal by constructed wetlands: role of biological volatilization. *Environ. Sci. Technol.* 32:591–97
53. Harms H, Bokern M, Kolb M, Bock C. 2003. Transformation of organic contaminants by different plant systems. In *Phytoremediation: Transformation and Control of Contaminants*, ed. SC McCutcheon, JL Schnoor, pp. 285–316. New York: Wiley
54. Hasegawa I, Terada E, Sunairi M, Wakita H, Shinmachi F, et al. 1997. Genetic improvement of heavy metal tolerance in plants by transfer of the yeast metallothionein gene (CUP1). *Plant Soil* 196:277–81
55. Hatton PJ, Dixon D, Cole DJ, Edwards R. 1996. Glutathione transferase activities and herbicide selectivity in maize and associated weed species. *Pestic. Sci.* 46:267–75
56. Hawkesford MJ. 2003. Transporter gene families in plants: the sulphate transporter gene family—redundancy or specialization? *Physiol. Plant* 117:155–63
57. Higuchi K, Suzuki K, Nakanishi H, Yamaguchi H, Nishizawa NK, Mori S. 1999. Cloning of nicotianamine synthase genes, novel genes involved in the biosynthesis of phytosiderophores. *Plant Physiol.* 119:471–79
58. Hirschi KD, Korenkov VD, Wilganowski NL, Wagner GJ. 2000. Expression of *Arabidopsis* CAX2 in tobacco. Altered metal accumulation and increased manganese tolerance. *Plant Physiol.* 124:125–33
59. Hong MS, Farmayan WF, Dortch IJ, Chiang CY, McMillan SK, Schnoor JL. 2001. Phytoremediation of MTBE from a groundwater plume. *Environ. Sci. Technol.* 35:1231–39
60. Horne AJ. 2000. Phytoremediation by constructed wetlands. In *Phytoremediation of Contaminated Soil and Water*, ed. N Terry, G Bañuelos, pp. 13–40. Boca Raton: Lewis
61. Hughes JB, Shanks J, Vanderford M, Lauritzen J, Bhadra R. 1997. Transformation of TNT by aquatic plants and plant tissue cultures. *Environ. Sci. Technol.* 31:266–71
62. Hutchinson SL, Schwab AP, Banks MK. 2003. Biodegradation of petroleum hydrocarbons in the rhizosphere. In *Phytoremediation: Transformation and Control of Contaminants*, ed. SC McCutcheon, JL Schnoor, pp. 355–86. New York: Wiley
63. Jacobson ME, Chiang SY, Gueriguian L, Westholm LR, Pierson J, et al. 2003. Transformation kinetics of trinitrotoluene conversion in aquatic plants. In *Phytoremediation: Transformation and Control of Contaminants*, ed. SC McCutcheon, JL Schnoor, 409–27. New York: Wiley
64. Jeffers PM, Liddy CD. 2003. Treatment of atmospheric halogenated hydrocarbons by plants and fungi. In *Phytoremediation: Transformation and Control of Contaminants*, ed. SC McCutcheon, JL Schnoor, pp. 787–804. New York: Wiley
65. Kapulnik Y. 1996. Plant growth promotion by rhizosphere bacteria. In *Plant Roots—The Hidden Half*, ed. Y Waisel, A Eshel, U Kaffkafi, pp. 769–81. New York: Marcel Dekker
66. Kirkwood NG. 2001. *Manufactured Sites. Rethinking the Post-Industrial Landscape*. New York: Spon. 256 pp.
67. Krämer U, Cotter-Howells JD, Charnock JM, Baker AJM, Smith JAC. 1996. Free histidine as a metal chelator in plants that accumulate nickel. *Nature* 379:635–38
68. Kreuz K, Tommasini R, Martinoia E. 1996. Old enzymes for a new job: herbicide detoxification in plants. *Plant Physiol.* 111:349–53
69. Küpper H, Zhao F, McGrath SP. 1999. Cellular compartmentation of zinc in leaves of the hyperaccumulator *Thlaspi caerulescens*. *Plant Physiol.* 119:305–11
70. Küpper H, Mijovilovich A, Meyer-Klaucke W, Kroneck MH. 2004. Tissue- and qge-dependent differences in the complexation of cadmium and zinc in the cadmium/zinc hyperaccumulator *Thlaspi caerulescens* (Ganges Ecotype) revealed by x-ray absorption spectroscopy. *Plant Physiol.* 134:748–57

71. LeDuc DL, Tarun AS, Montes-Bayon M, Meija J, Malit MF, et al. 2004. Overexpression of selenocysteine methyltransferase in Arabidopsis and Indian mustard increases selenium tolerance and accumulation. *Plant Physiol.* 135:377–83
72. Leigh MB, Fletcher JS, Fu X, Schmitz FJ. 2002. Root turnover: an important source of microbial substances in rhizosphere remediation of recalcitrant contaminants. *Environ. Sci. Technol.* 36:1579–83
73. Li Y-M, Chaney R, Brewer E, Roseberg R, Angle SJ, et al. 2003. Development of a technology for commercial phytoextraction of nickel: economic and technical considerations. *Plant Soil* 249:107–15
74. Lin Z-Q, Schemenauer RS, Cervinka V, Zayed A, Lee A, Terry N. 2000. Selenium volatilization from a soil-plant system for the remediation of contaminated water and soil in the San Joaquin Valley. *J. Environ. Qual.* 29:1048–56
75. Lu YP, Li ZS, Rea PA. 1997. AtMRP1 gene of *Arabidopsis* encodes a glutathione S-conjugate pump: isolation and functional definition of a plant ATP-binding cassette transporter gene. *Proc. Natl. Acad. Sci. USA* 94:8243–48
76. Lytle CM, Lytle FW, Yang N, Qian JH, Hansen D, Zayed A, Terry N. 1998. Reduction of Cr(VI) to Cr(III) by wetland plants: potential for in situ heavy metal detoxification. *Environ. Sci. Technol.* 32:3087–93
77. Ma LQ, Komar KM, Tu C. 2001. A fern that accumulates arsenic. *Nature* 409:579
78. Marrs KA. 1996. The functions and regulation of glutathione s-transferases in plants. *Annu. Rev. Plant Physiol. Plant. Mol. Biol.* 47:127–58
79. Marrs KA, Alfenito MR, Lloyd AM, Walbot VA. 1995. Glutathione s-transferase involved in vacuolar transfer encoded by the maize gene Bronze-2. *Nature* 375:397–400
80. Marschner H. 1995. *Mineral Nutrition of Higher Plants*. San Diego: Academic. 889 pp.
81. Martinoia E, Grill E, Tommasini R, Kreuz K, Amrehin N. 1993. ATP-dependent glutathione S-conjugate 'export' pump in the vacuolar membrane of plants. *Nature* 364:247–49
82. McCutcheon SC, Schnoor JL. 2003. Overview of phytotransformation and control of wastes. In *Phytoremediation: Transformation and Control of Contaminants*, ed. SC McCutcheon, JL Schnoor, pp. 3–58. New York: Wiley
83. McCutcheon SC, Medina VF, Larson SL. 2003. Proof of phytoremediation for explosives in water and soil. In *Phytoremediation: Transformation and Control of Contaminants*, ed. SC McCutcheon, JL Schnoor, pp. 429–80. New York: Wiley
84. McIntyre TC. 2003. Databases and protocol for plant and microorganism selection: hydrocarbons and metals. In *Phytoremediation: Transformation and Control of Contaminants*, ed. SC McCutcheon, JL Schnoor, pp. 887–904. New York: Wiley
85. Meagher RB, Rugh CL, Kandasamy MK, Gragson G, Wang NJ. 2000. Engineered phytoremediation of mercury pollution in soil and water using bacterial genes. In *Phytoremediation of Contaminated Soil and Water*, ed. N Terry, G Bañuelos, pp. 201–21. Boca Raton: Lewis
86. Morikawa H, Takahashi M, Kawamura Y. 2003. Metabolism and genetics of atmospheric nitrogen dioxide control using pollutant-philic plants. In *Phytoremediation: Transformation and Control of Contaminants*, ed. SC McCutcheon, JL Schnoor, pp. 765–86. New York: Wiley
87. Negri MC, Hinchman RR. 2000. The use of plants for the treatment of radionuclides. In *Phytoremediation of Toxic Metals. Using Plants to Clean up the Environment*, ed. I Raskin, BD Ensley, pp. 107–32. New York: Wiley
88. Negri MC, Gatliff EG, Quinn JJ, Hinchman RR. 2003. Root development and rooting at depths. In *Phytoremediation: Transformation and Control of Contaminants*, ed. SC McCutcheon, JL Schnoor, pp. 233–62. New York: Wiley

89. Neuefeind T, Reinemer P, Bieseler B. 1997. Plant glutathione S-transferases and herbicide detoxification. *Biol. Chem.* 378:199–205
90. Newman LA, Strand SE, Choe N, Duffy J, Ekuan G, et al. 1997. Uptake and biotransformation of trichloroethylene by hybrid poplars. *Environ. Sci. Technol.* 31:1062–67
91. Nriagu JO. 1979. Global inventory of natural and anthropogenic emissions of trace metals to the atmosphere. *Nature* 279:409–11
92. Nzengung VA, McCutcheon SC. 2003. Phytoremediation of perchlorate. In *Phytoremediation: Transformation and Control of Contaminants*, ed. SC McCutcheon, JL Schnoor, pp. 863–85. New York: Wiley
93. Olson PE, Reardon KF, Pilon-Smits EAH. 2003. Ecology of rhizosphere bioremediation. In *Phytoremediation: Transformation and Control of Contaminants*, ed. SC McCutcheon, JL Schnoor, pp. 317–54. New York: Wiley
94. Pence NS, Larsen PB, Ebbs SD, Letham DLD, Lasat MM, Garvin DF, et al. 2000. The molecular physiology of heavy metal transport in the Zn/Cd hyperaccumulator *Thlaspi caerulescens*. *Proc. Natl. Acad. Sci. USA* 97:4956–60
95. Pickering IJ, Prince RC, George MJ, Smith RD, George GN, Salt DE. 2000. Reduction and coordination of arsenic in Indian mustard. *Plant Physiol.* 122:1171–77
96. Pilon-Smits EAH, Hwang SB, Lytle CM, Zhu YL, Tai JC, et al. 1999. Overexpression of ATP sulfurylase in *Brassica juncea* leads to increased selenate uptake, reduction and tolerance. *Plant Physiol.* 119:123–32
97. Pilon-Smits EAH, de Souza MP, Hong G, Amini A, Bravo RC, et al. 1999. Selenium volatilization and accumulation by twenty aquatic plant species. *J. Environ. Qual.* 28:1011–17
98. Prescott AG. 1996. Dioxygenases: molecular structure and role in plant metabolism. *Annu. Rev. Plant Physiol. Plant Mol. Biol.* 47:247–71
99. Raskin I, Kumar PBAN, Dushenkov S, Salt DE. 1994. Bioconcentration of heavy metals by plants. *Curr. Opin. Biotechnol.* 5:285–90
100. Raskin I, Smith RD, Salt DE. 1997. Phytoremediation of metals: using plants to remove pollutants from the environment. *Curr. Opin. Biotechnol.* 8:221–26
101. Rock SA. 2003. Field evaluations of phytotechnologies. In *Phytoremediation: Transformation and Control of Contaminants*, ed. SC McCutcheon, JL Schnoor, pp. 905–24. New York: Wiley
102. Rogers EE, Eide DJ, Guerinot ML. 2000. Altered selectivity in an *Arabidopsis* metal transporter. *Proc. Natl. Acad. Sci. USA* 97:12356–60

This shows potential for tailoring transporters to specifically take up metals of interest while excluding other metals.

103. Ross SM. 1994. Toxic metals in soil-plant systems. Chichester, England: Wiley. 459 pp.
104. Rufyikiri G, Declerck S, Dufey JE, Delvaux B. 2000. Arbuscular mycorrhizal fungi might alleviate aluminum toxicity in banana plants. *New Phytol.* 148:343–52
105. Rugh CL, Wilde HD, Stack NM, Thompson DM, Summers AO, Meagher RB. 1996. Mercuric ion reduction and resistance in transgenic *Arabidopsis thaliana* plants expressing a modified bacterial *merA* gene. *Proc. Natl. Acad. Sci. USA* 93:3182–87

Introduction of the bacterial *MerA* gene resulted in enhanced Hg tolerance in *Arabidopsis*.

106. Ruiz ON, Hussein HS, Terry N, Daniell H. 2003. Phytoremediation of organomercurial compounds via chloroplast genetic engineering. *Plant Physiol.* 132:1344–52

When the bacterial *MerA* and *MerB* genes were integrated into the chloroplast genome, this significantly enhanced plant Hg tolerance.

107. Salt DE, Blaylock M, Kumar NPBA, Dushenkov V, Ensley BD, et al. 1995. Phytoremediation: a novel strategy for the removal of toxic metals from the environment using plants. *Biotechnology* 13:468–74
108. Salt DE, Smith RD, Raskin I. 1998. Phytoremediation. *Annu. Rev. Plant Physiol. Plant Mol. Biol.* 49:643–68
109. Sandermann H. 1994. Higher plant metabolism of xenobiotics: the “green liver” concept. *Pharmacogenetics* 4:225–41

110. Schnoor JL, Licht LA, McCutcheon SC, Wolfe NL, Carreira LH. 1995. Phytoremediation of organic and nutrient contaminants. *Environ. Sci. Technol.* 29:318A–23A
111. Shang TQ, Newman LA, Gordon MP. 2003. Fate of tricholorethylene in terrestrial plants. In *Phytoremediation: Transformation and Control of Contaminants*, ed. SC McCutcheon, JL Schnoor, pp. 529–60. New York: Wiley
112. Shibagaki N, Rose A, McDermott J, Fujiwara T, Hayashi H, et al. 2002. Selenate-resistant mutants of *Arabidopsis thaliana* identify Sultr1;2, a sulfate transporter required for efficient transport of sulfate into roots. *Plant J.* 29:475–86
113. Siciliano SD, Germida JJ. 1998. Mechanisms of phytoremediation: biochemical and ecological interactions between plants and bacteria. *Environ. Rev.* 6:65–79
114. Song W, Sohn EJ, Martinoia E, Lee YJ, Yang YY, et al. 2003. Engineering tolerance and accumulation of lead and cadmium in transgenic plants. *Nat. Biotechnol.* 21:914–19
115. Stephan UW, Schmidke I, Stephan VW, Scholz G. 1996. The nicotianamine molecule is made-to-measure for complexation of metal micronutrients in plants. *Biometals* 9:84–90
116. Taiz L, Zeiger E. 2002. *Plant Physiology*. Sunderland, MA: Sinauer. 690 pp.
117. Terry N, Carlson C, Raab TK, Zayed AM. 1992. Rates of selenium volatilization among crop species. *J. Environ. Qual.* 21:341–44
118. Terry N, Zayed A, Pilon-Smits E, Hansen D. 1995. Can plants solve the selenium problem? In *Proc. 14th Annu. Symp., Curr. Top. Plant Biochem., Physiol. Mol. Biol.: Will Plants Have a Role in Bioremediation?*, Univ. Missouri, Columbia, April 19–22, pp. 63–64
119. Terry N, Zayed AM, de Souza MP, Tarun AS. 2000. Selenium in higher plants. *Annu. Rev. Plant Physiol. Plant Mol. Biol.* 51:401–32
120. Theil EC. 1987. Ferritin: structure, gene regulation, and cellular function in animals, plants and microorganisms. *Annu. Rev. Biochem.* 56:289–315
121. Trapp S, McFarlane C, eds. 1995. Plant Contamination: Modeling and Simulation of Organic Processes. Boca Raton, FL: Lewis. 254 pp.
122. Tsao DT. 2003. Overview of phytotechnologies. *Adv. Biochem. Eng. Biotechnol.* 78:1–50
123. Van der Zaal BJ, Neuteboom LW, Pinas JE, Chardonnens AN, Schat H, et al. 1999. Overexpression of a novel *Arabidopsis* gene related to putative zinc-transporter genes from animals can lead to enhanced zinc resistance and accumulation. *Plant Physiol.* 119:1047–55
124. Van Huysen T, Abdel-Ghany S, Hale KL, LeDuc D, Terry N, Pilon-Smits EAH. 2003. Overexpression of cystathionine-γ-synthase enhances selenium volatilization in *Brassica juncea. Planta* 218:71–78
125. Van Huysen T, Terry N, Pilon-Smits EAH. 2004. Exploring the selenium phytoremediation potential of transgenic *Brassica juncea* overexpressing ATP sulfurylase or cystathionine-γ-synthase. *Intern. J. Phytorem.* 6:111–18
126. Volkering F, Breure AM, Rulkens WH. 1998. Microbiological aspects of surfactant use for biological soil remediation. *Biodegradation* 8:401–17
127. Von Wiren N, Klair S, Bansal S, Briat JF, Khodr H, Shiori T, et al. 1999. Nicotianamine chelates both FeIII and FeII. Implications for metal transport in plants. *Plant Physiol.* 119:1107–14
128. Winnike-McMillan SK, Zhang Q, Davis LC, Erickson LE, Schnoor JL. 2003. Phytoremediation of methyl tertiary-butyl ether. In *Phytoremediation: Transformation and Control of Contaminants*, ed. SC McCutcheon, JL Schnoor, pp. 805–28. New York: Wiley
129. Winter Sydnor ME, Redente EF. 2002. Reclamation of high-elevation, acidic mine waste with organic amendments and topsoil. *J. Environ. Qual.* 31:1528–37
130. Wolf AE, Dietz KJ, Schroder P. 1996. Degradation of glutathione s-conjugates by a carboxypeptidase in the plant vacuole. *FEBS Lett.* 384:31–34

131. Wolfe NL, Hoehamer CF. 2003. Enzymes used by plants and microorganisms to detoxify organic compounds. In *Phytoremediation: Transformation and Control of Contaminants*, ed. SC McCutcheon, JL Schnoor, pp. 159–87. New York: Wiley
132. Wullschleger S, Meinzer F, Vertessy RA. 1998. A review of whole-plant water use studies in trees. *Tree Physiol.* 18:499–512
133. Zayed A, Pilon-Smits E, deSouza M, Lin Z-Q, Terry N. 2000. Remediation of selenium polluted soils and waters by Phytovolatilization. In *Phytoremediation of Contaminated Soil and Water*, ed. N Terry, G Bañuelos, pp. 61–83. Boca Raton: Lewis
134. Zhu Y, Pilon-Smits EAH, Jouanin L, Terry N. 1999. Overexpression of glutathione synthetase in *Brassica juncea* enhances cadmium tolerance and accumulation. *Plant Physiol.* 119:73–79
135. Zhu Y, Pilon-Smits EAH, Tarun A, Weber SU, Jouanin L, Terry N. 1999. Cadmium tolerance and accumulation in Indian mustard is enhanced by overexpressing γ-glutamylcysteine synthetase. *Plant Physiol.* 121:1169–77

Calcium Oxalate in Plants: Formation and Function*

Vincent R. Franceschi[1] and Paul A. Nakata[2]

[1] School of Biological Sciences, Washington State University, Pullman, Washington 99164-4236; email: vfrances@wsu.edu

[2] Department of Pediatrics, USDA/ARS Children's Nutrition Research Center, Baylor College of Medicine, Houston, Texas 77030-2600; email: pnakata@bcm.tmc.edu

Annu. Rev. Plant Biol.
2005. 56:41–71

doi: 10.1146/
annurev.arplant.56.032604.144106

First published online as a
Review in Advance on
January 12, 2005

1543-5008/05/
0602-0041$20.00

Key Words

crystals, idioblast, morphology, oxalic acid, vacuole

Abstract

Calcium oxalate (CaOx) crystals are distributed among all taxonomic levels of photosynthetic organisms from small algae to angiosperms and giant gymnosperms. Accumulation of crystals by these organisms can be substantial. Major functions of CaOx crystal formation in plants include high-capacity calcium (Ca) regulation and protection against herbivory. Ultrastructural and developmental analyses have demonstrated that this biomineralization process is not a simple random physical-chemical precipitation of endogenously synthesized oxalic acid and environmentally derived Ca. Instead, crystals are formed in specific shapes and sizes. Genetic regulation of CaOx formation is indicated by constancy of crystal morphology within species, cell specialization, and the remarkable coordination of crystal growth and cell expansion. Using a variety of approaches, researchers have begun to unravel the exquisite control mechanisms exerted by cells specialized for CaOx formation that include the machinery for uptake and accumulation of Ca, oxalic acid biosynthetic pathways, and regulation of crystal growth.

Contents

Calcium oxalate: a highly insoluble crystalline salt of oxalic acid and calcium

Crystal cells: plant cells that form calcium oxalate crystals; they may be specialized for this (idioblast) or produce crystals in addition to normal functions in photosynthesis, assimilate storage, etc.

Crystal idioblast: a cell that is specialized for production of calcium oxalate crystals.

INTRODUCTION

The ubiquity of calcium oxalate (CaOx) crystals in higher plant families and the observation that they can account for a large amount of the total calcium (Ca) in a plant demonstrates that CaOx crystal formation is a basic and important process in many plant species. Plant CaOx crystals were first described by Leeuwenhoek in the late 1600s using a simple light microscope (as cited by 7), and are now known to occur in most plant families. CaOx has been observed in most plant tissues and organs as an intracellular or extracellular deposit. Intracellular crystals often occur within the vacuoles of cells specialized for crystal formation, called crystal idioblasts (35). Oxalate producing plants, which include many crop plants, accumulate oxalate in the range of 3% to 80% (w/w) of their dry weight (67, 110, 117, 119), with as much as 90% of the total Ca of a plant in the form of CaOx crystals (52, 53). Various functions have been ascribed to plant CaOx crystals and these diverse functions depend on crystal amount, distribution, and morphology as well as features of the cells that produce them. The great diversity in crystal morphologies, tissue, and cell types producing crystals, and specialized developmental features indicates that CaOx crystal formation has evolved numerous times in many plant lineages and with widely disparate functions.

A combination of genetic and environmental factors play a role in defining CaOx crystal amount, shape, and size, and thus function. Knowledge of the processes involved in CaOx formation is relevant to our basic understanding of Ca transport and regulation, oxalic acid biosynthesis, bulk Ca regulation in plant tissues and organs, and specialized defense mechanisms. Studies on CaOx formation and its regulation have also provided insights into the fascinating cellular specialization required for managing large fluxes of Ca across multiple compartments, and for controlling CaOx precipitation so that crystal growth does not cause unwanted damage to cells. Considering the complexity of crystal formation, regulation can occur at a number of steps. In this paper we present our current understanding of the structural, developmental, biochemical, and regulatory aspects of plant CaOx formation and place this knowledge into the context of its functional roles in plant growth and physiology.

CRYSTAL CHEMISTRY AND MORPHOLOGY

Chemistry

The major components of CaOx crystals are simple, but the resulting crystals can be complex in their morphology. Crystals in plants are formed from endogenously synthesized oxalic acid and Ca from the environment. Oxalic acid ($C_2H_2O_4$) is the simplest of the dicarboxylic acids and the most highly oxidized organic compound formed in plants. Oxalic acid is a strong organic acid with dissociation constants of $pK_1 = 1.46$ and $pK_2 = 4.40$. Oxalic acid can complex with Ca to form highly insoluble

Figure 1

Morphology of CaOx crystals. All are scanning electron micrographs. Insets are light micrographs. (*A*) Prismatic crystals in bean seed coat. Arrows indicate kinked twin crystals. (*B*) Crystal sand in sugar beet leaf cells. (*C*) Raphide crystals from a ruptured *Pistia* raphide idioblast. Note barbs on one end of the crystals (*arrows*). Inset shows large raphide idioblast in the leaf relative to much smaller adjacent mesophyll cells. (*D*) Developing druse crystals from *Pistia*. Only a few facets can be seen (*arrows*). (*E*) Isolated *Peperomia* druse crystal. Note the many facets radiating from a central core.

CaOx crystals (solubility product, K_{sp}, at 25°C of 2.32×10^{-9} for the monohydrate) with a striking range of morphologies (**Figure 1**). Recent evidence also demonstrates that plant CaOx crystals show distinct chirality, which is relevant to understanding how the crystals are formed (1, 104). Considerable work has been conducted on CaOx formation in vitro because it represents a major health problem with respect to kidney stone diseases. To form a crystal, the solution has to be in an unstable supersaturated state; however, agents in the environment can act as heterogeneous nucleators to lower the metastable limit and promote crystal formation (85). Various charged compounds, including organic acids, peptides, polysaccharides, glycoproteins, and lipids, have nucleating promoting or inhibiting properties in vitro (19, 20, 31, 73, 98, 99, 103, 150, 156, 157, 167). Adding these compounds to the solution from which CaOx is precipitated can change the physical-chemical dynamics and can affect the rate of formation, hydration state, morphology, and aggregation of crystals. Thus, although on the surface the

Oxalic acid: a strong and the simplest dicarboxylic acid, often thought of as a toxin or end product of metabolism, but which is also synthesized when calcium oxalate crystals are induced to form in plants

chemistry of CaOx precipitation is relatively simple, the addition of organic materials in the biological system complicates our understanding of the precipitation process.

Druse crystal: a multifaceted roughly spherical conglomerate calcium oxalate crystal commonly found as a single crystal, but also can be present as multiple crystals in cells of plants

Raphide crystal: a needle-shaped calcium oxalate crystal, often with grooves along its sides, that occurs as bundles of hundreds to thousands in the vacuole of a cell

Crystal chamber: a membranous sheath within the vacuole of crystal cells that defines the space in which calcium oxalate is precipitated

Morphology

CaOx crystals can occur in a wide range of morphologies (7, 24, 42, 190). The most common morphologies of CaOx crystals in plants are shown in **Figure 1**. These morphologies include block-like rhombohedral or prismatic crystals present as single or multiple crystals per cell, large elongate rectangular styloids that occur as single crystals per cell, bundles of needle-shaped (acicular) raphide crystals, masses of small angular crystals referred to as crystal sand, and multifaceted conglomerate crystals called druses (often single but also multiple per cell). Crystals found in cell walls are of the rhombohedral or prismatic shape, whereas the crystals found within cells can be any of these morphologies. Interpenetrant crystals and twinning (4, 8, 57) can lead to further variations of these basic forms (see **Figure 1**). In addition, crystals are often modified in a way that is not consistent with physical-chemical precipitation dynamics; for example, some crystals may have grooves or barbs, as shown in **Figures 1** and **3** (159). Crystal size can also vary tremendously. This is partly a function of the cell type in which the crystal is formed, the amount of available Ca (12, 123, 183), and other environmental factors (173).

Regulation of Morphology

A particular species will form only a certain crystal type or subset of crystal morphologies. This is significant because it indicates that the cells and the genetics of the particular species forming these crystals control the morphology. Features such as crystal hydration state, ratio of Ca to oxalate, presence of nucleating substances or contaminants (58, 139, 156, 180), and involvement of specialized cellular structures can all play a role in determining crystal morphology (reviewed by 7, 69, 79, 132, 190). CaOx can occur in various hydration states, with monohydrate (whewellite) and dihydrate (weddellite) being the dominant states in plants (3, 42, 49, 72). The hydration state is proposed to regulate morphology through coordination of the Ca and oxalate ions (49) and the hydration state itself can be affected by the presence of other organic acids (77). Hydration state has been determined for many different plant CaOx crystals, and although there is a good correlation between hydration and morphology for some crystal shapes, it is also clear that similar crystal morphologies, such as druses from different species, can be either monohydrate or dihydrate (42, 116, 128–130). It is unlikely that hydration state is the sole determining factor of complex crystal shapes (i.e., raphide and druses) although it may be a determining factor in the simple prismatic and wall crystals.

In vitro studies demonstrate that the relative amounts of Ca and oxalic acid in the solution from which the crystals are formed can affect crystal morphology (27, 81, 198). Unfortunately, it has not been possible to determine the concentrations of soluble oxalate and Ca during crystal formation in plants, particularly in the unique environment of the vacuole, where crystals are precipitated within a separate membrane-bound chamber. However, some plant crystal morphologies, such as raphides, have never been produced in vitro, regardless of relative substrate concentrations, nucleating agents, temperature, and a host of other physical-chemical parameters that have been explored. Thus, cell-directed morphology is implied with such crystal types.

Cellular compartments are almost certainly involved in determining crystal morphology. CaOx crystals are not formed free in solution in plants but are formed in association with cellular structures. With respect to crystal formation, membrane structures are perhaps among the most important. Intracellular crystals, and some extracellular crystals that have been studied, are formed within a membrane-lined "crystal chamber" (7, 42, 190; see Cell Specialization). It has been proposed that this chamber acts as a mold that determines the shape of the crystal. This is unlikely because crystals are not deposited into a chamber of mature size and shape, with the possible exception of prismatic crystals in soybean seed (75). The chamber is not rigid, but rather appears to be initiated as a small structure that grows as the crystal grows. In fact, raphides of *Pistia* grow in a bidirectional manner (91, 94) whereas druse crystals of the same species grow at all facets simultaneously (183). It is more likely that chemical features of the chamber directly affect crystallization dynamics, or that the chamber regulates the relative accumulation of oxalic acid and Ca, and probably other organic materials, into the crystallization space, thus affecting morphology. It is also well established that macromolecules are associated with mature crystals (5, 183, 191, 192). Recently, proteins were isolated from plant crystals and nucleated crystal formation (15, 107). Further studies should help determine the role of these proteins in regulating the growth and shape of the plant CaOx crystals.

DISTRIBUTION

CaOx can be found in all major groups of photosynthetic organisms including algae (50, 87, 101, 147, 148), lower vascular plants, gymnosperms, and angiosperms (7, 22, 23, 29, 42, 53, 110, 126, 201). CaOx crystals are particularly abundant in the angiosperms and gymnosperms, but not all plants produce CaOx. Production of crystals in plants is a normal physiological process. CaOx is also found in animals but in contrast to plants it is most commonly associated with the pathological condition of renal stone disease, although it occurs as a structural element in a few animals and as a potential defense in others (21).

Distribution of crystals within the plant is highly variable among species. CaOx may occur in a single tissue or it may be found in multiple tissues of an individual species. It is found in vegetative, reproductive, storage, and developing organs and also in photosynthetic and nonphotosynthetic tissues. It is striking that there are no generalities about the location where crystals can be formed. This is likely due to the variety of functions for CaOx formation and also the timing of when these functions need to be expressed with respect to development or physiological maturity.

The morphology of crystals produced may be of a single type throughout the plant, or multiple types with each specific for a certain organ, or multiple types that may exist within the same organ but different tissues or regions. Even with this great range of distribution among species, the morphology of crystal produced and its distribution is constant within a species. This constancy of crystal shape and distribution has found use as a taxonomic character (c.f. 102, 142, 145, 146, 205) and indicates tight genetic regulation of crystal deposition.

CaOx crystal deposition can occur within the vacuoles of cells or associated with the cell wall. The cells in which vacuolar crystals are deposited are often specialized for crystal formation and are referred to as crystal idioblasts (35). In some plants, crystals are deposited in the vacuoles of otherwise normal cell types such as storage parenchyma, bundle sheath cells, epidermal cells, or chlorenchyma. The deposition of CaOx into cell walls is common, especially in the gymnosperms, where there are large amounts of crystals (32, 42, 70, 74, 141). Deposition can be in almost any layer of the wall and in most cell types. However, as with intracellular deposits, a particular species will show deposition in selected cell types (74). In angiosperms, wall deposits of CaOx occur in specialized cells such as astrosclereids (7, 42, 70, 96). Ultrastructural observations indicate that, as with intracellular crystal formation, extracellular deposition is a very carefully regulated cellular process (96, 142, 144). Many fungi also produce cell wall CaOx crystals, through both intracellular (6) and extracellular processes (34). Extracellular deposits of CaOx are found in various lichens (34) and are due to secretion of oxalic acid and subsequent precipitation of Ca from the mineral substrate (30, 56). As an interesting aside, fungal and lichen production of CaOx is of considerable interest to art preservationists and archeologists, as it can cause damage to frescoes and stone structures, and ^{14}C ages of CaOx rock coatings have been used for archaeological and paleoclimate studies (10).

The distribution of CaOx among such a diverse group of organisms shows that it is a normal and important biomineralization function in plants, including some of the most ancient lineages. The huge variation in distribution among organs, tissues, cell types, and subcellular location among species suggests multiple independent origins of CaOx formation and its functions, and raises some interesting questions about the evolution of the mechanistic aspects of this process.

FUNCTION

Based on the diversity of crystal shapes and sizes, as well as their prevalence and spatial distribution, there are numerous hypotheses regarding crystal function in plants. These functions include calcium regulation, plant protection, detoxification (e.g., heavy metals or oxalic acid), ion balance, tissue support/plant rigidity, and even light gathering and reflection (40, 42, 165). Although there is a lack of evidence in support of some of these proposed functional roles, solid evidence is accumulating to support functions in calcium regulation, plant protection, and metal detoxification.

Calcium Regulation

There are a number of physiological and biochemical studies that demonstrate a role for CaOx production as part of a high-capacity mechanism for regulating bulk Ca levels in plant tissues and organs. This is necessary in plants that grow in environments where soluble Ca is relatively abundant and restriction of entry in the root zone is poor (39, 121, 183). Ca entry into many plants is related to substrate concentrations and the Ca permeability of the apoplastic pathway to the root xylem, and not necessarily metabolic requirements (88, 193, 194). One of the three recognized "physiotypes" of plants with respect to Ca metabolism is "oxalate plants," which precipitate Ca as crystals of CaOx (194). As water is lost from the surface of the plant through evaporation, the dissolved Ca is left behind and can build up to very high levels over time. Because Ca is involved in signal transduction pathways and in the regulation of other biochemical and cellular processes, careful control of Ca activity within the plant cell is critical. Large amounts of excess Ca can be precipitated as CaOx, which is physiologically and osmotically inactive. Studies in support of the function of CaOx in Ca regulation have shown, using a variety of plants, that the number and size of CaOx crystals are responsive to changes in the concentration of Ca (39, 41, 47, 123, 141, 142, 183, 200, 203). An example is illustrated in **Figure 2**, where a single mature soybean leaf was sampled successively over time. As tissue Ca levels increase in the mature leaf over time, CaOx crystals can be observed accumulating along the veins of the leaf. The only significant variable is the amount of Ca that was transported and accumulated in the leaf over a three-week differential. This pattern of accumulation along the veins is common in numerous species. One interpretation of the association of CaOx with veins is that because the site of entry of Ca is the xylem, precipitation in the cells surrounding the veins will prevent the Ca from accumulating around the chlorenchyma cells, which could affect cellular function (115).

Whereas CaOx is deposited in the mature organs in plants such as soybean, other plants deposit crystals in developing organs. Actively dividing cells and very young cells generally have little or no vacuolar compartment, and thus have a reduced capacity for dealing with Ca by intracellular sequestration. The cell wall has considerable exchange capacity for the binding of ionic Ca, but accumulation of Ca in the apoplast of young developing cells can interfere with the normal process of cell expansion by cross-linking acidic residues of cell wall polymers. Thus, a primary function of crystal idioblast formation in developing tissues may be to serve as a strong localized Ca sink to reduce the apoplastic Ca concentration around adjacent cells, allowing them to develop normally.

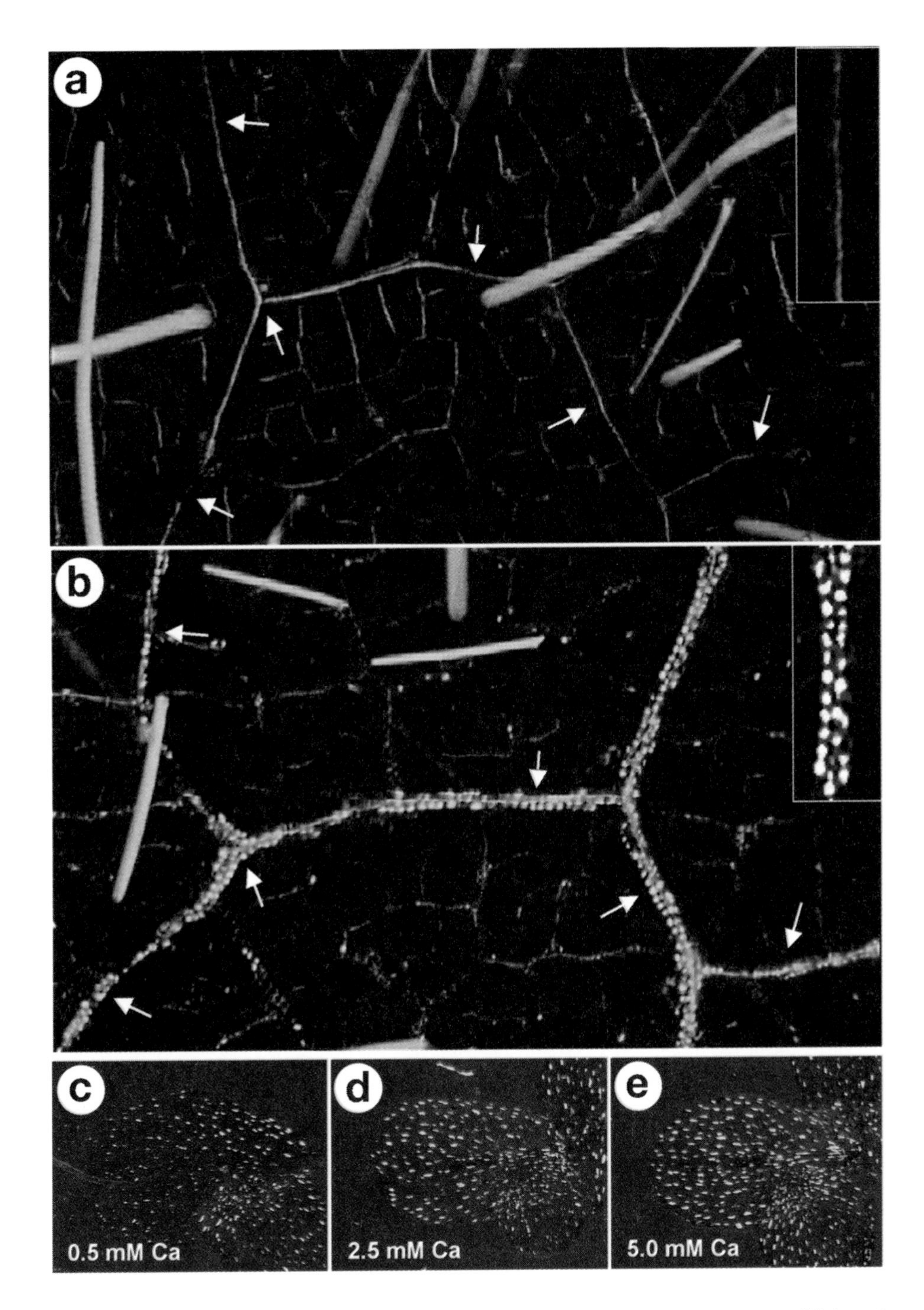

Figure 2

Accumulation of CaOx crystals in plants. (*A*, *B*) Clearings of samples taken from the same mature soybean leaf and viewed with crossed polarizing filters. B was taken 3 weeks after A and shows accumulation of CaOx crystals along the veins. (*C*, *D*, *E*) Clearings of whole *Lemna minor* fronds, demonstrating that the number and size of crystal idioblasts increases with increase of Ca in the medium.

Studies that provide direct support for this interpretation have demonstrated that as Ca in the nutrient solution is increased the number of crystal idioblasts increase, as shown in **Figure 2** for *Lemna minor* (12, 13, 28, 39, 41, 47, 97, 123, 143, 182, 183, 200, 202, 203). In addition, the spacing between the idioblasts decreases, indicating that each idioblast services a certain volume of tissue and as Ca levels increase, the limited capacity of the individual idioblast has to be supplemented by producing more idioblasts. Further evidence for a function in Ca regulation is indicated by the disappearance of such crystals under conditions of Ca deficiency (39, 183), developmental maturation (75, 76, 170), and where new growth is very active and Ca is limited, such as during the Spring flush, as observed in some perennials (9, 17, 18). Volk et al. (183) recently identified, using an oxalate oxidase antibody, a putative enzyme that could account for the degradation of oxalate liberated from crystal dissolution. This antigen was abundant in the crystal idioblast vacuole under plant growth conditions of Ca limitation where crystals were dissolving, but was absent under growth conditions of Ca availability.

There are now several examples illustrating a role of CaOx crystals in bulk Ca regulation in plants. However, this does not preclude other functions for CaOx crystals. In fact, it has been suggested that raphides in *Pistia* perform a dual function of, initially, Ca regulation and, in the mature state, a defense mechanism (183).

Plant Protection

A proposed role for CaOx crystals in protection or defense is based primarily on temporal, spatial, and/or morphological parameters of crystal formation in certain plants (29, 33, 74, 154, 155, 162, 176, 178, 183). CaOx crystals can have a passive mechanical role in defense or an active role as seen in a few well-known examples. One of the clearest and extreme examples of crystals as part of an active deterrent mechanism can be observed in the plant *Tragia ramosa* (177), which is covered with stinging hairs. These stinging hairs consist of an elongated stinging cell containing a large needle-shaped styloid crystal with a groove along one edge and a branched base. When an animal or human brushes up against these hairs, the tip of the cell ruptures, allowing the needle-shaped crystal to puncture the dermis of the animal. A toxin is then channeled along the groove in the crystal to the wound site, where it causes the dermal irritation or stinging sensation. Outbreaks of contact dermatitis have also been linked to CaOx crystals, and particularly raphides, among field workers and processors of the flower-growing (16, 80) and distillery industries (161), where an abundance of CaOx raphide crystals are liberated during the harvesting and processing of plant tissues. The agave juice that is used to make tequila contains as many as 6000 crystals/ml (161). A role of crystals in irritation due to contact with the sap of some common houseplants and other plants is also well documented (16, 54, 80, 95, 131, 151, 159, 160, 164, 174, 186). The crystals are mostly needle-shaped raphides, and in some cases (as seen in **Figure 3*D,E***) the crystal idioblast, which is pressurized, has a thin-walled end that is easily ruptured by contact resulting in the forcible expulsion of the crystals (151, 158). In other cases, the extremely large size

Figure 3

Examples of CaOx crystals as static or active defense structures in plants. (*A*) Cross section through a leaf of *Claoxylon sandwisence*, showing sections along two large styloid crystal idioblasts (*brackets*), which can be viewed as static defenses. Both almost span the entire width of the leaf and the enclosed crystals (C) can be envisioned as potentially wounding the mouth and soft tissues of a grazing animal. The smaller styloid crystal is still developing. (*B*) An example of another type of static crystal defense. These developing raspberry fruits have a distinct layer or sheath of prismatic crystals in the developing seed coat (*arrows*) that would provide a tough physical barrier. The crystals appear bright in this image taken with crossed polarizing filters. (*C*) Vascular bundles in the stem of a *Piper* sp., seen in cross section. The phloem (P) of the two bundles is surrounded by a sheath of druse crystals (*arrows*). The phloem bundle is also protected by a fiber cap at the top and the tough xylem (X) at the bottom. (*D*) A living raphide idioblast in the stem of *Pistia*. Note the tapered tip of the cell, which is much thinner than the general cell wall. (*E*) When the idioblast is subjected to mechanical pressure, the tip is ruptured and the crystals are forcibly expelled. (*F*) Scanning electron micrograph of the *Pistia* raphide crystals showing sharply pointed tips and grooves (*arrows*) along the edges that help channel toxins into wounds created by the crystals.

of crystals such as styloids, which may span the entire cross section of a leaf, can be a structural deterrent against grazing (**Figure 3*A***).

Not only the shape and size of crystals but also the placement and sheer number of crystals may prevent herbivory by large animals as well as insects. For example, collard peccaries, which graze on prickly pear, avoid those populations containing the most crystals (176). In addition, a correlation has been proposed between extensive crystal formation and resistance to bark-boring insects in conifers (74). It is also notable that the periderm in Spruce stems and many other conifers has layers of CaOx-encrusted

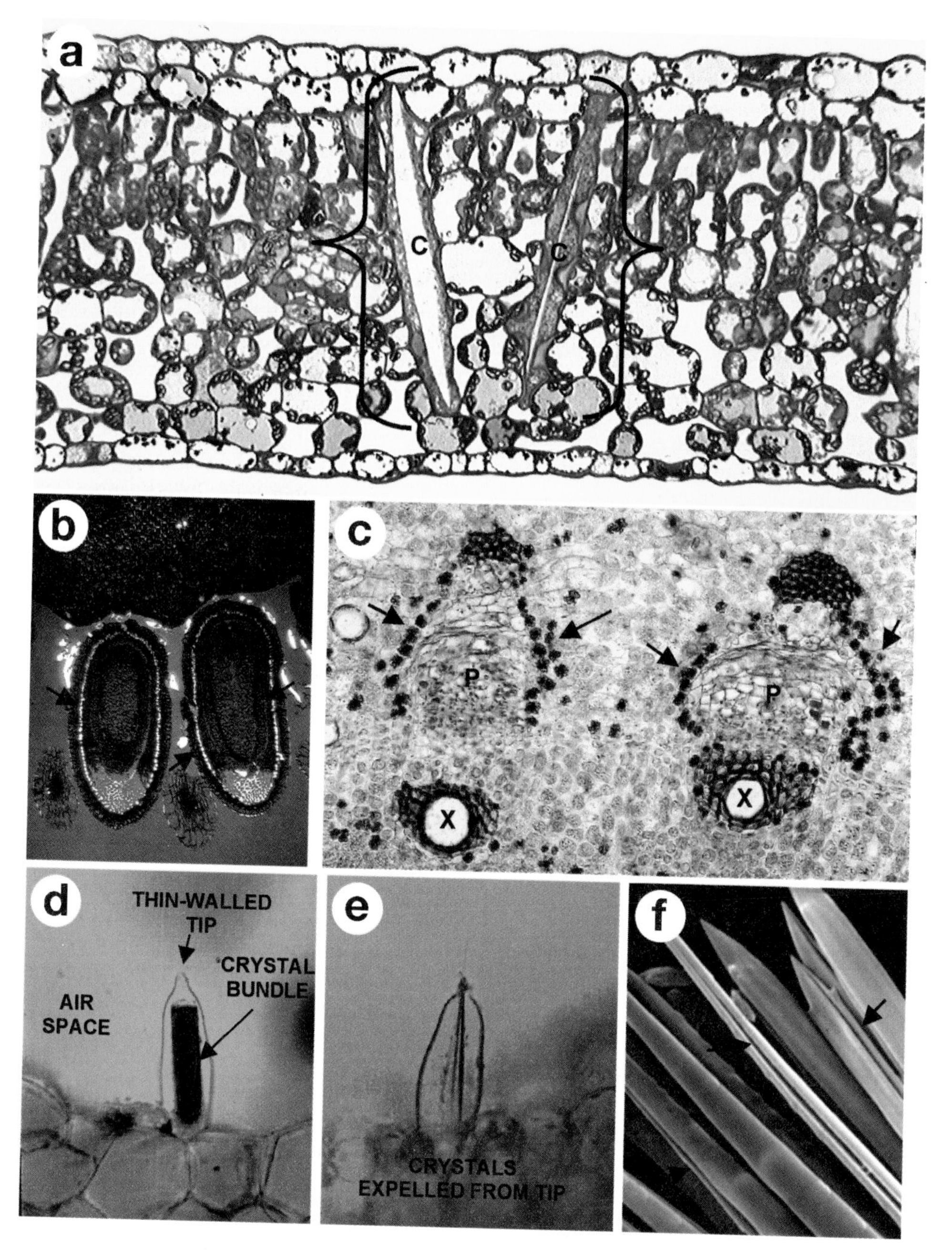

cells that make it more difficult for insects to chew through this outermost protective barrier (43). It is not uncommon to find layers or sheets of crystals surrounding an organ or tissue that form a physical barrier, as seen in the developing seeds in **Figure 3*B*** and the phloem in stem vascular bundles in **Figure 3*C***. They can easily be envisioned as protecting against grazing insects or animals, depending on the relative sizes. Related to this, in a gazelle herbivory study (162, 187), the authors noticed that only the leaf tips of desert lilies were consumed. Upon close examination of the leaves, they found that the tips were the only areas of the leaf devoid of raphide crystals (187). Ruiz et al. (154) compared the amount of CaOx the lily plants accumulated to the amount of herbivory that occurred at a particular location and found a correlation between the higher amounts of crystals and decreased grazing. Clipping or wounding the plant leaves, however, did not result in an increase in CaOx accumulation, indicating that crystal formation was not inducible and that the crystal formation was a developmentally programmed process and may have been under the selective pressure of herbivory (154). In contrast, CaOx crystal accumulation increased in leaves of *Sida rhombilfolia* (127) and seeds of Norway spruce (178) in response to artificial herbivory or tissue wounding, respectively. Thus, in some plants, CaOx formation appears to be an inducible defense response.

Oxalate itself, in the soluble form, can act as a toxic compound with consequences on grazing animals and sucking insects such as aphids and leaf hoppers (120, 197). For example, aversion to oxalate-containing plants was demonstrated with goats (51). Ingestion of soluble oxalates may interfere with Ca metabolism and can lead to CaOx precipitation in tissues and organs of the feeding organism. Ingestion of plants with high oxalic acid levels by grazing animals can have severe toxic consequences often attributable to renal failure (78, 184). From the 1940s to the 1970s the oxalate-accumulating plant *Halogeton* was considered one of the most economically destructive poisonous plants because consumption of large amounts of this plant during grazing resulted in significant losses of sheep and cattle. The cause of death was attributable to CaOx crystal formation in rumen walls, arteries, and kidneys (kidney stones). This problem can be extended to humans in the sense that oxalate load from consumed plant foods can have a similar effect if it is at a high enough level or if there is a genetic predisposition for oxalate kidney stone disease (66, 105, 117, 119).

Detoxification of Aluminum (and Other Heavy Metals)

In acid soils (approximately 40% of the world's arable land), aluminum (Al) toxicity is a major problem limiting crop production (36). Micromolar concentrations of Al can inhibit root growth and affect the acquisition of nutrients and water. This reduction in root function is a major issue in the production of many agriculturally important crop species (90, 114, 163). Some species of plants, however, can still thrive in Al-contaminated soils. Investigations into why these plants are resistant to Al toxicity revealed their use of organic acids such as citrate, malate, and/or oxalate. There are two different mechanisms of oxalate utilization for Al tolerance, known as the exclusion and internal mechanisms. The exclusion mechanism involves the excretion of oxalate into the environment by the roots and occurs in response to external Al stress (113). The internal tolerance mechanism involves the sequestration of Al in the nontoxic form of Al-oxalate within the aerial portion of the plant (112), and Al is also associated with CaOx crystals (121). Some plants, such as buckwheat, use both mechanisms. Oxalate may also be involved in the detoxification of other hazardous metals such as lead (122, 196), strontium (46, 122), cadmium (26, 122), and copper (122). In these cases the metal is incorporated into the oxalate crystals.

Oxalates and Human Health

Oxalate in plant foods can have a negative impact on human health (64, 110, 118, 138) by acting as an antinutrient, as a toxin, and in CaOx kidney stone formation, which is on the rise in humans (169). Recent studies have indicated that dietary oxalate contributes to increased urinary oxalate excretion to a much greater extent than previously thought (65, 66). Plant foods are the main source of dietary oxalate and thus efforts have been initiated to decrease the oxalate content in these foods. From a nutritional standpoint oxalate is also considered an antinutrient in that it renders Ca, and sometimes other minerals, unavailable for nutritional absorption. This issue of Ca bioavailability is important when one considers the reliance of different populations around the world on plant foods as their main source of Ca, as well as the failure of many in the United States to meet the recommended daily allowance (RDA) for Ca. Numerous studies using in vitro (11, 14, 82, 125, 195), rodent (189), and human (60, 61, 63, 188) systems have shown that Ca bound as the oxalate salt has reduced absorption. For example, in humans, Ca absorption from the high-oxalate vegetable spinach was only 5% (63) whereas absorption from the low-oxalate vegetable kale was 41% (61). However, the poor solubility of CaOx cannot account entirely for the poor absorption of Ca in all instances. Soybeans contain substantial amounts of CaOx yet have relatively good Ca bioavailability (62). There is also a report where Ca absorption from in vitro–generated CaOx crystals gave higher than expected values (60). Thus, a study using a single plant system that differs in only oxalate content is needed to clarify this issue by eliminating the variables introduced through the use of different plants. Such a system was recently identified and should allow direct assignment of Ca bioavailability to the presence or absence of oxalate (136).

Efforts continue in identifying germplasms (118), plant growth conditions (2, 153, 171), and breeding practices (108) that produce crops with lower oxalate levels in the edible tissues. Examples of some plant foods that contain substantial amounts of oxalate include spinach, peppers, rhubarb, nuts, plantains, dry beans, and that perennial favorite, chocolate. Recent evidence indicates that a genetic manipulation strategy to reduce the oxalate content in plant foods may be another feasible alternative (136). Thus, a better understanding of the mechanisms regulating CaOx formation in plants could lead to more targeted strategies to improve the nutritional composition of plant foods with respect to oxalate content.

OXALATE BIOSYNTHESIS

Accumulation of Ca and oxalic acid at the site of precipitation must be carefully coordinated to control crystal growth. Ca uptake from the substrate in which the plant is growing can be regulated by processes in the roots, or by the activity of pumps and channels in the crystal-producing cell. In contrast, the plant synthesizes oxalic acid and must coordinate the rate of synthesis with Ca accumulation and crystal development (106). There are a number of potential pathways for oxalic acid biosynthesis in plants (45). This organic acid can be formed through the oxidation of glycolate and glyoxylate by the activity of glycolate oxidase (152, 166). These potential substrates can be formed as a byproduct of photorespiration in photosynthetically active tissues, and glycolate oxidase, a peroxisomal enzyme, is fairly abundant in green tissues. Oxalate can also be produced by the activity of isocitrate lyase on isocitrate (55), and through oxidation of oxaloacetate, although the enzyme is not known and this has only been shown in a few tissues (59). Finally, L-ascorbic acid is a substrate for oxalate synthesis in a number of plant species (reviewed in 45, 111, 168), but the enzyme(s) responsible for this pathway has not been identified.

Which pathways are involved in oxalate synthesis used for crystal formation? The oxaloacetate pathway is unlikely because it has only been found in a few species, and the isocitrate pathway is also considered minor. The glycolate pathway is a strong possibility because it is common in tissues exhibiting C_3 photosynthesis. However, oxalate and CaOx are formed in abundance in many C_4 species (201), as well as in nonphotosynthetic tissues and organs. In addition, many crystal-forming cells lack photorespiration because they have weakly developed or no chloroplasts, and in fact often have specialized plastids called "crystalloplasts" (7, 42). Further, in two raphide-producing systems the crystal idioblasts did not have the enzyme glycolate oxidase (84, 106), and other studies have indicated that crystal formation is not directly associated with photorespiration (199, 204). So, at least in some systems, the glycolate pathway will not likely operate as a major source for oxalate synthesis used for crystal precipitation. Ascorbic acid has the potential to be a good substrate for oxalic acid formation because it is present in relatively high levels and is the most abundant of the possible precursor substrates. It is also found in all tissues, both green and nongreen (111, 168), which is also true for CaOx distribution. Early evidence for a role of ascorbic acid in crystal formation was indicated by studies that found ascorbic acid in the culture medium enhanced CaOx formation in a tissue culture system (41).

When considering oxalic acid biosynthesis it is worth noting that plants may accumulate large amounts of soluble oxalate, large amounts of CaOx, or both. The operational pathway for the production of oxalic acid used for crystal formation may not necessarily be the same as that used for accumulated soluble oxalate synthesis. Radiolabeling and microautoradiography can be used to determine pathways involved in oxalate production and in crystal formation. Radiolabeling studies have shown that in *Lemna minor* (39), *Yucca torreyii* (68), and *Pistia stratiotes* (86, 94), glycolate or glyoxylate are relatively poor precursors for synthesis of oxalate for CaOx crystal formation. All of these studies found that L-ascorbic acid gave rise to oxalic acid that was subsequently used for crystal formation. Using ^{14}C-labeled L-ascorbic acid the 1 and 2 carbons (38, 68, 86, 94) gave rise to the oxalic acid used for crystal formation. The 5-carbon ascorbate analog, erythorbic acid, and precursors of ascorbic acid labeled in the 1 carbon also gave rise to oxalic acid that was used for crystal formation (86, 94). These data strongly support a role of ascorbic acid as the primary substrate for oxalic acid synthesis used in crystal idioblasts for CaOx precipitation.

To gain a full understanding of this biomineralization process it is also important to determine the location of oxalate synthesis relative to crystal formation. As a strong dicarboxylic acid and chelator of Ca, accumulation of oxalate in the cytosol is unlikely. Therefore, the large-scale transfer of oxalic acid from adjoining cells is not considered a viable mechanism, even though plasmodesmata are present between idioblasts and adjoining cells (106). There is evidence indicating that oxalate synthesis from ascorbic acid occurs within individual crystal idioblasts. Crystal idioblasts in the few systems that have been studied lack glycolate oxidase, but some have L-galactono-gamma-lactone dehydrogenase, the enzyme responsible for the last step of ascorbate synthesis (93). Finally, Kostman et al. (94) demonstrated that isolated raphide crystal idioblasts from *Pistia* are capable of oxalic acid biosynthesis for use in crystal formation from ascorbic acid or its precursors. Related to this is the observation that increasing the Ca level often induces an increase in the amount of crystals formed in plants (as cited above). This finding suggests that the process of oxalate synthesis is driven or regulated by Ca levels and that crystal formation is not the end product of an attempt to sequester excess oxalic acid, as previously proposed (100). The picture that is developing is a semiautonomous nature of the crystal idioblasts with respect to synthesis and accumulation of the substrates needed for CaOx formation.

CELL SPECIALIZATIONS

CaOx crystal formation is a carefully regulated process, which is reflected in the features of the cells that form these crystals. We will focus on the crystal idioblasts that form internal crystals because they provide such an interesting model of the processes of mass Ca accumulation in a living plant cell. The ultrastructure of crystal idioblasts has been studied in a number of different plants and for a number of crystal types. Examining the cells revealed that they are specialized with respect to organelle amount and contain some unique features not seen in other cell types. Commonly observed features of crystal idioblasts include a very dense cytoplasm with an abundance of ER, golgi bodies, and small vesicles, presence of unusual plastids, and unique membrane structures associated with the crystals in the vacuole. Some of these features, such as the abundant golgi and mitochondria, are consistent with a very active, rapidly growing cell. Other ultrastructural features can be associated with processes that are necessary for CaOx formation within the cells. One example is the notable enrichment in ER (**Figure 4**). Although this ER may be linked to high metabolic activity, as indicated by rRNA content and rapid growth of the idioblasts (91), it is also proposed that the ER plays an important role in regulating cytosolic Ca activity through the buffering capacity of luminal proteins such as calreticulin. Calreticulin (133) and Ca (92, 123) accumulate in specialized subcompartments of the ER of raphide crystal idioblasts. The necessity for enhanced Ca buffering capacity is exemplified by the very rapid crystal formation, where under extreme conditions of elevated Ca the idioblast can be filled with crystals within one hour (39). This indicates the potential for large Ca fluxes from the apoplast to the cytosol. Thus, it is hypothesized that the abundance of ER is necessary to buffer this Ca flux (44, 92, 133). The large amount of ER provides a very large surface area for efficient uptake of Ca from the cytosol, and the enrichment of calreticulin provides enhanced capacity for buffering the Ca that is transported into the ER lumen.

One of the most unique and intriguing features of crystal idioblasts are the membranes that are associated with the developing and mature crystals. Intracellular crystals that have been examined, regardless of morphology, are produced in association with intravacuolar "membranes" that form a chamber or sheath around the crystal (7, 25, 48, 69, 83, 84, 100, 143, 146, 158, 179, 181, 185, 192). The chambers are formed de novo within the vacuole and are not an elaboration of the tonoplast (37, 69–71, 123, 183). During active crystal growth, addition of material to the surfaces of the chamber or elaborations of the surface occurred, as illustrated in **Figure 4** (39, 123, 183). The nature of the membranes has not been determined. In some species they clearly have a unit membrane structure whereas in others they look more like a protein precipitation layer (see reviews by 7, 70). Because they define the crystallization space their activity must affect CaOx precipitation. The chamber membrane could regulate crystal morphology by controlling the relative rate of transfer of Ca and oxalic acid into the crystallization space, and thus the ratio of these two ions, which can affect crystal shape (49). The mechanisms by which the ions are transferred across the membrane are not known. Energy-requiring transporters, or channels, could be used, but they might not be absolutely necessary. The chemical makeup of the membrane alone could provide for different reflection coefficients for the two ions, with the crystallization process providing the diffusion gradient to drive uptake across the membrane. Other potential mechanisms for transfer of Ca and oxalate to the chambers are suggested by images of developing crystals. For instance, images indicating fusion of small membrane vesicles with chambers have been published (37, 123), and recently it was shown that the chamber surfaces of developing but not mature druse crystals are covered with peg-like microstructures (**Figure 4**) that might channel ions to selected facets of the complex crystal conglomerate (183).

Additional vacuolar components such as nucleating agents are also likely involved in regulating crystal growth and shape. There is ample evidence for the presence of noncrystalline materials, called matrix materials, within the crystal structure (5, 7, 183, 191). Demineralizing isolated crystals leaves behind a matrix ghost, generally in the shape of the original crystal (107). This matrix promotes the formation of crystals of a similar form to the original from solutions of Ca and oxalic acid (107, 192). Part of the matrix is made of proteins, which are associated with isolated crystals from a number

ER: endoplasmic reticulum

Matrix protein: a protein that is proposed to play a role in nucleation or regulating calcium oxalate crystal growth and is incorporated into developing crystals to become a permanent component of the mature crystal

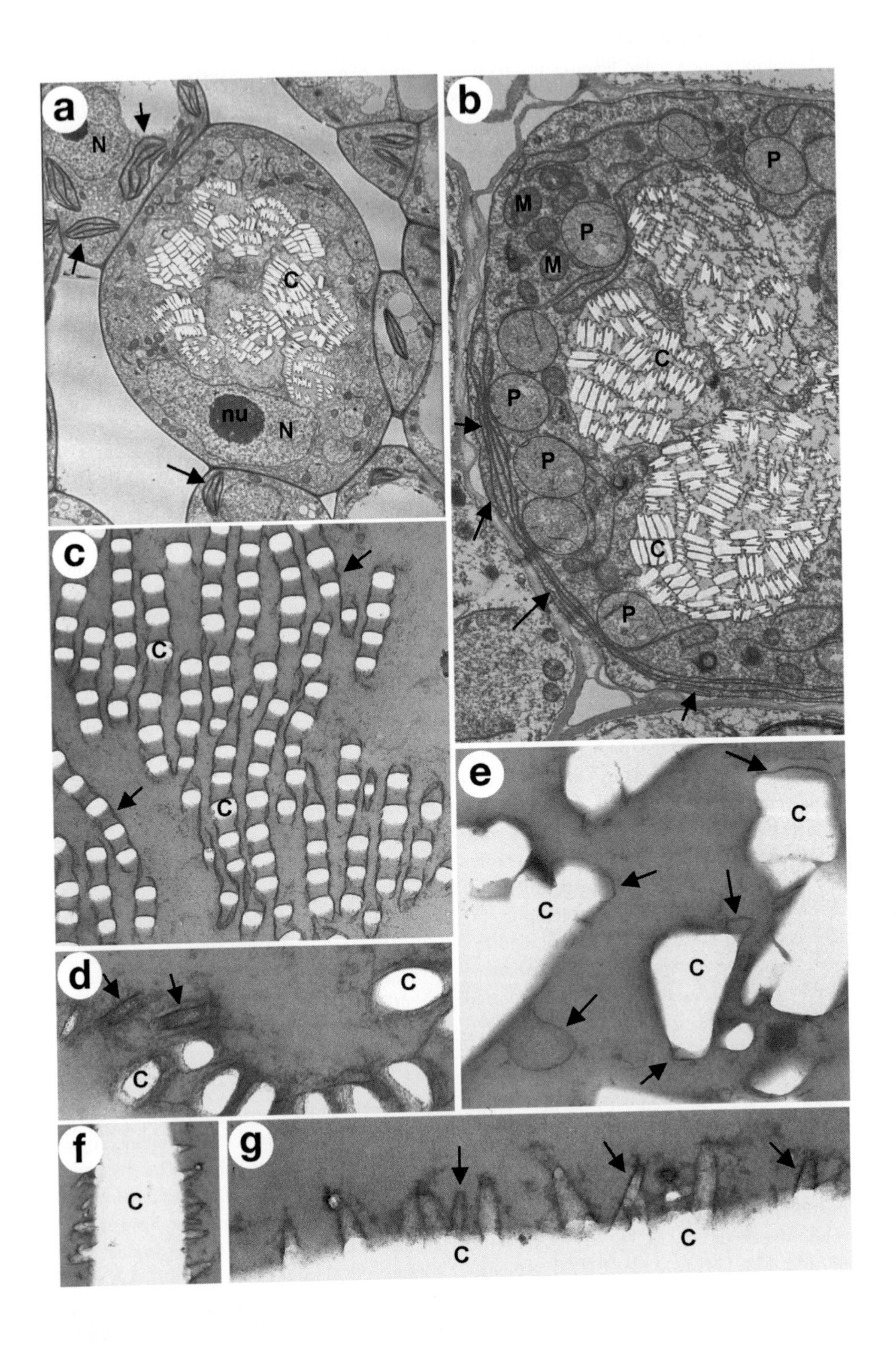
a
N
C
nu
N
b
P
M
P
M
C
P
P
C
P
c
C
C
d
C
C
e
C
C
C
f
C
g
C
C

of plants species (15, 107). Currently only one CaOx matrix protein has been isolated from crystals and partially characterized (107). This novel protein is specific to crystal idioblasts and makes up part of the matrix within the crystals. It has a considerable amount of acidic amino acid residues and strong Ca-binding properties. Further characterization of matrix proteins, or proteins associated with crystals or their surfaces, are important for understanding how crystal nucleation and growth is regulated within the chamber, and how morphology is determined.

Another interesting feature of many crystal idioblasts is the presence of plastids with an unusual structure (**Figure 4**). These generally lack thylakoids and grana, do not contain Rubisco (106), and are nonphotosynthetic, even when the cells next to them can be normal photosynthetic parenchyma. They are referred to as crystalloplastids (7). What is the function of these modified plastids? One can only speculate, but based on the extensive membrane systems formed in the cells, one important function is probably lipid biosynthesis. It is also possible that they are involved in the synthesis of oxalic acid. These modified plastids are abundant in idioblasts of species where ascorbic acid is a precursor of oxalate, and they could be a potential site of the cleavage in the conversion of ascorbate into oxalate.

The overall dynamics in play in forming a crystal idioblast and producing crystals with a defined shape are not simple and go well beyond transporting Ca and oxalate into the vacuole and allowing them to precipitate. Most crystal idioblasts show a considerable increase in size during crystal formation (**Figure 1**), presumably to increase the size of the sink for Ca accumulation while minimizing the number of cells that have to be modified for this process. Thus crystal growth and cell growth are tightly coordinated. This coordination is definitively shown in studies using colchicine to disrupt microtubules, and cell shape, during development in *Pistia stratiotes*. *Pistia* forms raphide crystals in elongated, football-shaped cells in the aerenchyma and druses in spherical cells in the compact chlorenchyma (91). In the presence of colchicine, druse and raphide idioblasts were still formed in their respective positions, but the raphide idioblasts had very irregular shapes. Even though the shape of the cell changed, often to a more spherical form, raphides, rather than druses, were still produced. The only change in the raphides was in length so that they never punctured the vacuoles of the much-shortened cells. This study demonstrates that crystal morphology is determined by mechanisms operating in the vacuoles, and that there are other mechanisms that ensure strict coordination between crystal growth and cell growth and expansion. Generally, crystal idioblasts reach a certain maximum size and crystal growth is terminated and the mature cell is a living deposit of mineralized Ca.

Figure 4

Specialized cellular features of crystal idioblasts. *A–D*, *F*, and *G* are from *Pistia stratiotes*. (*A*) Young raphide idioblast. Surrounding cells are forming chloroplasts (*arrows*) while the idioblast plastids are not developing thylakoids. Note enlarged nucleus and nucleolus, and bundle of crystals cut in cross section (*white rectangles*). (*B*) Idioblast fixed with $KMnO_4$ to emphasize membranes. Note the long endoplasmic reticulum (*arrows*), large crystalloplastids (P), and crystals (C). The H shape of some crystals is due to the groove along the sides. (*C*) Cross section through raphide crystals joined into files by parallel membranes. (*D*) Developing raphide crystals showing dark line of chamber (*arrow*) and matrix within the crystallization space. Older crystals appear as white profiles (C), where the crystal has been removed from the section. (*E*) Section through developing crystal sand in sugar beet leaf. Note the crystal chamber membrane (*arrows*), which sometimes appears as a protrusion from the surface. (*F*) Section through a facet of a growing druse crystal showing multiple projections on both surfaces. (*G*) Enlargement of a region of the facet in F, illustrating the surface of the growing druse crystal is covered with membrane-lined projections (*arrows*). These do not occur on mature crystals.

Without this coupling, crystals could grow too quickly or continue after cell growth ceases and disrupt the vacuole, which would then kill the cell. To our knowledge, this occurs only in one case where, as part of the development of stems in *Myriophyllum spicatum*, large druse crystals are produced in the end of a cell projecting into an air chamber and eventually grow large enough that they destroy the cell (41). This is a good illustration of the consequences of not coordinating cell and crystal growth.

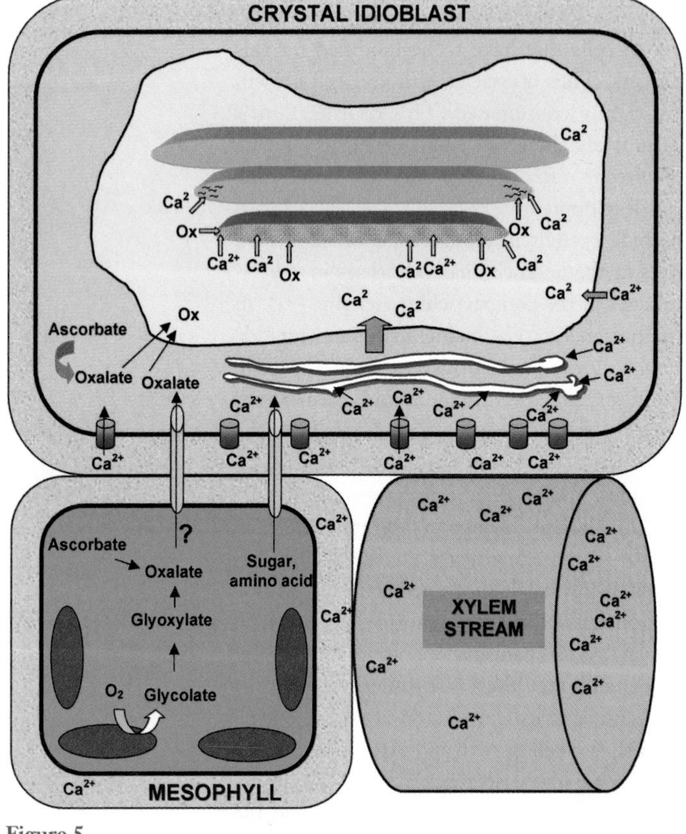

Figure 5

A simplified model of CaOx formation in crystals idioblasts. Calcium entering an organ, such as a leaf, with the xylem stream is distributed among cells via the apoplast (wall). Most cells regulate Ca levels by pumping it back out or by compartmentation, but crystal idioblasts accumulate large amounts of Ca via the activity of channels or pumps. Ca in the cytosol is taken up by the extensive endoplasmic reticulum and buffered by the activity of calreticulin. Oxalic acid can be synthesized in noncrystal cells; however, transport of this strong acid and chelator of Ca in the cytoplasm and via plasmodesmata is not likely, at least at levels that can support rapid growth of crystals. Oxalate can be generated from ascorbate in crystal idioblasts and transferred to the vacuole along with Ca. In this raphide crystal model, Ca and oxalate are transferred across the crystal chamber membrane and added along all the surfaces of young crystals, but as the crystal grows, they are added primarily to the ends. Mature crystals no longer add Ca and oxalate, although the cells are still living. Matrix proteins in the growing parts of the crystals may regulate precipitation or shape. Evidence for Ca channels, ER Ca accumulation, oxalate synthesis, matrix proteins, and the growth dynamics in crystal idioblasts has been published. Mechanisms for transfer of Ca and oxalate to the vacuole and into the crystals have not been identified.

CRYSTAL FORMATION: PUTTING THE PICTURE TOGETHER

Using the various and extensive information on crystal idioblast structure and development, and the biochemical and cell biology studies on *Pistia* raphide idioblasts, a summary of the processes and mechanisms operating in CaOx crystal formation can be composed. The process of CaOx formation requires coordinating a number of different events, as outlined in the model in **Figure 5**. These include uptake of Ca from the apoplast, transfer from the cytosol to the vacuole, and then transfer into the crystal chamber. Simultaneously, oxalic acid must be synthesized in the cytoplasm and transferred to the vacuole and chamber, but at a rate consistent with Ca delivery and crystal growth of a prescribed morphology. In many crystal idioblasts, cell growth occurs at the same time, and thus another level of control involves coordinating cell expansion and crystal growth, as well as directing cell expansion because different crystal morphologies require different cell shapes to accommodate them (see **Figures 1** and **3**).

Most of the physiological and biochemical studies on CaOx crystal idioblasts are concerned with Ca. There is an abundance of information that shows that crystal idioblast and crystal formation are Ca inducible. As Ca is increased in the medium, the amount of Ca in the apoplast will also increase and this is what elicits initiation of idioblast differentiation. Cells that are induced to begin differentiation will switch to a Ca-uptake mode from the general mode of preventing the accumulation of Ca by pumping Ca back out to the apoplast along with some intracellular compartmentation. As expected, idioblasts have a greater capacity for Ca uptake compared to adjacent nonidioblast cells (182).

There is good evidence to show that Ca uptake from the apoplast is mediated by activity of Ca channels, which would be the most energetically efficient mechanisms because they utilize the large inward directed electrochemical potential gradient for Ca. In support of this, Ca channel blockers inhibit CaOx crystal formation (39, 182). In addition, Volk et al. (182) found that fluorescently labeled channel blockers give rise to heavy labeling in developing idioblasts, and that a putative Ca channel protein was localized to developing crystal idioblasts. Most of the evidence indicates that a dihydropyridine type of Ca channel is operating in the crystal idioblasts, and that both plasma membrane and ER channels may be relatively abundant when compared to adjacent cells (182).

When Ca enters the developing idioblast its activity must be carefully regulated to avoid disrupting cell function. The abundance of ER accomplishes this by providing a large interface to the cytosol for rapid accumulation of the Ca. Ca channels may be involved in this but Ca ATPases, which operate in the ER, are also a likely mechanism for accumulation. Once in the ER lumen, the Ca is buffered through association with high-capacity Ca-binding proteins such as calreticulin. Crystal idioblasts of *Pistia* are enriched in this protein, and the ER even produces specialized subdomains where calreticulin and Ca accumulate (92, 133, 149). The process by which the Ca is transferred to the vacuole has not been determined. It is likely that some Ca is directly transported into the vacuole from the cytosol, and there is evidence that some Ca is transported via vesicle fusion to the vacuole during delivery of matrix protein (107, 123).

From the vacuolar solution, the Ca is transferred to the growing faces of the crystals (91). The simplest mechanism to accomplish this would be a mass transfer driven by a diffusion gradient generated by the precipitation of Ca at the surface of the crystals. Alternatively, the crystal chamber may have transporters that actively transfer Ca into the crystallization space. Because it is very difficult to isolate the chambers, little is known about their composition, activity, and actual role in crystal development. This remains an important area requiring intensive investigation, as it may be a key element in determining crystal morphology. It appears that surfaces of the crystals can become "poisoned" so that growth is in a preferential direction, as with elongate raphides and styloids, and this may also lead to shutting down growth of the crystals in mature idioblasts. This is supported by studies showing that mature crystals do not add Ca to their surfaces (91) nor are they easily solubilized during Ca starvation. In contrast, young developing crystals are capable of both of these processes (39, 183).

As Ca accumulates, oxalic acid must be synthesized and transported to the vacuole. Ascorbic acid is a major substrate for oxalate synthesis used for crystal formation and this biosynthetic pathway is present within developing idioblasts (94). The subcellular site of synthesis is unknown and could be either in the cytoplasm or vacuole, although it is more difficult to envision careful regulation of production in the vacuole. If synthesis were in the cytosol, oxalic acid would need to be rapidly transported to the vacuole to prevent accumulation of this strong acid in the cytoplasm. Channels and transporters for carboxylic acids are well known and it is proposed that a dicarboxylic or oxalate transporter is present in the idioblast tonoplast. Once in the vacuole, mechanisms for transfer to the crystal surface for precipitation presumably follow similar possibilities, as discussed for Ca.

Rate and direction of growth and crystal morphology are supposedly controlled by the chamber and included matrix materials. The role of the noncrystalline matrix in regulating crystal formation is only now being elucidated. Matrix protein components that are isolated from crystals are capable of promoting crystal precipitation and thus the matrix may act as the initial nucleating substance. Because the mature crystals have matrix protein incorporated throughout their structure (107), it may also have a function in directing growth or morphology of the crystals. This matrix is made up of a complex of different proteins and possibly other macromolecules, and until they are identified and characterized, we cannot fully understand how crystal growth is regulated.

As the biochemical mechanisms supporting crystal formation proceed, the process of crystal growth has to be coordinated with cell growth. Because crystals such as raphides and styloids can exhibit extensive elongation growth, the idioblast must undergo similarly extensive growth in a plane that mirrors what is occurring with the crystals. What is difficult to understand is how the process of cell elongation, known to be regulated by cytoskeleton arrangement (91), can be coupled to a vectorial growth process occurring in the vacuole that lacks any connection to the cytoskeleton. The initial disposition and orientation of the crystal chambers can be invoked as part of this mechanism, but how the chambers are oriented in the correct axis to begin with is difficult to envision. For raphide crystals, the parallel membranes that define files of crystals may be involved (see **Figure 4**), but such potential orienting structures have not been shown in other crystal types, such as styloids. This coordinated growth phenomenon is yet another reminder that CaOx formation is not a simple precipitation event.

GENETICS AND REGULATION

Considering the conservation of crystal type and location in various species, and the complex process of deposition, it has long been suspected that CaOx formation is under strict genetic control. One can hypothesize that major regulatory points may include oxalate synthesis, Ca uptake by the plant and accumulation by idioblasts, and crystal precipitation and growth. A few studies on plant cultivars suggest a genetic component to both oxalate accumulation and CaOx formation (89, 108, 109, 118), although the complexity of these traits has not been carefully evaluated. Genetic variation in oxalate content among various cultivars has been reported (89, 108, 109, 118, 172, 175). Particularly important to our understanding of crystal formation are recent genetic studies that have confirmed a genetic component of CaOx crystal morphology and formation (134). Mutant screens (124, 134) of an EMS (ethyl methyl sulfonate) population (140) of the model legume, *Medicago truncatula*, have led to the identification of several classes of mutants that display various crystal phenotypes. With the exception of roots, *M. truncatula* plants accumulate prismatic crystals along the vascular strands in all the different plant tissues. The mutant phenotypes include alterations in crystal nucleation, morphology, distribution, and/or amount (**Figure 6**). Although the specific genes that have been altered are not yet known, it is evident from the number of complementation groups that the control of crystal formation is complex and under strict genetic control (124, 134). As suggested by studies in other systems, mutations affecting protein, lipid, and polysaccharide function could all contribute to alterations in crystal size and shape. A genetic approach offers an additional tool to dissect the components involved in the regulation of crystal deposition and morphology as well as other critical features of this biomineralization process. A discussion of some of the results from this ongoing study are presented because they directly address the genetic components of crystal formation and offer a system that will enable the identification of various mechanisms required for the process of crystal development and control of crystal morphology.

Crystal Morphology Mutants

Isolating and characterizing a number of different crystal morphology defective (*cmd*) mutants revealed that a single-point mutation can drastically alter the size and morphology of the crystal (**Figure 6A**). Visual inspection of the different crystal morphologies that are present in various legumes to those present in the *cmd* population revealed striking similarities. Some of these mutations result in crystal shapes virtually identical to those observed in other legumes but not in wild-type *M. truncatula* (124). Based on these comparisons, over the course of evolution, such point mutations can account for, at least in part, the variations in crystal morphologies

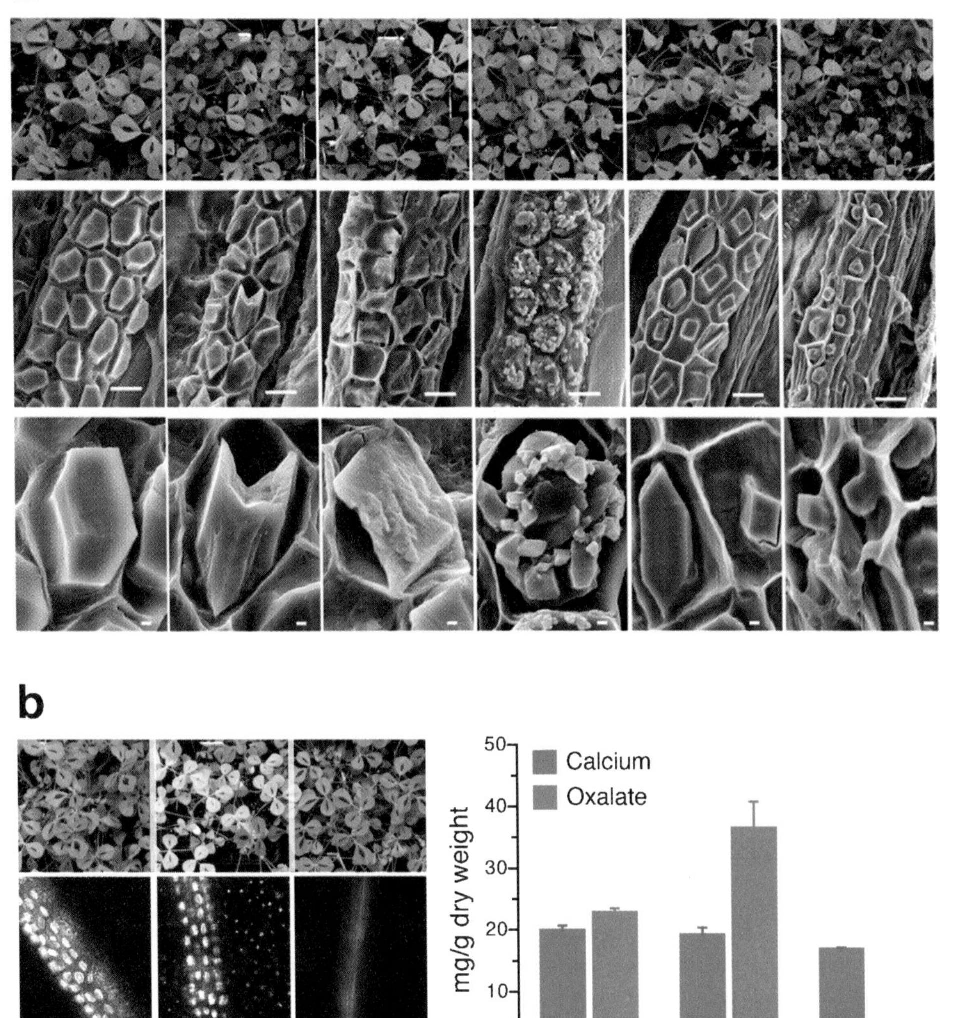

Figure 6

CaOx crystal mutants. (*A*) Plant and crystal phenotype comparisons of five ethyl methyl sulfonate (EMS)-generated crystal morphology *Medicago truncatula* mutants to wild type (*far left*). Note the differences in the crystal morphologies. All crystal images are scanning electron micrographs. Bar = 10 μm (*center panels*); Bar = 1 μm (*lower panels*). (*B*) Comparison of plant phenotypes (*left*) and calcium and oxalate content in mg/g dry weight (*right*) of two EMS-generated crystal content mutants compared to wild-type *Medicago truncatula*. Note the differences in the crystal content correlate with the differences in oxalate content. Crystal images are of whole-leaf clearings viewed between crossed polarizers. Bar = 10 μm.

commonly observed in these legumes today. Further characterization of these mutants should provide new insights into the regulation of crystal morphology and the genes involved.

Crystal Content Mutants

Several mutants with differences in the amount of CaOx accumulated were isolated from a mutagenized *M. truncatula* population (134). Two of the more extreme mutants in this regard were calcium oxalate defective 4 (*cod4*) and *cod5* (**Figure 6*B***). Compared to controls, *cod4* (137) accumulates more calcium oxalate and *cod5* (136) accumulates less. Measurements of oxalate content in the different *cod4* tissues showed elevated tissue oxalate levels as a result of druse crystal accumulation. The Ca content, however, did not increase in proportion to the increase in oxalate. Thus, more of the tissue Ca was partitioned into the crystalline form in the leaves and stems from *cod4* than from controls. The *cod4* mutation also resulted in an overall reduction in plant growth and chlorophyll content.

No prismatic crystals were detected in any of the different tissues of the *cod5* mutant, including leaves (**Figure 6*B***). Oxalate assays supported the observed crystal phenotype by showing low-oxalate levels compared to those of controls. Although compromised in its ability to accumulate prismatic crystals of CaOx, *cod5* exhibited growth that was similar to that of controls. Moreover, *cod5* and controls contained similar amounts of minerals, including Ca. These findings suggest that, in the case of *Medicago truncatula*, CaOx crystal formation is not essential for plant growth or development under greenhouse conditions.

Studies have been initiated to identify the molecular components involved in CaOx formation (135). The ability to genetically manipulate CaOx formation could prove fruitful in the development of strategies to improve the production and nutritional quality of plant foods. The rational design of such strategies would benefit from an understanding of the genetic mechanisms controlling CaOx crystal formation as well as its functional role in plants. The identified mutants in CaOx formation should prove useful in this endeavor.

CONCLUDING REMARKS

CaOx crystal formation is a fundamental part of the physiology of many plant species, and the taxonomic distribution among photosynthetic organisms indicates it is an ancient trait. Plant CaOx production appears to serve a primary function in bulk calcium regulation in tissues and organs and additional roles in defense and heavy metal tolerance. Functions in defense may have evolved secondarily, as they often involve dramatic changes in size, morphology or placement of crystals, or cellular modifications. Through the integration of ultrastructural, physiological, biochemical, and genetic approaches, researchers are beginning to identify the mechanisms responsible for this remarkable biomineralization process. L-ascorbic acid is the primary precursor for the biosynthesis of oxalic acid used for crystal formation in a number of species and the ascorbate utilized is produced directly within the crystal idioblast, indicating functional autonomy from surrounding cells. Some of the protein components of the Ca accumulation process and crystal-forming machinery are being identified and isolated; however, many features of crystal formation remain to be characterized. The recent establishment of a genetic model and mutants in crystal formation is a tremendous resource that provides additional powerful tools to help characterize this process. With these tools in hand, coupled with the developing *Medicago truncatula* insertional mutagenic lines, DNA arrays, and expected genome sequence, progress in deciphering the regulation of CaOx formation should be expedited in this as well as other plant systems in the coming years. Given the role of CaOx formation in Ca regulation and defense, and the antinutritive aspects of oxalates, a better understanding of the mechanisms operating in crystal idioblasts may lead to new strategies for improving productivity of crop plants and enhancing the nutritional quality of plant foods through manipulation of oxalate and CaOx crystal formation. Crystal-forming cells can also provide a model system for exploring Ca transport and sequestration processes and the mechanism for converting ascorbic acid to oxalic acid because they are enriched for these processes.

SUMMARY POINTS

1. Calcium oxalate is common in higher and lower vascular plants and algae, can account for most of the calcium in some species, and is a basic physiological process in plants.
2. Calcium oxalate occurs as crystals of various size and morphology, and as intracellular or extracellular deposits.

3. Crystal formation is not a simple physical-chemical process but is highly regulated by the cells producing the crystals.
4. High-capacity calcium regulation is one function demonstrated for calcium oxalate formation, and other functions include defense and detoxification of heavy metals.
5. Tight genetic regulation of crystal formation is illustrated by the constancy of morphology and rigid pattern of deposition within a particular tissue or organ.
6. Cells specialized for calcium oxalate formation, crystal idioblasts, show features that are necessary for dealing with mass accumulation of calcium and regulated control of crystal nucleation and growth.
7. Ascorbic acid is a major source of oxalic acid used for crystal formation, but the enzyme(s) catalyzing the cleavage between C2 and C3 of ascorbate has not been identified.
8. A genetic model system was established that will enable the characterization of the mechanisms and genes involved in regulating this complex process, including control of crystal shape and growth, oxalate synthesis, calcium accumulation and transfer between cellular compartments, and formation of unique crystal matrix proteins and intravacuolar membranes that are involved in calcium oxalate precipitation.

ACKNOWLEDGMENTS

Aspects of this work were supported by NSF grants MCB 96,32027 and MCB9904562 to V.R.F. and by the U.S. Department of Agriculture, Agricultural Research Service, under Cooperative Agreement number 58-6250-6-001 to P.A.N. We thank the WSU Electron Microscopy Center for access to their facilities. The contents of this publication do not necessarily reflect the views or policies of the U.S. Department of Agriculture, nor does mention of trade names, commercial products, or organizations imply endorsement by the U.S. Government.

LITERATURE CITED

1. Addadi L, Geva M. 2003. Molecular recognition at the interface between crystals and biology: generation, manifestation and detection of chirality at crystal surfaces. *Cryst. Eng. Comm.* 5:140–46
2. Ahmed AK, Johnson KA. 2000. The effect of the ammonium:nitrate nitrogen ration, total nitrogen, salinity (NaCl) and calcium on oxalate levels of *Tetragonia tetragonioides Pallas*. Kunz. *J. Hortic. Sci. Biotech.* 75:533–38
3. Al-Rais AH, Myers A, Watson L. 1971. The isolation and properties of oxalate crystals from plants. *Ann. Bot.* 35:1213–18
4. Arnott HJ. 1981. An SEM study of twinning in calcium oxalate crystals of plants. *Scan. Electron Microsc.* III:229–34
5. Arnott HJ. 1982. Three systems of biomineralization in plants with comments on the associated organic matrix. In *Biological Mineralization and Demineralization*, ed. GH Nancollas, pp. 199–218. Berlin: Springer Verlag
6. Arnott HJ. 1995. Calcium oxalate in fungi. In *Calcium Oxalate in Biological Systems*, ed. SR Khan, pp. 73–111. Boca Raton, FL: CRC

7. Arnott HJ, Pautard FGE. 1970. Calcification in plants. In *Biological Calcification: Cellular and Molecular Aspects*, ed. H Schraer, pp. 375–446. New York: Appleton-Century-Crofts
8. Arnott HJ, Webb MA. 2000. Twinned raphides of calcium oxalate in grape (*Vitis*): implications for crystal stability and function. *Int. J. Plant Sci.* 161:133–42
9. Assailly A. 1954. Sur les rapports de l'oxalate de chaux et de l'amidon. *C. R. Seances Acad. Sci. D.* 238:1902–4
10. Beazley MJ, Rickman RD, Ingram DK, Boutton TW, Russ J. 2002. Natural abundances of carbon isotopes (^{14}C, ^{13}C) in lichens and calcium oxalate pruina: implications for archaeological and paleoenvironmental studies. *Radiocarbon* 44:675–83
11. Benway DA, Weaver CM. 1993. Assessing chemical form of calcium in wheat, spinach, and kale. *J. Food Sci.* 58:605–8
12. Borchert R. 1985. Calcium-induced patterns of calcium-oxalate crystals in isolated leaflets of *Gleditsia triacanthos* L. and *Albizia julibrissin* Durazz. *Planta* 165:301–10
13. Borchert R. 1986. Calcium acetate induces calcium uptake and formation of calcium-oxalate crystals in isolated leaflets of *Gleditsia tracanthos* L. *Planta* 168:571–78
14. Bosscher D, Lu Z, Van Cauwenbergh R, Van Caillie-Bertrand M, Robberecht H, Deelstra H. 2001. A method for *in vitro* determination of calcium, iron and zinc availability from first-age infant formula and human milk. *Int. J. Food Sci. Nutr.* 52:173–82
15. Bouropoulos N, Weiner S, Addadi L. 2001. Calcium oxalate crystals in tomato and tobacco plants: morphology and in vitro interactions of crystal-associated macromolecules. *Chem. Eur. J.* 7:1881–88
16. Bruynzeel DB. 1997. Bulb dermatitis: dermatological problems in the flower bulb industries. *Contact Dermat.* 37:70–77
17. Calmes J. 1969. Contribution a l'etude du metabolisme de l'acide oxalique chez la Vigne vierge (*Parthenocissus tricuspidata* Planchon). *C. R. Seances Acad. Sci. D.* 269:704–7
18. Calmes J, Carles J. 1970. La repartition et l'evolution des cristaux d'oxalate de calcium dans les tissus de vigne vierge au cours d'un cycle de vegetation. *Soc. Bot. France Bull.* 117:189–98
19. Campbell AA, Ebrahimpour A, Perez L, Smesko SA, Nancollas GH. 1989. The dual role of polyelectrolytes and proteins as mineralization promotors and inhibitors of calcium oxalate monohydrate. *Calcif. Tissue Int.* 45:122–28
20. Campbell AA, Fryxell GE, Graff GL, Rieke PC, Tarasevich BJ. 1993. The nucleation and growth of Ca oxalate monohydrate on self-assembled monolayers (SAMS). *Scanning Microsc.* 7:423–29
21. Cerrano C, Bavestrello G, Arillo A, Benatti U, Bonpadre S, et al. 1999. Calcium oxalate production in the marine sponge *Chondrosia reniformis*. *Mar. Ecol.-Prog. Ser.* 179:297–300
22. Chattaway MM. 1955. Crystals in woody tissues I. *Trop. Woods* 102:55–74
23. Chattaway MM. 1956. Crystals in woody tissues II. *Trop. Woods* 104:100–24
24. Cheavin WHS. 1938. The crystals and cystoliths found in plant cells. Part 1: crystals. *Microscope* 2:155–58
25. Chiu MM, Falk RH. 1975. Ultrastructural study on *Lemna perpusilla*. *Cytologia (Tokyo).* 40:313–22
26. Choi Y-E, Harada E, Wada M, Tsuboi H, Morita Y, et al. 2001. Detoxification of cadmium in tobacco plants: formation and active excretion of crystals containing cadmium and calcium through trichomes. *Planta* 213:45–50
27. Cody AM, Horner HT. 1984. Crystallographic analysis of crystal images in scanning electron micrographs and their application to phytocrystalline studies. *Scan. Electron Microsc.* III:1451–60

28. De Silva DLR, Hetherington AM, Mansfield TA. 1996. Where does all the calcium go? Evidence of an important regulatory role for trichomes in two calcicoles. *Plant Cell Environ.* 19:880–86
29. Doaigey AR. 1991. Occurrence, type, and location of calcium oxalate crystals in leaves and stems of 16 species of poisonous plants. *Am. J. Bot.* 78:1608–16
30. Edwards HGM, Seaward MR, Attwood SJ, Little SJ, de Oliveira LFC, Tretiach M. 2003. FT-Raman spectroscopy of lichens on dolomitic rocks: an assessment of metal oxalate formation. *Analyst* 128:1218–21
31. Edyvane KA, Hibberd CM, Harnett RM, Marshall VR, Ryall RL. 1987. Macromolecules inhibit calcium oxalate crystal growth and aggregation in whole human urine. *Clin. Chim. Acta* 167:329–38
32. Fink S. 1991. The micromorphological distribution of bound calcium in needles of Norway spruce [*Picea abies* (L.) Karst.]. *N. Phytol.* 119:33–40
33. Finley DS. 1999. Patterns of calcium oxalate crystals in young tropical leaves: a possible role as an anti-herbivory defense. *Rev. Biol. Trop.* 47:27–31
34. Fomina M, Alexander IJ, Hillier S, Gadd GM. 2004. Zinc phosphate and pyromorphite solubilization by soil plant-symbiotic fungi. *Geomicrobiol. J.* 21:351–66
35. Foster AS. 1956. Plant idioblasts: remarkable examples of cell specialization. *Protoplasma* 46:184–93
36. Foy CD, Chaney RL, White MC. 1978. The physiology of metal toxicity in plants. *Annu. Rev. Plant Physiol.* 29:511–66
37. Franceschi VR. 1984. Developmental features of calcium oxalate crystal sand deposition in *Beta vulgaris* L leaves. *Protoplasma* 120:216–23
38. Franceschi VR. 1987. Oxalic-acid metabolism and calcium-oxalate formation in *Lemna minor* L. *Plant Cell Environ.* 10:397–406
39. Franceschi VR. 1989. Calcium oxalate formation is a rapid and reversible process in *Lemna minor* L. *Protoplasma* 148:130–37
40. Franceschi VR. 2001. Calcium oxalate in plants. *Trends Plant Sci.* 6:331
41. Franceschi VR, Horner HT Jr. 1979. Use of *Psychotria puncata* callus in study of calcium oxalate crystal idioblast formation. *Z. Pflanzenphysiol.* 67:61–75
42. Franceschi VR, Horner HT Jr. 1980. Calcium oxalate crystals in plants. *Bot. Rev.* 46:361–427
43. Franceschi VR, Krekling T, Christiansen E. 2002. Application of methyl jasmonate on *Picea abies* (Pinaceae) stems induces defense-related responses in phloem and xylem. *Am. J. Bot.* 89:578–86
44. Franceschi VR, Li X, Zhang D, Okita TW. 1993. Calsequestrin-like calcium binding protein is expressed in calcium accumulating cells of *Pistia stratiotes* L. *Proc. Natl. Acad. Sci. USA* 90:6986–90
45. Franceschi VR, Loewus FA. 1995. Oxalate biosynthesis and function in plants and fungi. In *Calcium Oxalate in Biological Systems*, ed. SR Khan, pp. 113–30. Boca Raton, FL: CRC
46. Franceschi VR, Schueren AM. 1986. Incorporation of strontium into plant calcium oxalate crystals. *Protoplasma* 130:199–205
47. Frank E. 1972. The formation of crystal idioblasts in *Canavalia ensiformis* DC. at different levels of calcium supply. *Z. Pflanzenphysiol.* 67:350–58
48. Frank E, Jensen WA. 1970. On the formation of the pattern of crystal idioblasts in *Canavalia ensiformis* DC. IV. The fine structure of the crystal cells. *Planta* 95:202–17
49. Frey-Wyssling A. 1981. Crystallography of the two hydrates of crystalline calcium oxalate in plants. *Am. J. Bot.* 68:130–41
50. Friedmann EI, Roth WC, Turner JB. 1972. Calcium oxalate crystals in the aragonite-producing green alga *Penicillus* and related genera. *Science* 177:891–93

51. Frutos P, Duncan AJ, Kyriazakis I, Gordon IJ. 1998. Learned aversion towards oxalic acid-containing foods by goats: Does rumen adaptation to oxalic acid influence diet choice? *J. Chem. Ecol.* 24:383–97
52. Gallaher RN, Jones JBJ. 1976. Total, extractable, and oxalate calcium and other elements in normal and mouse ear pecan tree tissues. *J. Am. Soc. Hortic. Sci.* 101:692–96
53. Gallaher RN, Perkins HF, Jones JBJ. 1975. Calcium concentration and distribution in healthy and decline peach tree tissues. *HortSci.* 10:134–37
54. Gardner DG. 1994. Injury to the oral mucous membranes caused by the common houseplant, *Dieffenbachia*. A review. *Oral Surg. Oral Med. Oral Pathol. Oral Radiol. Endod.* 78:631–33
55. Giachetti E, Pinzauti G, Bonaccorsi R, Vincenzini MT, Vanni P. 1987. Isocitrate lyase from higher plants. *Phytochemistry* 26:2439–46
56. Giordani P, Modenesi P, Tretiach M. 2003. Determinant factors for the formation of the calcium oxalate minerals, weddellite and whewellite, on the surface of foliose lichens. *Lichenologist* 35:255–70
57. Grimson MJ, Arnott HJ, Webb MA. 1982. A scanning electron microscopic study of winged twin crystals in the bean legume. *Scan. Electron Microsc.* III:1133–40
58. Guo SW, Ward MD, Wesson JA. 2002. Direct visualization of calcium oxalate monohydrate crystallization and dissolution with atomic force microscopy and the role of polymeric additives. *Langmuir* 18:4284–91
59. Hayaishi O, Shimazono H, Katagiri M, Saito Y. 1956. Enzymatic formation of oxalate and acetate from oxaloacetate. *J. Am. Chem. Soc.* 78:5126–27
60. Heaney RP, Weaver CM. 1989. Oxalate: effect on calcium absorbability. *Am. J. Clin. Nutr.* 50:830–32
61. Heaney RP, Weaver CM. 1990. Calcium absorption from kale. *Am. J. Clin. Nutr.* 51:656–57
62. Heaney RP, Weaver CM, Fitzsimmons ML. 1991. Soybean phytate content: effect on calcium absorption. *Am. J. Clin. Nutr.* 53:745–47
63. Heaney RP, Weaver CM, Recker RR. 1988. Calcium absorbability from spinach. *Am. J. Clin. Nutr.* 47:707–9
64. Hodgkinson A. 1977. *Oxalic Acid Biology and Medicine*. New York: Academic
65. Holmes RP, Goodman HO, Assimos DG. 1995. Dietary oxalate and its intestinal absorption. *Scanning Microsc.* 9:1109–20
66. Holmes RP, Goodman HO, Assimos DG. 2001. Contribution of dietary oxalate to urinary oxalate excretion. *Kidney Int.* 59:270–76
67. Holmes RP, Kennedy M. 2000. Estimation of the oxalate content of foods and daily oxalate intake. *Kidney Int.* 57:1662–67
68. Horner HT, Kausch AP, Wagner BL. 2000. Ascorbic acid: a precursor of oxalate in crystal idioblasts of *Yucca torreyi* in liquid root culture. *Int. J. Plant Sci.* 161:861–68
69. Horner HT, Wagner BL. 1980. The association of druse crystals with the developing stomium of *Capsicum annuum* (Solanaceae) anthers. *Am. J. Bot.* 67:1347–60
70. Horner HT, Wagner BL. 1995. Calcium oxalate formation in higher plants. In *Calcium Oxalate in Biological Systems*, ed. SR Khan, pp. 53–72. Boca Raton, FL: CRC
71. Horner HT, Whitmoyer RE. 1972. Raphide crystal cell development in leaves of *Psychotria punctata* (Rubiaceae). *J. Cell Sci.* 11:339–55
72. Horner HT, Zindler-Frank E. 1982. Histochemical, spectroscopic, and X-ray diffraction identifications of the two hydration forms of calcium oxalate crystals in three legumes and *Begonia*. *Can. J. Bot.* 60:1021–27
73. Hoyer JR, Asplin JR, Otvos L. 2001. Phosphorylated osteopontin peptides suppress crystallization by inhibiting the growth of calcium oxalate crystals. *Kidney Int.* 60:77–82

74. Hudgins JW, Krekling T, Franceschi VR. 2003. Distribution of calcium oxalate crystals in the secondary phloem of conifers: a constitutive defense mechanism? *N. Phytol.* 159:677–90
75. Ilarslan H, Palmer RG, Horner HT. 2001. Calcium oxalate crystals in developing seeds of soybean. *Ann. Bot.* 88:243–57
76. Ilarslan H, Palmer RG, Imsande J, Horner HT. 1997. Quantitative determination of calcium oxalate in developing seeds of soybean (*Leguminosae*). *Am. J. Bot.* 84:1042–46
77. Ishii Y. 1991. Three kinds of calcium-oxalate hydrates. *Nippon Kagaku Kaishi* 1:63–70
78. James LF. 1999. Halogenton poisoning in livestock. *J. Nat. Toxins* 8:395–403
79. Jauregui-Zuniga D, Reyes-Grajeda JP, Sepulveda-Sanchez JD, Whitaker JR, Moreno A. 2003. Crystallochemical characterization of calcium oxalate crystals isolated from seed coats of *Phaseolus vulgaris* and leaves of *Vitis vinifera*. *J. Plant Physiol.* 160:239–45
80. Julian CG, Bowers PW. 1997. The nature and distribution of daffodil picker's rash. *Contact Dermatitis* 37:259–62
81. Jung TS, Kim WS, Choi CK. 2004. Biomineralization of calcium oxalate for controlling crystal structure and morphology. *Mat. Sci. Eng. C-Bio. S.* 24:31–33
82. Kamchan A, Puwastien P, Sirichakwal PP, Kongkachuichai R. 2004. In vitro calcium bioavailability of vegetables, legumes and seeds. *J. Food Compos. Anal.* 17:311–20
83. Kausch AP, Horner HT. 1983. Development of syncytial raphide crystal idioblasts in the cortex of adventitious roots of *Vanilla planifolia* L. (Orchidiaceae). *Scan. Electron Microsc.* II:893–903
84. Kausch AP, Horner HT. 1985. Absence of $CeCl_3$-detectable peroxisomal glycolate-oxidase activity in developing raphide crystal idioblasts in leaves of *Psychotria punctata* Vatke and roots of *Yucca torreyi* L. *Planta* 164:35–43
85. Kavanagh JP. 1995. Calcium oxalate crystallization *in vitro*. In *Calcium Oxalate in Biological Systems*, ed. SR Khan, pp. 1–21. Boca Raton, FL: CRC
86. Keates SA, Tarlyn N, Loewus FA, Franceschi VR. 2000. L-Ascorbic acid and L-galactose are sources of oxalic acid and calcium oxalate in *Pistia stratiotes*. *Phytochemistry* 53:433–40
87. Kingsley RJ, Van Gilder R, LeGeros RZ, Watabe N. 2003. Multimineral calcareous deposits in the marine alga *Acetabularia acetabulum* (Chlorophyta; Dasycladaceae). *J. Phycol.* 39:937–47
88. Kirkby EA, Pilbeam DJ. 1984. Calcium as a plant nutrient. *Plant Cell Environ.* 7:397–405
89. Kitchen JW, Burns EE, Perry BA. 1964. Calcium oxalate content of spinach (*Spinacia oleracea* L.). *J. Am. Soc. Hortic. Sci.* 84:441–45
90. Kochian LV. 1995. Cellular mechanisms of aluminum toxicity and resistance in plants. *Annu. Rev. Plant Physiol. Plant Mol. Biol.* 46:237–60
91. Kostman TA, Franceschi VR. 2000. Cell and calcium oxalate crystal growth is coordinated to achieve high-capacity calcium regulation in plants. *Protoplasma* 214:166–79
92. Kostman TA, Franceschi VR, Nakata PA. 2003. Endoplasmic reticulum sub-compartments are involved in calcium sequestration within raphide crystal idioblasts of *Pistia stratiotes* L. *Plant Sci.* 165:205–12
93. Kostman TA, Koscher JR. 2003. L-galactono-gamma-lactone dehydrogenase is present in calcium oxalate crystal idioblasts of two plant species. *Plant Physiol. Biochem.* 41:201–6
94. Kostman TA, Tarlyn NM, Loewus FA, Franceschi VR. 2001. Biosynthesis of L-ascorbic acid and conversion of carbons 1 and 2 of L-ascorbic acid to oxalic acid occurs within individual calcium oxalate crystal idioblasts. *Plant Physiol.* 125:634–40
95. Kuballa B, Lugnier AAJ, Anton R. 1981. Study of *Dieffenbachia*-induced edema in mouse and rat hindpaw: respective role of oxalate needles and trypsin-like protease. *Toxicol. Appl. Pharmacol.* 58:444–51

96. Kuo-Huang LL, Chen SH, Chen SJ. 2000. Ultrastructural study on the formation of sclereids in the floating leaves of *Nymphoides coreana* and *Nuphar schimadai*. *Bot. Bull. Acad. Sinica* 41:283–91
97. Kuo-Huang LL, Zindler-Frank E. 1998. Structure of crystal cells and influences of leaf development on crystal cell development and vice versa in *Phaseolus vulgaris* (Leguminosae). *Bot. Acta* 111:337–45
98. Kurutz JW, Carvalho M, Nakagawa Y. 2003. Nephrocalcin isoforms coat crystal surfaces and differentially affect calcium oxalate monohydrate crystal morphology, growth, and aggregation. *J. Cryst. Growth* 255:392–402
99. Lanzalaco AC, Singh RP, Smesko SA, Nancollas GH, Sufrin G, et al. 1988. The influence of urinary macromolecules on calcium oxalate monohydrate crystal growth. *J. Urol.* 139:190–95
100. Ledbetter MC, Porter KR. 1970. *Introduction to the Fine Structure of Plant Cells*. New York: Springer-Verlag
101. Leliaert F, Coppejans E. 2004. Crystalline cell inclusions: a new diagnostic character in the Cladophorophyceae (Chlorophyta). *Phycologia* 43:189–203
102. Lersten NR, Horner HT. 2000. Calcium oxalate crystal types and trends in their distribution patterns in leaves of *Prunus* (Rosaceae : Prunoideae). *Plant Syst. Evol.* 224:83–96
103. Letellier SR, Lochhead MJ, Campbell AA, Vogel V. 1998. Oriented growth of calcium oxalate monohydrate crystals beneath phospholipid monolayers. *Biochim. Biophys. Acta* 1380:31–45
104. Levy-Lior A, Weiner S, Addadi L. 2003. Achiral calcium-oxalate crystals with chiral morphology from the leaves of some Solanacea plants. *Helv. Chim. Acta* 86:4007–17
105. Lewandowski S, Rodgers AL. 2004. Idiopathic calcium oxalate urolithiasis: risk factors and conservative treatment. *Clin. Chim. Acta* 345:17–34
106. Li XX, Franceschi VR. 1990. Distribution of peroxisomes and glycolate metabolism in relation to calcium oxalate formation in *Lemna minor* L. *Eur. J. Cell Biol.* 51:9–16
107. Li XX, Zhang DZ, Lynch-Holm VJ, Okita TW, Franceschi VR. 2003. Isolation of a crystal matrix protein associated with calcium oxalate precipitation in vacuoles of specialized cells. *Plant Physiol.* 133:549–59
108. Libert B. 1987. Breeding a low-oxalate rhubarb (Rheum sp. L.). *J. Hortic. Sci.* 62:523–29
109. Libert B, Creed C. 1985. Oxalate content of seventy-eight rhubarb cultivars and its relation to some other characters. *J. Hortic. Sci.* 60:257–61
110. Libert B, Franceschi VR. 1987. Oxalate in crop plants. *J. Agric. Food. Chem.* 35:926–38
111. Loewus F. 1999. Biosynthesis and metabolism of ascorbic acid in plants and of analogs of ascorbic acid in fungi. *Phytochemistry* 52:193–210
112. Ma JF, Hiradate S, Matsumoto H. 1998. High aluminum resistance in buckwheat. II. Oxalic acid detoxifies aluminum internally. *Plant Physiol.* 117:753–59
113. Ma JF, Hiradate S, Nomoto K, Iwashita T, Matsumoto H. 1997. Internal detoxification mechanism of Al in hydrangea. *Plant Physiol.* 113:1033–39
114. Ma JF, Ryan PR, Delhaize E. 2001. Aluminium tolerance in plants and the complexing role of organic acids. *Trends Plant Sci.* 6:273–78
115. Macnish AJ, Irving DE, Joyce DC, Vithanage V, Wearing AH, et al. 2003. Identification of intracellular calcium oxalate crystals in *Chamelaucium uncinatum* (Myrtaceae). *Aust. J. Bot.* 51:565–72
116. Malainine ME, Dufresne A, Dupeyre D, Vignon MR, Mahrouz M. 2003. First evidence for the presence of weddellite crystallites in *Opuntia ficus indica* parenchyma. Z. *Naturforsch. [C].* 58:812–16
117. Massey LK. 2003. Dietary influences on urinary oxalate and risk of kidney stones. *Front. Biosci.* 8:S584–S94

118. Massey LK, Palmer RG, Horner HT. 2001. Oxalate content of soybean seeds (*Glycine max:* Leguminosae), soyfoods, and other edible legumes. *J. Agric. Food Chem.* 49:4262–66
119. Massey LK, Roman-Smith H, Sutton RA. 1993. Effect of dietary oxalate and calcium on urinary oxalate and risk of formation of calcium oxalate kidney stones. *J. Am. Diet. Assoc.* 93:901–6
120. Massonie G. 1980. Elevage d'un biotype de *Myzus persicae* Sulzer sur milieu synthetique. V.- influence des acides oxalique et gentisique sur la valeur alimentaire d'un milieu synthetique. *Ann. Nutr. Aliment.* 34:139
121. Mazen AMA. 2004. Calcium oxalate deposits in leaves of *Corchorus olitotius* as related to accumulation of toxic metals. *Russ. J. Plant Physiol.* 51:281–85
122. Mazen AMA, El Maghraby OMO. 1998. Accumulation of cadmium, lead and strontium, and a role of calcium oxalate in water hyacinth tolerance. *Biol. Plant.* 40:411–17
123. Mazen AMA, Zhang DZ, Franceschi VR. 2004. Calcium oxalate formation in *Lemna minor*: physiological and ultrastructural aspects of high capacity calcium sequestration. *N. Phytol.* 161:435–48
124. McConn MM, Nakata PA. 2002. Calcium oxalate crystal morphology mutants from *Medicago truncatula*. *Planta* 215:380–86
125. McConn MM, Nakata PA. 2004. Oxalate reduces calcium availability in the pads of the prickly pear cactus through formation of calcium oxalate crystals. *J. Agric. Food Chem.* 52:1371–74
126. McNair JB. 1932. The intersection between substances in plants: essential oils and resins, cyanogen and oxalate. *Am. J. Bot.* 19:255–71
127. Molano-Flores B. 2001. Herbivory and calcium concentrations affect calcium oxalate crystal formation in leaves of *Sida* (Malvaceae). *Ann. Bot.* 88:387–91
128. Monje PV, Baran EJ. 1996. On the formation of weddellite in *Chamaecereus silvestrii*, a cactaceae from northern Argentina. *Z. Naturforsch.* [*C*]. 51:426–28
129. Monje PV, Baran EJ. 1997. On the formation of whewellite in the cactaceae species *Opuntia microdasys*. *Z. Naturforsch. [C].* 52:267–69
130. Monje PV, Baran EJ. 2002. Characterization of calcium oxalates generated as biominerals in cacti. *Plant Physiol.* 128:707–13
131. Muller N, Glaus T, Gardelle O. 1998. Extensive stomach ulcers due to *Dieffenbachia* intoxication in a cat. *Tierarztl. Prax. Ausg. K Klientiere. Heimtiere* 26:404–7
132. Nakata PA. 2003. Advances in our understanding of calcium oxalate crystal formation and function in plants. *Plant Sci.* 164:901–9
133. Nakata PA, Kostman TA, Franceschi VR. 2003. Calreticulin is enriched in the crystal idioblasts of *Pistia stratiotes*. *Plant Physiol. Biochem.* 41:425–30
134. Nakata PA, McConn MM. 2000. Isolation of *Medicago truncatula* mutants defective in calcium oxalate crystal formation. *Plant Physiol.* 124:1097–104
135. Nakata PA, McConn MM. 2002. Sequential subtractive approach facilitates identification of differentially expressed genes. *Plant Physiol. Biochem.* 40:307–12
136. Nakata PA, McConn MM. 2003. Calcium oxalate crystal formation is not essential for growth of *Medicago truncatula*. *Plant Physiol. Biochem.* 41:325–29
137. Nakata PA, McConn MM. 2003. Influence of the calcium oxalate defective 4 (*cod4*) mutation on the growth, oxalate content, and calcium content of *Medicago truncatula*. *Plant Sci.* 164:617–21
138. Noonan SC, Savage GP. 1999. Oxalate content of foods and its effect on humans. *Asia Pac. J. Clin. Nutr.* 8:64–74

139. Ouyang JM, Duan L, Tieke B. 2003. Effects of carboxylic acids on the crystal growth of calcium oxalate nanoparticles in lecithin-water liposome systems. *Langmuir* 19:8980–85

140. Penmetsa RV, Cook DR. 2000. Production and characterization of diverse developmental mutants of *Medicago truncatula*. *Plant Physiol.* 123:1387–97

141. Pennisi SV, McConnell DB. 2001. Inducible calcium sinks and preferential calcium allocation in leaf primordia of *Dracaena sanderiana* Hort. Sander ex M.T. Mast. (Dracaenaceae). *HortSci.* 36:1187–91

142. Pennisi SV, McConnell DB. 2001. Taxonomic relevance of calcium oxalate cuticular deposits in Dracaena vand. ex L. *Hort. Sci.* 36:1033–36

143. Pennisi SV, McConnell DB, Gower LB, Kane ME, Lucansky T. 2001. Intracellular calcium oxalate crystal structure in *Dracaena sanderiana*. *N. Phytol.* 150:111–20

144. Pennisi SV, McConnell DB, Gower LB, Kane ME, Lucansky T. 2001. Periplasmic cuticular calcium oxalate crystal deposition in *Dracaena sanderiana*. *N. Phytol.* 149:209–18

145. Prychid CJ, Furness CA, Rudall PJ. 2003. Systematic significance of cell inclusions in Haemodoraceae and allied families: silica bodies and tapetal raphides. *Ann. Bot.* 92:571–80

146. Prychid CJ, Rudall PJ. 1999. Calcium oxalate crystals in monocotyledons: a review of their structure and systematics. *Ann. Bot.* 84:725–39

147. Pueschel CM. 1995. Calcium oxalate crystals in the red alga *Antithamnion kylinii* (Ceramiales): cytoplasmic and limited to indeterminate axes. *Protoplasma* 189:73–80

148. Pueschel CM. 2001. Calcium oxalate crystals in the green alga *Spirogyra hatillensis* (Zygnematales, Chlorophyta). *Int. J. Plant Sci.* 162:1337–45

149. Quitadamo IJ, Kostman TA, Schelling ME, Franceschi VR. 2000. Magnetic bead purification as a rapid and efficient method for enhanced antibody specificity for plant sample immunoblotting and immunolocalization. *Plant Sci.* 153:7–14

150. Rao MV, Chhotray N, Agarwal JS, Kumar S. 1988. Calcium oxalate crystal growth studies in polyacrilamide gels: Part II. Influence of synthetic polypeptides and natural macromolecules on crystal aggregation. *Indian J. Exp. Biol.* 26:553–57

151. Rauber A. 1985. Observations on idioblasts of *Dieffenbachia*. *Clin. Toxicol.* 23:79–84

152. Richardson KE, Tolbert NE. 1961. Oxidation of glyoxylic acid to oxalic acid by glycolic acid oxidase. *J. Biol. Chem.* 236:1280–84

153. Rinallo C, Modi G. 2002. Content of oxalate in *Actinidia deliciosa* plants grown in nutrient solutions with different nitrogen forms. *Biol. Plant.* 45:137–39

154. Ruiz N, Ward D, Saltz S. 2002. Calcium oxalate crystals in leaves of *Pancratium sickenbergeri*: constitutive or induced defense? *Funct. Ecol.* 16:99–105

155. Ruiz N, Ward D, Saltz S. 2002. Responses of *Pancratium sickenbergeri* to simulated bulb herbivory: combining defence and tolerance strategies. *J. Ecol.* 90:472–79

156. Ryall RL, Fleming DE, Grover PK, Chauvet M, Dean CJ, Marshall VR. 2000. The hole truth: intracrystalline proteins and calcium oxalate kidney stones. *Mol. Urol.* 4:391–402

157. Ryall RL, Stapleton AMF. 1995. Urinary macromolecules in calcium oxalate stone and crystal matrix: good, bad, or indifferent? In *Calcium Oxalate in Biological Systems*, ed. SR Kahn, pp. 265–90. Boca Raton, FL: CRC

158. Sakai WS, Hanson M. 1974. Mature raphide and raphide idioblast structure in plants of the edible aroid genera *Colocasia*, *Alocasia* and *Xanthosoma*. *Ann. Bot.* 38:739–48

159. Sakai WS, Hanson M, Jones RC. 1972. Raphides with barbs and grooves in *Xanthosoma sagittifolium* (Aracea). *Science* 178:314–15

160. Sakai WS, Shiroma SS, Nagao MA. 1984. A study of raphide microstructure in relation to irritation. *Scan. Electron Microsc.* II:979–86

161. Salinas ML, Ogura T, Soffchi L. 2001. Irritant contact dermatitis caused by needle-like calcium oxalate crystals, raphides, in *Agave tequilana* among workers in tequila distilleries and agave plantations. *Contact Dermatitis* 44:94–96
162. Saltz S, Ward D. 2000. Responding to a three-pronged attack: desert lilies subject to herbivory by dorcas gazelles. *Plant Ecol.* 148:127–38
163. Samac DA, Tesfaye M. 2003. Plant improvement for tolerance to aluminum in acid soils—a review. *Plant Cell Tiss. Org.* 75:189–207
164. Schmidt RJ, Moult SP. 1983. The dermatitic properties of black bryony (*Tamus communis* L.). *Contact Dermatitis* 9:390–36
165. Schürhoff P. 1908. Ozellen und Lichtkondensoren bei einigen Peperomien. *Beihhefle zum Botanisches Zentralblatt* 23:14–26
166. Seal SN, Sen SP. 1970. The photosynthetic production of oxalic acid in *Oxalis corniculata*. *Plant Cell Physiol.* 11:119–28
167. Shiraga H, Min W, VanDusen WJ, Clayman MD, Miner D, et al. 1992. Inhibition of calcium oxalate crystal growth *in vitro* by uropontin: another member of the aspartic acid-rich protein superfamily. *Proc. Natl. Acad. Sci. USA.* 89:426–30
168. Smirnoff N, Conklin PL, Loewus FA. 2001. Biosynthesis of ascorbic acid in plants: a renaissance. *Annu. Rev. Plant Physiol. Plant Mol. Biol.* 52:437–67
169. Stamatelou KK, Francis ME, Jones CA, Nyberg LM Jr, Curhan GC. 2003. Time trends in reported prevalence of kidney stones in the United States: 1976–1994. *Kidney Int.* 63:1817–23
170. Storey R, Jones RGW, Schachtman DP, Treeby MT. 2003. Calcium-accumulating cells in the meristematic region of grapevine root apices. *Funct. Plant Biol.* 30:719–27
171. Sugiyama N, Okutani I. 1996. Relationship between nitrate reduction and oxalate synthesis in spinach leaves. *J. Plant Physiol.* 149:14–18
172. Tanaka M, Nakashima T, Mori K. 2003. Differences in density and size of idioblasts containing calcium oxalate crystals in petioles among cultivars of taro (*Colocasia esculenta* (L.) Schott. and *C. gigantea* Hook. f.). *J. Jpn. Soc. Hortic. Sci.* 72:551–56
173. Tanaka M, Nakashima T, Mori K. 2003. Effects of shading and soil moisture on the formation of idioblasts containing raphides in petioles of taro *Colocasia esculenta* (L.) Schott. *J. Jpn. Soc. Hortic. Sci.* 72:457–59
174. Tanaka M, Nakashima T, Mori K. 2003. Formation and distribution of calcium oxalate crystal idioblast in the tissues of taro (*Colocasia esculenta* (L.) Schott). *J. Jpn. Soc. Hortic. Sci.* 72:162–68
175. Tanaka M, Nakashima T, Mori K. 2003. Heterozygosity and inheritance pattern of characters of two taro *Colocasia esculenta* (L.) Schott. cultivars with edible petioles. *J. Jpn. Soc. Hortic. Sci.* 72:425–31
176. Theimer TC, Bateman GC. 1992. Patterns of prickly-pear herbivory by collared peccaries. *J. Wildl. Manage.* 56:234–40
177. Thurston EL. 1976. Morphology, fine structure and ontogeny of the stinging emergence of *Tragia ramosa* and *T. saxicola* (Euphorbiaceae). *Am. J. Bot.* 63:710–18
178. Tillman-Sutela E, Kauppi A. 1999. Calcium oxalate crystals in the mature seeds of Norway spruce, *Picea abies* (L.) Karst. *Trees* 13:131–37
179. Tilton VR, Horner HT. 1980. Calcium oxalate raphide crystals and crystalliferous idioblasts in the carpels of *Ornithogalum caudatum*. *Ann. Bot.* 46:533–39
180. Touryan LA, Lochhead MJ, Marquardt BJ, Vogel V. 2004. Sequential switch of biomineral crystal morphology using trivalent ions. *Nat. Mater.* 3:239–43
181. Vintejoux C, Shoar-Ghafari A. 1985. Repartition et ultrastructure compares des cellules oxaliferes en vie latente et en vie active de *Spirodela polyrrhiza* (Lemnaceae). *Soc. Bot. France Bull.* 132:52–39

182. Volk GM, Goss LJ, Franceschi VR. 2004. Calcium channels are involved in calcium oxalate crystal formation in specialized cells of *Pistia stratiotes* L. *Ann. Bot.* 93:741–53
183. Volk GM, Lynch-Holm VJ, Kostman TA, Goss LJ, Franceschi VR. 2002. The role of druse and raphide calcium oxalate crystals in tissue calcium regulation in *Pistia stratiotes* leaves. *Plant Biol.* 4:34–45
184. Von Burg R. 1994. Toxicology update. *J. Appl. Toxicol.* 14:233
185. Wagner BL. 1983. Genesis of the vacuolar apparatus responsible for druse formation in *Capsicum annuum* L. (Solanaceae) anthers. *Scan. Electron Microsc.* II:905–12
186. Walker S, Prescott J. 2003. Psychophysical properties of mechanical oral irritation. *J. Sens. Stud.* 18:325–45
187. Ward D, Spiegel M, Saltz S. 1997. Gazelle herbivory and interpopulation differences in calcium oxalate content of leaves of a desert lily. *J. Chem. Ecol.* 23:333–47
188. Weaver CM, Heaney RP, Nickel KP, Packard PI. 1997. Calcium bioavailability from high oxalate vegetables: Chinese vegetables, sweet potatoes and rhubarb. *J. Food Sci.* 62:524–25
189. Weaver CM, Martin BR, Ebner JS, Krueger CA. 1987. Oxalic acid decreases calcium absorption in rats. *J. Nutr.* 117:1903–6
190. Webb MA. 1999. Cell-mediated crystallization of calcium oxalate in plants. *Plant Cell* 11:751–61
191. Webb MA, Arnott HJ. 1983. Inside plant crystals: a study of the noncrystalline core in druses of *Vitis vinifera* endosperm. *Scan. Electron Microsc.* IV:1759–70
192. Webb MA, Cavaletto JM, Carpita NC, Lopez LE, Arnott HJ. 1995. The intravacuolar organic matrix associated with calcium oxalate crystals in leaves of *Vitis*. *Plant J.* 7:633–48
193. White PJ. 2000. Calcium channels in higher plants. *BBA-Biomembranes* 1465:171–89
194. White PJ, Broadley MR. 2003. Calcium in plants. *Ann. Bot.* 92:487–511
195. Wien EM, Schwartz R. 1983. Comparison of in vitro and in vivo measurements of dietary Ca exchangeability and bioavailability. *J. Nutr.* 113:388–93
196. Yang Y-Y, Jung J-Y, Song W-Y, Suh H-S, Lee Y. 2000. Identification of rice varieties with high tolerance or sensitivity to lead and characterization of the mechanism tolerance. *Plant Physiol.* 124:1019–26
197. Yoshihara T, Sogawa K, Pathak MD, Juliano BO, Sakamura S. 1980. Oxalic acid as a sucking inhibitor of the brown planthopper in rice (Delphacidae, Homoptera). *Entomol. Exp. Appl.* 27:149
198. Zhang DB, Qi LM, Ma JM, Cheng HM. 2002. Morphological control of calcium oxalate dihydrate by a double-hydrophilic block copolymer. *Chem. Mater.* 14:2450–57
199. Zindler-Frank E. 1974. The differentiation of crystal idioblasts in darkness and under inhibition of glycolic acid oxidase. *Z. Pflanzenphysiol.* 73:313–25
200. Zindler-Frank E. 1975. On the formation of the pattern of crystal idioblasts in *Canavalia ensiformis* D.C.: VII. Calcium and oxalate content of the leaves in dependence of calcium nutrition. *Z. Pflanzenphysiol.* 77:80–85
201. Zindler-Frank E. 1976. Oxalate biosynthesis in relation to photosynthetic pathways and plant productivity: a survey. *Z. Pflanzenphysiol.* 80:1–13
202. Zindler-Frank E. 1995. Calcium, calcium-oxalate crystals, and leaf differentiation in the common bean (*Phaseolus vulgaris* L). *Bot. Acta* 108:144–48
203. Zindler-Frank E, Honow R, Hesse A. 2001. Calcium and oxalate content of the leaves of *Phaseolus vulgaris* at different calcium supply in relation to calcium oxalate crystal formation. *J. Plant Physiol.* 158:139–44

204. Zindler-Frank E, Horner HT. 1985. Influence of humidity and atmospheres without O_2 or CO_2 on formation of calcium oxalate crystals in three legume taxa. *J. Plant Physiol.* 120:301–11
205. Zona S. 2004. Raphides in palm embryos and their systematic distribution. *Ann. Bot.* 93:415–21

Starch Degradation

Alison M. Smith,[1] Samuel C. Zeeman,[2] and Steven M. Smith[3,4]

[1]Department of Metabolic Biology, John Innes Centre, Norwich NR4 7UH, United Kingdom; email: alison.smith@bbsrc.ac.uk

[2]Institute of Plant Sciences, University of Bern, CH-3013 Bern, Switzerland; email: sam.zeeman@ips.unibe.ch

[3]Institute of Cell and Molecular Biology, University of Edinburgh, Edinburgh EH9 3JH, United Kingdom; email: s.smith@ed.ac.uk

[4]School of Biomedical, Biomolecular, and Chemical Sciences, University of Western Australia, Crawley, WA 6009 Australia; email: ssmith@cyllene.uwa.edu.au

Annu. Rev. Plant Biol. 2005. 56:73–98

doi: 10.1146/ annurev.arplant.56.032604.144257

First published online as a Review in Advance on January 13, 2005

1543-5008/05/0602-0073$20.00

Key Words

amylase, *Arabidopsis*, cereal endosperm, chloroplast, maltose

Abstract

Recent research reveals that starch degradation in *Arabidopsis* leaves at night is significantly different from the "textbook" version of this process. Although parts of the pathway are now understood, other parts remain to be discovered. Glucans derived from starch granules are hydrolyzed via β-amylase to maltose, which is exported from the chloroplast. In the cytosol maltose is the substrate for a transglucosylation reaction, producing glucose and a glucosylated acceptor molecule. The enzyme that attacks the starch granule to release glucans is not known, nor is the nature of the cytosolic acceptor molecule. An *Arabidopsis*-type pathway may operate in leaves of other species, and in nonphotosynthetic organs that accumulate starch transiently. However, in starch-storing organs such as cereal endosperms and legume seeds, the process differs from that in *Arabidopsis* and may more closely resemble the textbook pathway. We discuss the differences in relation to the biology of each system.

Contents

INTRODUCTION

Our aim in this article is to discuss new information about the pathway of starch degradation in *Arabidopsis* leaves at night, and to use it to reassess our understanding of starch degradation in other plant organs. Although starch degradation has been extensively studied in germinating cereal endosperm (8, 29, 76), the nature and regulation of the process in this and other plant organs is poorly understood. There is good a priori reason to think that the process in endosperms differs from that in other organs because the mature endosperm is not a living tissue whereas starch degradation in all other plant organs occurs within living cells. Biochemical analyses show that many plant organs possess a wealth of isoforms of several different types of enzymes capable of degrading starch and related glucans. However, discovering the roles and importance of each of these forms in catalyzing starch degradation in vivo has been hampered by a lack of tools for this purpose. In the last five years, the genetic and genomic resources available in *Arabidopsis* have facilitated new approaches to the pathway in leaves. We present below the picture that has emerged for *Arabidopsis* leaves, then discuss to what extent this is applicable to leaves of other species, and to other plant organs.

THE PATHWAY OF STARCH DEGRADATION IN *ARABIDOPSIS* LEAVES

During the day, starch and sucrose are synthesized together as the products of photosynthetic carbon assimilation in *Arabidopsis* leaves. Sucrose is exported to nonphotosynthetic parts of the plant, and starch accumulates in the chloroplasts. The ratio of starch-to-sucrose synthesis varies with environmental conditions, but in our "standard" growth conditions (12 h light, 20°C, and about 180-μmol quanta of photosynthetically active radiation m^{-2} s^{-1}) about half of the newly assimilated carbon is partitioned into starch, and the content at the end of the day is 10–15-mg g^{-1} fresh weight (14, 117). During the subsequent night, the starch is degraded to provide substrates for sucrose synthesis to allow continued export to nonphotosynthetic parts of the plant, and to provide carbon skeletons, energy, and reductant within the leaf cell. The supply of carbohydrate provided by nighttime starch degradation is essential for the normal

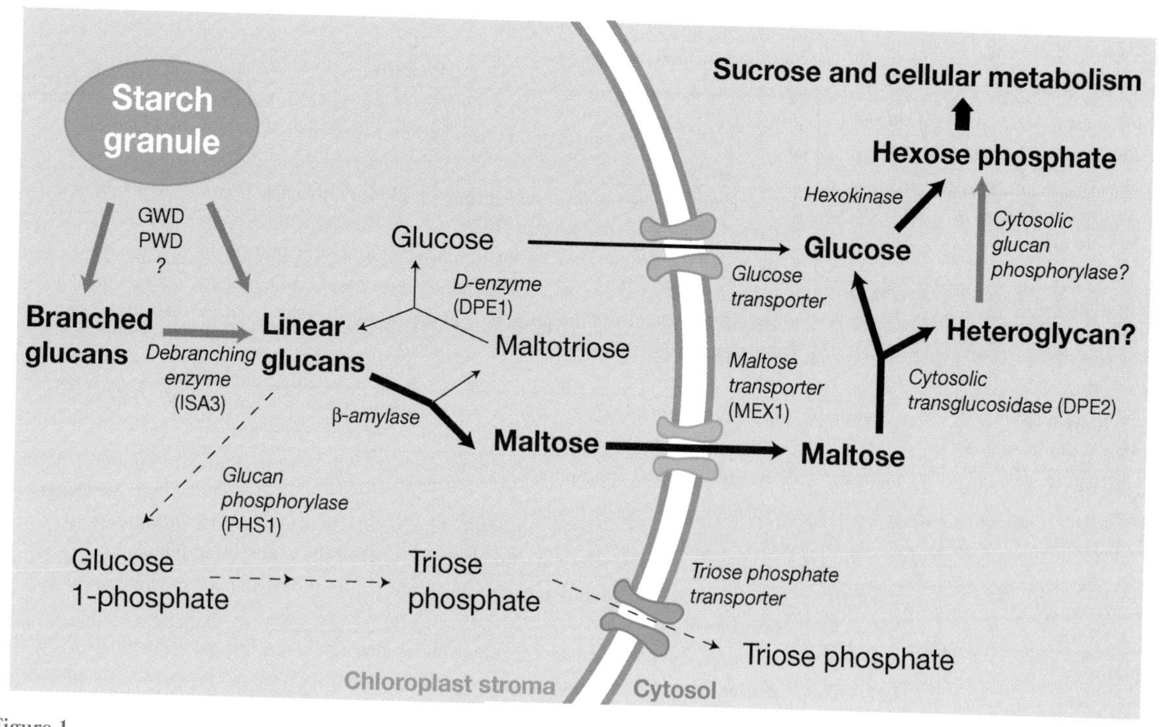

Figure 1

Proposed pathway of starch degradation in *Arabidopsis* leaves at night. Steps about which uncertainty remains are represented as stippled or dashed arrows, and with question marks. GWD is glucan, water dikinase and PWD is phosphoglucan, water dikinase. Further details of the enzymes and reactions involved are in **Table 1**.

growth of the plant. Mutants that synthesize less starch during the day, or have a reduced capacity to degrade it at night, have reduced growth rates under many environmental conditions (13, 85). In the following sections we discuss the steps involved in converting granular starch in the chloroplast into hexose phosphate in the cytosol. The scheme we describe is presented in **Figure 1**, and further details on the proteins involved are in **Table 1**.

The Attack on the Granule Surface

Starch in chloroplasts—like starch in other parts of plants—is in the form of granules composed of branched polymers of glucose. Most of these polymers are amylopectin—an α-1,4, α-1,6-linked polymer with a distinctive branching pattern that enables its organization into semicrystalline arrays (119, 120). The first step in the pathway of starch degradation must therefore be catalyzed by an enzyme capable of metabolizing polymers at the surface of a semicrystalline granule, rather than in a soluble form. Although several different types of enzymes are capable of releasing soluble glucans from purified starch granules in in vitro experiments (82, 95, 101), the only enzyme generally believed to do this in planta is the endoamylase α-amylase. In germinating cereal endosperm, α-amylases hydrolyze α-1,4 linkages within polymers exposed on the surface or in channels within granules, releasing soluble glucans that are the substrate for further degradation (see below).

Three proteins predicted to be α-amylases are encoded in the *Arabidopsis* genome, and one of these—AMY3 (**Table 1**)—is located in the chloroplast. It would seem reasonable to suppose that AMY3 catalyzes the attack on the granule surface in the chloroplast at night. However, surprisingly, none of these isoforms

TABLE 1 Starch-degrading enzymes in *Arabidopsis* leaves*

Enzyme	Locus	Reaction
Chloroplastic α-amylase; AMY3 EC 3.2.1.1	At1g69830	The enzyme is an endoamylase that hydrolyzes internal α-1,4 linkages of linear or branched glucans. It produces a mixture of linear and branched malto-oligosaccharides from amylopectin.
Glucan, water dikinase; GWD or GWD1 EC 2.7.9.4	At1g10760	The enzyme adds the β-phosphate group of ATP to either the 3- or the 6-carbon of a glucosyl residue of amylopectin (78). Glucan + ATP + water = Glucan-P + AMP + inorganic phosphate
Phosphoglucan, water dikinase; PWD or GWD3 EC 2.7.9.4	At4g24450	The *Arabidopsis* enzyme catalyzes the same reaction as GWD, but amylopectin must already be phosphorylated and the phosphate group is added to the 3- rather than the 6-position of glucosyl residues of amylopectin.[a]
Limit dextrinase EC 3.2.1.142	At5g04360	The enzyme hydrolyzes α-1,6 linkages of amylopectin and β-limit dextrin. The yeast α-1,4, α-1,6-linked glucan pullulan is a good substrate; glycogen is not.
Isoamylase; ISA1 and ISA2 EC 3.2.1.68	At2g39930 At1g03310	Recombinant ISA1 from potato is active; recombinant ISA2 from potato is not (12). Most of the isoamylase activity in potato (45) and *Arabidopsis*[b] is a function of a heterotetramer consisting of ISA1 and ISA2 proteins. The enzyme hydrolyzes α-1,6 linkages of amylopectin, β-limit dextrin, and glycogen, but not the yeast glucan pullulan.
Isoamylase; ISA3 EC3.2.1.68	At4g09020	Recombinant ISA3 from potato strongly prefers β-limit dextrin over amylopectin as a substrate (12).
Chloroplastic β-amylase EC 3.2.1.2	At3g23920 At4g00490 At4g17090 At5g55700	The enzyme is an exoamylase that hydrolyzes alternate α-1,4 linkages sequentially from the nonreducing end of a glucan chain, producing maltose. It cannot pass α-1,6 linkages, so the product of β-amylolysis of amylopectin is a dextrin with outer chains of two or three glucosyl units (a β-limit dextrin).
Chloroplastic glucan phosphorylase EC2.4.1.1	At3g29320	The enzyme catalyzes the conversion of the terminal glucosyl unit at the nonreducing end of glucan chains to glucose 1-phosphate, using inorganic phosphate. It cannot pass α-1,6 linkages. Chloroplastic isoforms prefer linear malto-oligosaccharides rather than large branched glucans as substrates (25, 64). Glucan(n) + inorganic phosphate = glucan(n-1) + glucose 1-phosphate.
Disproportionating enzyme; DPE1 2.4.1.25	At5g64860	The enzyme catalyzes the transfer of part of one α-1,4-linked glucan molecule (the donor) to another (the acceptor). The smallest donor molecule is maltotriose and the smallest acceptor is glucose, but the enzyme can use very large glucans as donors and acceptors. The preferred substrate is maltotriose, which is converted to glucose and maltopentaose (15).
Maltose transporter; MEX1	At5g17520	The *Arabidopsis* protein is located in the inner membrane of the chloroplast envelope and facilitates the transfer of maltose across the membrane (71). The mechanism is not known.
Cytosolic transglucosidase; DPE2	At2g40840	The *Arabidopsis* and potato enzymes catalyze the release of one of the glucosyl moieties of maltose and the transfer of the other to a glucan acceptor (14, 60). The nature of the endogenous acceptor is not known.
Cytosolic glucan phosphorylase; PHS2 EC 2.4.1.1	At3g46970	The enzyme catalyzes the same reaction as the chloroplastic isoform, but strongly prefers branched glucans rather than malto-oligosaccharides as substrates (25, 64).

*The table displays the EC numbers and chromosomal locations of *Arabidopsis* enzymes discussed in the text and shown in **Figure 1**, together with information about the reactions they catalyze. For amylases, phosphorylases, limit dextrinase, and disproportionating enzyme, catalytic properties have been characterized for several plant sources. In most of these cases the catalytic properties of the *Arabidopsis* enzymes have not been examined, but there is no reason to think that they will be substantially different from those of the enzymes from other plant species. For glucan, water dikinase, phosphoglucan, water dikinase, isoamylases, the maltose transporter, and the cytosolic transglucosidase, catalytic properties have been examined for the proteins from only one or two sources. The extent to which these properties will be shared by the proteins from other plant sources is not yet known.

[a]G. Ritte, personal communication.

[b]T. Delatte & S. Zeeman, unpublished data.

is necessary for starch degradation. T-DNA insertion mutants (referred to as knockout mutants) lacking either AMY1 (At4g25000), AMY2 (At1g76130), or AMY3 have normal rates of starch degradation in leaves at night. A triple mutant lacking all three α-amylases also has normal rates of starch degradation (128). It appears either that the initial attack on the granule surface does not require an endoamylase, or that *Arabidopsis* contains a novel endoamylase not recognizable from its primary amino acid sequence. These findings also leave open the role of AMY3. The AMY3 protein has α-amylase activity (128). It seems likely that it is involved in starch degradation, but—at least under the growth conditions we use—its absence can be compensated for by another enzyme.

The Importance of Starch Phosphorylating Enzymes

Whatever the mechanism of release of soluble glucans from the starch granule, it is clear that this process requires a newly discovered enzyme called glucan, water dikinase (GWD) (**Table 1**). Studies of the enzyme from potato show that GWD transfers the β-phosphate of ATP to either the 6- or the 3-position of glucosyl residues within amylopectin (67, 78). This phosphorylation probably occurs during both starch synthesis and starch degradation in vivo (70, 80, 112). Although phosphate groups occur at a low frequency in *Arabidopsis* leaf starch (about 1 in 2000 glucosyl residues is phosphorylated) (116), the presence of an active GWD appears to be essential for normal starch degradation. Mutations that eliminate the GWD protein or affect the dikinase domain of the enzyme dramatically reduce both the amount of phosphate in the amylopectin and the rate of starch degradation. Mature leaves of these mutants (*starch excess 1*, or *sex1*) accumulate amounts of starch up to seven times greater than those in wild-type leaves (13, 116, 117).

The simplest explanations for the involvement of GWD in starch degradation are that either the phosphate groups or the protein itself are necessary for the actions of the unknown enzyme(s) that attack the granule surface. The phosphate groups may influence the packing of the glucose polymers within the granule and hence the susceptibility of the granule surface to attack by enzymes (5, 6, 116). The GWD protein consists of both a dikinase domain and a large (approximately 120-kD) N-terminal domain of unknown function. This domain could be involved in promoting the activity of a granule-degrading enzyme, through an interaction with either the starch surface (see 79) or the enzyme itself.

Further complexity was recently added to this picture with the discovery that a second GWD-like enzyme [GWD3, or phosphoglucan, water dikinase (PWD)] (**Table 1**) is also required for normal starch metabolism (126, 127). Knockout mutants lacking this enzyme have increased amounts of leaf starch. However, unlike *sex1* mutants, the amount of phosphate in the starch of mutants lacking PWD is not dramatically affected. PWD is predicted to be chloroplastic. Although its C-terminal domain, like that of GWD, is closely related to those of other dikinases (67), its N-terminal domain is different from that of GWD. Recombinant PWD will phosphorylate amylopectin that already contains phosphate groups but, unlike GWD, it will not act on unphosphorylated glucans. This suggests that PWD action in vivo requires the presence of active GWD. It is possible to speculate that GWD and PWD actions together create a pattern of phosphorylation of amylopectin that makes it accessible to degradative enzymes at the granule surface, but much more information on these proteins and their products is necessary before their precise roles can be defined. There is also no information on the fate during starch degradation of the phosphate groups introduced by these enzymes.

Debranching

The *Arabidopsis* genome encodes four proteins predicted to catalyze the hydrolysis of α-1,6 linkages in glucan polymers (**Table 1**). One of these belongs to the limit dextrinase class and the remaining three to the isoamylase class of debranching enzymes (68). Limit dextrinase is present in high activity in germinating cereal endosperm, and is thought to be involved in the hydrolysis of the α-1,6 linkages during starch degradation in this organ (see below). However, a knockout mutant of *Arabidopsis* lacking limit dextrinase has normal rates of starch degradation in the leaf at night (T. Delatte & S. Zeeman, unpublished data). This indicates that one or more isoamylases is involved.

Understanding which debranching enzymes are involved in starch degradation is complicated by the fact that isoamylase also plays a role in starch synthesis. In *Arabidopsis* leaves, cereal endosperms, and potato tubers, reducing or eliminating measurable isoamylase activity causes abnormal starch synthesis. Starch granule numbers are increased, and at least some of the starch is replaced with a soluble α-1,4, α-1,6-linked glucan known as phytoglycogen (10, 12, 48, 122). The way in which isoamylase suppresses the initiation of starch granules and phytoglycogen particles in normal circumstances is not known. In *Arabidopsis*, mutations in either the *ISA1* or the *ISA2* (*DBE1*) gene cause the loss of detectable isoamylase activity and the disruption of normal starch structure (122; T. Delatte & S. Zeeman, unpublished data). In potato, the products of the *StISA1* and *StISA2* genes together form a heterotetramer responsible for the measurable isoamylase activity in the tuber (12, 45). It seems likely that the same is true for the enzyme in *Arabidopsis*.

Although ISA1 and ISA2 are indispensable for normal starch synthesis in *Arabidopsis* leaves, they are not necessary for hydrolysis of α-1,6 linkages during starch degradation at night. In *isa1* and *isa2* (*dbe1*) mutants, starch and phytoglycogen are both completely degraded at night (122; T. Delatte & S. Zeeman, unpublished data). In fact, given that phytoglycogen has a higher ratio of α-1,6 to α-1,4 linkages than amylopectin, the rate of degradation of α-1,6 linkages is probably faster than in wild-type leaves.

The lack of requirement for limit dextrinase, ISA1, or ISA2 in starch degradation suggests that ISA3 may be important in this process. Preliminary analyses of knockout mutants lacking ISA3 show that starch contents are higher than in wild-type leaves, consistent with a role for ISA3 in the hydrolysis of α-1,6 linkages at night (T. Delatte & S. Zeeman, unpublished data). The possibility that a further, unidentified activity is also involved cannot be ruled out at this stage.

Metabolism of Soluble, Linear Glucans

The initial stages of starch degradation proposed above will produce linear glucans that are soluble in the chloroplast stroma (**Figure 1**). Enzymes known to be present in the chloroplast can potentially catalyze two alternative pathways of further degradation. First, chloroplastic glucan phosphorylase (58, 118) (**Table 1**) can release glucose 1-phosphate, which can be converted to triose phosphate and exported from the chloroplast in exchange for inorganic phosphate via the triose-phosphate transporter (38). Second, β-amylase can catalyze the production of maltose from linear glucans (**Table 1**). Four of the nine β-amylases encoded in the *Arabidopsis* genome are predicted to be chloroplastic, and one of these proteins has been shown to be in the chloroplast (54).

Our recent results show that degradation of linear glucans in the *Arabidopsis* chloroplast usually proceeds via β-amylases rather than glucan phosphorylase. First, knockout mutants lacking chloroplastic glucan phosphorylase have normal rates of starch degradation under our growth conditions (121). Second, knockout mutants lacking one of the chloroplastic β-amylases have reduced rates of starch degradation (D. Fulton, H. Dunstan & S. Smith, unpublished data). It is not yet clear whether more than one β-amylase is required for normal rates of starch degradation. Third, plants lacking proteins required for maltose metabolism accumulate massive quantities of maltose in a pattern consistent with its production from starch, and have reduced rates of starch degradation (see below).

Fate of Maltose and Maltotriose

There are several ways in which maltose produced by β-amylolysis might be metabolized inside a chloroplast, but none of these appears to operate in *Arabidopsis*. For example, a maltose phosphorylase (EC 2.4.1.8) reported from pea leaf chloroplasts converted maltose to glucose 1-phosphate and glucose (52). However, this enzyme has not been reported from *Arabidopsis* and the genome does not encode a protein related to the maltose phosphorylase of bacteria (7). The genome encodes five enzymes classified as α-glucosidases (EC 3.2.1.20), at least some of which may be capable of hydrolyzing maltose to produce glucose, but none of these is predicted with confidence to be plastidial.

There is now strong evidence that maltose produced during starch degradation is exported to the cytosol via a specific transporter rather than metabolized inside *Arabidopsis* chloroplasts (**Figure 1**). Mutations at the *MEX1* locus cause accumulation of both starch and maltose in *Arabidopsis* leaves. Maltose levels are at least 40 times those of wild-type leaves. They rise at night and fall during the day, consistent with a block in the metabolism of maltose produced by β-amylolytic degradation of starch at night (71). *MEX1* encodes a protein located in the inner membrane of the chloroplast envelope (**Table 1**). When expressed in a mutant of *E. coli* lacking a component of the endogenous maltose transporter, MEX1 can restore the ability of this strain to grow on maltose (71). These data indicate that MEX1 is a maltose transporter responsible for the export to the cytosol of maltose produced during starch degradation in the chloroplast at night.

In addition to maltose, β-amylolytic degradation is also expected to produce a smaller amount of maltotriose because it is unable to act on chains of less than four glucosyl residues (e.g., 59). The only known maltotriose-metabolizing enzyme encoded by the genome and predicted to be chloroplastic is disproportionating enzyme, or α-1,4 glucanotransferase (DPE1) (**Table 1**), which can potentially convert two maltotriose molecules to one maltopentaose and one glucose. Evidence that this enzyme is responsible for maltotriose metabolism during starch degradation comes from a knockout mutant, which has a reduced rate of starch degradation and accumulates maltotriose at night to a far greater extent than wild-type leaves (15). We suggest that in the wild-type leaf the maltopentaose produced by disproportionating enzyme is acted on by β-amylase, and the glucose is exported from the chloroplast via the glucose transporter of the inner envelope (109) (**Figure 1**).

The Metabolism of Maltose in the Cytosol

The most obvious fate for maltose in the cytosol would be hydrolysis via an α-glucosidase, followed by conversion of the resulting glucose to glucose 6-phosphate via hexokinase. This is not, however, the pathway that appears to operate. Instead, maltose is metabolized via a transglucosylation reaction. Evidence for this fate for maltose comes from examining knockout mutants lacking a predicted transglucosidase (DPE2) (**Table 1**), structurally related to amylomaltases (7), which are involved in maltose metabolism in bacteria. Extracts of leaves of the mutants lack a transglucosidase activity, present in wild-type leaves, that can transfer one of the glucosyl moieties of maltose to branched polyglucans such as glycogen and release the other as glucose (14). The mutants have a phenotype similar to that of *mex1* mutants that lack the maltose transporter. Maltose levels are many times higher than those of normal plants, and starch degradation is inhibited (14, 62), suggesting that metabolism via DPE2 is the major or sole fate of maltose exported to the cytosol during starch degradation.

It is reasonable to assume that the free glucose released from maltose via DPE2 is converted to hexose phosphate via hexokinase. The fate of the second glucosyl moiety of maltose remains to be discovered. Presumably it is transferred by DPE2 to a cytosolic carbohydrate molecule, which is subsequently acted on by another enzyme that releases the glucosyl moiety either as glucose or as glucose phosphate (**Figure 1**, and see below).

Further Characterization of the Pathway

As described above, there are still major gaps in our understanding of the conversion of starch to hexose phosphate in *Arabidopsis* leaves. Here we suggest possible solutions to two of these outstanding problems and provide further support for novel parts of the pathway proposed above. The first outstanding problem is the nature of the enzyme responsible for attacking the starch granule. It remains entirely possible that the attack is catalyzed by an enzyme not recognizable as a α-1,4 endoglucanase from its predicted amino acid sequence. Continuing screens for mutant plants unable to degrade starch may lead to the identification of such an enzyme. An alternative possibility is that the attack on the granule is via β-amylase. β-amylase is an exoamylase that removes maltosyl units from the nonreducing ends of chains. It can act only on the outer chains of a branched glucan and cannot hydrolyze linkages beyond branch points. However, it could act in concert with a debranching enzyme, perhaps ISA3, to degrade starch granules progressively to maltose and maltotriose. The extent to which β-amylase can attack chains at the granule surface might be determined by the actions of GWD and PWD because the distribution of phosphate groups added to amylopectin by these enzymes affects the degree of crystalline packing of chains within the granule matrix (6). If this view of granule degradation is correct, the soluble linear and branched glucans postulated as the first products of starch degradation in **Figure 1** would not exist. Although unconventional, this scheme is appealing for its simplicity. Earlier studies concluded that β-amylase is not capable of degrading native starch granules (4, 24), but a chloroplastic β-amylase from potato leaves was recently shown to release malto-oligosaccharides from potato tuber starch granules (82).

The second major problem in understanding the pathway is the fate of the glucosyl moiety transferred by DPE2 from maltose to a carbohydrate acceptor. One possible candidate for the endogenous acceptor is a specific type of soluble arabinogalactan present in leaves of several species of plant (28, 114). This molecule is a mixed-linkage glycan consisting mainly of arabinose and galactose residues, with glucose, rhamnose, xylose, fucose, and mannose as minor components. Leaf and protoplast fractionation experiments indicate that it may be cytosolic: It is within the cell rather than associated with the wall, and is outside the chloroplast. Unlike cell-wall derived arabinogalactans, it is a good substrate in vitro for the ubiquitous cytosolic isoform of glucan phosphorylase (PHS2) (**Table 1**), which can transfer glucosyl moieties to the heteroglycan from glucose 1-phosphate. Thus, it is possible that glucosyl moieties from maltose are transferred to chains within this heteroglycan by DPE2 and released again as glucose 1-phosphate by PHS2 (**Figure 1**). This scheme is appealing because the PHS2 is likely readily reversible in vivo. PHS2 could potentially allow glucosyl moieties from the heteroglycan acceptor to be converted to hexose phosphate at a rate dependent on the rate of hexose phosphate utilization in the cytosol. The heteroglycan could thus act as a buffer between the rate of appearance in the cytosol of maltose produced in starch degradation and the rate of consumption of hexose phosphate in sucrose metabolism and cellular metabolism. We emphasize that this is speculation. Objective approaches are required to identify the acceptor and its further metabolism in vivo.

Although the first and latter stages of the pathway have yet to be discovered, there is good supporting evidence for the central portion in which chloroplastic maltose and maltotriose are converted to cytosolic hexose. In addition to the evidence discussed above, the creation of double mutants supports the view that the pathway presented in **Figure 1** accounts for most or all of the flux of carbon out of starch. Mutants lacking both MEX1 (so unable to metabolize maltose) and DPE1 (so unable to metabolize maltotriose) would be expected to be severely compromised, as no carbon from starch would be available for sucrose synthesis or cellular metabolism at night. The same should also be true of mutants that lack DPE1 and DPE2. Consistent with these expectations, *mex1/dpe1* (71) and *dpe2/dpe1* (T. Chia & A. Smith, unpublished data) mutants are extremely slow growing, much more so than their parental lines. In contrast, mutants lacking both MEX1 and DPE2 are very similar in appearance to their parental lines (T. Chia & A. Smith, unpublished data), consistent with the idea that MEX1 and DPE2 are consecutive steps on the same branch of the pathway.

CONTROL OF FLUX THROUGH THE PATHWAY IN *ARABIDOPSIS* LEAVES

Understanding the control of the pathway of starch degradation in *Arabidopsis* leaves presents two major challenges: switching on the pathway in response to darkness and other environmental signals, and controlling flux through the pathway at night. In standard growth conditions (see above) little or no starch degradation occurs while starch is being synthesized during the day. Pulse-chase experiments—in which $^{14}CO_2$ is supplied for a short period and then replaced with unlabeled carbon dioxide—indicate that carbon incorporated into granular starch is not released again until the light level falls below that required for net carbon dioxide assimilation (119). These experiments suffer from the technical difficulty that ^{14}C may be very rapidly buried by newly formed ^{12}C starch during the chase period, so that turnover of glucan chains at the surface is underestimated. However, release of ^{14}C was not observed even when light levels were lowered during the chase to a point only just sufficient for net assimilation, of which incorporation of ^{12}C into starch was minimal. Thus, under these growth conditions starch degradation is switched on by transition to conditions in which there is insufficient light for net carbon assimilation.

After the onset of the night, the rate of starch degradation increases over the first two hours and then remains relatively constant until the end of the night (117; authors' labs, unpublished data). At this point almost all of the starch in the leaf has been consumed. The rate of starch degradation is thus controlled such that all of the starch accumulated during the day is used to provide the cell with hexose phosphate at an almost constant rate throughout the night. This remarkable phenomenon occurs under a wide range of conditions. Mutations and environmental changes that decrease the accumulation of starch during the day result in a decrease in the rate of its consumption at night so that the supply of starch again allows a constant rate of hexose phosphate synthesis through the night (57). Except for photoperiods of more than 16 hours, alterations in the length of the photoperiod result in alterations of the rates of accumulation and degradation of starch so that the supply of starch again provides for a constant rate of hexose phosphate synthesis and is exhausted at the end of the night (34). These observations suggest that the control of the rate of degradation during the night is entrained to the amount of starch available for consumption and the expected length of the night. The importance of this control in the carbon economy of the plant is illustrated by experiments in which plants are subjected to a further period of darkness at the end of a normal 12-hour night. Within a few hours, massive changes in the transcriptome of the leaf indicate that the plant is entering carbon starvation. For example, expression of genes encoding enzymes involved in amino acid, lipid, and cell wall degradation increases (34, 106).

Regulation at the Level of Gene Expression

How are the switching-on and subsequent control of the rate of degradation achieved? The mechanisms are likely complex and are not yet understood, but some possibilities have emerged. The discovery that transcripts for several putative enzymes of starch degradation show strong diurnal rhythms and are under circadian control led to the suggestion that the diurnal control of starch degradation is primarily transcriptional (36). However, our examination of the diurnal pattern of transcript change for the enzymes in **Figure 1** casts strong doubt on the idea that transcription is of primary importance in switching on or controlling flux through the pathway (90).

First, diurnal patterns of transcript change are not the same as the pattern of change in the flux through the pathway of starch degradation. For several of the enzymes essential for starch degradation (DPE1, DPE2, ISA3, GWD, PWD), plus other glucan-metabolizing enzymes that may be involved (PHS1, PHS2, AMY3), there is a very similar pattern of transcript change in which levels are highest toward the end of the light period and lowest at the start of the light period (90). Although not mirroring changes in flux, this pattern provides high transcript levels at the time of day when starch degradation commences, and low levels at times when degradation does not usually occur. However, transcript levels for the four chloroplastic β-amylases and the maltose transporter MEX1 each show different diurnal patterns, none of which is obviously related to the pattern of starch degradation.

Second, there is no evidence so far that amounts of any protein necessary for starch degradation vary diurnally in *Arabidopsis* leaves in a manner consistent with a role in control of degradation. For several key enzymes with strong diurnal transcript changes, there is very little diurnal change in the amount of enzyme protein, and assayable activities remain high throughout the light period (90; authors' labs, unpublished data). A diurnally fluctuating endoamylolytic activity was reported from *Arabidopsis* leaves (50). However, neither this enzyme nor its subcellular location has been identified, and endoamylases appear to be unnecessary for normal rates of starch degradation in *Arabidopsis* leaves (see above). Levels of StDPE2 protein fluctuate diurnally in potato leaves grown in natural light, peaking in the first half of the night, but—in spite of strong diurnal changes in transcript levels—no equivalent fluctuation is seen in levels of DPE2 protein in *Arabidopsis* leaves grown in a controlled environment (60, 90).

Third, it is clear that the capacity for starch degradation remains high throughout the light as well as the dark period. Under appropriate conditions, high rates of starch degradation can be achieved in the middle of the light period. For example, if plants are transferred to low carbon dioxide and high oxygen concentrations (conditions that prevent net carbon assimilation and promote photorespiration) during the normal light period, starch degradation commences at once at a rate comparable with that in the dark period (S. Weise & T. Sharkey, personal communication).

Regulation of Enzyme Activity

pH, redox potential, and malto-oligosaccharide levels may all play a role in controlling starch degradation, via their effects on enzyme activities. All three of these factors undergo changes within the chloroplast during a light-dark transition. The pH of the stroma drops from about 8 to about 7 during a light-dark transition. This may promote activity of starch-degrading enzymes, but it is unlikely that it could switch on starch degradation or cause a major change in flux through the pathway (discussed in 98).

Redox potential presents interesting possibilities for flux control. Several chloroplast enzymes undergo very large changes in activity on light-dark transitions, mediated via oxidation and reduction of the sulfhydryl groups of cysteine residues. Reduction of the sulfhydryl groups is brought about by thioredoxin, which is reduced during photosynthesis by electrons from PSI, transferred via ferredoxin. Many enzymes interacting with thioredoxin were recently discovered, including a β-amylase from spinach (2). Recent work on ADPglucose pyrophosphorylase, the enzyme exerting most of the control over starch synthesis in *Arabidopsis* leaves, shows that it is activated by reduction (40). The level of activation is strongly influenced in a complex manner by sugars. It is tempting to suggest that starch degradation might be controlled in an analogous manner by oxidative activation of one or more key enzymes. Thus, degradation would be prevented during the day by reduction of the enzyme(s). Oxidation of this enzyme(s) at the onset of darkness would switch on the degradative process, and flux would be controlled by modulation of the redox state of the enzyme by sugars, linking the rate of degradation to the level of sucrose in the cytosol. The possibility that redox potential is involved in the control of starch degradation was dismissed in the past because phosphorolytic starch degradation in isolated chloroplasts is unaffected by dithiothreitol and because activities of chloroplastic α-amylase and phosphorylase are not modulated by thioredoxin or dithiothreitol (98). The discovery that neither α-amylase nor phosphorylase is necessary for starch degradation in vivo in *Arabidopsis* leaves reopens the possibility that redox potential may be involved in the control of the process.

Malto-oligosaccharide levels might act to control the rate of starch degradation by inhibiting a step in the pathway at or close to the attack on the granule, thus linking the rate of degradation of the granule to the rate of consumption of the products of degradation. We assume that this is what happens in the *dpe2*, *mex1*, and *dpe1* mutants, which accumulate abnormally high levels of maltose or maltotriose. The DPE1, MEX1, and DPE2 proteins do not attack the starch granule. However, losing any one of these enzymes causes a reduction in the rate of starch granule degradation (14, 15, 62, 71). It is likely that the malto-oligosaccharides inhibit an enzyme involved in the attack on the granule, perhaps by competing with granular starch for a starch-binding domain necessary for attack on the granule (113). Maltose also inhibits some β-amylases at high concentration (59).

Strong evidence for the involvement of protein phosphorylation in control of starch degradation was recently obtained from study of a starch-excess mutant *sex4*. We discovered that mutations at *sex4* lie in a gene encoding a protein phosphatase (authors' labs & J. Chen, unpublished data). This mutant accumulates up to three times more starch than the wild type under standard growth conditions because of a very low rate of degradation at night (117, 118). Amounts of the chloroplastic α-amylase AMY3 are reduced in this mutant (118). However, AMY3 is not encoded at the *sex4* locus and knockout mutants for AMY3 have normal rates of starch degradation (see above), so it is unlikely that the decrease in starch degradation in *sex4* is caused by its effect on AMY3 levels. The targets of this enzyme are not yet known, but it seems highly likely that either an enzyme of starch degradation or a protein controlling flux indirectly is a target for inactivation by phosphorylation. Whether this mechanism is of major importance in controlling flux in vivo, and whether it interacts with other possible regulatory mechanisms such as redox potential, remains to be established. However, discovering a protein phosphatase necessary for normal starch degradation presents the first direct evidence for a specific regulatory mechanism for this process and should allow further rapid process in this area.

The timing and pattern of starch degradation in plants growing in natural conditions is likely different from that in controlled environments. In some species there is good evidence that degradation of leaf starch can occur in the light when light levels are relatively low at the start and end of the day (e.g., 31). Starch degradation in *Arabidopsis* leaves is also switched on in the light by conditions that promote high rates of photorespiration (T. Sharkey, personal communication). Thus, the mechanisms that control the rate of degradation probably respond to complex and varying environmental and metabolic inputs in natural conditions. Data obtained from controlled environments will not necessarily reveal the full range of factors that can affect the onset and rate of degradation.

THE PATHWAY IN LEAVES OF OTHER SPECIES

Little direct evidence is available for any species other than *Arabidopsis* about which enzymes are important in leaf starch degradation. Activities of starch-degrading enzymes have been reported in leaves of numerous species, but most reports do not identify the isoforms that contribute to activity or their subcellular locations. Indirect evidence suggests that there may be variation in the pathway of starch degradation between species and with changing developmental and environmental conditions. Interspecific variations might be expected because there are enormous differences between species in the extent of accumulation of starch during photosynthesis and its importance in supplying carbon to the plant at night. Some species accumulate little or no starch in their leaves during the day and rely primarily on vacuolar sucrose and, in grasses, fructans for a continued carbohydrate supply at night. In other species the diurnal pattern of starch accumulation and degradation changes dramatically through leaf development. In tobacco, for example, there is a strong diurnal change in starch content at night in younger leaves whereas in older leaves progressively less of the starch is utilized at night and levels increase. During senescence all of this starch is degraded and the products are exported from the leaf (65, 66). Below we discuss likely similarities and differences between species in the pathway of degradation of leaf starch.

The Production of Soluble, Linear Glucans

There is no robust information about which enzyme attacks the starch granule in leaves of any plant species. Chloroplastic α-amylases similar to AtAMY3 are probably widespread. The three amylases of *Arabidopsis* are representatives of three distinct families of these enzymes found in higher plants. Evidence of expression of family 3 amylases—the family to which AMY3 belongs—has been reported for a wide range of dicots and for the gymnosperm *Pinus taeda*, and this family may also be represented in cereals (94). Endoamylase activity is present in chloroplasts of pea, spinach, sugar beet, and maize (bundle sheath), although the reported level of chloroplastic activity varies considerably between species and between different studies of single species (26, 56, 73, 74, 123). However, there is no direct evidence that endoamylases are necessary for starch degradation in leaves.

It seems likely that GWD is involved in starch degradation in leaves of a wide range of species. GWD protein and transcript have been reported for leaves of several species (77–79), and the amylopectin of leaf starches thus far examined contains phosphate groups (5, 116). For potato leaves, there is good evidence that GWD is necessary for normal rates of starch degradation, as in *Arabidopsis* leaves. Leaves of transgenic plants with reduced levels of GWD (formerly known as R1 protein) have a higher starch content than those of untransformed plants and fail to degrade all of their starch even after prolonged periods in the dark (61). The widespread occurrence of GWD may indicate that the nature and control of the attack on the starch granule is conserved between leaves of different species.

Proteins closely related to all four of the debranching enzymes present in *Arabidopsis* leaves are found in a wide range of plant species. Although the roles of isoamylases similar to AtISA1 and AtISA2 have not been examined in leaves other than those of *Arabidopsis*, it is likely that they are involved in starch synthesis rather than degradation in plant organs generally. StISA1 and StISA2 are necessary for normal starch synthesis in tubers of potato (12, 45), and ISA1-type isoamylases are involved in starch synthesis in cereal endosperms (11, 48, 53). Unlike *Arabidopsis*, limit dextrinase (pullulanase) is required in maize leaves for normal starch metabolism. The rates of both degradation and synthesis of starch are reduced in knockout maize plants lacking this enzyme (22). Nothing is yet known about the function of ISA3-type isoamylase in maize or in any species other than *Arabidopsis*. It may be that both ISA3 and limit dextrinase are capable of hydrolyzing α-1,6 linkages during starch degradation in leaves, and that their relative importance differs between species.

Hydrolytic versus Phosphorolytic Degradation

The debate about the relative importance in leaves of hydrolytic degradation via β-amylase versus phosphorolytic degradation via glucan phosphorylase (see **Figure 1**) is long standing. It is clear that chloroplasts from many species possess the capacity for both pathways. Work with transgenic and mutant plants has provided both direct and indirect evidence that degradation via β-amylase makes an important contribution to chloroplast starch degradation in a range of species. There is good direct evidence that β-amylase is necessary for normal starch degradation in leaves of potato. Transgenic plants with reduced activities of a chloroplastic isoform have reduced rates of starch degradation (82).

Good indirect evidence for the operation of a hydrolytic pathway of starch degradation comes from mutant and transgenic plants in which conversion of chloroplastic triose phosphate to cytosolic hexose phosphate is blocked, for example by lack of the triose phosphate transporter or cytosolic FBPase. Such plants have drastically reduced rates of export of carbon from the chloroplast during the day, and consequently accumulate much larger amounts of starch than wild-type plants. At night, all of this starch is degraded and the products are used for sucrose synthesis (38, 39, 87, 125). This implies that products of starch degradation are not exported from the chloroplast as triose phosphate at night but rather as hexose units and/or maltose. Nuclear magnetic resonance experiments also show that carbon for sucrose synthesis is exported from chloroplasts as hexose units or maltose, rather than triose phosphate, in intact bean and tomato leaves in the dark (83). There are strong indications that the chloroplast envelope has little or no capacity to transport hexose phosphate (30, 110). However, isolated chloroplasts from several sources produce maltose and glucose during starch degradation and can transport both of these compounds across the envelope (41, 75, 81, 86, 99, 111). Thus, these indirect lines of evidence are consistent with a predominantly hydrolytic route of starch degradation.

The involvement of β-amylase does not preclude the operation of a phosphorolytic pathway of degradation via a chloroplastic glucan phosphorylase. The activity of this enzyme in the *Arabidopsis* chloroplast is very low (58, 118) and probably insufficient to catalyze the observed rate of starch degradation, but in some other species phosphorylase activity in the chloroplast is much higher (for example, pea: 35, 73, 97). Isolated chloroplasts of spinach and pea are capable of phosphorolytic starch degradation (96, 98, 99). However, there is no evidence for a major role for glucan phosphorylase in starch degradation in leaves of any species. Carbohydrate metabolism was apparently normal in leaves of transgenic potato plants with no detectable chloroplastic glucan phosphorylase (91). It has been suggested that phosphorolytic starch degradation provides carbon for specific functions within the photosynthetic cell—for example for respiration at night (100), for the chloroplastic oxidative pentose phosphate pathway at night (121), or to maintain levels of Calvin cycle intermediates under photorespiratory conditions during the day (S. Weise & T. Sharkey, personal communication). The relative rates of phosphorolytic and hydrolytic degradation may thus vary with developmental stage and environmental conditions.

Maltose Metabolism

Consistent with the widespread involvement of β-amylase in leaf starch degradation, the proteins that act on its immediate products in *Arabidopsis* leaves—the maltose transporter MEX1, the chloroplastic disproportionating enzyme DPE1, and the cytosolic transglucosidase DPE2—are also of widespread occurrence. Genomic sequences or ESTs for proteins closely related to those in *Arabidopsis* have been reported from a wide range of species including both dicots and monocots (14, 62, 71, 104). The importance of MEX1 in starch degradation in leaves other than those of *Arabidopsis* has not yet been examined. Potato plants with drastically reduced levels of DPE1 grow more slowly than normal plants, suggesting a role for this enzyme in leaf starch metabolism (104). Leaves of potato plants with reduced levels of DPE2 have reduced rates of starch degradation, implying that this is a major, if not the sole, route of maltose metabolism (60). However, the DPE2 of potato is reported to be chloroplastic (60) whereas that in *Arabidopsis* is reported to be cytosolic (14, 62). A chloroplastic location for this transglucosidase implies a different route for starch degradation from that presented in **Figure 1**. This potentially major difference between potato and *Arabidopsis* is under further investigation.

THE PATHWAY IN OTHER PLANT ORGANS

A huge amount of information is available about the occurrence of starch-degrading enzymes in nonphotosynthetic organs of plants. For the most part, this information cannot be used to deduce the pathway of starch degradation because individual isoforms and their subcellular locations are not known, and the precise role and importance of enzymes cannot be assessed. Whereas in leaves starch degradation can be studied over a defined, short, and controllable period, the process in many other organs is much less amenable to study. Starch degradation may occur over long periods, during which there are profound changes in the developmental state of the organ concerned. Frequently, the extent and rate of degradation cannot be accurately measured because it may occur in parallel with starch synthesis and at different rates in different parts of the organ. Nonetheless, there are general reasons to believe that starch degradation in some nonphotosynthetic organs proceeds by a pathway different from that in *Arabidopsis* leaves. First, in some organs starch degradation does not occur inside a plastid. Second, the nature of the starch granule and its appearance during degradation differ substantially between plant organs. It is also clear that the control of degradation—if not the pathway—differs between organs. A host of developmental, metabolic, and environmental factors influence starch degradation in various organs. In a few instances control of expression of genes encoding specific enzymes of degradation has been studied in detail (e.g., 7, 51), but for the most part—reflecting the lack of knowledge of the pathway itself—the control of flux through the pathway as a whole is not understood. Rather than attempt to review the whole body of literature on the pathway of starch degradation and its control in nonphotosynthetic organs, we consider some examples for which there is sufficient information to make meaningful comparisons with the pathway presented in **Figure 1**. We deal first with starch-storing organs such as seeds and tubers in which large amounts of starch are stored over long periods (see **Figure 2**), then consider the pathway in cells in which starch forms a relatively transient store of carbon over a particular developmental period.

The Attack on the Granule in Storage Organs

For some starch-storing storage organs, there is good reason to think that the attack on the granule is catalyzed by α-amylase. The onset of starch degradation during germination of cereal seeds is accompanied by massive de novo synthesis of α-amylase. The control of this synthesis in the aleurone and scutellum, and the secretion of the enzyme into the starch-storing endosperm, is understood in detail at a molecular level (29, 76). The pattern of attack on the starch granule and the nature of the limit dextrins that appear in the endosperm as starch is degraded are both consistent with the idea that α-amylase participates in granule degradation (24). Experiments in vitro show that α-glucosidase from cereal endosperm can also attack cereal starch granules, and that this enzyme and α-amylase interact synergistically to promote granule degradation (88, 102). The extent of involvement of α-glucosidase in the attack on starch granules in vivo is not known.

The very substantial increases in α-amylase activity during the first few days of germination of starch-storing legume seeds are consistent with the idea that here too it is responsible for the attack on the starch granule (16, 105, 115). Changes in the properties of starch during degradation in germinating pea seeds also point to an endoamylolytic attack (49). An endoamylase is also implicated in the attack on the granule in one of the most spectacular examples of starch degradation in plants: the thermogenic spadix of *Arum maculatum*. In this organ—and those of other thermogenic members of the Araceae—enormous rates of degradation occur over just a few hours, fueling uncoupled respiration that heats the spadix to temperatures 10 or more degrees above ambient. Rates of starch degradation are typically more than 60 times greater on a fresh weight basis than those in *Arabidopsis* leaves under standard growth conditions at night (1). During spadix development prior to thermogenesis, the activity of an endoamylase increases dramatically, and the appearance of oligosaccharides in the spadix during starch degradation is consistent with the idea that this amylase is responsible for granule degradation (9).

For other types of starch-storing organs the nature of the attack on the starch granule is much less clear. In sprouting potato tubers, starch degradation is not initiated in a uniform manner across the tuber, and there are no consistent reports of accompanying substantial increases in either amylolytic or phosphorolytic starch-degrading enzymes (18–20). Starch degradation—accompanied by increases in sugars—also occurs when tubers are stored at low temperatures (cold-induced sweetening). A specific isoform of β-amylase increases in activity during this process, but its importance in degradation is not known (21, 69). The balance of information indicates that degradation during sprouting and cold-induced sweetening occurs within plastids. Although amyloplast membranes were reported to disappear during cold-induced sweetening (72), other reports, and the rapidly reversible nature of this process (21, 46, 47, 92), argue against loss of amyloplast integrity during starch degradation prior to tuber senescence.

An indication that starch degradation proceeds differently in cereal endosperms and potato tubers comes from the appearance of the granules during degradation. Granules from cereal endosperm have abundant channels leading from pores on the surface to the interior (3, 27, 42, 43). During degradation—both in vitro and in the germinating endosperm—they become deeply pitted, with loss of internal material surrounding the channels before much of the surface has been attacked. In contrast, granules from potato tubers have few if any pores or channels running inward from the surface. They are highly resistant to enzymic attack, and damage appears on the surface as degradation proceeds (23, 32, 33, 55). Because of these very different properties, it is tempting to suggest that the type of enzyme responsible for the attack on the granule in potato tubers may be different from that in cereal endosperm. Further information about the process in potato tubers is required to resolve this issue.

Starch in storage organs other than cereal endosperm is extensively phosphorylated, implying a possible role for GWD in controlling degradation. In mung bean seeds the level of phosphate in the starch is comparable with that in leaves, in potato tubers the level is up to 10 times that in *Arabidopsis* leaves, and in rhizomes of *Curcurma zedoaria* (ginger family) it is 3 times higher than that in potato tubers (5, 6). GWD homologues occur in a very wide range of species, and the enzyme may be ubiquitous in higher plants (78). For potato tuber, there is good indirect evidence that GWD is required for normal rates of starch degradation. Transgenic potatoes with greatly reduced levels of GWD had very low levels of phosphate in tuber starch and were less prone to starch loss and sugar accumulation when stored at low temperatures (cold-induced sweetening, see above), implying that mobilization of starch was

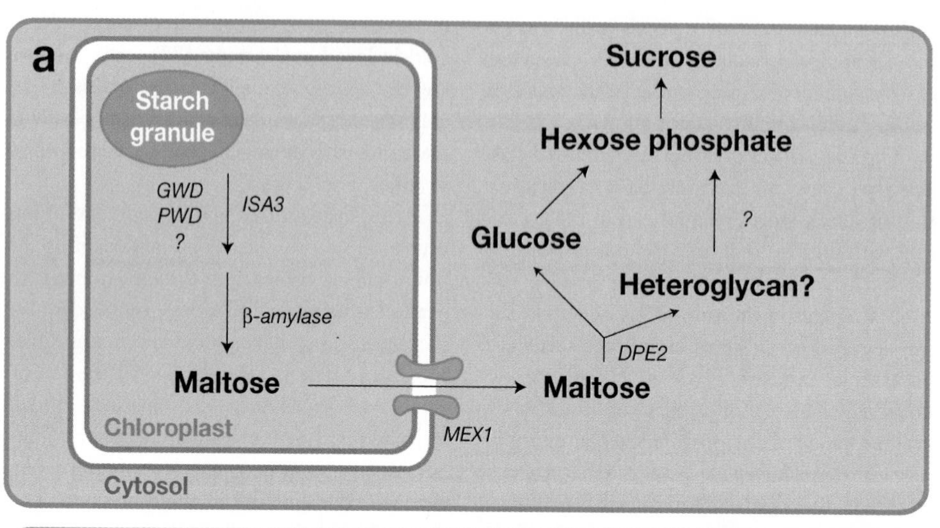
a
Starch granule
GWD
PWD
?
ISA3
β-amylase
Maltose
Chloroplast
Cytosol
MEX1
Maltose
DPE2
Glucose
Heteroglycan?
?
Hexose phosphate
Sucrose

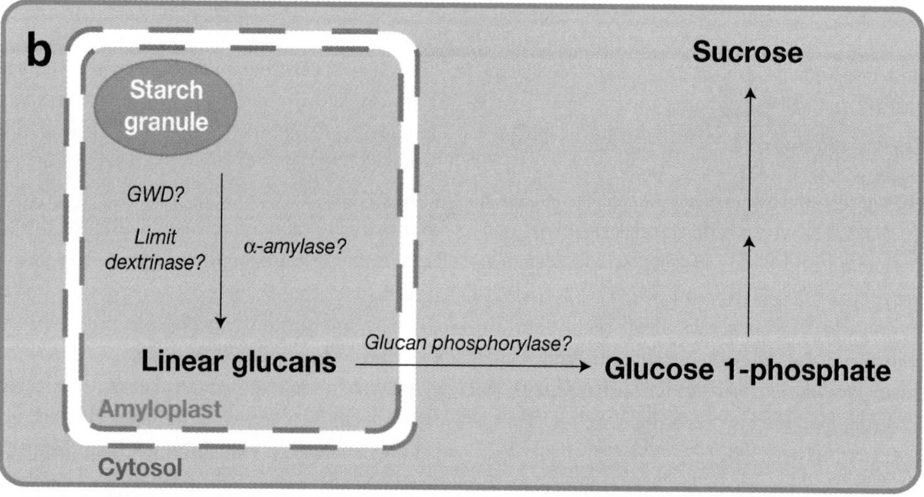
b
Starch granule
GWD?
Limit dextrinase?
α-amylase?
Linear glucans
Amyloplast
Cytosol
Glucan phosphorylase?
Glucose 1-phosphate
Sucrose

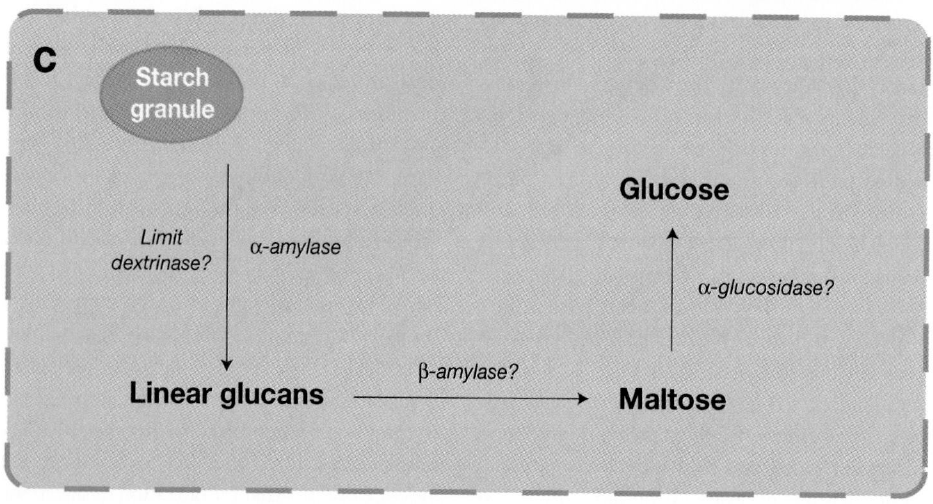
c
Starch granule
Limit dextrinase?
α-amylase
Linear glucans
β-amylase?
Maltose
α-glucosidase?
Glucose

impaired in these conditions (61). In contrast, the starch of most cereal endosperms contains almost undetectably low levels of phosphate. It is doubtful whether GWD plays a role in degradation in these organs (5, 6).

Degradation of Soluble Glucans in Storage Organs

The pathway of degradation of soluble glucans released from starch granules likely varies between organs because of the very different physical circumstances in which degradation occurs. In cereal endosperm, starch degradation takes place in a nonliving tissue—effectively in an acidic, apoplastic environment in which no intracellular or intercellular compartmentation remains. In starch-storing legume seeds, starch degradation takes place within living cells of the cotyledons, but probably not within the plastid in which the starch was synthesized. The plastid envelope is believed to disintegrate prior to germination so that degradation occurs within the cytosol (37, 108) (**Figure 2**). In potato tubers, the balance of evidence indicates that degradation takes place within intact plastids (see above).

In cereal endosperm, the degradation of glucans released from starch granules probably proceeds via limit dextrinase, α- and β-amylase, and α-glucosidase to maltose and glucose, which can enter the embryo (**Figure 2**). These enzymes are either synthesized within surrounding cell layers or mobilized within the endosperm as degradation proceeds. Study of a knockout mutant of maize shows that limit dextrinase is necessary for normal rates of starch degradation in the endosperm during the early stages of germination (up to seven days) but not thereafter (22). Questions about the importance of limit dextrinase in starch degradation in barley endosperm have been raised by the demonstration that newly synthesized enzyme in aleurone cells is targeted to the plastids and not to the secretory pathway. The extent to which the enzyme enters the endosperm and participates in starch degradation during germination is unclear (11, 84). The possibility remains open that another type of debranching enzyme also participates in starch degradation in cereal endosperm. A protein that binds to and strongly inhibits limit dextrinase in vitro (limit dextrinase inhibitor) is synthesized during endosperm development and present in the germinating seeds of some cereals (63, 93), but its importance in determining the course of starch degradation during germination is unclear. The extent to which β-amylase and α-glucosidase are necessary for starch degradation in cereal endosperm has been the subject of considerable debate (e.g., 17, 24, 103), and both this and the way in which the products of degradation in the endosperm are exported to the embryo (e.g., 44) remain unresolved.

The degradation of glucans produced from starch in germinating pea embryos is proposed to proceed via limit dextrinase and glucan phosphorylase in the cytosol (**Figure 2**), although

Figure 2

Comparison of the pathway of starch degradation in leaves with those proposed for germinating legume cotyledons and germinating cereal endosperm. (*a*) Major pathway by which starch is converted to sucrose in leaves (see **Figure 1** for more detail). Starch degradation occurs within the chloroplast, and maltose is exported to the cytosol for further metabolism to sucrose. (*b*) Proposed pathway by which starch is converted to sucrose in germinating legume cotyledons. The amyloplast membrane disintegrates so that starch degradation is catalyzed by cytosolic enzymes. Glucan phosphorylase is important in the degradation of soluble glucans. Sucrose synthesized from starch is exported from the starch-storing cells to the growing root and shoot of the seedling. (*c*) Proposed pathway by which starch is converted to maltose and glucose in germinating cereal endosperm. Both the amyloplast envelope and the plasma membrane disintegrate so that starch degradation takes place in a nonliving tissue. The granule is attacked by α-amylase. The precise roles and importance of α-amylase and other starch-degrading enzymes in metabolizing soluble glucans are not clear. Glucose and maltose produced in the endosperm are exported from the endosperm into the scutellum, then converted to sucrose for the growth of the embryo.

direct information about the importance of these enzymes is lacking. Limit dextrinase activity is present both inside and outside the plastid during embryo development. The activity of the extraplastidial form increases as the seed matures and remains high in the germinating embryo (124). Activity of a cytosolic isoform of glucan phosphorylase is low during seed development and then increases dramatically during the first five days of germination (64, 108).

The pathway of degradation of soluble glucans derived from starch in potato tubers remains unclear. Tubers of transgenic plants with reduced activity of cytosolic glucan phosphorylase have normal starch contents and actually sprout better than normal tubers, suggesting that this enzyme is not necessary for starch degradation (25). Tubers with reduced activities of disproportionating enzyme produce sprouts more slowly than normal tubers (104), and reductions in DPE2 have no effect on sprouting (60). Unfortunately these observations provide little information about the roles of the enzymes in tuber starch degradation: Activities of the enzymes have been reduced rather than eliminated in tubers and effects on starch degradation have not been directly assessed.

The Pathway in Vegetative Tissues

In many plant organs and cell types, starch accumulates and then disappears on a timescale of a few days (89). In *Arabidopsis*, for example, starch appears transiently during development of the embryo and testa (mature seeds contain little or no starch) and in the root cap prior to sloughing off. Although the pathway of degradation of these temporary starch reserves is not known, preliminary indications in *Arabidopsis* are that it may be the same as that presented for leaves in **Figure 1**. First, mutations that affect starch degradation in the leaf also affect the process in several other parts of the plant. In a study of the *sex1* mutant (lacking GWD) Caspar et al. (13) reported that starch accumulated to levels greater than those in the wild type in seeds, roots, petals, and anthers as well as leaves. Starch is present in mature seeds of the starch-excess mutants lacking disproportionating enzyme (*dpe1*), the chloroplastic maltose transporter (*mex1*), and the cytosolic transglucosidase (*dpe2*). Again consistent with observations in leaves, starch content is unaffected in developing seeds and other plant parts of mutants lacking plastidial glucan phosphorylase or the plastidial α-amylase AMY3 (authors' labs, unpublished data). Second, genes encoding enzymes important in starch degradation in leaves are widely expressed in other parts of the plant. For example, MEX1 was originally identified as a gene expressed in root caps (*RCP1* or root cap1) (107). We emphasize that these are preliminary observations rather than systematic analyses, but at present there is no good reason to believe that starch degradation proceeds via distinctly different pathways in different organs of the *Arabidopsis* plant.

CONCLUSION

Although we can present an experimentally based model of the pathway of starch degradation in *Arabidopsis* leaves, this is not possible at present for any other plant organ. However, the balance of evidence suggests that the pathway may be the same or similar in other leaves, and perhaps in nonphotosynthetic tissues in which starch is stored transiently. In some starch-storing organs in which starch is a major, long-term reserve compound, the pathway of degradation is different. These differences are probably related to the different subcellular locations in which the process is believed to occur (the apoplast in cereal endosperms, the cytosol in legume seeds, and the plastid in potato tubers) and the different ways in which the products of degradation are utilized in these organs.

We suggest that rapid progress in understanding this vital and commercially important pathway should now be possible. Information from *Arabidopsis* leaves, together with a wealth of existing descriptive and correlative observations for other species, allows hypotheses about the nature of the pathway to be formulated for other plant organs. The increasing availability of transformation, TILLING, and knockout technologies for crop and other model species should allow these hypotheses to be tested definitively in the near future. However, we emphasize that our understanding of the process in *Arabidopsis* is far from complete. Much further work is required to define this pathway fully, to test whether the present picture holds under natural environmental conditions, and in particular to understand how flux through the pathway is controlled.

SUMMARY POINTS

1. Forward and reverse genetic tools combined with biochemical approaches have produced a radically new picture of the pathway of starch degradation in *Arabidopsis* leaves at night.
2. The first steps in the pathway, in which glucans are released from the starch granule, remain unknown. Surprisingly, α-amylases are not required. A glucan water dikinase and a phosphoglucan water dikinase, which add phosphate groups to amylopectin, are necessary, but their influence on granule degradation is not understood.
3. Metabolism of glucans released from the starch granule is via β-amylase rather than glucan phosphorylase. The maltose product is exported from the chloroplast to the cytosol via a recently discovered maltose transporter.
4. Maltose in the cytosol is metabolized via a transglucosylation reaction. Details of this and downstream reactions in the conversion of maltose to hexose phosphate remain to be discovered.
5. Flux through this pathway is clearly subject to complex control, but the control mechanisms are poorly understood. Although levels of transcripts for several of the enzymes vary considerably on a diurnal basis, there are good reasons to think that control of flux is primarily posttranslational rather than transcriptional.
6. Information is sparse about starch degradation in leaves of other species of plant, and in nonphotosynthetic plant parts in which starch accumulates transiently. At present it seems likely that pathways similar to those in the *Arabidopsis* leaf may operate in these organs.
7. Knowledge of starch degradation in starch-storing organs such as cereal endosperm and legume seeds is likewise incomplete, but the process probably differs considerably between different types of storage organs and between storage organs and *Arabidopsis* leaves.

ACKNOWLEDGMENTS

We are very grateful to our colleague Professor Jychian Chen for many fruitful discussions and collaborations on *Arabidopsis* starch metabolism over several years. We also thank Professors Tom Sharkey, Mark Stitt, and Martin Steup and Drs. Lone Baunsgaard, James Lloyd, and Gerhard Ritte for making unpublished data and observations available to us, and members of our labs past and present for their hard work and insights.

LITERATURE CITED

1. ap Rees T, Wright BW, Fuller WA. 1977. Measurements of starch breakdown as estimates of glycolysis during thermogenesis by the spadix of *Arum maculatum*. *Planta* 134:53–56

2. Balmer Y, Koller A, del Val G, Manieri W, Schürmann P, Buchanan BB. 2003. Proteomics gives insights into the regulatory function of chloroplast thioredoxins. *Proc. Natl. Acad. Sci. USA* 100:370–75
3. Banks W, Muir DD. 1980. Structure and chemistry of the starch granule. In *The Biochemistry of Plants*, ed. J Preiss, 3:321–69. New York: Academic
4. Beck E, Ziegler P. 1989. Biosynthesis and degradation of starch in higher plants. *Annu. Rev. Plant Physiol. Plant Mol. Biol.* 40:95–117
5. Blennow A, Engelsen SB, Munck L, Møller BL. 2000. Starch molecular structure and phosphorylation investigated by a combined chromatographic and chemometric approach. *Carbohydr. Polymers* 41:163–74
6. Blennow A, Engelsen SB, Nielsen TH, Baunsgaard L, Mikkelsen R. 2002. Starch phosphorylation: a new front line in starch research. *Trends Plant Sci.* 7:445–50
7. Boos W, Schuman H. 1998. Maltose/maltodextrin system of *Escherichia coli*: transport, metabolism, and regulation. *Microbiol. Mol. Biol. Rev.* 62:204–29
8. Briggs DE. 1973. Hormones and carbohydrate metabolism in germinating cereal grains. In *Biosynthesis and its Control in Plants*, ed. BV Milborrow, pp. 219–77. London: Academic
9. Bulpin PV, ap Rees T. 1978. Starch breakdown in the spadix of *Arum maculatum*. *Phytochem.* 17:391–96
10. Burton R, Jenner H, Carrangis L, Fahy B, Fincher G, et al. 2002. Starch granule initiation and growth are altered in barley mutants that lack isoamylase activity. *Plant J.* 31:97–112
11. Burton RA, Zhang XQ, Hrmova M, Fincher GB. 1999. A single limit dextrinase gene is expressed both in the developing endosperm and in germinated grains of barley. *Plant Physiol.* 119:859–71
12. Bustos R, Fahy B, Hylton CM, Seale R, Nebane NM, et al. 2004. Starch granule initiation is controlled by a heteromultimeric isoamylase in potato tubers. *Proc. Natl. Acad. Sci. USA* 101:2215–20
13. Caspar T, Lin TP, Kakefuda G, Benbow L, Preiss J, Somerville C. 1991. Mutants of *Arabidopsis* with altered regulation of starch degradation. *Plant Physiol.* 95:1181–88
14. Chia T, Thorneycroft D, Chapple A, Messerli G, Chen J, et al. 2004. A cytosolic glucosyltransferase is required for conversion of starch to sucrose in *Arabidopsis* leaves at night. *Plant J.* 37:853–63
15. Critchley JH, Zeeman SC, Takaha T, Smith AM, Smith SM. 2001. A critical role for disproportionating enzyme in starch breakdown is revealed by a knock-out mutant in *Arabidopsis*. *Plant J.* 26:89–100
16. Dale JE. 1969. Gibberellins and early growth in seedlings of *Phaseolus vulgaris*. *Planta* 89:155–64
17. Daussant J, Zbaszyniak B, Sadowski J, Wiatroszak I. 1981. Cereal β-amylase: immunochemical study on two enzyme-deficient inbred lines of rye. *Planta* 151:176–79
18. Davies HV, Ross HA. 1984. The pattern of starch and protein degradation in tubers. *Potato Res.* 27:373–81
19. Davies HV, Ross HA. 1986. Hydrolytic and phosphorolytic enzyme activity and reserve mobilisation in sprouting tubers of potato (*Solanum tuberosum* L.). *J. Plant Physiol.* 126:379–86
20. Davies HV. 1990. Carbohydrate metabolism during sprouting. *Amer. Potato J.* 67:815–27
21. Deiting U, Zrenner R, Stitt M. 1998. Similar temperature requirement for sugar accumulation and for the induction of new forms of sucrose phosphate synthase and amylase in cold-stored potato tubers. *Plant Cell Env.* 21:127–38

22. Dinges JR, Colleoni C, James MG, Myers A. 2003. Mutational analysis of the pullulanase-type debranching enzyme of maize indicates multiple functions in starch metabolism. *Plant Cell* 15:666–80
23. Dronzek BL, Hwang P, Bushuk W. 1972. Scanning electron microscopy of starch from sprouted wheat. *Cereal Chem.* 49:232–39
24. Dunn G. 1974. A model for starch breakdown in higher plants. *Phytochem.* 13:1341–46
25. Duwenig E, Steup M, Willmitzer L, Kossmann J. 1997. Antisense inhibition of cytosolic phosphorylase in potato plants (*Solanum tuberosum* L.) affects tuber sprouting and flower formation with only little impact on carbohydrate metabolism. *Plant J.* 12:323–33
26. Echeverria E, Boyer CD. 1986. Localization of starch biosynthetic and degradative enzymes in maize leaves. *Amer. J. Bot.* 73:167–71
27. Fannon JE, Hauber RJ, BeMiller JN. 1992. Surface pores of starch granules. *Cereal Chem.* 69:284–88
28. Fettke J, Eckermann N, Poeste S, Pauly M, Steup M. 2004. The glycan substrate of the cytosolic (Pho 2) phosphorylase isozyme from *Pisum sativum* L.: identification, linkage analysis, and subcellular localisation. *Plant J.* 39:933–46
29. Fincher GB. 1989. Molecular and cellular biology associated with endosperm mobilization in germinating cereal grains. *Annu. Rev. Plant Physiol. Plant Mol. Biol.* 40:305–46
30. Flügge UI. 1999. Phosphate translocators in plastids. *Annu. Rev. Plant Physiol. Plant Mol. Biol.* 50:27–54
31. Fondy BR, Geiger DR, Servaites JC. 1989. Photosynthesis, carbohydrate metabolism, and export in *Beta vulgaris* L. and *Phaseolus vulgaris* L. during square and sinusoidal light regimes. *Plant Physiol.* 89:396–402
32. Fuwa H, Nakajima M, Hamada A, Glover DV. 1977. Comparative susceptibility to amylases of starches from different plant species and several single endosperm mutants and their double-mutant combinations with *opaque*-2 inbred Oh43 maize. *Cereal Chem.* 54:230–37
33. Gallant D, Mercier C, Guilbot A. 1972. Electron microscopy of starch granules modified by bacterial α-amylase. *Cereal Chem.* 49:354–65
34. Gibon Y, Bläsing OE, Palacios-Rojas N, Pankovic D, Hendriks JHM, et al. 2004. Adjustment of diurnal starch turnover to short days: depletion of sugar during the night leads to a temporary inhibition of carbohydrate utilization, accumulation of sugars and post-translational activation of ADP-glucose pyrophosphorylase in the following light period. *Plant J.* 39:847–62
35. Hammond JBW, Preiss J. 1983. Spinach leaf intra and extra chloroplast phosphorylase activities during growth. *Plant Physiol.* 73:709–12
36. Harmer SL, Hogenesch JB, Straume M, Chang HS, Han B, et al. 2000. Orchestrated transcription of key pathways in *Arabidopsis* by the circadian clock. *Science* 290:2110–13
37. Harris N. 1976. Starch grain breakdown in cotyledon cells of germinating mung bean seeds. *Planta* 129:271–72
38. Häusler RE, Schlieben NH, Schulz B, Flügge UI. 1998. Compensation of decreased triose phosphate/phosphate translocator activity by accelerated starch turnover and glucose transport in transgenic tobacco. *Planta* 204:366–76
39. Heineke D, Kruse A, Flügge UI, Frommer WB, Riesmeier JW, et al. 1994. Effect of antisense repression of the chloroplast triose-phosphate translocator on photosynthetic metabolism in transgenic potato plants. *Planta* 193:174–80
40. Hendricks JHM, Kolbe A, Gibon Y, Stitt M, Geigenberger P. 2003. ADP-glucose pyrophosphorylase is activated by post-translational redox-modification on response to light and sugars in leaves of *Arabidopsis* and other plant species. *Plant Physiol.* 133:838–49

41. Herold A, Leegood RC, McNeil PH, Robinson SP. 1981. Accumulation of maltose during photosynthesis in spinach protoplasts treated with mannose. *Plant Physiol.* 67:85–88
42. Huber KC, BeMiller JN. 1997. Visualization of channels and cavities of corn and sorghum starch granules. *Cereal Chem.* 74:537–41
43. Huber KC, BeMiller JN. 2000. Channels of maize and sorghum starch granules. *Carbohydr. Polymers* 41:269–76
44. Humphreys TE. 1975. Maltose uptake in the *Zea mays* scutellum. *Phytochem.* 14:333–40
45. Hussain H, Mant A, Seale R, Zeeman S, Hinchliffe E, et al. 2003. Three isoforms of isoamylase contribute different catalytic properties for the debranching of potato glucans. *Plant Cell* 15:133–49
46. Isherwood FA. 1973. Starch-sucrose interconversion in *Solanum tuberosum*. *Phytochem.* 12:2579–91
47. Isherwood FA. 1976. Mechanism of starch-sucrose interconversion in *Solanum tuberosum*. *Phytochem.* 15:33–41
48. James MG, Robertson DS, Myers AM. 1995. Characterization of the maize gene *sugary1*, a determinant of starch composition of kernels. *Plant Cell* 7:417–29
49. Juliano BO, Varner JE. 1969. Enzymic degradation of starch granules in the cotyledons of germinating peas. *Plant Physiol.* 44:886–92
50. Kakefuda G, Preiss J. 1997. Partial purification and characterization of a diurnally fluctuating novel endoamylase from *Arabidopsis thaliana* leaves. *Plant Physiol.* 35:907–13
51. Kaplan F, Guy CL. 2004. β-amylase induction and the protective role of maltose during temperature shock. *Plant Physiol.* 135:1674–84
52. Kruger NJ, ap Rees T. 1993. Maltose metabolism by pea chloroplasts. *Planta* 158:271–73
53. Kubo A, Fujita N, Harada K, Matsuda T, Satoh H, Nakamura Y. 1999. The starch-debranching enzymes isoamylase and pullulanase are both involved in amylopectin biosynthesis in rice endosperm. *Plant Physiol.* 121:399–409
54. Lao NT, Schoneveld O, Mould RM, Hibberd JM, Gray JC, Kavanagh TA. 1999. An *Arabidopsis* gene encoding a chloroplast-targeted β-amylase. *Plant J.* 20:519–27
55. Leach HW, Schoch TJ. 1961. Structure of the starch granule. II. Action of various amylases on granular starch. *Cereal Chem.* 38:34–46
56. Li B, Servaites JC, Geiger DR. 1992. Characterization and subcellular localization of debranching enzyme and endoamylase from leaves of sugar beet. *Plant Physiol.* 98:1277–84
57. Lin TP, Caspar T, Somerville C, Preiss J. 1988. A starch deficient mutant of *Arabidopsis thaliana* with low ADP glucose pyrophosphorylase activity lacks one of the two subunits of the enzyme. *Plant Physiol.* 88:1175–81
58. Lin TP, Spilatro SR, Preiss J. 1988. Subcellular localization and characterization of amylases in *Arabidopsis* leaf. *Plant Physiol.* 86:251–59
59. Lizotte PA, Henson CA, Duke SH. 1990. Purification and characterization of pea epicotyl β-amylase. *Plant Physiol.* 92:615–21
60. Lloyd JR, Blennow A, Burhenne K, Kossmann J. 2004. Repression of a novel isoform of disproportionating enzyme (stDPE2) in potato leads to inhibition of starch degradation in leaves but not tubers stored at low temperature. *Plant Physiol.* 134:1347–54
61. Lorberth R, Ritte G, Willmitzer L, Kossmann J. 1998. Inhibition of a starch-granule-bound protein leads to modified starch and repression of cold-induced sweetening. *Nature Biotech.* 16:473–77
62. Lu Y, Sharkey T. 2004. The role of amylomaltase in maltose metabolism in the cytosol of photosynthetic cells. *Planta* 218:466–73
63. MacGregor EA. 2004. The proteinaceous inhibitor of limit dextrinase in barley and malt. *Biochim. Biophys. Acta* 1696:165–70

64. Matheson NK, Richardson RH. 1976. Starch phosphorylase enzymes in developing and germinating pea seeds. *Phytochem.* 15:887–92
65. Matheson NK, Wheatley JM. 1962. Starch changes in developing and senescing tobacco leaves. *Austr. J. Biol. Sci.* 15:445–58
66. Matheson NK, Wheatley JM. 1963. Diurnal-nocturnal changes in the starch of tobacco leaves. *Austr. J. Biol. Sci.* 16:70–76
67. Mikkelsen R, Baunsgaard L, Blennow A. 2004. Functional characterisation of α-glucan, water dikinase, the starch phosphorylating enzyme. *Biochem. J.* 377:525–32
68. Nakamura Y. 1996. Some properties of starch debranching enzymes and their possible role in amylopectin biosynthesis. *Plant Sci.* 121:1–18
69. Nielsen TH, Deiting U, Stitt M. 1997. A β-amylase in potato tubers is induced by storage at low temperature. *Plant Physiol.* 113:503–10
70. Nielsen TH, Wischmann B, Enevoldsen K, Møller BL. 1994. Starch phosphorylation in potato tubers proceeds concurrently with *de novo* biosynthesis of starch. *Plant Physiol.* 105:111–17
71. Niittylä T, Messerli G, Trevisan M, Chen J, Smith AM, Zeeman SC. 2004. A previously unknown maltose transporter essential for starch degradation in leaves. *Science* 303:87–89
72. Ohad I, Friedberg I, Ne'eman Z, Schramm M. 1971. Biogenesis and degradation of starch. I. The fate of the amyloplast membranes during maturation and storage of potato tubers. *Plant Physiol.* 47:465–77
73. Okita TW, Greenberg E, Kuhn DN, Preiss J. 1979. Subcellular localization of the starch degradative and biosynthetic enzymes of spinach leaves. *Plant Physiol.* 64:187–92
74. Okita TW, Preiss J. 1980. Starch degradation in spinach leaves. Isolation and characterization of the amylases and R-enzyme of spinach leaves. *Plant Physiol.* 66:870–76
75. Peavey DG, Steup M, Gibbs M. 1977. Characterization of starch breakdown in the intact spinach chloroplast. *Plant Physiol.* 60:305–08
76. Ritchie S, Swanson SJ, Gilroy S. 2000. Physiology of the aleurone layer and starchy endosperm during grain development and early seedling growth: new insights from cell and molecular biology. *Seed Sci. Res.* 10:193–212
77. Ritte G, Eckermann N, Häbel S, Lorberth R, Steup M. 2002. Compartmentation of the starch-related R1 protein in higher plants. *Starch* 52:179–85
78. Ritte G, Lloyd JR, Eckermann N, Rottmann A, Kossmann J, Steup M. 2002. The starch-related R1 protein is an α-glucan, water dikinase. *Proc. Natl. Acad. Sci. USA* 99:7166–71
79. Ritte G, Lorberth R, Steup M. 2002. Reversible binding of the starch-related R1 protein to the surface of transitory starch granules. *Plant J.* 21:387–91
80. Ritte G, Scharf A, Eckermann N, Häbel S, Steup M. 2004. Phosphorylation of transitory starch is increased during degradation. *Plant Physiol.* 135:2068–77
81. Rost S, Frank C, Beck E. 1996. The chloroplast envelope is permeable for maltose, but not for maltodextrins. *Biochim. Biophys. Acta* 1291:221–27
82. Scheidig A, Frölich A, Schulze S, Lloyd JR, Kossmann J. 2002. Down-regulation of a chloroplast-targeted β-amylase leads to a starch-excess phenotype in leaves. *Plant J.* 30:581–91
83. Schleucher J, Vanderveer PJ, Sharkey TD. 1998. Export of carbon from chloroplasts at night. *Plant Physiol.* 118:1439–45
84. Schroder SW, MacGregor AW. 1998. Synthesis of limit dextrinase in germinated barley kernels and aleurone tissues. *J. Am. Soc. Brew. Chem.* 56:32–37
85. Schulze W, Stitt M, Schulze ED, Neuhaus HE, Fichtner K. 1991. A quantification of the significance of assimilatory starch for growth of *Arabidopsis thaliana* L. Heynh. *Plant Physiol.* 95:890–95

86. Servaites JC, Geiger DR. 2002. Kinetic characteristics of chloroplast glucose transport. *J. Exp. Bot.* 53:1581–91
87. Sharkey TD, Savitch LV, Vanderveer PJ, Micallef BJ. 1992. Carbon partitioning in a *Flaveria linearis* mutant with reduced cytosolic fructose bisphosphatase. *Plant Physiol.* 100:210–15
88. Sissons MJ, MacGregor AW. 1994. Hydrolysis of barley starch granules by alpha-glucosidases from malt. *J. Cereal Sci.* 19:161–69
89. Smith AM, Martin C. 1993. Starch biosynthesis and the potential for its manipulation. In *Biosynthesis and Manipulation of Plant Products*, ed. D Grierson, pp. 1–54. London: Blackie
90. Smith SM, Fulton DC, Chia T, Thorneycroft D, Chapple A, et al. 2004. Diurnal changes in the transcriptome encoding enzymes of starch metabolism provide evidence for both transcriptional and post-transcriptional regulation. *Plant Physiol.* 135:2687–99
91. Sonnewald U, Basner A, Greve B, Steup M. 1995. A second L-type isozyme of potato glucan phosphorylase: cloning, antisense inhibition and expression analysis. *Plant Mol. Biol.* 27:567–76
92. Sowokinos JR, Lulai EC, Knoper JA. 1985. Translucent tissue defects in *Solanum tuberosum* L. I. Alterations in amyloplast membrane integrity, enzyme activities, sugars, and starch content. *Plant Physiol.* 78:489–94
93. Stahl Y, Coates S, Bryce JH, Morris PC. 2004. Antisense downregulation of the barley limit dextrinase inhibitor modulates starch granule size distribution, starch composition and amylopectin structure. *Plant J.* 36:616–28
94. Stanley D, Fitzgerald AM, Farnden KJF, MacRae EA. 2002. Characterisation of putative α-amylases from apple (*Malus domestica*) and *Arabidopsis thaliana*. *Biologia* 57:137–48
95. Steup M, Robenek H, Melkonian M. 1983. In-vitro degradation of starch granules isolated from spinach chloroplasts. *Planta* 158:428–36
96. Stitt M, ap Rees T. 1980. Carbohydrate breakdown by the chloroplasts of *Pisum sativum*. *Biochim. Biophys. Acta* 627:131–43
97. Stitt M, Bulpin PV, ap Rees T. 1978. Pathway of starch breakdown in photosynthetic tissues of *Pisum sativum*. *Biochim. Biophys. Acta* 544:200–14
98. Stitt M, Heldt HW. 1981. Simultaneous synthesis and degradation of starch in spinach chloroplasts in the light. *Biochim. Biophys. Acta* 638:1–11
99. Stitt M, Heldt HW. 1981. Physiological rates of starch breakdown in isolated intact spinach chloroplasts. *Plant Physiol.* 68:755–61
100. Stitt M, Wirtz W, Gerhardt R, Heldt HW, Spencer C, et al. 1985. A comparative study of metabolite levels in plant leaf material in the dark. *Planta* 166:354–64
101. Sun Z, Duke SH, Henson CA. 1995. The role of pea chloroplast α-glucosidase in transitory starch degradation. *Plant Physiol.* 108:211–17
102. Sun Z, Henson CA. 1990. Degradation of native starch granules by barley α-glucosidases. *Plant Physiol.* 94:320–27
103. Sun Z, Henson CA. 1991. A quantitative assessment of the importance of barley seed α-amylase, β-amylase, debranching enzyme, and α-glucosidase in starch degradation. *Arch. Biochem. Biophys.* 234:298–305
104. Takaha T, Critchley J, Okada S, Smith SM. 1998. Normal starch content and composition in tubers of antisense potato plants lacking D-enzyme (4-α-glucanotransferase). *Planta* 205:445–51
105. Tarrágo JF, Nicolás G. 1976. Starch degradation in the cotyledons of germinating lentils. *Plant Physiol.* 58:618–21
106. Thimm O, Bläsing OE, Gibon Y, Nagel A, Meyer S, et al. 2004. Map-man: a user-driven tool to display genomics datasets onto diagrams of metabolic pathways and other biological processes. *Plant J.* 37:914–99

107. Tsugeki R, Fedoroff NV. 1999. Genetic ablation of root cap cells in *Arabidopsis*. *Proc. Natl. Acad. Sci. USA* 96:12941–46
108. van Berkel J, Conrads-Strauch J, Steup M. 1991. Glucan-phosphorylase forms in cotyledons of *Pisum sativum* L.: localization, developmental change, in-vitro translation, and processing. *Planta* 185:432–39
109. Weber A, Servaites JC, Geiger DR, Kofler H, Hille D, et al. 2000. Identification, purification, and molecular cloning of a putative plastidic glucose transporter. *Plant Cell* 12:787–801
110. Weber APM. 2004. Solute transporters as connecting elements between cytosol and plastid stroma. *Curr. Opinion Plant Biol.* 7:247–53
111. Weise S, Weber APM, Sharkey TD. 2004. Maltose is the major form of carbon exported from the chloroplast at night. *Planta* 218:474–82
112. Wischmann B, Nielsen TH, Møller BL. 1999. In vitro biosynthesis of phosphorylated starch in intact potato amyloplasts. *Plant Physiol.* 119:1–8
113. Witt W, Sauter JJ. 1996. Purification and properties of the starch granule-degrading α-amylase from potato tubers. *J. Exp. Bot.* 47:1789–95
114. Yang Y, Steup M. 1990. Polysaccharide fraction from higher plants which strongly interacts with the cytosolic phosphorylase isozyme. *Plant Physiol.* 94:960–69
115. Yomo H, Varner JE. 1973. Control of the formation of amylases and proteases in the cotyledons of germinating peas. *Plant Physiol.* 51:708–13
116. Yu TS, Kofler H, Häusler RE, Hille D, Flügge UI, et al. 2001. The Arabidopsis *sex1* mutant is defective in the R1 protein, a general regulator of starch degradation in plants, and not in the chloroplast hexose transporter. *Plant Cell* 13:1907–18
117. Zeeman SC, ap Rees T. 1999. Changes in carbohydrate metabolism and assimilate export in starch-excess mutants of *Arabidopsis*. *Plant Cell Env.* 22:1445–53
118. Zeeman SC, Northrop F, Smith AM, ap Rees T. 1998. A starch-accumulating mutant of *Arabidopsis thaliana* deficient in a chloroplastic starch hydrolyzing enzyme. *Plant J.* 15:357–65
119. Zeeman SC, Pilling E, Tiessen A, Kato L, Donald AM, Smith AM. 2002. Starch synthesis in *Arabidopsis*. Granule synthesis, composition, and structure. *Plant Physiol.* 129:516–29
120. Zeeman SC, Smith SM, Smith AM. 2004. The breakdown of starch in leaves. *N. Phytol.* 163:247–61
121. Zeeman SC, Thorneycroft D, Schupp N, Chapple A, Weck M, et al. 2004. Plastidial α-glucan phosphorylase is not required for starch degradation in *Arabidopsis* leaves but has a role in the tolerance of abiotic stress. *Plant Physiol.* 135:849–58
122. Zeeman SC, Umemoto T, Lue WL, Au-Yeung P, Martin C, et al. 1998. A mutant of Arabidopsis lacking a chloroplastic isoamylase accumulates both starch and phytoglycogen. *Plant Cell* 10:1699–711
123. Zeigler P. 1988. Partial purification and characterization of the major endoamylase of developing pea leaves. *Plant Physiol.* 86:659–66
124. Zhu ZP, Hylton CM, Rössner U, Smith AM. 1998. Characterization of starch-debranching enzymes in pea embryos. *Plant Physiol.* 118:581–90
125. Zrenner R, Krause KP, Apel P, Sonnewald U. 1996. Reduction of the cytosolic fructose –1,6-bisphosphatase in transgenic potato plants limits photosynthetic sucrose biosynthesis with no impact on plant growth and tuber yield. *Plant J.* 9:671–81
126. Baunsgaard L, Lütken H, Mikkelsen R, Glaring MA, Pham TT, Blennow A. 2005. A novel isoform of glucan, water dikinase phosphorylates pre-phosphorylated alpha-glucans and is involved in starch degradation in *Arabidopsis*. *Plant J.* 41:595–605

127. Kötting O, Pusch K, Tiessen A, Geigenberger P, Steup M, Ritte G. 2005. Identification of a novel enzyme required for starch metabolism in *Arabidopsis* leaves. The phosphoglucan, water dikinase. *Plant Physiol.* 137:242–52
128. Yu TS, Zeeman SC, Thorneycroft D, Fulton DC, Dunstan H, et al. 2005. alpha-amylase is not required for breakdown of transitory starch in *Arabidopsis* leaves. *J. Biol. Chem.* In press

CO_2 Concentrating Mechanisms in Algae: Mechanisms, Environmental Modulation, and Evolution

Mario Giordano,[1] John Beardall,[2] and John A. Raven[3]

[1]Department of Marine Sciences, Università Politecnica delle Marche, 60121 Ancona, Italy; email: m.giordano@univpm.it

[2]School of Biological Sciences, Monash University, Clayton, Australia 3800; email: John.Beardall@sci.monash.edu.au

[3]University of Dundee at the Scottish Crop Research Institute, Invergowrie, Dundee DD2 5DA, United Kingdom; email: j.a.raven@dundee.ac.uk

Annu. Rev. Plant Biol. 2005. 56:99–131

doi: 10.1146/annurev.arplant.56.032604.144052

First published online as a Review in Advance on January 14, 2005

1543-5008/05/0602-0099$20.00

Key Words

carbonic anhydrase, carboxysome, cyanobacteria, inorganic carbon, photosynthesis, pyrenoid

Abstract

The evolution of organisms capable of oxygenic photosynthesis paralleled a long-term reduction in atmospheric CO_2 and the increase in O_2. Consequently, the competition between O_2 and CO_2 for the active sites of RUBISCO became more and more restrictive to the rate of photosynthesis. In coping with this situation, many algae and some higher plants acquired mechanisms that use energy to increase the CO_2 concentrations (CO_2 concentrating mechanisms, CCMs) in the proximity of RUBISCO. A number of CCM variants are now found among the different groups of algae. Modulating the CCMs may be crucial in the energetic and nutritional budgets of a cell, and a multitude of environmental factors can exert regulatory effects on the expression of the CCM components. We discuss the diversity of CCMs, their evolutionary origins, and the role of the environment in CCM modulation.

Contents

DIC: dissolved inorganic carbon

RUBISCO: ribulose bisphosphate carboxylase/oxygenase

GENERAL INTRODUCTION

Why Do Some Algae Have CO_2 Concentrating Mechanisms?

Algal photosynthesis accounts for a large proportion (~50%) of the 111–117 Pg C y^{-1} global primary productivity (23). Although very few algae assimilate DIC from the environment via alternative pathways, most of this large carbon flux involves species using the C_3 pathway (the photosynthetic carbon reduction cycle or Calvin cycle) for DIC acquisition, fixing DIC directly via ribulose bisphosphate carboxylase oxygenase (Rubisco, Equation 1). However, RUBISCO has a relatively low affinity for CO_2 and, for most species, is consequently less than half saturated under current CO_2 levels. The poor performance of RUBISCO as a CO_2-fixing enzyme is aggravated by its dual role as an oxygenase (Equation 2).

$$\text{Ribulose-1,5-bisphosphate} + CO_2 + H_2O \rightarrow 2 \times \text{glycerate-3-P} \quad (1)$$

$$\text{Ribulose-1,5-bisphosphate} + O_2 \rightarrow \text{glycerate-3-P} + \text{glycolate-2-P} \quad (2)$$

The phosphoglycolate produced by the oxygenase activity of RUBISCO inhibits RUBISCO carboxylase activity, but this inhibition is alleviated by the dephosphorylation of phosphoglycolate via the enzyme phosphoglycolate phosphatase. The glycolate is then available for further metabolism via photorespiration (19, 159) or can be lost from algal cells by excretion. Decarboxylation of intermediates during photorespiration adds to the overall inefficiency of net carbon assimilation based on RUBISCO (19, 159).

All known RUBISCOs have competitive carboxylase and oxygenase functions and have relatively low substrate-saturated carboxylase activities on a protein mass basis. The extent to which the two competitive reactions of RUBISCO occur in autotrophic cells depends on the O_2 and CO_2 concentrations at the RUBISCO active site and the molecular nature of the RUBISCO molecule involved. Equation 3 gives the selectivity factor defining the relative rates of carboxylase and oxygenase reactions,

$$S_{rel} = \frac{K_{0.5}(O_2) \cdot k_{cat}(CO_2)}{K_{0.5}(CO_2) \cdot k_{cat}(O_2)} \quad (3)$$

where $K_{0.5}(CO_2)$ and $K_{0.5}(O_2)$ are the half-saturation constants for the carboxylase and oxygenase functions, respectively, and $k_{cat}(CO_2)$ and $k_{cat}(O_2)$ are the corresponding substrate-saturated rates of catalysis. Different species of autotrophs possess different forms of RUBISCO. Thus, green algae, β-cyanobacteria, and higher plants have Form 1B, with S_{rel} values ranging from 35 to 90; rhodophytes, cryptophytes, heterokonts, haptophytes, and α/β-proteobacteria have Form 1D RUBISCO (153), with much higher S_{rel}; and the dinophytes and proteobacteria exhibit Form II RUBISCOs with low S_{rel} values of 9–15 for proteobacteria and ~30 for dinophytes. The dinophytes are the only eukaryotes with Form II RUBISCO, this having arisen by lateral gene transfer from a δ-proteobacterium (7, 9, 146–148, 153). In general, a low $K_{1/2}(CO_2)$ and a high S_{rel} correlate with a low $k_{cat}(CO_2)$, and vice versa (153).

These biochemical properties of RUBISCO mean that for autotrophs dependent on diffusive CO_2 entry, the physiology of CO_2 assimilation shows inherent inefficiencies, such as significant inhibition of CO_2 fixation by oxygen, high CO_2 compensation points, and low affinities for external CO_2. However, all cyanobacteria examined, most algae and many aquatic plants have mechanisms that overcome the deficiencies of RUBISCO currently operating in what is essentially, in geological terms at least (20), a low-CO_2 environment. Collectively, these are referred to as CCMs but, as detailed below, the mechanisms involved are diverse.

CCM: CO_2 concentrating mechanism

CAM: crassulacean acid metabolism

Relatively few of the ~1500 described species of cyanobacteria, or ~53,000 described species of eukaryotic algae, have been examined for occurrence of a CCM. In the discussion below, we deal with each of the major lines (cyanobacteria, green algae, and red/brown algae) and consider what is known about inorganic C transport and CCM in each of them.

Evidence for, and Mechanisms of, CCMs

Table 1 summarizes the mechanisms by which algae can accumulate CO_2. These range from biochemical C_4 and CAM mechanisms involving additional DIC fixation prior to that by RUBISCO, to biophysical processes involving either localized enhancement of external CO_2 concentration by acidification of the external medium, or the active transport of DIC across one or more cellular membranes.

Some of these mechanisms involve the primary use of HCO_3^- ions whereas others involve CO_2 uptake. Determining whether algae use HCO_3^- or CO_2 can be based on a number of approaches. (*a*) pH drift experiments are perhaps the simplest approach, with HCO_3^--using algae capable of raising the pH in the surrounding medium (as a function of decreasing dissolved CO_2 to the compensation point) to values in excess of those attained by species only able to use CO_2 (155, 156, 168). (*b*) Comparing the pH dependence of $K_{0.5}(HCO_3^-)$ or $K_{0.5}(CO_2)$ for photosynthesis is another option, with CO_2 users showing pH-independent values for $K_{0.5}(CO_2)$ and, conversely, HCO_3^- users showing pH-independent values for $K_{0.5}(HCO_3^-)$ (32, 212). (*c*) Demonstrating that the rate of photosynthesis can exceed the uncatalyzed rate of conversion of HCO_3^- to CO_2 in the medium indicates that cells can use HCO_3^- (111, 112, 155, 156, 168, 212). (*d*) Isotope disequilibrium

Table 1 The major categories of CCM in terrestrial and aquatic phototrophs, their need for an energy input, and their necessity to elevate intracellular or intracompartmental DIC above extracellular levels

Mechanism	Energy input	Necessity for mean DIC_i or CO_{2i} to exceed DIC_o or CO_{2o}	Reference
C_4: inorganic C + C_3 → C_4 dicarboxylate in the cytosol → C_3 + CO_2 in plastid containing RUBISCO	In generation of C_3 acceptor (PEP)	Depends on relative volume of RUBISCO containing high-CO_2 compartment	(81, 87, 120, 163, 164)
CAM: inorganic C + C_3 →C_4 dicarboxylate in the cytosol at night; C_4 stored in vacuole until next day, released and decarboxylated with minimal CO_2 leakage (stomata closed in land plants)	In generation of C_3 acceptor and its conversion during decarboxylation to stored products. Also in transport of C_4 dicarboxylate to vacuole	Yes, at least in terrestrial CAM in decarboxylation phase	(86, 87)
HCO_3^- active influx, conversion to CO_2 by CA at RUBISCO site (often in carboxysome or pyrenoid)	In active influx of HCO_3^- at plasmalemma and/or plastid envelope	Yes for DIC_i, if active transport is at the plasmalemma	(7, 9, 11, 83)
CO_2 active influx	In active influx of CO_2 at plasmalemma and/or plastid envelope	Yes, for CO_2, unless the compartment in which CO_2 is accumulated is relatively small	(7, 37, 83)
CO_2 passive influx at plasmalemma of cyanobacteria with conversion of CO_2 to HCO_3^- by NADHdh, then conversion to CO_2 by CA in carboxysome	In NADHdh, bringing about the unidirectional CA conversion of CO_2 to HCO_3^-	Yes for DIC_i	(9, 11, 83)
Acidified compartment to which HCO_3^- has access; conversion of HCO_3^- (using CA) to give high equilibrium level of CO_2, CO_2 diffusion to RUBISCO compartment	In producing and maintaining a low-pH compartment using H^+ pumps at plasmalemma, thylakoid, and/or other (?) membranes	Yes for CO_2 if the compartment generating CO_2 and adjacent compartments are relatively large	(139–141, 145–147, 200, 205)

DIC, dissolved inorganic carbon
Subscripts i and o, respectively, refer to the inside and the outside of the cell.

CA: carbonic anhydrase

CA_{ext}: external (periplasmic) CA

techniques, following the kinetics of assimilation of inorganic carbon following supply of radioactively labeled HCO_3^- or CO_2, have also been applied to this question (e.g., 45, 89). The above techniques give no indication as to how HCO_3^- is utilized or whether CA_{ext} is involved. Careful use of CA inhibitors can, however, often elucidate this. (*e*) Recent developments using membrane-inlet mass spectroscopy have allowed major advances in determining the inorganic carbon species used by algae, with direct measurements of CO_2 transport (10); this is now a commonly used technique (2, 31, 40, 72, 170, 188, 190). Applying these various techniques shows that most algae examined can take up both HCO_3^- and CO_2, although, as detailed below, there are some chlorophytes and dinoflagellates and an eustigmatophyte, *Monodus subterraneus*, that will only take up CO_2 (37). In contrast, there is some evidence that two eustigmatophyte species can only actively transport HCO_3^- (37, 75).

Turning from evidence for use of different DIC species to evidence for possession of CCMs, all CCM mechanisms (**Table 1**) have the same physiological outcome, i.e., negligible inhibition of CO_2 fixation by oxygen, low CO_2 compensation points, and high affinities for external CO_2—features that are not characteristic of isolated RUBISCOs or intact cells of species that rely solely on CO_2 diffusion. Evidence for CCMs is thus based on physiological measurements of these parameters compared to those of the isolated RUBISCO and/or direct measurements of internal CO_2 pools using mass spectrometric or radioisotopic techniques (153). In addition, stable isotope measurements of the $^{13}C/^{12}C$ ratio of organic cellular material (relative to source DIC) can be used to indicate the presence of the CCM capacity of algae (153, 158). Isolated eukaryote RUBISCOs discriminate against ^{13}C to the extent of ~30‰, so species without CCMs may show isotope discrimination ratios approaching this value. CCMs tend to reduce discrimination, so lower discrimination values suggest CCM activity. However, note that $^{13}C/^{12}C$ ratios are indicative rather than definitive (153, 158).

CCMs Based on $C_3 + C_1$ Carboxylations (C_4 and CAM Metabolism)

The first two mechanisms for CO_2 accumulation in **Table 1** involve an additional carboxylation step, prior to that catalyzed by RUBISCO, which sees inorganic carbon derived from the external environment added to a C_3 carrier to form a C_4 intermediate that is decarboxylated at the site of RUBISCO, providing CO_2 to that enzyme's active site (145, 146). The role of such a C_4 dicarboxylic acid intermediate is a biochemical transporter of DIC from a site with access to exogenous inorganic C to the site where RUBISCO is active. This is C_4 photosynthesis. In CAM the primary ($C_3 + C_1$) carboxylation occurs at night and the ($C_4 - C_3$) decarboxylation occurs during the day, so the dicarboxylate residence time is about 12 hours.

The presence of C_4 (or C_4-like) metabolism in some algae has been suggested in several instances, although in most cases the evidence is weak (18, 33, 69, 125, 145, 163). The evidence for C_4 metabolism comes from the time course of ^{14}C incorporation into acid-stable compounds and the activity and location of (C_3 + C_1) carboxylases and (C_4 − C_1) decarboxylases. C_4-like metabolism is indicated if the first acid-stable product of ^{14}C-inorganic C assimilation is a C_4 dicarboxylic acid rather than 3-phosphoglycerate, and especially if pulse-chase experiments show a transfer of label from a C_4 acid to phosphoglycerate. In terms of enzyme activity and enzyme compartmentation C_4 metabolism requires a ($C_3 + C_1$) carboxylase in a compartment accessible to external DIC and a ($C_4 - C_1$) decarboxylase in the compartment containing RUBISCO. In individual algal cells the cytosol is the potential carboxylation site and the plastid stroma is the decarboxylation site (164). The cytosolic carboxylase could be PEPc or PEPck, and the chloroplastic decarboxylase could be PEPck or NAD^+ (or $NADP^+$) malic enzyme (ME) (153). If PEPck is to be used as both carboxylase and decarboxylase then it must be regulated such that the carboxylase activity is favored in the cytosol but the decarboxylase activity is favored in the plastid. To date there have been convincing cases for C_4 photosynthesis made for only two algae, the green ulvophycean benthic macroalga *Udotea flabellum* (165; see 145) and the planktonic diatom *Thalassiosira weissflogii* grown under inorganic C-limited conditions (120, 163, 164), although for the latter some of the data are contentious (81, 153).

PEPc: phosphenolpyruvate carboxylase

PEPck: phosphoenolpyruvate carboxykinase

CAM has also been proposed as a contributor to photosynthetic inorganic C assimilation in brown macroalgae, albeit providing less than 10% of the total organic C (80, 146). The evidence here comes from the high-PEPck activity in brown algae and the overnight increase and daytime decrease in titratable acidity and malate measurable in algal homogenates (80, 146). However, as for some data on C_4 metabolism, the evidence is equivocal (e.g., see 86).

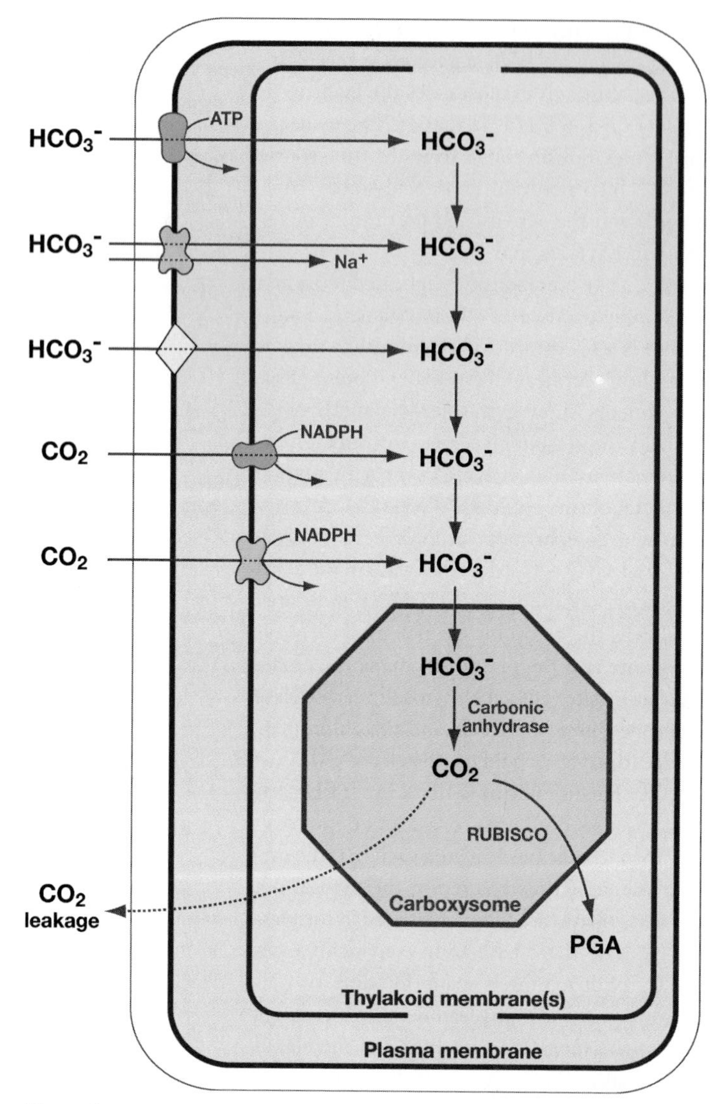

Figure 1

A schematic model for inorganic carbon transport and CO_2 accumulation in cyanobacterial cells. The model shown here incorporates low-affinity transport systems (shown in *gray*) and high-affinity systems (shown in *black*) at the plasmalemma and/or thylakoid membrane. Transporters whose characteristics are unknown are shown in white. Redrawn after Price et al. (138).

CCMs Based on Active Transport of Inorganic C Species

The mechanisms discussed above for DIC concentration at the RUBISCO active site can be thought of as biochemical CO_2 pumps. In contrast, the other mechanisms outlined in **Table 1** represent biophysical CO_2 pumps. Most CCMs in cyanobacteria and algae are based on active transport of HCO_3^- and/or CO_2 across one or more of the membranes separating the bulk medium from RUBISCO. This requires that the membrane across which active transport occurs has a low permeability to the DIC species delivered to the side of the membrane closest to RUBISCO, otherwise active transport is short-circuited (153). This rules out the outer of the two plastid envelope membranes in eukaryotes as well as the gram-negative outer membrane of cyanobacteria because these have high densities of porins with no selectivity for molecules of M_r less than ~800.

For cyanobacteria the CCM can be based on either CO_2 or HCO_3^- transport, either at the plasmalemma or thylakoid membrane (83, 88, 132, 153, 167). The various transporters deliver HCO_3^- to the cytosol regardless of the DIC species (CO_2 or HCO_3^-) removed from the periplasm (**Figure 1**). The HCO_3^- then diffuses into the carboxysomes, which show the only CA activity in the cytosol (93, 138, 181, 187). The CO_2 generated by this CA builds up to a higher steady-state concentration in the carboxysomes than in the bulk medium, thus strongly favoring the carboxylase over the oxygenase activity of RUBISCO (83, 100).

For eukaryotic algae with CCMs active transport mechanisms for DIC could be based on the plasma membrane or the inner plastid envelope membrane or both (2, 83, 124, 153, 202, 212) (**Figure 2**). Equilibration between CO_2 and HCO_3^- in the various compartments (periplasmic space, chloroplast stroma, thylakoid lumen) of eukaryotic algal cells can involve a range of CAs (4, 119, 123, 202). Although it is tempting to draw correlations between the function of carboxysomes in cyanobacterial CCMs and the role of pyrenoids (regions of the plastid stroma where most, if not all, of the cellular complement of RUBISCO is localized in eukaryotes), the evidence is weak. Although all pyrenoid-containing algae have CCMs (7, 146, 147, 153), not all algae capable of expressing CCMs have pyrenoids (7, 121, 122, 146, 147, 153, 157, 158).

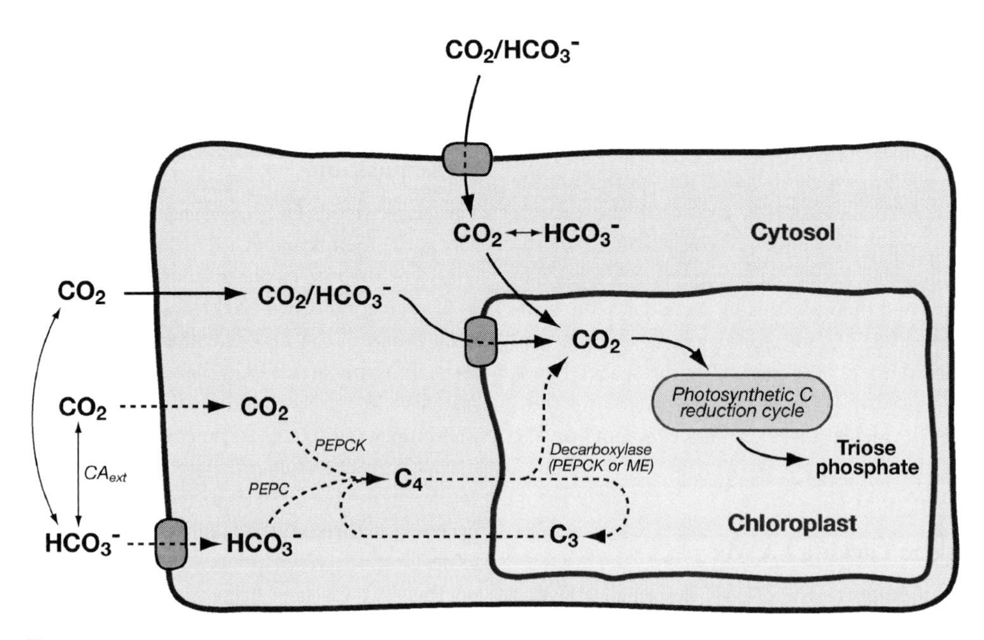

Figure 2

A schematic model for inorganic carbon transport and CO_2 accumulation processes in eukaryotic algal cells. The model incorporates the possibilities for DIC transport at the plasmalemma and/or chloroplast envelope as well as a putative C_4-like mechanism. CO_2 crosses membranes by diffusion, whereas active transport (shown by the *shaded boxes*) can be of CO_2 or HCO_3^-. No attempt has been made to show the roles of the various internal CAs in the different compartments. For this the reader is referred to the text and Badger (6). Redrawn after Sültemeyer (187) and Raven & Beardall (153).

CCMs Based on Enhancement of CO_2 Concentration Following Acidification in a Compartment Adjacent to RUBISCO

A final category of CCM concerns those species where acidification of a compartment containing high-HCO_3^- concentrations leads to enhanced levels of CO_2, which can then diffuse to the active site of RUBISCO at concentrations higher than would be expected by diffusion from the external bulk medium. The essential feature of this type of mechanism is that HCO_3^- from an alkaline medium/compartment is transported to a compartment that is maintained at a low pH by a proton pump (153). Here the equilibrium CO_2: HCO_3^- ratio is much higher than that in the first compartment. Raven (145) gives details of how this equilibrium may be maintained, but it is proposed to involve a relatively acid-stable CA or, in larger compartments, a proton-driven catalysis of the HCO_3^- to CO_2 conversion. The CO_2 produced in this acidic compartment then diffuses to RUBISCO in an adjacent, more alkaline (pH 7.5–8.0) compartment. Such mechanisms work most effectively if there is no CA in the compartment containing RUBISCO, at least in the cases where the RUBISCO compartment is closer to the medium than is the acidic compartment (e.g., a vacuole) (145). Walker et al. (205) first suggested and quantified, in its simplest forms, this model for the acid zones on the surface of internodal cells of characean freshwater macroalgae (see also 1, 53). A similar model for CO_2 concentration using the thylakoid lumen as the acidic compartment was later proposed by Pronina & Semenenko (141) and modified by Raven (145, 147).

For the variants of this mechanism that involve intracellular acid compartments there is a requirement for HCO_3^- transport from an adjacent alkaline (pH 7.0–8.0) compartment to the acid compartment. The alkaline compartment from which bicarbonate is transported is supplied with bicarbonate from the medium via the plasmalemma and/or, when the acid compartment is the thylakoid, the chloroplast envelope membranes (see 145–147). The compartments into which HCO_3^- is transported are the thylakoid lumen (141, 145–147), the vacuole (145–147), the aqueous compartment in the chloroplast endoplasmic reticulum of algae that obtained their plastids by secondary endosymbiosis (98, 99), and, for algae endosymbiotic in invertebrates, the perisymbiont space. Raven & Beardall (153) discuss in detail the (limited) evidence for these latter variants on the mechanism.

Algae Lacking CCMs

In addition to the CCMs described above, a number of algal species do not possess any means of concentrating CO_2 at the RUBISCO active site and rely solely on CO_2 diffusion from the external environment. Species with low-S_{rel} RUBISCOs would find it difficult to carry out net C assimilation under current CO_2 and O_2 levels without recourse to a CCM (7, 145, 146, 153). However, in at least some conditions, organisms with RUBISCOs with higher S_{rel} values could rely on diffusive CO_2 supply to RUBISCO. Red algae lacking CCMs grow in freshwater with CO_2 levels several times that at air equilibrium (102, 152, 154, 203) and often in fast-flowing conditions, which minimize the diffusion boundary layer. Similarly, almost all freshwater chrysophyte and synurophyte (heterokont) algae lack CCMs (12). Other species that rely on diffusive CO_2 supply are certain lichenized green algae (*Coccomyxa*: 134) and some marine red algae, which mainly live at low light levels where a smaller CO_2 flux is required to satisfy the assimilatory needs of the cells (145, 157, 158, 174).

CCMs IN CYANOBACTERIA

Several exhaustive reviews on cyanobacterial CCMs have been published recently (e.g., 11, 83); therefore, we refer the reader to these articles for more details. In the following paragraphs we describe the path of DIC toward the site of its fixation.

C_i Acquisition

The uncharged CO_2 molecule can diffuse across the membrane at a rate sufficient to account for the rate of photosynthesis with CO_2 as C source; no active CO_2 transporter has yet been found in the plasmalemma of cyanobacteria. It is possible (192) that, as with other systems (28, 127), including some photosynthetic ones (197), CO_2 diffusion into the cells occurs through aquaporins that are possibly similar to the bacterial Aqpz (35). Bicarbonate permeation through the plasmalemma is negligible. Cyanobacteria, however, can actively take up this DIC species from the medium. The HCO_3^- pumps afford intracellular concentrations up to three orders of magnitudes higher than those in the external medium (11, 83). Omata et al. (132) thoroughly characterized a low CO_2-induced ABC-type (68) HCO_3^- uniporter, named BCT1, with a high affinity for HCO_3^-. This transporter comprises four proteins encoded by the *cmp*ABCD operon (104, 105). The genes encoding BCT1 are exclusive to freshwater β-cyanobacteria and are absent from all sequenced genomes of marine α- and β-cyanobacteria (11). This may reflect variable, and frequently low, availability of DIC in freshwater (e.g., 101). In *Synechocystis* 6803, inactivation of the *cmp*ABCD operon does not block HCO_3^- transport; this is due to the presence of a Na^+-dependent HCO_3^- transporter with a lower affinity for HCO_3^- than BCT1 (175). This alternative HCO_3^- transport system requires Na^+ for a HCO_3^-/Na^+ symport, probably energized by an ATP-powered Na^+/H^+ antiport. The Na^+ dependence of this transporter may, at least in part, explain the Na^+ requirement of many cyanobacterial strains, especially at high pH (116). Two genes are necessary for expressing this HCO_3^- transport system: *sbtA* (or *slr 1512*) and *ntpJ* (175, 177). *SbtA* encodes the transmembrane protein responsible for the translocation of HCO_3^-, and its expression is triggered by exposure of cells to low CO_2, even if a small amount of transcript is also present in high-CO_2-grown cells (175). Two types of *sbtA* products, differing for the number of amino acids constituting the polypeptides, are present in cyanobacteria, with *Anabaena* PCC7120 possessing both of them (175). Homologues of *sbtA* have also been found in several β-cyanobacteria, but not in *Thermosynechococcus elongatus* and *Trichodesmium erythreum* (11 and references therein); more distantly related homologues are also present in the marine α-cyanobacteria (9, 138, 175). The *ntpJ* gene is present in all organisms possessing *sbtA*; the exact function of the product of this gene is not known, but it has been proposed that it is involved in the primary Na^+ pump (175). This Na^+-dependent HCO_3^- transport may have great ecological relevance because it appears to be the only DIC-active uptake system present in α-cyanobacteria-like *Prochlorococcus*, possibly among the most abundant marine phytoplankters (9, 175). In these organisms, this HCO_3^- transport system is probably constitutive (11). In *P. marinus* PCC 9511, high- and low-HCO_3^- uptake systems have been inferred from kinetic studies (133). The lack of molecular evidence for the presence of these systems, however, may be due to the fact that a biphasic system is encoded by a single gene (38). The constitutive nature of the HCO_3^- transport system may be related to the fact that *P. marinus* growth is presumably controlled by reduced N (NH_4^+ and amino acids) availability (43, 169), whose intracellular concentrations could be maintained at relatively high levels by transporters such as those for NH_4^+ present in *P. marinus* SS120 (43) and in *P. marinus* MIT9313 and MED4 (169). The fact that *P. marinus* cells can maintain relatively high-NH_4^+ concentrations is also confirmed by the $K_m(NH_4^+)$ of their glutamine synthetase, which is one order of magnitude higher than those of other cyanobacteria such us *Synechocystis* PCC 6301, *Calothrix* PCC 7601, and *Nostoc* PCC 7120 (44). In the presence of a control over growth exerted mostly by N, and of relatively constant intracellular N concentration, modulating C acquisition may not be necessary and the control mechanisms may have been "sacrificed" in the gene elimination process that presumably characterized the evolutionary path of *Prochlorococcus* (43). Because the N cell quota (26, 66) is probably highly regulated in the *Prochlorochoccus* strains exclusively using NH_4^+ and organic N (213), there is less need for a modulation in DIC acquisition to maintain a (near) optimal C/N ratio. The relative constancy of the C/N ratio in *Prochlorococcus* is also suggested by the nonphosphorylated state of the P_{II} protein in *P. marinus* PCC9511 (133); phosphorylation and dephosphorylation, mediated by a protein serine kinase and a phosphatase (49, 78) and by the redox state of the cell (71), generally control the activity of the P_{II} protein in cyanobacteria (49, 78, 96, 97, 126). The putatively permanently nonphosphorylated state of PII in *P. marinus* PCC9511 may betray the absence of a need to coordinate C and N metabolism in response to changes in availability of these two nutrients.

Cytosolic DIC

The form of DIC that accumulates in the cytosol is HCO_3^-, irrespective of the form of DIC entering the cells. The conversion of HCO_3^- to CO_2 in the cytosol is slow owing to the absence of a cytosolic CA in cyanobacteria (204). Accumulation of HCO_3^- in the cytosol is confirmed by the finding that, if equilibration of HCO_3^- to CO_2 in the cytosol is accelerated by the expression of human CA, *Synechococcus* PCC7942 cells release CO_2 and cannot maintain a functional CCM (137).

Because CO_2 influx across the external membrane occurs by passive diffusion, there is no CO_2 pump in the sense of a mechanism leading to a higher concentration (a surrogate for free energy for a neutral molecule) on the *trans* side of the plasmalemma than on the *cis* side, as a result of energizing the transmembrane flux of CO_2. An energized conversion of CO_2 to HCO_3^-, however, occurs on the cytosolic side of the thylakoid membrane (103, 130, 131, 176). This conversion is energized by the electron transfer through NAD(P)H dehydrogenase

(NADH-1 complex). Speculative modes of action for this system were illustrated by Kaplan & Reinhold (83) and Price et al. (138). According to Price et al. (138), the reduced intermediate generated within the NADH-1 complex by the electrons donated by NAD(P)H or ferredoxin converts Zn-H_2O to Zn-OH at the active site of the complex. As in CA (180), the Zn-OH is then involved in hydrating CO_2 to HCO_3^-. The residual proton is subsequently translocated into the lumen via proton channels of the complex. At least two systems have been described. One, named NADH-1_3 complex, is inducible, contains proteins encoded by the genes *ndhF3*, *ndhD3*, and *chpY*, and is characterized by a high affinity for CO_2; the other is the NADH-1_4 complex, which is presumably constitutive, with a much lower affinity for CO_2, and is encoded by the *ndhF4*, *ndhD4*, and *chpX* genes (103, 176). It was proposed that the chp proteins are involved in the CA-like reaction of CO_2 hydration (138), although these proteins have no sequence similarity with known CAs. The high-affinity CO_2 uptake system is absent from α-cyanobacteria genomes, and the low-affinity system is also absent in the species of *Prochlorococcus* sequenced, and a single copy of *ndhD4*, *ndhF4*, and *chpX* was found in *Synechococcus* WH8102. On the contrary, both high- and low-affinity systems are present in β-cyanobacteria, with the sole exception of *Trichodesmium erythraeum*, *which* only has the low-affinity CO_2 uptake system (11 and references therein).

CO_2 Production and Assimilation in the Carboxysome

Because the intracellular movement of DIC is by diffusion, the concentration of HCO_3^- in the carboxysome is presumably slightly lower than at the internal face of plasmalemma and the external face of the thylakoid. The HCO_3^- accumulated in the cytosol then serves as the substrate for a fixation trap, where, with the help of a CA, CO_2 is locally produced and used by the proximally located primary photosynthetic carboxylase, RUBISCO. The trap can be physically identified with the carboxysome, where RUBISCO and CA are usually confined (11, 36). It has been suggested that the protein shell of the carboxysome represents a barrier to CO_2 back diffusion, facilitating carboxylation by RUBISCO and reducing short circuiting of the CCMs. However, little evidence is available on the existence of such a barrier to CO_2 diffusion. The closest analogue to the 3–4-nm thick protein shell of the carboxysome is possibly the protein envelope of cyanobacterial gas vacuoles (207). The gas vesicles of *Anabaena* are permeable to CO_2 and in general to gas molecules smaller than perfluorocyclobutane (C_4F_8), whose collision diameter is 0.63 mm (206). Thus, to efficiently act as a barrier to CO_2 diffusion, the maximum pore size in the carboxysome envelope should be substantially smaller than that in gas vesicles. It has been suggested that O_2 readily diffuses into the carboxysomes (179); if this is true, in the absence of presently undiscovered active selection mechanisms, the pores of the protein shell of carboxysomes would be larger than the collision diameter of O_2, but smaller than that of CO_2. However, Marcus et al. (107) reported that, in *Synechocystis* PCC6803, RUBISCO activity is less sensitive to O_2 in intact carboxysomes than in ruptured carboxysomes, suggesting that the carboxysome envelope is more permeable to CO_2 than O_2. In the absence of confirmation of the "intracarboxysomal" kinetics of RUBISCO reported by Marcus et al. (107), further investigations on the physical chemistry of CO_2 permeation across protein envelopes are required to determine the nature of the hypothesized carboxysomal diffusion barrier.

The traditional view of the CO_2 production in the carboxysome requires a carboxysomal CA, which would accelerate HCO_3^- conversion to CO_2, providing the substrate for RUBISCO carboxylation at a relatively high rate. It has also been proposed that CA is located at the center of the carboxysome, and that the CO_2 that it generates is used up before it can diffuse across the thick intracarboxysonal protein arrangement (36). A carboxysomal CA is present in a number of β-cyanobacteria (6, 11), and a new type of CA (ε-CA) was identified in carboxysomes of the chemoautotroph *Halothiobacillus neapolitanus* (182). The gene encoding this ε-CA, named *csoS3*, has homologues in the α-cyanobacteria *Prochlorococcus* sp. MED4 and MIT9313 and *Synechococcus* WH8102 (182). The presence of a carboxysomal CA has not been confirmed for all cyanobacteria: The genes encoding specific carboxysome CAs are lacking from the genomes of the β-cyanobacteria *Trichodesmium erythreum* and *Thermosynechococcus elongatus* (6, 9, 11). Thus, if all cyanobacteria carboxysomes are involved in CCMs, the presence of a CA (at least from a known CA family) in the carboxysome is not an absolute requirement for efficient CO_2 fixation. Shively et al. (179) point out that producing protons by the carboxylation reaction of RUBISCO, especially in a closed compartment packed with the enzyme, may be sufficient to drive the dehydration of HCO_3^- at a rate sufficient to minimize the oxygenase reaction.

The CO_2 generated in the carboxysome of cyanobacteria serves as the substrate for Form 1A or 1B RUBISCOs. Their distribution strictly parallels the type of carboxysome: α-cyanobacteria contain α-carboxysomes with Form 1A RUBISCO; β-cyanobacteria contain β-carboxysomes with Form 1B RUBISCO (9). Whether each carboxysome-RUBISCO pair has evolved in parallel or is the result of a single lateral gene transfer is a matter of dispute (9). What advantages, disadvantages, or differences in activity are related to either type of carboxysome and RUBISCO are not yet clear due to the absence of adequate knowledge of the kinetics of Form 1A RUBISCO. Regardless of the type of RUBISCO, packing RUBISCO in carboxysomes is crucial for the efficiency of cyanobacterial photosynthesis (36), and packing may increase the RUBISCO activation state (173).

CCMs IN THE CHLOROPHYTA, CHLORARACHNIOPHYTA, AND EUGLENOPHYTA

In this section we consider the evidence for CCMs in the chlorophyte algae, including the euglenoids and chlorarachniophytes, which arose from chlorophytes following an endosymbiotic event with a flagellated protozoan and consequently have one and two additional chloroplast envelope membranes, respectively.

All chlorophytes examined have Form 1B RUBISCOs, with S_{rel} values in the range of 54 to 83; the highest value is for the lichen alga *Coccomyxa* sp. and it is similar to values for C_3 higher plants. **Figure 2** shows a model for CCMs in eukaryotes, based mainly on experiments with chlorophytes such as *Chlamydomonas reinhardtii*.

DIC Entry

On the basis of evidence from isotope disequilibrium and mass spectrometric approaches, most chlorophyte algae can use both HCO_3^- and CO_2 as DIC sources (37). Notable exceptions are *Nannochloris atomus* and *N. maculata* (74, 76), which are only capable of using CO_2. *Coccomyxa* is unusual in that it shows no CCM activity and CO_2 entry is via diffusion (134).

Many green algae can use HCO_3^-. However, although direct uptake of HCO_3^- was demonstrated by mass spectrometry (2, 136), in some cases concurrently with active CO_2 transport (189), in other instances it appears that HCO_3^- use involves the activity of a CA_{ext}. This enzyme activity is responsible for dehydrating HCO_3^- in the periplasmic space, increasing the potential for CO_2 uptake. In *Chlamydomonas reinhardtii*, in which both CO_2 and HCO_3^- are actively transported across the plasmalemma (although the former is the preferred species), there are two isozymes of CA_{ext} (50, 51, 162). These are pCA1 (the gene product of *Cah1*) and pCA2 (the gene product of *Cah2*). The former is highly expressed and inducible by limiting CO_2 and the latter is of unknown function and is repressed under low-CO_2 conditions and only weakly expressed under high CO_2 (186). A periplasmic CA was also reported in the charophytes *Chara zeylanica* and *Mougeotia* sp. (4).

Although pCA1 in *Chlamydomonas* certainly enhances the use of external HCO_3^- at neutral and alkaline pH, evidence from a null mutant

AE: Anion exchange

DIDS: 4,4′-diisothicyantostilbene-2,2′-disulfonate

of *Cah1* (*cah1-1*) suggests that pCA1 activity is not absolutely essential for growth in low CO_2 or for operation of the CCM (186, 201). Other species of microalgae have CCMs CA_{ext} activity. In these cases, the major flux is likely HCO_3^-, except for the *Nannochloris* species mentioned above, where CO_2 is the only DIC species actively transported (37).

In some ulvophyte macroalgae, HCO_3^- use based on CA_{ext} and entry of CO_2 is supplemented, when the algae are exposed to the very low CO_2 levels associated with high-pH environments, by induction of an AE HCO_3^- transporter. This system is characteristically inhibited by AE inhibitors such as DIDS, and under high-pH conditions both CA_{ext} and DIDS-sensitive systems co-occur, although in normal environmental conditions the CA_{ext} mechanism predominates. Not all green macroalgae show DIDS-sensitive HCO_3^- use and other factors are required before the system is induced (95). Furthermore, much care is needed in the use and interpretation of data obtained using AE inhibitors such as DIDS (212), and the occurrence of an AE exchange HCO_3^- transporter in macroalgae needs additional verification. DIDS-sensitive HCO_3^- transport occurs in chlorophyte microalgae, although at least in *Dunaliella tertiolecta*, this is much less significant than the CA_{ext} system is for HCO_3^- use (212).

DIC in the Cytosol and Plastid

The DIC delivered from the external environment across the plasmalemma can be acted on in numerous ways. In contrast to the cyanobacteria, chlorophytes and other eukaryotic algae have the added complication of additional compartments, specifically the chloroplast and the included pyrenoid, which can be involved in CO_2 accumulation. Within the cytosol, DIC is present as both CO_2 and HCO_3^-. As yet, there is no evidence for a putative cytosolic CA except in *Coccomyxa*, which lacks a CCM (70), but chloroplasts of *Chlamydomonas* contain a stromal β-CA (Cah6) and a thylakoid lumen α-CA (Cah3), which catalyze the CO_2/HCO_3^- equilbrium within those compartments (84, 119). How this stromal CA fits into models of CO_2 accumulation is unclear, although the model proposed by Mitra et al. (119) invokes not only active transport of DIC into the chloroplast but its transfer to the pyrenoid and the concentration of CO_2 based on localized acidification within the thylakoid lumen (see discussion below and figure 10 in 119).

There is considerable evidence in chlorophyte algae for a role for the plastid envelope in CCMs based on active transport of DIC. Photosynthetically active chloroplasts from high and low CO_2-grown chlorophyte cells possess low- and high-affinity DIC uptake systems respectively, as do intact cells (2, 10, 136). Such data are consistent with a major role for plastid envelope–based DIC transport in CCMs in these organisms. Soupene et al. (184) presented evidence that *Chlamydomonas reinhardtii* cells grown at high CO_2 show expression of the Rh1 (Rhesus) protein usually associated with red blood cells. Expression of *rh1* is suppressed under low CO_2. In high-grown CO_2 cells, Rh1 is expressed as a CO_2 channel in the plastid envelope, maximizing CO_2 influx in the absence of active DIC transport systems. Its suppression under low-CO_2 conditions minimizes CO_2 leakage and short circuiting of CCMs (184; see also 82). To date, active DIC uptake (as both CO_2 and HCO_3^-) and CO_2 accumulation have been demonstrated in isolated chloroplasts of *Chlamydomonas reinhardtii* and *Dunaliella tertiolecta* (2) and in *Tetraedon minimum* and *Chlamydomonas noctigama* (198). In contrast, intact cells of *Chlorella ellipsoidea* show a CCM, but isolated chloroplasts from this species are incapable of active CO_2 uptake and have very limited capacity for HCO_3^- use, suggesting that the major site of DIC transport in this species is the plasmalemma (171, 172). Note that the possession of active transport of DIC at the chloroplast envelope does not rule out the option of active DIC transport at the plasmalemma (198).

The role of the pyrenoid in chlorophyte CCMs is unclear. Morita et al. (121) showed that species of *Chloromonas* can accumulate CO_2 and express a CCM even though they do not possess pyrenoids. However, in a later paper Morita et al. (122) showed that, for a range of *Chlamydomonas* and *Chloromonas* species, the size of the DIC pool was associated with pyrenoid form. Species that form typical spheroidal, electron dense pyrenoids with strong RUBISCO immuno-labeling had higher DIC pool sizes (150–250 μM), whereas species with atypical pyrenoids had lower DIC pools (13–31 μM) (121, 122). Given the significant number of algae of other lineages (see below) that lack pyrenoids yet express CCMs, the exact role of pyrenoids in CCM function requires further investigation.

Evidence for Other Forms of CCM in Chlorophyte Algae

Evidence for a CCM based on thylakoid acidification. The role of thylakoid acidification in CCM function derives from earlier work by Pronina and coworkers (139–141) and was developed by Raven (147). The model is based on the occurrence of an α-carbonic anhydrase (Cah3; 84) on the inner side of the thylakoid membrane. In *Chlamydomonas*, this carbonic anhydrase is necessary for growth of ambient CO_2 (84, 199). Following active transport of HCO_3^- to the compartment containing RUBISCO, a passive uniport involving HCO_3^- channels would transport HCO_3^- to the thylakoid lumen. Here it reacts with H^+, accumulated in the lumen as a consequence of light-driven electron transport, and in the presence of Cah3 leads to localized elevated CO_2 concentrations in the lumen. The CO_2 then diffuses out of the lumen to the site of RUBISCO in the chloroplast stroma (or pyrenoid, where present). However, van Hunnik & Sültemeyer (200) were unable to demonstrate HCO_3^- influx (or CO_2 efflux) in isolated thylakoids of green algae. More work, including a search for bicarbonate channels or their functional equivalents, is needed to test this model. Involvement of Cah3 in CO_2 generation from HCO_3^- as part of the reaction sequence between exogenous inorganic C and RUBISCO does not preclude a role for the enzyme in photoactivation and function of the water-oxidizing complex of photosystem II (3).

However, it is clear that such a mechanism can account for cases where gas exchange physiology suggests the presence of a CCM yet the average inorganic C concentration in the cells is not greater than that in the medium. Furthermore, the compartmentation of well-established carbonic anhydrases in *Chlamydomonas reinhardtii* is consistent with Raven's hypotheses (119, 145–147, 149).

A final role for carbonic anhydrase in CCMs in chlorophyte algae relates to the β-CA associated with mitochondria (187). This mtCA is proposed to function to convert CO_2, produced in the mitochondrial matrix by the TCA cycle and photorespiratory glycine decarboxylase activity, to HCO_3^-, limiting the potential for CO_2 leakage through the plasmalemma (149). Giordano et al. (60) proposed a role for mtCA in the supply of HCO_3^- for anaplerotic carbon assimilation; mtCA is not only inducible under low CO_2 (46) but also by C:N ratio (60).

C_4 mechanisms in the chlorophyta. There is evidence for C_4 photosynthesis only in a single ulvophyte, the macroalga *Udotea*. The evidence is based on the kinetics of isotope labeling, including pulse-chase experiments and measurements of enzyme activity and localization. *Udotea* uses a cytosolic PEPck as the ($C_3 + C_1$) carboxylase, and, possibly, NAD ME as the stromal decarboxylase (146, 165). The kinetic data from labeling by $^{14}C_i$ show transfer of ^{14}C from malate to sugar phosphates during the chase period. Furthermore, the ^{14}C label in malate relative to phosphoglycerate during the pulse labeling is higher than expected for C_3 biochemistry (165, 166). However, keep in mind that the shortest pulse-labeling times were 10 sec (165, 166) and Johnston (80) showed that labeling times as short as 1 sec were needed to show that phosphoglycerate, rather than a C_4 acid, was the initial product of photosynthetic DIC fixation in the brown macroalga *Ascophyllum*. It would be especially useful to have more very short-term pulse-label data not only for *Udotea* but also for other algae that are presently believed to have C_3 biochemistry. Remember that CCMs based on active DIC (above) or H^+ (below) transport would, like C_4-like metabolism, account for C_4-like gas exchange physiology (high CO_2 affinity; O_2 insensitivity), and that active DIC transport could make pulse-chase experiments more difficult to interpret because any $^{14}CO_2$ released from ^{14}C-dicarboxylates would be less likely to be "chased" out of the cell and more likely to be refixed by RUBISCO even if this is not an obligate pathway from external DIC to RUBISCO. A low $^{13}C/^{12}C$ ratio in the organic C of the organism relative to source CO_2 can also be explained by a CCM based on DIC or H^+ active transport as well as by CCMs based on C_4-like metabolism using PEPc as the (C_3 + C_1) carboxylase, and more readily than if PEPck is the carboxylase (146).

There is no evidence for CAM metabolism in green algae.

CCMs IN RED AND CHROMIST ALGAE AND DINOFLAGELLATES

All of the algae considered here have red algal plastids. The plastids are either within red algae or were transferred, by secondary endosymbiosis, to other unicellular heterotrophs. The secondary endosymbiotic event gave rise to the photosynthetic members of the Chromista (i.e., photosynthetic cryptophytes, haptophytes, and heterokonts) and of the Alveolata (i.e., photosynthetic dinoflagellates) (47). The ancestral RUBISCO in this group is Form 1D, which has the highest known selectivity for CO_2 over O_2. This Form 1D RUBISCO is found in all red algae, cryptophytes, haptophytes, and heterokonts that have been examined, but most dinoflagellates have a Form II RUBISCO with a much lower selectivity for CO_2 over O_2 (48, 153). A number of dinoflagellates have plastids from tertiary endosymbioses, and they variously have Form 1B (green symbiont) or Form 1D (chromistan symbiont); these organisms have not yet been examined with respect to inorganic carbon acquisition. Also not examined for DIC acquisition are the glaucocystophytes, a small group of freshwater unicells with plastids from the primary endosymbiosis. This is also the case for the related red algae and green algae plus higher plants, but they have retention of the Form 1B RUBISCO from the cyanobacterial plastid ancestor, as do the green algae and higher plants. These algae retain the peptidoglycan wall from the plastid ancestor around their plastid and have structures in the stroma that have been variously interpreted as carboxysomes and as pyrenoids, allowing speculation as to their DIC acquisition mechanism (150).

Much of the effort in investigating carbon acquisition in these organisms focused on diatoms (heterokonts) and coccolithophores and other haptophytes, which are important members of the freshwater (diatoms only) and marine phytoplankton, with less work on the experimentally less tractable dinoflagellates. There has also been considerable work on red and brown (heterokont) macroalgae, as well as on red microalgae. The evidence considered below indicates that the majority of the organisms examined have CCMs (153) but that, for various reasons, there is less mechanistic evidence on CCMs than on green algae and, especially, cyanobacteria. The availability of complete genome sequences for the marine diatom *Thalassiosira pseudonana* and for the thermoacidophilic red alga *Cyanidioschyzon merolae* (113) should help in interpreting physiological and biochemical data on these and related organisms.

Algae Apparently Lacking a CCM

The CCM-less organisms include some red macroalgae, which grow in freshwater, some subtidal and low-intertidal marine red macroalgae, some marine red macroalgae, which grow at or above the high tide level or on the shaded side of intertidal rocks (157, 158), and almost all of the freshwater chrysophyte (heterokont) algae examined (12).

Algae with a CCM: Inorganic C Entry

The generation of a higher internal than external DIC, and of higher CO_2 concentration in cells than in the medium, has been demonstrated for one or more species of red alga, diatom, haptophyte, tribophyte, dinoflagellate, and eustigmatophyte (37, 153). The mass spectrometric method of Badger et al. (10) and the isotope disequilibrium technique show that many of the microscopic algae examined can take up both CO_2 and HCO_3^- (27, 31, 37, 111, 153, 170). The organisms concerned are a red alga, diatoms, haptophytes, and some eustigmatophytes. Other eustigmatophytes can take up either CO_2 or HCO_3^-, whereas dinoflagellates can only take up CO_2 (see table 2 in 37, 40). Some eustigmatophytes (*Nannochloropsis* spp.; marine) can take up HCO_3^-, whereas *Monodus subterraneus* (freshwater/soil) only takes up CO_2 (37).

For the marine organisms with the capacity to take up both HCO_3^- and CO_2, growth at decreasing concentrations of CO_2 (and hence a lower CO_2:HCO_3^- ratio) in equilibrium with the seawater generally decreased the ratio of uptake of CO_2 to that of HCO_3^- in diatoms, but was less obvious in the haptophytes *Phaeocystis globosa* and, especially, *Emiliania huxleyii* (31, 111, 170). These experiments were carried out in the presence of dextran-bound sulfanilamide to inhibit extracellular CA activity. CO_2 is often the dominant form of DIC taken up under these conditions in air-equilibrated seawater, and would be even more important as the form entering the cells when CA_{ext} is not inhibited. The 5.4-fold slower rate of uncatalyzed dehydration of HCO_3^- in seawater than in freshwater (79, 111) means that extracellular CA activity is even more important in making the dominant (HCO_3^-) form of DIC in seawater available to the CO_2 transport mechanism than is the case in fresh waters. In addition to its role as the major, or only, form of DIC entering the cell in most microalgal cells, CO_2 is the form of DIC whose concentration regulates the expression of the CCM components in some organisms, e.g., the marine diatom *Phaeodactylum tricornutum* (111).

Algae with a CCM: Inorganic C in the Cytosol

The red alga *Porphyridium* and the diatom *Phaeodactylum* have intracellular β-CAs that may be cytosolic but have leader sequences consistent with a plastid location. If they are cytosolic then they would equilibrate cytosolic DIC species, facilitating uptake of DIC by plastids if the DIC species taken up by the plastids was not the one that was delivered to the cytosol (see below). Lack of understanding of the full-range plastid targeting sequences in diatoms means that there is uncertainty as to where other CAs detected in the *Thalassiosira pseudonana* genome (**http://genome.jgi-psf.org.thaps1/thaps1.home.html**), including γ-CAs, are expressed.

For the C_4-like photosynthetic C assimilation proposed by Reinfelder et al. (163, 164) and Morel et al. (120, cf. 80), for C_i- or Zn-limited cells of the diatom *Thalassiosira weissflogii* there is a clear role for cytosolic CA if CO_2 is the inorganic C species delivered to the cytosol. As in terrestrial C_4 plants, cytosolic CA is required to convert CO_2 into HCO_3^-, the substrate for PEPc. The product of this reaction, oxaloacetate, is reduced to malate, which is transported into the plastids. There are still uncertainties in this proposed C_4-like pathway in a diatom, e.g., definitive data on the location of two of the enzymes involved, PEPc and PEPck. However, inhibitor evidence is consistent with a role for PEPc, and feeding of the proposed intermediates oxaloacetate and malate restricts the fixation of exogenous DIC while maintaining the rate of O_2 evolution (164). There is also evidence from silicone oil centrifugation experiments with a range of extraction-trapping solutions for a large fraction of the $^{14}C_i$ taken up in 10 sec occurring in an organic acid rather than in DIC (164).

Algae with a CCM: Role of the Plastid Envelope

Plastids isolated from the giant-celled marine red alga *Griffithsia*, and from two species of diatom, can photosynthesize at close to the in vivo rates on a chorophyll basis, but there have been no studies of the mechanism(s) of DIC entry (145, 153). The isolated diatom plastids lack the two "chloroplast endoplasmic reticulum" membranes, as is typically the case for kleptoplastids from the tribophyte *Vaucheria litorea*, which can photosynthesize for many months in the saccoglossan mollusk *Elysia chlorotica* (63, 151, 161). The proposed C_4-like photosynthesis in diatoms requires importing malate into the plastids and exporting phosphoenolpyruvate or pyruvate to the cytosol.

Algae with a CCM: Reactions in the Stroma

Many of the algae with CCMs have pyrenoids containing much of the plastid complement of RUBISCO, although a significant number lack pyrenoids. The common assumption is that the pyrenoid also contains a CA, perhaps the β-CA, which may be targeted to the plastid in *Porphyridium* and *Phaeodactylum*. The most effective use of such a CA would be to produce CO_2 at the site of RUBISCO activity from HCO_3^- transported into the stroma. There is no obvious role for such a CA, or the pyrenoids, if CO_2 is the form of DIC that is concentrated in the plastid by a plasmalemma and/or a plastid envelope–located CO_2 pump, or for a C_4 acid decarboxylase such as PEPck in the case of the proposed C_4-like photosynthesis. However, such a CA activity would not be as damaging to the effective operation of the C_4 mechanism proposed by Reinfelder et al. (163, 164) as cytosolic CA expression in the bundle sheaths of terrestrial C_4 plants would be, unless there were HCO_3^--conducting channels on the plastid envelope membranes leaking HCO_3^- back to the cytosol.

ENVIRONMENTAL MODULATION OF CCMs

There have been numerous roles suggested for CCMs, namely (*a*) improving CO_2 supply and providing a competitive advantage when the DIC and/or CO_2 in the environment is decreased, (*b*) improving resource-use efficiency when nutrients such as N, P, Fe, S, etc. are in short supply, and (*c*) acting as a means of energy dissipation (14). Consequently, it is not surprising that environmental factors such as nutrient availability, energy supply, and CO_2 availability play a significant role in modulating CCM activity. A full discussion of environmental regulation of CCMs can be found in Beardall & Giordano (14).

Inorganic Carbon Supply

DIC transport and CCM capacity are downregulated by increasing CO_2 concentrations in the gas phase in equilibrium with the external medium. In *Chlamydomonas*, at least, high CO_2 suppresses expression of a high-affinity DIC state, but CO_2 accumulation can still occur as a low-affinity, constitutive mechanism (2, 136, 188, 194).

There appears to be a continuum in the degree to which the CCM is expressed in response to external DIC concentration, with higher concentrations leading to a greater degree of suppression of CCM activity (8, 14, 114, 117, 178). Many genes associated with aspects of CCM function are, in *Chlamydomonas reinhardtii*, controlled by DIC supply. These include the α-CA in the periplasmic space (*Cah1*), the β-CA associated with mitochondria (*Mca1*, *Mca2*), which are upregulated by low CO_2, and the other periplasmic α-CA (*Cah2*), which is downregulated. Low CO_2 also modulates the activity of a number of enzymes associated with photorespiratory and nitrogen metabolism (see 186 for a detailed review). Low-CO_2-grown cells of *Chlamydomonas reinhardtii* also show substantial changes in ultrastructure, including an increase in the starch sheath associated with the pyrenoid (92, 142) and shifts in mitochondrial distribution from the center of the cell to the periphery, between the plasmalemma and the chloroplast envelope (56).

In cyanobacteria, the HCO_3^- concentration in the external medium is the controlling signal governing CCM expression (114). However, in some eukaryotes, e.g., *Peridinium gatunense*

(25), *Chlorella ellipsoidea* (109, 110), *C. kessleri* (29), and *Phaeodactylum tricornutum* (111), CO_2 is the DIC species that controls CCM activity. Fukuzawa et al. (52) and Xiang et al. (210) identified a gene (named *ccm1* or *cia5*, respectively) that appears to code for a zinc finger transcription factor. Cia5 is clearly a key component in the regulation of genes controlled by low CO_2, although its precise mode of action is subject to debate (186).

Any environmental factor that affects inorganic carbon levels in the bulk medium around algal cells can thus affect the phenotypic expression of CCMs. In addition to the direct effect of levels of CO_2 in equilibrium with the bulk medium, factors such as pH, temperature, and salinity also alter CCM activity (14, 17).

Light Availability

C_i transport and accumulation is an active process, with evidence from several groups suggesting that the ATP necessary for carbon transport is derived from electron flow associated with PSI (128, 129, 135, 185), although some eustigmatophyte algae are unusual in having a CCM driven by respiratory ATP (73). Given that CO_2 fixation coupled to operation of a CCM is energetically costly (17), it might be expected, a priori, that limitations on energy supply (i.e., photon flux) could affect the activity of CCMs. In a number of species there is evidence that rates of inorganic carbon transport and CCM activity are greatest under high-photon flux densities (13), and acclimation to low-photon flux results in a decreased capacity and/or affinity for DIC transport in numerous algae (14, 90, 91, 178). However, in *Anabaena variabilis*, a decrease in CCM capacity was only found under severe light limitation (13).

In both *Chlamydomonas reinhardtii* (77) and *Synechocystis* sp. PCC6803 (115), high light regulates expression of a number of genes, including some of those that are low-CO_2 inducible and under the control of the *Cia5* gene product (77), suggesting that any signaling mechanism modulating expression of CCM activity is not based solely on external DIC concentration (115).

Other Nutrients

In most cases, CCM function depends on photosynthesis, and light energy harvesting can be compromised by macronutrient and iron deficiency (54, 55, 64). It is therefore important to understand the interactions between CCM activity, light, and nutrient availability. CCM function requires synthesis of specific proteins (106, 186, 193), which represents a demand for cellular nitrogen. The amount of resources that a cell invests in acquiring carbon through a CCM is likely to be coupled to the availability of other nutrients. Controlling cellular elemental ratios is fundamentally important for optimizing resources and maintaining the operational enzymic machinery necessary for survival and growth. Thus, it is expected that DIC acquisition interacts with nutrient acquisition, although, as we shall see, such interactions are complex.

For nutrients such as phosphorus and sulfur, uptake and assimilation mechanisms are induced by limitation by these elements (41, 59, 61). Unless limitation is mild, acclimation strategies are not sufficient to restore optimal growth and are employed in parallel with a downregulation of cell activities not related to acquiring the limiting nutrient (41, 61). For instance, because of the dependency of DIC acquisition on ATP supply, it seems possible that phosphorus limitation would affect CCMs (14, 17). There are some limited data for *Chlorella emersonii* that CCMs are downregulated under severe P limitation (21a).

Little information is available on the effect of sulfur limitation on regulating CCMs. In general terms, under sulfur limitation cells tend to reallocate their resources so that synthesis of low-sulfur proteins and degradation of high-sulfur proteins are favored (61). RUBISCO is a major reservoir of sulfur in the cells and therefore it is not surprising that sulfur limitation characteristically leads to a conspicuous reduction in RUBISCO. This substantially reduces photosynthetic capacity (61) and also seems to impact the photosynthetic C-use efficiency (14). In *Chlamydomonas*, the reduced photosynthetic performance is a specific response controlled by the *sac1* gene product. *Sac1* mutants are unable to downregulate photosynthesis and rapidly die in the light when sulfur limited (42). However, in *Dunaliella salina*, periplasmic CA is not downregulated under sulfur-limited conditions (61), despite a reduction in affinity for inorganic carbon.

For nitrogen the situation is more complex. In *Chlorella emersonii* (15, 21) and *Dunaliella tertiolecta*, N limitation leads to an enhanced affinity for CO_2 in whole cells (211a). In contrast, mildly N-limited cells of *Chlamydomonas reinhardtii* show a downregulation of the CCM and of mitochondrial β-CA (60). In *Chlamydomonas reinhardtii*, the photosynthetic CO_2 use efficiency increases with increasing NH_4^+ (14). The stimulation of photosynthetic performance is thus somehow related to the concentration of N and might reflect a demand for carbon to maintain a correct C:N ratio, thereby ensuring optimal operational capacity for the cell. In species that show upregulation of CCMs under very low N availability, it has been suggested that this is a response to enhance the N-use efficiency of the cells (15, 21, 211a). If a possession of a CCM substantially favors the carboxylase function of RUBISCO, then the achieved rate of CO_2 fixation per unit nitrogen in RUBISCO increases. Therefore, under N limitation, CCMs will increase the nitrogen efficiency of growth (rate of biomass production per unit algal nitrogen), provided the nitrogen allocated to the CCM components does not offset the nitrogen savings permitted by action of the CCM (160). At the same time, by possibly reducing losses of nitrogen through photorespiration, an active CCM in starved cells may limit the increase in the cell C:N ratio, somewhat moderating the decrease in size and functionality of the enzymatic machinery (14).

Cells of *Dunaliella salina* and *D. parva* grown on NH_4^+ have a higher affinity for CO_2 than those grown on NO_3^- (14, 57, 58). This effect is largely independent of the external CO_2 availability and seems to provide additional evidence that CCMs may be involved in the control of cellular elemental ratios (and specifically the C:N ratio) rather than simply in DIC acquisition (14). In this respect it is worth noting that Giordano et al. (60) showed that induction of mtCA is regulated as much by C:N ratio as by low CO_2.

Iron availability, like that of nitrogen, can often compromise light-harvesting processes and photosynthesis (54, 55, 64). Large areas of the world's oceans, such as the equatorial Pacific Ocean (22) and the Southern Ocean (195), are potentially iron limited. Nonetheless, there is little information available on the interaction between CCMs and iron limitation. In iron-limited, as in N-limited, chemostat cultures of *Dunaliella tertiolecta*, the affinity of cells for DIC was enhanced at lower growth rates (211, 211a), reflecting an increased investment in a CCM under these conditions. Maintaining high-affinity DIC acquisition under both Fe- and N-limited conditions in *Dunaliella tertiolecta* (211, 211a) may confer improved energy and resource use efficiency when photosynthetic energy-harvesting capacity is thus impaired. These findings suggest that maintaining efficient DIC-uptake kinetics could be a general response to energy limitation imposed by nutrient deficiency in this alga.

Zinc is used as the metal cofactor of α-, β-, and γ-CAs and is directly involved in the catalytic mechanism of these enzymes (123 and references therein). Thus, it can be expected that a reduction in Zn availability would affect CCMs and, in diatoms, the level of induction of Zn-CA would depend on the availability of Zn (93). However, Zn can be effectively substituted with cobalt (94). There are also indications of Cd-CA distinct from Zn-CA in the diatom *Thalassiosira weissglogii* (39), which is more abundant when cells are exposed to low-CO_2 and -Zn concentrations. However, little is known of the physiological role of this Cd-CA.

Other Environmental Factors

Temperature has been proposed to regulate the CCM capacity of cells. At low temperatures, the need for a CCM is diminished due to the increased solubility of CO_2, CO_2:O_2 ratio, and pKa and the consequent greater availability of CO_2 (143, 157). However, Mitchell & Beardall (118) and Roberts & Beardall (168) showed that Antarctic micro- and macroalgae have similar DIC-acquisition capacities to their respective temperate-water counterparts. Changes in CA activity with temperature are inconsistent. Thus, CA_{ext} activity in *D. salina* is inversely related to temperature with a twofold decrease in CA_{ext} activity in cells grown at 25°C compared to cells grown at 10°C (B. Bernacchia & M. Giordano, unpublished), whereas the same authors observed a ~20% increase in CA_{ext} activity, for a similar temperature range, in the freshwater diatom *Asterionella formosa* and the marine coccolithophore *Emiliania huxleii* (14).

Although there are many reports on the effects of UV-B radiation on algal photosynthesis, there is virtually no information on the likely consequences of enhanced UV-B for CCM activity. In *Dunaliella tertiolecta*, UV-B inhibits carbon fixation but not transport, thereby causing intracellular DIC pools to be slightly elevated (16) and enhanced cellular affinity for CO_2 in photosynthesis. *Nannochloropsis gaditana*, which has active bicarbonate influx, showed more sensitivity to UV-B when grown under air levels of CO_2 than when the CCM was repressed by growth with 1% CO_2 in air (183). In contrast, the CO_2-transporting species *Nannochloris atomus* showed no difference in UV-B sensitivity between treatments. This implies that the bicarbonate transporter may be UV-B sensitive but that the CO_2 transporter may be resistant. However, this requires further investigation.

EVOLUTION

Here we briefly consider evidence from extant organisms as to the phylogeny of CCMs, and also evidence from the Earth sciences on the occurrence of cyanobacteria and algae and their impact on the environment over the last four Gyr. For cyanobacteria, the two categories (α and β) identified by carboxysome proteomics, the occurrence of the 1A or the 1B form of RUBISCO, and the membrane-associated CCM components map reasonably well onto the phylogeny deduced from other molecular genetic markers (11). Thus, the α-cyanobacteria *Prochlorococcus marinus* and *Synechococcus* sp. WH8102 (and other marine strains of *Synechococcus*) are a relatively derived (apomorphic) clade of cyanobacteria (169). Because cyanobacteria diversified prior to the endosymbiotic event leading to plastids (108, 196) it could be argued that at least β-carboxysomes and the associated components of the CCM evolved prior to the endosymbiotic origin of plastids. Badger & Price (11) suggest a later origin of CCMs and point out that the occurrence of both α- and β-carboxysomes in other kinds of bacteria means the evolution of cyanobacterial CCMs must take other bacteria, and the possibility of horizontal gene transfer, into account.

Gyr: billion years

Hypotheses as to the origin(s) of CCMs in eukaryotes should not be divorced from the occurrence (or not) of a CCM in the β-cyanobacterial plastid ancestor (11). If the plastid ancestor had a CCM, then the presence of a CCM would have been the plesiomorphic state for photosynthetic eukaryotes, with absence of CCMs as derived states. A similar argument can be applied to the subsequent secondary and tertiary endosymbioses. The membrane-associated components of the CCM would have been initially located on membranes derived from the plasmalemma or thylakoids of the plastid ancestor, although the subsequent transfer of genes to the host nucleus and retargeting of their products is a common feature of the evolution of photosynthetic eukaryotes (108). This hypothesis, and others arguing for the plesiomorphy of eukaryote CCMs, does not accord with the absence of reported homologues of cyanobacterial membrane-associated CCM components and the absence or rare occurrence of carboxysomes in eukaryotes.

A further problem is that phylogenetic analyses suggest that reacquisition of CCMs

Myr: million years

after loss may have occurred quite frequently, although much of the evidence favoring this view comes from the occurrence of pyrenoids (157), and although pyrenoids are good indicators of the capacity to express a CCM in a (wild-type) organism, the reverse is not true. In the absence of genetic data on (a) component(s) of CCM, the various possibilities cannot be distinguished.

In considering how the evolution of CCMs relates to their environment, a key factor is the kinetics of RUBISCO. The ancestral RUBISCO was derived from an enzyme in a bacterial methionine salvage pathway (5) and probably had a low CO_2 affinity and a low selectivity for CO_2 over O_2. At this time (perhaps ~3.8 Gyr ago), a greater greenhouse effect was needed to account for the presence of liquid water on an Earth orbiting the "faint young sun," which was only providing ~75% as much radiant energy as the sun does today (85). Mineralogical proxies for a 25°C surface temperature put a lower limit on CO_2 partial pressure of 250 pascal (Pa) at 3.2 Gyr ago (67), and an upper limit of 1 kPa at 2.8 Gyr ago (85); the latter value requires another greenhouse gas (e.g., CH_4) to account for Earth's surface temperature. O_2 was very low at this time, and for more than 100 Myr after the origin of oxygenic photosynthesis, not later than 2.7 Gyr ago (24, 30, 85, 191), such a diffusive supply of CO_2 to a phytoplankton cell would have been adequate even if its RUBISCO had a low CO_2 affinity and a low CO_2/O_2 selectivity. However, microbial mats and stromatolites (widespread from ~2.3 Gyr ago, at about the time that global oxygenation began) provided a high density of benthic (i.e., growing on a surface rather than free-floating) cells and could have caused a localized drawdown of CO_2 and accumulation of O_2.

Assuming a photosynthetic rate similar to extant microbial mats of 2 μmol m^{-2} s^{-1}, and a bulk-phase seawater CO_2 concentration (85) of 0.45 mM, a mean 0.4-mm diffusion distance to cyanobacterial cells yields an intracellular CO_2 of 50 μM and O_2 of 400 μM (cf. 144, which assumed higher CO_2 levels). This would constitute a selective factor favoring high-CO_2 affinity and higher CO_2/O_2 selectivity of RUBISCO, and/or a CCM, in the benthic cyanobacteria. These effects would have been exacerbated by the rise (via surface seawater O_2) in atmospheric O_2, which started later than 2.45 Gyr ago and had reached >0.02 Pa not later than 2.32 Gyr ago (24). The O_2 increase probably occurred at a similar time to that of the origin of eukaryotes (30), and is within the time frame of the Huronian glaciations (24). Increased O_2 would have converted the more potent greenhouse gas CH_4 into the less potent CO_2, so glaciations do not necessarily mean lower CO_2 (24, 85; figure 4 in 47).

These arguments suggesting that there could have been selective pressures for CCMs as early as 2.3 Gyr ago do not necessarily apply to planktonic cyanobacteria, so further evolution of cyanobacteria, including the primary endosymbiotic event to yield photosynthetic eukaryotes, may not have involved cyanobacteria and photosynthetic eukaryotes with CCMs. Later, there were the glaciations, possibly as extreme as a "Snowball Earth," in the Cryogenian/Ediacaran at 0.75 and 0.6 Gyr ago (85, 208). By 1.2 Gyr there were multicellular photosynthetic eukaryotes such as the red alga *Bangiomorpha* (34), with extant relatives that presumably survived the Snowball Earth in restricted, unfrozen areas of the ocean. These glaciations also could have involved very low CO_2 levels, and thus have been a time favorable to the evolution of CCMs (157). Badger & Price (11) suggest the origin of cyanobacterial and algal CCMs in the Carboniferous-Permian glacial episode was ~300 Myr ago. This period was characterized by low (~40 Pa) atmospheric CO_2 and high- (up to 35 kPa) atmospheric O_2 (87, 209); this was probably the time of origin of at least the submerged freshwater aquatic variant of CAM in vascular plants, although terrestrial C_4 metabolism probably did not originate until 20–30 Myr ago during a more recent decline in atmospheric CO_2 (87). Some examples of CCMs in algae also could have evolved at this time. The increase in the $^{13}C/^{12}C$ ratio of organic sediments in the ocean over the last 30 Myr needs investigation in this respect, although many factors other than the occurrence of CCMs influence this isotope ratio (65).

This brief survey shows that there were four times in Earth history (~2.3 Gyr ago; 0.75, 0.6 Gyr ago; ~300 Myr ago; from 30 Myr ago onward) that global conditions might have yielded selection pressures favoring CCMs. The earlier the suggested time of origin, the more problems there would have been in retaining the capacity for expressing CCMs during the intervening periods where only localized habitats would have favored the retention and expression of CCMs. Further advances in fixing the timing(s) of the origin(s) of CCMs may owe more to Earth sciences than to molecular clock (62) inputs.

CONCLUSIONS AND PROSPECTS

Since CCMs were last reviewed in this journal in 1999 there have been significant developments in our knowledge and understanding. There have been major advances in our understanding, at the molecular, mechanistic, and regulatory level, of the CCMs of β-cyanobacteria. However, complete genome sequences for several α-cyanobacteria have shown that these organisms from the oligotrophic ocean lack many of the components of the β-cyanobacterial CCMs without flagging up alternatives. The β-cyanobacterial genomic data have also not helped significantly in establishing the molecular basis for eukaryote CCMs, e.g., in *Chlamydomonas reinhardtii*. Another recent significant advance is the revival, with better evidence, of the hypothesis that diatoms have a CCM resembling the C_4 pathway of higher plants. This work on *Thalassiosira weissflogii* has not benefited as much as it might have from findings from the *T. pseudonana* genome project because we do not have a complete understanding of targeting sequences in diatoms to help establish the location of, for example, PEPck. Without playing down the advances that have been made, it is clear that much remains to be done to establish the mechanism(s), and regulation, of CCMs in many ecologically important groups of algae, such as the dinoflagellates.

As for environmental modulation, much has been achieved at the level of phenomena, but deeper understanding requires more knowledge of mechanisms and of regulation at the cellular and molecular level. This is an important area in view of the increasing CO_2 concentration, temperature, and acidity of the surface ocean and of inland water bodies. Increasing understanding of the mechanisms of CCMs, especially at the molecular level, will help our understanding of their evolution. Perhaps advances in our understanding of the phylogeny of algae might have predictive value in terms of the mechanisms of the CCM(s) of algae that have not previously been examined in detail. However, experience with α- and β-cyanobacteria suggests that caution is needed in the application of phylogeny as a predictive tool.

SUMMARY POINTS

1. CCMs evolved as a means of counteracting the effect of the increasing O_2/CO_2 ratio in the atmosphere on CO_2 fixation via RUBISCO.
2. The evolution of CCM is probably a polyphyletic process elicited by the selection pressure exerted by the increase in O_2/CO_2. Variations in this ratio occurred several times in the earth's history. It is possible that the prokaryotic CCM evolved in β-cyanobacteria; however, the possibility of lateral gene transfer makes it difficult to ascertain this. If the evolution of CCMs occurred before the primary endosymbiotic event, the presence of CCMs would be the plesiomorphic (ancestral) state for photosynthetic eukaryotes, some of which lost CCM at a later stage. If the plastid ancestor did not have a CCM, eukaryotes acquired CCMs polyphyletically.

3. Largely as a result of the presence of CCMs, the growth of most photolithotrophs present in today's aquatic environments is not limited by inorganic carbon availability.
4. CCMs can be based on biochemical mechanisms, such as C_4 photosynthesis and CAM, or on biophysical processes involving either localized enhancement of external CO_2 concentration by acidification of an external or internal compartment, or on the active transport of DIC across membranes. Different types of CCMs are present in almost all algal groups.
5. The carboxysome in prokaryotes and the pyrenoid in eukaryotes seem to play an important role in CCM (even though the absence of pyrenoids does not necessarily imply the absence of a CCM).
6. CCM activity can be modulated by environmental factors such as macro- and micronutrient supply, photosynthetically active and UV radiation, and temperature. CCMs likely play an important role in coupling DIC acquisition with the availability of energy and nutritional resources.

ACKNOWLEDGMENTS

John Beardall's work on inorganic carbon acquisition is supported by the Australian Research Council. John Raven's work on inorganic carbon acquisition is supported by the Natural Environment Research Council, United Kingdom. We are grateful to Ms. S. Stojkovic for her thorough and patient assistance in preparation of the manuscript.

LITERATURE CITED

1. Allemand D, Furia P, Bénazet-Tambutté S. 1998. Mechanisms of carbon acquisition for endosymbiont photosynthesis in Anthozoa. *Can. J. Bot.* 76:925–41
2. Amoroso G, Sültemeyer DF, Thyssen C, Fock HP. 1998. Uptake of HCO_3^- and CO_2 in cells and chloroplasts from the microalgae *Chlamydomonas reinhardtii* and *Dunaliella tertiolecta*. *Plant Physiol.* 116:193–201
3. Ananyev GM, Zoltsman L, Vasko C, Dismukes GC. 2001. The inorganic biochemistry of photosynthetic oxygen evolution/water oxidation. *Biochim. Biophys. Acta* 1503:52–68
4. Arancibia-Avila P, Coleman JR, Russin WA, Graham JM, Graham LE. 2001. Carbonic anhydrase localization in charophycean green algae: ecological and evolutionary significance. *Int. J. Plant. Sci.* 162:127–35
5. Ashida H, Saito Y, Kojima C, Kobayashi K, Ogasawara N, et al. 2003. A functional link between RuBisCO-like protein of *Bacillus* and photosynthetic RuBisCO. *Science* 302:286–90
6. Badger MR. 2003. The roles of carbonic anhydrases in photosynthetic CO_2 concentrating mechanisms. *Photosynth. Res.* 77:83–94
7. Badger MR, Andrews TJ, Whitney SM, Ludwig M, Yellowlees DC, et al. 1998. The diversity and co-evolution of Rubisco, plastids, pyrenoids and chloroplast-based CO_2-concentrating mechanisms in the algae. *Can. J. Bot.* 76:1052–71
8. Badger MR, Gallagher A. 1987. Adaptation of photosynthetic CO_2 and HCO_3^- accumulation by the cyanobacterium *Synechococcus* PCC6301 to growth at different inorganic carbon concentrations. *Aust. J. Plant Physiol.* 14:189–201

9. Badger MR, Hanson D, Price GD. 2002. Evolution and diversity of CO_2 concentrating mechanisms in cyanobacteria. *Funct. Plant Biol.* 29:161–73

10. Badger MR, Palmqvist K, Yu JW. 1994. Measurement of CO_2 and HCO_3^- fluxes in cyanobacteria and microalgae during steady-state photosynthesis. *Physiol. Plant.* 90:529–36

11. **Badger MR, Price GD. 2003. CO_2 concentrating mechanisms in cyanobacteria: molecular components, their diversity and evolution. *J. Exp. Bot.* 54:609–22**

12. Ball LA. 2003. *Carbon acquisition in the chrysophyte algae*. PhD thesis. Univ. Dundee, xv + 270 pp.

13. Beardall J. 1991. Effects of photon flux density on the "CO_2 concentrating mechanism" of the cyanobacterium *Anabaena variabilis*. *J. Plankton Res.* 13:133–41

14. **Beardall J, Giordano M. 2002. Ecological implications of microalgal and cyanobacterial CCMs and their regulation. *Funct. Plant Biol.* 29:335–47**

15. Beardall J, Griffiths H, Raven JA. 1982. Carbon isotope discrimination and the CO_2 accumulating mechanism in *Chlorella emersonii*. *J. Exp. Bot.* 33:729–37

16. Beardall J, Heraud P, Roberts S, Shelly K, Stojkovic S. 2002. Effects of UV-B radiation on inorganic carbon acquisition by the marine microalga *Dunaliella tertiolecta* (Chlorophyceae). *Phycologia* 41:268–72

17. Beardall J, Johnston AM, Raven JA. 1998. Environmental regulation of the CO_2 concentrating mechanism in cyanobacteria and microalgae. *Can. J. Bot.* 76:1010–17

18. Beardall J, Mukerji D, Glover HE, Morris I. 1976. The path of carbon in photosynthesis by marine phytoplankton. *J. Phycol.* 12:409–17

19. Beardall J, Quigg AS, Raven JA. 2003. Oxygen consumption: photorespiration and chlororespiration. In *Photosynthesis in Algae*, ed. AWD Larkum, SE Douglas, JA Raven, pp. 157–81. Dordrecht, The Netherlands: Kluwer

20. Beardall J, Raven JA. 2004. The potential effects of global climate change on microalgal photosynthesis, growth and ecology. *Phycologia* 43:31–45

21. Beardall J, Roberts S, Millhouse J. 1991. Effects of nitrogen limitation on inorganic carbon uptake and specific activity of ribulose-1,5-P_2 carboxylase in green microalgae. *Can. J. Bot.* 69:1146–50

21a. Beardall J, Roberts S, Raven JA. 2005. Regulation of inorganic carbon acquisition by phosphorus limitation in the green alga *Chlorella emersonii*. *Can. J. Bot.* In press

22. Behrenfeld MJ, Bale A, Kolber ZS, Aiken J, Falkowski PG. 1996. Confirmation of iron limitation of phytoplankton photosynthesis in the equatorial Pacific Ocean. *Nature* 383:508–11

23. Behrenfeld MJ, Randerson JT, McClain CR, Feldman GC, Los SO, et al. 2001. Biospheric primary production during an ENSO transition. *Science* 291:2594–97

24. Bekker A, Holland HD, Wang PL, Rumble D, Stein HJ, et al. 2004. Dating the rise of atmospheric oxygen. *Nature* 427:117–20

25. **Berman-Frank I, Kaplan A, Zohary T, Dubinsky Z.1995. Carbonic anhydrase activity in the bloom-forming dinoflagellate *Peridinium gatunense*. *J. Phycol.* 31:906–13**

26. Bertilsson S, Berglund O, Karl DM, Chisholm SW. 2003. Elemental composition of marine *Prochlorococcus* and *Synechococcus*: implications for the ecological stoichiometry of the sea. *Limnol. Oceanogr.* 48:1721–31

27. Bhatti S, Huertas IE, Colman B. 2002. Acquisition of inorganic carbon by the marine haptophyte *Isochrysis galbana* (Prymnesiophyceae). *J. Phycol.* 38:914–21

28. Blank ME, Ehmke H. 2003. Aquaporin-1 and HCO_3^--Cl^- transporter-mediated transport of CO_2 across the human erythrocyte membrane. *J. Physiol.* 550:419–29

This paper analyzes the occurrence of CCM components in relation to the phylogeny and evolution of cyanobacteria.

This is a comprehensive review of environmental factors modulating CCM activity in cyanobacteria and eucaryotic microalgae.

This is one of the few papers describing the induction of external CA activity (one of the components of CCM activity) in a developing natural population as well as in laboratory cultures.

29. Bozzo GG, Colman B, Matsuda Y. 2000. Active transport of CO_2 and bicarbonate is induced in response to external CO_2 concentration in the green alga *Chlorella kessleri*. *J. Exp. Bot.* 51:1341–48
30. Brocks JJ, Logan GA, Buick R, Summons RE. 1999. Archean molecular fossils and the early rise of eukaryotes. *Science* 285:1033–36
31. Burkhardt S, Amoroso G, Riebesell U, Sultemeyer D. 2001. CO_2 and HCO_3^- uptake in diatoms acclimated to different CO_2 concentrations. *Limnol. Oceanogr.* 46:1378–91
32. Burns DB, Beardall J. 1987. Utilization of inorganic carbon by marine microalgae. *J. Exp. Mar. Biol. Ecol.* 107:75–86
33. Busch S, Schmidt R. 2001. Enzymes associated with the β-carboxylation pathway in *Ectocarpus siliculosus* (Phaeophyceae): Are they involved in net carbon acquisition? *Eur. J. Phycol.* 36:61–70
34. Butterfield NJ. 2000. *Bangiomorpha pubescens* n. gen., n. sp.: implications for the evolution of sex, multicellularity, and the mesoproterozoic/neoproterozoic radiation of eukaryotes. *Palaeobiology* 26:386–404
35. Calamita G, Kempf B, Rudd KE, Bonhivers M, Kneip S, et al. 1997. The aquaporin-Z water channel gene of *Escherichia coli*: structure, organization and phylogeny. *Biol. Cell* 85:321–29
36. Cannon GC, Bradburne CE, Aldrich HC, Baker SH, Heinhorst S, Shively JM. 2001. Microcompartments in prokaryotes: carboxysomes and related polyhedra. *Appl. Environ. Microbiol.* 67:5351–61
37. **Colman B, Huertas IE, Bhatti S, Dason JS. 2002. The diversity of inorganic carbon acquisition mechanisms in eukaryotic microalgae. *Funct. Plant Biol.* 29:261–70**

This shows the range of microalgae with CCMs and the common occurrence of influx of both CO_2 and HCO_3^-.

38. Cram WJ. 1974. Influx isotherms—their interpretation and use. In *Membrane Transport in Plants*, ed. U Zimmermann, J Dainty, pp. 334–37. Berlin: Springer-Verlag
39. Cullen JT, Lane TW, Morel FMM, Sherrel RM. 1999. Modulation of cadmium uptake in phytoplankton by seawater CO_2 concentration. *Nature* 402:165–67
40. Dason JS, Huertas IE, Colman B. 2004. Source of inorganic carbon for photosynthesis in two marine dinoflagellates. *J. Phycol.* 40:285–92
41. Davies JP, Grossman AR. 1998. Responses to deficiencies in macronutrients. In *The Molecular Biology of Chloroplasts and Mitochondria in Chlamydomonas*. ed. J-D Rochaix, M Golschmidt-Clermont, S Merchant, pp. 613–33. Amsterdam: Kluwer
42. Davies JP, Yildiz FH, Grossman AR. 1996. *Sac1*, a putative regulator that is critical to survival of *Chlamydomonas reinhardtii* during sulfur deprivation. *EMBO J.* 15:2150–59
43. Dufresne A, Salanoubat M, Partensky F, Artiguenave F, Axmann IM, et al. 2003. Genome sequence of the cyanobacterium *Prochlorococcus marinus* SS120, a nearly minimal oxyphototrophic genome. *Proc. Nat. Acad. Sci. USA* 100:10020–25
44. El Alaoui S, Diez J, Toribio F, Gomez-Baena G, Dufresne A, Garcia-Fernandez JM. 2003. Glutamine synthetase from the marine cyanobacteria *Prochlorococcus* spp.: characterization, phylogeny and response to nutrient limitation. *Environ. Microbiol.* 5:412–23
45. Elzenega JTM, Prins HBA, Stefels J. 2000. The role of extracellular carbonic anhydrase activity in inorganic carbon utilization of *Phaeocystis globosa* (Prymnesiophyceae): a comparison with other marine algae using the isotopic disequilibrium technique. *Limnol. Oceanogr.* 45:372–80
46. Eriksson M, Karlsson J, Ramazanov Z, Gardesrtröm P, Samuelsson G. 1998. Discovery of an algal mitochondrial carbonic anhydrase: molecular cloning and characterization of a low-CO_2-induced polypeptide in *Chlamydomonas reinhardtii*. *Proc. Natl. Acad. Sci. USA* 93:12031–34
47. Falkowski PG, Katz ME, Knoll AH, Quigg A, Raven JA, et al. 2004. The evolution of modern eukaryotic phytoplankton. *Science* 305:354–60

48. Falkowski PG, Raven JA. 1997. *Aquatic Photosynthesis*. Malden, MA: Blackwell Sci.
49. Forschammer K. 1999. The P_{II} protein in *Synechococcus* PCC 7942 senses and signals 2-oxoglutarate under ATP-replete conditions. In *The Phototrophic Prokaryotes*, ed. W Löffelhardt, G Schmetterer, GA Peschek, pp. 549–53. New York: Kluwer Plenum
50. Fujiwara S, Fukuzawa H, Tachiki A, Miyachi S. 1990. Structure and differential expression of two genes encoding carbonic anhydrase in *Chlamydomonas reinhardtii*. *Proc. Natl. Acad. Sci. USA* 87:9779–83
51. Fukuzawa H, Fujiwara S, Yamamoyo Y, Dionisio-Sese ML, Miyachi S. 1990. cDNA cloning, sequence, and expression of carbonic anhydrase in *Chlamydomonas reinhardtii*: regulation by environmental CO_2 concentration. *Proc. Natl. Acad. Sci. USA* 87:4383–87
52. **Fukuzawa H, Miura K, Ishizaki K, Kucho K-I, Saito T, et al. 2001. *Ccm1*, a regulatory gene controlling the induction of a carbon concentrating mechanism in *Chlamydomonas reinhardtii* by sensing CO_2 availability. *Proc. Natl. Acad. Sci. USA* 98:5347–52**

This paper, with Reference 210, describes what is currently the best candidate for a component of the regulatory cascade for CCMs in eukaryotes.

53. Furla P, Bénazet-Tambatté S, Allemand D. 1998. Functional polarity of the tentacle of the sea anemone *Anemonia viridis*: role in inorganic carbon acquisition. *Am. J. Physiol.* 274:R303–10
54. Geider RJ, LaRoche J, Greene RM, Olaizola M. 1993. Response of the photosynthetic apparatus of *Phaeodactylum tricornutum* (Bacillariophyceae) to nitrate, phosphate, or iron starvation. *J. Phycol.* 29:755–66
55. Geider RJ, MacIntyre HL, Graziano LM, McKay RML. 1998. Responses of the photosynthetic apparatus of *Dunaliella tertiolecta* (Chlorophyceae) to nitrogen and phosphorus limitation. *Eur. J. Phycol.* 33:315–32
56. Geraghty AM, Spalding MH. 1996. Molecular and structural changes in *Chlamydomonas reinhardtii* under limiting CO_2: a possible mitochondrial role in adaptation. *Plant Physiol.* 111:1339–47
57. Giordano M. 2001. Interactions between C and N metabolism in *Dunaliella salina* cells cultured at elevated CO_2 and high N concentrations. *J. Plant Physiol.* 158 (5):577–81
58. Giordano M, Bowes G. 1997. Gas exchanges, metabolism, and morphology of *Dunaliella salina* in response to the CO_2 concentration and nitrogen source used for growth. *Plant Physiol.* 115:1049–56
59. Giordano M, Hell R. 2001. Mineral nutrition in photolithotrophs: cellular mechanisms controlling growth in terrestrial and aquatic habitats. *Recent Res. Dev. Plant Physiol.* 2:95–123
60. Giordano M, Norici A, Forssen M, Eriksson M, Raven JA. 2003. An anaplerotic role for mitochondrial carbonic anhydrase in *Chlamydomonas reinhardtii*. *Plant Physiol.* 132:2126–34
61. Giordano M, Pezzoni V, Hell R. 2000. Strategies for the allocation of resources under sulfur limitation in the green alga *Dunaliella salina*. *Plant Physiol.* 124:857–64
62. Graur D, Martin W. 2004. Reading the entrails of chickens: molecular timescales of evolution and the illusion of precision. *Trends Genet.* 20:80–86
63. Green BJ, Li W-Y, Manhart JR, Fox TC, Summer EJ, et al. 2000. Mollusc-algal chloroplast endosymnbiosis. Photosynthesis, thylakoid maintenance, and chloroplast gene expression continues for many months in the absence of the algal nucleus. *Plant Physiol.* 124:331–42
64. Greene RM, Geider RJ, Kolber Z, Falkowski PG. 1992. Iron-induced changes in light harvesting and photochemical energy conversion processes in eukaryotic marine algae. *Plant Physiol.* 100:565–75
65. Hayes JM, Strauss H, Kaufman AJ. 1999. The abundance of ^{13}C in marine organic matter and isotopic fractionation in the global biogeochemical cycle of carbon during the last 800 Ma. *Chem. Geol.* 161:103–25

86. Keeley JE. 1996. Aquatic photosynthesis. In *Crassulacean Acid Metabolism: Biochemistry, Ecophysiology and Evolution*. ed. K Winter, JAC Smith, pp. 281–95. Berlin: Springer-Verlag
87. Keeley JE, Rundel PW. 2003. Evolution of CAM and C_4 carbon-concentrating mechanisms. *Int. J. Plant Sci.* 164:S55–S77
88. Klughammer B, Sültemeyer D, Badger MR, Price GD. 1999. The involvement of NAD(P)H dehydrogenase subunits, NdhD3 and NdhF3, in high-affinity CO_2 uptake in *Synechoccus* sp. PCC7002 gives evidence for multiple NDH-1 complexes with specific roles in cyanobacteria. *Mol. Microbiol.* 32:1305–15
89. Korb RE, Saville PJ, Johnston AM, Raven JA. 1997. Sources of inorganic carbon for photosynthesis by three species of marine diatom. *J. Phycol.* 33:433–40
90. Kübler JE, Raven JA. 1994. Consequences of light-limitation for carbon acquisition in three rhodophytes. *Mar. Ecol. Pogr. Ser.* 110:203–8
91. Kübler JE, Raven JA. 1995. The interaction between inorganic carbon supply and light supply in *Palmaria palmata* (Rhodophyta). *J. Phycol.* 31:369–75
92. Kuchitsu TM, Tsuzuki M, Miyachi S. 1988. Changes of starch localization within the chloroplast induced by changes in CO_2 concentration during growth of *Chlamydomonas reinhardtii*: independent regulation of pyrenoid starch and stroma starch. *Plant Cell Physiol.* 29:1269–78
93. Lane TW, Morel FMM. 2000. Regulation of carbonic anhydrase expression by zinc, cobalt and carbon dioxide in the marine diatom *Thalassiosira weissflogii*. *Plant Physiol.* 123:345–52
94. Lane TW, Morel FMM. 2000. A biological function for cadmium in marine diatoms. *Proc. Natl. Acad. Sci. USA* 97:4627–31
95. Larsson C, Axelsson L.1999. Bicarbonate uptake and utilization in marine macroalgae. *Eur. J. Phycol.*34:79–86
96. Lee H-M, Flores E, Herrero A, Houmard J, Tandeau de Marsac N. 1998. A role for the signal transduction protein PII in the control of nitrate/nitrite uptake in a cyanobacterium. *FEBS Lett.* 427:291–95
97. Lee H-M, Vasquez-Bermudez MF, Tandeau de Marsac N. 1999. The global nitrogen regulator NtcA regulates transcription of the signal transducer P_{II} (GlnB) and influences its phosphorylation level in response to nitrogen and carbon supplies in the cyanobacterium *Synechococcus* sp. strain PCC 7942. *J. Bacteriol.* 181:2697–702
98. Lee RE, Kugrens PA. 1998. Hypothesis: the ecological advantage of chloroplast ER—The ability to outcompete at low dissolved CO_2 concentrations. *Protist* 149:341– 45
99. Lee RE, Kugrens PA. 2000. Ancient atmospheric CO_2 and the timing of evolution of secondary endosymbioses. *Phycologia* 39:167–72
100. Ludwig M, Sültemeyer D, Price GD. 2000. Isolation of *ccmKLMN* genes from the marine cyanobacterium *Synechococcus* sp. PCC7002 (cyanobacteria), and evidence that *ccmM* is essential for carboxysome assembly. *J. Phycol.* 36:1109–18
101. Maberly SC. 1996. Diel, episodic and seasonal changes in pH and concentrations of inorganic carbon in a productive lake. *Freshwater Biol.* 35:579–98
102. MacFarlane JJ, Raven JA. 1990. C, N and P nutrition of *Lemanea mammilosa* Kütz. (Batrachospermales, Rhodophyta) in the Dighty Burn, Angus, Scotland. *Plant Cell Environ.* 13:1–13
103. Maeda S, Badger MR, Price GD. 2002. Novel gene products associated with NdhD3/D4-containing NDH-1 complexes are involved in photosynthetic CO_2 hydration in the cyanobacterium *Synechococcus* sp. PCC7942. *Mol. Microbiol.* 43:425–36
104. Maeda S, Price GD, Badger MR, Enomoto C, Omata T. 2000. Bicarbonate binding activity of the CmpA protein of the cyanbacterium *Synechococcus* sp strain PCC 7942 involved in active transport of bicarbonate. *J. Biol. Chem.* 275:20551–55

105. Maeda S, Omata T. 1997. Substrate-binding lipoprotein of the cyanobacterium *Synechococcus* sp. Strain PCC 7942 involved in the transport of nitrate and nitrite. *J. Biol. Chem.* 272:3036–41
106. Manuel LJ, Moroney JV. 1988. Inorganic carbon accumulation in *Chlamydomonas reinhardtii*: new proteins are made during adaptation to low CO_2. *Plant Physiol.* 88:491–96
107. Marcus Y, Berry JA, Pierce J. 1992. Photosynthesis and photorespiration in a mutant of the cyanobacterium *Synechocystis* PCC 6803 lacking carboxysomes. *Planta* 187:511–16
108. Martin W, Rujan T, Richley E, Hansen A, Cornelson S, et al. 2002. Evolutionary analysis of *Arabidopsis*, cyanobacterial and chloroplast genomes reveals plastid phylogeny and thousands of cyanobacterial genes in the nucleus. *Proc. Natl. Acad. Sci. USA* 99:12246–51
109. Matsuda Y, Colman B. 1995. Induction of CO_2 and bicarbonate transport in the green alga *Chlorella ellipsoidea*. II. Evidence for induction in response to external CO_2 concentration. *Plant Physiol.* 108:253–60
110. Matsuda Y, Colman B. 1996. Active uptake of inorganic carbon by *Chlorella saccharophila* is not repressed by growth in high CO_2. *J. Exp. Bot.* 47:1951–56
111. Matsuda Y, Hara T, Colman B. 2001. Regulation of the induction of bicarbonate uptake by dissolved CO_2 in the marine alga *Phaeodactylum tricornutum*. *Plant Cell Environ.* 24:611–20
112. Matsuda Y, Satoh K, Harada H, Satoh D, Hiraoka Y, Hara T. 2002. Regulation of the expressions of HCO_3^- uptake and intracellular carbonic anhydrase in response to CO_2 concentration in the diatom *Phaeodactylum* sp. *Funct. Plant Biol.* 29:279–87
113. Matsuzaki M, Misumu O, Shin-I T, Maruyama M, Miyagishima SY, et al. 2004. Genome sequence of the ultrasmall unicellular red alga *Cyanidioschyzon merolae* 10D. *Nature* 428:653–57
114. Mayo WP, Williams TG, Birch DG, Turpin DH. 1986. Photosynthetic adaptation by *Synechococcus leopoliensis* in response to exogenous dissolved inorganic carbon. *Plant Physiol.* 80:1038–40
115. McGinn PJ, Price GD, Badger MR. 2004. High light enhances the expression of low-CO_2-inducible transcripts involved in the CO_2-concentrating mechanism in Synechocystis sp. PCC6803. *Plant Cell Environ.* 27:615–26

This paper describes the regulation, by high light, of genes involved in the CCM and suggests that any signaling mechanism modulating expression of CCM activity need not be based solely on external DIC concentration.

116. Miller AG, Turpin DH, Canvin DT. 1984. Na^+ requirement for growth, photosynthesis, and pH regulation in the alkalotolerant cyanobacterium *Synechococcus leopoliensis*. *J. Bacteriol.* 159:100–6
117. Miller AG, Turpin DH, Canvin DT. 1984. Growth and photosynthesis of the cyanobacterium *Synechococcus leopoliensis* in HCO_3^--limited chemostats. *Plant Physiol.* 75:1064–70
118. Mitchell C, Beardall J. 1996. Inorganic carbon uptake by an Antarctic sea-ice diatom *Nitzchia frigida*. *Polar Biol.* 16:95–99
119. Mitra M, Lato SM, Ynalvez RA, Xiao Y, Moroney JV. 2004. Identification of a new chloroplast carbonic anhydrase in *Chlamydomonas reinhardtii*. *Plant Physiol.* 135:173–82
120. Morel FMM, Cox EH, Kraepiel AML, Lane TW, Milligan AJ, et al. 2002. Acquisition of inorganic carbon by the marine diatom *Thalassiosira weissflogii*. *Funct. Plant Biol.* 29:301–8
121. Morita E, Abe T, Tsuzuki M, Fujiwana S, Sato N, et al. 1998. Presence of the CO_2-concentrating mechanism in some species of the pyrenoid-less free-living algal genus *Chloromonas* (Volvocales, Chlorophyta). *Planta* 204:269–76
122. Morita E, Abe T, Tsuzuki M, Fujiwana S, Sato N, et al. 1999. Role of pyrenoids in the CO_2-concentrating mechanism: comparative morphology, physiology and molecular phylogenetic analysis of closely related strains of *Chlamydomonas* and *Chloromonas*. *Planta* 208:365–72
123. Moroney JV, Bartlett SG, Samuelsson G. 2001. Carbonic anhydrases in plants and algae. *Plant Cell Environ.* 24:141–53

124. Moroney JV, Chen ZY. 1998. The role of the chloroplast in inorganic carbon uptake by eukaryotic algae. *Can. J. Bot.* 76:1025–34
125. Morris I, Beardall J, Mukerji D. 1978. The mechanisms of carbon fixation in phytoplankton. *Mitt. Internat. Verein. Limnol.* 21:174–83
126. Muro-Pastor MI, Reyes JC, Florencio FJ. 2001. Cyanobacteria perceive nitrogen status by sensing intracellular 2-oxoglutarate. *J. Biol. Chem.* 276:38320–28
127. Nakhoul NL, Davis BA, Romero MF, Boron WF. 1998. Effect of expressing the water channel aquaporin-1 on the CO_2 permeability of *Xenopus* oocyte? *Am. J. Physiol.* 274:C543–48
128. Ogawa T, Miyano A, Inoue Y. 1985. Photosystem-I-driven inorganic carbon transport in the cyanobacterium, *Anacystis nidulans. Biochim. Biophys. Acta.* 808:74–75
129. Ogawa T, Ogren WL. 1985. Action spectra for accumulation of inorganic carbon in the cyanobacterium, *Anabaena variabilis. Photochem. Photobiol.* 41:583–87
130. Ohkawa H, Pakrasi HB, Ogawa T. 2000. Two types of functionally distinct NAD(P)H dehydrogenases in *Synechocystis* sp strain PCC6803. *J. Biol. Chem.* 275:31630–34
131. Ohkawa H, Price GD, Badger MR, Ogawa T. 2000. Mutation of ndh genes leads to inhibition of CO_2 uptake rather than HCO_3^- uptake in *Synechocystis* sp strain PCC 6803. *J. Bacteriol.* 182:2591–96
132. Omata T, Price GD, Badger MR, Okamura M, Gohta S, Ogawa T. 1999. Identification of an ATP-binding cassette transporter involved in bicarbonate uptake in the cyanobacterium *Synechococcus* sp. Strain. *Proc. Natl. Acad. Sci. USA* 96:13571–756
133. Palinska KA, Laloui W, Bèdu S, Loiseaux-de Goër S, Castets AM, et al. 2002. The signal transducer PII and bicarbonate acquisition in *Prochlorococcus marinus* PCC 9511, a marine cyanobacterium naturally deficient in nitrate and nitrite assimilation. *Microbiol.* 148:2405–12
134. Palmqvist K. 2000. Carbon economy in lichens. *New Phytol.* 148:11–36
135. Palmqvist K, Sundblad L-G, Wingsle G, Samuelsson G. 1990. Acclimation of photosynthetic light reactions during induction of inorganic carbon accumulation in the green alga *Chlamydomonas reinhardtii. Plant Physiol.* 94:357–66
136. Palmqvist K, Yu J-W, Badger MR. 1994. Carbonic anhydrase activity and inorganic carbon fluxes in low- and high-Ci cells of *Chlamydomonas reinhardtii* and *Scenedesmus obliquus. Physiol. Plant.* 90:537–47
137. Price GD, Badger MR. 1989. Expression of human carbonic anhydrase in the cyanobacterium *Synechococcus* PCC7942 creates a high CO_2-requiring phenotype. Evidence for a central role for carboxysomes in the CO_2 concentrating mechanism. *Plant Physiol.* 91:505–13
138. Price GD, Maeda S, Omata T, Badger M. 2002. Modes of active inorganic carbon uptake in the cyanobacterium *Synechococcus* sp. PCC7942. *Funct. Plant Biol.* 29:131–49
139. Pronina NA, Borodin VV. 1993. CO_2 stress and CO_2 concentration mechanism: investigation by means of photosystem-deficient and carbonic anhydrase-deficient mutants of *Chlamydomonas reinhardtii. Photosynthetica* 28:515–22
140. Pronina NA, Semenenko VE. 1984. Localisation of membrane bound and soluble carbonic anhydrase in the *Chlorella* cell. *Fiziologia Rastenii* 31:241–51
141. Pronina NA, Semenenko VE. 1992. Role of the pyrenoid in concentration, generation and fixation of CO_2 in the chloroplast of microalgae. *Sov. Plant Physiol.* 39:470–76
142. Ramazanov Z, Rawat M, Henk MC, Mason CB, Matthews SW, Moroney JV. 1994. The induction of the CO_2-concentrating mechanism is correlated with the formation of the starch sheath around the pyrenoid of *Chlamydomonas reinhardtii. Planta* 195:210–16

143. Raven JA. 1991. Physiology of inorganic carbon acquisition and implications for resource use efficiency by marine phytoplankton: relation to increased CO_2 and temperature. *Plant Cell Environ.* 14:779–94
144. Raven JA. 1991. Plant responses to high O_2 concentrations: relevance to previous high O_2 episodes. *Palaeogeogr. Palaeoclimatol. Palaeoecol.* (*Glob. Planet. Chang. Sect.*) 97:19–38
145. Raven JA. 1997. Putting the C in phycology. *Eur. J. Phycol.* 32:319–33
146. Raven JA. 1997. Inorganic carbon acquisition by marine autotrophs. *Adv. Bot. Res.* 27:85–209
147. Raven JA. 1997. CO_2 concentrating mechanisms: a role for thylakoid lumen acidification? *Plant Cell Environ.* 20:147–54
148. Raven JA. 2000. Land plant biochemistry. *Phil. Trans. Roy. Soc. Lond. B.* 355:833–45
149. Raven JA. 2001. A role for mitochondrial carbonic anhydrase in limiting CO_2 leakage from low CO_2-grown cells of *Chlamydomonas reinhardtii*. *Plant Cell Environ.* 24:261–65
150. Raven JA. 2003. Carboxysomes and peptidoglycan walls of cyanelles: possible physiological functions. *Eur. J. Phycol.* 38:47–53
151. Raven JA. 2003. Inorganic carbon concentrating mechanisms in relation to the biology of algae. *Photosynth. Res.* 77:155–71
152. Raven JA, Beardall J. 1981. Carbon dioxide as the exogenous inorganic carbon source for *Batrachospermum* and *Lemanea*. *Br. Phycol. J.* 16:165–75
153. Raven JA, Beardall J. 2003. Carbon acquisition mechanisms in algae: carbon dioxide diffusion and carbon dioxide concentrating mechanisms. In *Photosynthesis in Algae*, ed. AWD Larkum, SE Douglas, JA Raven, pp. 225–44. Dordrecht, The Netherlands: Kluwer
154. Raven JA, Beardall J, Griffiths H. 1982. Inorganic C sources for *Lemanea*, *Cladophora* and *Ranunculus* in a fast-flowing stream: measurements of gas exchange and of carbon isotope ratio and their ecological implications. *Oecologia* 53:68–78
155. Raven JA, Beardall J, Johnston AM, Kübler J, Geoghegan I. 1995. Inorganic carbon acquisition by *Hormosira banksii* (Phaeophyta: Fucales) and its epiphyte *Notheia anomala* (Phaeophyta: Fucales). *Phycologia* 34:267–77
156. Raven JA, Beardall J, Johnston AM, Kübler JE, McInroy S. 1996. Inorganic carbon acquisition by *Xiphophora chondrophylla* (Phaeophyta: Fucales). *Phycologia.* 35:83–89
157. Raven JA, Johnston AM, Kübler JE, Korb J, McInroy SG, et al. 2002. Seaweeds in cold seas: evolution and carbon acquisition. *Ann. Bot.* 90:525–36
158. Raven JA, Johnston AM, Kübler JE, Korb RE, McInroy SG, et al. 2002. Mechanistic interpretation of carbon isotope discrimination by marine macroalgae and seagrasses. *Funct. Plant Biol.* 29:355–78
159. Raven JA, Kübler JE, Beardall J. 2000. Put out the light and then put out the light. *J. Mar. Biol. Assoc. UK* 80:1–25
160. Raven JA, Osborne A, Johnston AM. 1985. Uptake of CO_2 by aquatic vegetation. *Plant Cell Environ.* 8:417–25
161. Raven JA, Walker DI, Jensen KR, Handley LL, Scrimgeour CM, et al. 2001. What fraction of the organic carbon in saccoglosan is obtained from photosynthesis by kleptoplastids? An investigation using the natural abundance of stable isotopes. *Mar. Biol.* 138:537–45
162. Rawat M, Moroney JV. 1991. Partial characterization of a new isozyme of carbonic anhydrase isolated from *Chlamydomonas reinhardtii*. *J. Biol. Chem.* 266:9719–23
163. Reinfelder JR, Kraepiel AML, Morel FMM. 2000. Unicellular C_4 photosynthesis in a marine diatom. *Nature* 407:996–99
164. Reinfelder JR, Milligan AJ, Morel FMM. 2004. The role of C_4 photosynthesis in carbon accumulation and fixation in a marine diatom. *Plant Physiol.* 135:2106–11

Despite the data reported in this paper, the evidence for C_4 photosynthesis in a diatom is still incomplete.

165. Reiskind JB, Bowes G.1991. The role of phosphoenol pyruvate carboxykinase in a marine macroalga with C_4-like photosynthetic characteristics. *Proc. Natl. Acad. Sci. USA* 88:2883–87

166. Reiskind JB, Seaman PT, Bowes G. 1988. Alternative methods of photosynthetic carbon assimilation in marine macroalgae. *Plant Physiol.* 87:686–92

167. Ritchie RJ, Nadolny C, Larkum AWD. 1996. Driving forces for bicarbonate transport in the cyanobacterium *Synechococcus* R-Z (PCC 7942). *Plant Physiol.* 112:1573–84

168. Roberts S, Beardall J. 1999. Inorganic carbon acquisition by two species of Antarctic macroalgae: *Porphyra endivifolium* (Rhodophyta: Bangiales) and *Palmaria decipiens* (Rhodophyta: Palmariales). *Polar Biol.* 21:310–15

169. Rocap G, Larimer FW, Lamerdin J, Malfatti S, Chain P, et al. 2003. Genome divergence in two *Prochlorococcus* ecotypes reflects niche differentiation. *Nature* 424:1042–47

170. Rost B, Riebesell U, Burkhardt S, Sültemeyer D. 2003. Carbon acquisition of bloom-forming marine phytoplankton. *Limnol. Oceanogr.* 48:55–67

171. Rotatore C, Colman B. 1990. Uptake of inorganic carbon by isolated chloroplasts of the unicellular green alga *Chlorella ellipsoidea*. *Plant Physiol.* 93:1597–600

172. Rotatore C, Colman B. 1991. The localization of active carbon transport at the plasma membrane in *Chlorella ellipsoidea*. *Can. J. Bot.* 69:1025–31

173. Schwarz R, Reinhold L, Kaplan A. 1995. Low activation state of ribulose-1,5-bisphosphate carboxylase/oxygenase in carboxysome-defective *Synechococcus* mutants. *Plant Physiol.* 108:183–90

174. Sherlock DJ, Raven JA. 2001. Interactions between carbon dioxide and oxygen in the photosynthesis of three species of marine red algae. *Bot. J. Scotl.* 53:33–43

175. Shibata M, Katoh H, Sonoda M, Ohkawa H, Shimoyama M, et al. 2002. Genes essential to sodium-dependent bicarbonate transport in cyanobacteria. Function and phylogenetic analysis. *J. Biol. Chem.* 277:18658–64

176. Shibata M, Ohkawa H, Kaneko T, Fukuzawa H, Tabata S, et al. 2001. Distinct constitutive and low-CO_2-induced CO_2 uptake systems in cyanobacteria: genes involved and their phylogenetic relationship with homologous genes in other organisms. *Proc. Nat. Acad. Sci. USA* 98:11789–94

177. Shibata M, Ohkawa H, Katoh H, Shimoyama M, Ogawa T. 2002. Two CO_2-uptake systems: four systems for inorganic carbon acquisition in *Synechocystis* sp. strain PCC6803. *Funct. Plant Biol.* 29:123–29

178. Shiraiwa Y, Miyachi S. 1985. Effects of temperature and CO_2 concentration on induction of carbonic anhydrase and changes in efficiency of photosynthesis in *Chlorella vulgaris* 11h. *Plant Cell Physiol.* 26:543–59

179. Shively JM, van Keulen G, Meijer WG. 1998. Something from almost nothing: carbon dioxide fixation in chemoautotrophs. *Annu. Rev. Microbiol.* 52:191–230

180. Silverman DN. 2000. Marcus rate theory applied to enzymatic proton transfer. *Biochim. Biophys. Acta Bioenergetics* 1458:88–103

181. Smith KS, Ferry JG. 2000. Prokaryotic carbonic anhydrases. *FEMS Microbiol. Revs.* 24:335–66

182. So AK-C, Espie GS, Williams EB, Shively JM, Heinhorst S, Cannon GC. 2004. A novel evolutionary lineage of CAs (ε class) is a component of the carboxysome shell. *J. Bacteriol.* 186 : 623–30

183. Sobrino C, Neale PJ, Lubián LM. 2001. Effects of UV-radiation and CO_2 concentration on photosynthesis of two marine microalgae with different carbon concentrating mechanisms. *Phycologia* 40(Suppl.): 92–93

This contains some of the earliest evidence for a biochemical CCM in algae suggesting that even a simple anatomy is compatible with C_4 metabolism, and that the separation between β-carboxylation and decarboxylation can occur within the same cell.

184. Soupene E, Inwood W, Kustu S. 2004. Lack of the Rhesus protein Rh1 impairs growth of the green alga *Chlamydomonas reinhardtii* at high CO_2. *Proc. Natl. Acad. Sci. USA* 101:7787–92
185. Spalding MH, Critchley C, Govindjee, Ogren WL. 1984. Influence of carbon dioxide concentration during growth on fluorescence induction characteristics of the green alga *Chlamydomonas reinhardtii*. *Photosynth. Res.* 5:169–76
186. Spalding MH, Van K, Wang Y, Nakamura Y. 2002. Acclimation of *Chlamydomonas* to changing carbon availability. *Funct. Plant Biol.* 29:221–30
187. Sültemeyer D. 1998. Carbonic anhydrase in eukaryotic algae: characterization, regulation and possible functions during photosynthesis. *Can. J. Bot.* 76:962–72
188. Sültemeyer D, Amoroso G, Fock HP. 1995. Induction of intracellular carbonic anhydrases during the adaptation to low inorganic carbon concentrations in wild-type and *ca-1* mutant cells of *Chlamydomonas reinhardtii*. *Planta* 196:217–24
189. Sültemeyer DF, Fock HP, Canvin DT. 1991. Active uptake of inorganic carbon by *Chlamydomonas reinhardtii*: evidence for simultaneous transport of HCO_3^- and CO_2 and characterization of active transport. *Can. J. Bot.* 69:995–1002
190. Sültemeyer D, Klughammer B, Badger MR, Price GD. 1998. Fast induction of high-affinity HCO_3- transport in cyanobacteria. *Plant Physiol.* 116:183–92
191. Summons RE, Jahnke LL, Hope JM, Logan GA. 1999. 2-Methylhopanoids as biomarkers for cyanobacterial oxygenic photosynthesis. *Nature* 400:554–57
192. Tchernov D, Helman Y, Keren N, Luz B, Ohad I, et al. 2001. Passive entry of CO_2 and its energy-dependent intracellular conversion to HCO_3^- in cyanobacteria are driven by a photosystem I-generated $\Delta\mu H^+$. *J. Biol. Chem.* 276:23450–55
193. Thielmann J, Goyal A, Tolbert NE. 1992. Two polypeptides in the inner chloroplast envelope of *Dunaliella tertiolecta* induced by low CO_2. *Plant Physiol.* 100:2113–15
194. Thyssen C, van Hunnik E, Navarro MT, Fernández E, Galván A, et al. 2002. Analysis of *Chlamydomonas* mutants with abnormal expression of CO_2 and HCO_3^- uptake systems. *Funct. Plant Biol.* 29:251–60
195. Timmermans KR, van Leeuwe MA, de Jong JTM, McKay RML, Nolting RF, et al. 1998. Iron stress in the Pacific region of the Southern Ocean: evidence from enrichment bioassays. *Mar. Ecol. Prog. Ser.* 166:27–41
196. Tomitani A, Okada K, Miyashita H, Matthijs HCP, Ohno T, et al. 1999. Chlorophyll *b* and phycobilins in the common ancestor of cyanobacteria and chloroplasts. *Nature* 400:159–62
197. **Uehlein N, Lovisolo C, Siefritz F, Kaldenhoff R. 2003. The tobacco aquaporin NtAQP1 is a membrane CO_2 pore with physiological functions. *Nature* 425:734–37**

This paper provides the first molecular genetic evidence of a role for an aquaporin in CO_2 transport in photosynthetic organisms, providing a new perspective in the investigations for the pathway for CO_2.

198. van Hunnik E, Amoroso G, Sültemeyer D. 2002. Uptake of CO_2 and bicarbonate by intact cells and chloroplasts of *Tetraedon minimum* and *Chlamydomonas noctigama*. *Planta* 215:763–69
199. van Hunnik E, Livne A, Pogenberg V, Spijkerman E, van den Ende H, et al. 2001. Identification and localisation of a thylakoid-bound carbonic anhydrase from the green algae *Tetraedon minimum* (Chlorophyta) and *Chlamydomonas noctigama* (Chlorophyta). *Planta* 212:454–59
200. van Hunnik E, Sültemeyer D. 2002. A possible role for carbonic anhydrase in the lumen of chloroplast thylakoids in green algae. *Funct. Plant Biol.* 29:243–49
201. Van K, Spalding MH. 1999. Periplasmic carbonic anhydrase (*cah1*) structural gene mutant in *Chlamydomonas reinhardtii*. *Plant Physiol.* 120:757–64
202. Villarejo A, Rolland N, Martinez F, Sültemeyer DF. 2001. A new chloroplast envelope carbonic anhydrase activity is induced during acclimation to low inorganic carbon concentrations in *Chlamydomonas reinhardtii*. *Planta* 213:286–95

203. Vis ML, Entwisle TJ. 2000. Insights into the phylogeny of the Batrachospermales (Rhodophyta) with *rbcl* sequence data of Australian taxa. *J. Phycol.* 36:1175–82
204. Volokita M, Zenvirth D, Kaplan A, Reinhold L. 1984. Nature of the inorganic carbon species actively taken up by the cyanobacterium *Anabaena variabilis*. *Plant Physiol.* 76:599–602
205. Walker NA, Smith FA, Cathers IR. 1980. Bicarbonate assimilation by freshwater charophytes and higher plants. I. Membrane transport of bicarbonate is not proven. *J. Membr. Biol.* 57:51–58
206. Walsby AE. 1971. The pressure relationship of gas vacuoles. *Proc. Royal Soc. London B.* 178:301–26
207. Walsby AE. 1994. Gas vesicles. *Microbiol. Rev.* 58:94–144
208. Whitfield J. 2004. Geology: Time lords. *Nature* 429:124–25
209. Wildman RA, Hickey LJ, Dickinson MB, Berner RA, Robinson JM, et al. 2004. Burning of forest materials under late Palaeozoic oxygen levels. *Geology* 32:457–60
210. Xiang YB, Zhang J, Weeks DP. 2001. The *cia5* gene controls formation of the carbon concentrating mechanism in *Chlamydomonas*. *Proc. Natl. Acad. Sci USA.* 98:5341–46
211. Young EB. 1999. Interactions between photosynthetic carbon and nitrogen acquisition in the marine microalga *Dunaliella tertiolecta* Butcher. PhD thesis, Monash Univ., Melbourne, Australia. 228 pp.
211a. Young EB, Beardall J. 2005. Modulation of photosynthesis and inorganic carbon acquisition in a marine microalga by nitrogen, iron and light availability. *Can. J. Bot.* In press
212. Young EB, Beardall J, Giordano M. 2001. Investigation of inorganic carbon acquisition by *Dunaliella tertiolecta* (Chlorophyta) using inhibitors of putative HCO_3^- utilization pathways. *Eur. J. Phycol.* 36:81–88
213. Zubkov MV, Fuchs BM, Tarran GA, Burkill PH, Amann R. 2003. High rate of uptake of organic nitrogen compounds by *Prochlorococcus* cyanobacteria as a key to their dominance in oligotrophic oceanic waters. *App. Environ. Microbiol.* 69:1299–304

Solute Transporters of the Plastid Envelope Membrane

Andreas P.M. Weber,[1] Rainer Schwacke,[2] and Ulf-Ingo Flügge[2]

[1]Department of Plant Biology, Michigan State University, East Lansing, Michigan 48824-1312; email: aweber@msu.edu

[2]Lehrstuhl Botanik II, Universität zu Köln, 50931 Köln, Germany; email: ui.fluegge@uni-koeln.de

Annu. Rev. Plant Biol. 2005. 56:133–64

doi: 10.1146/annurev.arplant.56.032604.144228

First published online as a Review in Advance on January 14, 2005

1543-5008/05/0602-0133$20.00

Key Words

transport, metabolism, genomics, bioinformatics

Abstract

Plastids are metabolically extraordinarily active and versatile organelles that are found in all plant cells with the exception of angiosperm pollen grains. Many of the plastid-localized biochemical pathways depend on precursors from the cytosol and, in turn, many cytosolic pathways depend on the supply of precursor molecules from the plastid stroma. Hence, a massive traffic of metabolites occurs across the permeability barrier between plastids and cytosol that is called the plastid envelope membrane. Many of the known plastid envelope solute transporters have been identified by biochemical purification and peptide sequencing. This approach is of limited use for less abundant proteins and for proteins of plastid subtypes that are difficult to isolate in preparative amounts. Hence, the majority of plastid envelope membrane transporters are not yet identified at the molecular level. The availability of fully sequenced plant genomes, the progress in bioinformatics to predict membrane transporters localized in plastids, and the development of highly sensitive proteomics techniques open new avenues toward identifying additional, to date unknown, plastid envelope membrane transporters.

Contents

Plastids: semi-autonomous, membrane-bound organelles of plant cells that carry out a large number of biosynthetic and other functions such as photosynthesis (chloroplasts), ammonia and sulfur assimilation, and starch storage (amyloplasts)

INTRODUCTION

Plastids are essential to plants and are vital for life on Earth. These versatile plant organelles perform many specialized functions that are crucial for plant growth and development such as photosynthesis, nitrogen and sulphur assimilation, synthesis of amino acids and fatty acids, storage of carbohydrates and lipids, and formation of secondary compounds (**Figure 1**). Many of the plastid-synthesized products are further metabolized in other cell compartments and plastid metabolism depends on precursor supply from the cytosol. Therefore, the functioning of plant cells strongly relies on a complex metabolic network that intertwines reactions in the plastids with those in extraplastidial compartments by the exchange of metabolic intermediates. The plastid envelope represents the interface between both compartments and contains numerous transport proteins that mediate the flux of metabolites and ions across the envelope membrane(s).

It has been more than five years since the last review on metabolite transporters in the plastid envelope membrane was published in this series (24) and many novel plastid transporters have been discovered since then. Several recent reviews have addressed the physiology, biochemistry, and molecular biology of known plastid envelope transporters (23, 25, 76, 143, 144, 146). Therefore, we only briefly review these established transporter systems and summarize the most recent developments. The main focus is on the identification of novel transporters (i.e., transporters of unknown function) and on the impact of the completed genome sequence of *Arabidopsis thaliana* (125) on the field of plastid membrane transport. In addition, we analyze recent findings from proteomics and bioinformatics studies and discuss novel strategies for identifying plastid envelope membrane transporters. Finally, we give an outlook on future challenges, such as the elucidation of the structures of plastid envelope transporters.

BIOINFORMATICS APPROACHES TOWARD IDENTIFYING PLASTID ENVELOPE MEMBRANE TRANSPORTERS

The completely sequenced *Arabidopsis thaliana* genome (125) offered for the first time unique possibilities for a global analysis of plant proteins as a whole or as a subset of proteins, e.g., plant membrane proteins. Transmembrane proteins contain in general from one to over 14 transmembrane spans in the form of α-helices.

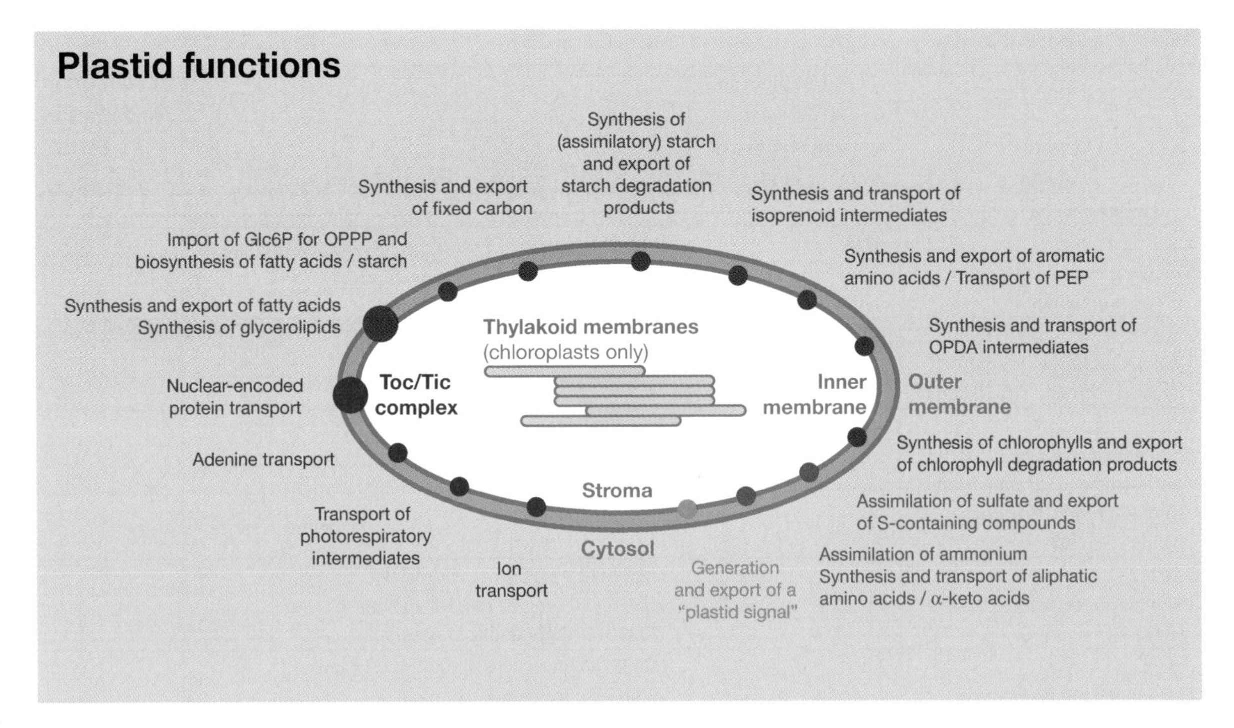

Figure 1

Schematic representation of plastid functions. All functions shown in purple, red, and orange also take place in the stroma. Glc6P, glucose 6-phosphate; OPDA, oxophytodienoic acid; OPPP, oxidative pentose phosphate pathway; PEP, phosphoenolpyruvate.

There are also β-barrel proteins with an even number of β-strands found in the outer membrane of bacteria, mitochondria, and plastids. A novel plant membrane protein database, ARAMEMNON, was recently established (110) (**http://aramemnon.botanik.uni-koeln.de/**), offering unique features such as calculation of consensus transmembrane topologies from numerous individual prediction programs, the collection of subcellular targeting predictions from several programs, and the display of paralogs and orthologs of any transmembrane protein in the *Arabidopsis* and rice proteomes, and of orthologs in cyanobacterial proteomes.

The predicted number of transmembrane proteins present in the genome of *A. thaliana* amounts to approximately 5800 proteins containing one or more transmembrane spans, corresponding to 22% of the 26,200 proteins encoded by the *Arabidopsis* genome. Approximately 660 of these 5800 transmembrane proteins are probably targeted to plastids because they possess predicted putative N-terminal targeting sequences that could direct the adjacent proteins to plastids. It is likely that not all of these 660 proteins are membrane transporters; most of these proteins contain only one or two predicted transmembrane domains, and transporters are polytopic membrane proteins that frequently have four or more transmembrane domains.

Table 1 is an attempt to categorize these putative plastidic transmembrane proteins encoded by the *A. thaliana* genome according to the TC system (6, 7). The TC system is analogous to the EC system for the classification of enzymes, except that the TC system considers both functional and phylogenetic information, whereas the EC system is based solely on the biochemical reaction catalyzed. The TC classification uses a five-digit system where the first

Plastid envelope membrane: Plastids of green plants are surrounded by two biomembranes, the inner and the outer plastid envelope membrane. This membrane system represents the permeability barrier between plastid stroma and surrounding cytosol. In some algae, the plastid envelope membrane may consist of up to four membrane layers.

Proteomics: systematic, comprehensive analysis of the full set of proteins (the proteome) in an organism, a specific cell or tissue, or a cellular fraction such as internal membrane systems

Table 1 Classification of putative envelope transport proteins according to the Transport Classification (TC) System[a]

TC number TC name	Total number in *A. thaliana* genome/estimated number in plastids	AGI numbers, (putative) function	Localization confirmed by proteomic data
1. Channels/pores			
1.A.1 Voltage-gated ion channel (VIC) superfamily	41/5	*At4g32650, At1g19780, At2g46430, At3g17690, At3g17700*	
1.A.18 Chloroplast envelope anion channel-forming Tic110 family	1/1	At1g06950 (atTic110; Reference 133)	Yes (51)
1.A.20 Gp91phox phagocyte NADPH oxidase-associated cytochrome b558 (Cyt b) H^+-channel family	18/2	*At5g49740* *At4g11230*	
1.A.26 Plant plasmodesmata (PPD) family	115/6	*At4g34440, At2g18470, At1g52290, At1g70460, At3g24550, At5g38560* (Putative receptor-like kinases)	
1.A.35 CorA metal ion transporter (MIT) family	11/3	*At4g28580* (putative magnesium transporter AtMGT5; Reference 61) *At5g22830* (putative magnesium transporter AtMGT10; Reference 61) *At3g58970* (putative magnesium transporter AtMGT6; Reference 61)	
2. Electrochemical potential-driven transporters			
2.A.1 Major facilitator superfamily (MFS)	76/10	*At4g04750* (putative sugar transporter) *At5g59250* (putative sugar transporter) *At1g05030* (putative sugar transporter) *At5g16150* (putative glucose translocator AtpGlcT1; Reference 142) *At3g20660* (putative organic cation transporter; Reference 82) *At2g29650* (putative anion transporter ANTR1; Reference 98) *At4g00370* (putative anion transporter ANTR2; Reference 98) *At2g38060* (putative anion transporter ANTR3) *At2g13100* (putative glycerol 3-phosphate transporter) *At4g30580*	Yes (20, 21, 28, 51, 97) [At5g59250] Yes (20, 21, 28, 97) [At1g05030] Yes (97) [At4g00370]
2.A.2 Glycoside-pentoside-hexuronide (GPH):cation symporter family	9/1	*At5g43610* (putative sucrose-proton symporter AtSUC6; pseudogene coding for an aberrant protein; Reference 104)	
2.A.3 Amino acid-polyamine-organocation (APC) family	17/3	*At5g04770, At1g05940* (putative cationic amino acid transporters AtCAT6 and AtCAT9) *At5g05630* (putative amino acid transporter AtLAT1)	

(*Continued*)

Table 1 (*Continued*)

TC number TC name	Total number in *A. thaliana* genome/estimated number in plastids	AGI numbers, (putative) function	Localization confirmed by proteomic data
2.A.4 Cation diffusion facilitator (CDF) family	8/2	*At2g04620*, *At2g47830*	
2.A.7 Drug/metabolite transporter (DMT) superfamily	117/9	At5g46110 (AtTPT; Reference 53) At3g01550 (AtPP2; References 53, 54) At5g33320 (AtPPT1; References 53, 54) At5g54800 (AtGPT1; (53)) At1g61800 (AtGPT2; (53)) At5g17630 (AtXPT; (17, 53)) *At1g12500* (phosphate translocator homolog; Referecne 53) *At3g10290* (phosphate translocator homolog; Reference 53) At3g10290	Yes (20, 21, 28, 51, 97) Yes (21, 51, 97)
2.A.9 Cytochrome oxidase biogenesis (Oxa1) family	6/1	At2g28800 (ALBINO3, localized to thylakoids; Reference 124)	
2.A.12 ATP:ADP antiporter (AAA) family	2/2	At1g80300 (ATP/ADP transporter AtNTT1; References 50, 70, 75) At1g15500 (ATP/ADP transporter AtNTT2; References 50, 70, 75)	Yes (28) Yes (21, 97)
2.A.18 Amino acid/auxin permease (AAAP) family	46/2	*At1g47670*, *At2g40420*	
2.A.28 Bile acid:Na^+ symporter (BASS) family	5/5	*At4g12030* *At4g22840* *At1g78560* *At2g26900* *At3g25410*	 Yes (20, 21, 97)
2.A.29 Mitochondrial carrier (MC) family	59/6	*At5g01500* *At4g39460* *At5g42130*, *At3g48850* *At3g51870* *At2g35800*	Yes (20, 97) Yes (20, 21, 97) Yes (51) Yes (20, 97)
2.A.37 Monovalent cation:proton antiporter-2 (CPA2) family	32/4	At1g05580 [$Na^+(K^+)H^+$ exchanger AtCHX23; Reference 119] *At5g37060* (putative cation proton exchanger AtCHX24; Reference 66) *At5g58460* (putative cation-proton exchanger AtCHX25; Reference 66) *At1g01790* (putative potassium efflux antiporter AtKEA; Reference 66)	 Yes (20, 51, 97)
2.A.39 Nucleobase:cation symporter-1 (NCS1) family	1/1	*At5g03555*	

(*Continued*)

Table 1 (*Continued*)

TC number TC name	Total number in *A. thaliana* genome/estimated number in plastids	AGI numbers, (putative) function	Localization confirmed by proteomic data
2.A.40 Nucleobase:cation symporter-2 (NCS2) family	12/2	*At4g38050* (AtNAT11) *At2g27810* (AtNAT12)	
2.A.47 Divalent anion:Na^+ symporter (DASS) family	4/3	At5g12860 (AtDiT1, AtpOMT1; References 89, 129, 140) At5g64280 (AtDiT2.2; References 89, 129, 140) At5g64290 (AtDiT2.1; References 89, 129, 140)	Yes (20, 21, 28, 51, 97) Yes (20, 21, 28, 51)
2.A.49 Ammonium transporter (Amt) family	6/1	*At1g64780* (AtAMT1.2; Reference 31)	
2.A.53 Sulfate permease (SulP) family	12/3	*At5g13550* (AtSulP, AtSultr4.1; References 5, 115, 126) *At5g19600* (AtSultr3.5; References 5, 115) *At4g08620* (AtSultr1.1; Reference 151)	
2.A.64 Twin arginine targeting (Tat) family	4/3	At2g01110 (cpTatC, localized to thylakoids; Reference 72) *At5g28750* (presumably localized to thylakoids) *At5g52440* (presumably localized to thylakoids)	 Yes (28) Yes (51)
2.A.66 Multidrug/oligosaccharidyl-lipid/polysaccharide (MOP) flippase superfamily	58/3	*At4g38380* (putative MATE-related efflux carrier AtDTX45; Reference 60) *At2g21340* (putative MATE-related efflux carrier AtDTX46; Reference 60) *At2g38330* (putative MATE-related efflux carrier AtDTX44M; Reference 60)	
2.A.67 Oligopeptide transporter (OPT) family	9/1	*At5g45450*	
2.A.71 Folate-biopterin transporter (FBT) family	35/2	*At1g04570*, *At2g32040*	
2.A.84 Chloroplast Maltose Exporter (MEX) Family	1/1	At5g17520 (AtMEX1; Reference 77)	Yes (20, 97)
3. Primary active transporters			
3.A.1 ATP-binding cassette (ABC) superfamily	115/19	Full-size transporters: *At3g21250* (AtMRP6), *At3g62700* (AtMRP10) *At1g15210* (AtPDR7) *At1g59870* (AtPDR8) *At1g66950* (AtPDR13) Half-transporters: *At1g54350* (AtPMP1), *At2g28070* (AtWBC3), *At4g25750* (AtWBC4), *At5g52860* (AtWBC8), *At5g19410* (AtWBC24), *At3g52310* (AtWBC28), *At1g70610* (AtTAP1), *At2g39190* (AtATH8), *At5g03990* (AtATH12)	 Yes (51) Yes (51) Yes (28)

(*Continued*)

Table 1 (*Continued*)

TC number TC name	Total number in *A. thaliana* genome/estimated number in plastids	AGI numbers, (putative) function	Localization confirmed by proteomic data
		Nontransport ABC systems: *At5g60790* (AtGCN1), *At5g09930* (AtGCN2), At4g04770 (AtNAP1), *At1g65410* (AtNAP11), *At4g25450* (AtNAP8)	Yes (20, 97)
3.A.2 H⁺- or Na⁺-translocating F-type, V-type and A-type ATPase (F-ATPase) superfamily	9/3 2 plastom-encoded proteins	*At5g08670* *At5g08680* *At5g08690* AtCg00120 (ATPase α-subunit, localized to thylakoids) AtCg00480 (ATPase β-subunit, localized to thylakoids)	Yes (51)
3.A.3 P-type ATPase (P-ATPase) superfamily (3)	48/4	*At4g33520* (putative P1B-type ATPase AtHMA6) *At4g37270* (putative P1B-type ATPase AtHMA1) *At5g21930* (putative P1B-type ATPase AtHMA8) *At1g27770* (putative P2B-type ATPase AtPEA1)	Yes (20, 97)
3.A.5 General secretory pathway (Sec) family	2/2	At2g18710 (cpSecY, localized to thylakoids; Reference 47) *At2g31530* (presumably localized to thylakoids)	Yes (97) Yes (97)
3.A.9 Chloroplast envelope protein translocase (CEPT or Tic-Toc) family	40/7	At4g03320 (AtTiC20-IV; Reference 46) *At4g25650* At5g16620 (AtTic40; References 15, 46) *At5g55710* At1g04940 (AtTiC20-I; Reference 46) At2g24820 (AtTic55; Reference 46) *At2g47840*	Yes (28, 51, 97) Yes (20, 51, 97) Yes (20, 28, 97) Yes (20, 51, 97) Yes (20, 51, 97)
3.D.1 Proton or sodium ion-translocating NADH dehydrogenase (NDH) family	5 plastom-encoded proteins	AtCg00890 (NADH Dehydrogenase ND2) AtCg01250 (NADH Dehydrogenase ND2) AtCg01050 (NADH Dehydrogenase ND5) AtCg01050 (NADH Dehydrogenase ND4) AtCg00440 (NADH Dehydrogenase D3) (All localized to thylakoids)	Not analyzed
3.D.3 Proton-translocating quinol:cytochrome c reductase (QCR) superfamily	6/1 2 plastom-encoded proteins	At4g03280 (Rieske FeS protein, localized to thylakoids; References 27, 55, 65) AtCg00720 (cytochrome b6, localized to thylakoids) AtCg00540 (cytochrome f, localized to thylakoids)	Yes (51)
3.E.2 Photosynthetic reaction center (PRC) family	2 plastom-encoded proteins	AtCg00270 (Photosystem II D2 protein, localized to thylakoids) AtCg00020 (Photosystem II D1 protein, localized to thylakoids)	Not analyzed
9. Incompletely characterized transport systems			
9.A.1 Phospholipid importer (PLI) family	5/3	*At1g54320*, *At1g79450*, *At3g12740*	

(*Continued*)

Table 1 (*Continued*)

TC number TC name	Total number in *A. thaliana* genome/estimated number in plastids	AGI numbers, (putative) function	Localization confirmed by proteomic data
9.A.12 Copper transporter-2 (Ctr2) family	6/2	At5g59030 (AtCOPT1; References 49, 99, 100) At5g59040 (AtCOPT3; Reference 100)	
9.B.17 Putative fatty acid transporter (FAT) family	45/8	*At3g23790, At4g14070, At4g19010, At1g20500, At1g20510, At5g38120, At1g77240, At3g05970*	
9.A.23 Ferroportin (FP) family	3/1	*At5g26820*	
9.B.26 PF27 (PF27) family	5/2	*At4g13590* *At1g64150*	Yes (20, 97)
Total number	995/137		

[a]For classification, all Arabidopsis membrane protein sequences are aligned to reference proteins (Transport Classification Database, http://tcdb.ucsd.edu/tcdb). All significantly similar sequences for each class are included (second column). Proteins with a plastid targeting sequence either predicted by at least three out of five prediction programs (ARAMEMNON database, http://aramemnon.botanik.uni-koeln.de/) or by experimental evidence (see references) are listed as putative plastidic proteins.

Bioinformatics: research, development, or application of computational tools and approaches to explore biological data such as genome sequences or proteomics data

ARAMEMNON: a database of transmembrane proteins from *Arabidopsis* and rice that stores genomic, cDNA, and protein sequence data, subcellular targeting information and transmembrane region prediction, and reference information (http://aramemnon.botanik.uni-koeln.de/)

TC: Transport Classification

EC: Enzyme Commission

digit designates the transporter class (e.g., primary active transporter) and the second digit corresponds to the subclass, with both digits referring to the mechanism of translocation and the source of energy used for the process. The next two digits refer to the transporter family and subfamily and the last digit corresponds to the substrates transported.

Table 1 shows those plastid-targeted membrane proteins that can be classified by the TC system down to the transporter family level (third digit of the TC system). Of the 5800 *A. thaliana* membrane proteins, 137 plastidic membrane proteins can be classified by the TC system. The total number of proteins, which fall in TC categories possessing plastidic representatives, amounts to 995. The remainder of the proteins possessing a predicted plastid targeting sequence do not fit into one of the currently established (as of August 2004) TC categories. Not all of these membrane proteins reside in the plastid envelope membrane—several (might) actually represent proteins of the thylakoid membrane and are indicated in **Table 1**, if known. The first column in **Table 1** refers to the TC number and TC name, the second column refers to the total number of proteins found for a specific subclass of transporters in the genome of *A. thaliana*, and the second and third columns give the number of putative plastidic transporters, the corresponding AGI numbers, and the putative functions. If a transporter function could not be experimentally assigned to a given protein or a member of a protein subfamily, the corresponding AGI numbers are italicized. In the case where the plastid localization of the membrane protein was confirmed by proteomics studies, this is indicated in column four of **Table 1**. **Table 1** also includes 11 plastom-encoded plastidic membrane proteins that can be classified by the TC system but that are involved in electron and ion transport rather than metabolite transport. Based on the bioinformatics analysis of the predicted *Arabidopsis* proteome, we estimate the number of metabolite transporters in the inner plastid envelope membrane in the range of 100 to 150 distinct proteins. This figure is derived from the 137 above-mentioned proteins and is corrected for those proteins potentially targeted to thylakoids, those proteins most likely involved in protein or electron transport, and the many proteins that currently cannot be classified by the TC system. Major limitations of the bioinformatics approach are potentially incorrect predictions of N termini in conceptual translations of genome sequences. In these cases the prediction programs will not correctly predict a plastidic localization of the protein. In addition, recent evidence shows that some envelope proteins (and also stroma-targeted proteins) do not possess cleavable targeting sequences that would be recognized by prediction programs (69, 74). Hence, 100 to 150 potential plastid envelope membrane transporters might represent an underestimation of the actual number.

PROTEOMICS STUDIES OF THE PLASTID ENVELOPE MEMBRANE

The recent developments of highly sensitive proteomics techniques have enabled the medium- to high-throughput analysis of notoriously difficult to handle membrane proteins (34, 103). Whereas van Wijk's laboratory has focused on the proteomes of the thylakoid membrane and the stroma (27, 81, 153), four recent studies concentrated on chloroplast envelope membranes (20, 21, 28, 97), and one group attempted a comprehensive approach toward the complete chloroplast proteome (51). Together, the latter five studies have identified 45 putative metabolite transporters (i.e., hydrophobic proteins with three or more putative transmembrane domains that do not show similarities to proteins involved in protein or electron transport). Twenty-two of these proteins could be assigned with one of the functional groups represented in **Table 1** (see column 4, **Table 1**). **Table 2** shows that 30 of the 45 putative transporters that were identified by proteomics could also be predicted to be plastid-localized by at least one of the prediction programs ChloroP (19), PCLR (105), or TargetP (18), and 25 were predicted to be plastid-targeted by two or more prediction programs (marked in gray in **Table 2**). Some of the proteins that were assigned to the plastid envelope by proteomics but were not categorized as plastidic proteins by the bioinformatics approach might represent contaminations of the plastidic fraction with proteins of the plasma membrane or other cellular membranes (e.g., plasma membrane or tonoplast membrane intrinsic proteins, see **Table 2**), whereas only a minor fraction might in fact belong to the plastid envelope membrane. The subcellular localization of these disputable candidates needs to be further examined by, e.g., GFP-fusion studies. Based on this evaluation, the proteomics approaches have revealed limited novel information on plastid envelope membrane transporters in comparison to the in silicio approaches. Nevertheless, these studies are important because they confirm the validity of the bioinformatics approach and they provide valuable information about the plastid envelope proteome of the tissue from which the plastids have been isolated. The in silicio approach tells us only what could be there, whereas the proteomics studies tell us what actually is there. The question arises why many of the computer-predicted envelope membrane proteins have not yet been confirmed by proteomics studies. It is likely that many of these proteins are not expressed in photosynthetically active tissues, which were exclusively used for the proteomic studies so far, but in tissues from which plastids cannot be easily isolated in preparative amounts. Future proteomics studies of, e.g., root plastids are necessary to complete the dataset and to gain information on plastid-type specific protein subsets. In addition, some proteins might only be present in minute amounts and may therefore be underrepresented in proteomics studies.

ABC TRANSPORTERS

ABC transporters (TC family 3.A.1) form a large family of proteins responsible for the translocation of many compounds across biological membranes. All ABC transporters share a common structural organization with two TMDs forming the translocation path and two cytosolic ABCs (NBDs), which couple ATP

Table 2 Putative plastid envelope membrane transporters as identified by proteomics[a]

AGI-number	Description	ChloroP	PCLR	TargetP	References	TM Domains (Aramemnon consensus)
At1g01620	Plasma membrane intrinsic protein 1c (AtPIP1.3/AtTMP-B)	0.105	0.021	0.066	(51)	6
At1g01790	Putative potassium efflux antiporter, CPA2 subfamily (AtKEA1)	0.19	0.004	0.01	(20, 51, 97)	13
At1g08890	Sugar transporter family	0.495	0.759	0.481	(51)	13
At1g15210	Pleiotropic drug resistance ABC transporter (AtPDR7)	0.555	0.375	0.642	(51)	14
At1g15500	AtNTT2, ATP carrier	0.515	0.323	0.279	(20, 97)	11
At1g15690	Putative vacuolar pyrophosphatase	0.175	0.107	0.016	(51)	16
At1g32080	Hypothetical protein	0.785	0.83	0.93	(20, 21, 97)	12
At1g59870	Pleiotropic drug resistance ABC transporter (AtPDR8)	0.515	0.386	0.778	(51)	13
At1g60160	Putative potassium transporter (AtKUP12/AtHAK12/AtKT12)	0.12	0.019	0.126	(51)	13
At1g78620	Integral membrane protein family	0.91	0.979	0.909	(20)	6
At1g80300	AtNTT1, ATP/ADP transporter	0.51	0.46	0.542	(28)	11
At2g26900	Bile acid:Na^+ symporter (BASS)	0.715	0.721	0.732	(20, 21, 97)	9
At2g35800	Mitochondrial carrier protein family	0.155	0.014	0.105	(20, 97)	2
At2g45690	Putative peroxisome formation protein (AtPex16/AtSSE1)	0.175	0.2	0.025	(51)	2
At2g48020	Sugar transporter, putative	0.465	0.469	0.074	(51)	12
At3g08580	Putative mitochondrial adenylate translocator (AtAAC1/AtANT1)	0.7	0.269	0.146	(51)	3
At3g16240	Putative tonoplast intrinsic protein 1 delta (AtTIP2.1)	0.305	0.023	0.006	(21, 51)	6
At3g26570	AtPHT2;1	0.51	0.384	0.156	(20, 97)	12
At3g45060	Putative high-affinity nitrate transporter (AtNRT2.6)	0.29	0.118	0.145	(51)	11
At3g51870	Putative mitochondrial carrier	0.55	0.18	0.513	(51)	2
At3g53420	Putative plasma membrane intrinsic protein 2a (AtPIP2.1)	0.13	0.003	0.0016	(51)	6
At3g57280	Expressed protein	0.87	0.871	0.89	(20, 97)	4
At3g60590	Hypothetical protein	0.295	0.114	0.001	(20, 97)	5
At4g00370	Putative anion transporter (AtANTR2)	0.465	0.184	0.684	(97)	10
At4g13590	PF27 Protein	0.82	0.979	0.738	(20, 97)	6
At4g17340	Tonoplast intrinsic protein 2 delta (AtTIP2.2)	0.35	0.036	0.005	(51)	6
At4g25450	Nonintrinsic ABC protein (AtNAP8)	0.745	0.745	0.539	(20, 97)	5
At4g37270	Putative $Zn_2{}^+/Co_2{}^+/Cd_2{}^+/Pb_2{}^+$-transporting P1B-type ATPase (AtHMA1)	0.675	0.919	0.659	(20, 97)	6
At4g39460	Mitochondrial carrier-like protein	0.72	0.369	0.626	(20, 21, 97)	5
At5g01500	Putative mitochondrial carrier	0.555	0.516	0.764	(20, 97)	3
At5g02180	Amino acid transporter family	0.045	0	0.032	(51)	10

(*Continued*)

Table 2 (*Continued*)

AGI-number	Description	ChloroP	PCLR	TargetP	References	TM Domains (Aramemnon consensus)
At5g03910	ABC-transporter like AtATH12	0.805	0.98	0.734	(28)	6
At5g12860	2-oxoglutarate/malate translocator DiT1	0.935	0.994	0.937	(20, 21, 28, 51, 97)	14
At5g13490	Putative mitochondrial adenylate translocator (AtANT2/AtAAC2)	0.44	0.339	0.132	(51)	3
At5g14040	Mitochondrial phosphate transporter (AtPHT3-1)	0.855	0.996	0.943	(51)	3
At5g16150	Putative glucose transporter pGlcT	0.83	0.473	0.674	(20, 21, 28, 97)	12
At5g17520	AtMEX1, maltose transporter	0.84	0.677	0.94	(20, 97)	9
At5g19760	Mitochondrial dicarboxylate/tricarboxylate carrier	0.25	0.105	0.051	(28, 51)	2
At5g33320	Phosphoenolpyruvate/phosphate translocator PPT1	0.835	0.993	0.975	(20, 51, 97)	6
At5g46110	Triose phosphate/phosphate translocator TPT	0.445	0.548	0.316	(20, 21, 28, 51, 97)	8
At5g52540	Hypothetical protein	0.86	0.976	0.956	(20, 97)	9
At5g59250	Putative xylose-H^+-symporter	0.755	0.847	0.907	(20, 21, 28, 51, 97)	11
At5g64290	Glutamate/malate translocator DiT2.1	0.795	0.972	0.819	(20, 21, 28, 51, 97)	10
At5g64940	ABC-transporter like AtATH13	0.765	0.956	0.942	(28)	11

[a]For each protein, the corresponding prediction scores of the programs ChloroP (19), PCLR(105), and TargetP (18) are listed. Prediction scores indicating plastidial localization are highlighted in gray.

binding to the translocation of the substrate. Some ABC transporters have only one TMD fused to one NBD and such a transporter is thus designated a half-transporter, which probably dimerizes to form a full-size transporter. In addition, there are soluble ABC proteins (nontransport ABC systems) lacking contiguous transmembrane domains.

The ABC transporter family in *A. thaliana* comprises a surprisingly high number of more than 100 members compared to the human genome, which has only about 50 members (101, 102). According to structural features, the ABC proteins can be grouped into different subfamilies. Most of the subfamilies also comprise members, which are putatively targeted to plastids. A total of 19 members represent the largest group of transporters present in the envelope membrane (**Table 1**; five full-size transporters, nine half-size transporters, and five nontransport ABC proteins). AtATH13, which also is putatively targeted to plastids, does not represent an ABC transporter (80). One of the largest ABC transporter subfamilies, the MDR subfamily with 23 members, does not appear to possess a plastidic representative. The function of most of the plastidic ABC transporters remains elusive to date. Only one plastidic nontransport ABC protein (AtNAP1) could be assigned with a function in light signaling (71).

TMD: transmembrane domain

NBD: nucleotide-binding domain

THE PLASTIDIC PHOSPHATE TRANSLOCATOR FAMILY

The genome of *A. thaliana* encodes four different subfamilies of plastidic phosphate

PT: phosphate translocator

Phosphate translocators: a class of plastidic solute transporters that catalyze the strict 1:1 counter exchange (antiport) of inorganic phosphate with a variety of phosphorylated carbon compounds such as triose phosphates, hexose phosphates, and pentose phosphates

TPT: triose phosphate/phosphate translocator

HMWP: high molecular weight polysaccharide

antiport systems, which exchange various phosphorylated C3-, C5-, and C6-compounds for inorganic phosphate between plastids and the cytosol. These PTs, all belonging to the Drug/Metabolite Transporter Family (TC DMT-family 2.A.7), comprise the best-characterized transporter family of plastids; the reader is referred to recent reviews dealing with the characterization of these transporters (23–25, 143, 146). Here, we summarize the impact the PTs have on plant metabolism and performance and focus on new emerging aspects.

The Triose Phosphate/Phosphate Translocator and Pathways of Carbon Allocation

The TPT represents the "day path of carbon allocation" from chloroplasts and exports the fixed carbon in the form of triose phosphates (trioseP). Triose phosphates are generated by the reductive pentose phosphate cycle (Calvin-Benson Cycle) at the expense of photosynthetically generated energy and reducing equivalents (ATP and NADPH). In most plants, sucrose formed in the cytosol from four molecules of trioseP is the main product allocated to heterotrophic plant organs for further metabolism or conversion into storage products, e.g., starch. In turn, the inorganic P_i released in the cytosol during the biosynthetic processes is transported back into chloroplasts where it is used for the synthesis and replenishment of ATP to sustain the Calvin-Benson Cycle and photosynthetic electron transport. TrioseP can be also used to produce transitory starch within the chloroplast, which is mobilized during the following night and exported from the source tissue to ensure a continuous supply of heterotrophic tissues with photoassimilates in the absence of photosynthesis.

Transformants with a reduced activity of the TPT or mutants defective in TPT function do not show a substantial growth phenotype but a severe alteration in carbon metabolism (38–40, 93, 107). Only plants in which trioseP export and starch biosynthesis are inhibited simultaneously show a severe growth phenotype (36, 107). Double mutants defective in both TPT function and starch mobilization degradation due to lack of the starch phosphorylating enzyme glucan, water dikinase (GWD) (95, 152) were less compromised, probably because of being able to synthesize and to turn over HMWPs (107). Forming these HMWPs is an additional mechanism for compensating photoassimilate export in case starch degradation is impaired.

Mutants defective in the TPT activity use a different path of photoassimilate allocation from the chloroplast to the cytosol to compensate for the deficiency in TPT activity. This "night path of carbon allocation" involves transporters different from the TPT, includes starch breakdown preferentially using an amylolytic pathway that leads to the formation and export of free sugars (33, 106), and results in the formation of maltose as the main starch breakdown product (147). Maltose is exported from the chloroplasts by the recently discovered maltose transporter MEX1 (TC number 2.A.84.1.1) (77). The current picture for assimilatory starch breakdown and export of starch breakdown products in the model plant *A. thaliana* (see 154 for a recent review) includes the action of β-amylase, an exoamylase that releases mainly maltose units and, to a minor extent, also maltotriose from linear α-1,4-glucan chains, whereas starch phosphorylase is not involved in transitory starch breakdown (155). Maltotriose is the substrate of a stromal disproportionating enyme (DPE1, D-enzyme, α glucanotransferase) releasing glucose, which can be exported by the putative glucose translocator pGlcT1 (TC family 2.A.1) (142). The further metabolism of the exported maltose in the cytosol comprises the action of a cytosolic disproportionating enyme (DPE2) producing glucose and transferring the second glucose moiety to a so far unknown acceptor (14, 63) from which it is released by an also so far unknown mechanism (see 112 for a discussion of possible mechanisms). Both glucose residues derived from cytosolic maltose and glucose exported by pGlcT1 are finally fed into the hexose phosphate pool. Together, after many years of research on assimilatory starch metabolism, the emerging picture of starch breakdown in *Arabidopsis* is almost complete. In the future it has to be shown how far this picture can extend to other plants.

The Glucose 6-Phoshate/Phosphate Translocator is Essential for Early Plant Development

Plastids of nongreen tissues can import carbon in the form of glucose 6-phosphate by the GPT (48). The genome of *Arabidopsis* harbors two paralogous GPT genes, AtGPT1 and AtGPT2 (53). *AtGPT1* transcripts accumulate ubiquitously during plant development, whereas *AtGPT2* expression is restricted to flowers, siliques, and senescent leaves. Expression of these genes in yeast cells reveals that both proteins represent functional GPTs. When ectopically expressed in a plastidic phosphoglucoisomerase-deficient mutant line, both *Arabidopsis* GPTs complemented the low-starch leaf phenotype of this mutant, indicating that both GPTs are functional also in planta (P. Niewiadomski, S. Knappe, S. Geimer, K. Fischer, B. Schulz, U.S. Unte, N. Strizhov, P. Ache, U.I. Flügge & A. Schneider, manuscript in preparation). By reverse genetics two mutants, *gpt1-1* and *gpt1-2*, were isolated, harboring T-DNA insertions in the GPT1 gene. In the homozygous state both mutations resulted in lethality. In both mutant lines a distorted segregation ratio together with a reduced male and female transmission efficiency and an arrest in pollen and ovule development indicated profound defects on gametogenesis. The mutant female gametophyte development was arrested at a stage before the polar nuclei fuse. The mutant pollen development was associated with less-lipid body formation and a disintegration of the membrane system. In addition, an effect on the developing sporophyte was indicated by the complete lack of homozygous adult mutant plants and the occasional occurrence of aborted embryos. The pleiotropic effect on gametophyte development could be fully reversed by introducing the *AtGPT1* gene under transcriptional control of its authentic promoter in the *gpt1-2* mutant. On the other hand, disruption of *GPT2* had no obvious effect on growth and development under greenhouse conditions (P. Niewiadomski, S. Knappe, S. Geimer, K. Fischer, B. Schulz, U.S. Unte, N. Strizhov, P. Ache, U.I. Flügge & A. Schneider, manuscript in preparation). Taken together, these results implicate that import of glucose 6-phosphate by AtGPT1 into nongreen plastids is crucial for pollen maturation and female gametophyte development. It is likely that glucose 6-phosphate serves as a substrate for the oxidative pentose phosphate cycle to create reducing power, which is required, e.g., for fatty acid biosynthesis.

GPT: glucose 6-phosphate/phosphate translocator

PEP: phosphoenolpyruvate

ME: malic enzyme

CAM: Crassulacean acid metabolism

PPT: PEP/phosphate translocator

The Phosphoenolpyruvate/Phosphate Translocator AtPPT1 is Essential for the Development of Mesophyll Tissues

Plastids of both C_4- and C_3-plants rely on the transport of PEP. In C_4-plants of the NADP ME type such as maize the necessity of high PEP transport rates across the envelope of mesophyll chloroplasts is obvious. Pyruvate produced by ME in the bundle sheath has to be converted to PEP by the chloroplast-localized pyruvate, phosphate dikinase (PPDK). Thus, in C_4-metabolism PEP is exported from the stroma to the cytosol, where it is required by PEP carboxylase for primary carbon fixation. There is a similar scenario in ME-type CAM plants during the day, when malic acid, which accumulates during the night, is decarboxylated by a cytosolic NADP ME and or mitochondrial NAD ME. Again, pyruvate has to enter the chloroplasts to be converted into PEP by PPDK. Because *M. crystallinum* chloroplasts lack a complete glycolytic/gluconeogenetic pathway (43), PEP is exported from the stroma via a PPT and acts as a precursor for the biosynthesis of storage carbohydrates. This function was reinforced by investigations on the expression of the PPT gene isolated from the inducible CAM plant *M. crystallinum*. The PPT is induced several-fold during CAM induction and PPT mRNA is more

MYB: myeloblastosis-type

DCG: dehydrodiconiferylalcohol glycoside

abundant during the light period. Transport activities determined during the diurnal CAM cycle followed a similar pattern (37).

The function of the PPT in C_3-plants or nongreen tissues of C_4-plants is less obvious. The proposed function of the PPT in these tissues is the import rather than the export of PEP (22). Together with erythrose 4P, an intermediate of the Calvin-Benson cycle and the oxidative pentose phosphate pathway, PEP is required as a precursor for the plastid-localized shikimate pathway, from which aromatic amino acids and an array of secondary plant products such as anthocyanins, flavonoids, or lignin derive. Chloroplasts and most nongreen plastids appear to lack the glycolytic sequence leading from 3-phosphoglycerate to PEP [i.e., phosphoglyceromutase and enolase (109, 121)]. Thus, the provision of cytosolic PEP to the stroma via the PPT would be essential for the shikimate pathway. This proposed role of the PPT was enforced following the identification of the mutated gene responsible for the severe reticulate phenotype of the *Arabidopsis cue1* [*c*hlorophyll a/b (CAB)-binding protein *u*nder*e*xpressed] mutant, which was isolated in a screen for mutants with a lowered CAB expression during de-etiolation (59). *Cue1* is defective in a PPT (122) and is impaired in mesophyll development combined with smaller and fewer chloroplasts therein, whereas cells adjacent to the vasculature of the rosette leaves show wild-type-like characteristics. The mutant phenotype could be rescued by feeding aromatic amino acids to the roots or by ectopic overexpression of a heterologous PPT and also of a plastid-targeted PPDK, which converts pyruvate to PEP (136). These experiments provide strong evidence that the lack of PEP and of aromatic compounds derived thereof in plastids is responsible for the *cue1* phenotype. A detailed analysis of amino acid contents as well as secondary plant products revealed characteristic tissue and cell-specific changes in their abundance between the *cue1* mutant and the wild type. Secondary plant products were only selectively affected in the *cue1* mutant and contents of aromatic amino acids remained fairly unchanged compared to the wild-type, of which observations threw some doubt on the proposed general involvement of the PPT in the shikimate pathway. In addition, antisense inhibition of transketolase (producing erythrose 4P as the second immediate entry substrate of the shikimate pathway) in transgenic tobacco plants resulted in a dramatic decline in aromatic amino acids and secondary metabolites contents (41). The general involvement of the *PPT* defective in *cue1* was further challenged by the isolation of a second functional *PPT* gene from *Arabidopsis*, *AtPPT2* (54). *AtPPT2* is expressed ubiquitously in the leaf blade, but is absent from the roots in contrast to *AtPPT1*, which is expressed mainly in certain tissues of the root, such as the root tip (most pronounced in the columella), lateral root tips, and in the central cylinder. In leaves, *AtPPT1* promoter activity is restricted to the vasculature of younger leaves and is absent from the mesophyll. Thus, in leaves AtPPT1 expression is restricted to zones that appear like the wild type in the *cue1* mutant. Despite the fact that AtPPT2 is expressed ubiquitously in the leaf blade, it is not able to compensate for a deficiency in AtPPT1 activity in the vasculature.

It is likely that AtPPT2 fulfills housekeeping functions in the provision of PEP for the shikimate pathway and the subsequent biosynthesis of aromatic amino acids and derived secondary components. By contrast, AtPPT1 might be involved in the generation of a signal molecule, which has the potential to trigger the correct development of the mesophyll. This hypothesis is based partially on the observation that a MYB transcription factor of *Anthirrinum majus*, when overexpressed in transgenic tobacco plants, competes for binding at target sites of endogenous MYB factors that trigger the biosynthesis of phenylpropanoids (127, 128). The transgenic tobacco lines have a similar reticulate leaf phenotype combined with aberrant-shaped mesophyll cells. These plants appear to have severely lowered contents of DCG, a substance that is derived from phenylpropanoid metabolism and has hormonal functions (132). Applying the aglycon or DCG to tobacco mesophyll cell cultures generated from the transgenic plants could rescue the aberrant shape of the cells, thus supporting the idea that in tobacco this substance can trigger developmental programs impaired in the transgenics. It is conceivable that a similar scenario occurs in the mutant. If, for instance, DCGs were formed only in those cell types where AtPPT1 is present, the production of this substance would be restricted in the *cue1* mutant. A lack of DCG production might then lead to the developmental constraints observed in the mutant. To fully understand the function of the PPTs in plant development and metabolism, an analysis of spatial and temporal distribution of enzymes involved in the proliferation of PEP and compounds derived thereof appears indispensable.

Other Plastic Phosphate Translocators of the DMT-Family

The recently discovered pentose phosphate translocator (XPT) represents the fourth member of the plastidic PT family (17). At least in *Arabidopsis*, the cytosolic oxidative pentose phosphate pathway proceeds only to the stage of interconvertible pentose phosphates but not further due to the absence of cytosolic isoforms of transketolase and transaldolase (17, see 56 for a recent review). The XPT exchanges pentose phosphates between the cytosol and the plastids in which pentose phosphates are intermediates of both the oxidative pentose phosphate pathway and the Calvin-Benson cycle.

The genome of *A. thaliana* contains numerous membrane proteins of the DMT-family, sharing significant similarities with the plastidic PTs (53). Two of them (At1g12500, At3g10290) contain N-terminal plastidic targeting sequences presumably directing these proteins to plastids. So far, the functions of all PT-paralogs remain elusive.

TRANSPORTERS OF THE OUTER ENVELOPE MEMBRANE

The vast majority of plastid proteins are encoded by the nuclear genome, synthesized as precursors in the cytosol, and imported post-translationally into the plastids. The precursor proteins are recognized and transported across both the outer and inner envelope membrane by components of the Toc and Tic machineries [TC family 3.A.9 (see 116 for a recent review)]. In addition, the outer envelope membrane contains various β-sheet outer envelope proteins (OEPs; OEP24, OEP21, OEP16) that possess channel-like activities and may act as selectivity filters for transporting metabolites including amino acids, inorganic phosphate, trioseP, and hexose phosphates (4, 84, 85, 120). Electrophysiological measurements clearly demonstrated the potential of the OEPs to regulate solute flow through the outer envelope membrane. However, there is no experimental evidence for a restriction of solute transport across envelope membranes exerted by these proteins in vivo.

The function of OEP16, which had been proposed to possess a selective permeability to amino acids, was challenged by recent studies demonstrating that OEP16 is involved in importing the precursor of NADPH: protochlorophyllide oxidoreductase A into chloroplasts (87, 88). These experiments were performed using chloroplasts of barley and it remains to be shown whether or not OEP16 can act as a general precursor translocase. Interestingly, OEP16 shares sequence similarities with components of the mitochondrial protein import machinery, constituting a family of PRATs (86).

XPT: xylulose 5P/phosphate translocator

PRAT: preprotein and amino acid transporter

OTHER TRANSPORT SYSTEMS

Owing to space constraints, a comprehensive treatise of all plastidic transport systems characterized to date is beyond the scope of this review. Transport systems not discussed in detail include the plastidic dicarboxylate transporter family (89, 129, 130, 141), carbohydrate transporters such as the maltose transporter MEX1 (77) and the putative glucose transporter pGlcT (142), the adenylate translocators (NTTs; 75), the low-affinity phosphate transporter PHT2;1 (135), and the ADP-glucose transporter Bt1 (123). For further information and for a classification of these transport systems, the reader is referred to several recent reviews (23–25, 76, 140, 143, 144, 146) and to **Table 1**.

STRATEGIES TO IDENTIFY NOVEL TRANSPORTERS

Although the bioinformatics and proteomics approaches provided a relatively comprehensive list of putative envelope membrane metabolite transporters, only a minor portion of these proteins is functionally characterized to date. Whereas less than 20 transporters are assigned with a function, probably more than 100 transporters are still awaiting a functional assignment. Strategies to identify transport systems in plants were recently reviewed (2); therefore, we limit this chapter to a brief overview and discuss novel model systems for characterizing plant membrane transporters.

In some cases, hypotheses about possible functions can be derived from homologies to proteins of known functions. These hypotheses can then be tested by producing recombinant protein in a heterologous host and functional analysis after reconstitution into liposomes. This approach, however, is limited by the availability of radioactively labeled tracer substrates used in these transport assays and because many of the antiporters catalyze an electro-neutral exchange of metabolites that cannot be assessed by electrophysiological methods. Nevertheless, some of the putative amino acid transporters, monosaccharide transporters, and organic anion transporters are promising candidates for this strategy.

Many transporters of the plasma membrane have been functionally characterized by complementation of yeast mutants defective in the uptake of a particular metabolite (9, 10, 32, 78, 90, 94, 111), and several plant mitochondrial transporters have been identified by complementation of the corresponding yeast mutants (9, 10). Surprisingly, no plastid envelope membrane transporters have been identified by this strategy. This may be mainly due to the fact that plastidial transporters are not targeted to the plasma membrane but to the endomembrane system when expressed in yeast cells, even in the absence of a targeting peptide (62), but this explanation may not hold in all cases. For example, it was demonstrated that the plastidic phosphate transporter PHT2;1 was efficiently targeted to the plasma membrane in yeast when the transit peptide was removed and the mature PHT2;1 protein was able to functionally complement a phosphate-uptake-deficient yeast mutant strain (135). This finding encourages the development of advanced complementation screens, for example the construction of cDNA libraries in which plant cDNAs are fused to sequences encoding targeting signals that route membrane proteins to the yeast plasma membrane. In addition, it might be possible to functionally complement yeast mutants defective in mitochondrial transporters with transporters of the plastid envelope membrane. For example, the chloroplast outer envelope protein OEP24 could functionally replace a mitochondrial voltage-gated anion channel in yeast (96), and similar strategies might also work for inner membrane proteins.

Additional promising strategies are forward and reverse genetics approaches. Several recent studies demonstrate the power of traditional forward genetic screens for identifying plastid envelope membrane transporters. For example, the maltose transporter AtMEX1 was identified in a screen for maltose-accumulating *Arabidopsis* mutants (77), the plastidic glutamate/malate translocator AtDiT2.1 was identified by map-based cloning of the defective gene in the photorespiratory mutant *dct* (89, 117, 118), and the putative lipid permease TGD1 was identified in a high-throughput screen for *Arabidopsis* mutants with altered lipid metabolism (150). In theory, it should be possible to identify additional plastidic membrane transporters such as the glycolate and the glycerate transporters by screening for additional photorespiratory mutants. An important example for a reverse genetic approach is the recent unraveling of the role of the $Na^+(K^+)H^+$ exchanger AtCHX23 (TC family 2.A.37) in regulating the stromal pH, an important factor for high photosynthetic activity (119). Additional examples can be found in a recent review (146). The many available T-DNA insertion lines for *Arabidopsis* (57) and efficient RNAi-mediated gene-silencing techniques (134, 148) will be instrumental for the functional identification of novel plastid envelope membrane transporters.

Although this review focuses on *Arabidopsis thaliana*, the exciting potential of unicellular algae as model systems (35, 42) should not be neglected. The genomes of the unicellular green alga *Chlamydomonas reinhardtii* and of the unicellular thermo-acidophilic red alga *Cyanidioschyzon merolae* were recently finished or will soon be finished (67, 114), and additional algal genomes will become available soon. Both algae are amenable to genetic modification by transformation (68, 113) and there is evidence for targeted gene disruption by homologous recombination in *C. merolae* (68). Traditional microbiological techniques such as replica plating can be applied to these organisms and mutant screens can be handled by laboratory robots, allowing high-throughput mutant screens (see 92 for a recent example). Large mutant collections are available for *C. reinhardtii*, map-based cloning of mutated genes is usually faster than with *Arabidopsis*, and additional mutant libraries can be conveniently generated by random DNA insertion techniques. In addition, mutated genes can be rapidly identified by complementation cloning (11, 92). Particularly promising are screens for resistance to toxic substrates and substrate analoga, and for auxotrophic mutants, for example mutants in amino acid biosynthesis pathways. Such mutants are potentially useful in identifying plastidic amino acid transporters. However, findings obtained with unicellular algae cannot necessarily be transferred to vascular plants. For example, red algae do not seem to possess plastidic monosaccharide and maltose transporters or plastidic dicarboxylate transporters (145). Chlamydomonas possesses a plastidic ABC-transporter-type sulfate permease that does not seem to exist in angiosperms or gymnosperms (12, 13), a CO_2 concentrating mechanism that involves plastid envelope membrane proteins (69a, 97a), and a plastidic nitrite transport system that does not find related proteins in the *Arabidopsis* or rice genomes (91). Nevertheless, these currently underused resources deserve further exploitation.

MFS: major facilitator superfamily

Solute transporter: integral membrane proteins that catalyze the transport of solutes across biological lipid-bilayer membranes

STRUCTURE OF PLASTID ENVELOPE MEMBRANE SOLUTE TRANSPORTERS

There has recently been exciting progress in the determination of the tertiary and quaternary structures of solute transporters. For example, the atomic resolution crystal structures of the mitochondrial ADP/ATP carrier from beef heart mitochondria (79) and the glycerol permease GlpF (29), the lactose permease LacY (1), and the glycerol-3-phosphate transporter GlpT (44) from *E. coli* were recently elucidated. LacY and GlpT are both secondary active transporters belonging to the MFS (6). They reside in the plasma membrane and have twelve α-helical transmembrane domains. The ADP/ATP carrier belongs to the mitochondrial transporter family, it resides in the mitochondrial inner membrane, and the active protein is a homodimer consisting of two identical protomers that each have six alpha-helical transmembrane domains (52, 58, 79, 83). Hence, atomic resolution structures of membrane transporters with 12 and 6 transmembrane helix topologies, respectively, have now been solved and can be considered representatives for their respective protein classes. In addition, the 3D structures of primary active transporters of the ABC-transporter superfamily, for example the multidrug exporter AcrB from *E. coli* (73), were also elucidated recently.

The plastid envelope membrane harbors numerous transporters that belong to the above-mentioned classes. For example, several plastid envelope membrane solute transporters, such as the putative monosaccharide transporter pGlcT (142), the dicarboxylate transporters DiT1 and DiT2 (89, 129, 130, 141), the ATP/ADP transporter (75), and the phosphate transporter PHT2;1 (135), are members of the 12-transmembrane helix MFS class of transporters. Others, such as the putative

ADP-glucose transporter BT1 (8, 123), are related to mitochondrial transporters that most likely have 6 transmembrane domains, and several members of ABC-transporter family, such as the lipid permease TGD1 (150) and the chloroplast envelope membrane sulphate permease SulP from *C. reinhardtii*, were recently described (12, 13). However, several important plastid envelope membrane transporters do not belong to these classes. For example, the maltose transporter AtMEX1 likely has 9 α-helical transmembrane domains and does not belong to any of the previously established classes of membrane transporters (77). Likewise, members of the plastidic PT family cannot be assigned with a 6- or 12-helix topology. The TPT is the best-studied member of the PT family and it is the most abundant protein of the chloroplast envelope membrane. The transmembrane topology of the TPT represents the plastidic PT family, and, potentially, also other PT-related proteins in plants, animals, and fungi (53). Therefore, we discuss possible topology models for the PT-family in more detail in the following paragraphs.

Several different transmembrane topologies have been proposed for the TPT: six (139), seven (53, 149) (**Figure 2**), and eight transmembrane helices (108) (**Figure 3**). In addition, a nine-helix topology was proposed for the TPT-related yeast protein SLY41 (16). These predictions (with the exception of 53) were based on single-protein sequences and are therefore not very reliable. The primary structures of the all PT-protein family members from *Arabidopsis* and of many PT-related proteins from other species are known. This enabled an improved topology prediction by aligning the transmembrane domain predictions for several individual protein family members and generating a consensus balance by accumulating the scores of all consensus predictions for the selected proteins. In other words, the quality of transmembrane topology prediction was improved by generating a consensus topology from several related protein sequences. This approach was applied to the six members of the plastidic PT family from *Arabidopsis* (53) using the corresponding algorithms of ARAMEMNON (110). The resulting consensus topology consists of nine transmembrane domains, implying that N and C termini of the proteins would reside on opposite sites of the inner plastid envelope membrane. Seven of the nine predicted helices have relatively high scores (between 60% and 100% of the maximal score) and can thus be considered certain. Helices 5 and 8 (**Figure 4**), however, have relatively low scores of 48% and 31%, respectively. Based on this computer-based prediction, a topology with less than seven helices is unlikely.

The localization of at least one terminus of TPT in the intermembrane space is in concordance with preliminary experimental results: When intact, [^{3}H]-H_2-DIDS-labeled spinach chloroplasts were incubated with a combination of phospholipase C and trypsin for 1.5 h at room temperature, two characteristic [^{3}H]-H_2-DIDS-labeled fragments of TPT with apparent molecular masses of 27.5 kDa and 26.3 kDa were observed (A. Weber & U.-I. Flügge, unpublished). The intact TPT protein from spinach has an apparent molecular mass of 29 kDa in the SDS-PAGE. It can be hypothesized that the DIDS-labeled peptides of 27.5 kDa and 26.3 kDa represent fragments of the intact TPT protein with one or both termini, respectively, removed by trypsin. This hypothesis implies that both termini are facing the intermembrane space because the protease trypsin is unable to penetrate the chloroplast inner envelope membrane (45, 64). To test this hypothesis, the DIDS-labeled 26.3 kDa fragment was isolated and sequenced, and the resulting peptide sequence (T G F L E K Y P A) was identical to amino acids 11–19 of the mature spinach TPT (26). Hence, the N terminus of spinach TPT is susceptible to proteolytic cleaveage by trypsin in intact chloroplast and therefore likely faces the intermembrane space. No sequence information could be obtained from the 27.5-kDa fragment, leading to the hypothesis that this fragment represents a tryptic TPT fragment lacking its C terminus. This hypothesis is supported by the fact that the N terminus of intact mature spinach TPT is not accessible to

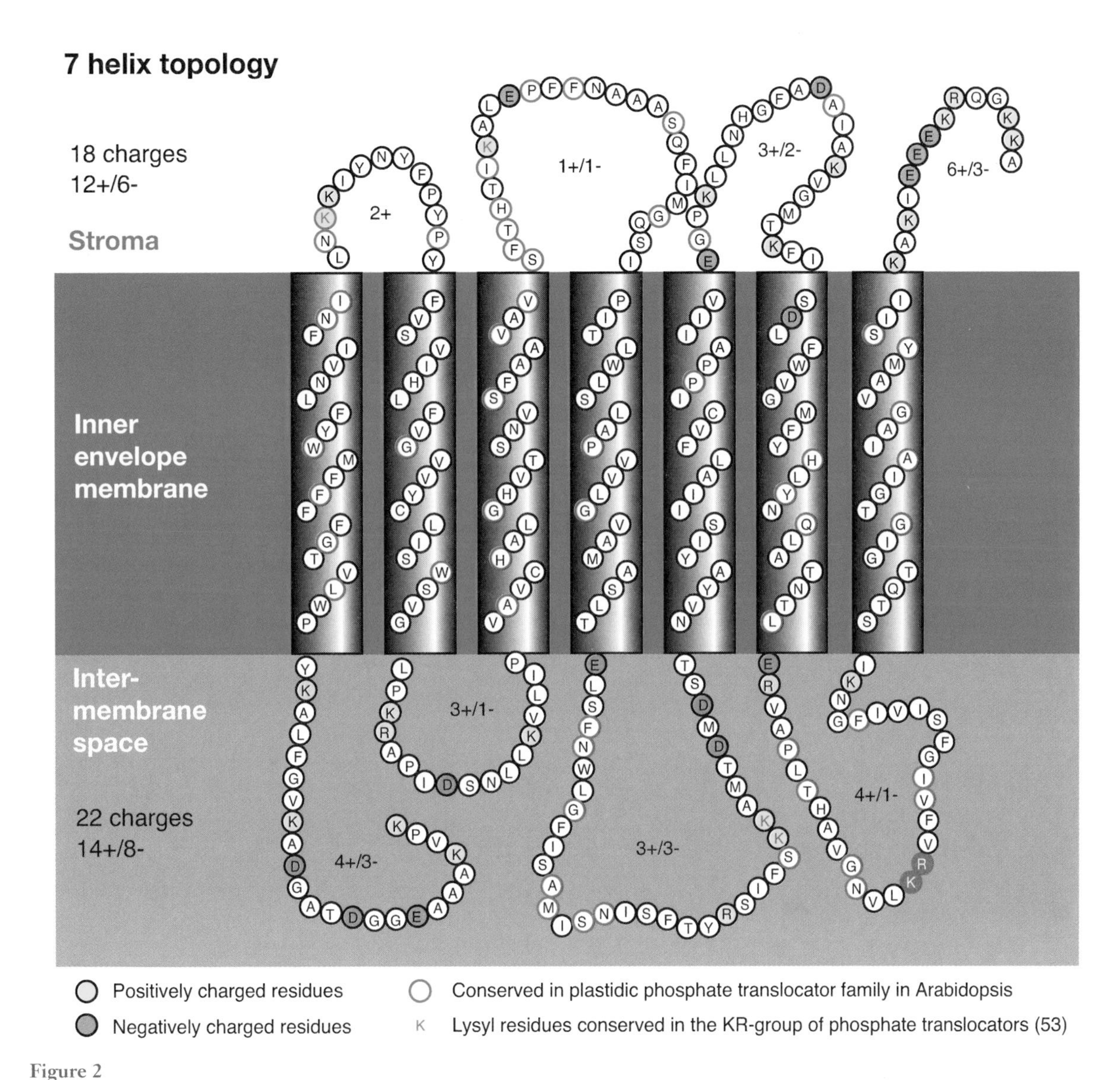

Figure 2

A possible 7-helix transmembrane topology of the plastidic triose phosphate/phosphate translocator TPT.

Edman degradation (149). Therefore N-terminal amino acid sequence information could not be expected from a tryptic fragment lacking part of its C terminus, but with intact N terminus.

Although further experimental support is required, the results of the proteolysis experiments indicate that both termini of TPT are facing the intermembrane space of the plastid envelope membrane and the protein must therefore have an even number of transmembrane domains. Hence, topologies with seven or nine helices are unlikely. Both odd-numbered topologies also do not agree with the "positive-inside" rule (137, 138). This rule observes that regions of polytopic membrane proteins facing the cytoplasm are generally enriched in positively charged lysyl and arginyl residues, whereas the translocated regions are largely devoid of these regions. However, the odd-numbered topologies are characterized by

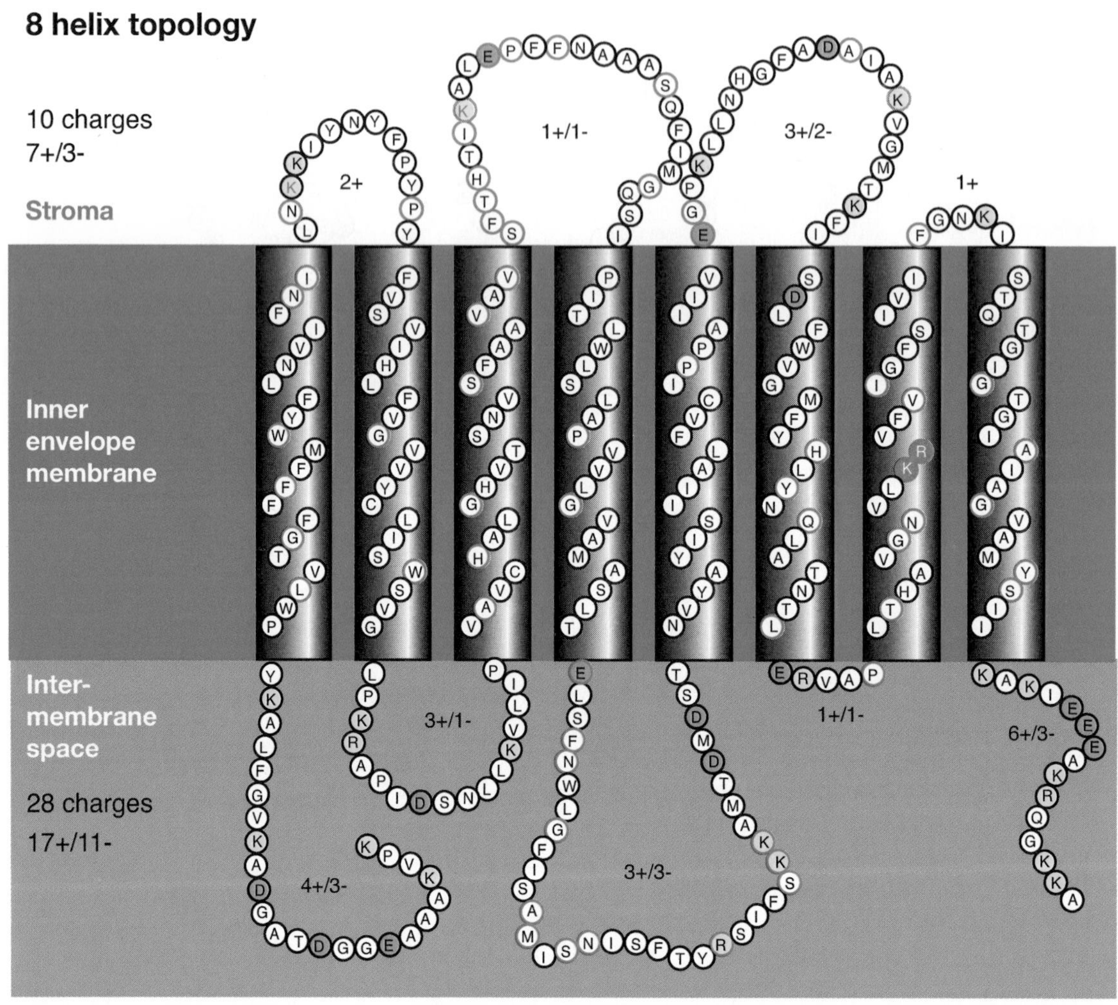

Figure 3

A possible 8-helix transmembrane topology of the plastidic triose phosphate/phosphate translocator TPT.

similar frequencies of positively charged residues (marked in blue in **Figures 2–4**) on the *cis*- and *trans*-sides of the inner envelope membrane. The eight-helix topology, however, is characterized by a 2.5-fold lower frequency of arginyl and lysyl residues on the stroma side of the inner envelope membrane compared to the cytosolic side. This topology shows a 3.6-fold higher frequency of negatively charged residues (marked in pink in **Figures 2–4**) and a 2.8-fold higher frequency of total charges on the cytosolic side of the inner membrane. A similar bias against the translocation of negatively charged residues was previously observed for nuclear-encoded proteins of the inner mitochondrial membrane (30).

Additional arguments favor the eight-helix topology: An amino acid alignment of all six members of the plastidic PT family in *Arabidopsis* reveals several regions of high similarity

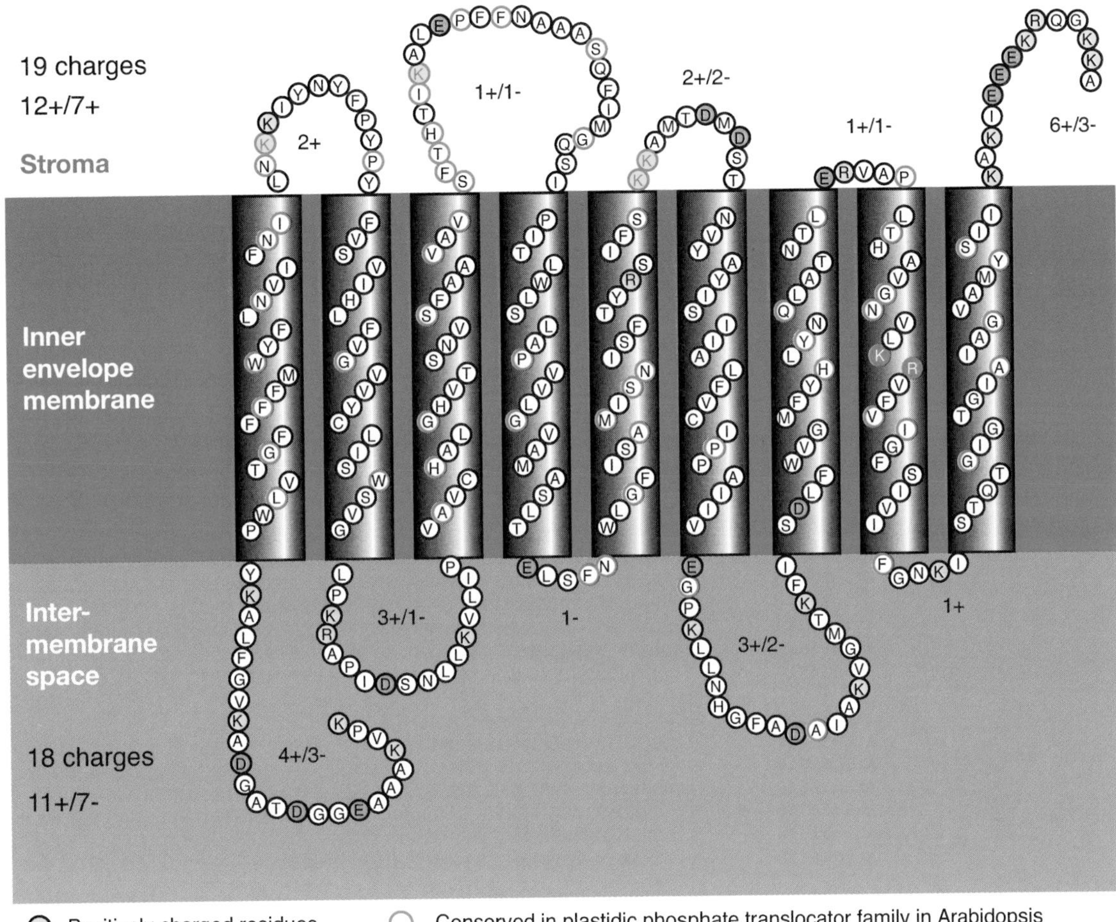

Figure 4

A possible 9-helix transmembrane topology of the plastidic triose phosphate/phosphate translocator TPT.

between the proteins (**Figure 5**; red-circled residues in **Figures 2–4**) and it is tempting to hypothesize that these conserved residues could be involved in the formation of hydrophilic pockets that are required for substrate binding and recognition on both sides of the membrane. In a nine-helix topology, however, most of the conserved residues are found on the stroma side of the inner membrane (**Figure 4**), whereas both seven- and eight-helix topologies lead to a balanced distribution of conserved regions on both sides of the membrane (**Figures 3,4**). In summary, computational and experimental evidence favors an eight-helix topology of plastidic PTs and argues against a nine-helix topology, whereas a seven-helix topology cannot unequivocally be ruled out. Further structural studies arerequired to settle this question.

Plastidic PTs represent particularly interesting targets for structural studies because these polytopic membrane proteins do not belong to any of the structural classes of membrane proteins for which atomic resolution structures have been solved. In contrast to

```
At5g54800      ----MVLSVKQTLS-------------PKIGLFRRNPSSSLGRSPVSLSFPST-ELPKR
At1g61800      ----MLSSIKPSSSS----------FSTAISGSVRRSIPTKLKFSPLLIIKNCHNQSFNA
At5g17630      ---MISLNLSPSLNPGLLHKTRTCQQPTRLSALLVTNPKPFNHRHPLGLSPIPNLQIRDV
At5g33320      MQSSAVFSLSPSLPL-----LKPRRLSLRHHPITTAASSSDLNVSPNVVSIPSLSRRSWR
At3g01550      MFALTFLNPNPRLP----------------SPLFLAKSTPES----------ALSRRSRA
At5g46110      MESRVLLRATANVVG--------IPKLRRPIGAIHRQFSTASSSSFSVKPIGGIGEGANL
                        .                             .              .

At5g54800      TVLAVSKPLHLSSSLRAKSPVVRCEAYEADRSEPHPIGDDAAAAETKSEAAKKLKIGIYF
At1g61800      NVVSHQKPLHISSASNFKR-EVKVEAYEADRSRPLDINIELPDEQ----SAQKLKIGIYF
At5g17630      SAKPLLSLTNPESSSGFSRKPRSIAAVGSSDSNPDEK-SDLGEAEKKEKKAKTLQLGIVF
At5g33320      LASSDSPLRAWSGVPSPISHSLDTNRFRTAATAVPES--AEEGDN-SGKLTKVLELGLLF
At3g01550      FSSSNS---------YPWRPNLRFNGFKLKSATVPEN--VEGGDLESGSLVKGLKLGGMF
At5g46110      ISGRQLRPILLLDSSAINGGEKREILKPVKAAAAEGGDTAGDAKVGFLAKYPWLVTGFFF
                           .                 :     .               *  *  *

At5g54800      ATWWALNVVFNIYNKKVLNAYPYPWLTSTLSLAAGSLMMLISWAVGIVETPKTDFDFWKT
At1g61800      ATWWALNVVFNIYNKKVLNAFPYPWLTSTLSLACGSLMMLVSWATRIADAPKTDLEFWKT
At5g17630      GLWYFQNIVFNIFNKKALNVFPYPWLLASFQLFAGSIWMLVLWSFKLYPCPKISKPFIIA
At5g33320      AMWYLFNIYFNIYNKQVLKALHAPMTVTLVQFAVGSVLITIMWVLNLYKRPKISGAQLAA
At3g01550      GVWYLLNIYYNIFNKQVLRVYPYPATVTAFQLGCGTLMIAIMWLLKLHPRPKFSPSQFTV
At5g46110      FMWYFLNVIFNILNKKIYNYFPYPYFVSVIHLFVGVVYCLISWSVGLPKRAPIDSNLLKV
                 *:  *: :** **:  .    *   : . :  *:    : *   :   .  .     .

At5g54800      LFPVAVAHTIGHVAATVSMSKVAVSFTHIIKSGEPAFSVLVSRFILGETFPTSVYLSLIP
At1g61800      LFPVAVAHTIGHVAATVSMSKVAVSFTHIIKSGEPAFSVLVSRFFMGETFPLPVYLSLLP
At5g17630      LLGPALFHTIGHISACVSFSKVAVSFTHVIKSAEPVFSVIFS-SLLGDSYPLAVWLSILP
At5g33320      ILPLAVVHTLGNLFTNMSLGKVSVSFTHTIKAMEPFFSVLLSAMFLGEKPTPWVLGAIVP
At3g01550      IVQLAVAHTLGNLLTNVSLGRVNVSFTHTIKAMEPFFTVLLSVLLLGEWPSLWIVCSLLP
At5g46110      LIPVAVCHALGHVTSNVSFAAVAVSFTHTIKALEPFFNAAASQFIMGQSIPITLWLSLAP
               :.  *: *::*:: : :*:. * ***** **: ** *..  *  ::*:  .  :  :: *

At5g54800      IIGGCALSALTELNFNMIGFMGAMISNLAFVFRNIFSKK-GMKG-KSVSGMNYYACLSML
At1g61800      IIGGCALAAITELNFNITGFMGAMISNLAFVFRNIFSKK-GMKG-KSVSGMNYYACLSMM
At5g17630      IVMGCSLAAVTEVSFNLGGLSGAMISNVGFVLRNIYSKR-SLQSFKEIDGLNLYGCISIL
At5g33320      IVGGVALASISEVSFNWAGFSSAMASNLTNQSRNVLSKKVMVKKDDSLDNITLFSIITLM
At3g01550      IVAGVSLASFTEASFNWIGFCSAMASNVTNQSRNVLSKKFMVGKD-ALDNINLFSIITII
At5g46110      VVLGVAMASLTELSFNWLGFISAMISNISFTYRSIFSKK----AMTDMDSTNVYAYISII
               :: * ::::.:* .**  *: .** **:    *.: **:        :.. . :. ::::

At5g54800      SLLILTPFAIAVEGPQMWVDGWQTALATVG--PQFVWWVVAQSVFYHLYNQVSYMSLDQI
At1g61800      SLVILTPFSIAVEGPQMWAAGWQNAVSQVG--PNFVWWVVAQSVFYHLYNQVSYMSLDQI
At5g17630      SLLYLFPVAIFVEG-SHWVPGYHKAIASVGTPSTFYFWVLLSGVFYHLYNQSSYQALDEI
At5g33320      SLVLMAPVTFFTEGIKFTPSYIQSA---GVNVKQIYTKSLIAALCFHAYQQVSYMILARV
At3g01550      SFILLVPLAILIDGFKVTPSHLQVATSQGLSVKEFCIMSLLAGVCLHSYQQVSYMILEMV
At5g46110      ALFVCIPPAIIVEGPKLLNHGFADAIAKVG-MTKFISDLFWVGMFYHLYNQLATNTLERV
               ::.   * ::  :* .        * :       :    .  .:  * *:* :   *  :

At5g54800      SPLTFSVGNTMKRISVIVSSIIIFRTPVQPVNALGAAIAILGTFLYSQAKL---------
At1g61800      SPLTFSIGNTMKRISVIVASIIIFHTPIQPVNALGAAIAIFGTFLYSQAKQ---------
At5g17630      SPLTFSVGNTMKRVVVIISTVLVFRNPVRPLNALGSAIAIFGTFLYSQATAKKKKIEVGG
At5g33320      SPVTHSVGNCVKRVVVIVSSVIFFKTPVSPVNAFGTGIALAGVFLYSRVK--GIKPKPKT
At3g01550      SPVTHSVGNCVKRVVVITSSILFFKTPVSPLNSIGTATALAGVYLYSRAKRVQVKPNPKM
At5g46110      APLTHAVGNVLKRVFVIGFSIVIFGNKISTQTGIGTGIAIAGVAMYSIIKAKIEEEKRQG
               :*:*.::** :**: **  :::.* . : . ..:*:. *: *. :**  .

At5g54800      ----
At1g61800      ----
At5g17630      DKKN
At5g33320      A---
At3g01550      S---
At5g46110      KKA-
```

Figure 5

Amino acid alignment of all six members of the plastidic phosphate translocator family in *Arabidopsis*. Asterisks indicate amino acids that are identical in all six proteins.

most other members of the DMT superfamily, pPTs are functionally characterized, they can be expressed in relatively large amounts in recombinant hosts such as yeast cells, and recombinant, hexahistidine-tagged transporter protein can be purified to homogeneity by metal affinity chromatography (17, 22, 48, 62). A highly specific inihibitor for PTs is available and the use of inhibitors to rigidify the structure of transport proteins during crystallization (131) has been successfully applied to the mitochondrial ADP/ATP carrier (79). We recently identified a cDNA encoding the plastidic PT from the thermophilic red alga *Galdieria sulphuraria* (145). This protein likely shows greater thermal stability than the corresponding proteins from mesophilic organisms. Initial studies indicate that this protein could be produced in the active state in large amounts in recombinant yeast cells (M. Zimmermann, A. Jamai & A. Weber, unpublished) and thus represents a promising target for structural studies.

CONCLUSIONS AND PERSPECTIVES

Thanks to recent progress in genome sequencing, bioinformatics, and proteomics, we are now in the fortunate situation to have a fairly comprehensive list of putative plastid inner envelope membrane transporters. Less is known about the outer envelope membrane, mainly because a reliable prediction of protein targeting to the outer envelope is not yet feasible. The recent discovery of a novel targeting pathway to the inner envelope membrane that does not rely on a cleavable transit peptide (74) will eventually require a reassessment of genome sequences for additional inner envelope membrane proteins that are routed through this novel pathway. Hence, the list of inner envelope membrane proteins will likely increase in the future. A major challenge remains the assignment of functions to the vast majority of envelope metabolite transporters. To this end, we will need to extend the choice of model systems for the study of plastid transporters beyond *Arabidopsis* to include, for example, unicellular algae and nonvascular plants with less redundant genomes and simpler cell biology. Structural studies of plastid envelope membrane transporters are another emerging frontier. The plastid envelope membrane harbors membrane transporters with different transmembrane topologies that belong to various structural classes and several of these proteins were the first members of their protein families that were functionally characterized and for which efficient recombinant expression systems were established. Recently, orthologous proteins were identified from thermophilic unicellular algae and these potentially thermo-stable membrane proteins represent interesting candidates for structural studies.

SUMMARY POINTS

1. Bioinformatics analysis shows that the *Arabidopsis* genome encodes at least 150 plastid envelope membrane transporters.
2. Proteomics of the plastid envelope confirms many of the predicted transporters and adds information on tissue-specific protein expression.
3. Proteomics identifies additional putative transporters and provides evidence for noncanonical protein import pathways into plastids. Potentially, the envelope membrane might contain many more proteins, as predicted by bioinformatics.
4. The plastid envelope harbors a diverse array of transporters that belong to different structural and functional categories.

5. The vast majority of putative plastid envelope membrane transporters have not yet been assigned a function.
6. Additional, currently underused model systems such as unicellular algae are potentially useful for identifying novel plastid transporters because they allow the use of traditional microbiological techniques in combination with high-throughput robotic screening methods.
7. Plastid envelope membrane transporters are interesting candidates for structural studies because they can be considered proxies for large protein families and can be produced in large amounts in recombinant systems.
8. We currently see only the tip of the plastid membrane transporter iceberg.

ACKNOWLEDGMENTS

The authors gratefully acknowledge financial support by the National Science Foundation (awards MCB-03,48074 and EF-03,32882), the U.S. Department of Energy (DE-FG02-04ER15562), the MSU Center for Plant Products and Technologies, MSU-IRGP, the Deutsche Forschungsgemeinschaft, the Bundesministerium für Bildung und Forschung, and the Fonds der Chemischen Industrie. We thank Norbert Sauer and Tasios Melis for providing manuscripts prior to publication.

LITERATURE CITED

1. Abramson J, Smirnova I, Kasho V, Verner G, Iwata S, Kaback HR. 2003. The lactose permease of *Escherichia coli*: overall structure, the sugar-binding site and the alternating access model for transport. *FEBS Lett.* 555:96–101
2. Barbier-Brygoo H, Gaymard F, Rolland N, Joyard J. 2001. Strategies to identify transport systems in plants. *Trends Plant Sci.* 6:577–85
3. Baxter I, Tchieu J, Sussman MR, Boutry M, Palmgren MG, et al. 2003. Genomic comparison of P-type ATPase ion pumps in Arabidopsis and rice. *Plant Physiol.* 132:618–28
4. Bölter B, Soll J, Hill K, Hemmler R, Wagner R. 1999. A rectifying ATP-regulated solute channel in the chloroplastic outer envelope from pea. *EMBO J.* 18:5505–16
5. Buchner P, Takahashi H, Hawkesford MJ. 2004. Plant sulphate transporters: co-ordination of uptake, intracellular and long-distance transport. *J. Exp. Bot.* 55:1765–73
6. Busch W, Saier MH Jr. 2002. The transporter classification (TC) system, 2002. *Crit. Rev. Biochem. Mol. Biol.* 37:287–337
7. Busch W, Saier MH Jr. 2003. The IUBMB-endorsed transporter classification system. *Methods Mol. Biol.* 227:21–36
8. Cao HP, Sullivan TD, Boyer CD, Shannon JC. 1995. Bt1, a structural gene for the major 39–44 kda amyloplast membrane polypeptides. *Physiol. Plant.* 95:176–86
9. Catoni E, Desimone M, Hilpert M, Wipf D, Kunze R, et al. 2003. Expression pattern of a nuclear encoded mitochondrial arginine-ornithine translocator gene from Arabidopsis. *BMC Plant Biol.* 3:1
10. Catoni E, Schwab R, Hilpert M, Desimone M, Schwacke R, et al. 2003. Identification of an Arabidopsis mitochondrial succinate-fumarate translocator. *FEBS Lett.* 534:87–92
11. Chang M, Li F, Odom OW, Lee J, Herrin DL. 2003. A cosmid vector containing a dominant selectable marker for cloning Chlamydomonas genes by complementation. *Plasmid* 49:75–78

12. Chen HC, Melis A. 2004. Localization and function of SulP, a nuclear-encoded chloroplast sulfate permease in *Chlamydomonas reinhardtii Planta*. doi: 10.1007/s00425-004-1331-5
13. Chen HC, Yokthongwattana K, Newton AJ, Melis A. 2003. SulP, a nuclear gene encoding a putative chloroplast-targeted sulfate permease in *Chlamydomonas reinhardtii. Planta* 218:98–106
14. Chia T, Thorneycroft D, Chapple A, Messerli G, Chen J, et al. 2004. A cytosolic glucosyltransferase is required for conversion of starch to sucrose in *Arabidopsis* leaves at night. *Plant J.* 37:853–63
15. Chou ML, Fitzpatrick LM, Tu SL, Budziszewski G, Potter-Lewis S, et al. 2003. Tic40, a membrane-anchored co-chaperone homolog in the chloroplast protein translocon. *EMBO J.* 22:2970–80
16. Dascher C, Ossig R, Gallwitz D, Schmitt HD. 1991. Identification and structure of four yeast genes (SLY) that are able to suppress the functional loss of YPT1, a member of the RAS superfamily. *Mol. Cell Biol.* 11:872–85
17. Eicks M, Maurino V, Knappe S, Flügge UI, Fischer K. 2002. The plastidic pentose phosphate translocator represents a link between the cytosolic and the plastidic pentose phosphate pathways in plants. *Plant Physiol.* 128:512–22
18. Emanuelsson O, Nielsen H, Brunak S, von Heijne G. 2000. Predicting subcellular localization of proteins based on their N- terminal amino acid sequence. *J. Mol. Biol.* 300:1005–16
19. Emanuelsson O, Nielsen H, von Heijne G. 1999. ChloroP, a neural network-based method for predicting chloroplast transit peptides and their cleavage sites. *Protein Sci.* 8:978–84
20. Ferro M, Salvi D, Brugiere S, Miras S, Kowalski S, et al. 2003. Proteomics of the chloroplast envelope membranes from *Arabidopsis thaliana*. *Mol. Cell Proteomics* 2:325–45
21. Ferro M, Salvi D, Riviere-Rolland H, Vermat T, Seigneurin-Berny D, et al. 2002. Integral membrane proteins of the chloroplast envelope: identification and subcellular localization of new transporters. *Proc. Natl. Acad. Sci. USA* 99:11487–92
22. Fischer K, Kammerer N, Gutensohn M, Arbinger B, Weber A, et al. 1997. A new class of plastidic phosphate translocators: a putative link between primary and secondary metabolism by the phosphoenolpyruvate/phosphate antiporter. *Plant Cell* 9:453–62
23. Fischer K, Weber A. 2002. Transport of carbon in non-green plastids. *Trends Plant Sci.* 7:345–51
24. Flügge UI. 1999. Phosphate translocators in plastids. *Annu. Rev. Plant Physiol. Plant Mol. Biol.* 50:27–45
25. Flügge UI, Häusler RE, Ludewig F, Fischer K. 2003. Functional genomics of phosphate antiport systems of plastids. *Physiol. Plant.* 118:475–82
26. Flügge UI, Weber A, Fischer K, Lottspeich F, Eckerskorn C, et al. 1991. The major chloroplast envelope polypeptide is the phosphate translocator and not the protein import receptor. *Nature* 353:364–67
27. Friso G, Giacomelli L, Ytterberg AJ, Peltier JB, Rudella A, et al. 2004. In-depth analysis of the thylakoid membrane proteome of *Arabidopsis thaliana* chloroplasts: new proteins, new functions, and a plastid proteome database. *Plant Cell* 16:478–99
28. Froehlich JE, Wilkerson CG, Ray WK, McAndrew RS, Osteryoung KW, et al. 2003. Proteomic study of the *Arabidopsis thaliana* chloroplastic envelope membrane utilizing alternatives to traditional two-dimensional electrophoresis. *J. Proteome Res.* 2:413–25
29. Fu D, Libson A, Miercke LJ, Weitzman C, Nollert P, et al. 2000. Structure of a glycerol-conducting channel and the basis for its selectivity. *Science* 290:481–86

The strategy described in Reference 21 was extended to chloroplast envelope membrane proteins from *Arabidopsis* and a number of additional putative membrane transporters were identified.

The first of a series of papers on proteomics studies of the chloroplast envelope membranes. A specific enrichment method for membrane proteins was used to increase the representation of spinach chloroplast envelope membrane proteins in the proteomics approach.

This group used several overlapping strategies such as MudPIT to increase the representation of membrane proteins in a proteomics study of *Arabidopsis* chloroplast envelope membranes.

30. Gavel Y, von Heijne G. 1992. The distribution of charged amino acids in mitochondrial inner-membrane proteins suggests different modes of membrane integration for nuclearly and mitochondrially encoded proteins. *Eur. J. Biochem.* 205:1207–15
31. Gazzarrini S, Lejay L, Gojon A, Ninnemann O, Frommer WB, von Wiren N. 1999. Three functional transporters for constitutive, diurnally regulated, and starvation-induced uptake of ammonium into Arabidopsis roots. *Plant Cell* 11:937–48
32. Gillissen B, Bürkle L, André B, Kühn C, Rentsch D, et al. 2000. A new familiy of high-affinity transporters for adenine, cytosine, and purine derivatives in Arabidopsis. *Plant Cell* 12:291–300
33. Gleixner G, Scrimgeour C, Schmidt HL, Viola R. 1998. Stable isotope distribution in the major metabolites of source and sink organs of *Solanum tuberosum* L.: a powerful tool in the study of metabolic partitioning in intact plants. *Planta* 207:241–45
34. Gygi SP, Aebersold R. 2000. Mass spectrometry and proteomics. *Curr. Opin. Chem. Biol.* 4:489–94
35. Harris EH. 2001. *Chlamydomonas* as a model organism. *Annu. Rev. Plant Physiol. Plant Mol. Biol.* 52:363–406
36. Hattenbach A, Müller-Röber B, Nast G, Heineke D. 1997. Antisense repression of both ADP-glucose pyrophosphorylase and triose phosphate translocator modifies carbohydrate partitioning in potato leaves. *Plant Physiol.* 115:471–75
37. Häusler RE, Baur B, Scharte J, Teichmann T, Eicks M, et al. 2000. Plastidic metabolite transporters and their physiological functions in the inducible crassulacean acid metabolism plant Mesembryanthemum crystallinum. *Plant J.* 24:285–96
38. Häusler RE, Schlieben NH, Flügge UI. 2000. Control of carbon partitioning and photosynthesis by the triose phosphate/phosphate translocator in transgenic tobacco plants (*Nicotiana tabacum*). II. Assessment of control coefficients of the triose phosphate/phosphate translocator. *Planta* 210:383–90
39. Häusler RE, Schlieben NH, Nicolay P, Fischer K, Fischer KL, Flügge UI. 2000. Control of carbon partitioning and photosynthesis by the triose phosphate/phosphate translocator in transgenic tobacco plants (*Nicotiana tabacum* L.). I. Comparative physiological analysis of tobacco plants with antisense repression and overexpression of the triose phosphate/phosphate translocator. *Planta* 210:371–82
40. Heineke D, Kruse A, Flügge UI, Frommer WB, Riesmeier JW, et al. 1994. Effect of antisense repression of the chloroplast triose-phosphate translocator on photosynthetic metabolism in transgenic potato plants. *Planta* 193:174–80
41. Henkes S, Sonnewald U, Badur R, Flachmann R, Stitt M. 2001. A small decrease of plastid transketolase activity in antisense tobacco transformants has dramatic effects on photosynthesis and phenylpropanoid metabolism. *Plant Cell* 13:535–51
42. Hicks GR, Hironaka CM, Dauvillee D, Funke RP, D'Hulst C, et al. 2001. When simpler is better. Unicellular green algae for discovering new genes and functions in carbohydrate metabolism. *Plant Physiol.* 127:1334–38
43. Holtum JAM, Osmond CB. 1981. The gluconeogenetic metabolism of pyruvate during deacification in plants with crassulacean acid metabolism. *Aust. J. Plant Physiol.* 8:31–44
44. Huang Y, Lemieux MJ, Song J, Auer M, Wang DN. 2003. Structure and mechanism of the glycerol-3-phosphate transporter from *Escherichia coli*. *Science* 301:616–20
45. Jackson DT, Froehlich JE, Keegstra K. 1998. The hydrophilic domain of Tic110, an inner envelope membrane component of the chloroplastic protein translocation apparatus, faces the stromal compartment. *J. Biol. Chem.* 273:16583–88
46. Jackson-Constan D, Keegstra K. 2001. Arabidopsis genes encoding components of the chloroplastic protein import apparatus. *Plant Physiol.* 125:1567–76

47. Jonas-Straube E, Hutin C, Hoffman NE, Schünemann D. 2001. Functional analysis of the protein-interacting domains of chloroplast SRP43. *J. Biol. Chem.* 276:24654–60
48. Kammerer B, Fischer K, Hilpert B, Schubert S, Gutensohn M, et al. 1998. Molecular characterization of a carbon transporter in plastids from heterotrophic tissues: the glucose6-phosphate/phosphate antiporter. *Plant Cell* 10:105–17
49. Kampfenkel K, Kushnir S, Babiychuk E, Inze D, Van Montagu M. 1995. Molecular characterization of a putative *Arabidopsis thaliana* copper transporter and its yeast homologue. *J. Biol. Chem.* 270:28479–86
50. Kampfenkel KH, Möhlmann T, Batz O, van Montague M, Inzé D, Neuhaus HE. 1995. Molecular characterization of an *Arabidopsis thaliana* cDNA encoding a novel putative adenylate translocator of higher plants. *FEBS Lett.* 374:351–55
51. **Kleffmann T, Russenberger D, von Zychlinski A, Christopher W, Sjolander K, et al. 2004. The *Arabidopsis thaliana* chloroplast proteome reveals pathway abundance and novel protein functions. *Curr. Biol.* 14:354–62**

Of the 690 different proteins from isolated *Arabidopsis* chloroplast, a number of putative membrane transporters were identified by shotgun proteomics. Surprisingly, many of the found proteins were not previously predicted to localize to the chloroplast and do not seem to possess plastid-targeting signals.

52. Klingenberg M. 1981. Membrane protein oligomeric structure and transport function. *Nature* 290:449–54
53. Knappe S, Flügge UI, Fischer K. 2003. Analysis of the plastidic phosphate translocator gene family in Arabidopsis and identification of new phosphate translocator-homologous transporters, classified by their putative substrate-binding site. *Plant Physiol.* 131:1178–90
54. Knappe S, Löttgert T, Schneider A, Voll L, Flügge UI, Fischer K. 2003. Characterization of two functional phosphoenolpyruvate/phosphate translocator (PPT) genes in Arabidopsis–AtPPT1 may be involved in the provision of signals for correct mesophyll development. *Plant J.* 36:411–20
55. Knight JS, Duckett CM, Sullivan JA, Walker AR, Gray JC. 2002. Tissue-specific, light-regulated and plastid-regulated expression of the single-copy nuclear gene encoding the chloroplast Rieske FeS protein of Arabidopsis thaliana. *Plant Cell Physiol.* 43:522–31
56. Kruger NJ, von Schaewen A. 2003. The oxidative pentose phosphate pathway: structure and organisation. *Curr. Opin. Plant Biol.* 6:236–46
57. Krysan PJ, Young JC, Sussman MR. 1999. T-DNA as an insertional mutagen in Arabidopsis. *Plant Cell* 11:2283–90
58. Laloi M. 1999. Plant mitochondrial carriers: an overview. *Cell Mol. Life Sci.* 56:918–44
59. Li H-m, Culligan K, Dixon RA, Chory J. 1995. *Cue1:* a mesophyll cell-specific positive regulator of light-controlled gene expression in Arabidopsis. *Plant Cell* 7:1599–610
60. Li L, He Z, Pandey GK, Tsuchiya T, Luan S. 2002. Functional cloning and characterization of a plant efflux carrier for multidrug and heavy metal detoxification. *J. Biol. Chem.* 277:5360–68
61. Li L, Tutone AF, Drummond RS, Gardner RC, Luan S. 2001. A novel family of magnesium transport genes in Arabidopsis. *Plant Cell* 13:2761–75
62. Loddenkötter B, Kammerer B, Fischer K, Flügge UI. 1993. Expression of the functional mature chloroplast triose phosphate translocator in yeast internal membranes and purification of the histidine-tagged protein by a single metal-affinity chromatography step. *Proc. Natl. Acad. Sci. USA* 90:2155–59
63. Lu Y, Sharkey TD. 2004. The role of amylomaltase in maltose metabolism in the cytosol of photosynthetic cells. *Planta* 218:466–73
64. Lübeck J, Soll J, Akita M, Nielsen E, Keegstra K. 1996. Topology of IEP110, a component of the chloroplastic protein import machinery present in the inner envelope membrane. *EMBO J.* 15:4230–38
65. Marmagne A, Rouet MA, Ferro M, Rolland N, Alcon C, et al. 2004. Identification of new intrinsic proteins in Arabidopsis plasma membrane proteome. *Mol. Cell Proteomics* 3:675–91

66. Maser P, Thomine S, Schroeder JI, Ward JM, Hirschi K, et al. 2001. Phylogenetic relationships within cation transporter families of Arabidopsis. *Plant Physiol.* 126:1646–67
67. Matsuzaki M, Misumi O, Shin IT, Maruyama S, Takahara M, et al. 2004. Genome sequence of the ultrasmall unicellular red alga *Cyanidioschyzon merolae* 10D. *Nature* 428:653–57
68. Minoda A, Sakagami R, Yagisawa F, Kuroiwa T, Tanaka K. 2004. Improvement of culture conditions and evidence for nuclear transformation by homologous recombination in a red alga, *Cyanidioschyzon merolae* 10D. *Plant Cell Physiol.* 45:667–71
69. **Miras S, Salvi D, Ferro M, Grunwald D, Garin J, et al. 2002. Non-canonical transit peptide for import into the chloroplast. *J. Biol. Chem.* 277:47770–78**

For the first time convincing evidence for targeting a protein (ceQORH; Chloroplast Envelope Quinone OxidoReductase Homologue) to the inner chloroplast envelope membrane in the absence of a cleavable, N-terminal targeting sequence is presented.

69a. Miura K, Yamano T, Yoshioka S, Kohinata T, Inoue Y, et al. 2004. Expression profiling-based identification of CO_2-responsive genes regulated by CCM1 controlling a carbon-concentrating mechanism in *Chlamydomonas reinhardtii. Plant Physiol.* 135:1595–607
70. Möhlmann T, Tjaden J, Schwoppe C, Winkler HH, Kampfenkel K, Neuhaus HE. 1998. Occurrence of two plastidic ATP/ADP transporters in *Arabidopsis thaliana* L.–molecular characterisation and comparative structural analysis of similar ATP/ADP translocators from plastids and *Rickettsia prowazekii. Eur. J. Biochem.* 252:353–59
71. Møller SG, Kunkel T, Chua NH. 2001. A plastidic ABC protein involved in intercompartmental communication of light signaling. *Genes Dev.* 15:90–103
72. Motohashi R, Nagata N, Ito T, Takahashi S, Hobo T, et al. 2001. An essential role of a TatC homologue of a Delta pH-dependent protein transporter in thylakoid membrane formation during chloroplast development in *Arabidopsis thaliana. Proc. Natl. Acad. Sci. USA* 98:10499–504
73. Murakami S, Nakashima R, Yamashita E, Yamaguchi A. 2002. Crystal structure of bacterial multidrug efflux transporter AcrB. *Nature* 419:587–93
74. **Nada A, Soll J. 2004. Inner envelope protein 32 is imported into chloroplasts by a novel pathway. *J. Cell Sci.* 117:3975–82**

Demonstration of a novel protein import pathway into chloroplasts.

75. Neuhaus HE, Thom E, Möhlmann T, Steup M, Kampfenkel K. 1997. Characterization of a novel eukaryotic ATP/ADP translocator located in the plastid envelope of *Arabidopsis thaliana* L. *Plant J.* 11:73–82
76. Neuhaus HE, Wagner R. 2000. Solute pores, ion channels, and metabolite transporters in the outer and inner envelope membranes of higher plant plastids. *Biochim. Biophys. Acta* 1465:307–23
77. **Niittyla T, Messerli G, Trevisan M, Chen J, Smith AM, Zeeman SC. 2004. A previously unknown maltose transporter essential for starch degradation in leaves. *Science* 303:87–89**

This paper demonstrates that maltose is the major carbohydrate exported from chloroplasts at night. It provides an excellent example of the power of forward genetics for identifying membrane transporters.

78. Ninnemann O, Jauniaux J-C, Frommer WB. 1994. Identification of a high affinity NH_4^+ transporter from plants. *EMBO J.* 13:3464–71
79. Pebay-Peyroula E, Dahout-Gonzalez C, Kahn R, Trezeguet V, Lauquin GJ, Brandolin G. 2003. Structure of mitochondrial ADP/ATP carrier in complex with carboxyatractyloside. *Nature* 426:39–44
80. Peelman F, Labeur C, Vanloo B, Roosbeek S, Devaud C, et al. 2003. Characterization of the ABCA transporter subfamily: identification of prokaryotic and eukaryotic members, phylogeny and topology. *J. Mol. Biol.* 325:259–74
81. Peltier JB, Friso G, Kalume DE, Roepstorff P, Nilsson F, et al. 2000. Proteomics of the chloroplast: systematic identification and targeting analysis of lumenal and peripheral thylakoid proteins. *Plant Cell* 12:319–41
82. Peltier JB, Ripoll DR, Friso G, Rudella A, Cai Y, et al. 2004. Clp protease complexes from photosynthetic and non-photosynthetic plastids and mitochondria of plants, their predicted three-dimensional structures, and functional implications. *J. Biol. Chem.* 279:4768–81

83. Picault N, Hodges M, Palmieri L, Palmieri F. 2004. The growing family of mitochondrial carriers in Arabidopsis. *Trends Plant Sci.* 9:138–46
84. Pohlmeyer K, Soll J, Grimm R, Hill K, Wagner R. 1998. A high-conductance solute channel in the chloroplastic outer envelope from Pea. *Plant Cell* 10:1207–16
85. Pohlmeyer K, Soll J, Steinkamp T, Hinnah S, Wagner R. 1997. Isolation and characterization of an amino acid-selective channel protein present in the chloroplastic outer envelope membrane. *Proc. Natl. Acad. Sci. USA* 94:9504–9
86. Rassow J, Dekker PJ, van Wilpe S, Meijer M, Soll J. 1999. The preprotein translocase of the mitochondrial inner membrane: function and evolution. *J. Mol. Biol.* 286:105–20
87. Reinbothe S, Quigley F, Gray J, Schemenewitz A, Reinbothe C. 2004. Identification of plastid envelope proteins required for import of protochlorophyllide oxidoreductase A into the chloroplast of barley. *Proc. Natl. Acad. Sci. USA* 101:2197–202
88. Reinbothe S, Quigley F, Springer A, Schemenewitz A, Reinbothe C. 2004. The outer plastid envelope protein OEP16: role as precursor translocase in import of protochlorophyllide oxidoreductase A. *Proc. Natl. Acad. Sci. USA* 101:2203–8
89. Renné P, Dreßen U, Hebbeker U, Hille D, Flügge UI, et al. 2003. The *Arabidopsis* mutant *dct* is deficient in the plastidic glutamate/malate translocator DiT2. *Plant J.* 35:316–31
90. Rentsch D, Laloi M, Rouhara I, Schmelzer E, Delrot S, Frommer WB. 1995. NTR1 encodes a high affinity oligopeptide transporter in *Arabidopsis*. *FEBS Lett.* 370:264–68
91. Rexach J, Fernández E, Galván A. 2000. The *Chlamydomonas reinhardtii* nar1 gene encodes a chloroplast membrane protein involved in nitrite transport. *Plant Cell* 12:1441–53
92. Riekhof WR, Ruckle ME, Lydic TA, Sears BB, Benning C. 2003. The sulfolipids 2′-O-acyl-sulfoquinovosyldiacylglycerol and sulfoquinovosyldiacylglycerol are absent from a *Chlamydomonas reinhardtii* mutant deleted in SQD1. *Plant Physiol.* 133:864–74
93. Riesmeier JW, Flügge UI, Schulz B, Heineke D, Heldt HW, et al. 1993. Antisense repression of the chloroplast triose phosphate translocator affects carbon partitioning in transgenic potato plants. *Proc. Natl. Acad. Sci. USA* 90:6160–64
94. Riesmeier JW, Willmitzer L, Frommer WB. 1992. Isolation and characterization of a sucrose carrier cDNA from spinach by functional expression in yeast. *EMBO J.* 11:4705–13
95. Ritte G, Lloyd JR, Eckermann N, Rottmann A, Kossmann J, Steup M. 2002. The starch-related R1 protein is an alpha -glucan, water dikinase. *Proc. Natl. Acad. Sci. USA* 99:7166–71
96. Rohl T, Motzkus M, Soll J. 1999. The outer envelope protein OEP24 from pea chloroplasts can functionally replace the mitochondrial VDAC in yeast. *FEBS Lett.* 460:491–94
97. Rolland N, Ferro M, Seigneurin-Berny D, Garin J, Douce R, Joyard J. 2003. Proteomics of chloroplast envelope membranes. *Photosynth. Res.* 78:205–30
97a. Rolland N, Dorne AJ, Amoroso G, Sültemeyer DF, Joyard J, Rochaix JD. 1997. Disruption of the plastid ycf10 open reading frame affects uptake of inorganic carbon in the chloroplast of *Chlamydomonas*. *EMBO J.* 16:6713–26
98. Roth C, Menzel G, Petetot JM, Rochat-Hacker S, Poirier Y. 2004. Characterization of a protein of the plastid inner envelope having homology to animal inorganic phosphate, chloride and organic-anion transporters. *Planta* 218:406–16
99. Sancenon V, Puig S, Mateu-Andres I, Dorcey E, Thiele DJ, Penarrubia L. 2004. The Arabidopsis copper transporter COPT1 functions in root elongation and pollen development. *J. Biol. Chem.* 279:15348–55
100. Sancenon V, Puig S, Mira H, Thiele DJ, Penarrubia L. 2003. Identification of a copper transporter family in *Arabidopsis thaliana*. *Plant Mol. Biol.* 51:577–87
101. Sanchez-Fernandez R, Davies TG, Coleman JO, Rea PA. 2001. The *Arabidopsis thaliana* ABC protein superfamily, a complete inventory. *J. Biol. Chem.* 276:30231–44

102. Sanchez-Fernandez R, Rea PA, Davies TG, Coleman JO. 2001. Do plants have more genes than humans? Yes, when it comes to ABC proteins. *Trends Plant Sci.* 6:347–48
103. Santoni V, Molloy M, Rabilloud T. 2000. Membrane proteins and proteomics: un amour impossible? *Electrophoresis* 21:1054–70
104. Sauer N, Ludwig A, Knoblauch A, Rothe P, Gahrtz M, Klebl F. 2004. *AtSUC8* and *AtSUC9* encode functional sucrose transporters, but the closely related *AtSUC6* and *AtSUC7* genes encode aberrant proteins in different Arabidopsis ecotypes. *Plant J.* 40:120–30
105. Schein AI, Kissinger JC, Ungar LH. 2001. Chloroplast transit peptide prediction: a peek inside the black box. *Nucleic Acids Res.* 29:E82
106. Schleucher J, Vanderveer PJ, Sharkey TD. 1998. Export of carbon from chloroplasts at night. *Plant Physiol.* 118:1439–45
107. Schneider A, Häusler RE, Kolukisaoglu U, Kunze R, van der Graaff E, et al. 2002. An *Arabidopsis thaliana* knock-out mutant of the chloroplast triose phosphate/phosphate translocator is severely compromised only when starch synthesis, but not starch mobilisation is abolished. *Plant J.* 32:685–99
108. Schnell DJ, Blobel G, Pain D. 1990. The chloroplast import receptor is an integral membrane protein of the chloroplast envelope contact sites. *J. Cell Biol.* 111:1825–38
109. Schulze-Siebert D, Heineke D, Scharf H, Schulz G. 1984. Pyruvate-derived amino acids in spinach chloroplasts: synthesis and regulation during photosynthetic carbon metabolism. *Plant Physiol.* 76:465–71
110. Schwacke R, Schneider A, van der Graaff E, Fischer K, Catoni E, et al. 2003. ARAMEMNON, a novel database for Arabidopsis integral membrane proteins. *Plant Physiol* 131:16–26

A comprehensive analysis of the Arabidopsis proteome for putative membrane proteins. The membrane proteome can be searched for putative plastidic membrane proteins based on the consensus of several prediction programs.

111. Sentenac H, Bonneaud N, Minet M, Lacroute F, Salmon J-M, et al. 1992. Cloning and expression in yeast of plant potassium ion transport system. *Science* 256:663–65
112. Sharkey TD, Laporte M, Lu Y, Weise S, Weber AP. 2004. Engineering plants for elevated CO_2: a relationship between starch degradation and sugar sensing. *Plant Biol.* (*Stuttg*) 6:280–88
113. Shimogawara K, Fujiwara S, Grossman A, Usuda H. 1998. High-efficiency transformation of *Chlamydomonas reinhardtii* by electroporation. *Genetics* 148:1821–28
114. Shrager J, Hauser C, Chang CW, Harris EH, Davies J, et al. 2003. *Chlamydomonas reinhardtii* genome project. A guide to the generation and use of the cDNA information. *Plant Physiol.* 131:401–8
115. Smith FW, Rae AL, Hawkesford MJ. 2000. Molecular mechanisms of phosphate and sulphate transport in plants. *Biochim. Biophys. Acta* 1465:236–45
116. Soll J, Schleiff E. 2004. Protein import into chloroplasts. *Nat. Rev. Mol. Cell Biol.* 5:198–208
117. Somerville SC, Ogren WL. 1983. An *Arabidopsis thaliana* mutant defective in chloroplast dicarboxylate transport. *Proc. Natl. Acad. Sci. USA* 80:1290–94
118. Somerville SC, Somerville CR. 1985. A mutant of *Arabidopsis* deficient in chloroplast dicarboxylate transport is missing an envelope protein. *Plant Sci.* 37:217–20
119. Song CP, Guo Y, Qiu Q, Lambert G, Galbraith DW, et al. 2004. A probable Na+(K+)/H+ exchanger on the chloroplast envelope functions in pH homeostasis and chloroplast development in *Arabidopsis thaliana*. *Proc. Natl. Acad. Sci. USA* 101:10211–16
120. Steinkamp T, Hill K, Hinnah SC, Wagner R, Rohl T, et al. 2000. Identification of the pore-forming region of the outer chloroplast envelope protein OEP16. *J. Biol. Chem.* 275:11758–64
121. Stitt M, ap Rees T. 1979. Capacities of pea chloroplasts to catalyse the oxidative pentose phosphate and glycolysis. *Phytochemistry* 18:1905–11

122. Streatfield SJ, Weber A, Kinsman EA, Häusler RE, Li JM, et al. 1999. The phosphoenolpyruvate/phosphate translocator is required for phenolic metabolism, palisade cell development, and plastid- dependent nuclear gene expression. *Plant Cell* 11:1609–21
123. Sullivan TD, Kaneko Y. 1995. The maize *brittle1* gene encodes amyloplast membrane polypeptides. *Planta* 196:477–84
124. Sundberg E, Slagter JG, Fridborg I, Cleary SP, Robinson C, Coupland G. 1997. ALBINO3, an Arabidopsis nuclear gene essential for chloroplast differentiation, encodes a chloroplast protein that shows homology to proteins present in bacterial membranes and yeast mitochondria. *Plant Cell* 9:717–30
125. TAGI. 2000. Analysis of the genome sequence of the flowering plant *Arabidopsis thaliana*. *Nature* 408:796–815
126. Takahashi H, Asanuma W, Saito K. 1999. Cloning of an Arabidopsis cDNA encoding a chloroplast localizing sulphate transporter isoform. *J. Exp. Bot.* 50:1713–14
127. Tamagnone L, Merida A, Parr A, Mackay S, Culianez-Macia FA, et al. 1998. The AmMYB308 and AmMYB330 transcription factors from antirrhinum regulate phenylpropanoid and lignin biosynthesis in transgenic tobacco. *Plant Cell* 10:135–54
128. Tamagnone L, Merida A, Stacey N, Plaskitt K, Parr A, et al. 1998. Inhibition of phenolic acid metabolism results in precocious cell death and altered cell morphology in leaves of transgenic tobacco plants. *Plant Cell* 10:1801–16
129. Taniguchi M, Taniguchi Y, Kawasaki M, Takeda S, Kato T, et al. 2002. Identifying and characterizing plastidic 2-oxoglutarate/malate and dicarboxylate transporters in *Arabidopsis thaliana*. *Plant Cell Physiol.* 43:706–17
130. Taniguchi Y, Nagasaki J, Kawasaki M, Miyake H, Sugiyama T, Taniguchi M. 2004. Differentiation of dicarboxylate transporters in mesophyll and bundle sheath chloroplasts of maize. *Plant Cell Physiol.* 45:187–200
131. Tate CG. 2001. Overexpression of mammalian integral membrane proteins for structural studies. *FEBS Lett.* 504:94–98
132. Teutonico RA, Dudley MW, Orr JD, Lynn DG, Binns AN. 1991. Activity and accumulation of cell division-promoting phenolics in tobacco tissue-cultures. *Plant Physiol.* 97:288–97
133. van den Wijngaard PW, Vredenberg WJ. 1999. The envelope anion channel involved in chloroplast protein import is associated with Tic110. *J. Biol. Chem.* 274:25201–4
134. Vaucheret H, Beclin C, Fagard M. 2001. Post-transcriptional gene silencing in plants. *J. Cell Sci.* 114:3083–91
135. Versaw WK, Harrison MJ. 2002. A chloroplast phosphate transporter, PHT2;1, influences allocation of phosphate within the plant and phosphate-starvation responses. *Plant Cell* 14:1751–66
136. Voll L, Häusler RE, Hecker R, Weber A, Weissenbock G, et al. 2003. The phenotype of the Arabidopsis *cue1* mutant is not simply caused by a general restriction of the shikimate pathway. *Plant J.* 36:301–17
137. von Heijne G. 1986. The distribution of positively charged residues in bacterial inner membrane proteins correlates with the trans-membrane topology. *EMBO J.* 5:3021–27
138. von Heijne G, Gavel Y. 1988. Topogenic signals in integral membrane proteins. *Eur. J. Biochem.* 174:671–78
139. Wallmeier H, Weber A, Gross A, Flügge UI. 1992. Insights into the structure of the chloroplast phosphate translocator protein. In *Transport and Receptor Proteins of Plant Membranes*, ed. DT Clakson, DT Cooke, pp. 77–89. New York: Plenum
140. Weber A, Flügge UI. 2002. Interaction of cytosolic and plastidic nitrogen metabolism in plants. *J. Exp. Bot.* 53:865–74

141. Weber A, Menzlaff E, Arbinger B, Gutensohn M, Eckerskorn C, Flügge UI. 1995. The 2-oxoglutarate/malate translocator of chloroplast envelope membranes: molecular cloning of a transporter containing a 12-helix motif and expression of the functional protein in yeast cells. *Biochemistry* 34:2621–27
142. Weber A, Servaites JC, Geiger DR, Kofler H, Hille D, et al. 2000. Identification, purification, and molecular cloning of a putative plastidic glucose translocator. *Plant Cell* 12:787–801
143. Weber APM. 2004. Solute transporters as connecting elements between cytosol and plastid stroma. *Curr. Opin. Plant Biol.* 7:247–53
144. Weber APM. 2005. Synthesis, export, and partitioning of the end products of photosynthesis. In *The Structure and Function of Plastids*, ed. RR Wise, JK Hoober. Dordrecht, Netherlands: Kluwer Academics
145. Weber APM, Oesterhelt C, Gross W, Bräutigam A, Imboden LA, et al. 2004. EST-analysis of the thermo-acidophilic red microalga *Galdieria sulphuraria* reveals potential for lipid A biosynthesis and unveils the pathway of carbon export from rhodoplasts. *Plant Mol. Biol.* 55:17–32
146. Weber APM, Schneidereit J, Voll LM. 2004. Using mutants to probe the in vivo function of plastid envelope membrane metabolite transporters. *J. Exp. Bot.* 55:1231–44
147. Weise SE, Weber APM, Sharkey TD. 2004. Maltose is the major form of carbon exported from the chloroplast at night. *Planta* 218:474–82
148. Wesley SV, Helliwell CA, Smith NA, Wang M, Rouse DT, et al. 2001. Construct design for efficient, effective and high-throughput gene silencing in plants. *Plant J.* 27:581–90
149. Willey DL, Fischer K, Wachter E, Link TA, Flügge UI. 1991. Molecular cloning and structural analysis of the phosphate translocator from pea chloroplasts and its comparison to the spinach phosphate translocator. *Planta* 183:451–61
150. Xu C, Fan J, Riekhof W, Froehlich JE, Benning C. 2003. A permease-like protein involved in ER to thylakoid lipid transfer in Arabidopsis. *EMBO J.* 22:2370–79
151. Yoshimoto N, Takahashi H, Smith FW, Yamaya T, Saito K. 2002. Two distinct high-affinity sulfate transporters with different inducibilities mediate uptake of sulfate in Arabidopsis roots. *Plant J.* 29:465–73
152. Yu TS, Kofler H, Häusler RE, Hille D, Flügge UI, et al. 2001. The Arabidopsis *sex1* mutant is defective in the R1 protein, a general regulator of starch degradation in plants, and not in the chloroplast hexose transporter. *Plant Cell* 13:1907–18
153. Zabrouskov V, Giacomelli L, van Wijk KJ, McLafferty FW. 2003. New approach for plant proteomics—characterization of chloroplast proteins of *Arabidopsis thaliana* by top-down mass spectrometry. *Mol. Cell. Proteomics* 2:1253–60
154. Zeeman SC, Smith SM, Smith AM. 2004. The breakdown of starch in leaves. *New. Phytol.* 163:247–61
155. Zeeman SC, Thorneycroft D, Schupp N, Chapple A, Weck M, et al. 2004. Plastidial alpha-glucan phosphorylase is not required for starch degradation in Arabidopsis leaves but has a role in the tolerance of abiotic stress. *Plant Physiol.* 135:849–58

An excellent review on our current understanding of transitory starch degradation that integrates plastid envelope transport processes with biochemical pathways on both sides of the plastid envelope membrane.

Abscisic Acid Biosynthesis and Catabolism

Eiji Nambara[1] and Annie Marion-Poll[2]

[1]Laboratory for Reproductive Growth Regulation, Plant Science Center, RIKEN, Yokohama, 230-0045, Japan; email: nambara@postman.riken.go.jp

[2]Seed Biology Laboratory, UMR 204 INRA-INAPG, Jean-Pierre Bourgin Institute, 78026 Versailles cedex, France; email: annie.marion-poll@versailles.inra.fr

Annu. Rev. Plant Biol.
2005. 56:165–85

doi: 10.1146/
annurev.arplant.56.032604.144046

1543-5008/05/0602-
0165$20.00

Key Words

ABA conjugation, ABA hydroxylation, carotenoid, seed, stress adaptation

Abstract

The level of abscisic acid (ABA) in any particular tissue in a plant is determined by the rate of biosynthesis and catabolism of the hormone. Therefore, identifying all the genes involved in the metabolism is essential for a complete understanding of how this hormone directs plant growth and development. To date, almost all the biosynthetic genes have been identified through the isolation of auxotrophic mutants. On the other hand, among several ABA catabolic pathways, current genomic approaches revealed that *Arabidopsis* CYP707A genes encode ABA 8′-hydroxylases, which catalyze the first committed step in the predominant ABA catabolic pathway. Identification of ABA metabolic genes has revealed that multiple metabolic steps are differentially regulated to fine-tune the ABA level at both transcriptional and post-transcriptional levels. Furthermore, recent ongoing studies have given new insights into the regulation and site of ABA metabolism in relation to its physiological roles.

Contents

INTRODUCTION

ABA belongs to a class of metabolites known as isoprenoids, also called terpenoids. They derive from a common five-carbon (C_5) precursor, isopentenyl (IDP). Until recently, it was thought that all isoprenoids were synthesized from MVA. However, recently, an alternative pathway to synthesize IDP was discovered, first in certain eubacteria and then in higher plants (65a, 89a). Plastidic isoprenoids, including carotenoids, originate from IDP synthesized from this MVA-independent pathway, called the 2-C-methyl-D-erythritol-4-phosphate (MEP) pathway (24, 89).

ABA: abscisic acid

MVA: mevalonic acid

MEP pathway: Isopentenyl diphosphate derives in plastids from pyruvate and glyceraldehyde 3-phosphate via the formation of 2-C-methyl-D-erythritol-4-phosphate (MEP). In higher plants, both MVA and MEP pathways coexist, in contrast to many eubacteria and green algae, in which only the MEP pathway is present.

Although ABA contains 15 carbon atoms, in plants it is not derived directly from the C_{15} sesquiterpene precursor, farnesyl diphosphate (FDP), but is rather formed by cleavage of C_{40} carotenoids originating from the MEP pathway (47, 56, 72). Evidence for ABA synthesis from carotenoids has been obtained by ^{18}O labeling experiments, molecular genetic analysis of auxotrophs, and biochemical studies. The milestones of the discovery of this "indirect pathway" are described in detail in the previous review on ABA metabolism in this series (128).

ABA BIOSYNTHETIC AND CATABOLIC PATHWAYS

The molecular basis of ABA metabolism was established by genetic approaches. Most of *viviparous* mutants in maize are defective in carotenoid biosynthesis (68). These mutants showed an albino phenotype with a reduced ABA level. In contrast, in a variety of plant species, phenotypes of mutants defective in downstream of xanthophyll cycle are most likely due to ABA deficiency, which is characterized by a wilty plant and production of nondormant seeds. Several recent reviews described the upstream of ABA biosynthesis, particularly MEP and carotenoid pathways (24, 28, 89). Therefore, we focus on the current advances in ABA biosynthetic steps following xanthophyll formation and on the catabolic pathway. Several other reviews describing the ABA metabolic pathway were recently published (20, 98, 101, 120).

ABA Biosynthesis

Epoxy-carotenoid synthesis. Zeaxanthin is produced as a *trans*-isomer after cyclization and hydroxylation of all-*trans*-lycopene via ß-carotene. The following steps consist of the synthesis of *cis*-isomers of violaxanthin and neoxanthin that will be cleaved to form a C_{15} precursor of ABA (**Figure 1**).

Conversion of zeaxanthin to violaxanthin is catalysed by zeaxanthin epoxidase (ZEP) via the intermediate antheraxanthin. The *ZEP* gene, which was first cloned in *Nicotiana plumbaginifolia* by insertional mutagenesis, encodes a protein with sequence similarities to FAD-binding monooxygenases that requires ferredoxin (14, 67). Mutants impaired in ZEP have been isolated in several species, including *Arabidopsis* (60, 69, 77, 119), *N. plumbaginifolia* (67), and rice (1). They accumulate zeaxanthin and show a severe reduction in ABA content, which

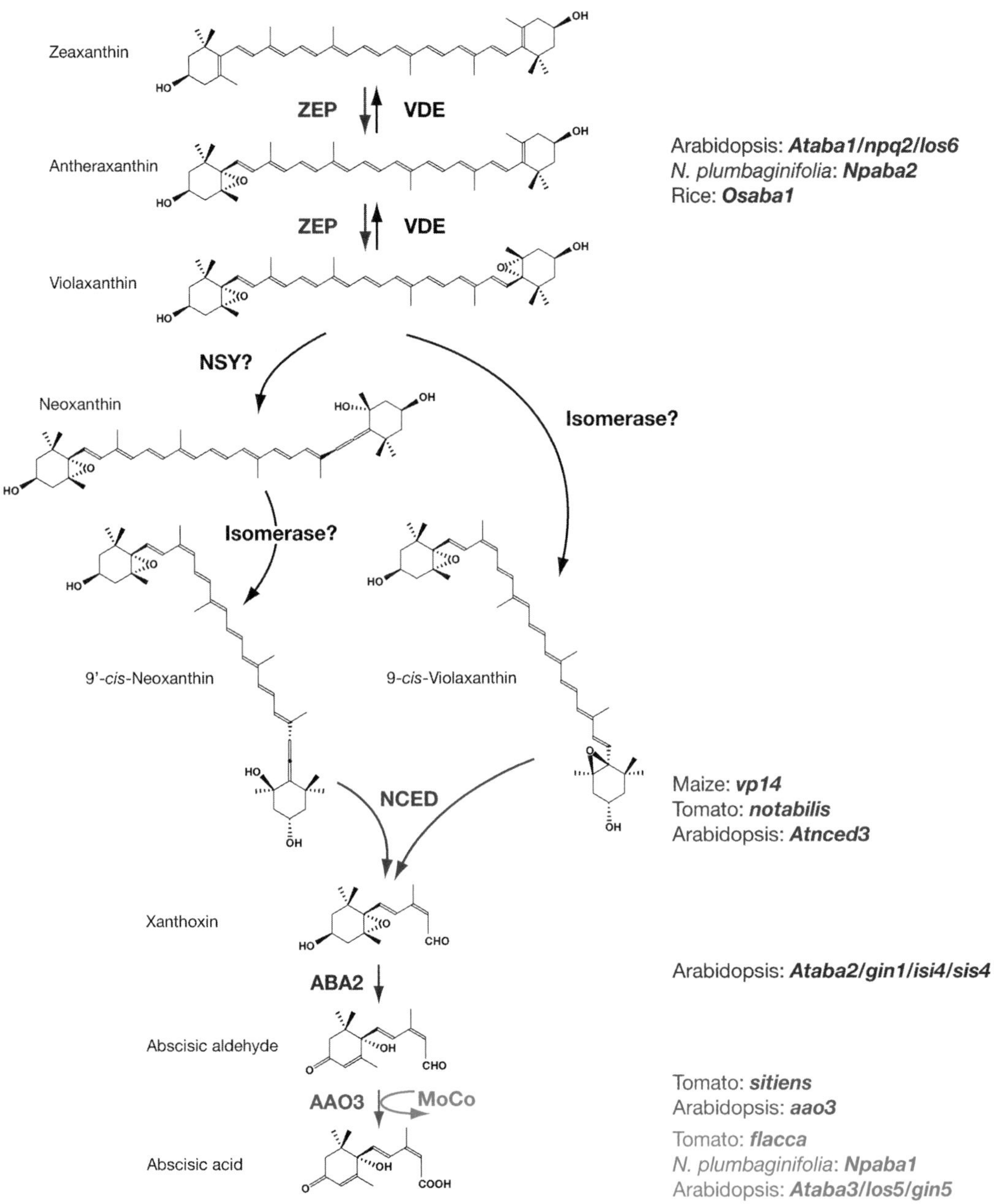

Zeaxanthin

ZEP VDE

Antheraxanthin

ZEP VDE

Violaxanthin

Arabidopsis: ***Ataba1/npq2/los6***
N. plumbaginifolia: ***Npaba2***
Rice: ***Osaba1***

NSY?

Neoxanthin

Isomerase?

Isomerase?

9'-*cis*-Neoxanthin

9-*cis*-Violaxanthin

NCED

Maize: ***vp14***
Tomato: ***notabilis***
Arabidopsis: ***Atnced3***

Xanthoxin

ABA2

Arabidopsis: ***Ataba2/gin1/isi4/sis4***

Abscisic aldehyde

AAO3 MoCo

Tomato: ***sitiens***
Arabidopsis: ***aao3***

Abscisic acid

Tomato: ***flacca***
N. plumbaginifolia: ***Npaba1***
Arabidopsis: ***Ataba3/los5/gin5***

Figure 1

ABA biosynthetic pathway. Synthesis of violaxanthin is catalyzed by zeaxanthin epoxidase (ZEP). A reverse reaction occurs in chloroplasts in high light conditions catalysed by violaxanthin de-epoxidase (VDE). The formation of *cis*-isomers of violaxanthin and neoxanthin may require two enzymes, a neoxanthin synthase (NSY) and an isomerase. Cleavage of *cis*-xanthophylls is catalysed by a family of 9-*cis*-epoxycarotenoid dioxygenases (NCED). Xanthoxin is then converted by a short-chain alcohol dehydrogenase (ABA2) into abscisic aldehyde, which is oxidized into ABA by an abscisic aldehyde oxidase (AAO3). AAO3 protein contains a molydenum cofactor activated by a MoCo sulfurase. A list of defective mutants, which have been named separately depending on species or selective screens, is given on the right side of each enzymatic step.

SDR: short-chain dehydrogenase/reductase

leads to a wilty phenotype and production of nondormant seeds. In *Arabidopsis*, mutations causing amino acid substitutions in the monooxygenase domain impair enzyme function, indicating that this domain might be important for activity (69, 119).

Synthesis of neoxanthin from violaxanthin is not fully elucidated. By homology to lycopene β-cyclase (*LCYB*) and capsanthin capsorubin synthase from pepper, putative neoxanthin synthase (*NSY*) genes of tomato and potato have been isolated (2, 13). However, no *NSY* homologous gene could be found in the *Arabidopsis* genome that contains a unique *LCYB* gene (28). Furthermore, mutations in the putative tomato *NSY* gene were later found to affect ß-carotene synthesis from lycopene, therefore proving that this gene encoded a LCYB isoform (90). Recently, mutants lacking neoxanthin isomers were identified in *Arabidopsis* (H. North & A. Marion-Poll, unpublished results) and tomato (J. Hirchberg, personal communication). The *Arabidopsis* gene has been cloned (H. North & A. Marion-Poll, unpublished) and further biochemical analysis will indicate whether the encoded protein exhibits NSY activity and produces only all-*trans* neoxanthin or both neoxanthin isomers. The gene encoding a *trans-cis* isomerase has not yet been found.

Xanthophyll cleavage. Nine-*cis*-epoxycarotenoid dioxygenase (NCED) enzymes cleave the *cis*-isomers of violaxanthin and neoxanthin to a C_{15} product, xanthoxin, and a C_{25} metabolite (98). The first *NCED* gene (*VP14*) was cloned in maize by insertional mutagenesis (99, 109). Maize VP14 recombinant protein was able to cleave 9-*cis*-violaxanthin and 9′-*cis*-neoxanthin but not *trans*-xanthophyll isomers (99). Enzyme activity requires iron and oxygen to form a *cis*-isomer of xanthoxin (99). In all plant species analyzed, NCED genes belong to a multigene family. In accordance, *nced* mutants, such as *vp14* of maize and *notabilis* of tomato, exhibit mild ABA-deficient phenotypes due to gene redundancy (17, 109). In *Arabidopsis*, nine NCED-related sequences have been identified, and the sequence and functional analyses indicate that five of them (AtNCED2, 3, 5, 6, and 9) are most probably involved in ABA biosynthesis (50, 98). Recently, a new leading compound for ABA biosynthesis inhibitors targeting the NCED was developed (39). This inhibitor might facilitate the study of ABA-mediated physiology in many plant species for which genetic approaches are not available.

As is the case for other carotenoid biosynthesis enzymes, NCED proteins from various species are chloroplast-targeted (51, 83, 107, 108). Because the following enzymatic reaction takes place in the cytosol (18), xanthoxin is presumed to migrate from plastid to cytosol by an unknown mechanism.

C_{15} cytosolic pathway. ABA, the biologically active form, is produced from *cis*-xanthoxin by two enzymatic steps via the intermediate abscisic aldehyde (**Figure 1**). To date, genes encoding these enzymes have been identified only from *Arabidopsis*. The conversion of xanthoxin to abscisic aldehyde is catalysed by AtABA2, belonging to the SDR family. This gene was identified by map-based cloning (18, 35) after the isolation of numerous *Arabidopsis* mutant alleles from various genetic screens (35, 65, 69, 74, 85, 91). AtABA2 protein is encoded by a single gene in the *Arabidopsis* genome; therefore, loss-of-function of this gene leads to a severe ABA deficiency. Mutations have been identified in putative functional domains (NAD binding domain, catalytic center, subunit interacting helix, and substrate binding site) that affect ABA production, indicating the importance of these domains for enzyme activity (35). Furthermore, intragenic complementation between mutant alleles suggests that AtABA2 might have a multimeric structure in accordance with the dimeric or tetrameric structure for most SDR proteins from various organisms (54, 69, 91).

The oxidation of the abscisic aldehyde to the carboxylic acid is the final step in ABA biosynthesis, catalyzed by an abscisic aldehyde oxidase. Among four *Arabidopsis* aldehyde

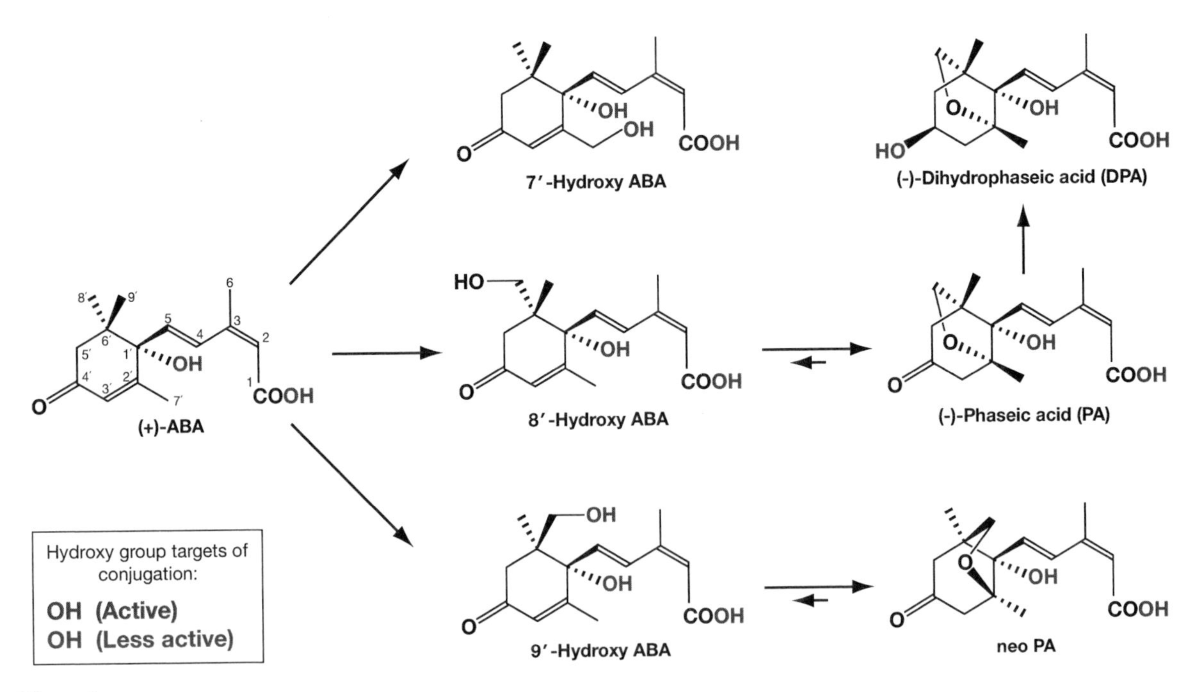

Figure 2

ABA catabolic pathways. Three different hydroxylation pathways are shown. The 8′-hydroxylation is thought to be the predominant pathway for ABA catabolism. Asterisks indicate targets for the conjugation. Red and blue asterisks indicate active and less-active hydroxy groups for conjugation, respectively.

oxidases (AAOs), *AAO3* encodes an enzyme active on abscisic aldehyde (102). The *aao3-1* mutant, containing a mutation in an intron-splicing site, showed a wilty phenotype but only a minor reduction in seed dormancy, compared to other *Arabidopsis aba* mutants affected on unique genes. This mild seed phenotype was thought to be attributed to gene redundancy and it had been postulated that other AAO genes might also be involved in ABA biosynthesis (100). However, identification of null *aao3* alleles exhibiting significant ABA-deficient phenotypes in seeds indicated that *AAO3* is most likely the only *AAO* gene involved in ABA synthesis (34, 99a).

Aldehyde oxidase requires a molybdenum cofactor (MoCo) for its catalytic activity. Therefore, mutations in the genes for MoCo biosynthesis lead to ABA deficiency. Consistent with this, mutations in the *FLACCA* in tomato (92) and *AtABA3* in *Arabidopsis* (10, 122) encoding a MoCo sulfurase confer the expected ABA-deficient phenotypes.

PA: phaseic acid

DPA: dihydrophaseic acid

ABA Catabolism

ABA catabolism is largely categorized into two types of reactions, hydroxylation and conjugation (**Figure 2**). There are three different ABA hydroxylation pathways that oxidize one of the methyl groups of the ring structure (C-7′, C-8′, and C-9′). Three forms of hydroxylated ABA contain substantial biological activities (130, 131), but hydroxylation triggers further inactivation steps. The hydroxylation at C-8′ position is commonly thought to be the predominant ABA catabolic pathway (20, 128). In addition to hydroxylation pathways, ABA and its hydroxylated catabolites [8′-hydroxy ABA, PA, DPA, and *epi*-DPA] are conjugated to glucose (15, 45). A minor inactive form, 2-*trans*-ABA, was also identified. The *cis-trans* isomerization is a photo-permissive equilibrium reaction and is not an enzymatic conversion in tomato (70).

P450: Cytochrome P450 monooxygenases are heme-containing enzymes that catalyze the oxidative reaction of diverse organic compounds by utilizing atmospheric O_2.

EST: expressed sequence tag

ABA hydroxylation. Among the ABA catabolic pathways, the 8′-hydroxylation is reportedly the major regulatory step in many physiological events controlled by ABA. In accordance, PA and DPA are the most widespread and abundant ABA catabolites (20, 128). In addition, ABA analogs modified at the C-8′ methyl group that are resistant to the hydroxylation exhibit stronger ABA-like activities compared to other substitutions (21, 113, 115). ABA 8′-hydroxylation is catalyzed by a cytochrome P450 monooxygenase (P450) and 8′-hydroxy ABA is then isomerized spontaneously to PA (32, 61). Ninety-eight percent of 8′-hydroxy ABA exist as PA at the equilibrium under normal laboratory conditions (114). Although this isomerization occurs quickly in vitro, this reaction is thought to be catalyzed enzymatically in vivo (71). PA is further catabolized to DPA by a soluble reductase (32).

ABA is biologically inactivated in a stepwise manner during the course of catabolism. The 8′-hydroxy ABA contains substantial biological activity (4, 131). Spontaneous cyclization to form PA causes a significant reduction in biological activity, although the degree of reduction varies among bioassays (8, 9, 38, 44, 88, 131). Recent reports showed that the ABA-binding proteins from apple fruit and barley aleurone layers are unable to bind to PA (87, 129), suggesting that PA is an inactive catabolite at least for some physiological processes. DPA is inactive in various bioassays; therefore ABA inactivation is complete by this stage (116).

The 7′-hydroxy ABA is found in a variety of plant species as the minor catabolite (116, 128), and 9′-hydroxy ABA and its isomer neoPA were recently identified as abundant ABA catabolites in *Brassica napus* immature seed (130). In addition, this 9′-hydroxylated product appears to exist also in other plant species such as pea, orange, barley, and *Arabidopsis* (130). Further investigation of this catabolic route should elucidate new aspects of ABA catabolism.

Identification of CYP707A genes encoding ABA 8′-hydroxylase. Recently, *Arabidopsis* P450 CYP707A genes were identified by the reverse genetic approach to encode ABA 8′-hydroxylases (62, 93). Biochemical analysis showed that the recombinant CYP707A protein converts ABA to PA in vitro, but none of the other hydroxylated catabolites (such as 7′-hydroxy or 9′-hydroxy ABA) were produced. CYP707A does not appear to be involved in cyclization of 8′-hydroxy ABA to PA because ABA is primarily converted to 8′-hydroxy ABA in a short incubation period and then 8′-hydroxy ABA is autoisomerized to PA (93). The activity of CYP707A was inhibited by a P450 inhibitor tetcyclasis, which was originally developed as an inhibitor of GA biosynthesis, but not by another P450 inhibitor metyrapone (62). This indicates that CYP707A discriminates between two different known P450 inhibitors. Therefore, it might be possible to develop a specific inhibitor of this enzyme in the future.

CYP707A appears to be widespread in many plant species. CYP707A-related sequences are found in rice genome and among ESTs from tomato, soybean, and maize (**http://drnelson.utmem.edu/CytochromeP450.html**). *CYP 707A* sequences are also identified in lettuce and wheat (T. Toyomasu, N. Kawakami & E. Nambara, unpublished results).

ABA conjugation. The carboxyl (at the C-1) and hydroxyl groups of ABA and its oxidative catabolites are the potential targets for conjugation with glucose (**Figure 2**). ABA glucosyl ester (ABA-GE) is the most widespread conjugate (15). In addition to the glucosyl esters, other conjugates with the hydroxyl groups of ABA and its hydroxylated catabolites are also reported. ABA conjugates had been thought to be physiologically inactive and accumulate in vacuoles during aging (16, 63). However, recently ABA-GE was proposed to be involved in long-distance transport of ABA (42, 118). ABA-GE was identified as an allelopathic substance of *Citrus junis* (57), and soil in agricultural fields contains higher concentrations of

ABA-GE (up to 30 nM) than ABA. It has been hypothesized that ABA-GE is taken up by the root (96). Furthermore, ABA-GE is the most abundant catabolite in the sunflower xylem sap (40). ß-D-glucosidase releases ABA from ABA-GE in wheat, barley, and sunflower (23, 64, 95). The ß-D-glucosidase activity is enhanced by salinity and is inhibited competitively by ABA-GE or zeatin riboside. However, because ABA-GE cannot migrate passively through the plasma membrane, the molecular mechanism underlying the transport of ABA or its conjugates remains unclear.

Identification of the AOG gene encoding ABA glucosyltransferase. The *AOG* gene encoding ABA glucosyltransferase was identified from adzuki bean as the first reported gene for ABA catabolism (123). The AOG recombinant protein can conjugate ABA with UDP-D-glucose. AOG exhibits a broad substrate specificity compared to other ABA catabolic enzyme CYP707As. AOG catalyzes the conjugation of 2-*trans*-ABA to glucose more efficiently than natural 2-*cis*-ABA, consistent with previous feeding experiments in tomato (70). AOG can also use an ABA analog (-)-*R*-ABA or cinnamic acid as substrates, but not the immediate ABA catabolite PA. Therefore, PA and DPA glucosylation might be catalyzed by different enzymes (123).

REGULATION OF ABA METABOLISM IN RELATION TO ITS PHYSIOLOGICAL ROLES

Regulatory Steps, Factors, and Levels

The endogenous ABA level is modulated by the precise balance between biosynthesis and catabolism of this hormone. With regard to ABA biosynthesis, NCED has been proposed to be the regulatory enzyme because its expression is well correlated to endogenous ABA content (98) and its overexpression confers a significant ABA accumulation (50, 84, 111). On the other hand, ABA 8′-hydroxylase is most likely the major regulatory enzyme in many physiological processes, as described below (20, 62, 93, 128).

Aside from these two main regulatory steps in the ABA metabolic pathway, metabolic steps upstream of ABA metabolism also contribute to determining the ABA level. Overexpression of genes encoding regulatory enzymes for the MEP pathway (1-deoxy-D-xylulose 5-phosphate synthase), carotenoid biosynthesis (phytoene synthase), and xanthophyll cycle (*ZEP*) causes an enhanced accumulation of ABA in seeds or seedlings (25, 29, 66). Taken together, this indicates that the regulation of ABA metabolism is not merely restricted to specific steps in ABA metabolism (i.e., NCED and CYP707A), but is also coordinated with the upstream metabolism.

To date, the regulation of ABA metabolism has been studied mostly at the transcription level. This process is differentially regulated by external and endogenous signals. In particular, the expression of *AtNCED3*, *AAO3*, *AtABA3*, and *AtZEP*, but not *AtABA2*, genes are induced by dehydration in *Arabidopsis*, as detailed below, whereas the expression of *AtABA2*, *AtZEP*, and *AAO3*, but not *AtNCED3*, are induced by application of glucose that induces ABA accumulation (18). Moreover, all four *Arabidopsis* *CYP707A* genes are induced by osmotic stresses (62). In addition to external signals, the expression of the *CYP707A3* is positively regulated by gibberellin (GA) and brassinolide (93), indicating that CYP707A genes function as the node of hormone interactions. Aside from these interactions, several reports indicate that many biosynthetic and catabolic genes are also upregulated by the application of ABA, suggesting that ABA might regulate its own accumulation (18, 62, 93, 119, 121, 123).

Genetic analysis of the *sad1* (supersensitive to ABA and drought) mutant of *Arabidopsis* indicated that ABA biosynthesis is also regulated at the level of mRNA stability. The *SAD1* locus encodes a peptide similar to multifunctional Sm-like snRNP proteins required for mRNA processing (121). The *sad1* mutant shows reduced levels of ABA and PA, and expression

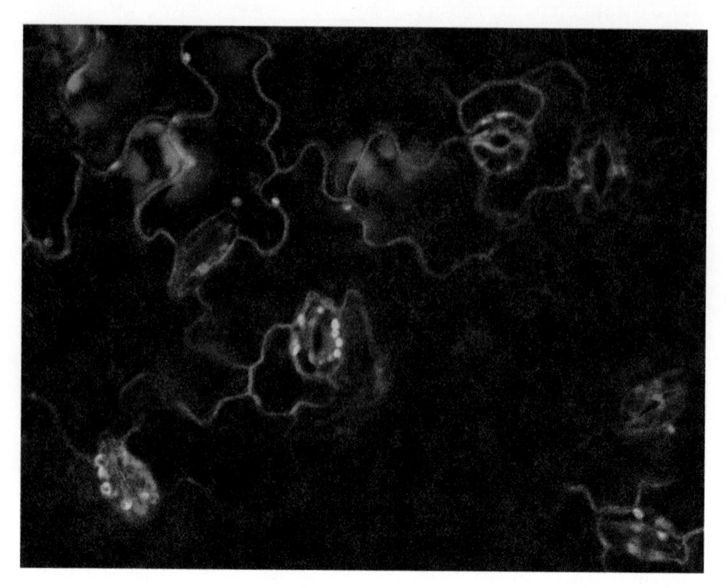

Figure 3

Localization of the pAAO3::AAO3-GFP protein in guard cells. A representative view of turgid transgenic *Arabidopsis* carrying pAAO3::AAO3-GFP is shown.

and feeding analyses demonstrated that SAD1 is a positive regulator of *AAO3* and *AtABA3*. Although it is unclear how SAD1 functions in regulating feedback on ABA metabolism and signaling, the identification of several RNA processing genes through genetic screens suggests that RNA processing is closely tied to the regulation of these processes.

The Sites of ABA Biosynthesis

Study on the site of ABA biosynthesis is essential to link the understanding of the regulation of ABA metabolism to physiology and development. Especially, in contrast to stress-induced ABA accumulation, little is known about the role and function of ABA in plant growth and development under unstressed conditions. Identification of ABA metabolic genes enables the study of where and when these genes are expressed in the plant life cycle.

In turgid tissues the expression of *AtNCEDs*, *AtABA2*, and *AAO3* genes is observed in vascular bundles (18, 59, 108). Koiwai et al. (2004) reported that the AAO3 protein is abundantly localized in phloem companion cells and xylem parenchyma cells of turgid plants (59). Therefore, vascular tissues are probably the main site of ABA biosynthesis in unstressed plants and ABA and its precursors might be synthesized in vascular tissues and transported to target cells such as stomata.

The localization and regulation of the expression of biosynthetic genes in guard cells is particularly interesting with regard to the role of ABA in stomatal closure. Although many studies have given evidence for the transport of ABA to target cells (94), recent data indicate that ABA synthesis is also active in guard cells. Through reporter gene analysis *AtNCED2* and *AtNCED3* transcripts were expressed in guard cells of senescent leaves and cotyledons, respectively (108). In addition, the *AAO3* encoding the enzyme for the final step in ABA biosynthesis is induced in guard cells upon stress. In the same study, by immuno-fluorescence and expression studies using a reporter-fused protein, the AAO3 protein was also present in guard cells (59) (**Figure 3**). It is, therefore, probable that ABA synthesized inside guard cells, in concert with transported ABA, triggers the downstream signaling cascade leading to stomatal closure.

Nevertheless, it remains to be elucidated whether ABA synthesis is still restricted to the same sites or becomes activated in other plant tissues under stress conditions.

Regulation of ABA metabolism depends on internal and external signals, as well as developmental stages, organs, or tissues. This is illustrated in the two sections below, which describe in more detail the regulation of ABA metabolism genes in two physiological processes highly controlled by the hormone, i.e., seed physiology and stress tolerance.

ABA Metabolism in Seeds

Embryogenesis and seed maturation. ABA has a dual role in embryo growth during seed development, as deduced from the physiological analysis of ABA-deficient mutants. In early embryogenesis, ABA prevents seed abortion and promotes embryo growth (18, 30). In contrast, during late embryogenesis when the hormone level increases, ABA blocks the embryo growth by counteracting the action of GA to promote germination (86, 117). Recent studies show that the transcription factor FUS3 prevents vivipary in *Arabidopsis* seeds by positively regulating ABA levels and downregulating GA synthesis (31, 73). In addition, maternal ABA can inhibit viviparous germination in *fus3* mutants defective in embryo growth arrest (86). It will be interesting to see if other developmental mutants that affect embryogenesis impinge on ABA metabolism.

Despite the low levels of ABA generally detected during early embryogenesis, the ABA biosynthetic pathway is apparently active at this stage. *AtZEP*, *AtNCED5*, and *AtNCED6* transcripts have been detected in *Arabidopsis* young embryos (7, 108). In maternal tissues, the *AtZEP* gene was expressed in testa, *AtNCED3* and *AtABA2* in funicules, and *AtNCED5* and *AtNCED6* in maternal nucellar tissues of newly fertilized ovules (18, 108). These tissues might provide ABA or its precursors to the embryo, and in agreement, high ABA levels have been found in the pedicel/placento-chalazal complex of maize kernels (53).

A major increase in ABA levels occurs during the maturation phase in relation to the positive regulation of a number of genes for seed reserves (27). Carotenoid precursors accumulate in seeds of most plant species and their synthesis is expected to precede their cleavage into xanthoxin. Consistent with this, maximal *ZEP* gene expression in *N. plumbaginifolia* appears to peak earlier (6) than that found in *Arabidopsis NCED* genes. Indeed, *AtNCED5* and *AtNCED6* show the strongest expression in *Arabidopsis* embryos at mid- to late-developmental stages (108). A decrease in the ABA level during the desiccation phase is expected to result from decreased ABA synthesis, as evidenced by very low *ZEP* transcript levels at this stage (6).

Although ABA catabolites have been detected in developing seeds and reproductive organs (19, 103), expression studies of catabolic genes are still limited. *CYP707A1* and *CYP707A3* are expressed abundantly during the mid-stage of *Arabidopsis* seed development and are downregulated during late embryo development (62).

Seed dormancy and germination. Aside from its role in embryogenesis and seed maturation, ABA is absolutely required to induce seed dormancy during late embryogenesis. Genetic studies show that ABA produced by the embryo itself, and not maternal ABA, is necessary to impose dormancy (37, 55). Besides induction of dormancy in developing seeds, ABA is involved in maintaining dormancy during imbibition (3, 22, 36). Germination is preceded by a decrease in ABA levels resulting from both the suppression of de novo synthesis and the activation of catabolism (26, 62). In contrast, dormant seeds generally maintain endogenous ABA at the high levels, and dormancy is effectively released by the application of fluridone, which blocks the synthesis of carotenoid precursors of ABA. When ABA levels decrease during seed imbibition, concomitant increases in PA/DPA levels were observed in barley (52), lettuce (33), yellow-cedar (97), white pine (26), and *Arabidopsis* (62). In high-temperature-induced dormant lettuce seeds, ABA catabolism is positively regulated by GA because PA/DPA accumulation is accelerated by GA application (33).

In *Arabidopsis*, the catabolic enzyme CYP707A2 plays a major role in the rapid decrease in ABA levels during early seed imbibition (62). *CYP707A2* transcripts accumulate to a high level in dry seed, whereas other CYP707A transcripts are scarce. *CYP707A2* transcript levels increase within six hours of imbibition and decrease thereafter. The *cyp707a2* seeds exhibit hyperdormancy when sown without stratification. Furthermore, *cyp707a2* dry seeds accumulate sixfold more ABA than wild type and this high ABA level is maintained during seed imbibition (62). In addition, gene expression analysis suggests that *CYP707A1* and *CYP707A3* are possibly involved in seed germination and early seedling development because their transcripts are gradually accumulated after 12 hours of imbibition.

ABA Metabolism in Abiotic Stress Adaptation

Transcript levels of several ABA biosynthetic genes are upregulated by osmotic stresses. The expression of the *ZEP* gene is induced by both rapid or progressive drought stress in roots of *N. plumbaginifolia*, tomato, and *Arabidopsis* (6, 7, 110). In leaves, the high carotenoid levels are not likely to contribute positively to ABA synthesis even under stress conditions. In agreement, no upregulation of the *ZEP* gene has been reported in *Arabidopsis* or in *N. plumbaginifolia* and tomato (6, 7, 110). Moreover, *N. plumbaginifolia* transgenic plants overexpressing *ZEP* transcripts do not exhibit higher ABA levels and water stress tolerance compared to wild type (12). However, opposite results were also reported in *Arabidopsis* leaves, in which overexpression of the *ZEP* gene upregulates the expression of genes induced by drought, salt, and osmotic stress (119).

The induction of *NCED* gene expression has been observed in several species, both in roots and in leaves (51, 83, 108, 109, 111). Detailed studies with the *PvNCED1* gene from *Phaseolus vulgaris* provide evidence that the oxidative cleavage of xanthophylls is a major regulatory step of ABA accumulation under drought stress. Water stress–induced ABA accumulation is preceded by large increases in both *PvNCED1* transcript and protein levels in leaves and roots (83). In *Arabidopsis*, among the five *NCED* genes involved in ABA biosynthesis, only *AtNCED3* is highly induced by dehydration, although a positive but minor regulation of the other *NCED* genes was also observed (50, 108). In addition, *AtNCED3* overexpression in transgenic *Arabidopsis* plants increases both ABA levels and desiccation tolerance. This result is also seen in tomato and *N. plumbaginifolia* after transformation with the *LeNCED1* and *PvNCED1* genes, respectively (84, 111). Interestingly, the induction of *AtNCED3* under stress conditions is reduced in carotenoid-deficient mutants, suggesting that the expression level of this gene is correlated with the levels of its substrates and/or ABA (112). Furthermore, the induction of this gene in response to exogenous ABA is highly enhanced in ABA-deficient mutant backgrounds (119).

Biochemical studies indicate that the activity of the last two biosynthetic enzymes is constitutive (105) and *AtABA2* transcripts are not induced upon osmotic stress in *Arabidopsis* (18). However, the expression of the two other genes, *AAO3* and *AtABA3*, involved in the conversion of abscisic aldehyde into ABA, is upregulated under osmotic stresses (122).

Although it is clear that ABA biosynthesis is responsive to stress conditions, it is becoming evident that the catabolism is also required to determine the ABA level in response to environmental conditions. PA levels and occasionally DPA levels increase following the increase in ABA content. Furthermore, the PA level continues to increase even after ABA levels reach the plateau. However, when dehydrated plants are subsequently rehydrated, the ABA level decreases and a concomitant increase in the PA level is observed in the *P. vulgaris*, *Xanthium strumarium* and *Arabidopsis* (41, 62, 126). In *Arabidopsis*, multiple *CYP707A* genes are expressed in most organs and the ratio of *CYP707A* transcripts varies among tissues. The expression of all *CYP707A* genes is induced by dehydration, although the induction is slower than that of *AtNCED3*, which encodes the key biosynthetic enzyme under dehydration conditions in *Arabidopsis* (62). A significant increase in all *CYP707A* transcripts is observed upon rehydration, in accordance with the increase in PA levels. *CYP707A* are also upregulated by salinity and osmotic stresses (93). In contrast to PA and DPA, ABA conjugate levels do not always vary in parallel to the change in ABA levels, suggesting that the conjugation is regulated in particular tissues and conditions (127). In agreement with this, *AOG* gene expression is significantly induced by dehydration and wounding in adzuki bean hypocotyls, but not in leaves (123).

EVOLUTION OF ABA METABOLISM

ABA Metabolism in Fungi

Some phytopathogens are known to synthesize ABA (81). The ABA level is thought to determine a plant's susceptibility to these fungi by negatively regulating the salicylic acid–dependent defense pathway (5). Studies on fungal ABA biosynthesis have been mostly conducted in *Cercospora rosicola*, *Cercospora cruenta*, *Cercospora pini-densiflorae*, and *Botrytis cinerea*. Feeding experiments using [1-^{13}C]-D-glucose clearly demonstrate that fungal ABA is derived from the MVA pathway (47, 124), in contrast to plant ABA, which originates from the MEP pathway.

ABA biosynthesis in fungi can be divided largely into two parts (**Figure 4**). The isomers of ionylideneethanol and/or ionylideneacetate have been identified from several fungi (81). Therefore, the early steps seem to convert FDP derivatives to ionylideneacetate, and the latter steps in their oxidation at C-1′ and C-4′ to produce ABA. Feeding experiments show that similar but distinct intermediates are identified among fungal genera (81, 128). This suggests that ABA biosynthetic pathway and/or its regulation might be different among these fungi.

In early steps, MVA is converted into ionylideneacetoaldehyde via cyclization and isomerization of FDP derivatives. Direct cyclization of the sesquiterpene was proposed in fungal ABA biosynthesis, although experimental evidence is still missing. $^{18}O_2$ labeling experiments demonstrate that the oxygen atom at C-1 of ABA is derived from molecular oxygen in *C. cruenta* and *B. cinerea* (49, 124). This opens two possibilities. One is that dephosphorylation and reduction of FDP occurs to produce allofarnesene prior to the cyclization, which is then oxidized to ionylideneacetoaldehyde. Alternatively, ionylideneacetoaldehyde is produced via the cleavage of C_{40} carotenoids. Some fungi, such as *C. rosicola* and *C. cruenta*, produce β-carotene and other carotenoids (78, 124), but carotenoids (except for phytoene) are not found in other ABA-synthesizing fungus *B. cinerea* (49, 49a). Recently in *B. cinerea* and *C. cruenta*, allofarnesene and ionylideneethane were shown to be endogenous compounds that were able to convert to ABA (49, 49a). This indicates that ABA is synthesized by the direct pathway via the cyclization of allofarnesene and oxidation of ionylideneethane in this fungus.

The latter steps involve the oxidation of C-1′ and C-4′ of ionylideneacetate. α-Ionylideneethanol/ionylideneacetate were converted to ABA and 1′-deoxy ABA in *C. rosicola* (75). 1′-Deoxy ABA is thought to be the precursor of ABA in this fungus, because it is oxidized stereoselectively to ABA (75). On one hand, in *B. cinerea* and *C. pini-densiflorae*, 1′,4′-*trans*-diol ABA is likely the predominant precursor whose endogenous levels are correlated with ABA synthesis (46, 80). On the other hand, 1′,4′-*trans*-dihydro-γ-ionylideneacetoaldehyde is thought to be the intermediate of ABA biosynthesis in *C. cruenta* (82). Three oxygen atoms at C-1, C-1′, and C-4′ derive from atmospheric oxygen in *C. cruenta* and *B. cinerea* (49, 124). In addition, it has been reported that several P450 inhibitors block the ABA synthesis in *C. rosicola* (79), indicating that P450 is most likely involved in these oxidations. This was recently proven by the genomic approach, which showed that targeted inactivation of P450 oxidoreductase reduced ABA production in *B. cinerea* (104). Furthermore, loss-of-function of a P450 gene, *BcABA1*, whose expression is associated with the ABA production, abolished the accumulation of ABA in this fungus (104). The *BcABA1* gene will likely be the first ABA biosynthetic gene identified from fungi. Because functionally related genes are often clustered in fungal genomes, the molecular basis of fungal ABA biosynthesis will be elucidated in the near future.

MVA pathway: Isopentenyl diphosphate derives in the cytosol from acetyl-coenzymeA via the formation of MVA. It is the only pathway for isoprenoid synthesis in archaebacteria, fungi, and animals.

ABA Metabolism in Lower Plants

In addition to higher plants and fungi, ABA is synthesized in moss, fern, and algae. ABA is found in all divisions and classes of algae, including colorless species (48). In green algae

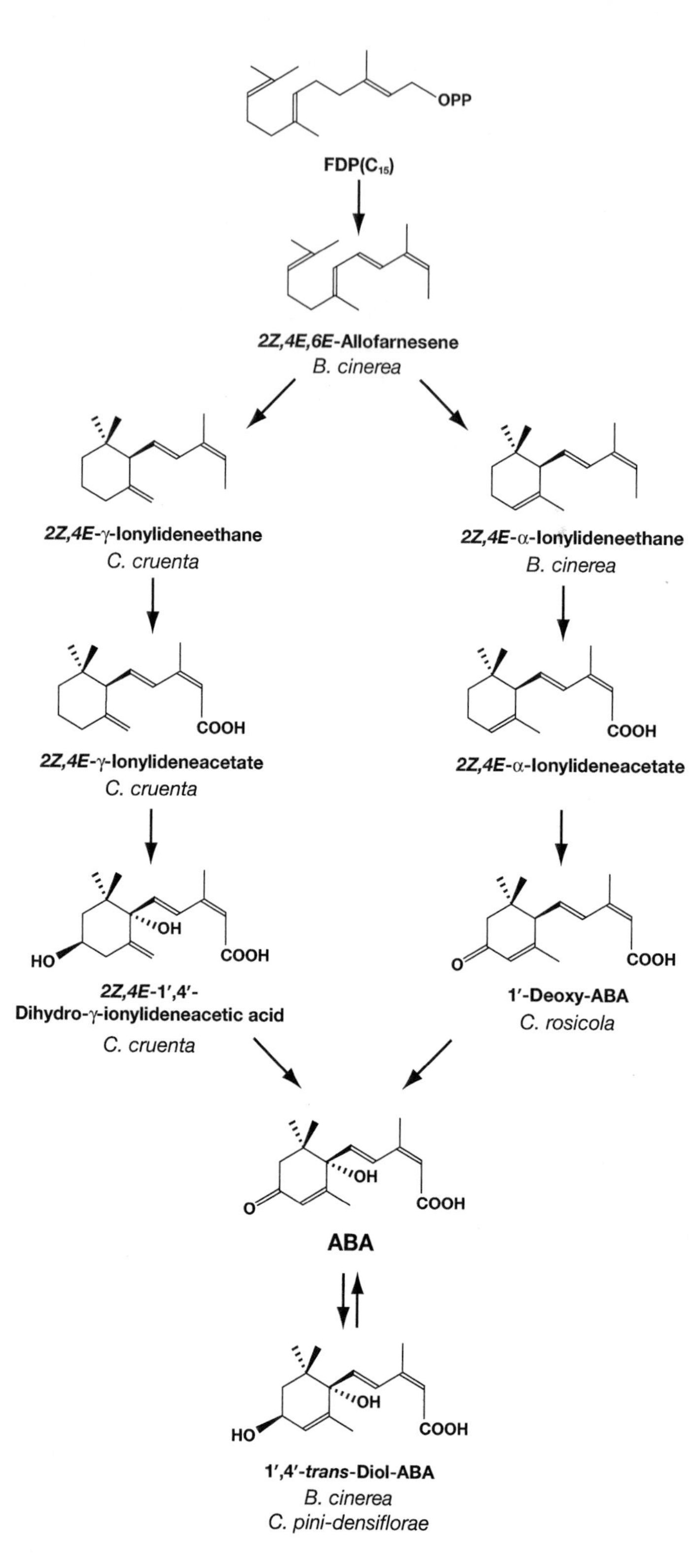

Chlamydomonas reinhardtii, application of ABA enhances the resistance to the oxidative stress (125). A study using carotenoid biosynthesis inhibitors suggests the direct pathway is proposed to be active in green algae *Dunaliella* (11). However, this issue is still in debate about whether these organisms synthesize ABA through a direct pathway or a carotenoid-derived indirect pathway.

In *Riccia fluitans*, ABA and its catabolite content fluctuates in response to water status (43). Accumulation of ABA in *Riccia* thalli is inhibited by the application of fluridone, suggesting that ABA is synthesized via carotenoids. Feeding experiments using radio-labeled ABA show PA and DPA as major catabolites, and minor catabolites include ABA-GE. PA/DPA production is inhibited by tetcyclacis, a P450 inhibitor. Therefore, it is likely that *Riccia* synthesizes and catabolizes ABA through the same metabolic pathway as that in higher plants. Interestingly, the same ABA responsive element as in higher plants acts on ABA-mediated transcription in the moss *Physcomitrella patens* (58).

Current genome sequencing and EST projects in many organisms facilitate the understanding of the evolutionary aspects of the ABA metabolic pathway. The EST project of the moss *Physcomitrella patens* reveals genes highly similar to *NCED* and *CYP707A* (76). This indicates that this moss synthesizes and catabolizes ABA through pathways similar to those in higher plants. *Chlamydomonas* contains a gene highly homologous to *ZEP*, consistent with the identification of endogenous xanthophylls (such as 9′-*cis*-neoxanthin) (106). In addition, *Chlamydomonas* contains several genes belonging to the RPE65 family, although protein sequences are more related to the carotenoid-cleaving dioxygenases acting at the 15′ position rather than NCED, which cleaves at the

Figure 4

Proposed ABA biosynthetic pathways in fungi. Direct (cyclization of C15 terpenoid) is proposed in fungal ABA biosynthesis. Identified potential intermediates with fungal species are shown.

9′ or 9 positions of xanthophylls. Interestingly, *Chlamydomonas* also contains a P450 gene that is related to CYP707 (85-clan). Functional analyses of these genes will elucidate the nature of evolution of ABA and carotenoid metabolic pathways in lower plants.

SUMMARY POINTS

1. Both forward and reverse genetic analyses have been instrumental in identifying ABA biosynthetic genes, and most of the genes have been identified in higher plants.
2. *Arabidopsis* CYP707A genes were recently shown to encode ABA 8′-hydroxylases, which catalyze the committed step in the predominant ABA catabolic pathway.
3. Localization analyses of biosynthetic enzymes and their transcripts indicate that the vascular bundles are the active site of ABA synthesis in turgid plants.
4. Expression of ABA metabolism genes is temporarily and spatially regulated during seed development and germination.
5. Transcriptional and post-transcriptional gene regulation plays a crucial role in the ABA accumulation in response to osmotic stresses.
6. ABA biosynthesis in fungi is thought to occur through the direct cyclization of FDP derivatives, and allofarnesene and ionylideneethane are proposed to be intermediates in *Botrytis cinerea*. Furthermore, the first fungal ABA biosynthetic gene, *BcABA1*, was recently identified in *B. cinerea*.

FUTURE ISSUES TO BE RESOLVED

1. In higher plants, the enzymes catalysing the isomerization of xanthophylls are the only biosynthetic enzymes remaining to be identified. Furthermore, recent isolation of genes encoding two major catabolic enzymes encourages the identification of other genes involved in the diverse catabolic pathways.
2. A large research field needs to be explored to unravel the signaling pathways regulating the metabolic genes and their interactions with endogenous and environmental signals.
3. In other organisms, such as fungi, algae, or mosses, the ABA biosynthetic and catabolic pathways require further investigation to identify all the genes, enzymes, precursors, and catabolites. The regulation of ABA produced by phytopathogenic fungi might be of particular interest to study plant-pathogen interaction.

ACKNOWLEDGMENTS

The authors acknowledge Drs. Peter McCourt, Tetsuo Kushiro, Nobuhiro Hirai, and Hiroshi Kawaide for critical reading of the manuscript. The authors also thank Drs. Yossi Hirschberg, Naoto Kawakami, and Tomonobu Toyomasu for sharing unpublished results, and Drs. Masanori Okamoto, Hanae Koiwai, Mitsunori Seo, and Tomokazu Koshiba for providing photographs.

LITERATURE CITED

1. Agrawal GK, Yamazaki M, Kobayashi M, Hirochika R, Miyao A, et al. 2001. Screening of the rice viviparous mutants generated by endogenous retrotransposon *Tos17* insertion. Tagging of a zeaxanthin epoxidase gene and a novel *OsTATC* gene. *Plant Physiol.* 125:1248–57
2. Al-Babili S, Hugueney P, Schledz M, Welsch R, Frohnmeyer H, et al. 2000. Identification of a novel gene coding for neoxanthin synthase from *Solanum tuberosum*. *FEBS Lett.* 485:168–72
3. Ali-Rachedi S, Bouinot D, Wagner MH, Bonnet M, Sotta B, et al. 2004. Changes in abscisic acid levels during dormancy release and maintenance of mature seeds: studies with the Cape Verde Islands ecotype, the dormant model of *Arabidopsis thaliana*. *Planta* 219:479–88
4. Arai S, Todoroki Y, Ibaraki S, Naoe Y, Hirai N, et al. 1999. Synthesis and biological activity of 3′-chloro, -bromo, and -iodoabscisic acids, and biological activity of 3′-fluoro-8′-hydroxyabscisic acid. *Phytochemistry* 52:1185–93
5. Audenaert K, De Meyer GB, Höfte MM. 2002. Abscisic acid determines basal susceptibility of tomato to *Botrytis cinerea* and suppresses salicylic acid-dependent signaling mechanisms. *Plant Physiol.* 128:491–501
6. Audran C, Borel C, Frey A, Sotta B, Meyer C, et al. 1998. Expression studies of the zeaxanthin epoxidase gene in *Nicotiana plumbaginifolia*. *Plant Physiol.* 118:1021–28
7. Audran C, Liotenberg S, Gonneau M, North H, Frey A, et al. 2001. Localisation and expression of zeaxanthin epoxidase mRNA in Arabidopsis in response to drought stress and during seed development. *Aust. J. Plant Physiol.* 28:1161–73
8. Balsevich JJ, Abrams SR, Lamb N, Konig WA. 1994. Identification of unnatural phaseic acid as a metabolite derived from exogenously added (-)-abscisic acid in a maize cell-suspension culture. *Phytochemistry* 36:647–50
9. Balsevich JJ, Cutler AJ, Lamb N, Friesen LJ, Kurz EU, et al. 1994. Response of cultured maize cells to (+)-abscisic acid, (−)-abscisic acid, and their metabolites. *Plant Physiol.* 106:135–42
10. Bittner F, Oreb M, Mendel RR. 2001. ABA3 is a molybdenum cofactor sulfurase required for activation of aldehyde oxidase and xanthine dehydrogenase in *Arabidopsis thaliana*. *J. Biol. Chem.* 276:40381–84
11. Bopp-Buhler ML, Wabra P, Hartung W, Gimmler H. 1991. Evidence for direct ABA synthesis in *Dunaliella* (Volvocales). *Crypt. Bot.* 2/3:192–200
12. Borel C, Audran C, Frey A, Marion-Poll A, Tardieu F, et al. 2001. *N. plumbaginifolia* zeaxanthin epoxidase transgenic lines have unaltered baseline ABA accumulations in roots and xylem sap, but contrasting sensitivities of ABA accumulation to water deficit. *J. Exp. Bot.* 52:427–34
13. Bouvier F, d'Harlingue A, Backhaus RA, Kumagai MH, Camara B. 2000. Identification of neoxanthin synthase as a carotenoid cyclase paralog. *Eur. J. Biochem.* 267:6346–52
14. Bouvier F, d'Harlingue A, Hugueney P, Marin E, Marion-Poll A, et al. 1996. Xanthophyll biosynthesis. Cloning, expression, functional reconstitution, and regulation of β-cyclohexenyl carotenoid epoxidase from pepper (*Capsicum annuum*). *J. Biol. Chem.* 271:28861–67
15. Boyer GL, Zeevaart JAD. 1982. Isolation and quantitation of β-D-glucopyranosyl abscisate from leaves of *Xanthium* and spinach. *Plant Physiol.* 70:227–31
16. Bray EA, Zeevaart JAD. 1985. The compartmentation of abscisic acid and β-D-glucopyranosyl abscisate in mesophyll cells. *Plant Physiol.* 79:719–22
17. Burbidge A, Grieve TM, Jackson A, Thompson A, McCarty DR, et al. 1999. Characterization of the ABA-deficient tomato mutant *notabilis* and its relationship with maize *Vp14*. *Plant J.* 17:427–31

18. Cheng WH, Endo A, Zhou L, Penney J, Chen HC, et al. 2002. A unique short-chain dehydrogenase/reductase in Arabidopsis glucose signaling and abscisic acid biosynthesis and functions. *Plant Cell* 14:2723–43
19. Chichowa S, von Aderkas P. 2002. Endogenous levels of free and conjugated forms of auxin, cytokinins and abscisic acid during seed development in Douglas fir. *Plant Growth Regul.* 36:191–200
20. Cutler AJ, Krochko JE. 1999. Formation and breakdown of ABA. *Trends Plant Sci.* 4:472–78
21. Cutler AJ, Rose PA, Squires TM, Loewen MK, Shaw AC, et al. 2000. Inhibitors of abscisic acid 8′-hydroxylase. *Biochemistry* 39:13614–24
22. Debeaujon I, Koornneef M. 2000. Gibberellin requirement for Arabidopsis seed germination is determined both by testa characteristics and embryonic abscisic acid. *Plant Physiol.* 122:415–24
23. Dietz KJ, Sauter A, Wichert K, Messdaghi D, Hartung W. 2000. Extracellular β-glucosidase activity in barley involved in the hydrolysis of ABA glucose conjugate in leaves. *J. Exp. Bot.* 51:937–44
24. Eisenreich W, Bacher A, Arigoni D, Rohdich F. 2004. Biosynthesis of isoprenoids via the non-mevalonate pathway. *Cell. Mol. Life Sci.* 61:1401–26
25. Estévez JM, Cantero A, Reindl A, Reichler S, Leon P. 2001. 1-deoxy-D-xylulose-5-phosphate synthase, a limiting enzyme for plastidic isoprenoid biosynthesis in plants. *J. Biol. Chem.* 276:22901–9
26. Feurtado JA, Ambrose SJ, Cutler AJ, Ross ARS, Abrams SR, et al. 2004. Dormancy termination of western white pine (*Pinus monticola* Dougl. Ex D. Don) seeds is associated with changes in abscisic acid metabolism. *Planta* 218:630–39
27. Finkelstein RR, Gampala SSL, Rock CD. 2002. Abscisic acid signaling in seeds and seedlings. *Plant Cell* 14:S15–S45
28. Fraser PD, Bramley PM. 2004. The biosynthesis and nutritional uses of carotenoids. *Prog. Lipid Res.* 43:228–65
29. Frey A, Audran C, Marin E, Sotta B, Marion-Poll A. 1999. Engineering seed dormancy by the modification of zeaxanthin epoxidase gene expression. *Plant Mol. Biol.* 39:1267–74
30. Frey A, Godin B, Bonnet M, Sotta B, Marion-Poll A. 2004. Maternal synthesis of abscisic acid controls seed development and yield in *Nicotiana plumbaginifolia*. *Planta* 218:958–64
31. **Gazzarrini S, Tsuchiya Y, Lumba S, Okamoto M, McCourt P. 2004. The transcription factor FUSCA3 controls developmental timing in Arabidopsis through the action of the hormones gibberellins and abscisic acid. *Dev. Cell* 7:73–85**

This article reported that the developmental regulator FUS3 modulates ABA biosynthesis in seed development.

32. Gillard DF, Walton DC. 1976. Abscisic acid metabolism by a cell-free preparation from *Echinocystic lobata* liquid endosperm. *Plant Physiol.* 58:790–95
33. Gonai T, Kawahara S, Tougou M, Satoh S, Hashiba T, et al. 2004. Abscisic acid in the thermoinhibition of lettuce seed germination and enhancement of its catabolism by gibberellin. *J. Exp. Bot.* 55:111–18
34. González-Guzman M, Abia D, Salinas J, Serrano R, Rodríguez PL. 2004. Two new alleles of the *abscisic aldehyde oxidase 3* gene reveal its role in abscisic acid biosynthesis in seeds. *Plant Physiol.* 135:325–33
35. González-Guzmán M, Apostolova N, Bellés JM, Barrero JM, Piqueras P, et al. 2002. The short-chain alcohol dehydrogenase ABA2 catalyzes the conversion of xanthoxin to abscisic aldehyde. *Plant Cell* 14:1833–46
36. Grappin P, Bouinot D, Sotta B, Miginiac E, Jullien M. 2000. Control of seed dormancy in *Nicotiana plumbaginifolia*: post-imbibition abscisic acid synthesis imposes dormancy maintenance. *Planta* 210:279–85

37. Groot SPC, Van Yperen I, Karssen CM. 1991. Strongly reduced levels of endogenous abscisic acid in developing seeds of the tomato mutant *sitiens* do not influence in vivo accumulation of dry matter and storage proteins. *Physiol. Plant* 81:73–78
38. Gusta LV, Ewan B, Reaney MJT, Abrams SR. 1992. The effect of abscisic acid and abscisic acid metabolites on the germination of cress seed. *Can. J. Bot.* 70:1550–55
39. Han S, Kitahata N, Saito T, Kobayashi M, Shinozaki K, et al. 2004. A new leading compound for abscisic acid biosynthesis inhibitor targeting 9-*cis*-epoxycarotenoid dioxygenase. *Bioorg. Med. Chem. Lett.* 14:3033–36
40. Hansen H, Dörffling K. 1999. Changes of free and conjugated abscisic acid and phaseic acid in xylem sap of drought-stressed sunflower plants. *J. Exp. Bot.* 50:1599–605
41. Harrison MA, Walton DC. 1975. Abscisic acid metabolism in water-stressed bean leaves. *Plant Physiol.* 56:250–54
42. Hartung W, Sauter A, Hose E. 2002. Abscisic acid in the xylem: Where does it come from, where does it go to? *J. Exp. Bot.* 53:27–32
43. Hellwege EM, Hartung W. 1997. Synthesis, metabolism and compartmentation of abscisic acid in *Riccia fluitans* L. *J. Plant Physiol.* 150:287–91
44. Hill RD, Liu JH, Durnin D, Lamb N, Shaw A, et al. 1995. Abscisic acid structure-activity relationships in barley aleurone layers and protoplasts—Biological activity of optically active, oxygenated abscisic acid analogs. *Plant Physiol.* 108:573–79
45. Hirai N, Kondo S, Ohigashi H. 2003. Deuterium-labeled phaseic acid and dihydrophaseic acids for internal standards. *Biosci. Biotechnol. Biochem.* 67:2408–15
46. Hirai N, Okamoto M, Koshimizu K. 1986. The 1′,4′-*trans*-diol of abscisic acid, a possible precursor of abscisic acid in *Botrytis cinerea*. *Phytochemistry* 25:1865–68
47. Hirai N, Yoshida R, Todoroki Y, Ohigashi H. 2000. Biosynthesis of abscisic acid by the non-mevalonate pathway in plants, and by the mevalonate pathway in fungi. *Biosci. Biotechnol. Biochem.* 64:1448–58
48. Hirsch R, Hartung W, Gimmler H. 1989. Abscisic acid content of algae under stress. *Bot. Acta.* 102:326–34
49. Inomata M, Hirai N, Yoshida R, Ohigashi H. 2004. Biosynthesis of abscisic acid by the direct pathway *via* ionylideneethane in a fungus, *Cercospora cruenta*. *Biosci. Biotechnol. Biochem.* In press
49a. Inomata M, Hirai N, Yoshida R, Ohigashi H. 2004. The biosynthetic pathway to abscisic acid *via* ionylideneethane in the fungus *Botrytis cinerea*. *Phytochemistry* 65:2667–78

This article reported the identification of allofarnesene and ionylideneethane as the endogenous compounds of *B. cinerea*, suggesting that ABA is synthesized via direct pathway in this fungus.

50. Iuchi S, Kobayashi M, Taji T, Naramoto M, Seki M, et al. 2001. Regulation of drought tolerance by gene manipulation of 9-*cis*-epoxycarotenoid dioxygenase, a key enzyme in abscisic acid biosynthesis in *Arabidopsis*. *Plant J.* 27:325–33
51. Iuchi S, Kobayashi M, Yamaguchi-Shinozaki K, Shinozaki K. 2000. A stress-inducible gene for 9-*cis*-epoxycarotenoid dioxygenase involved in abscisic acid biosynthesis under water stress in drought-tolerant cowpea. *Plant Physiol.* 123:553–62
52. Jacobsen JV, Pearce DW, Poole AT, Pharis RP, Mander LN. 2002. Abscisic acid, phaseic acid and gibberellin contents associated with dormancy and germination in barley. *Physiol. Plant* 115:428–41
53. Jones RJ, Brenner ML. 1987. Distribution of abscisic acid in maize kernels during grain filling. *Plant Physiol.* 83:905–9
54. Jornvall H, Persson B, Krook M, Atrian S, Gonzalez-Duarte R, et al. 1995. Short-chain dehydrogenases/reductases (SDR). *Biochemistry* 34:6003–13

55. Karssen CM, Brinkhorst-van der Swan DLC, Breekland AE, Koornneef M. 1983. Induction of dormancy during seed development by endogenous abscisic acid: Studies on abscisic acid deficient genotypes of *Arabidopsis thaliana* (L) Heynh. *Planta* 157:158–65
56. Kasahara H, Takei K, Ueda N, Hishiyama S, Yamaya T, et al. 2004. Distinct isoprenoid origins of *cis*- and *trans*-zeatin biosyntheses in *Arabidopsis*. *J. Biol. Chem.* 279:14049–54
57. Kato-Noguchi H, Tanaka Y. 2003. Allelopathic potential of citrus fruit peel and abscisic acid-glucose ester. *Plant Growth Regul.* 40:117–20
58. Knight CD, Sehgal A, Atwal K, Wallace JC, Cove DJ, et al. 1995. Molecular responses to abscisic acid and stress are conserved between moss and cereals. *Plant Cell* 7:499–506
59. **Koiwai H, Nakaminami K, Seo M, Mitsuhashi W, Toyomasu T, et al. 2004. Tissue-specific localization of an abscisic acid biosynthetic enzyme, AAO3, in Arabidopsis. *Plant Physiol.* 134:1697–707**

This article reported the localization of AAO3, which catalyzes the final step in ABA biosynthesis.

60. Koornneef M, Jorna ML, Brinkhorst-van der Swan DLC, Karssen CM. 1982. The isolation of abscisic acid (ABA) deficient mutants by selection of induced revertants in non-germinating gibberellin sensitive lines of *Arabidopsis thaliana* (L.) Heynh. *Theor. Appl. Genet.* 61:385–93
61. Krochko JE, Abrams GD, Loewen MK, Abrams SR, Cutler AJ. 1998. (+)-abscisic acid 8′-hydroxylase is a cytochrome P450 monooxygenase. *Plant Physiol.* 118:849–60
62. **Kushiro T, Okamoto M, Nakabayashi K, Yamagishi K, Kitamura S, et al. 2004. The Arabidopsis cytochrome P450 CYP707A encodes ABA 8′-hydroxylases: key enzymes in ABA catabolism. *EMBO J.* 23:1647–56**

This article reported the first identification and characterization of CYP707A genes encoding ABA 8′-hydroxylases.

63. Lehmann H, Glund K. 1986. Abscisic acid metabolism-vascular/extravascular distribution of metabolites. *Planta* 168:559–62
64. Lehmann H, Vlasov P. 1988. Plant growth and stress—the enzymic hydrolysis of abscisic acid conjugate. *J. Plant Physiol.* 132:98–101
65. Léon-Kloosterziel KM, Gil MA, Ruijs GJ, Jacobsen SE, Olszewski NE, et al. 1996. Isolation and characterization abscisic acid-deficient *Arabidopsis* mutants at two new loci. *Plant J.* 10:655–61

65a. Lichtenthaler HK. 1999. The 1-deoxy-D-xylulose-5-phosphate pathway of isoprenoid biosynthesis in plants. *Annu. Rev. Plant Physiol. Plant Mol. Biol.* 50:47–65

66. Lindgren LO, Stalberg KG, Hoglund AS. 2003. Seed-specific overexpression of an endogenous Arabidopsis phytoene synthase gene results in delayed germination and increased levels of carotenoids, chlorophyll, and abscisic acid. *Plant Physiol.* 132:779–85
67. Marin E, Nussaume L, Quesada A, Gonneau M, Sotta B, et al. 1996. Molecular identification of zeaxanthin epoxidase of *Nicotiana plumbaginifolia*, a gene involved in abscisic acid biosynthesis and corresponding to the ABA locus of *Arabidopsis thaliana*. *EMBO J.* 15:2331–42
68. McCarty DR. 1995. Genetic control and integration of maturation and germination pathways in seed development. *Annu. Rev. Plant Physiol. Plant Mol. Biol.* 46:71–93
69. Merlot S, Mustilli AC, Genty B, North H, Lefebvre V, et al. 2002. Use of infrared thermal imaging to isolate Arabidopsis mutants defective in stomatal regulation. *Plant J.* 30:601–9
70. Milborrow BV. 1970. The metabolism of abscisic acid. *J. Exp. Bot.* 21:17–29
71. Milborrow BV, Carrington NJ, Vaughan GT. 1988. The cyclization of 8′-hydroxy abscisic acid to phaseic acid in vivo. *Phytochemistry* 27:757–59
72. Milborrow BV, Lee HS. 1998. Endogenous biosynthetic precursors of (+)-abscisic acid. VI—Carotenoids and ABA are formed by the 'non-mevalonate' triose-pyruvate pathway in chloroplasts. *Aust. J. Plant Physiol.* 25:507–12
73. Nambara E, Hayama R, Tsuchiya Y, Nishimura M, Kawaide H, et al. 2000. The role of *ABI3* and *FUS3* loci in *Arabidopsis thaliana* on phase transition from late embryo development to germination. *Dev. Biol.* 220:412–23

74. Nambara E, Kawaide H, Kamiya Y, Naito S. 1998. Characterization of an *Arabidopsis thaliana* mutant that has a defect in ABA accumulation: ABA-dependent and ABA-independent accumulation of free amino acids during dehydration. *Plant Cell. Physiol.* 39:853–58
75. Neill SJ, Horgan R, Walton DC, Mercer CAM. 1987. The metabolism of α-ionylidene compounds by *Cercospora rosicola*. *Phytochemistry* 26:2515–19
76. Nishiyama T, Fujita T, Shin-I T, Seki M, Nishide H, et al. 2003. Comparative genomics of *Physcomitrella patens* gametophytic transcriptome and *Arabidopsis thaliana*: implication for land plant evolution. *Proc. Natl. Acad. Sci. USA* 100:8007–12
77. Niyogi KK, Grossman AR, Bjorkman O. 1998. Arabidopsis mutants define a central role for the xanthophyll cycle in the regulation of photosynthetic energy conversion. *Plant Cell* 10:1121–34
78. Norman SM. 1991. β-Carotene from the abscisic acid producing strain of *Cercospora rosicola*. *Plant Growth Regul.* 10:103–8
79. Norman SM, Bennett RD, Poling SM, Maier VP, Nelson MD. 1986. Paclobutrazol inhibits abscisic acid biosynthesis in *Cercospora rosicola*. *Plant Physiol.* 80:122–25
80. Okamoto M, Hirai N, Koshimizu K. 1988. Biosynthesis of abscisic acid in *Cercospora pini-densiflorae*. *Phytochemistry* 27:2099–103
81. Oritani T, Kiyota H. 2003. Biosynthesis and metabolism of abscisic acid and related compounds. *Nat. Prod. Rep.* 20:414–25
82. Oritani T, Yamashita K. 1985. Biosynthesis of (+)-abscisic acid in *Cercospora cruenta*. *Agric. Biol. Chem.* 49:245–49
83. Qin XQ, Zeevaart JAD. 1999. The 9-*cis*-epoxycarotenoid cleavage reaction is the key regulatory step of abscisic acid biosynthesis in water-stressed bean. *Proc. Natl. Acad. Sci. USA* 96:15354–61
84. Qin XQ, Zeevaart JAD. 2002. Overexpression of a 9-*cis*-epoxycarotenoid dioxygenase gene in *Nicotiana plumbaginifolia* increases abscisic acid and phaseic acid levels and enhances drought tolerance. *Plant Physiol.* 128:544–51
85. Quesada V, Ponce MR, Micol JL. 2000. Genetic analysis of salt-tolerant mutants in *Arabidopsis thaliana*. *Genetics* 154:421–36
86. Raz V, Bergervoet JHW, Koornneef M. 2001. Sequential steps for developmental arrest in Arabidopsis seeds. *Development* 128:243–52
87. Razem FA, Luo M, Liu JH, Abrams SR, Hill RD. 2004. Purification and characterization of a barley aleurone abscisic acid-binding protein. *J. Biol. Chem.* 279:9922–29
88. Robertson AJ, Reaney MJT, Wilen RW, Lamb N, Abrams SR, et al. 1994. Effects of abscisic acid metabolites and analogs on freezing tolerance and gene expression in Bromegrass (*Bromus inermis* Leyss) cell cultures. *Plant Physiol.* 105:823–30
89. Rodríguez-Concepción M, Boronat A. 2002. Elucidation of the methylerythritol phosphate pathway for isoprenoid biosynthesis in bacteria and plastids. A metabolic milestone achieved through genomics. *Plant Physiol.* 130:1079–89
89a. Rohmer M, Knani M, Simonin P, Sutter B, Sahm H. 1993. Isoprenoid biosynthesis in bacteria: a novel pathway for early steps leading to isopentenyl diphosphate. *Biochem. J.* 295:517–24
90. Ronen G, Carmel-Goren L, Zamir D, Hirschberg J. 2000. An alternative pathway to β-carotene formation in plant chromoplasts discovered by map-based cloning of *beta* and *old-gold* color mutations in tomato. *Proc. Natl. Acad. Sci. USA* 97:11102–7
91. Rook F, Corke F, Card R, Munz G, Smith C, et al. 2001. Impaired sucrose-induction mutants reveal the modulation of sugar-induced starch biosynthetic gene expression by abscisic acid signalling. *Plant J.* 26:421–33

92. Sagi M, Scazzocchio C, Fluhr R. 2002. The absence of molybdenum cofactor sulfuration is the primary cause of the *flacca* phenotype in tomato plants. *Plant J.* 31:305–17

93. Saito S, Hirai N, Matsumoto C, Ohigashi H, Ohta D, et al. 2004. Arabidopsis CYP707As encode (+)-abscisic acid 8′-hydroxylase, a key enzyme in the oxidative catabolism of abscisic acid. *Plant Physiol.* 134:1439–49

This article reported the first identification and characterization of CYP707A genes encoding ABA 8′-hydroxylases.

94. Sauter A, Davies WJ, Hartung W. 2001. The long-distance abscisic acid signal in the droughted plant: the fate of the hormone on its way from root to shoot. *J. Exp. Bot.* 52:1991–97

95. Sauter A, Dietz KJ, Hartung W. 2002. A possible stress physiological role of abscisic acid conjugates in root-to-shoot signalling. *Plant Cell Environ.* 25:223–28

96. Sauter A, Hartung W. 2000. Radial transport of abscisic acid conjugates in maize roots: its implication for long distance stress signals. *J. Exp. Bot.* 51:929–35

97. Schmitz N, Abrams SR, Kermode AR. 2002. Changes in ABA turnover and sensitivity that accompany dormancy termination of yellow-cedar (*Chamaecyparis nootkatensis*) seeds. *J. Exp. Bot.* 53:89–101

98. Schwartz SH, Qin X, Zeevaart JA. 2003. Elucidation of the indirect pathway of abscisic acid biosynthesis by mutants, genes, and enzymes. *Plant Physiol.* 131:1591–601

99. Schwartz SH, Tan BC, Gage DA, Zeevaart JA, McCarty DR. 1997. Specific oxidative cleavage of carotenoids by VP14 of maize. *Science* 276:1872–74

99a. Seo M, Aoki H, Koiwai H, Kamiya Y, Nambara E, Koshiba T. 2004. Comparative studies on the *Arabidopsis* aldehyde oxidase (*AAO*) gene family revealed a major role of *AAO3* in ABA biosynthesis in seeds. *Plant Cell Physiol.* 45:1694–703

100. Seo M, Koiwai H, Akaba S, Komano T, Oritani T, et al. 2000. Abscisic aldehyde oxidase in leaves of *Arabidopsis thaliana*. *Plant J.* 23:481–88

101. Seo M, Koshiba T. 2002. Complex regulation of ABA biosynthesis in plants. *Trends Plant Sci.* 7:41–48

102. Seo M, Peeters AJM, Koiwai H, Oritani T, Marion-Poll A, et al. 2000. The *Arabidopsis aldehyde oxidase 3* (*AAO3*) gene product catalyzes the final step in abscisic acid biosynthesis in leaves. *Proc. Natl. Acad. Sci. USA* 97:12908–13

103. Setha S, Kondo S, Hirai N, Ohigashi H. 2004. Xanthoxin, abscisic acid and its metabolite levels associated with apple fruit development. *Plant Sci.* 166:493–99

104. Siewers V, Smedsgaard J, Tudzynski P. 2004. The P450 monooxygenase BcABA1 is essential for abscisic acid biosynthesis in *Botrytis cinerea*. *Appl. Environ. Microb.* 70:3868–76

This article reported the first identification of fungal ABA biosynthetic gene from *B. cinerea*.

105. Sindhu RK, Walton DC. 1987. Conversion of xanthoxin to abscisic acid by cell-free preparations from bean leaves. *Plant Physiol.* 85:916–21

106. Takaichi S, Mimuro M. 1998. Distribution and geometric isomerism of neoxanthin in oxygenic phototrophs: 9′-*Cis*, a sole molecular form. *Plant Cell. Physiol.* 39:968–77

107. Tan BC, Cline K, McCarty DR. 2001. Localization and targeting of the VP14 epoxycarotenoid dioxygenase to chloroplast membranes. *Plant J.* 27:373–82

108. Tan BC, Joseph LM, Deng WT, Liu LJ, Li QB, et al. 2003. Molecular characterization of the *Arabidopsis* 9-*cis* epoxycarotenoid dioxygenase gene family. *Plant J.* 35:44–56

109. Tan BC, Schwartz SH, Zeevaart JA, McCarty DR. 1997. Genetic control of abscisic acid biosynthesis in maize. *Proc. Natl. Acad. Sci. USA* 94:12235–40

110. Thompson AJ, Jackson AC, Parker RA, Morpeth DR, Burbidge A, et al. 2000. Abscisic acid biosynthesis in tomato: regulation of zeaxanthin epoxidase and 9-*cis*-epoxycarotenoid dioxygenase mRNAs by light/dark cycles, water stress and abscisic acid. *Plant Mol. Biol.* 42:833–45

111. Thompson AJ, Jackson AC, Symonds RC, Mulholland BJ, Dadswell AR, et al. 2000. Ectopic expression of a tomato 9-*cis*-epoxycarotenoid dioxygenase gene causes over-production of abscisic acid. *Plant J.* 23:363–74
112. Tian L, DellaPenna D, Zeevaart JAD. 2004. Effects of hydroxylated carotenoid deficiency on ABA accumulation in *Arabidopsis*. *Physiol Plant* 122:314–20
113. Todoroki Y, Hirai N, Koshimizu K. 1995. 8′,8′-Difluoroabscisic acid and 8′,8′,8′-trifluoroabscisic acid as highly potent, long-lasting analogs of abscisic acid. *Phytochemistry* 38:561–68
114. Todoroki Y, Hirai N, Ohigashi H. 2000. Analysis of isomerization process of 8′-hydroxy-abscisic acid and its 3′-fluorinated analog in aqueous solutions. *Tetrahedron* 56:1649–53
115. Todoroki Y, Sawada M, Matsumoto M, Tsukada S, Ueno K, et al. 2004. Metabolism of 5′ alpha,8′-cycloabscisic acid, a highly potent and long-lasting abscisic acid analogue, in radish seedlings. *Bioorgan. Med. Chem.* 12:363–70
116. Walton DC, Li Y. 1995. Abscisic acid biosynthesis and metabolism. In *Plant Hormones*, ed. PJ Davis, pp. 140–57. Dordrecht: Kluwer Acad.
117. White CN, Proebsting WM, Hedden P, Rivin CJ. 2000. Gibberellins and seed development in maize. I. Evidence that gibberellin/abscisic acid balance governs germination versus maturation pathways. *Plant Physiol.* 122:1081–88
118. Wilkinson S, Davies WJ. 2002. ABA-based chemical signalling: the co-ordination of responses to stress in plants. *Plant Cell Environ.* 25:195–210
119. Xiong L, Lee H, Ishitani M, Zhu JK. 2002. Regulation of osmotic stress-responsive gene expression by the *LOS6/ABA1* locus in *Arabidopsis*. *J. Biol. Chem.* 277:8588–96
120. Xiong L, Zhu JK. 2003. Regulation of abscisic acid biosynthesis. *Plant Physiol.* 133:29–36
121. **Xiong LM, Gong ZZ, Rock CD, Subramanian S, Guo Y, et al. 2001. Modulation of abscisic acid signal transduction and biosynthesis by an Sm-like protein in Arabidopsis. *Dev. Cell* 1:771–81**

This article reported that the expression of *AAO3* and *AtABA3* genes is regulated by SAD1, an Sm-like RNA binding protein.

122. Xiong LM, Ishitani M, Lee H, Zhu JK. 2001. The Arabidopsis *LOS5/ABA3* locus encodes a molybdenum cofactor sulfurase and modulates cold stress- and osmotic stress-responsive gene expression. *Plant Cell* 13:2063–83
123. Xu ZJ, Nakajima M, Suzuki Y, Yamaguchi I. 2002. Cloning and characterization of the abscisic acid-specific glucosyltransferase gene from adzuki bean seedlings. *Plant Physiol.* 129:1285–95
124. Yamamoto H, Inomata M, Tsuchiya S, Nakamura M, Uchiyama T, et al. 2000. Early biosynthetic pathway to abscisic acid in *Cercospora cruenta*. *Biosci. Biotechnol. Biochem.* 64:2075–82
125. Yoshida K, Igarashi E, Mukai M, Hirata K, Miyamoto K. 2003. Induction of tolerance to oxidative stress in the green alga, *Chlamydomonas reinhardtii*, by abscisic acid. *Plant Cell Environ.* 26:451–57
126. Zeevaart JAD. 1980. Changes in the levels of abscisic acid and its metabolites in excised leaf blades of *Xanthium strumarium* during and after water stress. *Plant Physiol.* 66:672–78
127. Zeevaart JAD. 1999. Abscisic acid metabolism and its regulation. In *Biochemistry and Molecular Biology of Plant Hormones*, ed. PJJ Hooykaas, MA Hall, KP Libbenga, 8:189–207. Amsterdam: Elsevier Science
128. Zeevaart JAD, Creelman RA. 1988. Metabolism and physiology of abscisic acid. *Annu. Rev. Plant Physiol. Plant Mol. Biol.* 39:439–73
129. Zhang DP, Chen SW, Peng YB, Shen YY. 2001. Abscisic acid-specific binding sites in the flesh of developing apple fruit. *J. Exp. Bot.* 52:2097–103

130. Zhou R, Cutler AJ, Ambrose SJ, Galka MM, Nelson KM, et al. 2004. A new abscisic acid catabolic pathway. *Plant Physiol.* 134:361–69
131. Zou JT, Abrams GD, Barton DL, Taylor DC, Pomeroy MK, et al. 1995. Induction of lipid and oleosin biosynthesis by (+)-abscisic acid and its metabolites in microspore-derived embryos of *Brassica napus* L. Cv Reston—Biological responses in the presence of 8′-hydroxyabscisic acid. *Plant Physiol.* 108:563–71

Redox Regulation: A Broadening Horizon

Bob B. Buchanan and Yves Balmer

Department of Plant and Microbial Biology, University of California, Berkeley, California 94720; email: view@nature.berkeley.edu

Annu. Rev. Plant Biol.
2005. 56:187–220

doi: 10.1146/
annurev.arplant.56.032604.144246

First published online as a Review in Advance on January 27, 2005

1543-5008/05/0602-0187$20.00

Key Words

thioredoxin, glutaredoxin, redox signaling

Abstract

Initially discovered in the context of photosynthesis, regulation by change in the redox state of thiol groups (S–S ↔ 2SH) is now known to occur throughout biology. Several systems, each linking a hydrogen donor to an intermediary disulfide protein, act to effect changes that alter the activity of target proteins: the ferredoxin/thioredoxin system, comprised of reduced ferredoxin, a thioredoxin, and the enzyme, ferredoxin-thioredoxin reductase; the NADP/thioredoxin system, including NADPH, a thioredoxin, and NADP-thioredoxin reductase; and the glutathione/glutaredoxin system, composed of reduced glutathione and a glutaredoxin. A related disulfide protein, protein disulfide isomerase (PDI) acts in protein assembly. Regulation linked to plastoquinone and signaling induced by reactive oxygen species (ROS) and other agents are also being actively investigated. Progress made on these systems has linked redox to the regulation of an increasing number of processes not only in plants, but in other types of organisms as well. Research in areas currently under exploration promises to provide a fuller understanding of the role redox plays in cellular processes, and to further the application of this knowledge to technology and medicine.

Contents

INTRODUCTION

Twenty-five years ago, one of us (BBB) wrote a review that described the role of light in the regulation of chloroplast enzymes (27). At that time, the disulfide protein thioredoxin had recently been found to function in enzyme regulation, initially with chloroplasts (199). Thioredoxin that had been reduced photochemically with ferredoxin via the enzyme ferredoxin-thioredoxin reductase (FTR) was found to activate (or deactivate) chloroplast enzymes by reduction of disulfide bonds—elements then considered to play a structural or a catalytic role, but not known to function in regulation. Originally demonstrated to target fructose-1,6-bisphosphatase and other regulatory members of the Calvin cycle, thioredoxin had then been linked to ATP synthesis and the oxidative pentose phosphate cycle. It was clear that thioredoxin enabled the chloroplast to minimize so-called futile cycling and maximize use of available energy resources.

The 1980 review covered basic principles of a new regulatory mechanism (called the "ferredoxin-thioredoxin system") and its coordination with other light-actuated reactions of chloroplasts that collectively functioned in modulating enzyme activity, i.e., changes in pH and in the concentration of Mg^{2+} and metabolites. In extending the role of light, this review helped set in motion change in a longstanding concept—namely, that the sole contribution of light in photosynthesis was to provide the reductant (NADPH) and energy (ATP) for the CO_2 assimilation reactions. Ensuing developments led to a revision in widely used terminology and replacement of the term "dark reactions" by "carbon reactions" in the scientific literature and to an increasing extent in textbooks, which, hopefully, will eventually include those used in high schools (125).

When the 1980 review was written, it seemed likely that thioredoxin would play a significant role in regulating the Calvin cycle and related processes, but it was unclear whether its function would extend beyond chloroplasts. The NADP-linked thioredoxin system had been described for a plant source, flour from wheat endosperm. However, as in other organisms, the evidence suggested that the NADP-linked mechanism functioned primarily as an enzyme substrate—i.e., in the reduction of ribonucleotides (183). It was, therefore, uncertain whether the horizon for redox regulation would remain relatively narrow and be confined to photosynthesis, or whether it would extend to other processes of the cell.

The redox regulation field has expanded and its horizon has broadened immeasurably since 1980. Regulation by redox now embraces virtually every type of living organism. Perhaps the greatest strides have been made with animal systems where an exponentially expanding literature links redox not only to the expression of key genes (6), but also to a spectrum of human diseases (73, 139). Impressive progress has also been made with plants. A number of timely reviews summarizing progress on the regulatory role of redox, especially with respect to plant thioredoxins, were written in the period leading up to 2000, when an excellent update listed the major reviews published up to that time (172). More recently, progress made on plant thioredoxins is described in a number of pieces on individual topics (1, 9, 16, 24, 35, 40, 44, 46, 47, 55, 64, 88, 103, 117, 120, 133, 150, 151, 167, 170, 171, 188, 194) and an historical account has been published (32). The reviews covered major recent achievements at that time. Especially noteworthy was the elucidation of the tertiary structure of the specific members of the ferredoxin/thioredoxin system: FTR (44–46); thioredoxins *f* (34) and *m* (34, 112); and three target enzymes: fructose bisphosphatase (38, 192), NADP-malate dehydrogenase (36, 93), and NADP-glyceraldehyde 3-phosphate dehydrogenase (213). The determination of the redox potentials of the active disulfide groups of participating proteins also strengthened our understanding by giving a physical basis for the mechanism of enzyme regulation by the process now known as thiol/disulfide exchange or thiol modulation (80, 103, 170, 177). Structures of members of the NADP/thioredoxin system—NADP-thioredoxin reductase (NTR) (43) and thioredoxin *h* (134)-have also been elucidated.

A number of developments have, however, taken place since a broad review was written. The growth of the thioredoxin family has continued and now includes thioredoxins *x*, *y* and *o* in addition to the longer known *f*, *m*, and *h* types. One long-awaited result was genetic evidence for the requirement of the ferredoxin/thioredoxin system in plants. Functions of thioredoxin-like proteins have also surfaced in unexpected places. Furthermore, although described for bacteria and animals, our knowledge of glutaredoxins in plants was scant until recently (116, 163). Finally, in the first 25 years following their description in chloroplasts, the function of thioredoxins appeared to be linked to a limited number of major proteins. The advent of proteomics has broadened their role to include processes that were beyond imagination in 1980. Totally new research areas have also developed such that redox now seems to intervene in the regulation of multiple processes functional at virtually every stage of plant development.

The purpose of the present account is to build on the 2000 review (172) and relate themes stemming from the early thioredoxin research on photosynthesis to new developments implicating a general role for redox in regulating plant processes. In describing this work, we will first provide background information on the interacting disulfide groups and participating proteins that link redox to the regulation of myriad enzymatic reactions throughout the plant—i.e., glutaredoxins, thioredoxins, thioredoxin-related proteins, and protein disulfide isomerases (PDIs). We will then describe regulation associated with other types of redox changes. Rather than writing an all-inclusive survey, we will highlight key events in the developing areas in an attempt to provide a synoptic overview of the role of redox in regulation throughout the plant.

Thioredoxin: a small protein, reduced enzymatically by NADPH or ferredoxin, that is active in thiol/disulfide exchange and results in regulation or substrate conversion

Redox regulation: a reversible posttranslational alteration in the properties of a protein, typically the activity of an enzyme, as a result of change in its oxidation state. Regulation is independent of terminal oxidation—an irreversible reaction that marks proteins for degradation.

Glutaredoxin: a small protein, reduced by glutathione, that is active in thiol/disulfide exchange and results in regulation or substrate conversion

NATURE AND FUNCTION OF DISULFIDES

Types of Disulfide Groups

Disulfide bonds are of three types, depending of the origin of the two participating cysteines: intra, inter, and mixed. A major S–S group is the intramolecular type formed between two cysteines of the same polypeptide. The two interacting amino acids can be positioned either nearby in the primary sequence or far apart. In the latter case, participating amino acids brought together by secondary and tertiary folding of the protein can form a disulfide bridge. Another linkage, interdisulfide, is located between thiols of two different polypeptides (in most cases identical subunits), which yields a covalent dimer. The third type, a variant of the interdisulfide category, is a mixed disulfide where a cysteine reacts with its counterpart of an oxidized glutathione molecule. Known as glutathionylation, this mechanism has been extensively studied in animal systems, but is only now being investigated in plants.

Function of Disulfides

Covalent postranslational modification, such as phosphorylation, usually results in a change in enzyme activity or in signaling (101). By contrast, formation of a disulfide bond may be purely structural and act to stabilize the protein by maintaining its tertiary structure. In plants, this stabilizing type of covalent modification is especially prominent in secretory and storage proteins, where it helps prevent denaturation and decrease susceptibility to proteolytic degradation.

In addition to stabilizing structure, a disulfide bond can be reversibly oxidized (S–S) and reduced (–SH HS–). This change in redox state may result in either a catalytic or regulatory change (27, 82, 207). Catalytic disulfides are often formed between two cysteines separated by one or two amino acids. This redox active site functions either as an electron (hydrogen) donor or acceptor. Though it took longer to recognize, the interconversion of thiols to a disulfide equally provides a mechanism for the regulation of catalytic activity. In this case, formation of a disulfide can physically block the enzyme active site as for chloroplast phosphoribulokinase (23). Alternatively, if the regulatory disulfide is far from the catalytic site, reduction may produce a small structural shift to affect the catalytic properties of the enzyme as exemplified by chloroplast fructose-1,6-bisphosphatase (38) or NADP-malate dehydrogenase (107). Here, the regulatory disulfide is located more than 20 Å from the active site on the opposite side of the protein.

Finally, a third function of the S–S bond is the capability for glutathionylation, even if, in some cases, the formation of an intra- or interdisulfide could theoretically achieve the same result, i.e., the protection of SH groups. Indeed, thiols can be oxidized to sulfinic (SO_2H) and cysteic acids (SO_3H) as a result of oxidative stress in reactions that appear irreversible. Damage of this type, which generally results in proteolysis of the affected protein, can be prevented by the formation of a mixed disulfide with glutathione. It is noteworthy that stress-induced disulfide formation also appears to play a role in enzyme regulation of an oxidative type, as recently proposed for chloroplasts (10).

PROTEINS ACTIVE IN FORMATION AND CLEAVAGE OF DISULFIDES

Thioredoxins

Plant thioredoxins are the subject of numerous recent reviews and, thus, only relevant aspects will be discussed in this article (9, 16, 32, 64, 88, 172). Thioredoxins are small proteins with a characteristic structural motif—4 alpha helixes surrounding a beta sheet composed of 5 strands—common to most, if not all, protein disulfide oxidoreductase family members. Thioredoxins appear to be present in all living cells with the exception of a fastidious human pathogen (155). In plants, unlike bacteria and animals, a large number of genes encode thioredoxins—19 different thioredoxins have been identified in the genome of *Arabidopsis thaliana* (**Figure 1**) that can be grouped in 6 subfamilies (117, 118, 132). Chloroplasts contain 4 types of thioredoxin—*f*, *m*, *x*, and *y*—whereas thioredoxin *o* is located in mitochondria. The *h* representatives are distributed in multiple cell compartments: cytosol, nucleus, endoplasmic

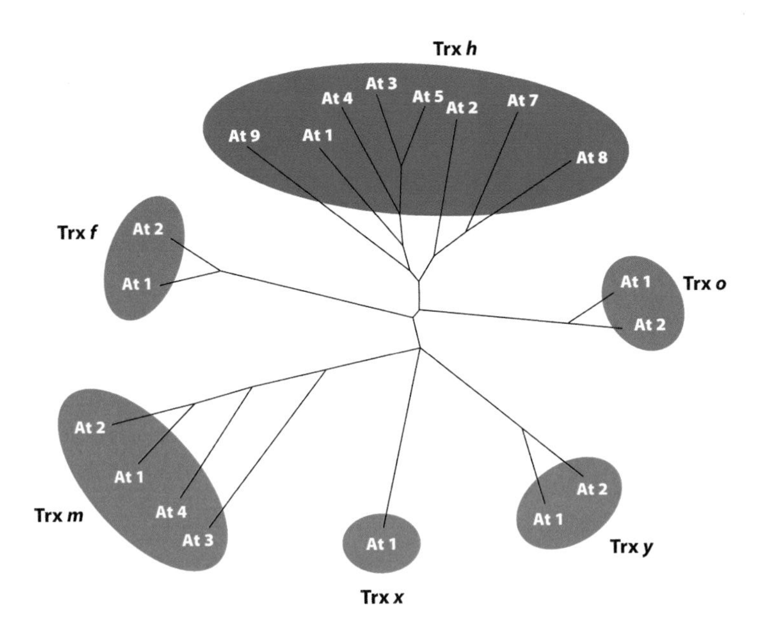

Figure 1

Patterns of sequence similarity showing the relationship of members of the thioredoxin family of *Arabidopsis thaliana*. Thioredoxins *f*, *m*, *x*, and *y* are found in the chloroplast, *o* and *h* types At7 and At8 in the mitochondrion. Other forms of thioredoxin *h* have been localized in the cytosol and ER. The SwissProt ID numbers are: Trx *h*: At1, P29448; At2, Q38879; At3, Q42403; At4, Q39239; At5, Q39241; At7, Q9XIF4; At8, Q9CAS1; At9, Q9C9Y6. Trx *o*: At1, O64764; At2, Q39VQ9. Trx *y*: At1, Q9SRD7; At2, Q8L7S9. Trx *x*: At1, Q9SEU5. Trx *m*: At1, O48737; At2, Q9SEU8; At3, Q9SEU7; At4, 9SEU6. Trx *f*: At1, Q9XFH8; At2, Q9XFH9. Adapted from Reference 120.

reticulum, as well as mitochondria (57, 62, 64, 132, 174, 175). The evidence suggests that individual organs express characteristic members of the thioredoxin *h* family (154). Thioredoxins contain a conserved redox active site with the sequence WC[G/P]PC located at the periphery of the protein. In plants, the disulfide formed between the two cysteines is reduced in chloroplasts by ferredoxin and FTR and in the other cell compartments by NADPH via a flavin enzyme, NTR. By reducing disulfide groups, thioredoxins, in turn, function as electron (hydrogen) donors for the regulation of enzymes, such as those of chloroplasts, or for the reduction of chemical substrates, such as ribonucleotides or methionine sulfoxide.

Glutaredoxins

Glutaredoxins are small proteins, closely related to thioredoxins, that exhibit a similar overall 3D structure, but have a more positive redox potential and, unlike typical thioredoxins, are reduced by glutathione (116, 163). With 31 members, glutaredoxins represent a more heterogeneous group than thioredoxins (**Figure 2**). The plant glutaredoxins first described (here called "classical glutaredoxins") contain a redox active disulfide with the conserved sequence, CPYC. However, analysis of the *Arabidopsis thaliana* genome reveals a number of glutaredoxins and glutaredoxin-like proteins that show variation in their active site, in some cases with replacement of the second cysteine by a serine (116, 163). Phylogenic analysis of the *Arabidopsis* glutaredoxin representatives reveals three clusters with distinct active sites: a classical type with the typical CPYC disulfide, a CC type with two contiguous cysteines in the motif CCxC/S/G, and finally a C type with a single cysteine in the conserved sequence CGFS (**Figure 2**) (116, 163). Recent

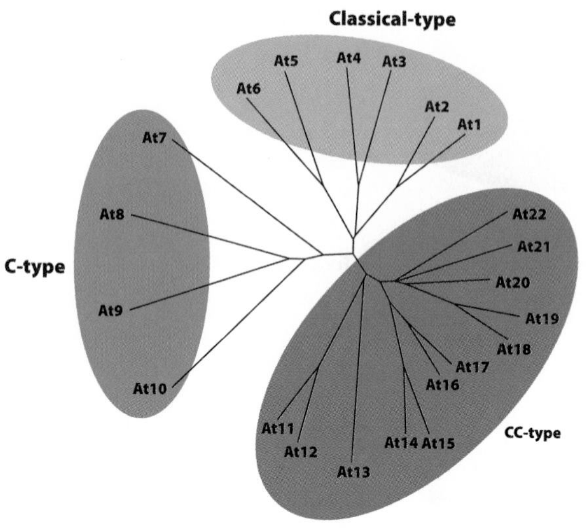

Figure 2

Patterns of sequence similarity of members of the glutaredoxin family of *Arabidopsis thaliana*. Classical (CPYC), C, and CC types are shown. The protein sequences fall into 22 clusters. Consistent with the proteins originally described for *E. coli* and humans (193), it is assumed that plant glutaredoxins have a low molecular weight (ca. 12 to 15 kDa) and contain the first cysteine of the active site. The SwissProt ID numbers are: At1, Q8RXH1; At2, Q9FNE2; At3, Q9SK75; At4, Q8GWS0; At5, Q9M457; At6, Q9FVX1; At7, O80451; At8, Q9SV38; At9, Q9LW13; At10, Q9ZPH2; At11, Q8L9S3; At12, Q9LYI0; At13, Q9SGP6; At14, Q9SA68; AT15, Q9LYC8; At16, Q8L8Z8; At17, O23420; At18, Q9S7X8; At19, Q9FMU3; At20, Q9LIF1; At21, O82254; At22, O82255. Adapted from References 116, 120, 163.

experiments indicate that only the first cysteine of these latter representatives is essential for catalytic activity with peroxiredoxin (161, 162). Additionally, data showed that a yeast glutaredoxin with a CGFS motif exhibits a redox active disulfide formed with a second cysteine located 50 amino acids upstream of the active site (185).

Peroxiredoxin: a small protein linked to reduced thioredoxin or glutaredoxin that functions as a peroxidase to remove hydrogen peroxide

In plants, relatively little is known of the large glutaredoxin family of protein disulfide oxidoreductases. In addition to interacting with peroxiredoxins, one of the proposed functions is the cleavage of mixed disulfides formed between cysteines of a given protein and glutathione. Additional studies are required to clarify the respective specific functions and redundancy of the thioredoxin and glutaredoxin systems as well as the interplay between the two. In the latter context, a poplar thioredoxin *h* (related to *Arabidopsis* thioredoxin *h9*) was reduced by glutathione and glutaredoxin rather than NADPH and NTR (63). It has long been known that, while preferring glutaredoxin, ribonucleotide reductase can use either glutaredoxin or thioredoxin as hydrogen donor (83)—a feature recently extended to a peroxiredoxin (164).

Protein Disulfide Isomerases

PDIs are a family of oxidoreductases typically containing two, or in some cases more, thioredoxin domains involved in the formation of S–S bonds (60, 88, 120). Analysis of the *Arabidopsis* genome reveals nine sequences with the conserved active site PWCGHC, four of which cluster in one major group (classical type) whereas the others form rather divergent subfamilies (**Figure 3**) (120). When the cysteines of the PDI active site are in the disulfide form, the enzyme may be involved in oxidizing target dithiols during protein folding. In its reduced form, PDI more likely functions in the reshuffling of incorrect S–S bridges (60). Thus, whereas thioredoxin and glutaredoxin seem to function mainly in the reduction of disulfides in regulation, PDI acts primarily as a dithiol oxidase in protein folding.

Other Disulfide Proteins

A variety of other proteins is likely involved in thiol-disulfide interconversion. A search of sequence databases reveals a large number of thioredoxin-like proteins (proteins that show some homology with thioredoxin) and polypeptides that contain a thioredoxin domain. An important case in point is a drought-induced chloroplast protein of 32 kDa (CDSP32) that is composed of two thioredoxin modules and is expressed under oxidative stress (26). Recently, an HSP70-interacting protein (HIP) that contains a functional thioredoxin domain was identified in *Arabidopsis* (191, 196). Additionally, bacteria possess a membrane protein at the cytoplasm/periplasm interface (DsbD) that transfers electrons from cytoplasmic thioredoxin to DsbC, a disulfide protein located in the periplasm, via a succession of thiol-disulfide exchanges (99, 158). DsbC appears to be involved in the formation of properly folded disulfide proteins by isomerization. Similarly, another protein at the same membrane interface (DsbB) transfers electrons from the dithiol of periplasmic DsbA to oxygen via ubiquinone and cytochrome oxidase. In this way, DsbA acts as an oxidant for sulfhydryl groups of other proteins, thereby facilitating the formation of disulfide bonds in the periplasmic space (98, 99). A protein with significant homology to DsbD (CcdA-like protein) was recently identified in the *Arabidopsis* genome (138, 144). The CcdA mechanism appears to represent a means whereby reducing equivalents are transferred from the chloroplast stroma to the thylakoid lumen—a research area that, as seen below, has become quite active. Functions of other thioredoxin-like proteins are beginning to surface in plants. HCF164 of the thylakoid lumen is required for the formation of the cytochrome b_6/f complex (121) and CITRX, considered of cytoplasmic origin, is implicated in viral resistance in tomato (159).

REVERSIBLE ACTIVATION/ DEACTIVATION OF ENZYMES IN RESPONSE TO LIGHT/DARK CONDITIONS

The reversible activation/deactivation of chloroplast enzymes and other proteins in response to light and dark are linked both to photosystem I (PSI) and photosystem II (PSII). The mechanisms linked to each of these systems are summarized below.

REGULATION LINKED TO PHOTOSYSTEM I

The mechanism of redox regulation first described, the ferredoxin/thioredoxin system, is linked to PSI (**Figure 4**, upper scheme). Since the first report almost 40 years ago (30, 32), an impressive body of biochemical and physiological evidence has accumulated for thioredoxin function in chloroplasts. On the other hand, although obtained early on with cyanobacteria (137, 141), supporting genetic evidence has been limited for plants. Quite recently, however, relevant results were obtained in two lines of research with *Arabidopsis*. First, inactivation of one of the two variable FTR subunit genes demonstrated that the system is required for growth when plants are oxidatively stressed, as well as for the *in planta* activation of a classical thioredoxin-linked enzyme, NADP-malate dehydrogenase, especially under high light or oxygen stress (102). Similar results were independently obtained with plants having a partially inactivated gene for the PSI-D subunit that serves as the PSI docking site for ferredoxin (75). These two studies add a genetic dimension to the other lines of evidence supporting a major role for the ferredoxin/thioredoxin system in chloroplasts. It should be mentioned that, although the mechanism by which the light signal is transmitted sequentially from ferredoxin to FTR and thioredoxin (44–46) and

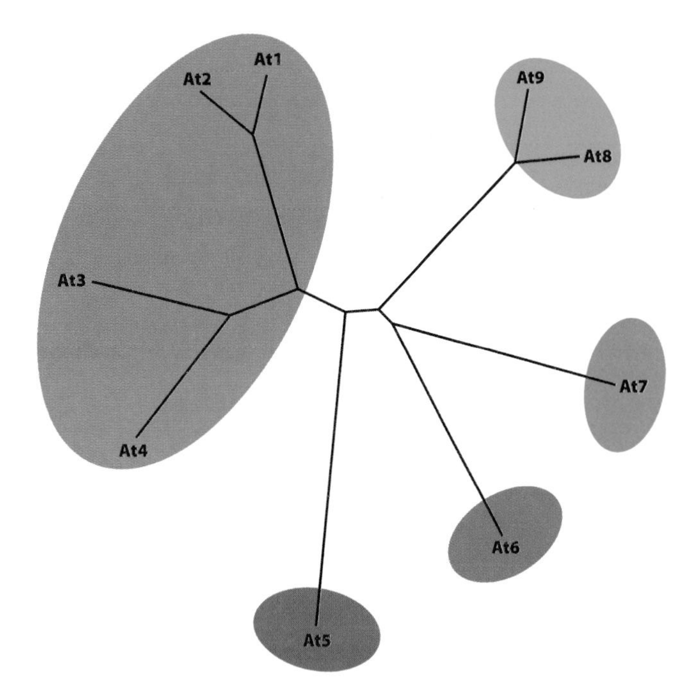

Figure 3

Patterns of sequence similarity of members of the protein disulfide isomerase family of *Arabidopsis thaliana*. Although all members show the conserved site, PWCGHC, their sequences fall into five clusters. The SwissProt ID numbers are: At1, Q9XI01; At2, Q9SRG3; At3, Q8LAM5; At4, Q8VX13; At5, Q9LQG5; At6, O22263; At7, O48773; At8, Q9MAU6; At9, Q8GYD1. Adapted from Reference 120.

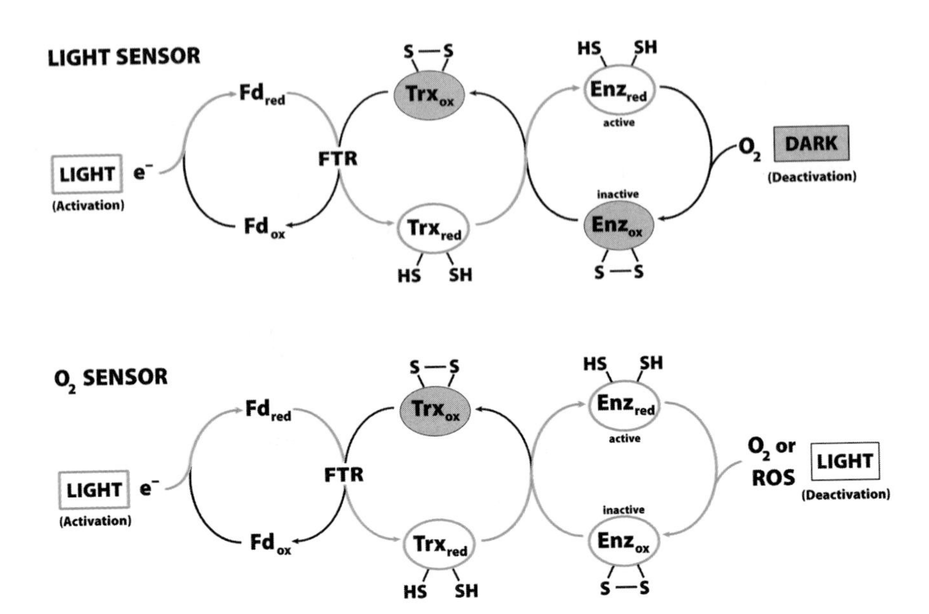

Figure 4

Role of the ferredoxin/thioredoxin system in chloroplasts. The classical role of thioredoxin as an "eye" in sensing light is shown in the upper part of the figure. The more recently recognized role in oxidative regulation is shown in the lower part. Under some conditions, oxidation could occur via a thioredoxin-linked peroxiredoxin.

then to classical target enzymes was described in a series of creative experiments (12, 23, 77, 107, 179, 211), details of how the activated enzymes are deactivated are not fully clear. Available evidence indicates that, in the main, the sulfhydryls formed on reduction of the regulatory disulfides are transferred to oxygen, either directly or, more efficiently, via the oxidized form of thioredoxin (169). In some cases, this oxidation may be linked to a peroxiredoxin.

The chloroplast thioredoxin system was long considered to be comprised of two thioredoxins (*f* and *m*) in addition to ferredoxin and FTR. Recent genome research has added two additional thioredoxin members (thioredoxins *x* and *y*) (118, 132) (**Figure 1**). The experimental evidence suggests that thioredoxins *f* and *m* represent the major chloroplast forms and *x* and *y*, whose functions are now being studied, are of lower abundance. As discussed elsewhere, *f* and *m* show some degree of specificity *in vitro*, *f* being linked to Calvin cycle enzymes and *m* to either glucose 6-phosphate dehydrogenase or, under certain conditions, to NADP-malate dehydrogenase (27, 28). Similarly, recent studies suggest some degree of specificity of thioredoxin *x* and *y* for peroxiredoxin (41; M. Miginiac-Maslow, personal communication). The finding of the two additional chloroplast thioredoxins raises new questions with respect both to function and specificity in vivo. The role of thioredoxin in chloroplasts is further enriched by the recent finding of a novel NTR that has an associated thioredoxin domain (176). Although the NTR and thioredoxin components show the respective independent activities, it appears that the two do not interact directly. The protein seems to function in oxidative stress.

A comment seems in order on the study with the inactivated FTR variable subunit gene in which mutant plants appeared normal until stressed (102). The results indicate that there is sufficient overlap in function of the two subunit isoforms for one to serve the needs of the plant under normal conditions. Thus, only when faced with adversity, i.e., oxidative stress, is the mutant plant with only one of the subunits unable to cope, with the result that a new phenotype develops. It will be of interest to find out if this response applies generally to studies on the inactivation of individual members of multigene families relating to thioredoxin.

Thioredoxin Targets

One limitation in our understanding of the chloroplast thioredoxin system has been knowledge of target enzymes. In the first 25 years following its discovery in chloroplasts, thioredoxin was found to target approximately 16 proteins functional in 8 different processes (10). The advent of proteomics and the capability to trap potential thioredoxin targets by mutant affinity chromatography (136) has led to prodigious advances in our understanding of thioredoxin function in land plants as well as algae (*Chlamydomonas*) and cyanobacteria (*Synechocystis*). In land plants, 27 potential new chloroplast targets were added, increasing the number from 16 to 43 (10, 136). The chloroplast proteins were found to function in 5 established processes (11 targets) and in 10 processes not previously known to be thioredoxin-linked (15 targets) (**Figure 5**). The new processes range from DNA replication, transcription, and plastid division to metabolism and biosynthesis, protein assembly/folding, and degradation, to HCO_3^-/CO_2 equilibration. Significantly, 9 of the 15 known chloroplast targets were captured by the affinity column. For reasons that are not fully clear, the 6 additional established chloroplast thioredoxin targets were not observed. It is possible that these proteins were not trapped on the column owing to the participation of an additional cysteine positioned such that it could displace the trapped member of the active site, as appears to happen with fructose bisphosphatase (13, 38). Alternatively, proteins could have been missed owing to lack of similarity of the spinach targets with proteins in the database.

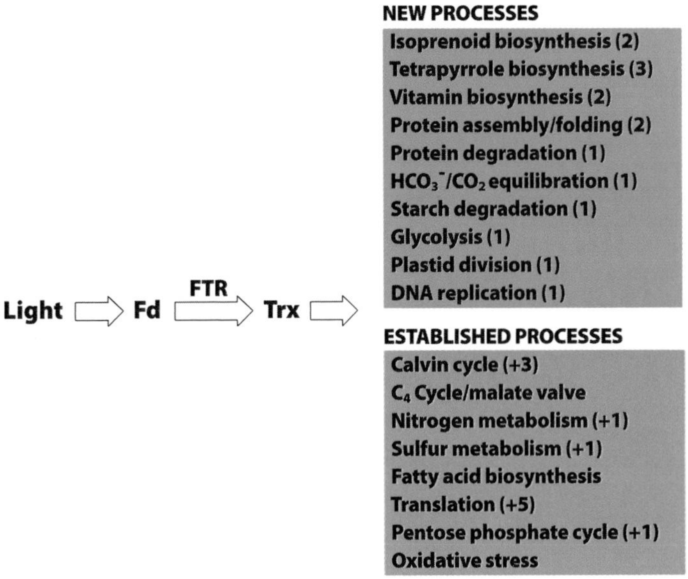

Figure 5

Chloroplast processes regulated by thioredoxin. The numbers in parentheses indicate the number of targets recently identified by the affinity chromatography/proteomic method.

Thioredoxin Specificity

An interesting point emerging from the proteomic studies is the apparent lack of specificity shown by the column-bound mutant *f*- and *m*-type thioredoxins (10, 136). Although many of the previously known enzyme targets identified are preferably regulated by thioredoxin *f* in vitro, specificity was not reflected in the studies summarized above with the mutated proteins. With few exceptions, no significant specificity for thioredoxin *f* or *m* was seen with either the newly identified or the previously known proteins. Such a lack of specificity can be explained by assuming that the replacement of one cysteine of the thioredoxin active site by either serine or alanine abolished specificity by inducing a slight change in the microenvironment. The absence of specificity could also be due to the high concentration of thioredoxins bound to the Sepharose column—a factor that would favor the formation of a heterodisulfide regardless of enzyme preference. The question of thioredoxin specificity emerges as a major challenge for the future, for compartments that contain either multiple types of thioredoxin (chloroplasts and mitochondria) or multiple forms of a single type (cytosol). The occurrence of multiple thioredoxins has made it difficult to assign a function to a particular member of this family, such as representatives of the *h* type occurring in the cytosol. Although specificity of chloroplast thioredoxins seems not to depend on redox potential, the different *h* isoforms have not been extensively studied in this respect.

Target Diversity

As noted above, preparations of an alga, *Chlamydomonas reinhardtii* (119), and a cyanobacterium, *Synechocystis* sp. PCC 6803 (123),

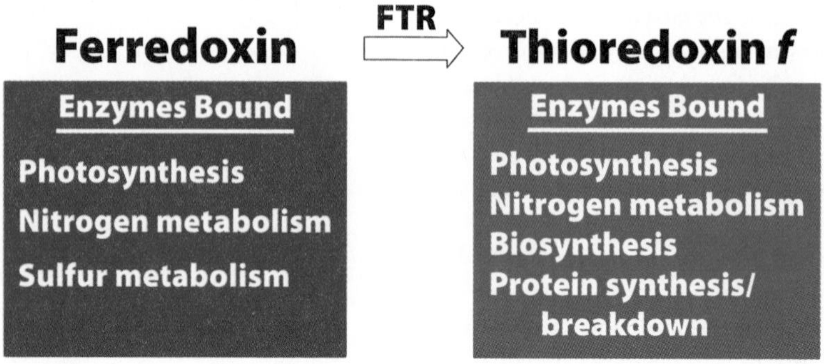

Ferredoxin —FTR→ Thioredoxin *f*

Enzymes Bound
Photosynthesis
Nitrogen metabolism
Sulfur metabolism

Enzymes Bound
Photosynthesis
Nitrogen metabolism
Biosynthesis
Protein synthesis/breakdown

Figure 6

Chloroplast enzymes that form an electrostatic complex with members of the ferredoxin/thioredoxin system. Interacting enzymes are associated with biochemical processes indicated.

have also been probed by applying an affinity column approach. Of the 27, 57, and 25 targets captured in spinach chloroplasts, *Chlamydomonas* and *Synechocystis* extracts, respectively, only one, elongation factor Tu, was found in all three sources; five additional targets were common to chloroplasts and *Chlamydomonas* (rubisco small subunit, transketolase, glutamine synthetase, 2-Cys peroxiredoxin, and thiazole biosynthetic protein) whereas no additional targets were uniquely common to *Chlamydomonas* and *Synechocystis*. Although the *Chlamydomonas* and cyanobacterial extracts—prepared from whole cells—contained nonphotosynthetic targets, the unexpected lack of similarity in the thioredoxin-linked proteins identified from the three sources is striking. The results suggest that fundamental biochemical processes have developed their own patterns of thioredoxin recognition in the three systems, seemingly in keeping with cell architecture and individual lifestyle.

Electrostatically Interacting Proteins

It has been known for many years that thioredoxin *f* forms a noncovalent complex with fructose bisphosphatase (42). Recent work reveals that numerous other proteins have such an interaction. In a study based on a combination of proteomics and an affinity column in which the wild-type protein replaced the mutant form, 27 chloroplast proteins interacted electrostatically with thioredoxin *f* in addition to fructose bisphosphatase (11). The proteins function in 10 processes, most of which also contain covalently interacting thioredoxin targets (**Figure 6**). The results suggest that an association with thioredoxin enables interacting proteins to achieve an optimal conformation, so as to promote: (*a*) the formation of multienzyme complexes—a function facilitated by CP12, an 8.5 kDa disulfide protein (197); (*b*) the transfer of reducing equivalents from the ferredoxin/FTR complex to a target protein; and (*c*) the channeling of metabolite substrates. A similar approach was independently successfully carried out with proteins of *E. coli* (108).

REGULATION LINKED TO PHOTOSYSTEM II

An extensive body of evidence indicates that electron flux through PSII is linked to a network whereby light plays a regulatory function in controlling the optimization of energy utilization through both a change in thylakoid structure (state transition) and gene expression [the historical development of the field is given in (3)]. Each of these topics is addressed below.

State Transition

LHC-II is a light-harvesting protein complex that functions as an auxiliary antenna for the efficient distribution of light energy between PSI and PSII. The spatial distribution of the two chloroplast photosystems is key to this energy distribution. PSII is found almost exclusively in oppressed grana membranes, and PSI is enriched in nonoppressed stromal thylakoid membranes. It has been known for many years that a small part of the total LHC-II undergoes reversible phosphorylation under low to medium light, thereby changing the surface such that the negatively charged proteins are displaced from the hydrophobic core of the granal stacks to the less hydrophobic exposed stromal region (2). The migration of a portion of the PSII antenna complex decreases light absorption by PSII in the granal membrane by decreasing the antennae associated with PSII and increasing energy available for PSI. This transition results in an increase in the rate of noncyclic electron transport.

The most widely accepted model for enabling chloroplasts to sense an imbalance in excitation is based on the phosphorylation of LHC-II by a redox-regulated kinase (61) (**Figure 7**). The kinase is activated when the pool of plastoquinone (PQ) is highly reduced, i.e., when PSII receives more light than PSI, resulting in movement of the LHC-II complex. It seems that activation is linked to the binding of PQH_2 to the Q_o pocket of the cytb_6/f complex. Further details of the mechanism of regulation of the kinase continue to unfold.

There is evidence that LHC-II is regulated by a PSI thiol-based system in addition to the PQ mechanism (7). Under increasing light intensity, when the formation of reducing equivalents exceeds the needs of the Calvin cycle and related reactions, thioredoxin, and possibly other thiol components of the stroma, become more highly reduced. Thioredoxin, in turn, then reduces a disulfide bridge located on the surface of the LHC-II kinase (converts S–S to SH) to yield an inhibited form of the enzyme. This inhibition takes place only when the kinase is in the fully oxidized state as determined by its association with the oxidized cytochrome b_6/f complex (157). When thioredoxin becomes more oxidized, LHC-II kinase returns to the deactivated state. Recent experiments with *Chlamydomonas* have led to an advance in the field through the identification of a LCH-II protein kinase (51). The work, which was based on the use of two mutants deficient in the *Stt7* gene, revealed the enzyme to be an 85-kDa serine-threonine-type protein kinase with homologs in land plants. Significantly, the LHC-II kinase showed a Cys motif in keeping with the potential to interact with thioredoxin. It remains to be determined whether the Stt7 kinase interacts directly with LHC-II or whether it is part of a regulatory cascade and is situated upstream. The family of thylakoid-associated kinases (TAKs) that also phosphorylate LHC-II are candidates that may act in such a system (5).

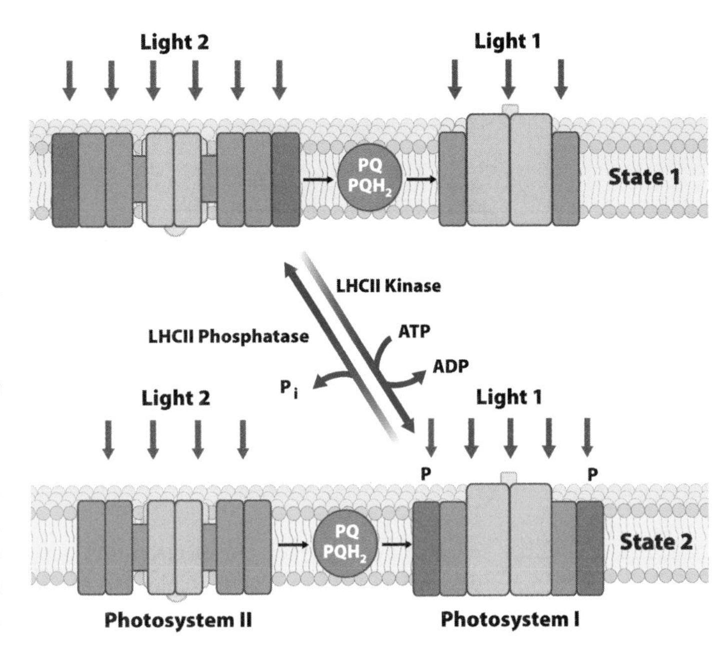

Figure 7

Proposed mechanism for regulating thylakoid LHC-II kinase by redox. Reduced plastoquinone (PQH_2) activates LHC-II kinase that, in turn, phosphorylates the accessory pigment, LHC-II, thereby leading to a redistribution of the PSII and PSI and optimization of light utilization. Adapted from Reference 4.

	High Light		Low Light	
LHC-II Kinase	$\xleftarrow{Trx_{red}}$	LHC-II Kinase	$\xrightarrow{PQH_2}$	LHC-II Kinase
Inhibited	$\xrightarrow[Trx_{ox}]{}$	Deactivated	$\xleftarrow[PQ]{}$	Activated
	PS i		**PS ii**	

Redox Signaling

The role of redox discussed so far relates to reversible posttranslational regulatory changes

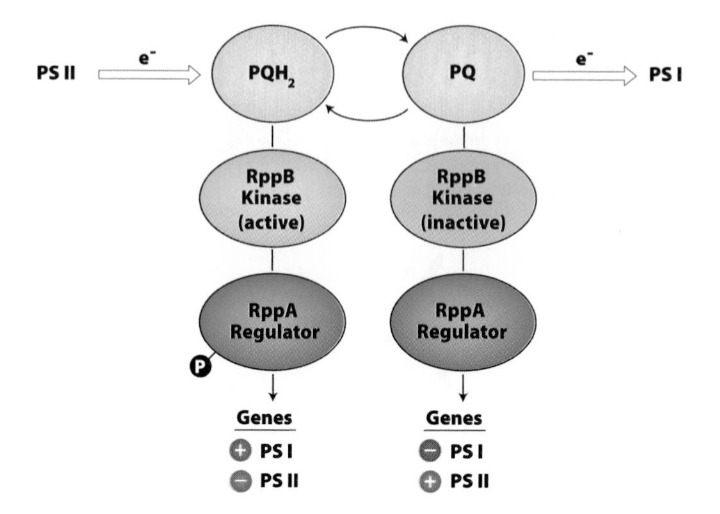

Figure 8

Model for control of gene expression by plastoquinone (PQ). According to this model, reduced plastoquinone (PQH_2) activates a specific protein kinase (RppB kinase) that, in turn, controls the expression of the indicated photosystem I (PSI) or II (PSII) genes. Adapted from Reference 147.

Redox signaling: a specific form of redox regulation linked to transcription, translation, and associated ROS-induced changes.

effected by photosynthetic electron transport, designated redox regulation. Chloroplasts also utilize light and redox to initiate processes at the transcriptional and translational levels. Consistent with current usage, we refer to these activities as a form of redox signaling—a special type of redox regulation. Both of these changes are independent of terminal oxidation—an irreversible reaction that marks proteins for degradation (67).

Transcription. Results obtained during the past decade have linked PSII to the regulation of selected genes integral to photosynthesis. Included are genetic elements transcribed in the chloroplast (e.g., the PSII reaction center gene, *psbA*) as well as the nucleus (e.g., PSII light-harvesting protein *Lhcb*). Although details are still evolving, the mechanism, like the one controlling state transition, centers on the redox state of the PQ pool (**Figure 8**) which, when reduced, activates a specific protein kinase of the histidine type (147). The activated RppB kinase, in turn, phosphorylates a response regulator protein (RppA) that enhances the expression of PSI genes and decreases the expression of PSII genes. Although the overall effect seems clear, numerous unanswered questions remain. For example, as with state transition, it is not known how the PQ redox intermediary leaves the thylakoid membrane and alters kinase activity. Further, the mechanism by which the response regulator exerts its effects on transcription remains to be elucidated. The regulatory system is complicated by evidence that the redox state of PQ also controls the expression of genes active in splicing the relevant structural protein genes. An RppB/RppA system originally described for cyanobacteria (122) has been renamed nrsS/nvsR after finding its involvement in nickel sensing (214).

In addition to transcription, light plays an increasingly important role in translation, as confirmed in recent proteomic studies (10). Much of the work in this area has been carried out with the gene encoding the D1 protein, *psbA*, using *Chlamydomonas*. Building on earlier results (131), the most recent evidence indicates that light acts via two signals in this connection: a PSII "priming" signal generated by the PQ pool and a PSI dithiol signal from thioredoxin (187). The priming signal initiates an as yet unidentified activity that oxidizes the protein complex, thus making it susceptible to reductive activation by thioredoxin (**Figure 9**). It seems that a balance between oxidized PQ and reduced thioredoxin controls the redox state of a poly(A)-binding protein (RB60, not shown in **Figure 9**), which through SH/S–S exchange regulates the translation of the D1 protein—a mechanism that also likely functions in land plants. It remains to be seen how the control of *psbA* transcription (**Figure 8**) and translation (**Figure 9**) are coordinated.

REGULATION LINKED TO NADP

Redox regulation has been invoked in a growing number of organelles and tissues during plant development. In a number of cases, such as auxin signalling, the overall effects seem clear, but details of the inner workings remain to be elucidated (89, 113, 148). Discoveries concerning the mechanisms involved have been most extensive with thioredoxin-utilizing pathways. Relevant findings are summarized below following a brief description of the responsible mechanism—the NADP/thioredoxin system. The system is comprised of the reductant, NADPH, a thioredoxin (*h* or *o* type in plants)

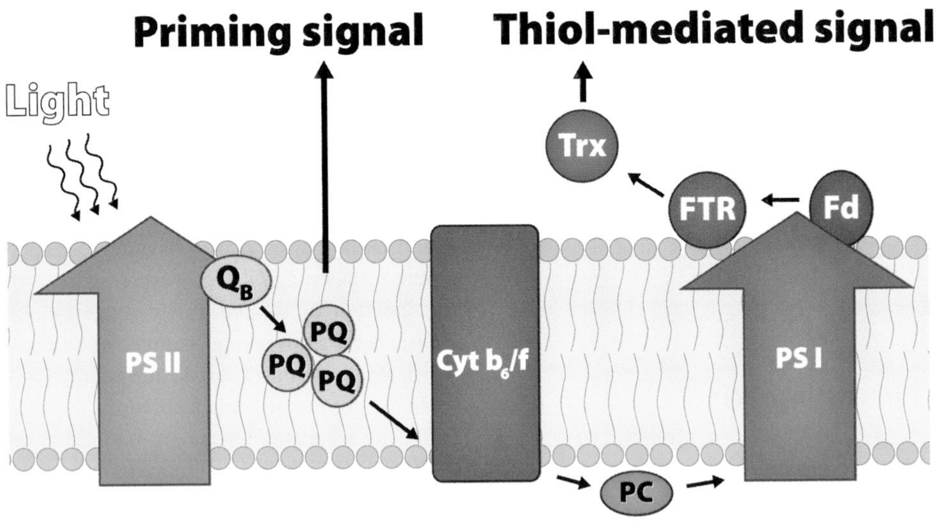

Figure 9

Role of light in regulating translation via redox. The results suggest that a PSII priming signal initiated by reduced PQ prepares the target, RB60, for subsequent regulation by PSI via reduced thioredoxin. Adapted from Reference 187.

and the flavin enzyme, NTR (15, 57, 58, 83, 94, 95). NTR catalyzes the transfer of reducing equivalents from NADPH to thioredoxin (83) (Equation 1).

$$\underset{}{\text{NADPH}} + \underset{(-\text{S}-\text{S}-)}{\text{Thioredoxin } h/o_{\text{ox}}} \xrightarrow{\text{NTR}} \underset{(-\text{SH HS}-)}{\text{Thioredoxin } h/o_{\text{red}}} + \text{NADP} \quad (1)$$

The reduced thioredoxin, in turn, spontaneously reduces a target protein (Equation 2).

$$\underset{(-\text{SH HS}-)}{\text{Thioredoxin } h/o_{\text{red}}} + \underset{(-\text{S}-\text{S}-)}{\text{Target Protein}_{\text{ox}}} \rightarrow \underset{(-\text{SH HS}-)}{\text{Target Protein}_{\text{red}}} + \underset{(-\text{S}-\text{S}-)}{\text{Thioredoxin } h/o_{\text{ox}}} \quad (2)$$

Unlike its ferredoxin-linked counterparts, which are confined to plastids, thioredoxins linked to NADP are multicompartmental and, so far, have been observed in the cytosol, mitochondria, nucleus, and ER. Although knowledge of the function of the NADP/thioredoxin system is still in its infancy, it appears to be connected to the regulation of a diversity of processes. In this section, we discuss the role of this system in an organelle (the mitochondrion) and in two different organs, seed and pollen.

Mitochondrial Processes

Evidence that mitochondria, from plants as well as animals, contain thioredoxin was reported in 1989 (21). Shortly thereafter, a thioredoxin purified from spinach roots, earlier designated thioredoxin *h* (94, 95), was found to represent a new type (128). As seen in **Figure 1**, thioredoxin *h* belongs to a large family, seemingly present in most, if not all, plant tissues. In the 1991 study, cell fractionation experiments revealed that, in addition to cytosol and ER, thioredoxin *h* was present in mitochondria—a conclusion recently confirmed and extended in an investigation with poplar (62). It has also been found that mitochondria contain a characteristic type of thioredoxin, designated "*o*" (**Figure 1**). A mitochondrial NTR was also identified in that study (111).

Until quite recently, little was known about mitochondrial thioredoxin function. By analogy to the animal and yeast organelles, a role was proposed early in antioxidant defense and in regulating cyanide-resistant respiration (9, 111, 172). This latter function was recently shown to be effected by a form of thioredoxin *h* specific to mitochondria (62). The advent of proteomic capability has expanded the role of redox and tied thioredoxin to a range of fundamental processes of mitochondria (14). Application of the mutant affinity column approach led

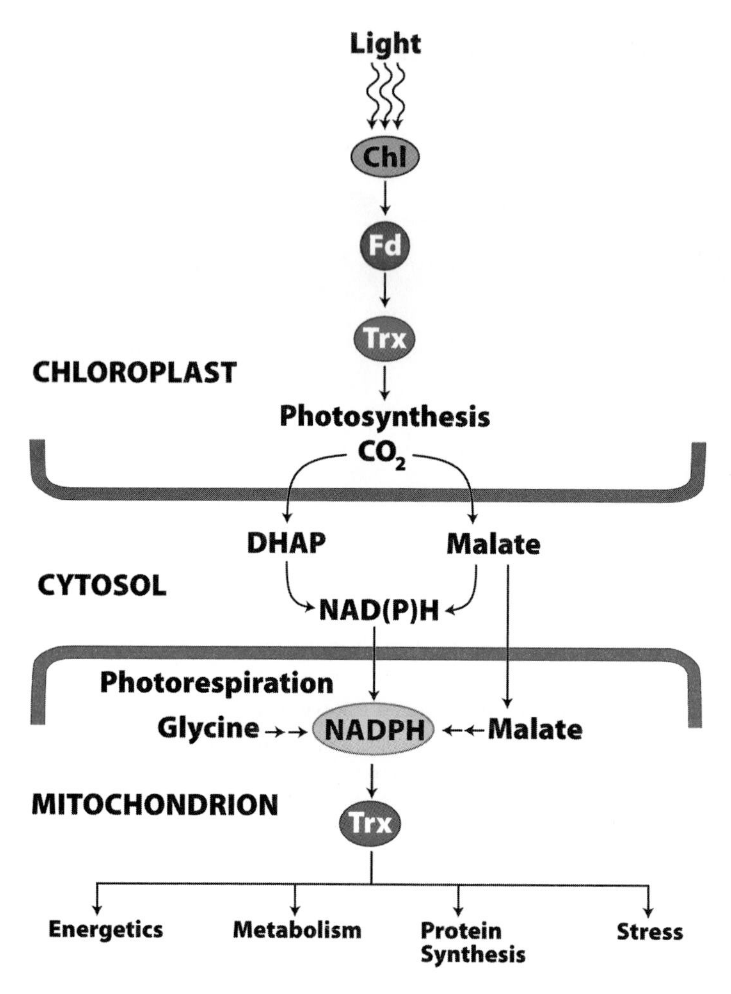

Figure 10

Role of thioredoxin as a regulatory link between chloroplasts and mitochondria. The light signal is transmitted from chloroplast thioredoxin to a mitochondrial counterpart either directly by NAD(P)H or indirectly by transported metabolites (malate or the glycine precursor, glycolate). Adapted from Reference 14.

to the identification of 50 potential thioredoxin-linked proteins functional in 12 processes: photorespiration, citric acid cycle and associated reactions, lipid metabolism, electron transport, ATP synthesis/transformation, membrane transport, translation, protein assembly/folding, nitrogen metabolism, sulfur metabolism, hormone synthesis, and stress-related reactions. Almost all of these same targets were confirmed by a fluorescent gel electrophoresis procedure in which reduction by thioredoxin was observed directly (more below). The results suggest that thioredoxin enables mitochondria to adjust key reactions in accord with prevailing redox state. These and earlier findings further indicate that, by sensing redox in chloroplasts and mitochondria, thioredoxin enables the two organelles of photosynthetic tissues to communicate via a network of transportable metabolites such as dihydroxyacetone phosphate, malate, and glycolate (**Figure 10**). In this way, light absorbed and processed via chlorophyll can be perceived and function in regulating fundamental mitochondrial processes akin to its mode of action in chloroplasts. It will be of interest to learn the function of thioredoxin in mitochondria from other organisms, e.g., mammals and yeast, which also contain a thioredoxin system.

In the proteomic study, thioredoxin was also linked to a voltage-dependent anion channel (VDAC) porin functional in programmed cell death (69), and to aldehyde dehydrogenase, a restoration factor in cytoplasmic male sterility in maize (168).

Seed Development and Germination

Redox changes. Evidence obtained during the past 15 years has established that seed proteins containing a disulfide (S–S) group, including insoluble storage family members of cereals, undergo redox change during development and germination (48, 68, 105, 126, 129, 156, 208). The proteins are synthesized in the reduced (SH) state and become oxidized to a more stable disulfide (S–S) state during maturation and drying. Upon germination, the proteins are converted back to the reduced (SH) state to facilitate mobilization (105, 126, 127, 129, 175, 202, 204, 208).

Developing →	Mature →	Germinating
Seed	Seed	Seed
–SH	S–S	–SH

Reduction effects an increase in (*a*) solubility (202, 204), (*b*) susceptibility to proteases (84, 91, 92, 165), and (*c*) heat sensitivity (91, 92). The reduced form also shows (*d*) a change in enzyme activity either directly by reduction or indirectly by inactivation of specific inhibitor proteins (17, 91, 92, 203). In some cases, there is also (*e*) a mitigation of allergenicity (29, 49).

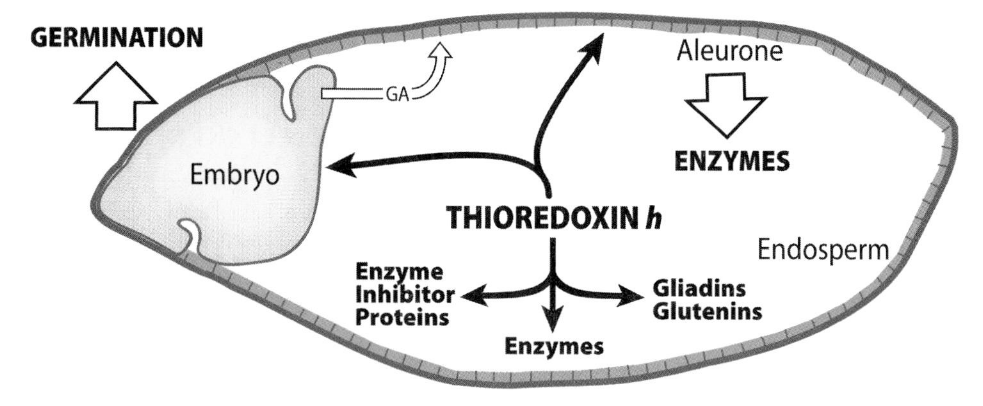

Figure 11

Role of thioredoxin *h* in barley seeds. Thioredoxin acts in the endosperm to facilitate the mobilization of nitrogen and carbon by (*a*) reducing gliadin and glutenin storage proteins in the endosperm, thereby leading to increased solubility and susceptibility to proteolysis; (*b*) inactivating small proteins that inhibit enzymes of starch degradation; and (*c*) directly reductively activating individual enzymes (*lower arrows*). Thioredoxin also seems to function as a member of a communication network linking the endosperm to the embryo and aleurone (*upper arrows*). The increase in GA effected by thioredoxin is partly, but not fully, responsible for the enhanced production of hydrolytic enzymes by the aleurone. Adapted from Reference 204.

Role of thioredoxin. Early experiments with wheat provided in vitro evidence that thioredoxin *h* facilitated the protein redox changes observed during germination (105, 126). The evidence was consistent with the view that thioredoxin *h*, reduced with metabolically generated NADPH, acted early in the imbibed seed to initiate the mobilization of nitrogen and carbon in the cereal starchy endosperm, the major repository for these reserves. The NADPH needed for this reduction could be generated enzymatically from stored carbohydrate via glucose 6-phosphate and 6-phosphogluconate dehydrogenases (126).

Through the reduction of intramolecular disulfide bonds (Equation 2), thioredoxin promoted the degradation of major storage proteins, the inactivation of small proteins that inhibit amylolytic enzymes, and the activation of a novel calcium-dependent substrate-specific protease, thiocalsin (18, 105, 126). Work with transgenic barley supported this conclusion in revealing that thioredoxin *h* overexpressed in the endosperm enhanced the activity of starch debranching enzyme (39) and accelerated the appearance of α-amylase as well as the rate of germination (204). The results provide evidence that thioredoxin *h* acts as a wake-up call in germination and seedling development by catalyzing reductive reactions in the endosperm and acting as a member of a transmembrane communication network that seems to involve gibberellic acid (**Figure 11**).

Recent investigations made possible with proteomics have lent further support to this conclusion. In the original seed proteomic experiments, carried out with peanut, a gel procedure was devised in which thioredoxin targets could be visualized directly on gels following labeling of extracts with the fluorescent probe, monobromobimane (mBBr), and excised for proteomic analysis (209). Five potential target proteins, including three allergens, were initially identified in peanut seeds (209) and six in barley embryo (129). The mBBr procedure was then applied to wheat endosperm (200) and barley grain (127), resulting in the identification of 23 and 5 potential targets, respectively. In parallel experiments, barley grain was analyzed with the more sensitive fluorescent probe, Cy5 maleimide, in addition to mBBr, resulting in the identification of approximately 34 potential targets (127). The latter authors pointed out

ROS: reactive oxygen species

Oxidative regulation: posttranslational regulatory changes resulting from oxidation by molecular oxygen or ROS such as hydrogen peroxide

that, although it was advantageous to use both fluorescent probes, Cy5 proved more effective in their hands.

In more recent work, the mBBr and mutant column affinity procedures were compared in analyzing thioredoxin targets in wheat endosperm (201). The procedures were complementary: of the total targets identified, one third were observed with both procedures and one third were unique to each. Altogether, 68 potential targets were identified in this study, including 40 targets not previously observed in seeds. Overall, the proteomic studies have collectively led to the identification of approximately 70 potential thioredoxin targets in seeds (peanut, barley, wheat). In a related study (127), 15 putative targets, 12 of which were detected in dry grain, were identified in germinating barley. Many of these targets were identified independently in *Arabidopsis* seedlings by Yamazaki et al. (206).

Self-Incompatability

A role for thioredoxin recently surfaced in an unsuspected place: self-incompatibility, a widespread mechanism that prevents self-fertilization in flowering plants. The pheonomenon has fascinated generations of biologists, among them Charles Darwin. With the advent of molecular genetics, our understanding of self-incompatility has made impressive strides. Progress has been notable with the *Brassica* system. In brief, a receptor kinase on the pistil stigma (SRK, or S-locus receptor kinase) undergoes autophosphorylation after coming in contact with a surface protein of incompatibile pollen (SCR, or S-locus cysteine-rich protein) (71, 140). Phosphorylation of the SR kinase initiates a signaling cascade that leads to rejection of the pollen. According to a growing body of evidence, thioredoxin *h* bound to the pistil acts as a negative regulator of the SR kinase and blocks autophosphorylation in the absence of incompatible pollen (22, 33). Thioredoxin seems to act as a guard in protecting the pistil against false fertilization. It remains to be seen whether the participating thioredoxin undergoes redox change as a part of this function.

REGULATION LINKED TO OXYGEN

Two types of control linked to oxygen have been described: oxidative regulation and redox signaling. Oxidative regulation refers to posttranslational redox changes effected by oxygen by analogy to those long described for light and dark transition. Redox signaling, on the other hand, represents a special form of redox regulation linked to transcription, translation, and associated changes that are induced by ROS. These two regulatory systems are discussed in this section.

Oxidative Regulation

Current findings suggest that, in addition to classical light/dark modulation, thioredoxin functions in a previously unrecognized oxidative type of regulation in chloroplasts (10). The latter role provides a reversible mechanism for the deactivation of photosynthetic CO_2 assimilation in response to newly formed oxidants and for subsequent reactivation with the return of agreeable conditions, all in the light (**Figure 4**, lower scheme). This oxidant-induced regulation allows the chloroplast to slow down light-dependent biosynthesis and accelerate carbohydrate degradation under constant illumination conditions, thereby minimizing energy loss and marshalling available biochemical resources for antioxidant defense.

Such an oxidative control mechanism is compatible with the observation that carbon dioxide assimilation is among the first chloroplast processes affected when plants experience adverse conditions that result in the production of ROS (115, 198). Furthermore, several thioredoxin-linked enzymes of the Calvin cycle are sensitive to H_2O_2 (184), including transketolase (100), a pivotal enzyme in chloroplasts. Transketolase is a member of the Calvin as well as the oxidative pentose phosphate cycle. This enzyme also forms the substrates that are the starting point of several other pathways, including the one leading to the synthesis of shikimate. Its activity thus affects not only the Calvin cycle, where the enzyme may be rate-limiting, but also carbohydrate, amino acid, and natural product metabolism (79).

Other thioredoxin-linked enzymes are also candidates for oxidative regulation (10), including 5-adenylyl sulfate reductase (20) and phosphribulokinase (104). Furthermore, it is likely that the activity of each enzyme regulated by thiol-disulfide exchange will be affected by the production of ROS, either directly via the oxidant-induced formation of regulatory disulfide bonds or through the change in the ratio of reduced and oxidized thioredoxin (or other reducing agents). A small variation in the redox status of the thioredoxin pool can impede its ability to catalyze the cleavage of target disulfides efficiently.

Oxidative conditions also promote the formation of mixed disulfides between glutathione and cysteinyl residues. This reaction, glutathionylation, prevents the irreversible oxidation of SH groups and the subsequent degradation of the protein (67, 149). Similarly, in addition to the modulation of activity, the formation of disulfide bonds may play a role in protecting thiols and, as discussed above, in enhancing protein stability (14). Accordingly, disulfide linkages could assist in the adaptation of plants to temperature. Enzymes of psychrophilic organisms show high flexibility and low thermal stability whereas thermophilic enzymes display the reverse, i.e., low flexibility and high thermal stability (66). Plants could, therefore, cope with a pronounced shift in temperature via thiol-disulfide interconversion that would affect the physico-chemical properties of enzymes. The formation of disulfide bond would increase thermostability (153), whereas their subsequent reduction would increase flexibility and enhance catalytic activity at low temperature.

Redox Signaling

ROS are toxic oxygen-derived molecules that are generated by (*a*) external factors, (*b*) by-products of metabolism, or (*c*) dedicated specific enzymes. ROS induced by external stimuli are the result of an environmental change that either directly favors their generation or that creates an imbalance in the cellular machinery leading to their production. For example, external factors can affect the production of oxygen intermediates linked to the activity of highly reactive electron carriers in chloroplasts and mitochondria or can promote the photorespiratory generation of H_2O_2 in peroxisomes (**Figure 12**). However, even under normal conditions, ROS are an inevitable by-product of aerobic metabolism. Thus, like other organisms living in the presence of oxygen, plants possess an impressive array of protective mechanisms to remove these intermediates. Despite their deleterious effects, ROS also play a central role both as indicators of oxidative stress and as signaling molecules.

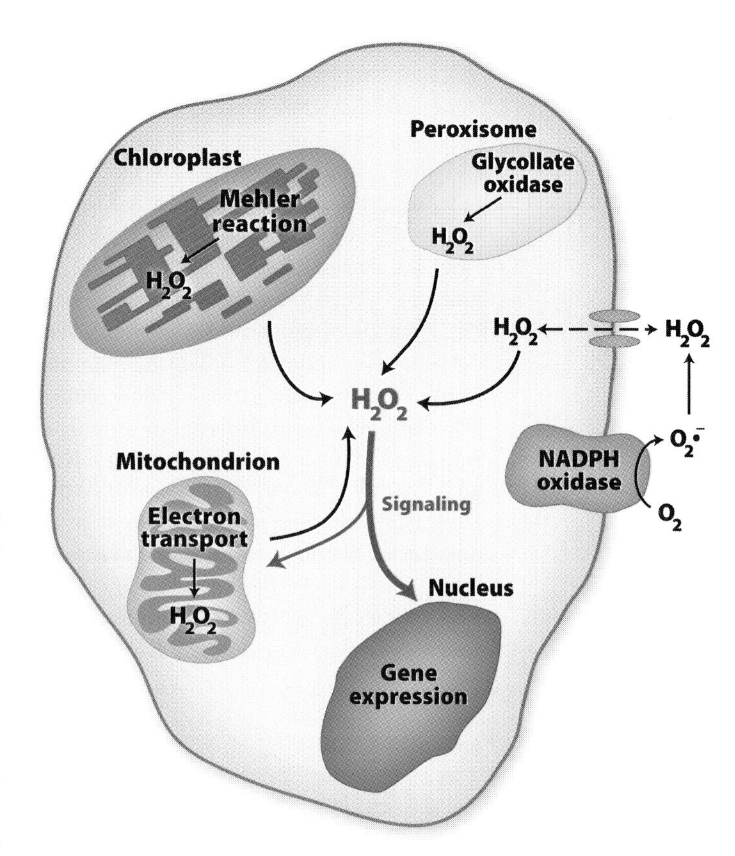

Figure 12

Hydrogen peroxide generation in a photosynthetic plant cell. Hydrogen peroxide produced by pseudocyclic phosphorylation (Mehler reaction) in chloroplasts, electron transport in mitochondria, photorespiration in peroxisomes, and specific enzymes such as NADPH oxidase, acts as a signaling molecule. The oxidant signal is transduced in the nucleus primarily for transcription and in the mitochondrion to initiate programmed cell death. Adapted from Reference 142.

Translation. Participants in translation—including initiation and elongation factors, ribosomal and RNA binding proteins—have been identified as potential thioredoxin targets and thus could play a role in modulating protein synthesis under oxidative stress [references in 10]. These proteins could sense ROS by activity changes accompanying the formation of disulfides, such as the increase in the synthesis of polypeptides of the NADH dehydrogenase complex following photooxidative stress or hydrogen peroxide treatment (37). This induction is believed to be initiated via a rapid translation of a pre-existing mRNA pool followed by an increase in transcript levels. Similarly, high light exposure induces a dramatic reduction in synthesis of the Rubisco large subunit in *Chlamydomonas*, concomitant with an increase in ROS level. In this case, it seems that, upon oxidation, the Rubisco large subunit exposes an RNA binding domain and mediates translational arrest by binding to its own mRNA (210).

A second type of redox control of translation is linked to RNA stability, as suggested by experiments with *Chlamydomonas* cells expressing a GUS:psaB mRNA fusion when subjected to different oxidizing and reducing conditions (166). Addition of a strong oxidizing agent such as diamide seemed to enhance mRNA stability, whereas DTT promoted degradation of the chloroplast transcripts (166). A similar situation could be present in chloroplasts of land plants, where mRNA stability is linked to external signals such as light (8).

Transcription. Hydrogen peroxide is one of the major molecules triggering response to oxidative stress in plants as well as animals and microorganisms (142). To gain insight into the molecular mechanisms involved in redox signaling, microarray analyses have been undertaken to monitor gene expression under H_2O_2 or high light treatment by several laboratories (52, 160, 190). Although the individual studies reveal different patterns, likely due to experimental approach and type of stress applied, certain general trends emerge. ROS production seems to induce a rapid increase in the expression of genes involved in stress signal transduction, antioxidants, and cell rescue machinery concomitant with repression of genes encoding photosynthetic components. Little information was gained, however, on the nature of the redox sensors involved.

Transcription factors have been described in animals, yeast, and bacteria that respond to H_2O_2 via the formation of disulfides, either directly or through thioredoxin and glutaredoxin, e.g., AP-1, Yap1, and OxyR (65, 173, 181, 212). Although the picture is less clear in plants, the mitogen-activated protein kinase (MAPK) pathway is one ROS-sensing mechanism that can result in the activation of transcription factors (109). The ROS sensor/initiator of this cascade is still elusive, but could be a redox-controlled protein tyrosine phosphatase (74). In addition, several potential redox-regulated transcription factors were recently identified in plants. Members of the homeodomain transcription factors form covalent dimers via conserved cysteines that can be reduced by the NADP/thioredoxin system, thereby increasing their DNA binding capacity (189). Furthermore, NPR1, a regulator of salicylic acid–dependent gene expression interacts differentially with the reduced and oxidized forms of the transcription factor TGA1, thereby altering its DNA binding activity (53). The formation of an intramolecular disulfide in TGA1 precludes an interaction with NPR1 to prevent the stimulatory effect of the cofactor (53). Additionally, results by others suggest that the formation of a disulfide in the maize P1 regulator of flavonoid biosynthesis, a member of the R2R3 MYB-domain transcription factor family, prevents its binding to DNA, a feature seemingly specific to plants (78).

Interestingly, the proteomic-based approaches used to identify potential thioredoxin targets described above failed to detect redox-sensing transcription factors in plant extracts. Assuming their existence, this lack of success could be due either to the relatively low abundance of these factors or to the fact that nuclear preparations were not specifically studied. Nonetheless, Ran, a protein involved in nuclear transport, was identified as a potential thioredoxin target in *Chlamydomonas* (119). Additionally, recent evidence shows that thioredoxin accumulates in the nucleus of developing wheat under oxidative stress (174) and a nucleoredoxin—a Zn finger-containing member of the thioredoxin superfamily that is proposed to act directly on the reduction state of transcription factors—was identified in the nucleus of maize kernels (114). Taken together, these data imply that redox plays a role in controlling transcription in plants.

Programmed cell death. There is a growing body of evidence that ROS are key elements in initiating programmed cell death. The results suggest that the balance between ROS production and removal is critical in initiating this process. In barley aleurone, programmed cell death induction by gibberellic acid is linked to a decrease in cellular antioxidant capacity (19, 56). In addition, incubation of cells with an oxidant or a pathogen induces an oxidative burst that is followed by cell death (50, 85). One recently identified pathway leading to programmed cell death centers on the formation of a permeability transition pore in mitochondria that releases cytochrome *c*. The cytochrome *c*, in turn, initiates a cascade leading to programmed cell death (81, 96, 130). ROS could act in the formation of the permeability transition pore and/or by directly activating the MAPK pathway (106, 186). As noted above, one of the proteins that seems critical to the process, the voltage-dependent anion channel protein (VDAC), has been identified as a thioredoxin target (14, 69).

Related processes. Evidence also associates hydrogen peroxide signaling with plant hormones. In the guard cell, abscisic acid (ABA) activates the synthesis of H_2O_2 that, in turn, induces the closure of the stomata (142, 145). Similarly, auxin and H_2O_2 jointly play a key role in root gravitropism among other processes (97). Finally, recent evidence suggests that both ethylene and ROS are required for the initiation of nodulation in the semiaquatic legume *Sesbania rostrata* (54).

In addition to translation and transcription, ROS emerge as central messengers in the control of a number of processes such as hypersensitive pathogen response, cell expansion, stomatal closure, and root gravitropism (109, 142). To this end, plants possess several enzymes, NADPH oxidase, xanthine oxidase, amine oxidase, and peroxidase, that generate ROS as signaling molecules catalytically. Although recent evidence suggests the pattern of the observed response may differ depending on the specific ROS intermediate (85, 143), H_2O_2, acting alone or in conjunction with other species, seems to be active in cell signaling related to these processes (142).

The high antioxidant capacity of the cell ensures that ROS signaling is a localized event under physiological conditions. Thus, even if H_2O_2 is able to move freely, the molecule has to interact with sensors in close proximity to the site of production. In its formation of superoxide, NADPH oxidase bound to the plasma membrane is one of the main sources of ROS. This enzyme plays a central role in the regulation of plant cell growth via the activation of Ca^{2+} channels by ROS (59). Moreover, following its conversion to H_2O_2, superoxide could be the substrate of the cell wall peroxidases, producing hydroxyl radicals responsible for wall loosening by cleavage of the polymers and, hence, a prerequisite for cell extension (124). In addition, NADPH oxidase seems to be involved in the increase of the antioxidant defense capacity induced by ABA or water (drought) stress (90) and in the response elicited by pathogens (178). It remains a mystery as to how the low concentrations of H_2O_2 effective in signaling achieve their effects (181).

Environmental and biological stresses. Due to an inability to move, plants have an added problem in coping with a range of deleterious stresses produced by sporadic changes in the environment—e.g., drought, ozone, heat, cold, UV, and high light—or by biological agents such as pathogens. Most of these stresses lead to a burst of ROS, notably H_2O_2, which, in the absence of deterrence, overwhelms the detoxification capacity of the cell (195). On encountering environmental stresses, plants respond by adapting to the new condition by transcriptional, translational, and posttranslational alterations to increase detoxification capacity and to decrease ROS production (76, 135). In this way cell death is minimized. In contrast, biologically induced stresses, such as pathogen attack, result in the formation of ROS that contribute to resistance against the intruder and ultimately to the initiation of programmed cell death of the infected cells (72, 85). In one case, infection with the bacterium *Pseudomonas syringe* induced the expression of an *h*-type thioredoxin, *Arabidopsis* TRXh5 (110), prior to hypersensitive cell death.

RESEARCH ON THE HORIZON

The results summarized above illustrate how a field can evolve and expand over a 25-year period. First linked to photosynthesis, regulation by redox is now associated with other major plant processes via mechanisms that continue to unfold. In reviewing the field, a number of areas stemming from recent progress stand out as timely areas for future investigation.

A lingering problem since the finding of multiple types of thioredoxins in chloroplasts, the issue of specificity has become more important as the number of plant thioredoxins has increased (**Figure 1**). Progress has been made with individual members of the *m*-, *x*- (41, 86), and *h*-families (25) using yeast complementation, but we have little understanding of the specificity and interaction of the different thioredoxins collectively present in an organelle. It is possible that RNA silencing technology will facilitate work in this challenging area that is becoming increasingly important in plants, as the number of target proteins (currently 180) continues to mount (**Table 1**) (10, 14, 127, 136, 200, 201, 206). Further studies on the individual thioredoxins should be coupled with work on glutaredoxins to deduce not only

TABLE 1 Established and potential thioredoxin targets identified by proteomics in plants

*Acetyl-CoA carboxylase
Acetyl-CoA synthetase
60S Acidic ribosomal protein P
Actin
Aconitase
Adenylate kinase
*5′-Adenylylsulfate reductase
S-Adenosylmethionine synthetase
ADP-glucose pyrophosphorylase
AIR synthetase
Alanine aminotransferase
Alcohol dehydrogenase
Aldehyde dehydrogenase
Aldehyde oxidase
Aldolase
Aldose reductase
Allene oxide cyclase
Allyl alcohol dehydrogenase
*α-Amylase
*α-Amylase inhibitors
*α-Amylase/chymtotrypsin inhibitor
*α-Amylase/subtilisin inhibitor
*α-Amylase/trypsin inhibitor
Argininosuccinate synthetase
Ascorbate peroxidase
Aspartate aminotransferase
ATP synthase α, β, δ su
ATP synthase, su α, chloroplast
ATP synthase, su α, vacuole
*ATP synthase, su γ, chloroplast
ATP-dependent clp protease
ATP-dependent DNA helicase
Auxin-induced protein
Avenins
Barley embryo globulin
Barwin
Branched-chain keto acid decarboxylase E1
Carbonic anhydrase
Carboxysomal protein (CcmM)
Catalase
Chaperonin 60 KD αβ su
Chitinase
CoA-thioester hydrolase
CP12
Cpn20
Cyclophilin
Cysteine synthase

(Continued)

TABLE 1 (*Continued*)

Cytochrome c oxidase
*DAHP synthase
Dessication-related protein
Diaminopimelate epimerase (Lys)
Dihydrolipoamide acetyltransferase
Dihydrolipoamide dehydrogenase
Dihydroxyacid dehydratase (Val, Ile)
DXP reductoisomerase
Elongation factor 1
Elongation factor 1b
Elongation factor 2
Elongation factor G
Elongation factor Tu
Elongation factor EF-2
Endochitinase
Enolase
ER membrane protein
Eukaryotic translation initiation factor eIF4A
Ferredoxin
Ferredoxin-GOGAT
Formate dehydrogenase
10-Formyltetrahydrofolate synthase
*Fructose bisphosphatase
FtsZ protein
GcpE protein
GDP-mannose dehydratase
Germin-like protein
Globulin
Glucan branching enzyme
6-P-Gluconate dehydrogenase
*Glucose 6-P dehydrogenase
β-Glucosidase
Glutamate dehydrogenase
*Glutamine synthetase
Glutaredoxin-like protein
Glutathione peroxidase
Glutathione S-transferase
Glyceraldehyde 3-P dehydrogenase (NAD)
*Glyceraldehyde 3-P dehydrogenase (NADP)
*Glycerate kinase
Glycine cleavage system (H, P, T-proteins)
Glycinin
Glycogen synthase
Glyoxalase
GSA aminomutase
GSH dehydroascorbate reductase
Heat shock protein 17.5 kDa class II
Heat shock protein 70 kDa
Inorganic pyrophosphatase

(*Continued*)

TABLE 1 (*Continued*)

Isocitrate dehydrogenase
Isocitrate lyase
3-Isopropylmalate dehydrogenase (Leu)
Isovaleryl-CoA dehydrogenase
α-Ketoglutarate dehydrogenase complex, su E2
Ketol-acid reductoisomerase
Leucine aminopeptidase
LMW glutenin subunit
LytB
Magnesium chelatase
Malate dehydrogenase (NAD)
*Malate dehydrogenase (NADP)
Malic enzyme
Malonyl CoA ACP transacylase
Mercaptopyruvate sulfurtransferase
*Methionine sulfoxide reductase
Methionine synthase
Methylmalonate-semialdehyde dehydrogenase
NADH-GOGAT
NADH-ubiquinone oxidoreductase
Non-specific lipid transfer protein
Nucleoside diphosphate kinase
OEE1
Peroxidase 1
*1-Cys Peroxiredoxin
*2-Cys Peroxiredoxin
*Peroxiredoxin type II
P-Glycerate dehydrogenase
Phosphoglucomutase
Phosphoglycerate kinase
Phosphoglycerate mutase
Phospholipid hydroperoxide GSH reductase
*Phosphoribulokinase
Phycobilisome linker (LCM)
Poly(A)-binding protein
Porin (VDAC)
Porphobilinogen synthase
*PPi-fructose-6-P 1-phosphotransferase
Profilin
26S Proteasome regulatory S12 subunit
Protein disulfide isomerase
2S Proteins
Pyruvate dehydrogenase E1
Pyruvate Pi dikinase
Ran
RB60 PDI
28 kDa Ribonucleoprotein
Ribose 5P isomerase
30S Ribosomal protein S1

(*Continued*)

TABLE 1 (*Continued*)

30S Ribosomal protein S3
Ribosomal protein S6 (PrpS6)
Ribosomal protein SR
Ribulose-P 3-epimerase
RNA binding protein
RNA polymerase subunit
RPN12
*Rubisco activase
Rubisco L, S su
*Sedoheptulose bisphosphatase
Serine hydroxymethyl transferase
Serine protease-like
Serpin
Soybean seed maturation protein
Stress-induced protein, sti-1 like
Subtilisin/chymotrypsin inhibitor
Succinate dehydrogenase
Succinyl-CoA ligase
Sugar-nucleotide epimerase
Sulfate adenylyltransferase
Sulfite reductase
Superoxide dismutase (Cu-Zn)
Superoxide dismutase Mn
Thaumatin-like protein
Thiamin biosynthesis protein
Thiazole biosynthetic enzyme
Threonine synthase (Thr)
Transketolase
Tricitin
Triosephosphate isomerase
Trypsin inhibitor
Tubulin
Ubiquinol-cytochrome c reductase
Ubiquitin conjugating enzyme E2
UDP-glucose Ppase
Uroporphyrinogen decarboxylase
Valyl-tRNA synthetase (ValS)

*Identified prior to proteomics.

new functions but also how the systems interact (cf., 63).

While progress has been made in identifying new thioredoxin targets in chloroplasts and mitochondria, our understanding of the role of redox signaling in processes such as transcription within these organelles or the nucleus is quite limited. Assuming that the situation is similar to that in animals and bacteria, the identification of redox-regulated transcription factors surfaces as an area awaiting discovery in plants. Studies should ideally focus on chloroplasts and mitochondria, as well as the nucleus where H_2O_2 seems to be important. Thioredoxin has been shown to be localized in the nucleus of wheat aleurone and scutellum (174, 175) and a thioredoxin-linked nuclear transport factor, Ran, has been identified in *Chlamydomonas* (119). It may be that, as indicated by the studies on state transition described above, protein phosphorylation serves as a link between redox and the control of gene expression in chloroplasts. Such a signaling link could extend to other processes, as recently illustrated by evidence that phosphoenolpyruvate carboxylase is controlled by both redox and phosphorylation (87). Most likely, as more thioredoxin-like proteins are functionally identified, redox signaling will continue to expand into new areas.

Advances made possible with proteomics indicate that the site of oxygen evolution in chloroplasts, the thylakoid lumen, is a rich source for discovery. The lumen is now known to contain a relatively large number of proteins, many of unknown function (146, 180, 182). Recent research on its structure has linked redox and possibly light to the regulation of a major resident of the lumen: FKBP13, an immunophilin that functions as a peptidyl-prolyl isomerase (70). The results are in accord with a previously unrecognized paradigm functional in linking light to redox regulation. Work that may be related has uncovered a fascinating mechanism by which reducing equivalents appear to be transferred from the stroma to the lumen (144). Further research will determine whether the two systems functionally interact.

A final area awaiting exploitation centers on the technological applications of thioredoxin. Largely because of public concern over genetically modified foods, progress in this promising area (31) has been sluggish and pales with the strides being made in the medical arena. The promising leads on applying thioredoxin, such as in the production of hypoallergenic foods (29, 49) and improved barley malt (204) and dough products (205), should be pursued. Applications stemming directly from research on plants may, however, find their way via unsuspected routes. For example, a recent report describes a dramatic increase in the solubility of proteins of sputem of patients suffering from cystic fibrosis following reduction by thioredoxin (152). If results hold up in clinical trials, thioredoxin treatment offers possible relief for patients who typically have difficulties expectoring the accumulated highly viscous mucous.

CONCLUDING REMARKS

Originating with a report on a single chloroplast enzyme published nearly 40 years ago (30), the field has grown to include mechanisms linking redox regulation and redox signaling to the activity of a spectrum of processes functional not only in plants, but throughout biology. In addition to serving important structural and catalytic functions, it is now recognized that, in many cases, disulfide bonds can be broken and reformed for regulation. The history of the development of the chloroplast thioredoxin system has been recorded (32) and selected aspects pertaining to plants have been described and updated in a number of outstanding reviews noted above. The present account represents an attempt to present a broad picture, of necessity abridged, describing current knowledge of mechanisms whereby regulatory disulfide proteins—thioredoxins, glutaredoxins, PDIs, and others still being discovered—link redox to the control of enzymes and proteins functional in different processes. We hope that this approach will enable students to gain an awareness of the regulatory role redox plays in the life of the plant. It is also our wish that this review will serve as a source of ideas to encourage newcomers to study redox regulation. In our view, the field is poised to undergo progress such that the horizon will achieve dimensions presently difficult to fathom. All evidence portends that discoveries with far-reaching consequences lie ahead.

SUMMARY POINTS

1. Thiol/disulfide exchange is a major focus of this review.
2. The review describes the differences between redox regulation and redox sensing/oxidative regulation.
3. The relative roles of thioredoxin and glutaredoxin in regulating enzymes are a major focus of the article.
4. Regulation linked to NADP and oxygen is discussed in addition to regulation linked to ferredoxin and light.
5. Signaling linked to ROS (redox signaling) is presented as a special form of redox regulation.
6. The role of redox in regulating photosynthesis, seed development, and germination is an important part of the review.
7. The developing area of control of gene expression by redox is presented as a part of the overall field of redox regulation.
8. Redox is discussed from a perspective of its function in regulating processes taking place in cellular organelles.

ACKNOWLEDGMENTS

We are indebted to Paul Bethke, Bill Hurkman, and Russell Jones for helpful comments, to Nick Cai for preparing the final figures, and to Heather Lee in finalizing the manuscript.

LITERATURE CITED

1. Aalen RB. 1999. Peroxiredoxin antioxidants in seed physiology. *Seed Sci. Res.* 9:285–95
2. Allen JF. 1992. Protein phosphorylation in regulation of photosynthesis. *Biochim. Biophys. Acta* 1098:275–335
3. Allen JF. 2002. Plastoquinone redox control of chloroplast thylakoid protein phosphorylation and distribution of excitation energy between photosystems: discovery, background, implications. *Photosynth. Res.* 73:139–48
4. Allen JF. 2003. Botany. State transitions–a question of balance. *Science* 299:1530–32
5. Allen JF, Race HL. 2002. Will the real LHC II kinase please step forward? *Science STKE*: PE43
6. Arnér ES, Holmgren A. 2000. Physiological functions of thioredoxin and thioredoxin reductase. *Eur. J. Biochem.* 267:6102–9
7. Aro EM, Ohad I. 2003. Redox regulation of thylakoid protein phosphorylation. *Antioxid. Redox Signal.* 5:55–67
8. Baginsky S, Gruissem W. 2002. Endonucleolytic activation directs dark-induced chloroplast mRNA degradation. *Nucleic Acids Res.* 30:4527–33
9. Balmer Y, Buchanan BB. 2002. Yet another plant thioredoxin. *Trends Plant Sci.* 7:191–93
10. **Balmer Y, Koller A, del Val G, Manieri W, Schürmann P, Buchanan BB. 2003. Proteomics gives insight into the regulatory function of chloroplast thioredoxins. *Proc. Natl. Acad. Sci. USA* 100:370–75**

In corroborating the strength of the approach described in Reference 136, the authors demonstrated that thioredoxin functions broadly in chloroplasts and paved the way for the identification of new target proteins in a number of systems.

11. Balmer Y, Koller A, del Val G, Schürmann P, Buchanan BB. 2004. Proteomics uncovers proteins interacting electrostatically with thioredoxin in chloroplasts. *Photosynth. Res.* 79:275–80
12. Balmer Y, Schürmann P. 2001. Heterodimer formation between thioredoxin *f* and fructose 1,6-bisphosphatase from spinach chloroplasts. *FEBS Lett.* 492:58–61
13. Balmer Y, Stritt-Etter AL, Hirasawa M, Jacquot J-P, Keryer E, et al. 2001. Oxidation-reduction and activation properties of chloroplast fructose 1,6-bisphosphatase with mutated regulatory site. *Biochemistry* 40:15444–50
14. Balmer Y, Vensel WH, Tanaka CK, Hurkman WJ, Gelhaye E, et al. 2004. Thioredoxin links redox to the regulation of fundamental processes of plant mitochondria. *Proc. Natl. Acad. Sci. USA* 101:2642–47
15. Banze M, Follmann H. 2000. Organelle-specific NADPH thioredoxin reductase in plant mitochondria. *J. Plant Physiol.* 156:126–29
16. Baumann U, Juttner J. 2002. Plant thioredoxins: the multiplicity conundrum. *Cell. Mol. Life Sci.* 59:1042–57
17. Besse I, Buchanan BB. 1997. Thioredoxin-linked plant and animal processes: the new generation. *Bot. Bull. Acad. Sinica (Taipei)* 38:1–11
18. Besse I, Wong JH, Kobrehel K, Buchanan BB. 1996. Thiocalsin: a thioredoxin-linked, substrate-specific protease dependent on calcium. *Proc. Natl. Acad. Sci. USA* 93:3169–75
19. Bethke PC, Jones RL. 2001. Cell death of barley aleurone protoplasts is mediated by reactive oxygen species. *Plant J.* 25:19–29
20. Bick JA, Setterdahl AT, Knaff DB, Chen Y, Pitcher LH, et al. 2001. Regulation of the plant-type 5′-adenylyl sulfate reductase by oxidative stress. *Biochemistry* 40:9040–48

21. Bodenstein-Lang J, Buch A, Follmann H. 1989. Animal and plant mitochondria contain specific thioredoxins. *FEBS Lett.* 258:22–26
22. Bower MS, Matias DD, Fernandes-Carvalho E, Mazzurco M, Gu T, et al. 1996. Two members of the thioredoxin-h family interact with the kinase domain of a Brassica S locus receptor kinase. *Plant Cell* 8:1641–50
23. Brandes HK, Larimer FW, Hartman FC. 1996. The molecular pathway for the regulation of phosphoribulokinase by thioredoxin *f*. *J. Biol. Chem.* 271:3333–35
24. Bréhélin C, Laloi C, Setterdahl AT, Knaff DB, Meyer Y. 2004. Cytosolic, mitochondrial thioredoxins and thioredoxin reductases in *Arabidopsis thaliana*. *Photosynth. Res.* 79:295–304
25. Bréhélin C, Mouaheb N, Verdoucq L, Lancelin J-M, Meyer Y. 2000. Characterization of determinants for the specificity of Arabidopsis thioredoxins *h* in yeast complementation. *J. Biol. Chem.* 275:31641–47
26. Broin M, Cuiné S, Peltier G, Rey P. 2000. Involvement of CDSP 32, a drought-induced thioredoxin, in the response to oxidative stress in potato plants. *FEBS Lett.* 467:245–48
27. **Buchanan BB. 1980. Role of light in the regulation of chloroplast enzymes. *Annu. Rev. Plant Physiol.* 31:341–74**

This review set the stage for the development of the redox regulation field by summarizing then recent evidence for the role of light and redox in the regulation of chloroplast enzymes.

28. Buchanan BB. 1991. Regulation of CO2 assimilation in oxygenic photosynthesis: the ferredoxin/thioredoxin system. Perspective on its discovery, present status and future development. *Arch. Biochem. Biophys.* 288:1–9
29. Buchanan BB, Adamidi C, Lozano RM, Yee BC, Momma M, et al. 1997. Thioredoxin-linked mitigation of allergic responses to wheat. *Proc. Natl. Acad. Sci. USA* 94:5372–77
30. Buchanan BB, Kalberer PP, Arnon DI. 1967. Ferredoxin-activated fructose diphosphatase in isolated chloroplasts. *Biochem. Biophys. Res. Commun.* 29:74–79
31. Buchanan BB, Schürmann P, Decottignies P, Lozano RM. 1994. Thioredoxin: a multifunctional regulatory protein with a bright future in technology and medicine. *Arch. Biochem. Biophys.* 314:257–60
32. Buchanan BB, Schürmann P, Wolosiuk RA, Jacquot J-P. 2002. The ferredoxin/thioredoxin system: from discovery to molecular structures and beyond. *Photosynth. Res.* 73:215–22
33. **Cabrillac D, Cock JM, Dumas C, Gaude T. 2001. The S-locus receptor kinase is inhibited by thioredoxins and activated by pollen coat proteins. *Nature* 410:220–23**

The authors characterized the role of thioredoxin in preventing self-fertilization in incompatible pollen—an historic problem of great interest.

34. Capitani G, Markovic-Housley Z, del Val G, Morris M, Jansonius JN, Schürmann P. 2000. Crystal structures of two functionally different thioredoxins in spinach chloroplasts. *J. Mol. Biol.* 302:135–54
35. Capitani G, Schürmann P. 2004. On the quaternary assembly of spinach chloroplast thioredoxin *m*. *Photosynth. Res.* 79:281–85
36. Carr PD, Verger D, Ashton AR, Ollis DL. 1999. Chloroplast NADP-malate dehydrogenase: structural basis of light-dependent regulation of activity by thiol oxidation and reduction. *Structure Fold. Des.* 7:461–75
37. Casano LM, Martin M, Sabater B. 2001. Hydrogen peroxide mediates the induction of chloroplastic Ndh complex under photooxidative stress in barley. *Plant Physiol.* 125:1450–58
38. Chiadmi M, Navaza A, Miginiac-Maslow M, Jacquot J-P, Cherfils J. 1999. Redox signalling in the chloroplast: structure of oxidized pea fructose-1,6-bisphosphate phosphatase. *EMBO J.* 18:6809–15
39. Cho M-J, Wong JH, Marx C, Jiang W, Lemaux PG, Buchanan BB. 1999. Overexpression of thioredoxin *h* leads to enhanced activity of starch debranching enzyme (pullulanase) in barley grain. *Proc. Natl. Acad. Sci. USA* 96:14641–46

40. Chueca A, Sahrawy M, Pagano EA, Gorge JL. 2002. Chloroplast fructose-1,6-bisphosphatase: structure and function. *Photosynth. Res.* 74:235–49
41. Collin V, Issakidis-Bourguet E, Marchand C, Hirasawa M, Lancelin JM, et al. 2003. The Arabidopsis plastidial thioredoxins: new functions and new insights into specificity. *J. Biol. Chem.* 278:23747–52
42. Crawford NA, Yee BC, Hutcheson SW, Wolosiuk RA, Buchanan BB. 1986. Enzyme regulation in C4 photosynthesis: purification, properties, and activities of thioredoxins from C4 and C3 plants. *Arch. Biochem. Biophys.* 244:1–15
43. Dai S, Saarinen M, Ramaswamy S, Meyer Y, Jacquot J-P, Eklund H. 1996. Crystal structure of Arabidopsis thaliana NADPH dependent thioredoxin reductase at 2.5 A resolution. *J. Mol. Biol.* 264:1044–57
44. Dai S, Schwendtmayer C, Johansson K, Ramaswamy S, Schürmann P, Eklund H. 2000. How does light regulate chloroplast enzymes? Structure-function studies of the ferredoxin/thioredoxin system. *Q. Rev. Biophys.* 33:67–108
45. **Dai S, Schwendtmayer C, Schürmann P, Ramaswamy S, Eklund H. 2000. Redox signaling in chloroplasts: cleavage of disulfides by an iron-sulfur cluster. *Science* 287:655–58**

This paper solved the structure of FTR and the mechanism by which electrons are transferred from photoreduced ferredoxin to thioredoxin and then to target enzymes—a long-standing challenging problem in the field.

46. Dai SD, Johansson K, Miginiac-Maslow M, Schürmann P, Eklund H. 2004. Structural basis of redox signaling in photosynthesis: structure and function of ferredoxin: thioredoxin reductase and target enzymes. *Photosynth. Res.* 79:233–48
47. Danon A. 2002. Redox reactions of regulatory proteins: do kinetics promote specificity? *Trends Biochem. Sci.* 27:197–203
48. De Gara L, de Pinto MC, Moliterni VMC, D'Egidio MG. 2003. Redox regulation and storage processes during maturation in kernels of *Triticum durum*. *J. Exp. Bot.* 54:249–58
49. del Val G, Yee BC, Lozano RM, Buchanan BB, Ermel RW, et al. 1999. Thioredoxin treatment increases digestibility and lowers allergenicity of milk. *J. Allergy Clin. Immunol.* 103:690–97
50. Delledonne M, Zeier J, Marocco A, Lamb C. 2001. Signal interactions between nitric oxide and reactive oxygen intermediates in the plant hypersensitive disease resistance response. *Proc. Natl. Acad. Sci. USA* 98:13454–59
51. **Depège N, Bellafiore S, Rochaix JD. 2003. Role of chloroplast protein kinase Stt7 in LHCII phosphorylation and state transition in Chlamydomonas. *Science* 299:1572–75**

The authors identified a protein kinase functional in thylakoid state transition and showed a potential site for its redox-linked regulation via thioredoxin.

52. Desikan R, A-H-Mackerness S, Hancock JT, Neill SJ. 2001. Regulation of the Arabidopsis transcriptome by oxidative stress. *Plant Physiol.* 127:159–72
53. Després C, Chubak C, Rochon A, Clark R, Bethune T, et al. 2003. The Arabidopsis NPR1 disease resistance protein is a novel cofactor that confers redox regulation of DNA binding activity to the basic domain/leucine zipper transcription factor TGA1. *Plant Cell* 15:2181–91
54. D'Haeze W, De Rycke R, Mathis R, Goormachtig S, Pagnotta S, et al. 2003. Reactive oxygen species and ethylene play a positive role in lateral root base nodulation of a semiaquatic legume. *Proc. Natl. Acad. Sci. USA* 100:11789–94
55. Dietz KJ. 2003. Plant peroxiredoxins. *Annu. Rev. Plant Biol.* 54:93–107
56. Fath A, Bethke PC, Jones RL. 2001. Enzymes that scavenge reactive oxygen species are down-regulated prior to gibberellic acid-induced programmed cell death in barley aleurone. *Plant Physiol.* 126:156–66
57. Florencio FJ, Yee BC, Johnson TC, Buchanan BB. 1988. An NADP/thioredoxin system in leaves: purification and characterization of NADP-thioredoxin reductase and thioredoxin h from spinach. *Arch. Biochem. Biophys.* 266:496–507
58. Follmann H, Haberlein I. 1995. Thioredoxins: universal, yet specific thiol-disulfide redox cofactors. *Biofactors* 5:147–56

59. Foreman J, Demidchik V, Bothwell JH, Mylona P, Miedema H, et al. 2003. Reactive oxygen species produced by NADPH oxidase regulate plant cell growth. *Nature* 422:442–46
60. Frand AR, Cuozzo JW, Kaiser CA. 2000. Pathways for protein disulphide bond formation. *Trends Cell Biol.* 10:203–10
61. Gal A, Zer H, Ohad I. 1997. Redox controlled thylakoid protein phosphorylation. News and views. *Physiol. Plantarum* 100:869–85
62. Gelhaye E, Rouhier N, Gérard J, Jolivet Y, Gualberto J, et al. 2004. Thioredoxin *h* regulates enzymes of plant mitochondria. *Proc. Natl. Acad. Sci. USA* 101:14545–50
63. Gelhaye E, Rouhier N, Jacquot J-P. 2003. Evidence for a subgroup of thioredoxin *h* that requires GSH/Grx for its reduction. *FEBS Lett.* 555:443–48
64. Gelhaye E, Rouhier N, Jacquot J-P. 2004. The thioredoxin *h* system of higher plants. *Plant Physiol. Biochem.* 42:265–71
65. Georgiou G. 2002. How to flip the (redox) switch. *Cell* 111:607–10
66. Georlette D, Damien B, Blaise V, Depiereux E, Uversky VN, et al. 2003. Structural and functional adaptations to extreme temperatures in psychrophilic, mesophilic, and thermophilic DNA ligases. *J. Biol. Chem.* 278:37015–23
67. Ghezzi P, Bonetto V. 2003. Redox proteomics: identification of oxidatively modified proteins. *Proteomics* 3:1145–53
68. Gobin P, Ng PKW, Buchanan BB, Kobrehel K. 1997. Sulfhydryl-disulfide changes in proteins of developing wheat grain. *Plant Physiol. Biochem.* 35:777–83
69. Godbole A, Varghese J, Sarin A, Mathew MK. 2003. VDAC is a conserved element of death pathways in plant and animal systems. *Biochim. Biophys. Acta* 1642:87–96
70. Gopalan G, He Z, Balmer Y, Romano P, Gupta R, et al. 2004. Structural analysis uncovers a role for redox in regulating FKBP13, an immunophilin of the chloroplast thylakoid lumen. *Proc. Natl. Acad. Sci. USA* 101:13945–50
71. Goring DR, Walker JC. 2004. Plant sciences. Self-rejection–a new kinase connection. *Science* 303:1474–75
72. Greenberg JT, Yao N. 2004. The role and regulation of programmed cell death in plant-pathogen interactions. *Cell Microbiol.* 6:201–11
73. Gromer S, Urig S, Becker K. 2004. The thioredoxin system–from science to clinic. *Med. Res. Rev.* 24:40–89
74. Gupta R, Luan S. 2003. Redox control of protein tyrosine phosphatases and mitogen-activated protein kinases in plants. *Plant Physiol.* 132:1149–52
75. Haldrup A, Lunde C, Scheller HV. 2003. *Arabidopsis thaliana* plants lacking the PSI-D subunit of photosystem I suffer severe photoinhibition, have unstable photosystem I complexes, and altered redox homeostasis in the chloroplast stroma. *J. Biol. Chem.* 278:33276–83
76. Hazen SP, Wu Y, Kreps JA. 2003. Gene expression profiling of plant responses to abiotic stress. *Funct. Integr. Genomics* 3:105–11
77. He X, Miginiac-Maslow M, Sigalat C, Keryer E, Haraux F. 2000. Mechanism of activation of the chloroplast ATP synthase. A kinetic study of the thiol modulation of isolated ATPase and membrane-bound ATP synthase from spinach by Eschericia coli thioredoxin. *J. Biol. Chem.* 275:13250–58
78. Heine GF, Hernandez MJ, Grotewold E. 2004. Two cysteines in plant R2R3 MYB domains participate in REDOX-dependent DNA binding. *J. Biol. Chem.* 279:37878–85
79. Henkes S, Sonnewald U, Badur R, Flachmann R, Stitt M. 2001. A small decrease of plastid transketolase activity in antisense tobacco transformants has dramatic effects on photosynthesis and phenylpropanoid metabolism. *Plant Cell* 13:535–51

This article showed that, as in animal cells, a membrane-bound NADPH oxidase forms the reactive oxygen species controling cell expansion via activation of the Ca^{2+} channel.

80. Hirasawa M, Brandes HK, Hartman FC, Knaff DB. 1998. Oxidation-reduction properties of the regulatory site of spinach phosphoribulokinase. *Arch. Biochem. Biophys.* 350:127–31
81. Hoeberichts FA, Woltering EJ. 2003. Multiple mediators of plant programmed cell death: interplay of conserved cell death mechanisms and plant-specific regulators. *Bioessays* 25:47–57
82. Hogg PJ. 2003. Disulfide bonds as switches for protein function. *Trends Biochem. Sci.* 28:210–14
83. Holmgren A. 1985. Thioredoxin. *Annu. Rev. Biochem.* 54:237–71
84. Huang DJ, Chen HJ, Hou WC, Lin CD, Lin YH. 2004. Active recombinant thioredoxin h protein with antioxidant activities from sweet potato (*Ipomoea batatas* [L.] Lam Tainong 57) storage roots. *J. Agric. Food Chem.* 52:4720–24
85. Hückelhoven R, Kogel KH. 2003. Reactive oxygen intermediates in plant-microbe interactions: who is who in powdery mildew resistance? *Planta* 216:891–902
86. Issakidis-Bourguet E, Mouaheb N, Meyer Y, Miginiac-Maslow M. 2001. Heterologous complementation of yeast reveals a new putative function for chloroplast *m*-type thioredoxin. *Plant J.* 25:127–35
87. Izui K, Matsumura H, Furumoto T, Kai Y. 2004. Phosphoenolpyruvate carboxylase: A new era of structural biology. *Annu. Rev. Plant Biol.* 55:69–84
88. Jacquot J-P, Gelhaye E, Rouhier N, Corbier C, Didierjean C, Aubry A. 2002. Thioredoxins and related proteins in photosynthetic organisms: molecular basis for thiol dependent regulation. *Biochem. Pharmacol.* 64:1065–69
89. Jiang K, Meng YL, Feldman LJ. 2003. Quiescent center formation in maize roots is associated with an auxin-regulated oxidizing environment. *Development* 130:1429–38
90. Jiang M, Zhang J. 2002. Involvement of plasma-membrane NADPH oxidase in abscisic acid- and water stress-induced antioxidant defense in leaves of maize seedlings. *Planta* 215:1022–30
91. Jiao J, Yee BC, Kobrehel K, Buchanan BB. 1992. Effect of thioredoxin-linked reduction on the activity and stability of the Kunitz and Bowman-Birk soybean trypsin inhibitor proteins. *J. Agr. Food Chem.* 40:2333–36
92. Jiao J-A, Yee BC, Wong JH, Kobrehel K, Buchanan BB. 1993. Thioredoxin-linked changes in regulatory properties of barley alpha-amylase/subtilisin inhibitor protein. *Plant Physiol. Biochem.* 31:799–804
93. Johansson K, Ramaswamy S, Saarinen M, Lemaire-Chamley M, Issakidis-Bourguet E, et al. 1999. Structural basis for light activation of a chloroplast enzyme: The structure of sorghum NADP-malate dehydrogenase in its oxidized form. *Biochemistry* 38:4319–26
94. Johnson TC, Cao RQ, Kung JE, Buchanan BB. 1987. Thioredoxin and NADP-thioredoxin reductase from cultured carrot cells. *Planta* 171:321–31
95. Johnson TC, Wada K, Buchanan BB, Holmgren A. 1987. Reduction of purothionin by the wheat seed thioredoxin system and potential function as a secondary thiol messenger in redox control. *Plant Physiol.* 85:446–51
96. Jones A. 2000. Does the plant mitochondrion integrate cellular stress and regulate programmed cell death? *Trends Plant Sci.* 5:225–30
97. Joo JH, Bae YS, Lee JS. 2001. Role of auxin-induced reactive oxygen species in root gravitropism. *Plant Physiol.* 126:1055–60
98. Kadokura H, Katzen F, Beckwith J. 2003. Protein disulfide bond formation in prokaryotes. *Annu. Rev. Biochem.* 72:111–35
99. Kadokura H, Tian H, Zander T, Bardwell JC, Beckwith J. 2004. Snapshots of DsbA in action: detection of proteins in the process of oxidative folding. *Science* 303:534–37

100. Kaiser W. 1976. The effect of hydrogen peroxide on CO2 fixation of isolated intact chloroplasts. *Biochim. Biophys. Acta* 440:476–82
101. Kalume DE, Molina H, Pandey A. 2003. Tackling the phosphoproteome: tools and strategies. *Curr. Opin. Chem. Biol.* 7:64–69
102. Keryer E, Collin V, Lavergne D, Lemaire S, Issakidis-Bourguet E. 2004. Characterization of Arabidopsis mutants for the variable subunit of ferredoxin: thioredoxin reductase. *Photosynth. Res.* 79:265–74
103. Knaff DB. 2000. Oxidation-reduction properties of thioredoxins and thioredoxin-regulated enzymes. *Physiol. Plantarum* 110:309–13
104. Kobayashi D, Tamoi M, Iwaki T, Shigeoka S, Wadano A. 2003. Molecular characterization and redox regulation of phosphoribulokinase from the cyanobacterium *Synechococcus sp.* PCC 7942. *Plant Cell Physiol.* 44:269–76
105. Kobrehel K, Wong JH, Balogh A, Kiss F, Yee BC, Buchanan BB. 1992. Specific reduction of wheat storage proteins by thioredoxin *h*. *Plant Physiol.* 99:919–24
106. Kovtun Y, Chiu WL, Tena G, Sheen J. 2000. Functional analysis of oxidative stress-activated mitogen-activated protein kinase cascade in plants. *Proc. Natl. Acad. Sci. USA* 97:2940–45
107. Krimm I, Goyer A, Issakidis-Bourguet E, Miginiac-Maslow M, Lancelin JM. 1999. Direct NMR observation of the thioredoxin-mediated reduction of the chloroplast NADP-malate dehydrogenase provides a structural basis for the relief of autoinhibition. *J. Biol. Chem.* 274:34539–42
108. Kumar JK, Tabor S, Richardson CC. 2004. Proteomic analysis of thioredoxin-targeted proteins in *Escherichia coli*. *Proc. Natl. Acad. Sci. USA* 101:3759–64
109. Laloi C, Apel K, Danon A. 2004. Reactive oxygen signalling: the latest news. *Curr. Opin. Plant Biol.* 7:323–28
110. Laloi C, Mestres-Ortega D, Marco Y, Meyer Y, Reichheld JP. 2004. The Arabidopsis cytosolic thioredoxin *h*5 gene induction by oxidative stress and its W-box-mediated response to pathogen elicitor. *Plant Physiol.* 134:1006–16
111. Laloi C, Rayapuram N, Chartier Y, Grienenberger J-M, Bonnard G, Meyer Y. 2001. Identification and characterization of a mitochondrial thioredoxin system in plants. *Proc. Natl. Acad. Sci. USA* 98:14144–49
112. Lancelin JM, Stein M, Jacquot J-P. 1993. Secondary structure and protein folding of recombinant chloroplastic thioredoxin Ch2 from the green alga *Chlamydomonas reinhardtii* as determined by 1H NMR. *J. Biochem. Tokyo* 114:421–31
113. Laskowski MJ, Dreher KA, Gehring MA, Abel S, Gensler AL, Sussex IM. 2002. FQR1, a novel primary auxin-response gene, encodes a flavin mononucleotide-binding quinone reductase. *Plant Physiol.* 128:578–90
114. Laughner BJ, Sehnke PC, Ferl RJ. 1998. A novel nuclear member of the thioredoxin superfamily. *Plant Physiol.* 118:987–96
115. Lawlor DW, Cornic G. 2002. Photosynthetic carbon assimilation and associated metabolism in relation to water deficits in higher plants. *Plant Cell. Environ.* 25:275–94
116. Lemaire SD. 2004. The glutaredoxin family in oxygenic photosynthetic organisms. *Photosynth. Res.* 79:305–18
117. Lemaire SD, Collin V, Keryer E, Issakidis-Bourguet E, Lavergne D, Miginiac-Maslow M. 2003. Chlamydomonas reinhardtii: a model organism for the study of the thioredoxin family. *Plant Physiol. Biochem.* 41:513–21
118. Lemaire SD, Collin V, Keryer E, Quesada A, Miginiac-Maslow M. 2003. Characterization of thioredoxin *y*, a new type of thioredoxin identified in the genome of *Chlamydomonas reinhardtii*. *FEBS Lett.* 543:87–92

119. Lemaire SD, Guillon B, Le Marechal P, Keryer E, Miginiac-Maslow M, Decottignies P. 2004. New thioredoxin targets in the unicellular photosynthetic eukaryote *Chlamydomonas reinhardtii*. *Proc. Natl. Acad. Sci. USA* 101:7475–80
120. Lemaire SD, Miginiac-Maslow M. 2004. The thioredoxin superfamily in *Chlamydomonas reinhardtii*. *Photosynth. Res.* 82:203–20
121. Lennartz K, Plucken H, Seidler A, Westhoff P, Bechtold N, Meierhoff K. 2001. HCF164 encodes a thioredoxin-like protein involved in the biogenesis of the cytochrome b(6)f complex in Arabidopsis. *Plant Cell* 13:2539–51
122. Li H, Sherman LA. 2000. A redox-responsive regulator of photosynthesis gene expression in the cyanobacterium *Synechocystis sp.* Strain PCC 6803. *J. Bacteriol.* 182:4268–77
123. Lindahl M, Florencio FJ. 2003. Thioredoxin-linked processes in cyanobacteria are as numerous as in chloroplasts, but targets are different. *Proc. Natl. Acad. Sci. USA* 16107-2
124. Liszkay A, Kenk B, Schopfer P. 2003. Evidence for the involvement of cell wall peroxidase in the generation of hydroxyl radicals mediating extension growth. *Planta* 217:658–67
125. Lonergan TA. 2000. The photosynthetic reactions do not operate in the dark. *Amer. Biol. Teacher* 62:166–70
126. Lozano RM, Wong JH, Yee BC, Peters A, Kobrehel K, Buchanan BB. 1996. New evidence for a role for thioredoxin *h* in germination and seedling development. *Planta* 200:100–6
127. Maeda K, Finnie C, Svensson B. 2004. Cy5 maleimide-labelling for sensitive detection of free thiols in native protein extracts: Identification of seed proteins targeted by barley thioredoxin *h* isoforms. *Biochem. J.* 378:497–507
128. Marcus F, Chamberlain S, Chu C, Masiarz F, Shin S, et al. 1991. Plant thioredoxin *h*: an animal-like thioredoxin occurring in multiple cell compartments. *Arch. Biochem. Biophys.* 287:195–98
129. Marx C, Wong JH, Buchanan BB. 2003. Thioredoxin and germinating barley: targets and protein redox changes. *Planta* 216:454–60
130. Mayer B, Oberbauer R. 2003. Mitochondrial regulation of apoptosis. *News Physiol. Sci.* 18:89–94
131. Mayfield SP, Yohn CB, Cohen A, Danon A. 1995. Regulation of chloroplast gene expression. *Annu. Rev. Plant Physiol. Plant Mol. Biol.* 46:147–66
132. Mestres-Ortega D, Meyer Y. 1999. The *Arabidopsis thaliana* genome encodes at least four thioredoxins *m* and a new prokaryotic-like thioredoxin. *Gene* 240:307–16
133. Miginiac-Maslow M, Johansson K, Ruelland E, Issakidis-Bourguet E, Schepens I, et al. 2000. Light-activation of NADP-malate dehydrogenase: a highly controlled process for an optimized function. *Physiol. Plantarum* 110:322–29
134. Mittard V, Blackledge MJ, Stein M, Jacquot J-P, Marion D, Lancelin JM. 1997. NMR solution structure of an oxidised thioredoxin *h* from the eukaryotic green alga *Chlamydomonas reinhardtii*. *Eur. J. Biochem.* 243:374–83
135. Mittler R. 2002. Oxidative stress, antioxidants and stress tolerance. *Trends Plant Sci.* 7:405–10
136. Motohashi K, Kondoh A, Stumpp MT, Hisabori T. 2001. Comprehensive survey of proteins targeted by chloroplast thioredoxin. *Proc. Natl. Acad. Sci. USA* 98:11224–29

This paper described a new approach for the identification of target proteins by a combination of thioredoxin-mutant affinity chromatography and proteomics.

137. Muller EG, Buchanan BB. 1989. Thioredoxin is essential for photosynthetic growth. The thioredoxin *m* gene of *Anacystis nidulans*. *J. Biol. Chem.* 264:4008–14
138. Nakamoto SS, Hamel P, Merchant S. 2000. Assembly of chloroplast cytochromes b and c. *Biochimie* 82:603–14

139. Nakamura H, Nakamura K, Yodoi J. 1997. Redox regulation of cellular activation. *Annu. Rev. Immunol.* 15:351–69
140. Nasrallah JB. 2000. Cell-cell signaling in the self-incompatibility response. *Curr. Opin. Plant Biol.* 3:368–73
141. Navarro F, Florencio FJ. 1996. The cyanobacterial thioredoxin gene is required for both photoautotrophic and heterotrophic growth. *Plant Physiol.* 111:1067–75
142. Neill S, Desikan R, Hancock J. 2002. Hydrogen peroxide signalling. *Curr. Opin. Plant Biol.* 5:388–95
143. op den Camp RG, Przybyla D, Ochsenbein C, Laloi C, Kim C, et al. 2003. Rapid induction of distinct stress responses after the release of singlet oxygen in Arabidopsis. *Plant Cell* 15:2320–32
144. Page ML, Hamel PP, Gabilly ST, Zegzouti H, Perea JV, et al. 2004. A homolog of prokaryotic thiol disulfide transporter CcdA is required for the assembly of the cytochrome b6f complex in Arabidopsis chloroplasts. *J. Biol. Chem.* 279:32474–82
145. Pei ZM, Ghassemian M, Kwak CM, McCourt P, Schroeder JI. 1998. Role of farnesyltransferase in ABA regulation of guard cell anion channels and plant water loss. *Science* 282:287–90
146. Peltier JB, Emanuelsson O, Kalume DE, Ytterberg J, Friso G, et al. 2002. Central functions of the lumenal and peripheral thylakoid proteome of Arabidopsis determined by experimentation and genome-wide prediction. *Plant Cell* 14:211–36
147. Pfannschmidt T. 2003. Chloroplast redox signals: how photosynthesis controls its own genes. *Trends Plant Sci.* 8:33–41

This recent review details the mechanism by which redox controls the expression of photosynthetic genes via the oxidation state of plastoquinone.

148. Pignocchi C, Fletcher JM, Wilkinson JE, Barnes JD, Foyer CH. 2003. The function of ascorbate oxidase in tobacco. *Plant Physiol.* 132:1631–41
149. Pompella A, Visvikis A, Paolicchi A, De Tata V, Casini AF. 2003. The changing faces of glutathione, a cellular protagonist. *Biochem. Pharmacol.* 66:1499–503
150. Portis AR Jr. 2003. Rubisco activase: Rubisco's catalytic chaperone. *Photosynth. Res.* 75:11–27
151. Raines CA. 2003. The Calvin cycle revisited. *Photosynth. Res.* 75:1–10
152. Rancourt RC, Tai S, King M, Heltshe SL, Penvari C, et al. 2004. Thioredoxin liquefies and decreases the viscoelasticity of cystic fibrosis sputum. *Am. J. Physiol. Lung Cell. Mol. Physiol.* 286:931–38
153. Reading NS, Aust SD. 2000. Engineering a disulfide bond in recombinant manganese peroxidase results in increased thermostability. *Biotechnol. Prog.* 16:326–33
154. Reichheld JP, Mestres-Ortega D, Laloi C, Meyer Y. 2002. The multigenic family of thioredoxin *h* in *Arabidopsis thaliana*: specific expression and stress response. *Plant Physiol. Biochem.* 40:685–90
155. Renesto P, Crapoulet N, Ogata H, La Scola B, Vestris G, et al. 2003. Genome-based design of a cell-free culture medium for *Tropheryma whipplei*. *Lancet* 362:447–49
156. Rhazi L, Cazalis R, Aussenac T. 2003. Sulfhydryl-disulfide changes in storage proteins of developing wheat grain: influence on the SDS-unextractable glutenin polymer formation. *J. Cereal Sci.* 38:3–13
157. Rintamäki E, Martinsuo P, Pursiheimo S, Aro EM. 2000. Cooperative regulation of light-harvesting complex II phosphorylation via the plastoquinol and ferredoxin-thioredoxin system in chloroplasts. *Proc. Natl. Acad. Sci. USA* 97:11644–49
158. Ritz D, Beckwith J. 2001. Roles of thiol-redox pathways in bacteria. *Annu. Rev. Microbiol.* 55:21–48
159. Rivas S, Rougon-Cardoso A, Smoker M, Schauser L, Yoshioka H, Jones JD. 2004. CITRX thioredoxin interacts with the tomato Cf-9 resistance protein and negatively regulates defence. *EMBO J.* 23:2156–65

160. Rossel JB, Wilson IW, Pogson BJ. 2002. Global changes in gene expression in response to high light in Arabidopsis. *Plant Physiol.* 130:1109–20
161. Rouhier N, Gelhaye E, Jacquot J-P. 2002. Exploring the active site of plant glutaredoxin by site-directed mutagenesis. *FEBS Lett.* 511:145–49
162. Rouhier N, Gelhaye E, Jacquot J-P. 2002. Glutaredoxin-dependent peroxiredoxin from poplar: protein-protein interaction and catalytic mechanism. *J. Biol. Chem.* 277:13609–14
163. **Rouhier N, Gelhaye E, Jacquot J-P. 2004. Plant glutaredoxins: still mysterious reducing systems. *Cell. Mol. Life Sci.* 61:1266–77**

This review presents a detailed analysis of members of the *Arabidopsis* glutaredoxin family and their function as disulfide oxidoreductases in plants.

164. Rouhier N, Gelhaye E, Sautière P-E, Brun A, Laurent P, et al. 2001. Isolation and characterization of a new peroxiredoxin from poplar sieve tubes that uses either glutaredoxin or thioredoxin as a proton donor. *Plant Physiol.* 127:1299–309
165. Roychaudhuri R, Sarath G, Zeece M, Markwell J. 2004. Stability of the allergenic soybean Kunitz trypsin inhibitor. *Biochim. Biophys. Acta* 1699:207–12
166. Salvador ML, Klein U. 1999. The redox state regulates RNA degradation in the chloroplast of *Chlamydomonas reinhardtii*. *Plant Physiol.* 121:1367–74
167. Scheibe R. 2004. Malate valves to balance cellular energy supply. *Physiol. Plantarum* 120:21–26
168. Schnable PS, Wise RP. 1998. The molecular basis of cytoplasmic male sterility and fertility restoration. *Trends Plant Sci.* 3:175–80
169. Schürmann P. 1983. The light activation of chloroplast enzymes through the ferredoxin/thioredoxin system. In *Photosynthesis and plant productivity*, ed. H Metzner, pp. 255–58. Stuttgart: Wissenschaftl. Verlagsgesellschaft mBH
170. Schürmann P. 2003. Redox signaling in the chloroplast: the ferredoxin/thioredoxin system. *Antioxid. Redox Signal.* 5:69–78
171. Schürmann P, Buchanan BB. 2001. The structure and function of the ferredoxin/thioredoxin system. In *Advances in Photosynthesis*, ed. B Andersson, EM Aro, pp. 331–61. Dordrecht, The Netherlands: Kluwer
172. **Schürmann P, Jacquot J-P. 2000. Plant thioredoxin systems revisited. *Annu. Rev. Plant Physiol. Plant Mol. Biol.* 51:371–400**

This account, which traces the development of the plant thioredoxin field during the 1980s and 1990s, was used as a point of reference in the current review.

173. Sen CK, Packer L. 1996. Antioxidant and redox regulation of gene transcription. *FASEB J.* 10:709–20
174. Serrato AJ, Cejudo FJ. 2003. Type-*h* thioredoxins accumulate in the nucleus of developing wheat seed tissues suffering oxidative stress. *Planta* 217:392–99
175. Serrato AJ, Crespo JL, Florencio FJ, Cejudo FJ. 2001. Characterization of two thioredoxins *h* with predominant localization in the nucleus of aleurone and scutellum cells of germinating wheat seeds. *Plant Mol. Biol.* 46:361–71
176. Serrato AJ, Perez-Ruiz JM, Spinola MC, Cejudo FJ. 2004. A novel NADPH thioredoxin reductase, localized in the chloroplast, which deficiency causes hypersensitivity to abiotic stress in *Arabidopsis thaliana*. *J. Biol. Chem.* 279:43821–27
177. Setterdahl AT, Chivers PT, Hirasawa M, Lemaire SD, Keryer E, et al. 2003. Effect of pH on the oxidation-reduction properties of thioredoxins. *Biochemistry* 42:14877–84
178. Simon-Plas F, Elmayan T, Blein JP. 2002. The plasma membrane oxidase NtrbohD is responsible for AOS production in elicited tobacco cells. *Plant J.* 31:137–47
179. Sparla F, Pupillo P, Trost P. 2002. The C-terminal extension of glyceraldehyde-3-phosphate dehydrogenase subunit B acts as an autoinhibitory domain regulated by thioredoxins and nicotinamide adenine dinucleotide. *J. Biol.Chem.* 277:44946–52

180. Spetea C, Hundal T, Lundin B, Heddad M, Adamska I, Andersson B. 2004. Multiple evidence for nucleotide metabolism in the chloroplast thylakoid lumen. *Proc. Natl. Acad. Sci. USA* 101:1409–14
181. Stone JR. 2004. An assessment of proposed mechanisms for sensing hydrogen peroxide in mammalian systems. *Arch. Biochem. Biophys.* 422:119–24
182. Sun Q, Emanuelsson O, van Wijk KJ. 2004. Analysis of curated and predicted plastid subproteomes of Arabidopsis. Subcellular compartmentalization leads to distinctive proteome properties. *Plant Physiol.* 135:723–34
183. Suske G, Wagner W, Follmann H. 1979. NADPH dependent thioredoxin reductase and a new thioredoxin from wheat. *Z. Naturforsch.* 34:214–21
184. Takeda T, Yokota A, Shigeoka S. 1995. Resistance of photosynthesis to hydrogen peroxide in algae. *Plant Cell Physiol.* 36:1089–95
185. Tamarit J, Belli G, Cabiscol E, Herrero E, Ros J. 2003. Biochemical characterization of yeast mitochondrial Grx5 monothiol glutaredoxin. *J. Biol. Chem.* 278:25745–51
186. Tiwari BS, Belenghi B, Levine A. 2002. Oxidative stress increased respiration and generation of reactive oxygen species, resulting in ATP depletion, opening of mitochondrial permeability transition, and programmed cell death. *Plant Physiol.* 128:1271–81
187. Trebitsh T, Danon A. 2001. Translation of chloroplast psbA mRNA is regulated by signals initiated by both photosystems II and I. *Proc. Natl. Acad. Sci. USA* 98:12289–94
188. Trivelli X, Bouillac S, Tsan P, Krimm I, Lancelin JM. 2004. NMR of redox proteins of plants, yeasts and photosynthetic bacteria. *Photosynth. Res.* 79:357–67
189. Tron AE, Bertoncini CW, Chan RL, Gonzalez DH. 2002. Redox regulation of plant homeodomain transcription factors. *J. Biol. Chem.* 277:34800–7
190. Vandenabeele S, Van Der Kelen K, Dat J, Gadjev I, Boonefaes T, et al. 2003. A comprehensive analysis of hydrogen peroxide-induced gene expression in tobacco. *Proc. Natl. Acad. Sci. USA* 100:16113–18
191. Vignols F, Mouaheb N, Thomas D, Meyer Y. 2003. Redox control of Hsp70-Co-chaperone interaction revealed by expression of a thioredoxin-like Arabidopsis protein. *J. Biol. Chem.* 278:4516–23
192. Villeret V, Huang S, Zhang Y, Xue Y, Lipscomb WN. 1995. Crystal structure of spinach chloroplast fructose-1,6-bisphosphatase at 2.8 Å resolution. *Biochemistry* 34:4299–306
193. Vlamis-Gardikas A, Holmgren A. 2002. Thioredoxin and glutaredoxin isoforms. *Methods Enzymol.* 347:286–96
194. Walters EM, Johnson MK. 2004. Ferredoxin: thioredoxin reductase: disulfide reduction catalyzed via novel site-specific [4Fe-4S] cluster chemistry. *Photosynth. Res.* 79:249–64
195. Wang W, Vinocur B, Altman A. 2003. Plant responses to drought, salinity and extreme temperatures: towards genetic engineering for stress tolerance. *Planta* 218:1–14
196. Webb MA, Cavaletto JM, Klanrit P, Thompson GA. 2001. Orthologs in *Arabidopsis thaliana* of the Hsp70 interacting protein Hip. *Cell Stress Chaperones* 6:247–55
197. Wedel N, Soll J, Paap BK. 1997. CP12 provides a new mode of light regulation of Calvin cycle activity in higher plants. *Proc. Natl. Acad. Sci. USA* 94:10479–84
198. Weis E. 1981. Reversible heat inactivation of the calvin cycle a possible mechanism of the temperature regulation of photosynthesis. *Planta* 151:33–39
199. Wolosiuk RA, Buchanan BB. 1977. Thioredoxin and glutathione regulate photosynthesis in chloroplasts. *Nature* 266:565–67
200. Wong JH, Balmer Y, Cai N, Tanaka CK, Vensel WH, et al. 2003. Unraveling thioredoxin-linked metabolic processes of cereal starchy endosperm using proteomics. *FEBS Lett.* 547:151–56

201. Wong JH, Cai N, Balmer Y, Tanaka CK, Vensel WH, et al. 2004. Thioredoxin targets of developing wheat seeds identified by complementary proteomic approaches. *Phytochemistry* 65:1629–40
202. Wong JH, Cai N, Tanaka CK, Vensel WH, Hurkman WJ, Buchanan BB. 2004. Thioredoxin reduction alters the solubility of proteins of wheat starchy endosperm: an early event in cereal germination. *Plant Cell Physiol.* 45:407–15
203. Wong JH, Jiao J-A, Kobrehel K, Buchanan BB. 1995. Thioredoxin-dependent deinhibition of pullulanase of barley malt by inactivation of a specific inhibitor protein. *Plant Physiol.* 108:67
204. Wong JH, Kim YB, Ren P-S, Cai N, Cho M-J, et al. 2002. Transgenic barley grain overexpressing thioredoxin shows evidence of communication between the endosperm and the embryo and aleurone. *Proc. Natl. Acad. Sci. USA* 99:16325–30
205. Wong JH, Kobrehel K, Nimbona C, Yee BC, Balogh A, et al. 1993. Thioredoxin and bread wheat. *Cereal Chem.* 70:113–14
206. Yamazaki D, Motohashi K, Kasama T, Hara Y, Hisabori T. 2004. Target proteins of the cytosolic thioredoxins in *Arabidopsis thaliana. Plant Cell Physiol.* 45:18–27
207. Yano H, Kuroda S, Buchanan BB. 2002. Disulfide proteome in the analysis of protein function and structure. *Proteomics* 2:1090–96
208. Yano H, Wong JH, Cho M-J, Buchanan BB. 2001. Redox changes accompanying the degradation of seed storage proteins in germinating rice. *Plant Cell Physiol.* 42:879–83
209. Yano H, Wong JH, Lee YM, Cho M-J, Buchanan BB. 2001. A strategy for the identification of proteins targeted by thioredoxin. *Proc. Natl. Acad. Sci. USA* 98:4794–99
210. Yosef I, Irihimovitch V, Knopf JA, Cohen I, Orr-Dahan I, et al. 2004. RNA binding activity of the ribulose-1,5-bisphosphate carboxylase/oxygenase large subunit from Chlamydomonas reinhardtii. *J. Biol. Chem.* 279:10148–56
211. Zhang N, Portis AR Jr. 1999. Mechanism of light regulation of Rubisco: a specific role for the larger Rubisco activase isoform involving reductive activation by thioredoxin-*f*. *Proc. Natl. Acad. Sci. USA* 96:9438–43
212. Zheng M, Storz G. 2000. Redox sensing by prokaryotic transcription factors. *Biochem. Pharmacol.* 59:1–6
213. Falini G, Fermani S, Ripamonti A, Sabatino P, Sparla F, et al. 2003. Dual coenzyme specificity of photosynthetic glyceraldehydes-3-phosphate dehydrogenase interpreted by the crystal structure of A4 isoform complexed with NAD. *Biochemistry* 42:4631–39
214. Lopez-Maury L, Garcia-Dominquez M, Florencio FJ, Reyes JC. 2002. A two-component signal transduction system involved in nickel sensing in the cyanobacterium *Synechocystis sp.* PCC 6803. *Mol. Microbiol.* 43:247–56

Endocytotic Cycling of PM Proteins

Angus S. Murphy,[1] Anindita Bandyopadhyay,[1] Susanne E. Holstein,[2] and Wendy A. Peer[1]

[1]Department of Horticulture, Purdue University, West Lafayette, Indiana 47907; email: murphy@purdue.edu

[2]Institute for Plant Sciences, University of Heidelberg, 69120, Heidelberg, Germany

Annu. Rev. Plant Biol. 2005. 56:221–51

doi: 10.1146/annurev.arplant.56.032604.144150

First published online as a Review in Advance on January 14, 2005

1543-5008/05/0602-0221$20.00

Key Words

endosomes, recycling, internalization, trafficking, auxin

Abstract

Plasma membrane protein internalization and recycling mechanisms in plants share many features with other eukaryotic organisms. However, functional and structural differences at the cellular and organismal level mandate specialized mechanisms for uptake, sorting, trafficking, and recycling in plants. Recent evidence of plasma membrane cycling of members of the PIN auxin efflux facilitator family and the KAT1 inwardly rectifying potassium channel demonstrates that endocytotic cycling of some form occurs in plants. However, the mechanisms underlying protein internalization and the signals that stimulate endocytosis of proteins from the cell-environment interface are poorly understood. Here we summarize what is known of endocytotic cycling in animals and compare those mechanisms with what is known in plants. We discuss plant orthologs of mammalian-trafficking proteins involved in endocytotic cycling. The use of the styryl dye FM4-64 to define the course of endocytotic uptake and the fungal toxin brefeldin A to dissect the internalization pathways are particularly emphasized. Additionally, we discuss progress in identifying distinct endosomal populations marked by the small GTPases Ara6 and Ara7 as well as recently described examples of apparent cycling of plasma membrane proteins.

Contents

INTRODUCTION

Eukaryotic cells respond to external signals by altering transcriptional output and regulating the abundance and distribution of PM proteins that comprise the functional interface with the external environment. In animals and yeast, PM protein removal and recompartmentalization is regulated by endocytosis and endocytotic cycling. As the mechanisms underlying endocytotic cycling have been elucidated, the expectation that similar mechanisms would be found in plants has increased. Recent evidence that the PM localization of the PIN1 auxin efflux carrier complex and the KAT1 potassium channel are dynamically regulated (44, 102) has intensified efforts to conclusively demonstrate endocytotic recycling in plants. In both cases, the proteins involved have been shown to respond to both signaling molecules and external environmental cues (12, 72, 123) although no apparent cell surface receptor is involved. Many efforts are now underway to fit plant vesicular cycling mechanisms into the context of mammalian models. However, as **Figure 1** shows, developmental and structural differences at both the cellular and organismal level preclude the wholesale application of animal models to plants. In this review, we attempt to summarize the current understanding of this area of plant cell biology.

PM: Plasma Membrane

Vectorial redirection: redistribution of a protein from the apical or basal plasma membrane to a lateral membrane

ENDOSOMAL CYCLING IN ANIMALS: AN OVERVIEW

Endocytosis is an essential eukaryotic phenomenon whereby cells take up extracellular substances and/or internalize PM proteins for transport to endosomes. Eukaryotic cellular membranes are highly dynamic structures; in a typical cell the entire surface area of the PM turns over completely on an hourly basis (171). Endocytotic cycling helps maintain homeostatic regulation primarily by increasing or decreasing the surface content of mediating molecules in response to extracellular cues. From the endosomes, PM proteins are either targeted into the lysosome/vacuole for degradation or recycled back to cell surface (116). Mammalian endocytosis regulates multiple physiological processes such as nutrient uptake, retrieval of exocytosed vesicle components, downregulation of signaling receptors, and the localization/abundance of membrane transporters (116).

In animals, endosomal cycling is an essential component of endocrine and cell-to-cell signaling. Although the cycling of membrane transport proteins in response to internal signals has been documented (93), the endocytotic cycling of PM receptors in a process known as RME has been more intensively studied. In RME, the binding of an extracellular ligand to a PM

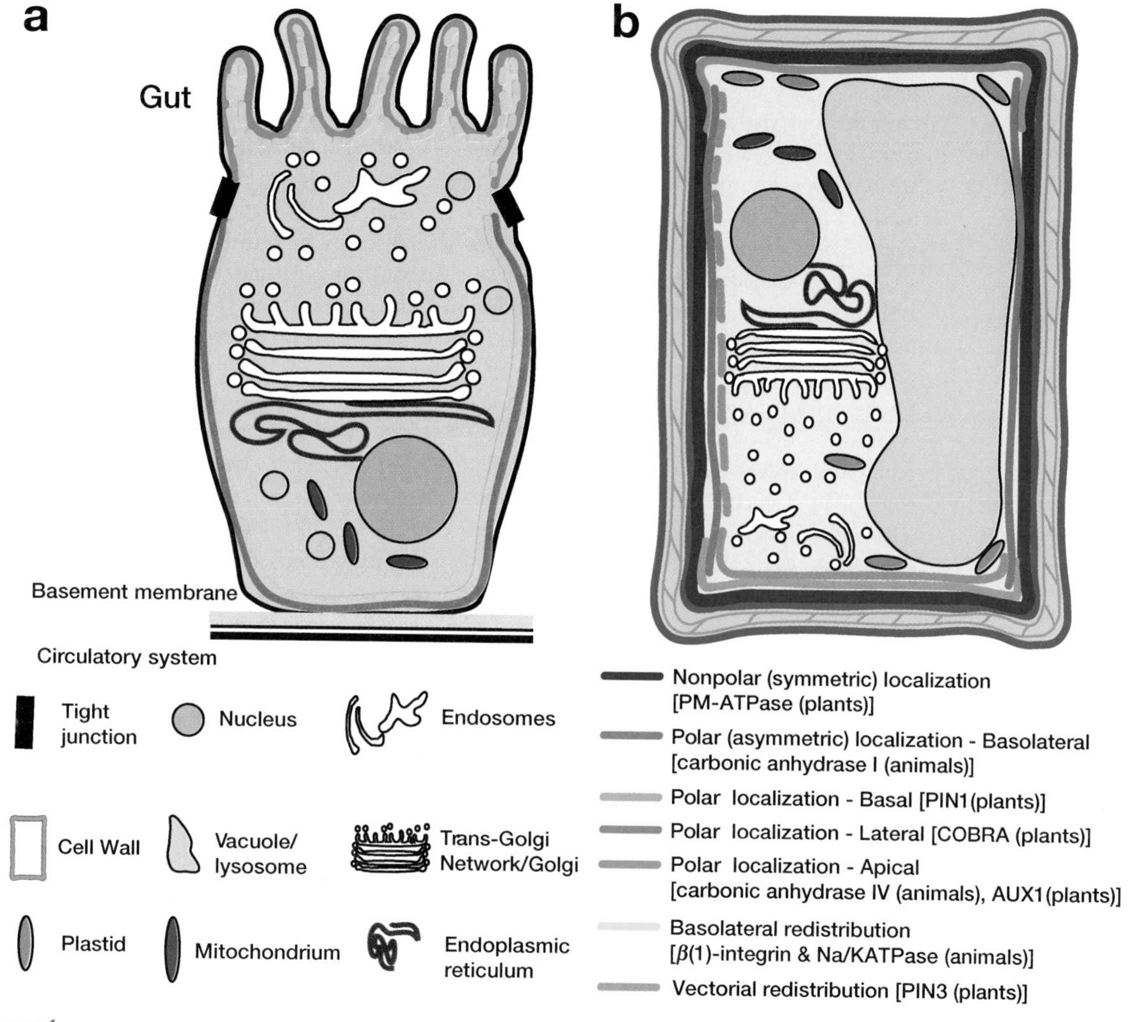

Figure 1

A comparison of plasma membrane (PM) localization in polarized animal and plant cells. Polarized animal cells have tight junctions that delineate the cell into basolateral and apical regions. As a result, animal cells exhibit targeting of PM proteins to basolateral or apical membranes (36, 85), basolateral redistribution of mistargeted proteins (186), and symmetric redistribution of proteins to all sides of the PM following loss of polarity, as seen when pathogenic *E. coli* infection disrupts tight junctions (115). Plant cells have cell walls and lack tight junctions to define the polarity of the cell. To date, four types of PM targeting have been documented in plants (basal, lateral, apical, and nonpolar) (28, 39, 145, 163, 165). Vectorial redistribution of PM proteins following a stimulus has been shown, as with the PIN3 auxin efflux facilitator that is redistributed from the basal to the lateral side of cells after tropic stimulus (39). (*A*) Generalized animal intestinal epithelial cell showing basolateral localization of carbonic anhydrase XII, apical localization of carbonic anhydrase IV, and basolateral redistribution of $\beta(1)$-integrin or Na^+/K^+ ATPase after a pathogen-induced change in cellular polarity. Dashed lines denote redistributed proteins. (*B*) Generalized plant cell showing nonpolar localization of the PM ATPase, basal localization of PIN1, lateral localization of COBRA, apical localization of AUX1, and vectorial redistribution of PIN3 after a gravitropic stimulus.

RME: receptor-mediated endocytosis, the process by which a ligand-bound receptor is internalized and sorted

ERC: endosomal recycling compartment, a network of tubules marked by the presence of LDL-R and iron-free Tf bound to the Tf-R. Proteins from the sorting endosomes are directed to the ERC for recycling to the plasma membrane or for sorting to the degradation pathway.

Rab: small GTPases implicated in several steps of endocytotis. Of the plant orthologs, Ara6 is unique to the kingdom.

receptor initiates the uptake of proteins or protein complexes from the cell surface. Of the various animal cell surface protein uptake mechanisms, clathrin-dependent endocytosis is by far the best investigated. Less is known about clathrin-independent processes, which are more difficult to study because they are not concentrative and often lack specific markers. Clathrin-independent pathways include caveolae/lipid raft-mediated endocytosis, fluid-phase endocytosis, and phagocytosis. PM lipid microdomains enriched in cholesterol and sphingolipids sequester membrane proteins and play a role in trafficking (59). Studies with recycling Shiga toxin and Glycosylphosphatidyl inositol (GPI)-anchored proteins such as the folate receptor suggest a role for lipid microdomains in sorting in the ERC. Fluid-phase endocytosis and phagocytosis have been documented in animal cells as molecular internalization mechanisms but it has not been demonstrated that molecules internalized by these processes are recycled to the PM.

In clathrin-mediated RME, ligands bound to their respective receptors and destined for degradation in lysosomes become internalized via clustering in clathrin-coated pits (103). As a result of recycling to the PM, some receptors are reused up to several hundred times. Endocytotic and biosynthetic recycling also maintain the identity of the respective organelles by retrieving specific sets of functional proteins (organelle homeostasis). RME is also essential to the basolateral redirection that maintains the specificity of apical and basolateral membranes in polarized cells (186). Because clathrin-mediated internalization is the most well understood endocytotic process, this overview focuses on recycling events in this pathway (**Figure 2**).

The first step in clathrin-mediated endocytosis is the recruitment of adaptor protein (AP) complexes and clathrin to the PM. Phosphoinositides initiate coat assembly by directly binding to endocytotic proteins such as the AP2 adaptor complex and AP-180 (88, 140). An ARF6 GTPase was also implicated in this initial recruitment step. APs link specific cargo to sites of coat assembly. The pinching off of the coated vesicle requires additional proteins such as dynamins, amphiphysin, and Eps15. The vesicles rapidly shed the coat proteins to undergo fusion with target endosomes and to allow recycling of the coat components. This uncoating process requires auxilin and the molecular chaperone Hsc70 (66, 175). The newly formed endosomes fuse with one another and with pre-existing sorting endosomes in a process that requires *N*-ethylmaleimide-sensitive factor adaptor protein receptor (SNARE) and Rab proteins.

The first stop for transmembrane proteins internalized by clathrin-mediated endocytosis is the early endosomes, which include the peripherally localized tubular-vesicular structures of the sorting endosomes as well as the ERC. Due to their intracellular localization, the sorting endosomes are the first organelles to receive PM-derived cargo such as the low-density lipoprotein (LDL) and transferrin (Tf). In the acidic lumen of the sorting endosome (pH 6.0) (75), the dissociation of ligands from their receptors occurs. The sorting endosome is also the first main junction in the RME pathway because two main destinations can be reached from there: the PM and the late endosomes. Extensive studies of Tf and its receptor (Tf-R) show that recycling to the PM can either occur directly from the sorting endosome or via the ERC (55, 154). For recycling the Tf-R and some fluorescent lipid analogs (e.g., C6-NBD-sphingomyelin) from the sorting endosome, the kinetics are fast, with a half-life ($t_{1/2}$) of about two minutes and with an efficiency of more than 95%.

The ERC is associated with microtubules and exhibits a variable distribution, although more specific distributions in the vicinity of the nucleus and microtubule-organizing centers have been reported (50). In contrast to the sorting endosome, the ERC does not contain ligands or receptors that are destined for degradation, such as lysosomal enzymes and LDL, but instead is marked by the presence of LDL-R and iron-free Tf bound to the

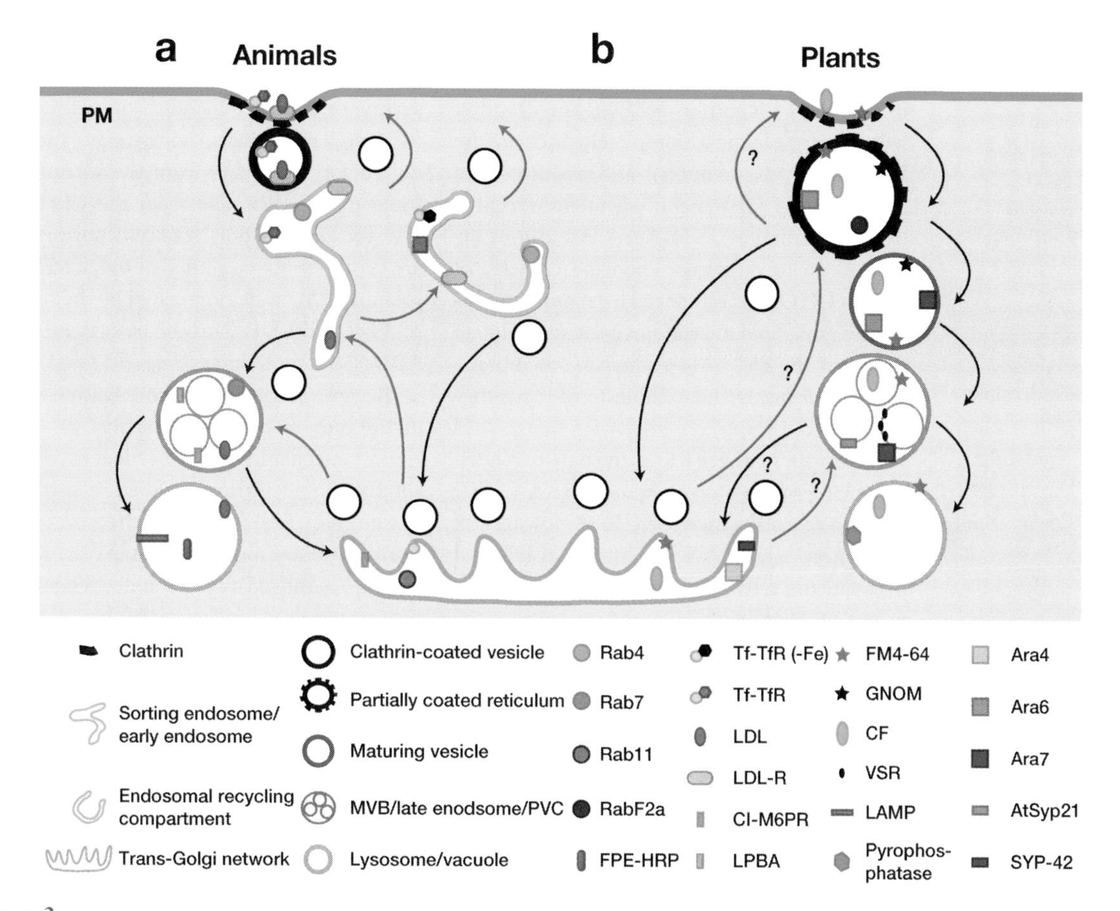

Figure 2

Endocytotic cycling in animals and plants. Endocytotic cycling is well studied in animal cells with several types of cycling known: clathrin-mediated, clathrin-coated receptor-mediated, nonclathrin-mediated (caveolae/lipid raft–mediated, fluid-phase endocytosis, phagocytosis). In contrast, little is known about endocytosis in plants and much has been extrapolated from animal literature. Two discrete endosomal compartments have been identified in plants, as well as a gradation of maturation from one to the other. Redirection of endocytosed PM proteins back to the PM has not been directly demonstrated but is strongly supported by indirect evidence (43, 44). Black arrows indicate the endocytotic pathway; red arrows indicate the secretory pathway. (*A*) In animals, clathrin-coated and receptor-mediated endocytosed vesicles from the plasma membrane (PM) are directed to the sorting endosome/early endosome [Rab4, LDL-receptor (LDL-R) and transferrin receptor (TfR) markers]. From there, cargo is transported to the PM, endosomal recycling compartment [Rab4, Rab11, LDL-R, transferrin bound to transferrin receptor (Tf-TfR)], or multivesicular body (MVB)/late endosome [Rab7, M6PR, lysobisphosphatidic acid (LPBA) (87)]. From the endosomal recycling compartment, cargo can traffic back to the PM or the trans-Golgi network (TGN) (Rab11). From the MVB/late endosome, cargo can travel to the TGN [cationic-independent mannose-6-phosphate receptor (CI-M6PR)] or the lysosome [fluid-phase endocytosed HRP (FPE-HRP), lysosome-associated membrane protein (LAMP) (24)]. Cargo can also travel from the TGN to the sorting endosome/early endosome and MVB/late endosome. (*B*) In plants, cargo and PM proteins/markers are endocytosed into the partially coated reticulum (PCR) [catonized ferritin (CF), GNOM, Ara6, FM4-64, RabF2a]. From the PCR, vesicle maturation results in direction of cargo to MVBs [CF, FM4-64, Ara7, AtSyp21, VSR proteins] through vesicle maturation with overlapping compartment markers [Ara6 and Ara7] or the TGN [CF, FM4-64, SYP-42, Ara4]. From the MVB/late endosome/prevacuolar compartment (PVC), cargo is trafficked to the vacuole [CF, FM4-64, pyrophosphatase (139)]; trafficking to the TGN is hypothesized. Trafficking from the TGN to the PCR or MVB has not been demonstrated.

IRAP: insulin-responsive aminopeptidase, a type II integral membrane protein that colocalizes with GLUT4 in specialized insulin-responsive compartments and exhibits similar trafficking dynamics as GLUT4

TGN: trans-golgi network

EGF: Epidermal Growth Factor

Tf-R (99, 111). Two kinetics have been described for ERC-to-PM recycling: a fast rate used by the Tf-R and C6-NBD-sphingomyelin, which recycle almost completely back to the PM ($t_{1/2}$ = 10 minutes), and a slower rate used by the GPI-anchored folate receptor ($t_{1/2}$ = 30 minutes). This latter rate can be increased up to threefold when the cholesterol level in CHO cells is reduced by about 40%, indicating that lipid microdomains may modulate sorting in the ERC (111). The ERC membrane is the main intracellular cholesterol repository (54).

The role of the ERC in insulin-regulated trafficking of the glucose-transporter GLUT4 was also recently described (18). At low insulin concentrations, GLUT4 is equally distributed between the ERC and another distinct compartment, characterized by the presence of the insulin-responsive aminopeptidase (IRAP) and the lack of Tf-R (18, 195) (**Figure 3**). In the absence of insulin stimulation, GLUT4 is constitutively transported from the internal compartments and cycles to the PM in a manner similar to Tf-R and other cell-surface proteins (141). However, in response to increased insulin concentrations, a rapid net translocation of GLUT4 from the IRAP internal storage compartments to the PM occurs, resulting in higher uptake of glucose in fat and muscle cells (195).

Unlike the sorting endosomes, which mature into late endosomes within 10 minutes, the ERC is a relatively long-lived organelle that requires vesicle coat proteins for recycling. The ERC uses clathrin-dynamin-coated vesicles for transport to the PM in nonpolarized and polarized cells (164, 177, 178). As such, the clathrin coat does not play a prominent role in sorting or recycling steps (11). It is also not required for concentrating the Tf-R because a truncated version of this protein that does not interact with APs recycles at the same rate as the wild-type receptor in nonpolarized cells (98). Rab11 and an Eps15-homology protein containing an EH domain (EHD1/Rme1) also specifically regulate transport from the ERC to the TGN and PM (92, 187). In addition, COPI coat proteins (2), sorting nexins, (191) and ARF6 (125) have been implicated in ERC sorting or ERC-mediated sorting of receptors. In nonpolarized cells, no sorting signal is required for trafficking from the sorting endosome to the ERC and from the ERC to the PM. However, in polarized cells, sorting motifs are required for targeting to the apical or basolateral PM, as demonstrated by the recycling of Tf-R and LDL-R to the basolateral PM via AP1/μ1B-containing vesicles (41).

Because sorting endosomes mature into late endosomes, no vesicular carriers or concentration mechanisms are needed to transport soluble ligands. However, before sorting endosomes mature into late endosomes, most transmembrane proteins destined to be recycled are removed rapidly and efficiently in a process described as "geometry-based" sorting. In this process, up to 80% of the membrane of sorting endosomes separates into narrow-diameter–tubules, which have a greater surface-area-to-volume ratio; removing these tubules results in separation of recycling receptors from soluble cargo (98, 111). Although this recycling step requires no sorting motifs, further sorting of the internalized transmembrane proteins into the degradation pathway (receptor downregulation) does.

Signaling receptors such as the G-protein-coupled receptors and the receptor–protein kinases contain an ubiquitin moiety at their cytoplasmic tails. Ubiquitination serves as a sorting signal for targeting into the invaginated membranes of the late endosome, a sorting process that simultaneously terminates their signaling function (81). Subsequently, an hepatocyte-growth-factor-regulated tyrosine kinase substrate (HRS) serves as a linker between the ubiquitinated receptors and the endosomal flat clathrin lattice, resulting in a bilayered clathrin coat that contains the mammalian equivalents to the yeast ESCRT (endosomal sorting complex required for transport) complexes (134, 143). For the EGF receptor protein kinase, the bulk of EGF receptors are bound to EGF and follow the process of receptor downregulation, but are also recycled from either the sorting

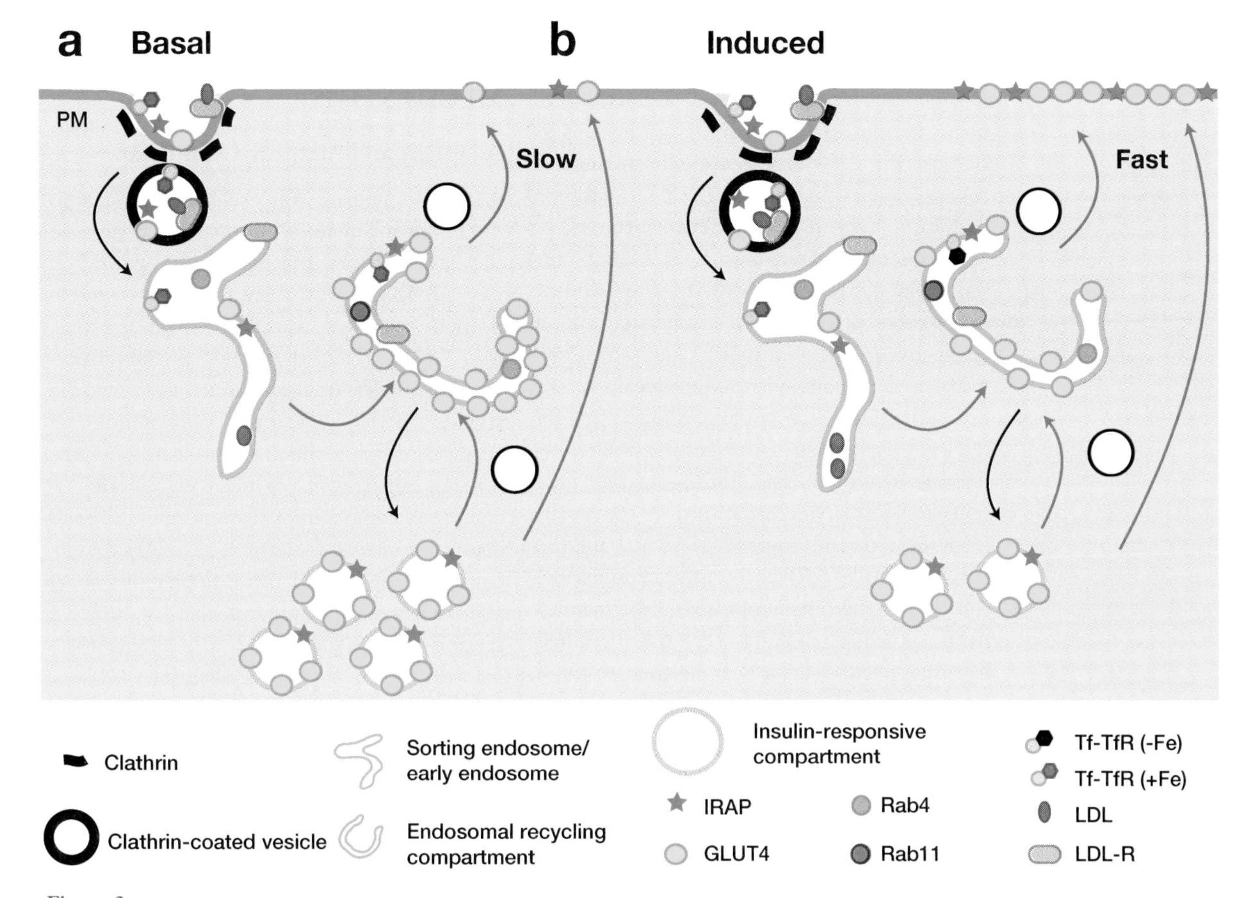

Figure 3

GLUT4/IRAP, a special case of endocytotic recycling in animals. The insulin-responsive compartment (IRC) is characterized by the presence of the GLUT4 glucose transporter, the insulin-responsive aminopeptidase (IRAP), and an absence of Tf-R. The IRC originates from the endosomal recycling compartment. Black arrows indicate the endocytotic pathway; red arrows indicate the secretory pathway. (*A*) Under basal insulin conditions, the trafficking of GLUT4/IRAP to the PM is constitutive. The presence of IRAP in the IRC has not been conclusively demonstrated. (*B*) Under high insulin conditions, induction occurs and trafficking of GLUT4/IRAP to the PM is ~sevenfold greater than that of the basal rate. When insulin levels return to the basal state, so does the trafficking rate.

endosome as free receptors or from the ERC after internalization of its ligand TGFα (161, 162). G-protein-coupled receptors follow the same recycling routes described for receptor protein kinases. The recycling of dephosphorylated G-protein-coupled receptors from the ERC via clathrin-coated vesicles containing βarrestin results in their rapid resensitization at the PM (46, 162). Because no direct input of PM vesicles to late endosomes has been demonstrated, late endosomes appear to function primarily as interfaces between endocytotic and biosynthetic pathways.

The cation-independent mannose 6-phosphate receptor (CI-M6PR) is an example of a more complex recycling pathway. Its early itinerary resembles that of TGN38 (45) but then resembles the itinerary of the transmembrane endoprotease furin in that it passes through the TGN and late endosomes en route to the PM. CI-M6PR is internalized via clathrin-coated pits from the PM and is delivered via sorting endosomes to the ERC. From there it cycles several times to the PM. During each recycling round, a fraction returns to the late endosome via the TGN. Because the CI-M6PR is also the sorting receptor for acid hydrolases from biosynthetic pathways, it is transported between the TGN and late endosomes, but it is clearly absent from lysosomes. TGN to late endosomal sorting of CI-M6PR is mediated via an acidic-cluster-dileucine motif in its carboxy terminal and a monomeric Golgi-localized, γ-ear-containing, ARF-binding adaptor protein (GGA2). Return to TGN from the late endosome involves TIP47 (tail-interacting protein) and Rab9, whereas return of CI-M6PR from the early endosome involves AP1 and the cytosolic PACS-1 APs. Recently, an additional retromer that is also implicated in the maintenance of cellular polarity (183) was identified in mammalian CI-M6PR endosomal retrieval to the Golgi (4, 149).

The small Rab GTPases are master regulators in vesicle targeting and fusion events and thus also regulate recycling (196). Although Rab5 regulates fusion of homotypic sorting endosomes, it is also required for heterotypic fusion of internalized clathrin-coated vesicles with sorting endosomes. This fusion requires an effector known as the early endosome antigen 1 (EEA1) (155). Rab4 and Rab11 are also implicated in Tf-R recycling because Tf-R sequentially moves through Rab4- (sorting endosome and ERC) and Rab11- (ERC, TGN) positive structures (196). Rab function is also regulated by interactions between proteins containing specific phospholipid interaction domains and localized endosomal membrane phospholipid concentrations (27). Because of this, Rab5, Rab4, and Rab11 are often distributed into distinct domains within the same endosome (160, 169). Proteins interacting with Rabs in this category are EEA1 and its associated endosomal SNARE protein, syntaxin13 (100), other PI3P-binding proteins such as the sorting nexin3 (193), and Rabip4′, which functions in endosome-to-PM recycling (37).

PROTEIN COMPONENTS OF ENDOCYTOTIC CYCLING MECHANISMS

Adaptins and Adaptor Complexes

Adaptors mediate the selection of cargo molecules for inclusion into coated vesicles in the late secretory and endocytotic pathways. One group of cargo adaptors are monomeric proteins such as AP180, the β-arrestins, the GGAs (Golgi-localized γ-adaptin-ear-homology-ARF-binding proteins), or the stonins. AP180 is the only plant monomeric adaptor ortholog identified to date (7). Another group consists of heterotetrameric AP complexes composed of adaptin subunits. Each complex contains two large ~100-kDa subunits consisting of one clathrin interactor (β_{1-4}) and another large subunit that is complex specific (α-adaptin/AP-2, γ-adaptin/AP-1, δadaptin/AP-3 and ε-adaptin/AP-4). A small ~20-kDa adaptin subunit (σ_{1-4}) and a ~50-kDa medium subunit t(μ_{1-4}) complete the complex (13, 84). In mammals, there are four AP complexes, but only the TGN-localized AP-1 and the endocytic AP-2 are connected with clathrin. *Saccharomyces cerevisiae*, *C. elegans*, and *Drosophila* all lack the AP-4 complex, but the total number of adaptin orthologs in *Arabidopsis* points to the existence of four AP complexes (13). Like mammals, the *Arabidopsis* genome contains several isoforms of some adaptins, namely two α-adaptin (7, 67), five β-adaptin (13), three γ-adaptin (13, 146), five μ-adaptin (56), and five σ-adaptin genes. In addition, δ- and ε-adaptins are coded for by single genes (13). Like mammalian β1- and β2-adaptins, the βB- and βC-adaptins from *Arabidopsis* are highly similar. Some plant adaptin homologs are referred to by letters because the composition of plant AP complexes has not been elucidated and assignment to particular complexes cannot be made on sequence similarity alone.

Adaptins have been identified in the rice genome and have been reported in other plant species, particularly a σ1-adaptin from the

66. Heldal M, Scanlan DJ, Norland S, Thingstad F, Mann NH. 2003. Elemental composition of single cells of various strains of marine *Prochlorococcus* and *Synechococcus* using X-ray microanalysis. *Limnol. Oceanogr.* 48:1732–43
67. Hessler AM, Lowe DR, Jones RL, Bird DK. 2004. A lower limit for atmospheric carbon dioxide levels 3.2 billion years ago. *Nature* 428:736–38
68. Higgins CF. 2001. ABC transporters: physiology, structure and mechanism—an overview. *Res. Microbiol.* 152:205–10
69. Hillrichs S, Schmid R. 2001. Activation by blue light of inorganic carbon acquisition for photosynthesis in *Ectocarpus siliculosus*: organic acid pools and short-term carbon fixation. *Eur. J. Phycol.* 36:71–79
70. Hiltonen T, Björkbacka H, Forsman C, Clarke AK, Samuelsson G. 1998. Intracellular β-carbonic anhydrase of the unicellular green alga *Coccomyxa*. Cloning of the cDNA and characterization of the functional enzyme overexpressed in *Escherichia coli*. *Plant Physiol.* 117:1341–49
71. Hisbergues M, Jeanjean R, Joset F, Tandeau de Marsac N, Bedu S. 1999. Protein PII regulates both inorganic carbon and nitrate uptake and is modified by a redox signal in *Synechocystis* PCC 6803. *FEBS Lett.* 463:216–20
72. Huertas IE, Bhatti S, Colman B. 2003. Inorganic carbon acquisition in two species of prymnesiophytes. *Eur. J. Phycol.* 38:181–89
73. Huertas IE, Colman B, Espie GS. 2002. Inorganic carbon acquisition and its energization in eustigmatophyte algae. *Funct. Plant Biol.* 29:271–77
74. Huertas IE, Colman B, Espie GS, Lubian LL. 2000. Active transport of CO_2 by three species of marine microalgae. *J. Phycol.* 36:314–20
75. Huertas IE, Espie GS, Colman B, Lubian LL. 2000. Light dependent bicarbonate transport and CO_2 efflux in the marine microalga *Nannochloropsis gaditana*. *Planta* 211:43–49
76. Huertas IE, Lubian LL. 1998. Comparative study of dissolved inorganic carbon utilization and photosynthetic responses in *Nannochloris* (Chlorophyceae) and *Nannochloropsis* (Eustigmatophyceae) species. *Can. J. Bot.* 76 : 1104–08
77. Im CS, Grossman AR. 2001. Identification and regulation of high light-induced genes in *Chlamydomonas reinhardtii*. *Plant J.* 30:301–13
78. Irmler A, Sanner S, Dierks H, Forschhammer K. 1997. Dephosphorylation of phosphoprotein P_{II} in *Synechococcus* PCC7942: identification of an ATP and 2-oxoglutarate-regulated phosphatase activity. *Mol. Microbiol.* 26:81–90
79. Johnson KS. 1981. Carbon dioxide hydration and dehydration kinetics in seawater. *Limnol. Oceanogr.* 27:849–55
80. Johnston AM. 1991. The acquisition of inorganic carbon by marine macroalgae. *Can. J. Bot.* 69:1123–32
81. Johnston AM, Raven JA, Beardall J, Leegood RC. 2001. C_4 photosynthesis in a marine diatom. *Nature* 412:40–41
82. Kaplan A, Lieman-Hurwitz J, Tchernov D. 2004. Resolving the biological role of the Rhesus (Rh) proteins of red blood cells with the aid of a green alga. *Proc. Natl. Acad. Sci. USA* 101:7497–98
83. Kaplan A, Reinhold L. 1999. CO_2 concentrating mechanisms in photosynthetic microorganisms. *Annu. Rev. Plant Physiol. Plant Mol. Biol.* 50:539–59
84. Karlsson L, Clarke AK, Chen Z-Y, Hugghins SY, Park YI. 1998. A novel α-type carbonic anhydrase associated with the thylakoid membrane in *Chlamydomonas reinhardtii* is required for growth at ambient CO_2. *EMBO J.* 17:1208–16
85. Kasting JF, Catling D. 2003. Evolution of a habitable planet. *Annu. Rev. Astron. Astrophys.* 41:429–63

Chinese medicinal tree (95), a σ2-adaptin from maize (138), and a zucchini CCV β-adaptin that was recognized by bovine $\beta 1/\beta 2$-specific antibodies (65) and tentatively localized at the PM (31).

Little is known about adaptin functions in plants. Only two adaptins and one monomeric adaptor homolog from *Arabidopsis* have been functionally characterized on the molecular level. The μA-adaptin AtμA-Ad was identified as a receptor-binding partner (56), and a plant ortholog of mammalian α-adaptins, which play a crucial role in endocytosis, was identified as a binding partner for network proteins (AtαC-Ad). AtμA-Ad and AtαC-Ad are found in high molecular weight AP-like complexes (7, 56). Consistent with a role in endocytosis, At-AP180 promoted the formation of clathrin cages of nearly uniform size, thus functioning as a plant clathrin-assembly protein (7). Thereby, the assembly of clathrin triskelia depends on a single-DLL clathrin-binding motif, a feature that has not been described for mammalian AP180 (7). AtμA-Ad is strongly implicated in clathrin-dependent vacuolar-trafficking events, and its receptor-binding domain exhibited tyrosine-dependent interactions with the tyrosine-based sorting motif (YXXØ) of the mammalian TGN38 protein and the (VSR-PS1) vacuolar sorting receptor from pea (56). Based on their conserved features, the other four plant μ-adaptins are also expected to function as receptor-binding partners of plant AP complexes because the YXXØ motif functions in endocytosis as well (15) and some plant receptor-like kinases contain this motif. At-αC-Ad functions as a network-protein-binding partner because its carboxy-terminal ear region interacted not only with the mammalian network proteins Eps15, AP180, and amphiphysin, but also with the plant monomeric At-AP180. Little is known about plant β- and γ-adaptins. β-adaptins were identified as interactors with PM complexes containing *Arabidopsis* auxin transport proteins and the *Arabidopsis* gluzincin membrane aminopeptidase and AtAPM1 (114), and, more recently, γ-adaptins were identified as interactors with ADL6 (DRP2) and the same complexes derived from microsomal membranes (A. Murphy, unpublished data). Yeast–two-hybrid studies confirm that an *Arabidopsis* γ-adaptin interacts with the dynamin-like protein ADL6 (DRP2), probably involved in Golgi-originating vesicle trafficking (90). Although adaptin function appears to be largely conserved between animals and plants, there are cases where plant function differs, as when plant adaptins other than the β subunit interacted with clathrin (S. Holstein, unpublished data).

ARF: ADP-ribosylation factor

GEF: guanine nucleotide exchange factor

ARF/GEF: ADP-ribosylation factor. Guanine nucleotide exchange factors are a family of small GTPases that regulate endocytotic cycling.

ARFs/ARF-GEFs

ARFs are a ubiquitous group of 20-kDa, ras-related GTP-binding proteins that maintain organellar integrity and regulate intracellular transport (21). Originally identified by their ability to stimulate the ADP-ribosyltransferase activity of cholera toxin (110), the six types of ARFs recruit vesicle coats crucial for vesicle budding and cargo selection (30). Under normal conditions, cytosolic GDP-bound ARFs are inactive (29). A vesicle-associated GTPase-activating protein (GAP) hydrolyzes GTP and releases the ARF protein, resulting in coat dissociation prior to vesicle fusion. GDP release requires an interaction with GEF proteins that contain a 200 amino acid sec domain required for GDP exchange activity. GEFs that contain sec7 are generally sensitive to Brefeldin A (BFA), a noncompetitive inhibitor that stabilizes an abortive ARF-GDP complex (126) and inhibits activation of ARF GTPases by blocking recruitment of vesicle coat components to membranes. However, a subset of this group contains a modified sec7 domain and is BFA resistant.

ARF6, the least conserved member of this family, appears to regulate endosome recycling and cortical actin reorganization (121). Studies in mammalian systems using Madin-Darby Canine Kidney cells indicate that ARF6 is present only on the apical surface of polarized epithelial cells (1). Overexpression of ARF6-Q67L, a GTP hydrolysis–deficient mutant, stimulates endocytosis at this surface. However, the

ARL: ARF-like

PH: pleckstrin homology domain

PRD: proline-rich domain

DRP: dynamin-related proteins in plants, large GTPases implicated in cycling and cell wall synthesis

ARF6-T27N that retains bound GDP also stimulates apical endocytosis to a lesser extent. A dominant negative form of dynamin or clathrin can inhibit this endocytosis. Overexpression of these mutant forms of ARF6 also leads to an increase in the number of clathrin-coated pits in the PM (1). This study indicates that ARF6 is an important modulator of clathrin-mediated endocytosis at the apical surface of the PM. In a similar study, when ARF6 and its mutant forms were expressed in Chinese hamster ovary cells, the protein either colocalized to a perinuclear recycling compartment or accumulated at the PM, depending on its nucleotide status. This suggests that the protein cycles between the PM and recycling endosomes and may be involved in the outward flow of recycling membranes (32). Franco et al. (38) showed that EFA6, a protein exchange factor for ARF6, regulates endosomal membrane recycling and promotes the redistribution of transferrin receptors to the cell surface. EFA6 stimulates ARF6 guanine nucleotide exchange and localizes to a dense matrix on the cytoplasmic face of PM invaginations. ARF6 is also required for cortical actin rearrangements (133) and may link membrane traffic with cytoskeleton organization.

The *Arabidopsis* genome contains 21 ARF GTPase family members genomes, with isoforms present in both ARF and ARL GTPase subfamilies. ARF6 localizes to the periphery of the cell where it has an essential role in endocytotic pathways (104). There are eight *Arabidopsis* proteins with sec7 domains, including GNOM, which is essential for recycling PM proteins from endosomes and is required for polarized distribution of the auxin efflux facilitator PIN1 (43, 44). However, the ARF substrate involved in GNOM-regulated endosomal trafficking has not been determined.

Dynamins

Dynamins constitute a family of large (100-kDa) GTP phosphohydrolases that carry out diverse roles in eukaryotic membrane cycling (61, 78). These proteins have a pH, which allows them to bind to phosphoinositides, and a PRD, which mediates binding to proteins with Src homology 3 (SH3) domains. A coiled-coil region (CC domain) serves as the GTPase effector domain (152). Instead of functioning as classic switch GTPases, dynamins participate in membrane scission events by self-assembling into multimeric ring structures (64). In vivo studies show interactions of dynamins with proteins such as endophilin1 that alter the geometry of lipid structures, resulting in the formation of vesicles (147). Association of dynamins with actin or actin-binding and actin-depolymerizing proteins suggests that dynamin-actin interactions may also contribute to vesicle formation (131, 190). Studies in mammals suggest that dynamins link cellular membranes to the actin cytoskeleton and thereby regulate endocytosis (120).

In animals, there is extensive evidence that dynamins are involved in clathrin-mediated endocytosis, internalization of caveolae, synaptic vesicle recycling, and vesicular trafficking to and from the Golgi complex (60, 118, 119, 137, 176). In cultured mammalian cells, mutations in the GTP-binding domain of dynamin resulted in markedly reduced endocytotic uptake of transferrin receptors (62). An analysis of intermediates in coated vesicle formation in these mutants indicated that endocytosis was blocked after the initiation of coat assembly and before the sequestration of ligands into deeply invaginated coated pits. ER-to-Golgi transport was unaffected in these mutants, suggesting that dynamins function in early endocytotic events (180). However, subsequent confocal imaging studies showed association of dynamin with the TGN and perinuclear late endosomes containing the CI-mannose-6-P receptor (118). GFP-dynamin chimeras localized in the clathrin-coated vesicles at the cell surface also localize to the TGN (76).

Six dynamin-like protein subfamilies have been identified in plants that include the 16 *Arabidopsis* dynamin-related proteins (DRPs) (70). Of these, only members of the DRP2 subgroup, which are implicated in TGN to vacuolar trafficking (74), contain all of the signature domains of classical mammalian dynamins (71). Members of DRP families 1 and 3 lack both pleckstrin and Pro-rich domains and are unique to plants (179). The DRP1 phragmoplastins constitute the largest plant-specific dynamin subfamily. In *Arabidopsis*, five dynamin-like protein gene families have been identified, with partially overlapping functions (78). DRP1-mediated membrane trafficking is essential for PM formation and recycling in plants (78, 79). A probable role in endocytotic events was suggested when DRP1 and several other proteins involved in endocytosis and vesicular trafficking copurified with auxin efflux protein complexes (114). More recently, the same complexes isolated from microsomal rather than PM fractions were found to be enriched in DRP2A bound to γ adaptin (A. Murphy, unpublished data; 114). Consistent with findings in animals, cellular localization of ADL1A (DRP1A) was disrupted by an actin antagonist, suggesting involvement of actin-dynamin interactions in endosomal regulation in *Arabidopsis* (78). Another yeast dynamin-like protein, VPS1p, is involved in protein transport from Golgi to an endosomal compartment (189), suggesting that the plant dynamin-like proteins could function in intracellular trafficking events other than endocytosis.

Rab GTPases

Rabs comprise a small family of ubiquitous proteins that are essential components of the membrane-trafficking machinery. Rabs are implicated in vesicular formation, loading, transport along cytoskeletal elements, and docking/fusion at target membranes (105, 109, 150, 167). Recent evidence shows that cargo proteins act as Rab effectors and regulate their own trafficking by direct interactions with the transport machinery (157). Like most GTPases, Rab proteins cycle between an inactive GDP-bound state and an active GTP-bound state. In the GTP-bound state, Rabs are associated with membranes and are attached to the cytoplasmic surface of compartments (106). The mechanism by which Rabs regulate endocytosis has not been elucidated, but it is hypothesized that the GTP-GDP cycle of Rabs might act as a timer switch to regulate the functionality of a membrane domain. It is also possible that Rabs act as GTP-activated switches that simply stabilize protein complexes required for transition events. However, the ability of Rabs to link membranes to cytoskeletal motor proteins (148) suggests that they can generate uniquely functional membrane subdomains.

In eukaryotes, specific Rab GTPases have been associated with the regulation of membrane-trafficking events in distinct compartments (184). In mammals Rab4, Rab5, Rab7, Rab9, and Rab11, are involved in endocytotic vesicular trafficking (127, 171). Rab4, Rab5, and Rab11 localize to early endocytic organelles where they regulate distinct events in the transferrin receptor pathway (106). Rab5 regulates docking and fusion of early endosomes and mediates interactions with multiple effectors. In its GTP-bound state, Rab5 builds an effector domain on the membrane, a process that requires the recruitment of PI3 kinases and PI3P-binding proteins that contain the FYVE motif (196). Rab4 and Rab11 are also involved in regulating protein recycling back to the PM. Rab4 regulates formation of endosomal recycling vesicles in ongoing cycles of association and dissociation from early endosomes (105). In mammals, human Rab11A and Rab11B isoforms have been localized in recycling endosomes (174), and in epithelial cells mammalian Rab11A is crucial for the exit of internalized proteins from apical recycling endosomes (33). Rab6, a mammalian homologue of yeast Ypt6p, initially thought to function only in *intra*-Golgi transport, is required for recycling proteins between endosomes and the TGN (96).

A large number of highly conserved Rab GTPases have been identified in plants, and their localization generally coincides with that of their eukaryotic homologs, particularly those functioning in the exocytotic pathway (129). However, unique isoforms of endocytosis-related Rabs have been identified in *Arabidopsis*, including two Rab5 orthologs, seven Rab7

PVC: prevacuolar compartment

orthologs, and Ara6, which is structurally dissimilar to all known GTPases (173). Orthologs of Rab4 and Rab9 are missing altogether in *Arabidopsis*. AtAra-4 and Pea Pra3 are similar to the mammalian Rab11 and localize to or function within the TGN (73, 172). A Rab6 ortholog in *Arabidopsis* complements a mutation in the yeast Ypt6 protein that was implicated in protein recycling from endosomes to the Golgi and from late to early Golgi (10, 94). AtRabA4b, which is similar to Rab11, labels a novel compartment that accumulates at the tips of root hair cells (129). The tobacco ortholog of the mammalian Rab2 that regulates membrane flow in Golgi intermediates localizes to Golgi bodies in pollen tubes (22, 167, 168). However, Rha1, which is associated with trafficking of cargo molecules to the vacuole through late endosomal/PVCs in *Arabidopsis*, is an ortholog of Rab5 (158), which regulates fusion of endosomal compartments in mammals. It is still unknown how Rabs are delivered to distinct subcellular compartments or what other effectors are involved.

SNAREs

N-ethylmaleimide-sensitive factor adaptor protein receptors (SNAREs) are essential components of vesicle-trafficking machinery where they play a central role in membrane fusion events. SNAREs comprise a family of small proteins that are associated with different intracellular membranes and, as such, define the interactions and dynamic structures involved in endocytotic cycling. The role of SNARES in the endosomal system has been studied using endosomal fusion in both cell-free and intact cell assays. In mammals, SNARES are involved in two distinct homotypic fusion steps involving early and late endosomes (52). The SNAREs VAMP-7, VAMP-8, and syntaxins 7, 8, and 13 have been implicated in both fusion events, whereas complexation of VAMP-8 with the syntaxins 7 and 8 and Vti1b are involved in the homotypic fusion of late endosomes (3). The v-SNARE cellubrevin has been implicated in the recycling of endocytic receptors back to the PM (101) and the t-SNARE Tlg2p in early stages of endocytotic internalization (151). More recently, direct transport from early/recycling endosomes to the TGN (bypassing the late endosome) was shown to be mediated by a complex formed between the TGN-localized syntaxin 6, syntaxin 16, and Vti1a and the EE/RE-localized v-SNARES VAMP 3 and VAMP4 (96). In a recent study, polarized recycling of T cell antigen receptors to the PM and their eventual fusion was shown to involve the t-SNARES syntaxin-4 and SNAP-23 (26).

In *Arabidopsis*, where studies of SNAREs have focused primarily on secretory and vacuolar targeting, two SNARES (AtVTI1a and AtVTI1b) are involved in trafficking to the PVC (198). In addition, the syntaxins SYP21, SYP22, SYP4, SYP5, SYP42, and SYP61 are associated with various endosomal compartments, including the PVC and TGN.

Other Cytoskeletal Interactions

All endocytotic pathways of membrane protein internalization depend on interaction with the cytoskeleton. Direct participation of F-actin in internalization events has been documented in animals and yeast (34). Actin dynamics are also crucial for lipid raft-mediated endocytosis (124). However, the molecular details of the interaction are not known and the specific steps of endocytosis where the interactions are involved have not been identified. Studies in *Arabidopsis* show that disrupting the actin cytoskeleton by latrunculin B or cytochalasin D inhibits the formation of BFA-induced endocytotic compartments involved in recycling PM proteins (44).

A number of proteins play an essential role in mediating interactions between vesicular structural components and the actin cytoskeleton in animals (34). Some of these molecular linkers have also been identified in plants. Myosin, molecular motors that traverse along actin filaments, have been implicated in the distribution of endocytotic vesicles and membrane-bound molecules in *Arabidopsis* (68). Components of the actin-related protein (ARP) 2/3 complex have also been implicated in the internalization step and in the organization of the actin cytoskeleton (144). These functions of the ARP2/3 proteins are mediated by their interactions with calmodulin. The ARP2/3 complex in *Arabidopsis* has a similar function (91).

Other Proteins Contributing to PM Protein Cycling

Rho GTPases are members of a small but diverse protein family (135, 199) involved in signaling and regulation of endocytotic traffic (130). Although Rho-GTPases are not found in plants, plant Rho-related GTPases (ROPs) comprise a plant-specific subfamily (194). ROPs have been localized to the tips of root hairs, have been implicated in polar growth (107), and play a key role in fusing docked vesicles and endocytosis in these growing tips (19). Immunophilins were originally discovered as receptors for a family of immunosuppressive drugs including cyclosporin A, FK506, and rapamycin. Later, when immunophilins were found in bacteria, yeast, animals, and higher plants, it became evident that immunophilin-like proteins were likely involved in fundamental cellular processes. Most immunophilins possess peptidyl prolyl cis/trans-isomerase activity required for protein folding and modification. The *Arabidopsis* genome contains 52 genes encoding putative immunophilins, including 23 putative FK506- and rapamycin-binding proteins (FKBPs) and 29 putative Cyclosporin As (CYPs) (57), many of which are localized to the chloroplast. Recent evidence suggests that a cyclophilin interacts with the ARF-GEF GNOM (49) and that an FKBP immunophilin, TWD1, may be involved in trafficking PM transport proteins (42). Fatty acyl CoA has been identified in cycling GLUT4 vesicles and has been implicated in membrane budding and fusion events (156). In another study, acyl CoA dehydrogenase (ACD) was found to mediate intracellular retention of GLUT4 vesicles via association with IRAP (80). A membrane-associated acyl CoA-binding protein that functions in intracellular lipid transfer from the ER and maintains a membrane-associated acyl pool at the PM was identified in *Arabidopsis* (47). AtAPM1, the IRAP homolog in *Arabidopsis*, was recently found to interact with ACDs (O. Lee & A. Murphy, unpublished data), suggesting a trafficking role for these proteins in plants.

ROP: Rho-related GTPases specific to plants, involved in endocytosis and fusion of docked vesicles

WHAT DO WE KNOW ABOUT ENDOCYTOTIC CYCLING IN PLANTS?

Endocytosis

Most of what is known about endocytosis comes from studies in yeast and mammalian cells. Until recently, endocytosis in plants was a questionable phenomenon because of the unfavorable energetics inherent in the endocytosis of vesicles in turgid cells with a rigid cell wall. Even though clathrin-coated pits were detected in the plant PM decades ago and polyvalent electron-dense tracers were subsequently shown to be internalized by protoplasts (136), this important phenomenon received little attention from plant biologists. However, recent studies utilizing styryl dyes, filipin-labeled plant sterols, reporter-tagged marker proteins characterizing endocytotic compartments, and the fluid-phase marker LuciferYellow indicate that endocytosis mediates the internalization and recycling of PM molecules, including membrane proteins and sterols in plants (5, 43, 44, 49, 72). Furthermore, because vigorous, constitutive endocytosis and recycling of the potassium channel KAT1 in turgid guard cells was recently demonstrated (102), turgidity can be ruled out as an inhibitory factor for endocytosis.

At least four forms of endocytosis operate in plants: clathrin independent and dependent, phagocytotic, fluid-phase and lipid raft-mediated endocytosis. The clathrin-independent pathways are mediated by actin polymerization and require small GTPases (20, 23). Rhizobia are internalized into root cells by a Rab 7-dependent phagocytotic process (159) and Baluska et al. (6) recently confirmed the occurrence of fluid-phase endocytosis in plants using the fluid-phase marker Lucifer Yellow. Mongrand (108) and collegues recently demonstrated the existence of lipid raft microdomains

PCR: Partially coated reticulum, an apparent plant analog of mammalian early endosomes.

MVBs: multivesicular bodies, also known as late endosomes or prevacuolar compartments

in tobacco PM. Many GPI-anchored proteins in animals cluster into sphingolipid- and sterol-enriched lipid raft microdomains (112). The structural sterols of plants, stigmasterol and sitosterol, form lipid rafts much more readily compared to cholesterol (192), suggesting the possibility of raft-mediated endocytosis in plants. Recent studies identified GPI-anchored proteins associated with lipid rafts in plants (17, 89). Other studies suggest that PINFORMED (PIN) auxin efflux proteins may depend on structural sterols and lipid rafts for their recycling in plants (49, 188). A GPI-anchored immunophilin (TWD1) and a fasciclin-like arabinogalactan protein (FAGP2) were recently shown to associate with auxin efflux proteins in *Arabidopsis* (42, 114). Although the direct involvement of clathrin-coated vesicles in regulated endocytosis has not yet been demonstrated in plants, clathrin-coated vesicles and associated APs mediating clathrin-dependent endocytosis have been found in *Arabidopsis* (7). To a large extent, studies of clathrin-mediated endocytosis are still predominantly exercises in comparison to mammalian processes (see Overview, above).

At present, molecular mechanisms underlying reversible membrane protein internalization in plants are unknown. However, polypeptide signals and receptors for plant defense, growth, and development have been identified (142), kinase-associated protein phosphatase (KAPP)-mediated endocytosis of AtSERK1 was recently documented (153), and many other plant membrane receptor kinases with phosphorylation- and tyrosine-based internalization motifs have been identified (67).

Endocytotic Compartments

Endosomal compartments have been characterized largely by studies tracing the course of endocytosed molecules through dynamically changing compartments. This makes the compartments difficult to define and identify because no molecule permanently resides in any compartment. Molecules may be transported into or out of the compartment through transport vesicles or, alternatively, the compartment itself may be modified and mature over time (98), as diagrammed in **Figure 2**.

Plant endocytotic compartments are not well characterized and the term endosome is often used for any compartment containing an endocytosed material. Styryl dyes such as FM4-64 are incorporated into the plant PM, subsequently endocytosed, and transported to vacuoles via transport intermediates (185) (**Figure 2**). These dyes are widely used in colocalization studies to ascertain the nature of intracellular vesicles that are putative endosomal compartments (14, 49, 173). Ultrastructural studies indicate that the PCR, a structure that was originally thought to arise from the TGN (63), is analogous to early/recycling endosomes of animals. Time-course studies tracing the path followed by cationized ferritin in soybean and white spruce protoplasts first labeled these compartments (40, 166). Molecule sorting occurs in the early endosomes, from where they are either recycled back to the PM or are transported to the Golgi apparatus or MVBs (8, 77). From the MVBs they are targeted to vacuoles for degradation. In a recent study using *N. tabacum* BY-2 cells, MVBs were identified as PVCs that are common to both secretory and endocytotic pathways (170).

In animals, proteins destined to be recycled back to the cell surface can be directly transported to the PM by the fission of tubules from sorting endosomes or can be transported via the ERCs. In plants there is still no clear evidence of sorting and recycling endosomes, but two distinct classes of early endosomes were identified in *Arabidopsis* using double-labeling experiments (173). One class of endosomes is characterized by Ara6, a unique plant Rab GTPase, whereas another is characterized by Ara7, the *Arabidopsis* homolog of mammalian Rab5. A small population of endosomes also contains both Ara6 and Ara7, suggesting endosomal maturation (**Figure 2**). However, although Ara6 and Ara7 both drive early endosomal fusion, it is unclear whether they mediate fusion of different endosomal populations. These two distinct early endosomal compartments could participate in the sorting events that target endocytosed proteins to the recycling or degradation pathways.

A number of plant proteins cycle between the PM and intracellular compartments. Pharmacological studies with the fungal toxin BFA, which targets the catalytic domains of ARF-GEFs (126), demonstrate that the PIN1 PM auxin efflux facilitator can be reversibly retained in intracellular compartments (44), diagrammed in **Figure 4**. The compartments are characterized by the presence of the endocytotic markers Ara6 and FM4-64. The ARF-GEF GNOM is a specific target of BFA (43), suggesting that similar compartments function in the normal cycling of PIN1 and other proteins. These results also suggest that GNOM is functionally distinct from similar ARF-GEFs in animals that mediate ER-Golgi or intra-Golgi trafficking (197).

Distinct compartments that are not labeled by endocytotic markers have also been identified both in animals and plants. In animals, the most striking of these is the compartments that house the GLUT4 glucose transporter in the absence of an insulin signal. These compartments lack transferrin receptors and are characterized by the presence of IRAP, which exhibits trafficking dynamics similar to GLUT4 (80). Comparable mechanisms in plants are suggested by the PM and intracellular localization of AtAPM1, an IRAP homolog in *Arabidopsis* (114).

Endosomal Sorting and Redirection

In animals the first branch point for endocytosed proteins is the sorting endosomes. Proteins in the sorting endosomes can have three distinct fates: they can be directly recycled back to the PM, transported to the ERC for eventual return to the PM, or they can end up in the late endosomes. The low luminal pH of the sorting endosomes underlies the first recycling step, dissociation of ligands from the receptors (128). In contrast to animals ligand-receptor binding in plants occurs in an acidic (pH ~5.5) extracellular environment. As such, it is likely that ligand dissociation in the early endosomes of plants requires either greater acidification than is seen in animal cells or a different sorting mechanism altogether. Studies with the vacuolar proton pump inhibitor bafilomycin A1 indicate that endosomal V-ATPase activity is essential for transporting proteins from late endosomes to lysosomes because, in its absence, receptor-ligand complexes continuously recycle between the PM and the early endosome (181). In *Arabidopsis*, acidification is also required for sorting, but may involve PM ATPase activity as well (9).

In mammals, phosphoinositol 3-phosphate (PI3P) and Rab5 recruit microtubule motors to endosomes, thereby establishing specific membrane domains (160). PI3P is required for two critical functions in early endosomes (51). It regulates the dynamics of the compartment and, subsequently, participates in the retention of proteins in domains that mature into late endosomes. PI3P fuses to FYVE domain from human EEA and tags to GFP colocalized with BP-80 in the PVC when transiently expressed. Overexpression of the chimeric protein inhibits targeting of Sporamin to the vacuole (83).

In mammals and yeast, ubiquitination of receptors and transporters serves as a signal for their internalization. Studies with the yeast endocytic cargo uracil permease (Fur4p) indicated that ubiquitination was required for its internalization but deubiquitination was not required for its recycling. Ubiquitinated early endosomal proteins are directed by the ESCRT I, II, and III complexes to the MVB. Upon fusion with the lysosome, the internal vesicles of the MVBs and their cargo are accessible to luminal lytic enzymes. Studies in *S. cerevisiae* lacking a functional ESCRT complex resulted in mistargeting of endocytotic and biosynthetic ubiquitinated cargoes. Deficiencies in this complex also lead to the accumulation of receptors and transporters at the PM in yeast and higher eukaryotes. Predicted proteins with sequence similarity to the Vps23 component of the ESCRT-I complex are present in the *Arabidopsis* genome, suggesting that a similar mechanism may exist in plants.

EVIDENCE FOR ENDOCYTOTIC RECYCLING IN PLANTS

Auxin Transport Proteins and GNOM

Establishing plant polarity depends on the polar transport of the growth hormone auxin. The directionality of auxin transport is maintained by a polar transport apparatus that requires asymmetrically localized transporters and regulators. The PIN proteins are the best characterized of such asymmetrically localized auxin efflux regulators. Pharmacological, physiological,

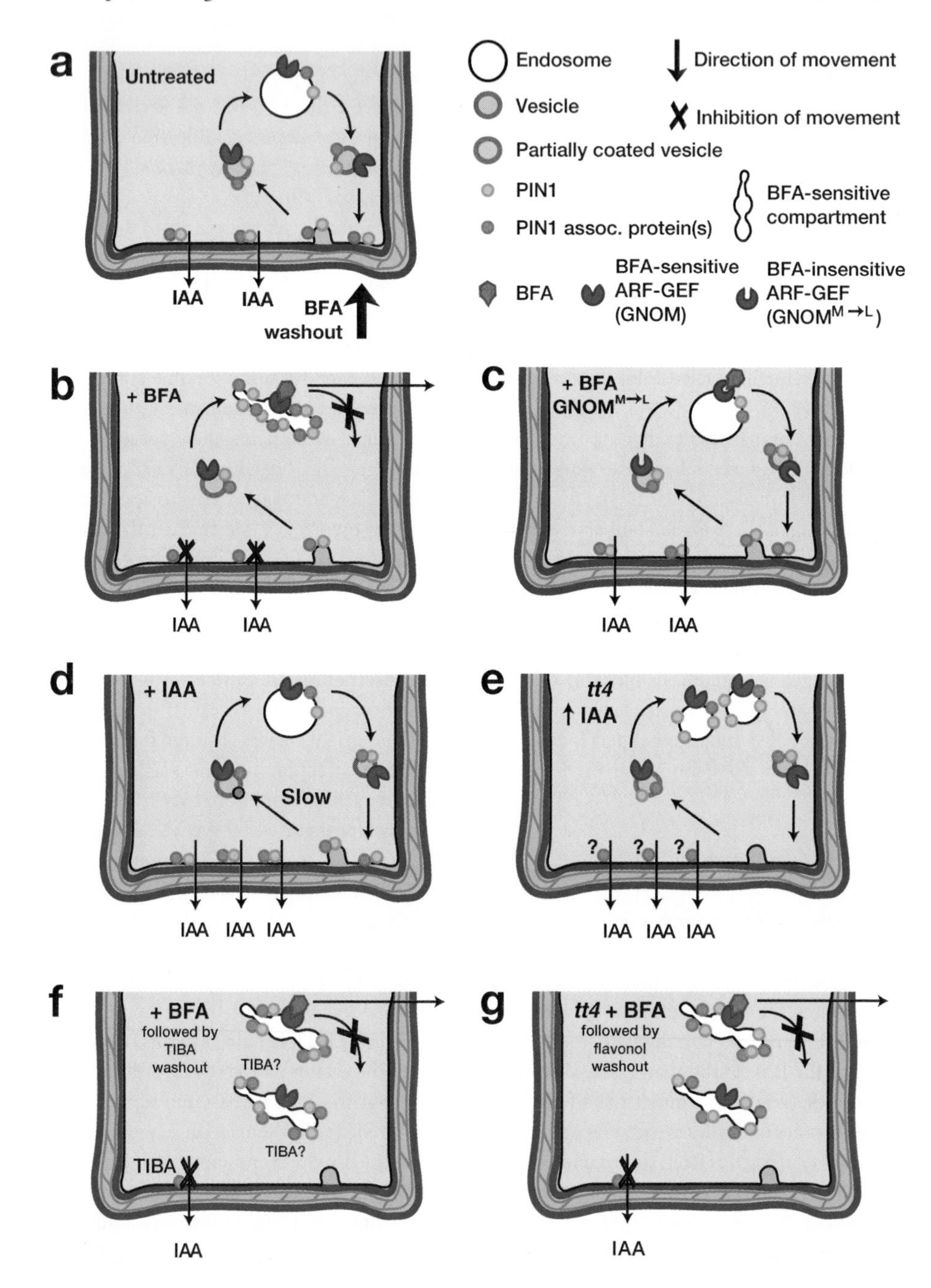

and molecular genetic studies suggest that activity, membrane localization, and vectorial realignment of PIN proteins is regulated by dynamic cycling between the PM and endosomal compartments (12, 39, 43, 44, 123).

PIN1 undergoes rapid actin-dependent cycling. When treated with BFA, PIN1 is still internalized but accumulates in unidentified, juxtanuclear compartments rather than returning to the PM (44). Although initially proposed to result from the fusion of endosomal and post-Golgi membranes (117), the BFA-induced bodies are surrounded by remnants of Golgi stacks and are actually aggregated endosomal compartments that accumulate the endocytotic tracer FM4-64 (43). Similar BFA-induced internalization occurs for other PM-ATPases (44), the PM marker Lti6a (49), the peripheral membrane protein ARG1 (16), and the putative auxin influx carrier AUX1 (48). Cell wall polymers such as pectins cross-linked with boron or calcium are also recycled by the same BFA-sensitive pathway in maize roots (5).

The BFA-induced compartments also accumulate other molecules involved in vesicular trafficking such as the cytokinesis-specific syntaxin KNOLLE and its interactor AtSNAP33, the small GTPase ARF1 and small GTPase Pra2 (25, 43, 44, 58, 73). This suggests that internalization and recycling mechanisms similar to those found in animals exist in plants. This cycling is regulated by extracellular signals such as hormones and flavonoids (123), as is the case with the mammalian glucose transporter GLUT4.

As diagrammed in **Figure 4**, the inhibition of PIN1 cycling by BFA is mediated by an ARF-GEF (43). In *gnom* mutant embryos, PIN1 localization is largely randomized, suggesting that GNOM is required for proper PIN1 localization on the PM (163). When plants are transformed with an engineered BFA-resistant version of GNOM, PIN1 is localized on the

Figure 4

PINFORMED (PIN) protein endocytosis and auxin transport in root tips. Indirect evidence suggests that the PIN components of the auxin efflux carrier complex undergo endocytosis and redistribution to the plasma membrane (PM). This was demonstrated with pharmacological studies utilizing the fungal exocytosis inhibitor brefeldin A (BFA), exogenous application of the transported hormone auxin (IAA), studies of mutants exhibiting altered auxin flux, and analysis of the effects of exogenous and endogenous auxin transport regulators (43, 44, 123). (*A*) In untreated WT root tip cells with normal auxin levels, PIN1 protein is detected on the PM. (*B*) Indirect evidence suggests that PIN1 undergoes endocytosis and redistribution to the PM. When root tips are treated with BFA, PIN1 is endocytosed into BFA-sensitive compartments and is no longer observed on the membrane. Following BFA washout, PIN1 is redistributed to the PM, as in (*A*). (*C*) BFA activity affecting PIN1 localization depends on binding to an ARF-GEF, GNOM. When GNOM is mutated (GNOM$^{M \rightarrow L}$) so that it can no longer bind BFA, then PIN1 is observed on the PM in BFA-treated cells. (*D*) When roots are treated with IAA, the rate of uptake of FM4-64 from the membrane is reduced (A. Murphy, unpublished data) and more PIN1 is observed on the PM. IAA also decreases the transcription of *PIN1*; therefore, no nascent PIN1 is delivered to the PM. If the cells are treated with the artificial auxin NAA plus BFA, then PIN1 is observed on the PM and not in BFA compartments (T. Paciorek & J. Friml, personal communication). (*E*) Flavonols are endogenous compounds that can modulate auxin efflux at the apices. In the flavonoid-deficient mutant *tt4*, auxin delivery to the root tip is greater than in WT, and PIN1 exhibits predominantly intracellular localization and is not found on the PM. *PIN1* transcription is also reduced in the mutant, apparently reflecting the effect of long-term IAA exposure on PIN1 localization at the root tip. (*F*) The competitive auxin efflux inhibitor triodobenzoic acid (TIBA) binds at the efflux site. When BFA is washed out with TIBA, PIN1 localization remains in BFA compartments and does not return to the PM. (*G*) When *tt4* is treated with BFA, PIN1 is observed in BFA compartments (as in BFA-treated WT, unlike untreated *tt4*). When BFA is washed out, then PIN1 is observed on the membrane, as in WT. In *tt4*, when BFA is washed out with flavonols, then PIN1 remains in BFA compartments; however, in WT, PIN1 returns to the membrane. This suggests that BFA has pleiotropic effects on cycling and does not inhibit exocytosis via ARF-GEFs only. This also suggests that flavonols interact with a protein that is required for PIN1 retention before sequestration in BFA compartments.

PM, but is not internalized in response to BFA treatment. Therefore, GNOM appears to be required for recycling of PIN1 from endosomes to the PM. However, cycling of other PM proteins such as PIN2, a PM-ATPase, and the syntaxin KNOLLE are not affected by GNOM, suggesting that specific ARF-GEFs might regulate the endocytotic recycling of discrete groups of PM proteins.

VDAC: voltage-dependent anion channel

In animals, endocytotic recycling also plays an essential role in maintaining the correct membrane protein composition in apical and basolateral membranes of polarized cells. However, in plants, although delocalization of PIN proteins from the PM (12) and vectorial relocalization of PIN3 in response to gravity (39) were recently demonstrated, no basolateral redirection has been observed. Recent evidence of polar reorientation of PIN proteins in *PID* overexpression transformants (39a) suggests that a PID-dependent regulatory mechanism controls the polarity of PIN proteins.

The KAT1 Inward-Rectifying Potassium Channel

The dogma that the high turgor of plant cells precludes endocytosis was recently challenged by evidence that the KAT1 K^+ inward-rectifying channel is constitutively endocytosed from the guard cell PMs against high turgor pressure (72) (**Figure 5**). In WT and KAT1-GFP transformed protoplasts, increases or decreases in PM surface area resulting from a change in cellular pressure led to the incorporation and withdrawal of vesicular membranes carrying an active K^+_{in} rectifier. The density of KAT1 was higher in vesicular membranes than in the PM, suggesting that the channels remained clustered while trafficking to and from the PM (72). The endocytosis process starts immediately following pressure stimulation, ruling out channel concentration prior to internalization. It is suggested that KAT1 channels form stable clusters and that membrane areas containing these clusters are retrieved preferentially during pressure-driven endocytosis. The KAT1::GFP chimera was later shown to be internalized and accumulated in small vesicles in the cortical regions of the cells. These bodies were also labeled with the styryl dye FM4-64 and were thus confirmed to be endocytosed vesicles (102).

Other Examples

Reports of environmentally responsive changes in the subcellular compartmentation of membrane proteins suggest that membrane protein internalization and recycling are part of the regulatory repertoire of plants. In mammals, hormonal signals trigger the cycling of aquaporins between the PM and internal vesicles (53, 86). Similar mechanisms may be at work in plants because the *Arabidopsis* aquaporins (PIPs) can localize both to the PM and vacuoles (132), and the *Mesembryanthemum crystallinum* aquaporins, *Mc*TIP1 and 2, are internalized from the PM and redistributed to the tonoplast and other membrane fractions in response to mannitol-induced water stress (182). *Mc*TIP1 and 2 were also localized to some unique multivesicular endosomal compartments and the observed redistribution was arrested when treated with BFA and other inhibitors of vesicular trafficking. The VDACs have a dual localization in plants as in animals, being localized to the mitochondrial outer membrane and PM (35, 97), which suggests cycling of these proteins between the PM and internal organelles. The recent finding that AtAPM1 interacts with VDACs (O. Lee & A. Murphy, unpublished data) suggests that VDACs may be trafficked by this mechanism. In another study focusing on membrane protein endocytosis, BFA induced internalization of the low-temperature-inducible protein Lti6a and resulted in its colocalization with FM4-64 in discrete subcellular compartments (49). A PM H^+-ATPase also accumulated within BFA compartments in an F-actin-dependent manner, but the redistribution did not require intact microtubules (49). In another recent study Kim et al. (82) reported that the metal tolerance protein (TgMTP1) can be localized to the PM and the

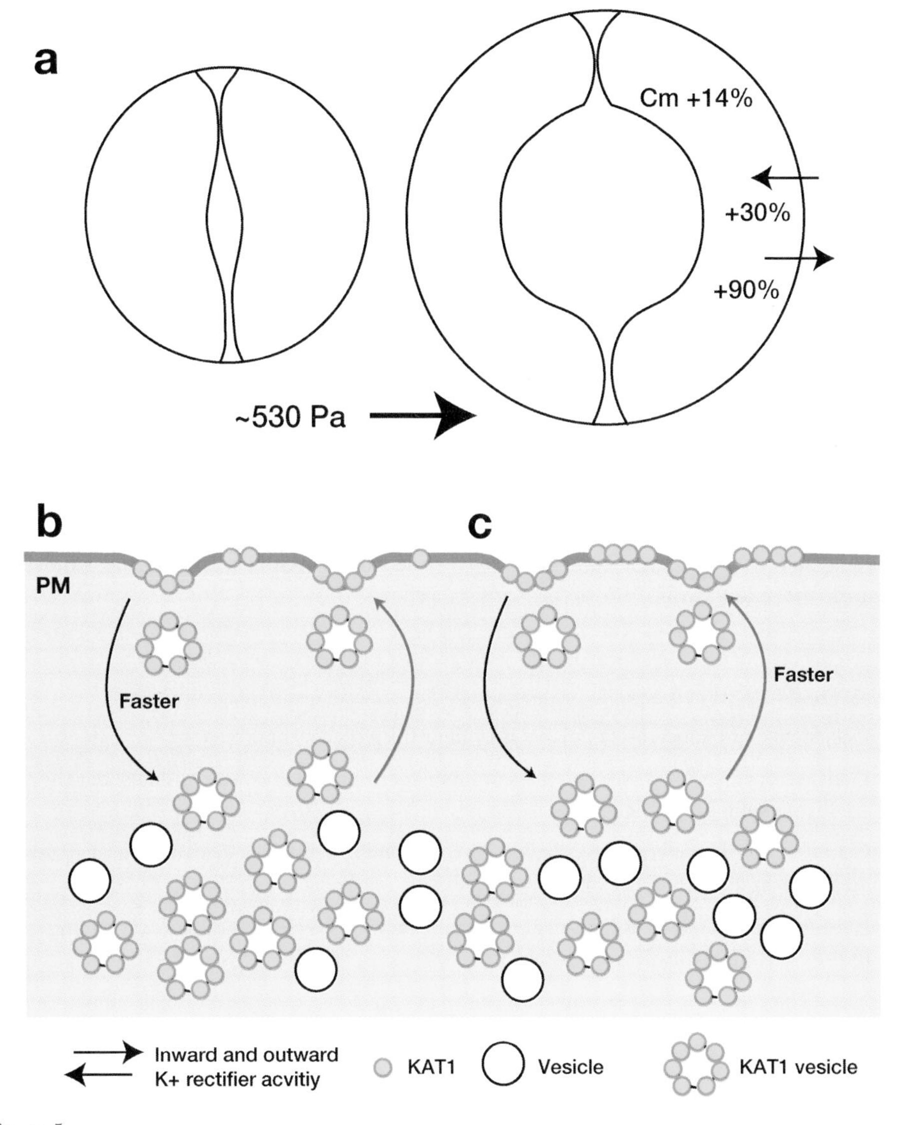

Figure 5

KAT1 recycling and a technique to quantitate endocytosis in plants. Endocytosis was thought not to occur in plants due to turgid cells and rigid cell walls. However, the variable surface area and capacitance of stomatal guard cells provide a method to quantitate endocytosis in plants (69, 72, 102). Black arrows indicate the endocytotic pathway; red arrows indicate the secretory pathway. (*A*) Guard cells swell and shrink to open and close the stomatal pores, thereby regulating the amount of gas exchanged with the environment. The activity of potassium channels, as indicated by membrane capacitance (Cm), correlates with the surface area of the guard cell protoplast. The surface area and the capacitance increase with hydrostatic pressure. The surface area increases up to 40%, and only a 3% to 4% increase of area can be accounted for by stretching the plasma membrane (PM); the capacitance increases by 14%. Therefore, recruitment of additional PM material and ion channels occurs during stomatal opening and retrieval of PM material and ion channels occurs as stomata are closing. (*B*) KAT1, an inward potassium channel rectifier, is found in greater abundance in the endomembrane system than on the PM. Negative hydrostatic pressure results in reduced cell area, lower capacitance, and faster and selective endocytotic retrieval of KAT1 from the PM. (*C*) Positive pressure results in greater cell surface area, higher capacitance, and faster and selective insertion of KAT1 into the PM.

vacuole, suggesting that its dynamic relocalization between the PM and various endomembrane systems could have a role in regulating metal hyperaccumulation. Similar endocytosis-mediated trafficking of metal transporters has been documented in mammals. The PM zinc transporter Dri 27/ZnT4 has been colocalized with the transferrin receptor and the clathrin adaptor complexes AP-1 and AP-2 in endosomal vesicles, suggesting that they are internalized in a clathrin-dependent manner from the PM (113).

CONCLUSION

Recent experimental evidence suggests that plant cells respond to developmental programming and environmental conditions by regulating the endocytosis of PM proteins. This evidence also suggests that some of the mechanisms underlying endocytotic cycling are conserved between plants and animals. However, as plant endosomal compartments and the proteins that characterize them are still poorly defined, the extent of that conservation has yet to be determined. Until recently, studies in plant cell biology focused primarily on secretory mechanisms at the expense of internalization pathways. Because of this, researchers should be cautious about overextrapolation of mammalian models to plant systems. The proliferation of molecular tools in *Arabidopsis*, the advances in imaging techniques, and the growing availability of organellar markers, dyes, and reporter proteins are expected to accelerate our understanding of endocytotic cycling in plants and animals as well.

SUMMARY POINTS

1. Occurrence of endocytosis in plants is evident from recent reports.
2. Plant endocytosis processes and players share common features with animals, but there are features unique to plants.
3. Clathrin and adaptor complexes have been identified in plants, but evidence for receptor-mediated endocytosis is lacking.
4. FM4-64 and BFA studies have made a significant contribution to the understanding of the endocytotic pathway in plants.
5. Endosomes are not well characterized in plants, but two early endosomal populations have been identified by double-labeling experiments; Ara6 is targeted to only one of them.
6. There is indirect evidence for cycling of PIN auxin efflux facilitators. The ARF-GEF GNOM is required for the observed cycling.
7. The KAT1 K^+ inward-rectifying channel is endocytosed against high-turgor pressure of guard cells, suggesting turgor is not an inhibitory factor for endocytosis.
8. Several other membrane transporters and channels are regulated by endocytotic recycling.

LITERATURE CITED

1. Altschuler Y, Liu S, Katz L, Tang K, Hardy S, et al. 1999. ADP-ribosylation factor 6 and endocytosis at the apical surface of Madin-Darby canine kidney cells. *J. Cell. Biol.* 147:7–12

2. Aniento F, Gu F, Parton RG, Gruenberg J. 1996. An endosomal beta COP is involved in the pH-dependent formation of transport vesicles destined for late endosomes. *J. Cell. Biol.* 133:29–41
3. Antonin W, Holroyd C, Tikkanen R, Honing S, Jahn R. 2000. The R-SNARE endobrevin/VAMP-8 mediates homotypic fusion of early endosomes and late endosomes. *Mol. Biol. Cell.* 11:3289–98
4. Arighi CN, Hartnell LM, Aguilar RC, Haft CR, Bonifacino JS. 2004. Role of the mammalian retromer in sorting of the cation-independent mannose 6-phosphate receptor. *J. Cell. Biol.* 165:123–33
5. Baluska F, Hlavacka A, Samaj J, Palme K, Robinson DG, et al. 2002. F-actin-dependent endocytosis of cell wall pectins in meristematic root cells. Insights from brefeldin A-induced compartments. *Plant Physiol.* 130:422–31
6. Baluska F, Samaj J, Hlavacka A, Kendrick-Jones J, Volkmann D. 2004. Actin-dependent fluid-phase endocytosis in inner cortex cells of maize root apices. *J. Exp. Bot.* 55:463–73
7. Barth M, Holstein SEH. 2004. Identification and functional characterization of *Arabidopsis* AP180, a binding partner of plant alpha C-adaptin. *J. Cell Sci.* 117:2051–62
8. Battey NH, James NC, Greenland AJ, Brownlee C. 1999. Exocytosis and endocytosis. *Plant Cell* 11:643–59
9. Baxter IR, Armstrong G, Foster N, Peer WA, Hazen SP, et al. 2005. A plasma membrane H+ ATPase is required for the formation of proanthocyanidins in the seed coat endothelium of *Arabidopsis* thaliana. *Proc. Natl. Acad. Sci. USA* 102:2649–54
10. Bednarek SY, Reynolds TL, Schroeder M, Grabowski R, Hengst L, et al. 1994. A small GTP-binding protein from *Arabidopsis* thaliana functionally complements the yeast YPT6 null mutant. *Plant Physiol.* 104:591–96
11. Bennett EM, Lin SX, Towler MC, Maxfield FR, Brodsky FM. 2001. Clathrin hub expression affects early endosome distribution with minimal impact on receptor sorting and recycling. *Mol. Biol. Cell.* 12:2790–99
12. Blakeslee JJ, Bandyopadhyay A, Peer WA, Makam SN, Murphy AS. 2004. Relocalization of the PIN1 auxin efflux facilitator plays a role in phototropic responses. *Plant Physiol.* 134:28–31
13. Boehm M, Bonifacino JS. 2001. Adaptins: the final recount. *Mol. Biol. Cell.* 12:2907–20
14. Bolte S, Talbot C, Boutte Y, Catrice O, Read ND, Satiat-Jeunemaitre B. 2004. FM-dyes as experimental probes for dissecting vesicle trafficking in living plant cells. *J. Microsc.* 214:159–73
15. Bonifacino JS, Traub LM. 2003. Signals for sorting of transmembrane proteins to endosomes and lysosomes. *Annu. Rev. Biochem.* 72:395–447
16. Boonsirichai K, Sedbrook JC, Chen RJ, Gilroy S, Masson PH. 2003. ALTERED RESPONSE TO GRAVITY is a peripheral membrane protein that modulates gravity-induced cytoplasmic alkalinization and lateral auxin transport in plant statocytes. *Plant Cell* 15:2612–25
17. Borner GHH, Lilley KS, Stevens TJ, Dupree P. 2003. Identification of glycosylphosphatidylinositol-anchored proteins in *Arabidopsis*. A proteomic and genomic analysis. *Plant Physiol.* 132:568–77
18. Bryant NJ, Govers R, James DE. 2002. Regulated transport of the glucose transporter GLUT4. *Nat. Rev. Mol. Cell. Biol.* 3:267–77
19. Camacho L, Malho R. 2003. Endo/exocytosis in the pollen tube apex is differentially regulated by Ca^{2+} and GTPases. *J. Exp. Bot.* 54:83–92

20. Caron E, Hall A. 1998. Identification of two distinct mechanisms of phagocytosis controlled by different Rho GTPases. *Science* 282:1717–21
21. Chavrier P, Goud B. 1999. The role of ARF and Rab GTPases in membrane transport. *Curr. Opin. Cell. Biol.* 11:466–75
22. Cheung AY, Chen CY, Glaven RH, de Graaf BH, Vidali L, et al. 2002. Rab2 GTPase regulates vesicle trafficking between the endoplasmic reticulum and the Golgi bodies and is important to pollen tube growth. *Plant Cell* 14:945–62
23. Chimini G, Chavrier P. 2000. Function of Rho family proteins in actin dynamics during phagocytosis and engulfment. *Nat. Cell. Biol.* 2:E191–96
24. Cook NR, Row PE, Davidson HW. 2004. Lysosome associated membrane protein 1 (Lamp1) traffics directly from the TGN to early endosomes. *Traffic* 5:685–99
25. Couchy I, Bolte S, Crosnier MT, Brown S, Satiat-Jeunemaitre A. 2003. Identification and localization of a beta-COP-like protein involved in the morphodynamics of the plant Golgi apparatus. *J. Exp. Bot.* 54:2053–63
26. Das V, Nal B, Dujeancourt A, Thoulouze MI, Galli T, et al. 2004. Activation-induced polarized recycling targets T cell antigen receptors to the immunological synapse; involvement of SNARE complexes. *Immunity* 20:577–88
27. De Matteis MA, Godi A. 2004. PI-loting membrane traffic. *Nat. Cell. Biol.* 6:487–92
28. DeWitt ND, Hong B, Sussman MR, Harper JF. 1996. Targeting of two *Arabidopsis* H(+)-ATPase isoforms to the plasma membrane. *Plant Physiol.* 112:833–44
29. Donaldson JG. 2003. Multiple roles for Arf6: sorting, structuring, and signaling at the plasma membrane. *J. Biol. Chem.* 278:41573–76
30. Donaldson JG, Jackson CL. 2000. Regulators and effectors of the ARF GTPases. *Cur. Opin. Cell Biol.* 12:475–82
31. Drucker M, Herkt B, Robinson DG. 1995. Demonstration of b-type adaptin at the plant PM. *Cell Biol. Int.* 19:191–201
32. D'Souza-Schorey C, van Donselaar E, Hsu VW, Yang C, Stahl PD, Peters PJ. 1998. ARF6 targets recycling vesicles to the plasma membrane: insights from an ultrastructural investigation. *J. Cell. Biol.* 140:603–16
33. Duman JG, Tyagarajan K, Kolsi MS, Moore HP, Forte JG. 1999. Expression of rab11a N124I in gastric parietal cells inhibits stimulatory recruitment of the H^+-K^+-ATPase. *Am. J. Physiol.* 277:C361–72
34. Engqvist-Goldstein AEY, Drubin DG. 2003. Actin assembly and endocytosis: from yeast to mammals. *Annu. Rev. Cell Dev. Biol.* 19:287–332
35. Fischer K, Weber A, Brink S, Arbinger B, Schunemann D, et al. 1994. Porins from plants. Molecular cloning and functional characterization of two new members of the porin family. *J. Biol. Chem.* 269:25754–60
36. Fleming RE, Parkkila S, Parkkila AK, Rajaniemi H, Waheed A, Sly WS. 1995. Carbonic anhydrase IV expression in rat and human gastrointestinal tract regional, cellular, and subcellular localization. *J. Clin. Invest.* 96:2907–13
37. Fouraux MA, Deneka M, Ivan V, van der Heijden A, Raymackers J, et al. 2004. Rabip4′ is an effector of rab5 and rab4 and regulates transport through early endosomes. *Mol. Biol. Cell.* 15:611–24
38. Franco M, Peters PJ, Boretto J, van Donselaar E, Neri A, et al. 1999. EFA6, a sec7 domain-containing exchange factor for ARF6, coordinates membrane recycling and actin cytoskeleton organization. *EMBO J.* 18:1480–91
39. Friml J, Wisniewska J, Benkova E, Mendgen K, Palme K. 2002. Lateral relocation of auxin efflux regulator PIN3 mediates tropism in *Arabidopsis*. *Nature* 415:806–9

39a. Friml J, Yang X, Michniewicz M, Weijers D, Quint A, et al. 2004. A pinoid-dependent binary switch in apical-basal pin polar targeting directs auxin efflux. *Science* 306:862–65
40. Galway ME, Rennie PJ, Fowke LC. 1993. Ultrastructure of the endocytotic pathway in glutaraldehyde-fixed and high-pressure frozen/freeze-substituted protoplasts of white spruce (Picea glauca). *J. Cell Sci.* 106 (Pt. 3): 847–58
41. Gan Y, McGraw TE, Rodriguez-Boulan E. 2002. The epithelial-specific adaptor AP1B mediates post-endocytic recycling to the basolateral membrane. *Nat. Cell. Biol.* 4: 605–9
42. Geisler M, Kolukisaoglu HU, Bouchard R, Billion K, Berger J, et al. 2003. TWISTED DWARF1, a unique plasma membrane-anchored immunophilin-like protein, interacts with *Arabidopsis* multidrug resistance-like transporters AtPGP1 and AtPGP19. *Mol. Biol. Cell* 14:4238–49
43. Geldner N, Anders N, Wolters H, Keicher J, Kornberger W, et al. 2003. The *Arabidopsis* GNOM ARF-GEF mediates endosomal recycling, auxin transport, and auxin-dependent plant growth. *Cell* 112:219–30
44. Geldner N, Friml J, Stierhof YD, Jurgens G, Palme K. 2001. Auxin transport inhibitors block PIN1 cycling and vesicle traficking. *Nature* 413:425–28
45. Ghosh P, Dahms NM, Kornfeld S. 2003. Mannose 6-phosphate receptors: new twists in the tale. *Nat. Rev. Mol. Cell. Biol.* 4:202–12
46. Goodman OB Jr, Krupnick JG, Santini F, Gurevich VV, Penn RB, et al. 1996. Beta-arrestin acts as a clathrin adaptor in endocytosis of the beta2-adrenergic receptor. *Nature* 383:447–50
47. Graham IA, Li Y, Larson TR. 2002. Acyl-CoA measurements in plants suggest a role in regulating various cellular processes. *Biochem. Soc. Trans.* 30:1095–99
48. Grebe M, Friml J, Swarup R, Ljung K, Sandberg G, et al. 2002. Cell polarity signaling in *Arabidopsis* involves a BFA-sensitive auxin influx pathway. *Curr. Biol.* 12:329–34
49. Grebe M, Xu J, Mobius W, Ueda T, Nakano A, et al. 2003. *Arabidopsis* sterol endocytosis involves actin-mediated trafficking via ARA6-positive early endosomes. *Curr. Biol.* 13:1378–87
50. Gruenberg J. 2001. The endocytic pathway: a mosaic of domains. *Nat. Rev. Mol. Cell. Biol.* 2:721–30
51. Gruenberg J. 2003. Lipids in endocytic membrane transport and sorting. *Cur. Opin. Cell Biol.* 15:382–88
52. Gruenberg J, Howell KE. 1989. Membrane traffic in endocytosis: insights from cell-free assays. *Annu. Rev. Cell Biol.* 5:453–81
53. Gustafson CE, Katsura T, McKee M, Bouley R, Casanova JE, Brown D. 2000. Recycling of AQP2 occurs through a temperature- and bafilomycin-sensitive trans-Golgi-associated compartment. *Am. J. Physiol. Renal. Physiol.* 278:F317–26
54. Hao M, Lin SX, Karylowski OJ, Wustner D, McGraw TE, Maxfield FR. 2002. Vesicular and non-vesicular sterol transport in living cells. The endocytic recycling compartment is a major sterol storage organelle. *J. Biol. Chem.* 277:609–17
55. Hao M, Maxfield FR. 2000. Characterization of rapid membrane internalization and recycling. *J. Biol. Chem.* 275:15279–86
56. Happel N, Honing S, Neuhaus JM, Paris N, Robinson DG, Holstein SE. 2004. *Arabidopsis* mu A-adaptin interacts with the tyrosine motif of the vacuolar sorting receptor VSR-PS1. *Plant J.* 37:678–93
57. He Z, Li L, Luan S. 2004. Immunophilins and parvulins. Superfamily of peptidyl prolyl isomerases in *Arabidopsis*. *Plant Physiol* 134:1248–67

58. Heese M, Gansel X, Sticher L, Wick P, Grebe M, et al. 2001. Functional characterization of the KNOLLE-interacting t-SNARE AtSNAP33 and its role in plant cytokinesis. *J. Cell Biol.* 155:239–49
59. Helms JB, Zurzolo C. 2004. Lipids as targeting signals: lipid rafts and intracellular trafficking. *Traffic* 5:247–54
60. Henley JR, Krueger EW, Oswald BJ, McNiven MA. 1998. Dynamin-mediated internalization of caveolae. *J. Cell. Biol.* 141:85–99
61. Henley JR, McNiven MA. 1996. Association of a dynamin-like protein with the Golgi apparatus in mammalian cells. *J. Cell. Biol.* 133:761–75
62. Herskovits JS, Burgess CC, Obar RA, Vallee RB. 1993. Effects of mutant rat dynamin on endocytosis. *J. Cell. Biol.* 122:565–78
63. Hillmer S, Freundt H, Robinson DG. 1988. The partially coated reticulum and its relationship to the golgi-apparatus in higher-plant cells. *Eur. J. Cell Biol.* 47:206–12
64. Hinshaw JE, Schmid SL. 1995. Dynamin self-assembles into rings suggesting a mechanism for coated vesicle budding. *Nature* 374:190–92
65. Holstein SE, Drucker M, Robinson DG. 1994. Identification of a beta-type adaptin in plant clathrin-coated vesicles. *J. Cell Sci.* 107 (Pt. 4): 945–53
66. Holstein SE, Ungewickell H, Ungewickell E. 1996. Mechanism of clathrin basket dissociation: separate functions of protein domains of the DnaJ homologue auxilin. *J. Cell. Biol.* 135:925–37
67. Holstein SEH. 2002. Clathrin and plant endocytosis. *Traffic* 3:614–20
68. Holweg C, Nick P. 2004. *Arabidopsis myosin XI* mutant is defective in organelle movement and polar auxin transport. *Proc. Natl. Acad. Sci. USA* 101:10488–93
69. Homann U, Thiel G. 2002. The number of K(+) channels in the plasma membrane of guard cell protoplasts changes in parallel with the surface area. *Proc. Natl. Acad. Sci. USA* 99:10215–20
70. Hong Z, Bednarek SY, Blumwald E, Hwang I, Jurgens G, et al. 2003. A unified nomenclature for *Arabidopsis* dynamin-related large GTPases based on homology and possible functions. *Plant Mol. Biol.* 53:261–65
71. Hong ZL, Geisler-Lee CJ, Zhang ZM, Verma DPS. 2003. Phragmoplastin dynamics: multiple forms, microtubule association and their roles in cell plate formation in plants. *Plant Mol. Biol.* 53:297–312
72. Hurst AC, Meckel T, Tayefeh S, Thiel G, Homann U. 2004. Trafficking of the plant potassium inward rectifier KAT1 in guard cell protoplasts of Vicia faba. *Plant J.* 37:391–97
73. Inaba T, Nagano Y, Nagasaki T, Sasaki Y. 2002. Distinct localization of two closely related Ypt3/Rab11 proteins on the trafficking pathway in higher plants. *J. Biol. Chem.* 277:9183–88
74. Jin JB, Kim YA, Kim SJ, Lee SH, Kim DH, et al. 2001. A new dynamin-like protein, ADL6, is involved in trafficking from the trans-Golgi network to the central vacuole in *Arabidopsis*. *Plant Cell* 13:1511–26
75. Johnson LS, Dunn KW, Pytowski B, McGraw TE. 1993. Endosome acidification and receptor trafficking: bafilomycin A1 slows receptor externalization by a mechanism involving the receptor's internalization motif. *Mol. Biol. Cell.* 4:1251–66
76. Jones SM, Howell KE, Henley JR, Cao H, McNiven MA. 1998. Role of dynamin in the formation of transport vesicles from the trans-Golgi network. *Science* 279:573–77
77. Juergens G. 2004. Membrane trafficking in plants. *Annu. Rev. Cell Dev. Biol.* 20:481–504
78. Kang BH, Busse JS, Bednarek SY. 2003. Members of the *Arabidopsis* dynamin-like gene family, ADL1, are essential for plant cytokinesis and polarized cell growth. *Plant Cell* 15:899–913

79. Kang BH, Busse JS, Dickey C, Rancour DM, Bednarek SY. 2001. The *Arabidopsis* cell plate-associated dynamin-like protein, ADL1Ap, is required for multiple stages of plant growth and development. *Plant Physiol.* 126:47–68
80. Katagiri H, Asano T, Yamada T, Aoyama T, Fukushima Y, et al. 2002. Acyl-coenzyme a dehydrogenases are localized on GLUT4-containing vesicles via association with insulin-regulated aminopeptidase in a manner dependent on its dileucine motif. *Mol. Endocrinol.* 16:1049–59
81. Katzmann DJ, Odorizzi G, Emr SD. 2002. Receptor downregulation and multivesicular-body sorting. *Nat. Rev. Mol. Cell Biol.* 3:893–905
82. Kim D, Gustin JL, Lahner B, Persans MW, Baek D, et al. 2004. The plant CDF family member TgMTP1 from the Ni/Zn hyperaccumulator Thlaspi goesingense acts to enhance efflux of Zn at the plasma membrane when expressed in Saccharomyces cerevisiae. *Plant J.* 39:237–51
83. Kim DH, Eu YJ, Yoo CM, Kim YW, Pih KT, et al. 2001. Trafficking of phosphatidylinositol 3-phosphate from the trans-Golgi network to the lumen of the central vacuole in plant cells. *Plant Cell* 13:287–301
84. Kirchhausen T. 1999. Adaptors for clathrin-mediated traffic. *Annu. Rev. Cell. Dev. Biol.* 15:705–32
85. Kivela A, Parkkila S, Saarnio J, Karttunen TJ, Kivela J, et al. 2000. Expression of a novel transmembrane carbonic anhydrase isozyme XII in normal human gut and colorectal tumors. *Am. J. Pathol.* 156:577–84
86. Klussmann E, Maric K, Rosenthal W. 2000. The mechanisms of aquaporin control in the renal collecting duct. *Rev. Physiol. Biochem. Pharmacol.* 141:33–95
87. Kobayashi T, Beuchat MH, Lindsay M, Frias S, Palmiter RD, et al. 1999. Late endosomal membranes rich in lysobisphosphatidic acid regulate cholesterol transport. *Nat. Cell. Biol.* 1:113–18
88. Krauss M, Kinuta M, Wenk MR, De Camilli P, Takei K, Haucke V. 2003. ARF6 stimulates clathrin/AP-2 recruitment to synaptic membranes by activating phosphatidylinositol phosphate kinase type I gamma. *J. Cell. Biol.* 162:113–24
89. Lalanne E, Honys D, Johnson A, Borner GHH, Lilley KS, et al. 2004. SETH1 and SETH2, two components of the glycosylphosphatidylinositol anchor biosynthetic pathway, are required for pollen germination and tube growth in *Arabidopsis*. *Plant Cell* 16:229–40
90. Lam BCH, Sage TL, Bianchi F, Blumwald E. 2002. Regulation of ADL6 activity by its associated molecular network. *Plant J.* 31:565–76
91. Li S, Blanchoin L, Yang Z, Lord EM. 2003. The putative *Arabidopsis* arp2/3 complex controls leaf cell morphogenesis. *Plant Physiol.* 132:2034–44
92. Lin SX, Grant B, Hirsh D, Maxfield FR. 2001. Rme-1 regulates the distribution and function of the endocytic recycling compartment in mammalian cells. *Nat. Cell. Biol.* 3:567–72
93. Loder MK, Melikian HE. 2003. The dopamine transporter constitutively internalizes and recycles in a protein kinase C-regulated manner in stably transfected PC12 cell lines. *J. Biol. Chem.* 278:22168–74
94. Luo Z, Gallwitz D. 2003. Biochemical and genetic evidence for the involvement of yeast Ypt6-GTPase in protein retrieval to different Golgi compartments. *J. Biol. Chem.* 278:791–99
95. Maldonado-Mendoza IE, Nessler CL. 1996. Cloning and expression of a plant homologue of the small subunit of the Golgi-associated clathrin assembly protein AP19 from *Camptotheca acuminata*. *Plant Mol. Biol.* 32:1149–53

96. Mallard F, Tang BL, Galli T, Tenza D, Saint-Pol A, et al. 2002. Early/recycling endosomes-to-TGN transport involves two SNARE complexes and a Rab6 isoform. *J. Cell. Biol.* 156:653–64
97. Marmagne A, Rouet MA, Ferro M, Rolland N, Alcon C, et al. 2004. Identification of new intrinsic proteins in *Arabidopsis* plasma membrane proteome. *Mol. Cell. Proteomics* 3:675–91
98. Maxfield FR, McGraw TE. 2004. Endocytic recycling. *Nat. Rev. Mol. Cell Biol.* 5:121–32
99. Mayor S, Presley JF, Maxfield FR. 1993. Sorting of membrane components from endosomes and subsequent recycling to the cell surface occurs by a bulk flow process. *J. Cell. Biol.* 121:1257–69
100. McBride HM, Rybin V, Murphy C, Giner A, Teasdale R, Zerial M. 1999. Oligomeric complexes link Rab5 effectors with NSF and drive membrane fusion via interactions between EEA1 and syntaxin 13. *Cell* 98:377–86
101. McMahon HT, Ushkaryov YA, Edelmann L, Link E, Binz T, et al. 1993. Cellubrevin is a ubiquitous tetanus-toxin substrate homologous to a putative synaptic vesicle fusion protein. *Nature* 364:346–49
102. Meckel T, Hurst AC, Thiel G, Homann U. 2004. Endocytosis against high turgor: intact guard cells of *Vicia faba* constitutively endocytose fluorescently labelled plasma membrane and GFP-tagged K^+-channel KAT1. *Plant J.* 39:182–93
103. Mellman I. 1996. Endocytosis and molecular sorting. *Annu. Rev. Cell. Dev. Biol.* 12:575–625
104. Menetrey J, Macia E, Pasqualato S, Franco M, Cherfils J. 2000. Structure of Arf6-GDP suggests a basis for guanine nucleotide exchange factors specificity. *Nat. Struct. Biol.* 7:466–69
105. Mohrmann K, Gerez L, Oorschot V, Klumperman J, van der Sluijs P. 2002. Rab4 function in membrane recycling from early endosomes depends on a membrane to cytoplasm cycle. *J. Biol. Chem.* 277:32029–35
106. Mohrmann K, van der Sluijs P. 1999. Regulation of membrane transport through the endocytic pathway by rabGTPases. *Mol. Membr. Biol.* 16:81–87
107. Molendijk AJ, Bischoff F, Rajendrakumar CS, Friml J, Braun M, et al. 2001. *Arabidopsis* thaliana Rop GTPases are localized to tips of root hairs and control polar growth. *EMBO J.* 20:2779–88
108. Mongrand S, Morel J, Laroche J, Claverol S, Carde JP, et al. 2004. Lipid rafts in higher plant cells: purification and characterization of Triton X-100-insoluble microdomains from tobacco plasma membrane. *J. Biol. Chem.* 279:36277–86
109. Moritz OL, Tam BM, Hurd LL, Peranen J, Deretic D, Papermaster DS. 2001. Mutant *rab8* impairs docking and fusion of rhodopsin-bearing post-Golgi membranes and causes cell death of transgenic Xenopus rods. *Mol. Biol. Cell.* 12:2341–51
110. Moss J, Vaughan M. 1998. Molecules in the ARF orbit. *J. Biol. Chem.* 273:21431–34
111. Mukherjee S, Ghosh RN, Maxfield FR. 1997. Endocytosis. *Physiol. Rev.* 77:759–803
112. Muniz M, Riezman H. 2000. Intracellular transport of GPI-anchored proteins. *EMBO J.* 19:10–15
113. Murgia C, Vespignani I, Cerase J, Nobili F, Perozzi G. 1999. Cloning, expression, and vesicular localization of zinc transporter Dri 27/ZnT4 in intestinal tissue and cells. *Am. J. Physiol.* 277:G1231–39
114. Murphy AS, Hoogner KR, Peer WA, Taiz L. 2002. Identification, purification, and molecular cloning of N-1-naphthylphthalmic acid-binding plasma membrane-associated aminopeptidases from *Arabidopsis*. *Plant Physiol.* 128:935–50

115. Muza-Moons MM, Koutsouris A, Hecht G. 2003. Disruption of cell polarity by enteropathogenic Escherichia coli enables basolateral membrane proteins to migrate apically and to potentiate physiological consequences. *Infect. Immun.* 71:7069–78
116. Naslavsky N, Boehm M, Backlund PS Jr, Caplan S. 2004. Rabenosyn-5 and EHD1 interact and sequentially regulate protein recycling to the plasma membrane. *Mol. Biol. Cell* 15:2410–22
117. Nebenfuhr A. 2002. Vesicle traffic in the endomembrane system: a tale of COPs, Rabs and SNAREs. *Curr. Opin. Plant Biol.* 5:507–12
118. Nicoziani P, Vilhardt F, Llorente A, Hilout L, Courtoy PJ, et al. 2000. Role for dynamin in late endosome dynamics and trafficking of the cation-independent mannose 6-phosphate receptor. *Mol. Biol. Cell.* 11:481–95
119. Oh P, McIntosh DP, Schnitzer JE. 1998. Dynamin at the neck of caveolae mediates their budding to form transport vesicles by GTP-driven fission from the plasma membrane of endothelium. *J. Cell. Biol.* 141:101–14
120. Orth JD, Krueger EW, Cao H, McNiven MA. 2002. The large GTPase dynamin regulates actin comet formation and movement in living cells. *Proc. Natl. Acad. Sci. USA* 99:167–72
121. Palacios F, Price L, Schweitzer J, Collard JG, D'Souza-Schorey C. 2001. An essential role for ARF6-regulated membrane traffic in adherens junction turnover and epithelial cell migration. *EMBO J.* 20:4973–86
122. Deleted in proof
123. Peer WA, Bandyopadhyay A, Blakeslee JJ, Makam SN, Chen RJ, et al. 2004. Variation in expression and protein localization of the PIN family of auxin efflux facilitator proteins in flavonoid mutants with altered auxin transport in *Arabidopsis* thaliana. *Plant Cell* 16:1898–911
124. Pelkmans L, Puntener D, Helenius A. 2002. Local actin polymerization and dynamin recruitment in SV40-induced internalization of caveolae. *Science* 296:535–39
125. Peters PJ, Gao M, Gaschet J, Ambach A, van Donselaar E, et al. 2001. Characterization of coated vesicles that participate in endocytic recycling. *Traffic* 2:885–95
126. Peyroche A, Antonny B, Robineau S, Acker J, Cherfils J, Jackson CL. 1999. Brefeldin A acts to stabilize an abortive ARF-GDP-Sec7 domain protein complex: involvement of specific residues of the Sec7 domain. *Mol. Cell* 3:275–85
127. Pfeffer SR. 2001. Rab GTPases: specifying and deciphering organelle identity and function. *Trends Cell Biol.* 11:487–91
128. Presley JF, Mayor S, McGraw TE, Dunn KW, Maxfield FR. 1997. Bafilomycin A1 treatment retards transferrin receptor recycling more than bulk membrane recycling. *J. Biol. Chem.* 272:13929–36
129. Preuss ML, Serna J, Falbel TG, Bednarek SY, Nielsen E. 2004. The *Arabidopsis* Rab GTPase RabA4b localizes to the tips of growing root hair cells. *Plant Cell* 16:1589–603
130. Qualmann B, Mellor H. 2003. Regulation of endocytic traffic by Rho GTPases. *Biochem. J.* 371:233–41
131. Qualmann B, Roos J, DiGregorio PJ, Kelly RB. 1999. Syndapin I, a synaptic dynamin-binding protein that associates with the neural Wiskott-Aldrich syndrome protein. *Mol. Biol. Cell.* 10:501–13
132. Quigley F, Rosenberg JM, Shachar-Hill Y, Bohnert HJ. 2002. From genome to function: the *Arabidopsis* aquaporins. *Genome Biol.* 3:RESEARCH0001
133. Radhakrishna H, Donaldson JG. 1997. ADP-ribosylation factor 6 regulates a novel plasma membrane recycling pathway. *J. Cell. Biol.* 139:49–61

134. Raiborg C, Bache KG, Gillooly DJ, Madshus IH, Stang E, Stenmark H. 2002. Hrs sorts ubiquitinated proteins into clathrin-coated microdomains of early endosomes. *Nat. Cell. Biol.* 4:394–98
135. Ridley A. 2000. Rho GTPases. Integrating integrin signaling. *J. Cell. Biol.* 150:F107–9
136. Robinson DG, Hillmer S. 1990. Endocytosis in plants. *Physiol. Planta.* 79:96–104
137. Robinson MS. 1994. The role of clathrin, adaptors and dynamin in endocytosis. *Curr. Opin. Cell. Biol.* 6:538–44
138. Roca R, Stiefel V, Puigdomenech P. 1998. Characterization of the sequence coding for the clathrin coat assembly protein AP17 (sigma2) associated with the plasma membrane from Zea mays and constitutive expression of its gene. *Gene* 208:67–72
139. Rocha Facanha A, de Meis L. 1998. Reversibility of H^+-ATPase and H^+-pyrophosphatase in tonoplast vesicles from maize coleoptiles and seeds. *Plant Physiol.* 116:1487–95
140. Rohde G, Wenzel D, Haucke V. 2002. A phosphatidylinositol (4,5)-bisphosphate binding site within mu2-adaptin regulates clathrin-mediated endocytosis. *J. Cell. Biol.* 158:209–14
141. Royle SJ, Murrell-Lagnado RD. 2003. Constitutive cycling: a general mechanism to regulate cell surface proteins. *Bioessays* 25:39–46
142. Ryan CA, Pearce G, Scheer J, Moura DS. 2002. Polypeptide hormones. *Plant Cell* 14(Suppl.): S251–64
143. Sachse M, Strous GJ, Klumperman J. 2004. ATPase-deficient hVPS4 impairs formation of internal endosomal vesicles and stabilizes bilayered clathrin coats on endosomal vacuoles. *J. Cell Sci.* 117:1699–708
144. Schaerer-Brodbeck C, Riezman H. 2000. Functional interactions between the p35 subunit of the Arp2/3 complex and calmodulin in yeast. *Mol. Biol. Cell.* 11:1113–27
145. Schindelman G, Morikami A, Jung J, Baskin TI, Carpita NC, et al. 2001. *COBRA* encodes a putative GPI-anchored protein, which is polarly localized and necessary for oriented cell expansion in *Arabidopsis*. *Genes Dev.* 15:1115–27
146. Schledzewski K, Brinkmann H, Mendel RR. 1999. Phylogenetic analysis of components of the eukaryotic vesicle transport system reveals a common origin of adaptor protein complexes 1, 2, and 3 and the F subcomplex of the coatomer COPI. *J. Mol. Evol.* 48:770–78
147. Schmidt A, Wolde M, Thiele C, Fest W, Kratzin H, et al. 1999. Endophilin I mediates synaptic vesicle formation by transfer of arachidonate to lysophosphatidic acid. *Nature* 401:133–41
148. Seabra MC, Coudrier E. 2004. Rab GTPases and myosin motors in organelle motility. *Traffic* 5:393–99
149. Seaman MN. 2004. Cargo-selective endosomal sorting for retrieval to the Golgi requires retromer. *J. Cell. Biol.* 165:111–22
150. Segev N. 2001. Ypt/rab gtpases: regulators of protein trafficking. *Sci. STKE* 2001:RE11
151. Seron K, Tieaho V, Prescianotto-Baschong C, Aust T, Blondel MO, et al. 1998. A yeast t-SNARE involved in endocytosis. *Mol. Biol. Cell.* 9:2873–89
152. Sever S, Muhlberg AB, Schmid SL. 1999. Impairment of dynamin's GAP domain stimulates receptor-mediated endocytosis. *Nature* 398:481–86
153. Shah K, Russinova E, Gadella TWJ, Willemse J, de Vries SC. 2002. The *Arabidopsis* kinase-associated protein phosphatase controls internalization of the somatic embryogenesis receptor kinase 1. *Genes Dev.* 16:1707–20
154. Sheff DR, Daro EA, Hull M, Mellman I. 1999. The receptor recycling pathway contains two distinct populations of early endosomes with different sorting functions. *J. Cell. Biol.* 145:123–39

155. Simonsen A, Lippe R, Christoforidis S, Gaullier JM, Brech A, et al. 1998. EEA1 links PI(3)K function to Rab5 regulation of endosome fusion. *Nature* 394:494–98
156. Sleeman MW, Donegan NP, Heller-Harrison R, Lane WS, Czech MP. 1998. Association of acyl-CoA synthetase-1 with GLUT4-containing vesicles. *J. Biol. Chem.* 273:3132–35
157. Smythe E. 2002. Direct interactions between rab GTPases and cargo. *Mol. Cell* 9:205–6
158. Sohn EJ, Kim ES, Zhao M, Kim SJ, Kim H, et al. 2003. Rha1, an *Arabidopsis* Rab5 homolog, plays a critical role in the vacuolar trafficking of soluble cargo proteins. *Plant Cell* 15:1057–70
159. Son O, Yang HS, Lee HJ, Lee MY, Shin KH, et al. 2003. Expression of *srab7* and *SCaM* genes required for endocytosis of *Rhizobium* in root nodules. *Plant Sci.* 165:1239–44
160. Sonnichsen B, De Renzis S, Nielsen E, Rietdorf J, Zerial M. 2000. Distinct membrane domains on endosomes in the recycling pathway visualized by multicolor imaging of Rab4, Rab5, and Rab11. *J. Cell. Biol.* 149:901–14
161. Sorkin A, Krolenko S, Kudrjavtceva N, Lazebnik J, Teslenko L, et al. 1991. Recycling of epidermal growth factor-receptor complexes in A431 cells: identification of dual pathways. *J. Cell. Biol.* 112:55–63
162. Sorkin A, von Zastrow M. 2002. Signal transduction and endocytosis: close encounters of many kinds. *Nat. Rev. Mol. Cell Biol.* 3:600–14
163. Steinmann T, Geldner N, Grebe M, Mangold S, Jackson CL, et al. 1999. Coordinated polar localization of auxin efflux carrier PIN1 by GNOM ARF GEF. *Science* 286:316–18
164. Stoorvogel W, Oorschot V, Geuze HJ. 1996. A novel class of clathrin-coated vesicles budding from endosomes. *J. Cell. Biol.* 132:21–33
165. Swarup R, Friml J, Marchant A, Ljung K, Sandberg G, et al. 2001. Localization of the auxin permease AUX1 suggests two functionally distinct hormone transport pathways operate in the *Arabidopsis* root apex. *Genes Dev.* 15:2648–53
166. Tanchak MA, Rennie PJ, Fowke LC. 1988. Ultrastructure of the partially coated reticulum and dictyosomes during endocytosis by soybean protoplasts. *Planta* 175:433–41
167. Tisdale EJ. 1999. A *Rab2* mutant with impaired GTPase activity stimulates vesicle formation from pre-Golgi intermediates. *Mol. Biol. Cell.* 10:1837–49
168. Tisdale EJ, Bourne JR, Khosravi-Far R, Der CJ, Balch WE. 1992. GTP-binding mutants of *rab1* and *rab2* are potent inhibitors of vesicular transport from the endoplasmic reticulum to the Golgi complex. *J. Cell. Biol.* 119:749–61
169. Trischler M, Stoorvogel W, Ullrich O. 1999. Biochemical analysis of distinct Rab5- and Rab11-positive endosomes along the transferrin pathway. *J. Cell Sci.* 112 (Pt. 24): 4773–83
170. Tse YC, Mo BX, Hillmer S, Zhao M, Lo SW, et al. 2004. Identification of multivesicular bodies as prevacuolar compartments in *Nicotiana tabacum* BY-2 cells. *Plant Cell* 16:672–93
171. Tuvim MJ, Adachi R, Hoffenberg S, Dickey BF. 2001. Traffic control: Rab GTPases and the regulation of interorganellar transport. *News Physiol. Sci.* 16:56–61
172. Ueda T, Anai T, Tsukaya H, Hirata A, Uchimiya H. 1996. Characterization and subcellular localization of a small GTP-binding protein (Ara-4) from *Arabidopsis*: conditional expression under control of the promoter of the gene for heat-shock protein HSP81-1. *Mol. Gen. Genet.* 250:533–39
173. Ueda T, Yamaguchi M, Uchimiya H, Nakano A. 2001. Ara6, a plant-unique novel type Rab GTPase, functions in the endocytic pathway of *Arabidopsis* thaliana. *EMBO J.* 20:4730–41
174. Ullrich O, Reinsch S, Urbe S, Zerial M, Parton RG. 1996. Rab11 regulates recycling through the pericentriolar recycling endosome. *J. Cell. Biol.* 135:913–24
175. Ungewickell E, Ungewickell H, Holstein SE, Lindner R, Prasad K, et al. 1995. Role of auxilin in uncoating clathrin-coated vesicles. *Nature* 378:632–35

176. Vallis Y, Wigge P, Marks B, Evans PR, McMahon HT. 1999. Importance of the pleckstrin homology domain of dynamin in clathrin-mediated endocytosis. *Curr. Biol.* 9:257–60
177. van Dam EM, Stoorvogel W. 2002. Dynamin-dependent transferrin receptor recycling by endosome-derived clathrin-coated vesicles. *Mol. Biol. Cell.* 13:169–82
178. van Dam EM, Ten Broeke T, Jansen K, Spijkers P, Stoorvogel W. 2002. Endocytosed transferrin receptors recycle via distinct dynamin and phosphatidylinositol 3-kinase-dependent pathways. *J. Biol. Chem.* 277:48876–83
179. van der Bliek AM. 1999. Functional diversity in the dynamin family. *Trends Cell Biol.* 9:96–102
180. van der Bliek AM, Redelmeier TE, Damke H, Tisdale EJ, Meyerowitz EM, Schmid SL. 1993. Mutations in human dynamin block an intermediate stage in coated vesicle formation. *J. Cell. Biol.* 122:553–63
181. van Weert AW, Dunn KW, Gueze HJ, Maxfield FR, Stoorvogel W. 1995. Transport from late endosomes to lysosomes, but not sorting of integral membrane proteins in endosomes, depends on the vacuolar proton pump. *J. Cell. Biol.* 130:821–34
182. Vera-Estrella R, Barkla BJ, Bohnert HJ, Pantoja O. 2004. Novel regulation of aquaporins during osmotic stress. *Plant Physiol.* 135:2318–29
183. Verges M, Luton F, Gruber C, Tiemann F, Reinders LG, et al. 2004. The mammalian retromer regulates transcytosis of the polymeric immunoglobulin receptor. *Nat. Cell. Biol.* 6:763–69
184. Vernoud V, Horton AC, Yang Z, Nielsen E. 2003. Analysis of the small GTPase gene superfamily of *Arabidopsis*. *Plant Physiol.* 131:1191–208
185. Vida TA, Emr SD. 1995. A new vital stain for visualizing vacuolar membrane dynamics and endocytosis in yeast. *J. Cell Biol.* 128:779–92
186. Wang E, Brown PS, Aroeti B, Chapin SJ, Mostov KE, Dunn KW. 2000. Apical and basolateral endocytic pathways of MDCK cells meet in acidic common endosomes distinct from a nearly-neutral apical recycling endosome. *Traffic* 1:480–93
187. Wilcke M, Johannes L, Galli T, Mayau V, Goud B, Salamero J. 2000. Rab11 regulates the compartmentalization of early endosomes required for efficient transport from early endosomes to the trans-golgi network. *J. Cell. Biol.* 151:1207–20
188. Willemsen V, Friml J, Grebe M, van den Toorn A, Palme K, Scheres B. 2003. Cell polarity and PIN protein positioning in *Arabidopsis* require STEROL METHYLTRANSFERASE1 function. *Plant Cell* 15:612–25
189. Wilsbach K, Payne GS. 1993. Vps1p, a member of the dynamin GTPase family, is necessary for Golgi membrane protein retention in Saccharomyces cerevisiae. *EMBO J.* 12:3049–59
190. Witke W, Podtelejnikov AV, Di Nardo A, Sutherland JD, Gurniak CB, et al. 1998. In mouse brain profilin I and profilin II associate with regulators of the endocytic pathway and actin assembly. *EMBO J.* 17:967–76
191. Worby CA, Dixon JE. 2002. Sorting out the cellular functions of sorting nexins. *Nat. Rev. Mol. Cell. Biol.* 3:919–31
192. Xu XL, Bittman R, Duportail G, Heissler D, Vilcheze C, London E. 2001. Effect of the structure of natural sterols and sphingolipids on the formation of ordered sphingolipid/sterol domains (rafts). *J. Biol. Chem.* 276:33540–46
193. Xu Y, Hortsman H, Seet L, Wong SH, Hong W. 2001. SNX3 regulates endosomal function through its PX-domain-mediated interaction with PtdIns(3)P. *Nat. Cell. Biol.* 3:658–66
194. Yang ZB. 2002. Small GTPases: versatile signaling switches in plants. *Plant Cell* 14:S375–S88

195. Zeigerer A, Lampson MA, Karylowski O, Sabatini DD, Adesnik M, et al. 2002. GLUT4 retention in adipocytes requires two intracellular insulin-regulated transport steps. *Mol. Biol. Cell* 13:2421–35
196. Zerial M, McBride H. 2001. Rab proteins as membrane organizers. *Nat. Rev. Mol. Cell Biol.* 2:107–17
197. Zhao X, Lasell TK, Melancon P. 2002. Localization of large ADP-ribosylation factor-guanine nucleotide exchange factors to different Golgi compartments: evidence for distinct functions in protein traffic. *Mol. Biol. Cell.* 13:119–33
198. Zheng H, von Mollard GF, Kovaleva V, Stevens TH, Raikhel NV. 1999. The plant vesicle-associated SNARE AtVTI1a likely mediates vesicle transport from the trans-Golgi network to the prevacuolar compartment. *Mol. Biol. Cell.* 10:2251–64
199. Zheng ZL, Yang Z. 2000. The Rrop GTPase switch turns on polar growth in pollen. *Trends Plant Sci.* 5:298–303

Molecular Physiology of Legume Seed Development

Hans Weber, Ljudmilla Borisjuk, and Ulrich Wobus

Institute of Plant Genetics and Crop Plant Research (IPK), D-06466 Gatersleben, Germany; email: weber@ipk-gatersleben.de, borisjuk@ipk-gatersleben.de, wobusu@ipk-gatersleben.de

Annu. Rev. Plant Biol. 2005. 56:253–79

doi: 10.1146/annurev.arplant.56.032604.144201

First published online as a Review in Advance on January 14, 2005

1543-5008/05/0602-0253$20.00

Key Words

seed maturation, seed metabolism, storage product synthesis, metabolic control, photoheterotrophic plastids

Abstract

Legume seed development is characterized by progressive differentiation of organs and tissues resulting in developmental gradients. The whole process is prone to metabolic control, and distinct metabolite profiles specify the differentiation state. Whereas early embryo growth is mainly maternally controlled, the transition into maturation implies a switch to filial control. A signaling network involving sugars, ABA, and SnRK1 kinases governs maturation. Processes of maturation are activated by changing oxygen/energy levels and/or a changing nutrient state, which trigger responses at the level of transcription and protein phosphorylation. This way seed metabolism becomes adapted to altering conditions. In maturing cotyledons photoheterotrophic metabolism improves internal oxygen supply and biosynthetic fluxes and influences assimilate partitioning. Transgenic legumes with changed metabolic pathways and seed composition provide suitable models to study pathway regulation and metabolic control. At the same time, desirable improvements of seed quality and yield may be achieved.

Contents

INTRODUCTION

Seed development is genetically programmed and correlated with changes on metabolite level. Differentiation occurs successively starting with the maternal and followed by the filial organs, which later become highly specialized storage tissues. A complex regulatory network triggers initiation of maturation and accumulation of storage products. This includes transcriptional and physiological reprogramming mediated by sugar and hormone-responsive pathways (43, 161). To study these processes the seeds of grain legumes, *Vicia* or pea, offer excellent models, which are accessible to a wide range of different methods due to their special morphology and large size. The aim is to understand how legume seed development is connected with growth, transport processes and the control of biosynthetic pathways (16, 18, 154, 155).

Most basic knowledge of the genetic background of growth regulation and organization of embryo tissue (77), cellular differentiation (45), and signal transduction (35, 125) comes from *Arabidopsis* seeds and is briefly quoted here. There is sufficient evidence to assume that the fundamental regulatory mechanisms are similar in legume seeds. However, we should be cautious with too much generalization because, unlike *Arabidopsis*, grain legumes are crop plants selected for high seed yield and characterized by high metabolic activity and fluxes in seeds.

Progress toward a better understanding of mechanisms controlling legume seed growth has been summarized in a number of overviews (16, 18, 57, 95, 96, 154, 155). In the present review we discuss recent research dealing with aspects of metabolic and hormonal regulation and highlight the role of regulatory networks coordinating seed development and metabolism.

SEED GROWTH AND DEVELOPMENTAL GRADIENTS

Seed Growth Characteristics

Growing seeds of legumes, consisting of the maternal seed coat and the filial endosperm and embryo, are genetically and physiologically heterogeneous. Development proceeds successively starting with the maternal organs. Careful studies in pea (summarized in 148) identified three rapid phases of seed growth separated by two lag phases. Whereas the first growth phase is confined to endosperm and seed coat the second one is associated with the embryo and continues until embryogenesis is complete. Up to this stage the embryo mainly grows by cell division. Another lag phase precedes the third growth period comprising the maturation phase and characterized by cell expansion (148). This discontinuous growth is accompanied by discontinuous changes in gene expression patterns (R. Radchuk & H. Weber, unpublished data). The described phases are at least partially identical with a general staging system that divides embryo or seed development into three parts: cell division or prestorage phase, maturation or storage phase, and desiccation phase (cf. 44 for soybean and **Figure 1** introducing an additional transition phase; see Transition Stage—Switch Toward Filial Control).

Developmental Gradients

Generally, cotyledon differentiation is a sequential process involving sucrose uptake, halted cell division, cell expansion and endopolyploidization, greening and gain of photosynthetic activity, accumulation of storage products and acquisition of desiccation tolerance, and dormancy (12, 18). The cotyledons reach physiological maturity when cell expansion and storage activity stop first in the center and then gradually in the more outer regions. Young legume cotyledons of *Vicia* or pea are highly mitotically active. During the transition stage the inner adaxial cells increase in size whereas in the outer abaxial region mitotic activity is maintained for longer. Thus, the differentiation process proceeds in a wave-like manner generating developmental gradients across the maturing embryo although the particular pattern can be species-specific (12, 53, 97).

For crop seeds the differentiation phase is of special interest because during this stage a regulatory network on different levels initiates the accumulation of storage products. The processes regulating the change from embryogenesis to maturation during the transition phase are currently not fully understood. Work on legume and in particular on *Arabidopsis* seeds suggests that metabolite- and hormone-responsive pathways are involved (35, 43, 161).

Transition phase: At transition stage (see **Figure 1**) the embryo switches from a meristem-like tissue into a differentiated storage organ when sugar and hormone-responsive signaling pathways trigger initiation of maturation.

Developmental gradient: A developmental gradient results from the gradual differentiation of different organs or within a single-seed organ and is reflected by heterogeneous populations of cells of different physiological age accumulating different amounts of substances, e.g., mRNAs, proteins, and starch.

CONTROL OF EARLY SEED DEVELOPMENT

The Role of the Seed Coat

Maternal seed coat, endosperm, and embryo physically interact. In pea seeds, the general pattern of seed development appears largely determined by the maternal parent, i.e., the seed, and final seed size is positively correlated with the maximum volume of the endosperm (148). Large-seeded genotypes of *Vicia faba* develop a larger seed resulting in a longer cell division period of the embryo (152). This is consistent with the observation that cell number of cotyledons is correlated with seed size, which is predominantly maternally determined (27). Because cell number is determined by cell division cycles during early growth, control of cell division is crucial. However, seed size control is complex. Studies in small- and large-seeded genotypes of *Vicia faba* suggest that seed coat–derived metabolic signals are critical (152). In *Arabidopsis*, maternal differences already affect processes at the level of the shoot meristem

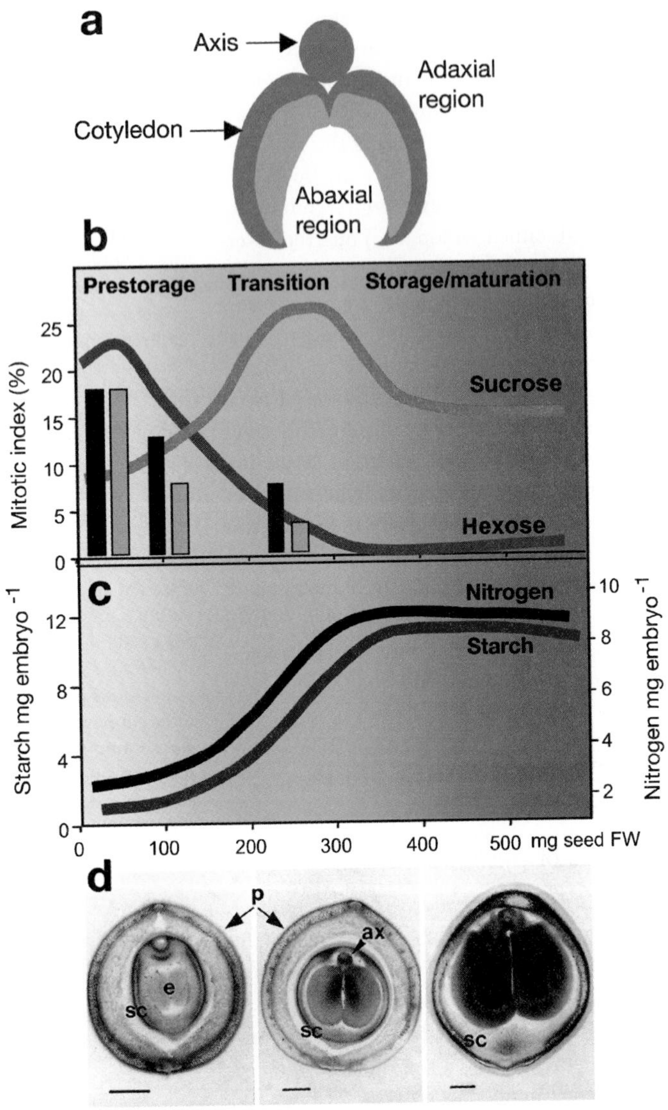

Figure 1

Physiological and biochemical characteristics of developing *Vicia faba* seeds. (*a*) Schematic view of a mid-maturation embryo, (*b*) changes in cotyledonary mitotic indices (*black columns*: adaxial region, *gray columns*: abaxial region) and sugars, (*c*) starch and nitrogen accumulation, (*d*) iodine starch staining in prestorage, transition, and storage phase seed sections. ax, embryo axis; e, embryo; p, pod; sc, seed coat; bars 1 mm.

that determine the number of flowers and seeds, characters associated with seed size (1). Also in *V. faba*, the genotypes differ already in the number of flowers per internode and ovaries per pod (L. Borisjuk, unpublished data).

ABA: abscisic acid

Young seed coats of *V. faba* and pea are transient storage organs, which accumulate starch and proteins before storage activity starts in the embryo (12, 28, 56, 105). When seed coat storage activity is affected by mutation, embryo growth is impeded. The pea *rb* mutation impairs growth rate and starch content of the seed coat by affecting ADP-glucose pyrophosphorylase (AGP) (61). Seed coat dry weight is reduced by 20% and, accordingly, the endospermal volume is decreased, implicating a physical restraint on the embryo (105). Reciprocal crosses between *rb*, with smaller, and *Rb*, with larger, seed coats reveal clear maternal effects on final seed weight. F1 seeds are smaller when the maternal genotype is *rb*. Interestingly, only embryo dry weight is reduced but not the proportion of starch, indicating that embryonic cell number is reduced (80). In transgenic seeds of *V. narbonensis*, where reduced AGP activity is restricted to the embryo, seed coat growth or starch content is not reduced and, consequently, final seed size is also not altered (157).

What determines growth rate and sink strength of the seed coat? In addition to enzymes related to sink strength like invertases, sucrose synthase, sucrose phosphate synthase (28), and transporters (153), hormonal control may also be important. In *Arabidopsis* seeds ABA, classically associated with seed maturation, is produced first in maternal tissues and later in the embryo (66). Maternal ABA, synthesized in the seed coat of *Arabidopsis* and *Nicotiana* and translocated to the embryo, promotes its growth and avoids abortion (39). Little is known about ABA functions in maternal organs. Seed coats pass through a developmental program and undergo a phase of maturation evidenced by the expression of storage protein genes (12), which may be upregulated by ABA. In this way, ABA synthesized in the seed coat could initiate its own maturation (39). In *V. faba*, delayed seed coat differentiation also delays embryo maturation and is accompanied by prolonged cell divisions first in the seed coat and then in the embryo (152). However, in legumes a direct effect of maternally derived ABA on the initiation of embryo differentiation is not proven. Pea and

V. faba embryo maturation is initiated at the inner adaxial region (12; L. Borisjuk, unpublished data). This spatial pattern is not consistent with maternally derived ABA diffusing into the embryo and initiating its differentiation. ABA synthesized within embryo axis and moving into cotyledons could play a role (L. Borisjuk, unpublished data). Maternal ABA might also control embryo development by regulating assimilate import (24, 39, 134).

Maternal Control Exerted by Invertases and Sugars

Generally, acid invertase is related to early development and sucrose synthase to maturation and storage (129, 150). Persistence of invertase at certain stages can delay differentiation (128). Conversely, differentiation can be induced when acid invertase activity decreases (123).

High activities of both vacuolar (soluble) invertase (vcINV) and cell wall–bound (insoluble) invertase (cwINV) are present in the maternal tissue of young seeds of legumes (150, 152), barley (159), and maize (3). In general, vcINV and cwINV are sequentially expressed (3). In response to drought stress soluble invertase decreased in young maize seeds, eventually leading to seed abortion (166). Stress-induced decrease of vcINV is a response to low source activity and reduces the capacity to use sucrose. It diminishes sink strength and disturbs the sucrose-to-hexose ratio in ovules, which disrupts further development at the level of the maternal tissue (3). Thus, the function of stress-regulated soluble invertases is adjusting seed numbers to the resources available to the plant during early reproductive development.

In contrast, cwINV is active in growing zones and expanding sink tissues and facilitates assimilate unloading by increasing the concentration gradient of sucrose. In *V. faba*, *VfcwINV1* is expressed in the unloading area of the seed coat, cleaving incoming sucrose within the apoplastic space separating maternal and filial tissues (150), thereby creating a high hexose environment at a stage when mitotic activity proceeds in the embryo. The high sugar status promotes embryo growth by cell divisions (13). Hexoses produced in the apoplastic space are probably taken up by monosaccharide transporters (153). In large-seeded genotypes of *V. faba*, cwINV is active for longer in the enlarged seed coat and, consequently, more cells are produced in the embryo (152). In maize the *miniature1*(*mn1*) mutation decreases invertase activity and impairs pedicel and endosperm development by reducing the cell number. The *Mn1*-locus determines an endosperm-specific cell wall–bound isoform, INCW2, which is critical to provide hexose sugars for mitoses and to generate appropriate sugar signals (23).

In the pea seed mutant *E2748*, in which embryo growth is strongly affected, the endospermal vacuole bearing either mutant or wild-type embryos contains similar concentrations of sugars. The characteristic change of the hexose-to-sucrose ratio during the transition phase within this vacuolar cavity was not different in the mutant and is, therefore, independent of normal embryo growth (15). Thus, in young seeds the seed coat probably determines, via invertases, both concentration and composition of sugars within the endospermal vacuole, which in turn affects embryo development.

Invertases are subject to inhibition by small proteins (100). Sucrose can protect cwINV (not vcINV) from inhibition in vitro (116) or release the inhibited enzyme (8). In young maize seeds invertase inhibitors are targeted to the apoplast within the embryo-surrounding region (8). In tobacco plants, invertase inhibitor mRNA is strongly induced by osmoticum and ABA and upon sugar depletion, e.g., under stress conditions (100). Thus, invertase inhibitors induced by ABA could control seed maturation by decreasing hexose levels, which in turn may initiate differentiation by terminating cell division.

Maternal or filial control: Early embryo growth is maternally controlled by physically restricting seed coat growth, activity of seed coat–borne invertases, and transient storage activity within the seed coat. During transition phase the formation of transfer cells, followed by increased sucrose uptake and generation of embryonic sink strength, marks the switch to filial control of further embryo/endosperm development.

Metabolic Control of Cell Cycle

The dependence of seed size on cell number implicates cell cycle control, which is the target for hormones and nutrients. For legume seeds it is unknown how a putative sugar signal generated in the maternal tissue is integrated in the cell division process of the embryo, but cyclin D induction could be important. D-type cyclins play an important role in cell cycle responses to nutrient signals (83). In *Arabidopsis* cyclin type D2 and D3 gene expression responds positively to sugar availability (103). *Physcomitrella CycD* knockouts confirmed the role of cycD in coupling developmental progression to sugar supply but failed to show a role in cell cycle regulation (81). Cell cycle progression in plant meristems is favored by elevated CO_2 (64, 68). The CO_2 effect could be explained by increased photosynthesis leading via increased sugar levels to cyclin D induction and to accelerated cell division.

Invertase control hypothesis: Cell wall–bound invertases in maternal seed tissues regulate assimilate unloading and elevate hexose levels. Therefore, sink strength is maintained at early stages and nature and amount of nutrients supplied to the embryo are controlled, thus regulating the cell division phase (see 154).

General Features of Maternal Control

Current knowledge shows that in legumes as well as in other seeds early embryo growth is subject to maternal control. This can be exerted by physical restriction, which determines how far the embryo can realize its growth potential later on (discussed in 148). Transient storage products within the seed coat could function as a buffer and are subsequently mobilized to supply the embryo and promote its growth (12, 28, 56, 105). In addition, the seed coat modifies and controls the nutrient supply to the embryo with seed coat–borne invertases as key players (150; see The Role of the Seed Coat). According to the invertase control hypothesis of seed development (154), invertases in young seeds have specific and important functions: Soluble invertases can act as stress sensors, thereby adjusting sink size, e.g., seed number, to the available resources. Cell wall–bound invertases can regulate assimilate unloading and elevate hexose levels. Therefore, sink strength is maintained at early growth stages and the nature and amount of sugars supplied to the embryo are controlled, thus regulating the cell division phase (see **Figure 1** for a schematic view).

TRANSITION STAGE—SWITCH TOWARD FILIAL CONTROL

During the transition stage the legume embryo develops from a meristem-like tissue into a highly differentiated storage organ. This switch is accompanied by multiple changes:

1) The initiation of a nutrient uptake system based on transfer cells formed by regional differentiation of the protoderm.
2) The subsequent strongly increased sucrose uptake leading to raising sucrose levels in the embryo and a switch from a high-hexose to a high-sucrose state.
3) The switch from mitotic growth to growth driven by cell expansion.
4) The gain of photosynthetic activity and the improvement of the energy state.
5) The induction of storage-associated gene expression leading to high metabolic fluxes into storage products.

Embryonic Transfer Cells

Transfer cells are ubiquitous in plants and important for nutrient transport. They are characterized by secondary wall ingrowths, resulting in an increased plasma membrane surface enriched in assimilate transporters (93). Transfer cell formation in seeds is an early differentiation event (138) and is found in close vicinity to the maternal unloading tissue in the cotyledonary epidermis of *V. faba* and pea (15, 135, 153), at the basal endosperm of maize (23, 138), and in barley endosperm cells facing the nucellar projection (158). Transfer cells represent an epidermal specification to improve embryo nutrition (15, 93, 138). In legume seeds transfer cells are formed at the early transition stage by regional differentiation of the protoderm.

Transfer cell formation is coupled to upregulated expression of transport-related genes (52, 135, 136, 153, 158). In *V. faba* cotyledons transfer cell-specific expression of the sucrose transporter VfSUT1 likely causes sucrose accumulation in underlying tissues (14),

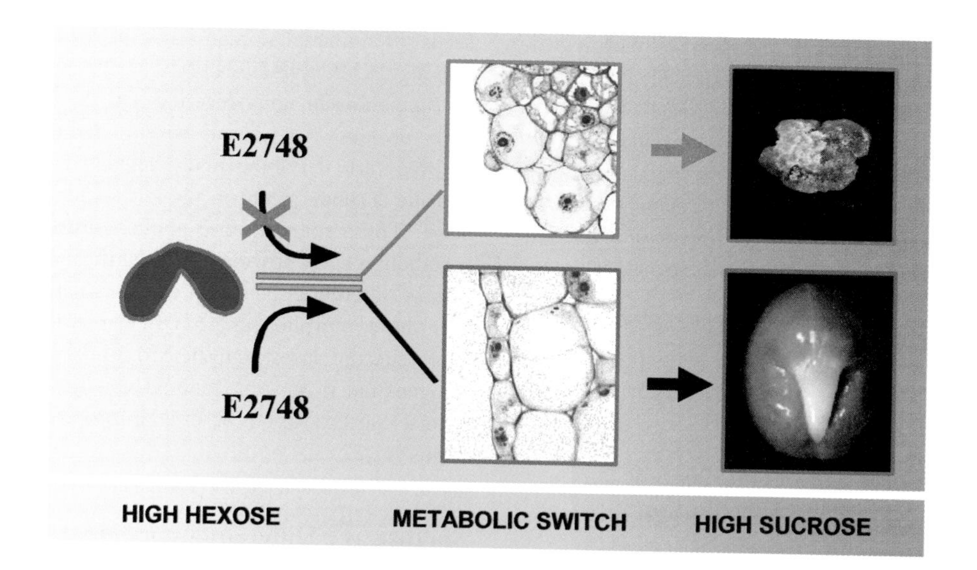

Figure 2

Schematic view of major events in the pea seed mutant *E2748* and wild-type embryos. Underlying color indicates the changing sugar status from high hexose to high sucrose. The micrographs during metabolic switch show mutant (*above*) and wild-type (*below*) embryo cell layers without and with a transfer cell layer, respectively.

and a high sucrose state is involved in triggering maturation (156). Thus, establishing the epidermis-localized sucrose uptake system during transfer cell formation is crucial for the initiation of the maturation phase, during which strong filial sink strength is established.

Light was shed on the essential role of transfer cells for legume embryos to proceed into maturation by mutant analysis. The pea *E2748* mutation blocks transfer cell formation in pea embryos and leads to the loss of epidermal identity (**Figure 2**). This has severe consequences for further embryo growth and is accompanied by lower starch levels within the embryo, unusual enlargement of the outer cell layers of the cotyledonary parenchyma, and restricted movement of a symplasmic tracer within the cotyledons. The lack of a proper epidermis changes the internal anatomy of the cotyledons as well as the gradient of developmental maturity and initiates callus-like growth (15). The morphological aberrations accompany the failure to reach a high sucrose state necessary for maturation initiation (156), which indicates that a functional epidermis is required to integrate metabolic signals. The E2748 gene product is still unknown but it is only required for transfer cell formation in the embryo epidermis. Homozygous mutants of *E2748* exclusively show the expected seed phenotype but no changes within the rest of the plant (15; L. Borisjuk, unpublished data).

Altogether, the formation of the transfer cell layer is an early event preceding maturation. It occurs during the transition stage and is important to render the filial tissues independent of maternal control.

Signaling Transfer Cell Formation

Embryonic transfer cells of *V. faba* form at late heart stage at regions where the cotyledons approach the seed coat. This zone enlarges and spreads along the outer surface down to the abaxial area. Epidermal cells covering the inner surface of the cotyledons and the hypocotyls do not form transfer cells. It is unclear what triggers transfer cell formation but the pattern is consistent with an apoplastic signal (153). Sucrose/hexoses and/or stresses such as energy or nutrient depletion may play a role (92, 93, 153). Recent studies in *Arabidopsis* show that FUS3, an essential regulator of seed development, is only required in the protoderm to carry out its function (142). However, whether FUS3 is involved in transfer cell formation is unknown. Several genes specifically expressed in transfer cells have been described but its function is unclear. In maize seeds the genes *BETL1-4*, which encode small cystein-rich cell wall proteins, are specifically expressed in basal endospermal transfer cells (60). At least BETL1 and 2 are trans-activated by ZmMRP-1, a MYB-related transcription factor. *ZmMRP-1* is first expressed in the nucleocytoplasmic domain of the multinuclear coenocyte and later throughout transfer cell formation and is regarded as a determinant of transfer cell–specific gene expression (48).

CONTROL OF EMBRYO MATURATION

Differentiation is Reflected by Metabolite Distribution

Legume embryo differentiation is evident on structural and morphological levels like differences in cell size, vacuolization, endopolyploidization, and accumulation of starch and storage proteins (12, 148). In addition, the metabolic state of a tissue can represent a characteristic feature of its differentiation stage. Using a bioluminescence-based imaging technique the spatial distribution of metabolites such as glucose, sucrose, and ATP was measured quantitatively in sections of *V. faba* cotyledons. The results revealed a spatial-temporal change of metabolite profiles (recently reviewed in 16).

During cotyledon differentiation gradients in metabolite concentration emerge. They are most pronounced during the transition stage and are related to particular cell types. Higher concentrations of glucose occur in nondifferentiated premature regions, indicating a direct correlation to mitotic activity. Mature starch-accumulating regions contain particularly low glucose levels (13). A strikingly different pattern was observed for sucrose. Before maturation, young embryos contain moderately low levels. At maturation actively elongating and starch-accumulating cells contain the highest sucrose concentrations, which are correlated with transcripts of storage-associated enzymes (14). ATP concentrations in young cotyledons are low. Levels increase toward maturation, starting from the abaxial region and moving toward the interior. In legumes as well as barley seeds differentiated, storage-active regions contain high ATP levels (17, 110).

Metabolite imaging revealed that the specific profiles that occur within maturing embryos are developmentally regulated and provide a fingerprint of the differentiation stage. These concentration gradient–forming metabolites likely provide signals for the induction of differentiation events and thus may be regarded as morphogens (162). In plants, auxin represents the closest equivalent to animal morphogens but it remains doubtful whether the animal concept should be directly adopted to plants (11).

Sucrose as a Differentiation Signal

Sucrose has a dual function as transport and nutrient sugar and as a signal molecule triggering storage-associated processes (70, 71, 125). The increasing sucrose concentration in *Vicia* cotyledons at the onset of maturation is mediated by sucrose transporter activity within epidermal transfer cells (153). Several lines of evidence suggest that sucrose can induce storage-associated gene expression. Sucrose acts at the transcriptional level causing upregulation of enzymes like sucrose synthase and ADPG-pyrophosphorylase (56, 155). The genes for the enzymes sucrose-phosphate synthase (SPS) and phospho*enol*pyruvate carboxylase (PEPC) are repressed and induced, respectively, in response to sucrose (46, 151). In vitro sucrose feeding disrupts the meristematic state, induces cell expansion and endopolyploidization in young explanted cotyledons (152), and promotes cotyledonary storage activity at the transcript level (2, 25). In potato tubers the sucrose-mediated upregulation of enzymes involves redox activation (41); the situation in seeds is unknown. Together, these findings suggest that sucrose signals the transition of the legume embryo into the storage mode.

Transgenic approaches have been performed that decreased sucrose levels in storage cells. *Vicia* embryos overexpressing a yeast-derived invertase (156) contain lower levels of sucrose but large amounts of hexoses. This potentially generates embryo lethality. Seeds with weaker phenotypes germinate. Analysis of transgenic seeds reveals that both transcript levels of AGP and sucrose synthase and starch accumulation correlate with sucrose but not with hexose concentrations in the cotyledons. A threshold level of approximately 20 mM sucrose is required for starch accumulation (156). In addition, cell differentiation is altered in the transgenic low-sucrose cotyledons. The cells appear developmentally younger and are strongly disturbed within the subcellular organization of the vacuolar system (90). Similar results were reported from nonseed-storage organs. In potato tubers, the expression of invertase (140) or sucrose phosphorylase (141) bypasses sucrose synthase and decreases sucrose levels. In these tubers starch accumulation is downregulated and respiration is largely stimulated.

Sucrose also has metabolic effects, which have to be distinguished from signaling. Sucrose cleavage by sucrose synthase is readily reversible (42), is inhibited by free hexoses (112, 151), and has a high K_M in *Vicia* seeds for sucrose (112). Thus, flux through sucrose synthase depends on high sucrose levels and removal of cleavage products and can be downregulated or even reversed at high hexose concentrations (56, 151).

The data indicate that impaired storage metabolism in seeds and tubers is due to decreased sucrose levels rather than to hexose accumulation. Sucrose potentially acts on transcriptional and posttranscriptional levels (58), thereby affecting carbon fluxes. It is a key player within the regulatory network controlling seed differentiation.

Trehalose may also be important for sugar-specific signaling, as with *Arabidopsis* seedlings (30). No data are available for legume seeds.

ABA as a Signal of Maturation

ABA regulates a wide range of developmental events and mediates responses to environmental stress conditions. The hormone is necessary to proceed through seed maturation and to acquire desiccation tolerance and dormancy. Based on genetic approaches ABA action and signaling networks have been studied preferably in *Arabidopsis* seeds (89, 106) and shown to be intimately connected to the transcriptional control of seed development governed by such major regulators as ABI3, FUS3, LEC1, and LEC2 (35, 89).

Increasing ABA levels during maturation induces expression of a cyclin-dependent kinase inhibitor leading to cell cycle arrest (147). Thus, embryo differentiation is partly controlled by ABA-regulated cell division. The *FUS3/LEC* genes are also involved because mutations in these genes cause prolonged cell divisions throughout seed maturation instead of growth arrest (102). Interestingly, the *FUS3* gene is only required in the epidermis to carry out its function, possibly by modulating the rate of cell division in these cells (142). The details of how FUS3 functions connect to the ABA signaling network is unknown, but epidermal cells may integrate FUS3-, ABA-, and sugar-derived signals to regulate seed maturation.

Little is known about the genes involved in legume seeds. An array-based gene expression analysis of pea seeds provides evidence for a developmental-specific regulation of ABA-mediated effects. The temporal expression pattern of the *FUS3/LEC* orthologous pea genes differs from that of *PsABI3*, as expected. The *PsFUS3/PsLEC1* genes are expressed at highest levels in young embryos at the transition stage, consistent with a major influence on early events. However, *PsABI3* is expressed later in accordance with storage protein synthesis (R. Radchuk & H. Weber, unpublished data).

Known antagonists of ABA function in seed development are GAs. GAs primarily promote germination-associated processes and seedling growth but are also required for early pea seed growth (133) and cell expansion in pea fruits (94). The *lh-2* locus, which encodes ent-kaurene oxidase, an enzyme of the GA biosynthetic pathway (26), is involved in seed

GA: gibberellic acid

development. Its mutation decreases GA1 and GA3 in filial organs, leading to retarded growth or abortion during the transition stage (133). The synthesis of GAs in the filial organs is necessary to initiate and/or promote cell expansion, and its effects are not mediated through ABA (9).

ABA and Sugar Signal Interaction

The analysis of ABA action in seeds has to consider the interactions with other signals (89). Transition of embryos into maturation is triggered by sugars and ABA and the interaction of both. Sugar responses are linked to that of ABA (21, 43). Mutants in the sugar-sensing pathway can germinate in the presence of sugars that are inhibitory to wild-type seeds. Some sugar-sensing mutations are allelic to known mutations in ABA synthesis or sensitivity (79). Because all known ABA and *ABI* mutants are also sensing mutants, sugar signaling requires an intact ABA transduction chain (125). It is possible that sucrose increases ABA sensitivity or levels (125). Alternatively, ABA could enhance the ability to respond to sugar signals (111). Interactions of ABA signals with other hormone-signaling pathways in *Arabidopsis* were recently reviewed (35).

ABA accumulation in seeds is regulated on spatial and temporal levels (89), but what triggers ABA synthesis is unknown. In legume embryos ABA levels are several-fold higher in the axis compared to the cotyledons (L. Borisjuk, unpublished data). Transcript levels of ABA biosynthesis genes in *Arabidopsis* leaves are raised by low concentrations of glucose, indicating a nutritional control on ABA biosynthesis (79). In addition, certain stress conditions can play a role. As already described, in legume seeds a switch of the principal sugars from hexoses to sucrose initiates embryo maturation. The switch could result in nutrient stress and/or energy limitation, which may stimulate ABA synthesis. However, direct experimental evidence is missing.

REGULATION OF MATURATION VIA PROTEIN PHOSPHORYLATION

Targets of Phosphorylation

The regulation of storage-associated processes in legume seeds involves protein phosphorylation (151; R. Radchuk & H. Weber, unpublished data). In *Vicia* cotyledons SPS is inactivated during the switch to the storage phase, shown by increased sensitivity to phosphate inhibition (151). PEPC, a storage-associated enzyme, is simultaneously activated upon phosphorylation, as seen by decreased sensitivity against malate inhibition (46). Activating PEPC specifically channels carbon into amino acid biosynthesis via the anaplerotic pathway (109), thereby promoting storage protein synthesis. In nonseed models phosphorylation of target enzymes, such as SPS, allows binding of 14-3-3 proteins, which causes inactivation (59). In the presence of high AMP, an indicator of a low energy state, the complex dissociates and inhibition is relieved (6). These mechanisms link enzyme activities to energy charge in the way that a low energy state downregulates metabolic fluxes. In legume seeds sugar signaling is part of that regulatory network. During the switch from high-hexose to high-sucrose levels in *V. faba* embryos SPS becomes inactivated by phosphorylation (151). This change is accompanied by an increased energy state and decreased AMP levels (17). Maize leaf sucrose synthase is phosphorylated by Ca^{2+}-dependent protein kinases, which activate the cleavage reaction (51). Interestingly, sucrose synthase phosphorylation can occur in response to low oxygen and/or energy stress (130). The results provide evidence that in seeds phosphorylation events are triggered by metabolic signals and/or by the oxygen or energy state. Phosphorylation/dephosphorylation events thus provide a fine-tuning, suitable way to rapidly adapt enzyme activities and metabolic fluxes to changing conditions.

The Role of Sucrose Nonfermenting-1-Related Protein Kinases

Sucrose nonfermenting-1-related protein kinase (SnRK1) is a major component of the sugar-sensing and response mechanism in yeast, animals, and plants (50). Plants contain several genes of each of the three components common to the yeast and mammalian SNF1/AMPK heterotrimeric complex: the kinase α or SNF1 subunit, the regulatory subunit γ or SNF4, and the β or GAL83/SIP subunit with adapter function (22, 125). In mammals the kinase is activated upon cellular stress such as heat, hypoxia, ATP depletion, or oxidative stress. It directly modulates the phosphorylation of some metabolic enzymes, leading to inactivation of ATP-consuming processes via phosphorylation and thus to energy preservation (49). In pea embryos SnRK1 (α subunit) is constitutively expressed whereas the mRNA level of two β and one γ subunit increases upon maturation (R. Radchuk & H. Weber, unpublished data). Similarly, in tomato seeds the α subunit is expressed constitutively whereas the β subunit is induced during maturation (20). Only the regulatory γ subunit (LeSNF4) is responsive to GA, ABA, and stress (20). In germinating *Medicago* seeds both the β and γ subunits are upregulated by starvation (22). SnRK1 proteins are closely related to calcium-dependent protein kinases (CPDK). In rice seeds, a CPDK isoform is required for storage product accumulation via phosphorylation of sucrose synthase (4). Overexpression of OsCDPK2 arrested seed development at an early stage (88). Thus, hormonal signals (possibly ABA) or certain stress conditions derived from carbohydrate and/or energy depletion activate SnRK1 kinases. However, a direct effect of sugars has not been shown in seeds.

Transgenic plants with altered gene expression of the α subunit provide more direct information about SnRK1 functions. In SnRK1 antisense barley, tobacco, and pea plants pollen development was arrested (165; R. Radchuk & H. Weber, unpublished data) due to impaired starch accumulation, possibly due to its inability to metabolize incoming sucrose. In transgenic pea with seed-specific antisense expression of SnRK1 the maturation phase is delayed, resulting in green seeds, which are occasionally viviparous. An array-based gene expression analysis revealed that expression of specific gene clusters, normally downregulated during the transition stage, remains high when SnRK1 kinase is reduced. Genes involved are associated with cell cycle regulation and cytosolic metabolism, indicating that SnRK1-deficient seeds have a prolonged prestorage phase. However, clusters related to storage protein synthesis, ABA-mediated signal transduction, and some other storage-associated events exhibit increased expression much later compared to the wild type (R. Radchuk & H. Weber, unpublished data). These results indicate that SnRK1 functions are cross-regulated with that of ABA and mediate responses related to differentiation at the level of transcription. A newly developed protein microarray method will allow a more direct identification of kinase protein targets (73).

SEED ENERGY METABOLISM

Seed Growth Occurs under Hypoxic Conditions

Maternal seed tissues are barely permeable for gases. The reason for the low gas permeability is unclear, but it may be required to prevent loss of CO_2 from the respiring embryos (36). Without retention and subsequent recycling of CO_2 about 20% of total carbon would be lost (36, 110). An alternative function could be that a CO_2-enriched environment within the endospermal vacuole promotes cell division activity (64, 68). Due to low gas diffusion the oxygen concentrations within growing seeds of *Brassica*, soybean, and *Vicia* are generally low (107, 145).

Growing embryos are heterotrophic. ATP is mainly produced by respiration and is imported into plastids via ATP/ADP translocators (101). ATP import into plastids is potentially rate limiting in storage tissues (139). Therefore, the low oxygen content of seeds can lead to energy

AEC: adenylate energy charge

Photoheterotrophic metabolism: Photoheterotrophic metabolism occurs in green seeds in which specific photoheterotrophic plastids import sugars, are photosynthetically active, and produce oxygen, thereby supporting respiration and overall metabolic activity.

depletion, a stress situation with physiological consequences. In *Vicia* and *Brassica* seeds, low oxygen levels under in vivo conditions limit respiration and thus lower the energy state of the seed cells (107, 145). A typical response to hypoxia is inducing fermentation, evidenced by ethanol and/or lactate production and increased alcohol dehydrogenase or lactate dehydrogenase activity. However, analyzing different developmental stages of *Vicia* seeds reveals that only the young premature embryos show such responses (107). Fermentative metabolism is not detectable during transition and maturation stages. Instead, maturing seed tissues in *Vicia* and barley seeds with high synthesis rates are characterized by high ATP and AEC levels (108, 110), although embryonic respiration is oxygen-limited during the entire storage phase, especially in the dark. In maturing embryos of *Vicia* and *Brassica* low oxygen restricts starch and lipid synthesis (107, 145). Despite low oxygen levels and impaired respiration, a high energy state in maturing legume embryos suggests an adaptation mechanism, which adjusts metabolic fluxes to prevent fermentation.

Adaptation to Low Oxygen

During transition stage legume embryos become adapted to oxygen availability. The metabolic and physiological adaptations to low-oxygen conditions are embedded in the embryos' differentiation program and are evident on different levels: First, respiration is more strongly inhibited by low oxygen during earlier than later stages, and overall rates decrease during embryo growth, indicating that during maturation embryonic respiration becomes tightly controlled and adapted to the low-oxygen conditions (108). There is a similar decrease of respiration from early cell division to maturation for wheat endosperm and barley (33, 110). Second, ATP concentrations and AEC are lowest at early stages but increase later on, indicating that embryos acquire the ability to elevate and stabilize ATP and AEC levels. During the transition to maturation, embryos switch from an invertase- and hexose-based metabolism to one that is sucrose based and controlled by a sucrose-synthase pathway (150, 151). Compared to invertase, sucrose synthase saves ATP. Flux through the enzyme is metabolically controlled and depends on high sucrose levels and removal of cleavage products. Thus, at maturation the embryonic metabolism becomes energetically more economically and metabolically controlled. Third, legume embryos become green during transition and acquire photosynthetic activity, which improves oxygen supply and energy state (17).

Low oxygen and/or energy activate specific responses. How this adaptation is programmed is not fully understood. However, ABA and sugars (67) and/or SnRK-like kinases may be involved. As a result, fermentation is prevented, metabolic fluxes become controlled, and energy state is increased to a constant high level. Our current knowledge of events during differentiation of seed-storage organs is summarized in **Figure 3**.

PHOTOHETEROTROPHIC SEED METABOLISM

Photosynthetic Oxygen Production

Maturing embryos of *V. faba*, pea, and soybean are green and photosynthetically active. In *Vicia* cotyledons the spatial pattern of photosynthesis measured as oxygen production upon illumination corresponds to the chlorophyll distribution (17). Similarly, in rapeseeds the chlorophyll content is correlated with photosynthesis-dependent oxygen evolution (32). Plastids in seeds are characterized by high rates of uncoupled electron transport and high photosystem II activity (5, 7). However, photosynthetic CO_2 fixation is low in embryos of canola (5), pea (36), and soybean (114). Rubisco activity in pea embryos is near detection limit and several magnitudes lower than in pods or leaves (55). The main CO_2-refixing enzyme in legume embryos is PEPC (46). It is likely that the primary effect of embryonic photosynthesis is to increase internal oxygen content. The measured rate of oxygen production in *Vicia*

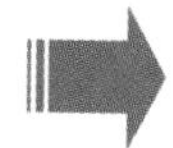

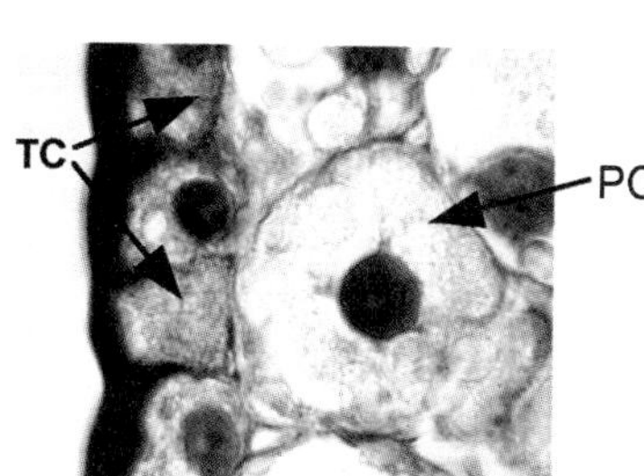

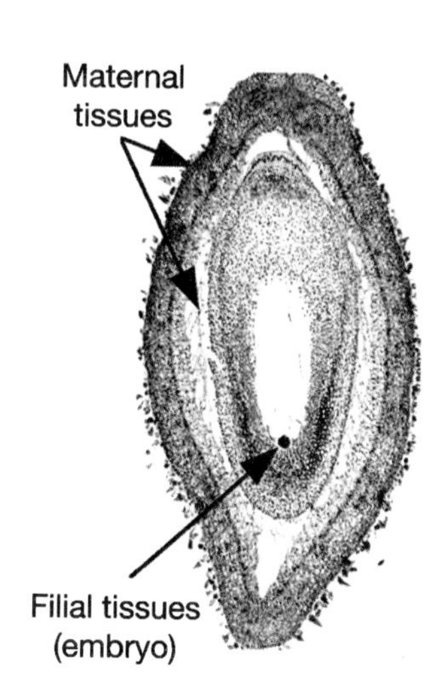

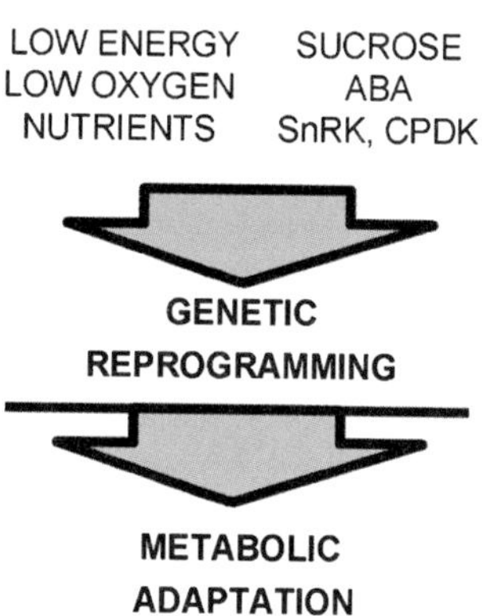

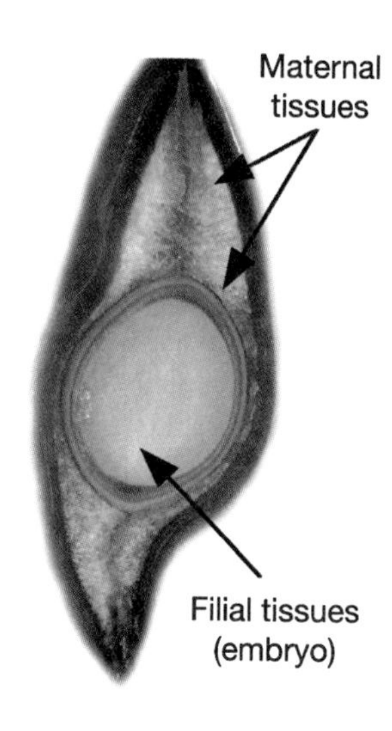

Figure 3

Sequence of events during differentiation of seed-storage organs. For more details see text.

cotyledons of 25 to 40 nmol g^{-1} min^{-1} should be sufficient to reach 50% of atmospheric oxygen concentration within the embryo within approximately 5 min (108).

Redox Signaling

The oxidation-reduction cascades can provide the driving force for metabolism and redox signals, which influence biosynthetic activity within seeds (38). The light- and photosynthesis-dependent increase of respiratory and biosynthetic fluxes in seeds could be achieved by redox signals coming from photosynthetic activity. In potato tubers the redox state regulates AGP activity (137). Accordingly, upon illumination in *Vicia* embryos, redox status immediately switches to a more reduced status (H. Rolletschek, unpublished data). Plastids of soybean embryos achieve a higher biosynthetic rate in the light, and products synthesized in the light are much more reduced (160). In chloroplasts fatty acid synthesis is redox-regulated because acetyl-CoA carboxylase (ACC), a key enzyme of fatty acid synthesis, is activated upon reduction, and a redox potential generated by photosynthesis is involved through thioredoxin (117). Thus, changes in plastid fatty acid synthesis during light-dark transitions may be controlled at the level of ACC (98) due to posttranslational redox modification (72, 117). However, ACC could also be the target of posttranslational inhibition by SnRK1 kinases (125). These kinases may be activated under energy-limiting conditions, e.g., darkness, as shown for yeast (163).

Seed Photosynthesis and Metabolism

Because plastids of heterotrophic tissues are generally energy-limited, seed photosynthesis can be important for storage activity (91). With improved oxygen supply or illumination, metabolic fluxes increase in embryos of *Vicia* and pea, whereas steady-state levels of metabolites are unchanged and thus reflect the developmental stage (108; H. Rolletschek, unpublished data). In rapeseed, lipid synthesis is stimulated at elevated oxygen concentrations, probably due to increased levels of ATP (145). Seed photosynthesis controls biosynthetic fluxes by providing oxygen and ATP, which is readily used for respiration and biosynthesis.

The energy state could also control partitioning of assimilates into different storage products. Energy demand is highest for lipids, followed by proteins, and lowest for starch. In *Vicia* embryos the pattern of starch accumulation during transition stage is inversely oriented to that of the storage protein legumin, and protein accumulation is initiated in regions enriched in ATP, whereas starch is detected first in regions with lower ATP levels (17). Starch synthesis in isolated cauliflower plastids and in rapeseed embryos is saturated at lower ATP concentrations than lipid synthesis (87, 145). In rapeseed, increased ATP-to-ADP ratios accompanies the developmental shift from starch to lipid storage at mid-maturation (145). The shift is related to increased availability of oxygen and ATP, probably derived from seed photosynthesis. Otherwise, stimulation of lipid synthesis could be due to increased sucrose unloading and/or increased levels of ATP or redox signals, which stimulate ACC (see above). In soybean embryos lipid biosynthesis starts when embryos become fully green (164). In *Arabidopsis* seeds maximum levels of photosynthesis-related mRNAs coincide with oil but not with starch synthesis (113). A similar association was observed with Rubisco activity in growing castor bean seeds (124). Therefore, the relative fluxes of substrates into different classes of storage products may be not only controlled by the genetically programmed developmental stage, but also by the energy state of the tissue prone to external control.

Photoheterotrophic Plastids

Seed embryos contain photoheterotrophic plastids. These are different from leaf chloroplasts with respect to morphology and physiology (5). Photoheterotrophic plastids import carbon, mainly Glc-6-P, PEP, and pyruvate, via specific translocators (37). In addition, photoheterotrophic plastids are photosynthetically active and have significant electron transport activity, high chlorophyll a-to-b ratios, and abundant proteins associated with photosystem II. However, the capacity for photosynthetic CO_2 fixation is low (5). Photoheterotrophic plastids perform cyclic electron transport via photosystem II at high rates.

The formation of photoheterotrophic plastids during transition stage could be triggered by sugars and is associated with the expression and activity of plastidial metabolite translocators. Sugar feeding of tobacco leaves induces Glc-6-P uptake into chloroplasts, which is a typical feature for photoheterotrophic plastids (99). The mRNA-encoding triose-P translocator protein is strongly decreased after feeding sucrose to tobacco seedlings (69). Canola embryos cultured on sucrose develop photoheterotrophic plastids functioning in storage. Without sucrose the embryos germinate and become photoautotrophic (65). *Vicia* embryos become green when cultured on sucrose but not on glucose (H. Weber, unpublished data). In accordance with the greening process during transition, the principal sugars change in *Vicia* embryos from high hexoses to high sucrose (13, 14, 150). In this context it is of interest that in transition-stage embryos highest sucrose concentrations are present within the abaxial half of the cotyledons, with much lower levels toward the interior (14). This pattern spatially and temporally coincides with that of ATP and chlorophyll distribution (17). For a schematic view of carbon fluxes within storage cells see **Figure 4**.

Integration of Plastidial and Cytosolic Metabolism

Storage metabolism in plant cells involves different compartments such as cytosol, plastids, and mitochondria. Pathways in these compartments interact (19, 119). Gene expression studies provide evidence that the central metabolism during *Arabidopsis* seed development is directed from the cytosol to the plastids (113). Plastidial biosynthesis in seeds depends on energy supply and the import of metabolic precursors (37, 91). Information about metabolite import for legume seeds is scarce, and most knowledge comes from *Brassica* (101). During maturation of soybean (164) and *Brassica* embryos (31) there is a switch from starch to lipid biosynthesis. The change is connected to the ability to import specific metabolites. *Brassica* embryos take up Glc-6-P during earlier stages when starch is accumulating. In the course of further growth pyruvate uptake activity increases more than 20-fold together with high lipid synthesis (31). PEP translocator activity also increases in *Brassica* embryos from pre- to mid-oil accumulation stages (74). This suggests that the relative fluxes into the different storage products are developmentally controlled and may depend on the activity of specific metabolite translocators. Alternatively, translocators found in any tissue may be flexible and reflect the metabolic demand (19).

Measuring metabolic fluxes in seeds is difficult. Most pathways are branched and there are alternative routes, futile cycles, and product turnover (118–120). To model metabolic fluxes, *Brassica* embryos are cultured in vitro with labeled substrates and then the distribution of label in different seed components is measured. There is a rapid exchange of metabolites between the cytosol and the plastids. For seed oil synthesis the net flux through the oxidative pentose phosphate pathway (OPPP) is approximately 10% of that through glycolysis. The OPPP provides ca. 44% of the required NADPH (119).

UNDERSTANDING STORAGE PROTEIN SYNTHESIS

Nitrogen Transport and Availability

Protein accumulation in legume seeds depends mainly on nitrogen uptake and availability (47, 85, 115). Glutamine and/or asparagine are imported from the phloem (84). Within the seed coat amino acids are metabolized and reconstructed (76, 104). Mainly glutamine, alanine, and threonine are released (76). At later stages asparagine is unloaded from the seed coat (104).

Efflux of amino acids (and sucrose) from pea seed coats is passive with linear kinetics, perhaps mediated by nonselective pores (29). Uptake into soybean and pea embryos occurs partly passively especially during early stages (10, 29). A saturable system, via H^+-amino acid cotransport, becomes important at later stages. The saturable system is induced by N starvation, indicating control by assimilate availability (10, 75). Much of the control of N uptake is embryonic (54). An amino acid transporter, VfAAP1, is expressed in *V. faba* cotyledon parenchyma cells but not in the epidermal cell layer, with maximum levels at the beginning of storage protein accumulation (85). By contrast, pea *PsAAP1* is localized in the seed cotyledonary epidermis (136). *PsAAP1* is not orthologous to the *Vicia AAP1* isoform but, aside from this, the differences need to be clarified by further studies.

In addition to amino acid transport, peptide transporters seem to be important in connection with protein mobilization predominantly, but not exclusively, during leaf senescence and germination (127). VfPTR1 of *V. faba* is a functional peptide transporter, complements a yeast mutant, and is expressed in growing cotyledons, most strongly during early germination but also in growing embryo axes and root hairs. Localization of transcripts suggests a temporal and spatial regulation during cotyledon and seedling development, perhaps by nutrient and/or senescence signals (86).

Seed-specific amino acid biosynthesis requires carbon skeletons via the anaplerotic reaction of PEPC (143). PEPC has been investigated in seeds of pea, soybean, *Vicia*, and wheat (36, 46, 55, 126). The enzyme refixes HCO_3^- from respiration in a reaction catalyzing the conversion of PEP. PEP is converted either to aspartate or to malate and other intermediates of the citric acid cycle. PEPC can control the anaplerotic carbon flow and potentially improve seed carbon economy. Correlative evidence points to a rate-limiting role of PEPC in *Vicia faba* storage protein synthesis (47). In soybean cultivars accumulating different amounts of protein PEPC activity correlates with seed protein content (126, 132). PEPC is a ubiquitous, highly regulated plant enzyme. Its allosteric properties are subject to opposite and antagonistic effects of metabolites such as malate and Glc-6-P. Phosphorylation modulates the metabolic regulation with respect to feedback inhibition by malate (46, 63).

Changing Sink Strength and Metabolic Pathways

Legume seeds provide important food and feed stuff. Therefore, increasing the protein content and/or improving nutritional quality of legume crop seeds is a desirable goal (149). Increasing the seed protein fraction or the protein-to-starch ratio is also interesting for biotechnological approaches aimed at producing valuable proteins (82) because its optimal production in transgenic seeds depends on its general capacity to synthesize proteins.

Besides directly introducing quality-relevant genes by genetic engineering it could be advantageous to manipulate metabolic pathways. For instance, seed sink strength may be elevated by changing gene expression of transporters. Likewise, assimilate flux into amino acid biosynthesis may be increased, or fluxes into starch biosynthesis decreased, with compensatory increases of competing pathways. Such transgenic models will improve our knowledge of pathway regulation and interaction.

To manipulate assimilate partitioning to elevate protein content in legume seeds, transgenic *Vicia* plants have been generated that specifically express a *Corynebacterium* PEPC in its seeds. The bacterial enzyme is not feedback inhibited by malate. Transgenic seeds show a higher $[^{14}C]$-CO_2 uptake and a threefold increased incorporation of labeled carbon into proteins. Changed metabolite profiles of maturing cotyledons indicate a shift of metabolic fluxes from sugars/starch into organic acids and free amino acids. These changes are consistent with an increased carbon flow through the anaplerotic pathway catalyzed by PEPC. Consequently, transgenic seeds accumulate up to 20% more protein per gram of seed dry weight. Additionally, seed dry weight is higher by 20% to 30% possibly due to improved carbon fixation. Thus, PEPC in seeds is a promising target for molecular plant breeding (109).

To improve nitrogen influx the amino acid transporter gene *VfAAP1* (85) was expressed in pea seeds. Overexpressers contain 10% to 25% more seed nitrogen. Extracting storage proteins from both the *PEPC* overexpressing *Vicia* seeds (109) and the *VfAAP1* expressing pea seeds (H. Rolletschek, F. Hosein, I. Saalbach & H. Weber, submitted) shows that the globulin content is increased, indicating a stimulation of globulin synthesis by amino acid availability.

Because *PEPC* and *VfAAP1* overexpression is restricted to seeds, a seed-specific signal has to be assumed that couples the higher demand of nitrogen of the embryo to an increased supply from the vegetative organs. Possible ways to meet the higher nitrogen demand are increased N translocation, higher uptake from the soil, and/or increased nodule fixation. Experiments where $^{15}NH_4Cl$ has been applied to the roots show that a higher proportion of the label is transported into the transgenic seeds than into the wild type (K.-P. Götz & H. Weber, unpublished data). This indicates that the higher demand of nitrogen in the transgenic seeds is covered by stimulation of root uptake, probably via whole-plant signaling of nitrogen demand (40). Studies with *Arabidopsis* (78) show that the nitrogen status of the whole plant controls uptake of mineral N via long-distance signaling.

FUTURE PERSPECTIVES

Large-seeded legumes contribute substantially to a better understanding of seed metabolism and the role of sugars as signals in developmental processes, as evidenced in this review. Nevertheless, the insights gained are insufficient and all too often we had to add nonseed or nonlegume data to passably complete a picture. Thus, many challenges are ahead. We need to further analyze regulatory networks of metabolic and hormonal signals in developing seeds as they relate to the genetic program and external signals. This demands an integrated experimental approach including genetical, biochemical, and histological methods, gene expression studies, and analysis of the spatial-temporal dynamics of metabolite levels and fluxes. It would be of great interest to determine metabolites in situ at the tissue, cellular, and subcellular levels. New techniques like NMR microimaging (121) and minimally invasive dynamic imaging (34) will provide major contributions. At the molecular level an attractive mechanism to look for is the direct metabolic regulation of gene expression at the mRNA level by riboswitches (146). Such metabolite-binding riboswitch elements in mRNAs have been well characterized in bacteria but are also found in plant mRNAs (131, 146) where its functional characterization has not yet been reported.

To bridge the gap between gene expression and yield-related seed characters a genetical-genomics approach (62) treating levels of RNA, proteins, and metabolites as quantitative traits promises new and integrated insights. Transgenic plants and induced mutants with seed

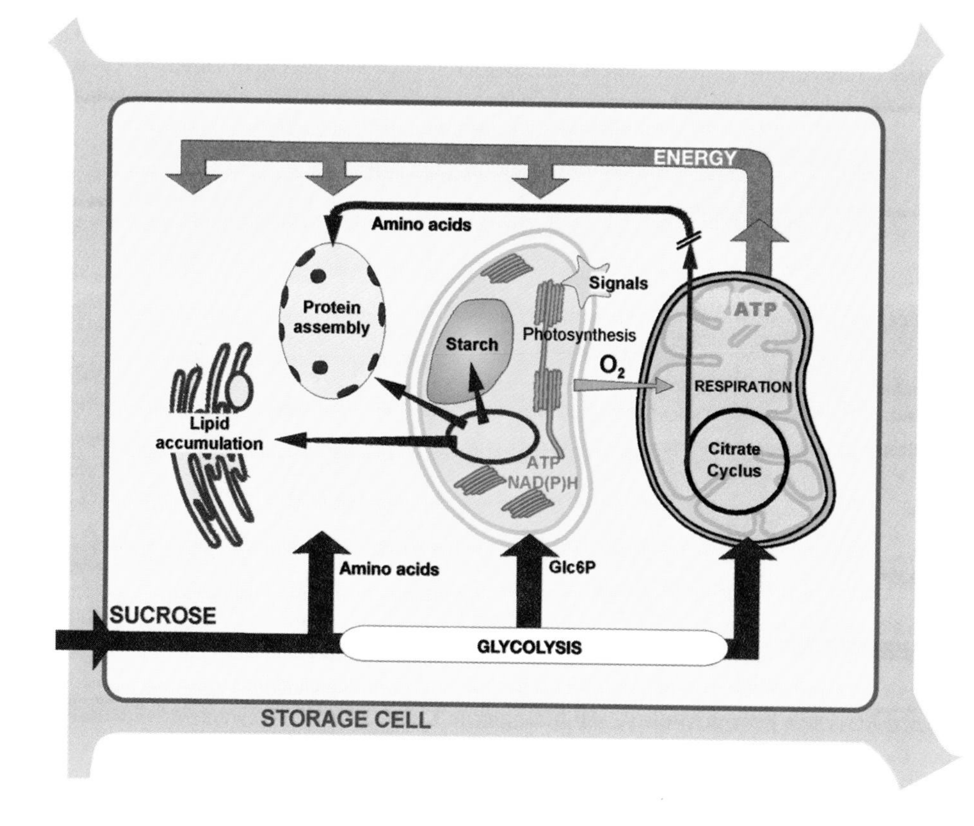

Figure 4
Model of photoheterotrophic seed metabolism. Carbon flux is indicated by brown-colored arrows. Photoheterotrophic plastids import sugars, are photosynthetically active, and produce oxygen (*blue arrow*) and signals derived from photosynthetic activity (*yellow star wheel*). ATP (*red arrows*) is supplied via mitochondrial respiration.

phenotypes are also indispensable for studying metabolic regulation. This is the only way to find out experimentally whether a particular gene product has really rate-limiting functions in vivo (122). It is important to study such plants more on the whole-plant level rather than to concentrate on seeds as an isolated system. Metabolic changes are often complex and affect development and cell structures (90). Thus, increased efforts are required to understand the intimate association of metabolism with development. "Omics" tools are very helpful when analyzing transgenic models and mutants but also represent a challenge in terms of interpreting the wealth of information. Only recently did such tools become available for legumes (144). However, omics-based analyses will not replace in-depth studies of single genes and proteins at the different levels of cell and tissue complexity. Taking advantage of the suitable model character, the large body of knowledge, and the powerful technologies available and under development, legume seeds will remain a favorable object to study.

SUMMARY POINTS

1. Legume seed development is characterized by a tissue-specific progression of differentiation and by the formation of gradients measured at the molecular and cellular levels.
2. During early legume seed growth the maternal seed coat modifies and controls the nutrient supply to the embryo, thus regulating the cell division phase. Cell wall–bound invertases play a major role by regulating assimilate unloading and elevating hexose levels.
3. With ceasing cell divisions the legume embryo enters a transition phase and develops from a meristem-like tissue into a highly differentiated storage organ. The process is controlled by a signaling network involving sugars, ABA, and SnRK1 kinases. Changing oxygen/energy or nutrient conditions are involved in its activation.
4. During transition embryos become green and photosynthetically active. Photosynthesis improves oxygen supply in a hypoxic seed environment. As an adaptation to low oxygen respiration decreases and energy states increase. Embryos switch from an invertase- and hexose-based to a sucrose-based metabolism controlled by a sucrose synthase.
5. Sucrose has a dual function as transport and nutrient sugar and as a signal molecule triggering storage-associated processes. Sucrose acts on transcriptional and posttranscriptional levels, thereby affecting carbon fluxes. It is a key player within the regulatory network controlling seed differentiation.
6. The described sequence of events is schematically depicted in **Figure 3**.

ACKNOWLEDGMENTS

We are grateful to our previous and present lab members for their committed work and stimulating discussions. We particularly thank Winfriede Weschke for critical reading of the manuscript. Our work has been funded by grants from Deutsche Forschungsgemeinschaft (DFG), the European Union, the German Federal Ministry of Education and Research (BMBF), Fonds der Chemischen Industrie, and IPK. We apologize to those of our collegues whose published work is not cited due to space limitations.

LITERATURE CITED

1. Alonso-Blanco C, Blankestijn-de Vries H, Hanhart CJ, Koornneef M. 1999. Natural allelic variation at seed size loci in relation to other life history traits of *Arabidopsis thaliana*. *Proc. Natl. Acad. Sci. USA* 13:4710–17
2. Ambrose MJ, Wang TL, Cook SK, Hedley CL. 1987. An analysis of seed development in *Pisum sativum* L. IV. Cotyledon cell population in vitro and in vivo. *J. Exp. Bot.* 38:1909–20
3. **Andersen MN, Asch F, Wu Y, Jensen CH, Naested, et al. 2002. Soluble invertase expression is an early target of drought stress during the critical, abortion-sensitive phase of young ovary development in maize. *Plant Physiol.* 130:591–604**
4. Asano T, Kunieda N, Omura Y, Ibe H, Kawasaki T, et al. 2002. Rice SPK, a calmodulin-like domain protein kinase, is required for storage product accumulation during seed development: phosphorylation of sucrose synthase is a possible factor. *Plant Cell* 14:619–28
5. Asokanthan P, Johnson RW, Griffith M, Krol M. 1997. The photosynthetic potential of canola embryos. *Physiol. Plant.* 101:353–60
6. Athwal GS, Huber JL, Huber SC. 1998. Phosphorylated nitrate reductase and 14-3-3 proteins. Site of interaction, effects of ions, and evidence for an amp-binding site on 14-3-3 proteins. *Plant Physiol.* 118:1041–48
7. Banerji D, Rauf A. 1979. Comparative growth and biochemical studies on seed development. 3. Chlorophyll development and Hill activity in developing seeds of *Pisum sativum* and *Vicia faba*. *Plant Biochem. J.* 6:31–35
8. Bate NJ, Niu X, Wang Y, Reimann KS, Helentjaris TG. 2004. An invertase inhibitor from maize localizes to the embryo surrounding region during early kernel development. *Plant Physiol.* 134:246–54
9. Batge SL, Ross JJ, Reid JB. 1999. Abscisic acid levels in seeds of the gibberellin-deficient mutant lh-2 of pea. *Physiol. Plant.* 105:485–90
10. Bennett AB, Spanswick RM. 1983. Derepression of amino acid-H+ cotransport in developing soybean embryos. *Plant Physiol.* 72:781–86
11. Bhalerao RP, Bennett MJ. 2003. The case for morphogens in plants. *Nat. Cell Biol.* 5:939–43
12. Borisjuk L, Weber H, Panitz R, Manteuffel R, Wobus U. 1995. Embryogenesis of *V. faba* histodifferentiation in relation to starch and storage protein synthesis. *J. Plant Physiol.* 147:203–18
13. **Borisjuk L, Walenta S, Weber H, Mueller-Klieser W, Wobus U. 1998. High resolution histographical mapping of glucose concentrations in developing cotyledons of *V. faba* in relation to mitotic activity and starch accumulation: glucose as a possible developmental trigger. *Plant J.* 15:583–91**
14. Borisjuk L, Walenta S, Rolletschek H, Mueller-Klieser W, Wobus U, et al. 2002. Spatial analysis of plant development: sucrose imaging within *Vicia faba* cotyledons reveals specific developmental patterns. *Plant J.* 29:521–30
15. Borisjuk L, Wang T, Rolletschek H, Wobus U, Weber H. 2002. A pea seed mutant affected in the differentiation of the embryonic epidermis leads to deregulated seed maturation and impaired embryo growth. *Development* 129:1595–607
16. Borisjuk L, Rolletschek H, Weber H, Wobus U, Weber H. 2003. Differentiation of legume cotyledons as related to metabolic gradients and assimilate transport into seeds. *J. Exp. Biol.* 54:503–12
17. **Borisjuk L, Rolletschek H, Walenta S, Panitz P, Wobus U, et al. 2003. Energy status and its control on embryogenesis of legumes: ATP distribution within *Vicia faba***

The paper highlights the role of soluble acid invertases in early maize seed development. Soluble invertase is responsive to drought stress, thereby adjusting seed sink size to the available source.

For the first time a sugar gradient in cotyledons is visualized and correlated to developmental processes.

This paper analyzes the role of energy state and seed photosynthesis in *Vicia* embryos as related to differentiation and presents a model of photoheterotrophic seed metabolism.

embryos is developmentally regulated and correlated with photosynthetic capacity. *Plant J.* 36:318–29

18. Borisjuk L, Rolletschek H, Radchuk R, Weschke E, Wobus U, et al. 2004. Seed development and differentiation: a role for metabolic regulation. *Plant Biol.* 6:375–86
19. Bowsher CG, Tobin AK. 2001. Compartmentation of metabolism within mitochondria and plastids. *J. Exp. Bot.* 52:513–27
20. Bradford KJ, Downie AB, Gee OH, Alvarado V, Yang H, et al. 2003. Abscisic acid and gibberellin differentially regulate expression of genes of the SNF1-related kinase complex in tomato seeds. *Plant Physiol.* 132:1560–76
21. Brocard-Gifford IM, Lynch TJ, Finkelstein RR. 2003. Regulatory networks in seeds integrating developmental, abscisic acid, sugar, and light signaling. *Plant Physiol.* 131:78–92
22. Buitink J, Thoma M, Gissot L, Leprince O. 2003. Starvation, osmotic stress and desiccation tolerance lead to expression of different genes of the regulatory γ and β subunits of the SnRK1 complex in germinating seeds of *Medicago truncatula*. *Plant Cell Environ.* 27:55–67
23. Cheng WH, Chourey PS. 1999. Genetic evidence that invertase-mediated release of hexoses is critical for appropriate carbon partitioning and normal seed development in maize. *Theor. Appl. Gen.* 98:485–95
24. Cheng WH, Endo A, Zhou L, Penney J, Chen HC, et al. 2002. A unique short-chain dehydrogenase/reductase in *Arabidopsis* glucose signaling and abscisic acid biosynthesis and functions. *Plant Cell* 14:2723–43
25. Corke FMK, Hedley CL, Wang TL. 1990. An analysis of seed development in *Pisum sativum* XI. Cellular development and the position of storage protein in immature embryos grown in vivo and in vitro. *Protoplasma* 155:127–35
26. Davidson SE, Smith JJ, Helliwell CA, Poole AT, Reid JB. 2004. The pea gene LH encodes ent-kaurene oxidase. *Plant Physiol.* 134:1123–34
27. Davies DR. 1975. Studies of seed development in *Pisum sativum* I. Seed size in reciprocal crosses. *Planta* 124:297–302
28. Déjardin A, Rochat C, Maugenest S, Boutin JP. 1997. Purification, characterization and physiological role of sucrose synthase in the pea seed coat (*Pisum sativum* L.). *Planta* 201:128–37
29. DeJong A, Koerselman-Kooij JW, Schuurmans JAMJ, Borstlap AC. 1997. The mechanism of amino acid efflux from seed coats of developing seeds as revealed by uptake experiments. *Plant Physiol.* 114:731–36
30. Eastmond PJ, Graham IA. 2003. Trehalose metabolism: a regulatory role for trehalose-6-phosphate? *Curr. Opin. Plant Biol.* 6:231–35
31. Eastmond PJ, Rawsthorne S. 2000. Coordinate changes in carbon partitioning and plastidial metabolism during the development of oilseed rape embryos. *Plant Physiol.* 122:767–74
32. Eastmond PJ, Kolacna L, Rawsthorne S. 1996. Photosynthesis by developing embryos of oilseed rape (*Brassica napus* L.). *J. Exp. Bot.* 47:1763–69
33. Emes MJ, Bowsher CG, Hedley C, Burrell MM, Scrase-Field ESF, et al. 2003. Starch synthesis and carbon partitioning in developing endosperm. *J. Exp. Bot.* 54:569–75
34. Fehr M, Ehrhardt DW, Lalonde S, Frommer WB. 2004. Minimally invasive dynamic imaging of ions and metabolites in living cells. *Curr. Opin. Plant Biol.* 7:345–51
35. Finkelstein RR, Gampala SS, Rock CD. 2002. Abscisic acid signaling in seeds and seedlings. *Plant Cell* 14 Suppl:S15–45
36. Flinn AM. 1985. Carbon dioxide fixation in developing seeds. In *The Pea Crop: A Basis for Improvement*, ed. PD Hebblethwaite, MC Heath, TCK Dawkins, pp. 349–58. London: Butterworths

37. Flügge UI. 1999. Phosphate translocators in plastids. *Annu. Rev. Plant Physiol. Plant Mol. Biol.* 50:27–45
38. Foyer CH, Noctor G. 2003. Redox sensing and signalling associated with reactive oxygen in chloroplasts, peroxisomes and mitochondria. *Physiol. Plant.* 119:355–64
39. Frey A, Godin B, Bonnet M, Sotta B, Marion-Poll A. 2004. Maternal synthesis of abscisic acid controls seed development and yield in *N. plumbaginifolia*. *Planta* 218:958–64
40. Gansel X, Muños S, Tillard P, Gojon A. 2001. Differential regulation of the NO3- and NH4+ transporter genes AtNrt2.1 and AtAmt1.1 in *Arabidopsis:* relation with long distance and local controls by N status of the plant. *Plant J.* 26:143–55
41. Geigenberger P. 2003. Regulation of sucrose to starch conversion in growing potato tubers. *J. Exp. Bot.* 54:457–65
42. Geigenberger P, Stitt M. 1993. Sucrose synthase catalyses a readily reversible reaction in vivo in developing potato tubers and other plant tissues. *Planta* 189:329–39
43. Gibson SI. 2004. Sugar and phytohormone response pathway: navigating a signalling network. *J. Exp. Bot.* 55:253–64
44. Goldberg RB, Barker SJ, Perez-Grau L. 1989. Regulation of gene expression during plant embryogenesis. *Cell* 56:149–60
45. Goldberg RB, DePaiva G, Yadegari R. 1994. Plant embryogenesis: zygote to seed. *Science* 266:605–14
46. Golombek S, Heim U, Horstmann C, Wobus U, Weber H. 1999. PEP-carboxylase in developing seeds of *Vicia faba*. Gene expression and metabolic regulation. *Planta* 208:66–72
47. Golombek S, Rolletschek H, Wobus U, Weber H. 2001. Control of storage protein accumulation during legume seed development. *J. Plant Physiol.* 158:457–64
48. Gomez E, Royo J, Thompson R, Hueros G. 2002. Establishment of cereal endosperm expression domains: identification and properties of a maize transfer cell-specific transcription factor, ZmMRP-1. *Plant Cell* 14:599–610
49. Halford NG, Paul MJ. 2003. Carbon metabolite signalling. *Plant Biotech. J.* 1:381–98
50. Hardie DG, Carling D, Carlson M. 1999. The AMP-activated/SNF1 protein kinase subfamily: metabolic sensors of the eukaryotic cell? *Annu. Rev. Biochem.* 67:821–55
51. Hardin SC, Winter H, Huber SC. 2004. Phosphorylation of the amino terminus of maize sucrose synthase in relation to membrane association and enzyme activity. *Plant Physiol.* 134:1427–38
52. Harrington GN, Franceschi VR, Offler CE, Patrick JW, Tegeder M, et al. 1997. Cell specific expression of three genes involved in plasma membrane sucrose transport in developing *Vicia faba* seed. *Protoplasma* 197:160–73
53. Hauxwell AJ, Corke FMK, Hedley CW, Wang TL. 1990. Storage protein gene expression is localised to regions lacking mitotic activity in developing pea embryos. An analysis of seed development in *Pisum sativum* XIV. *Development* 110:283–89
54. Hayati R, Egli DB, Crafts-Brandner SJ. 1996. Independence of nitrogen supplies and seed growth in soybean: studies using in vitro culture system. *J. Exp. Bot.* 47:33–40
55. Hedley CL, Harvey DM, Keely RJ. 1975. The role of PEP-carboxylase during seed development in *P. sativum*. *Nature* 258:352–54
56. Heim U, Weber H, Bäumlein H, Wobus U. 1993. A sucrose-synthase gene of *V. faba* L.: expression pattern in developing seeds in relation to starch synthesis and metabolic regulation. *Planta* 191:394–401
57. Hills MJ. 2004. Control of storage-product synthesis in seeds. *Curr. Opin. Plant Biol.* 7:302–8

58. Huber SC, Hardin SC. 2004. Numerous posttranslational modifications provide opportunities for the intricate regulation of metabolic enzymes at multiple levels. *Curr. Opin. Plant Biol.* 7:318–22
59. Huber SC, MacKintosh C, Kaiser WM. 2002. Metabolic enzymes as targets for 14-3-3-proteins. *Plant Mol. Biol.* 50:1053–63
60. Hueros G, Royo J, Maitz M, Salamini F, Thompson RD. 1999. Evidence for factors regulating transfer cell-specific expression in maize endosperm. *Plant Mol. Biol.* 41:403–14
61. Hylton C, Smith AM. 1992. The rb mutation of peas causes structural and regulatory changes in ADP glucose pyrophosphorylase from developing embryos. *Plant Physiol.* 99:1626–34
62. Jansen RC. 2003. Studying complex biological systems using multifactorial perturbation. *Nat. Rev. Genet.* 4:145–51
63. Jeanneau M, Vidal J, Gousset-Dupont A, Lebouteiller B, Hodges M, et al. 2002. Manipulating PEPC levels in plants. *J. Exp. Bot.* 53:1837–45
64. Jitla DS, Rogers GS, Seneweera SP, Basra AS, Oldfield RJ, et al. 1997. Accelerated early growth of rice at elevated CO_2 (Is it related to developmental changes in the shoot apex?). *Plant Physiol.* 115:15–22
65. **Johnson RW, Asokanthan PS, Griffith M. 1997. Water and sucrose regulate canola embryo development. *Physiol. Plant.* 101:361–66**

The authors demonstrate a change of embryonic developmental growth pattern from an autotrophic to a photoheterotrophic mode depending on sucrose availability.

66. Karssen C, Brinkhorst-van der Swan D, Breekland A, Koornneef M. 1983. Induction of dormancy during seed development by endogenous abscisic acid: studies of abscisic acid deficient genotypes of *Arabidopsis thaliana* (L.) Heynh. *Planta* 157:158–65
67. Kato-Noguchi H. 2000. Abscisic acid and hypoxic induction of anoxia tolerance in roots of lettuce seedlings. *J. Exp. Bot.* 51:1939–44
68. Kinsman EA, Lewis C, Davies MS, Young JE, Francis D, et al. 1997. Elevated CO_2 stimulates cells to divide in grass meristems: a differential effect in two natural populations of *Dactylis glomerata*. *Plant Cell Environ.* 20:1309–16
69. Knight JS, Gray JC. 1994. Expression of genes encoding the tobacco chloroplast phosphate translocator is not light regulated and is repressed by sucrose. *Mol. Gen. Genet.* 242:586–94
70. Koch KE. 1996. Carbohydrate-modulated gene expression in plants. *Annu. Rev. Plant Physiol. Plant Mol. Biol.* 47:509–40
71. Koch K. 2004. Sucrose metabolism: regulatory mechanisms and pivotal roles in sugar sensing and plant development. *Curr. Opin. Plant Biol.* 7:235–46
72. Kozaki A, Sasaki Y. 1999. Light-dependent changes in redox status of the plastidic acetyl-CoA carboxylase and its regulatory component. *Biochem. J.* 339:541–46
73. Kramer A, Feilner T, Possling A, Radchuk V, Weschke W, et al. 2004. Identification of barley CK2α targets by using the protein microarray technology. *Phytochem.* 65:1777–84
74. Kubis SE, Pike MJ, Everett CJ, Hill LM, Rawsthorne S. 2004. The import of phosphoenolpyruvate by plastids from developing embryos of oilseed rape, *Bassica napus* (L.) and its potential as a substrate for fatty acid synthesis. *J. Exp. Bot.* 55:1455–62
75. Lanfermeijer FC, Koerselman-Kooij JW, Borstlap AC. 1990. Changing kinetics of L-valine uptake by immature pea cotyledons during development. *Planta* 181:576–82
76. Lanfermeijer FC, van Oene MA, Borstlap AC. 1992. Compartmental analysis of amino-acid release from attached and detached pea seed coats. *Planta* 187:75–82
77. Laux T, Jürgens G. 1997. Embryogenesis: a new start in life. *Plant Cell* 9:989–1000
78. Lejay L, Tillard P, Lepetit M, Olive F, Filleur S, et al. 1999. Molecular and functional regulation of two NO3- uptake systems by N- and C-status of *Arabidopsis* plants. *Plant J.* 18:509–19

79. Léon P, Sheen J. 2003. Sugar and hormone connections. *Trends Plant Sci.* 8:110–16
80. Lloyd JR, Wang TL, Hedley CL. 1996. An analysis of seed development in *P. sativum* XIX. Effect of mutant alleles at the r and rb loci on starch grain size and on the content and composition of starch in developing pea seeds. *J. Exp. Bot.* 47:171–80
81. Lorenz S, Tintelnot S, Reski R, Decker EL. 2003. Cyclin D-knockout uncouples developmental progression from sugar availability. *Plant Mol. Biol.* 53:227–36
82. Ma JKC, Drake PMW, Christou P. 2003. The production of recombinant pharmaceutical proteins in plants. *Nat. Rev. Genet.* 4:794–805
83. Meijer M, Murray JAH. 2000. The role and regulation of D-type cyclins in the plant cell cycle. *Plant Mol. Biol.* 43:621–33
84. Miflin BJ, Lea PJ. 1977. Amino acid metabolism. *Annu. Rev. Plant Physiol.* 28:299–329
85. Miranda M, Borisjuk L, Tewes A, Heim U, Sauer N, et al. 2001. Amino acid permeases in developing seeds of *Vicia faba* L.: expression precedes storage protein synthesis and is regulated by amino acid supply. *Plant J.* 28:61–72
86. Miranda M, Borisjuk L, Tewes A, Dietrich D, Rentsch D, et al. 2003. Peptide and amino acid transporters are differentially regulated during seed development and germination in Faba bean. *Plant Physiol.* 132:1950–60
87. Möhlmann T, Scheibe R, Neuhaus HE. 1994. Interaction between starch synthesis and fatty-acid synthesis in isolated cauliflower-bud amyloplasts. *Planta* 194:492–97
88. Morello L, Frattini M, Giani S, Christou P, Breviario D. 2000. Overexpression of the calcium-dependent protein kinase OsCDPK2 in transgenic rice is repressed by light in leaves and disrupts seed development. *Transg. Res.* 9:453–62
89. Nambara E, Marion-Poll A. 2003. ABA action and interactions in seeds. *Trends Plant Sci.* 8:213–17
90. Neubohn B, Gubatz S, Wobus U, Weber H. 2000. Sugar levels altered by ectopic expression of a yeast-derived invertase affects cellular differentiation of developing cotyledons on *V. narbonensis*. *Planta* 211:325–34
91. Neuhaus HE, Emes MJ. 2000. Nonphotosynthetic metabolism in plastids. *Annu. Rev. Plant Physiol. Plant Mol. Biol.* 51:111–40
92. Offler CE, Liet E, Sutton EG. 1997. Transfer cell induction in cotyledons of *V. faba*. *Protoplasma* 200:51–64
93. Offler CE, McCurdy DW, Patrick JW, Talbot MJ. 2003. Transfer cells: cells specialized for a special purpose. *Annu. Rev. Plant Biol.* 54:431–54
94. Ozga JA, van Huizen R, Reinecke DM. 2002. Hormone and seed-specific regulation of pea fruit growth. *Plant Physiol.* 128:1379–89
95. Patrick JW, Offler CE. 1995. Post-sieve element transport of sucrose in developing seeds. *Aus. J. Plant Physiol.* 22:681–702
96. Patrick JW, Offler CE. 2001. Compartmentation of transport and transfer events in developing seeds. *J. Exp. Bot.* 52:551–64
97. Perez-Grau L, Goldberg RB. 1989. Soybean seed protein genes are regulated spatially during embryogenesis. *Plant Cell* 1:1095–109
98. Post-Beittenmiller D, Roughan G, Ohlrogge JB. 1992. Regulation of plant fatty acid synthesis: analysis of acyl-coenzyme A and acyl-acyl carrier protein substrate pools in spinach and pea chloroplasts. *Plant Physiol.* 100:923–30
99. Quick PW, Scheibe R, Neuhaus HE. 1995. Induction of a hexose-phosphate translocator activity in spinach chloroplasts. *Planta* 194:193–99
100. Rausch T, Greiner S. 2004. Plant protein inhibitors of invertases. *Biochim. Biophys. Acta* 1696:253–61

101. Rawsthorne S. 2002. Carbon flux and fatty acid synthesis in plants. *Prog. Lipid Res.* 41:182–96

102. Raz V, Bergervoet JHW, Koornneef M. 2001. Sequential steps for development arrest in *Arabidopsis* seeds. *Development* 128:243–52

103. Riou-Khamlichi C, Menges M, Healy JM, Murray JA. 2000. Sugar control of the plant cell cycle: differential regulation of *Arabidopsis* D-type cyclin gene expression. *Mol. Cell Biol.* 13:4513–21

104. Rochat C, Boutin JP. 1991. Metabolism of phloem-borne amino acids in maternal tissues of fruit of nodulated or nitrate-fed pea plants. *J. Exp. Bot.* 42:207–14

105. Rochat C, Wuilleme S, Boutin JP, Hedley CL. 1995. A mutation at the rb gene, lowering ADPGPPase activity, affects storage product metabolism of pea seed coats. *J. Exp. Bot.* 46:415–21

106. Rock CD. 2000. Pathways to abscisic acid-regulated gene expression. *N. Phytol.* 148:357–96

107. Rolletschek H, Borisjuk L, Koschorreck M, Wobus U, Weber H. 2002. Legume embryos develop in a hypoxic environment. *J. Exp. Bot.* 53:1099–107

For the first time topographic high-resolution oxygen profiles showing hypoxia were measured during seed development and related to the energy state.

108. Rolletschek H, Weber H, Borisjuk L. 2003. Energy status and its control on embryogenesis of legumes: embryo photosynthesis contributes to oxygen supply and is coupled to biosynthetic fluxes. *Plant Physiol.* 132:1196–206

109. Rolletschek H, Borisjuk L, Radchuk R, Miranda M, Heim U, et al. 2004. Seed-specific expression of a bacterial phosphoenolpyruvate carboxylase in *Vicia narbonensis* increases protein content and improves carbon economy. *Plant Biotech. J.* 2:211–20

110. Rolletschek H, Weschke W, Weber H, Wobus U, Borisjuk L. 2004. Energy state and its control on seed development: Starch accumulation is associated with high ATP and steep oxygen gradients within barley grains. *J. Exp. Bot.* 55:1351–59

111. Rook F, Corke F, Card R, Munz G, Smith C, et al. 2001. Impaired sucrose induction mutants reveal the modulation of sugar-induced starch biosynthetic gene expression by abscisic acid signalling. *Plant J.* 26:421–33

112. Ross HA, Davies HV. 1992. Purification and characterization of sucrose synthase from the cotyledons of *Vicia faba* L. *Plant Physiol.* 100:1008–13

113. Ruuska SA, Girke T, Benning C, Ohlrogge JB. 2002. Contrapunctal networks of gene expression during Arabidopsis seed filling. *Plant Cell* 14:1191–206

This paper describes the first extensive microarray-based transcriptome analyses of developing seeds, with several conclusions regarding molecular seed physiology.

114. Saito GY, Chang YC, Walling LL, Thompson WW. 1989. A correlation in plastid development and cytoplasmic ultrastructure with nuclear gene expression during seed ripening in soybean. *N. Phytol.* 113:459–69

115. Salon C, Munier-Jolain NG, Duc G, Voisin AS, Grandgirard D, et al. 2001. Grain legume seed filling in relation to nitrogen acquisition: a review and prospects with particular reference to pea. *Agronomie* 21:539–52

116. Sander A, Krausgrill S, Greiner S, Weil M, Rausch T. 1996. Sucrose protects cell wall invertase but not vacuolar invertase against proteinaceous inhibitions. *FEBS Lett.* 386:171–75

117. Sasaki Y, Kozaki A, Hatano M. 1997. Link between light and fatty acid synthesis: thioredoxin-linked reductive activation of plastidic acetyl-CoA carboxylase. *Proc. Natl. Acad. Sci. USA* 94:11096–101

118. Schwender J, Ohlrogge JB. 2002. Probing in vivo metabolism by stable isotope labeling of storage lipids and proteins in developing *Brassica napus* embryos. *Plant Physiol.* 130:347–61

Using an isotope-labeling technique the authors could quantitatively model metabolic fluxes in maturing *Brassica* embryos, an oilseed crop.

119. Schwender J, Ohlrogge JB, Shachar-Hill Y. 2003. A flux model of glycolysis and the oxidative pentosephosphate pathway in developing *Brassica napus* embryos. *J. Biol. Chem.* 278:29442–53
120. Schwender J, Ohlrogge J, Shachar-Hill Y. 2004. Understanding flux in plant metabolic networks. *Curr. Opin. Plant Biol.* 7:309–17
121. Shachar-Hill Y. 2002. Nuclear magnetic resonance and plant metabolic engineering. *Metab. Eng.* 4:90–97
122. Siedow JN, Stitt M. 1998. Plant metabolism: Where are all those pathways leading us? *Curr. Opin. Plant Biol.* 1:197–200
123. Silva MP, Ricardo CPP. 1992. ß-fructosidases and in vitro dedifferentiation-redifferentiation of carrot cells. *Phytochem.* 31:1507–11
124. Simcox PD, Garland W, Deluca V, Canvin DT, Dennis DT. 1979. Respiratory pathways and fat synthesis in the developing castor oil seed. *Can. J. Bot.* 57:1008–14
125. Smeekens S. 2000. Sugar-induced signal transduction in plants. *Annu. Rev. Plant Physiol. Plant Mol. Biol.* 51:49–81
126. Smith AJ, Rinne RW, Seif RD. 1989. PEP carboxylase and pyruvate kinase involvement in protein and oil biosynthesis during soybean seed development. *Crop Sci.* 29:349–53
127. Stacey G, Koh S, Granger C, Becker JM. 2002. Peptide transport in plants. *Trends Plant Sci.* 7:257–63
128. Sturm A. 1999. Invertases. Primary structure, functions, and roles in plant development and sucrose partitioning. *Plant Physiol.* 121:1–7
129. Sturm A, Tang GQ. 1999. The sucrose-cleaving enzymes of plants are crucial for development, growth and carbon partitioning. *Trends Plant Sci.* 4:401–7
130. Subbaiah CC, Sachs MM. 2001. Altered patterns of sucrose synthase phosphorylation and localization precede callose induction and root tip death in anoxic maize seedlings. *Plant Physiol.* 125:585–94
131. Sudarsan N, Barrick JE, Breaker RR. 2003 Metabolite-binding RNA domains are present in the genes of eukaryotes. *RNA* 9:644–47
132. Sugimoto T, Tanaka K, Monma M, Kawamura Y, Saio K. 1989. Phosphoenolpyruvate carboxylase level in soybean seed highly correlates to its contents of protein and lipid. *Agric. Biol. Chem.* 53:885–87
133. Swain SM, Ross JJ, Reid JB, Kamiya Y. 1995. Gibberellins and pea seed development. Expression of the lhi, ls and le5839 mutations. *Planta* 195:426–33
134. Tan BC, Joseph LM, Deng WT, Liu L, Li QB, et al. 2003. Molecular characterization of the *Arabidopsis* 9-cis epoxycarotenoid dioxygenase gene family. *Plant J.* 35:44–56
135. Tegeder M, Wang XD, Frommer WB, Offler EO, Patrick JW 1999. Sucrose transport into developing seeds of *Pisum sativum*. *Plant J.* 18:151–61
136. Tegeder M, Offler CE, Frommer WB, Patrick JW. 2000. Amino acid transporters are localized to transfer cells of developing pea seeds. *Plant Physiol.* 122:319–26
137. Tiessen A, Hendriks JHM, Stitt M, Branscheid A, Gibon Y, et al. 2002. Starch synthesis in potato tubers is regulated by post-translational redox modification of ADP-glucose pyrophosphorylase: a novel regulatory mechanism linking starch synthesis to the sucrose supply. *Plant Cell* 14:2191–213
138. Thompson RD, Hueros G, Becker HA, Maitz M. 2001. Development and functions of seed transfer cells. *Plant Sci.* 160:775–83
139. Tjaden J, Möhlmann T, Kampfenkel K, Henrichs G, Neuhaus HE. 1998. Altered plastidic ATP/ADP-transporter activity influences potato (*Solanum tuberosum* L.) tuber morphology, yield and composition of starch. *Plant J.* 16:531–40

140. Trethewey RN, Geigenberger P, Hajirezaei M, Sonnewald U, Stitt M, et al. 1998. Combined expression of glucokinase and invertase in potato tubers leads to a dramatic reduction in starch accumulation and a stimulation of glycolysis. *Plant J.* 15:109–18

141. Trethewey RN, Fernie AR, Bachmann A, Fleischer-Notter H, Geigenberger P, et al. 2001. Expression of a bacterial sucrose phosphorylase in potato tubers results in a glucose-independent induction of glycolysis. *Plant Cell Environ.* 24:357–65

142. Tsuchiya Y, Nambara E, Naito S, McCourt P. 2004 The FUS3 transcription factor functions through the epidermal regulator TTG1 during embryogenesis in *Arabidopsis*. *Plant J.* 37:73–81

143. Turpin DH, Weger HG. 1990. Interactions between photosynthesis, respiration and nitrogen assimilation. In *Plant Physiology, Biochemistry and Molecular Biology*, ed. DT Dennis, DH Turpin, pp. 422–33. Singapore: Longman Scientific

144. Van den Bosch KA, Stacey G. 2003. Summeries of legume genomics projects from around the globe. Community resources for crops and models. *Plant Physiol.* 131:840–65

145. Vigeolas H, van Dongen JT, Waldeck P, Hühn D, Geigenberger P. 2003. Lipid storage metabolism is limited by the prevailing low oxygen concentrations within developing seeds of oilseed rape. *Plant Physiol.* 133:2048–60

Increased ATP to ADP ratios accompanied the shift from starch to lipid storage in rapeseed embryos. The shift is related to increased availability of oxygen and ATP. Lipid synthesis is stimulated by improved oxygen supply.

146. Vitreschak AG, Rodionov DA, Mironov AA, Gelfand MS. 2004. Riboswitches: the oldest mechanism for the regulation of gene expression. *Trends Genet.* 20:1–44

147. Wang H, Qi Q, Schorr P, Cutler AJ, Crosby WL, et al. 1998. ICK1, a cyclin-dependent protein kinase inhibitor from *Arabidopsis thaliana* interacts with both Cdc2a and CycD3, and its expression is induced by abscisic acid. *Plant J.* 15:501–10

148. Wang TL, Hedley CL. 1993. Genetic and developmental analysis of the seed. In *Peas: Genetics, Molecular Biology and Biochemistry*, ed. R Casey, DR Davies, pp. 83–120. Cambridge, UK: CAB Intl.

149. Wang TL, Domoney C, Hedley CL, Casey R, Grusak MA. 2003. Can we improve the nutritional quality of legume seeds? *Plant Physiol.* 131:886–91

150. Weber H, Borisjuk L, Heim U, Buchner P, Wobus U. 1995. Seed coat associated invertases of Fava bean control both unloading and storage functions: cloning of cDNAs and cell type-specific expression. *Plant Cell* 7:1835–46

151. Weber H, Buchner P, Borisjuk L, Wobus U. 1996. Sucrose metabolism during cotyledon development of *Vicia faba* L. is controlled by the concerted action of both sucrose-phosphate synthase and sucrose synthase: expression patterns, metabolic regulation and implications on seed development. *Plant J.* 9:841–50

152. Weber H, Borisjuk L, Wobus U. 1996. Controlling seed development and seed size in *Vicia faba*: a role for seed coat-associated invertases and carbohydrate state. *Plant J.* 10:823–34

153. Weber H, Borisjuk L, Heim U, Sauer N, Wobus U. 1997. A role for sugar transporters during seed development: molecular characterization of a hexose and a sucrose carrier in faba bean seeds. *Plant Cell* 9:895–908

154. Weber H, Borisjuk L, Wobus U. 1997. Sugar import and metabolism during seed development. *Trends Plant Sci.* 22:169–74

155. Weber H, Heim U, Golombek S, Borisjuk L, Wobus U. 1998. Assimilate uptake and the regulation of seed development. *Seed Sci. Res.* 8:331–45

156. Weber H, Heim U, Golombek S, Borisjuk L, Manteuffel R, et al. 1998. Expression of a yeast-derived invertase in developing cotyledons of *Vicia narbonensis* alters the carbohydrate state and affects storage functions. *Plant J.* 16:163–72

157. Weber H, Golombek S, Heim U, Rolletschek H, Gubatz S, et al. 2000. Antisense-inhibition of ADP-glucose pyrophosphorylase in developing seeds of *Vicia narbonensis* moderately decreases starch but increases protein content and affects seed maturation. *Plant J.* 24:33–43
158. Weschke W, Panitz R, Sauer N, Wang Q, Neubohn B, et al. 2000. Sucrose transport into barley seeds: molecular characterisation of two transporters and implications for seed development and starch accumulation. *Plant J.* 21:455–67
159. Weschke W, Panitz R, Gubatz S, Wang Q, Sreenivasulu N, et al. 2003. The role of invertases and hexose transporters in controlling sugar ratios in maternal and filial tissues of barley caryopses during early development. *Plant J.* 33:395–411
160. Willms JR, Salon C, Layzell DB. 1999. Evidence for light-stimulated fatty acid synthesis in soybean fruit. *Plant Physiol.* 120:1117–28
161. Wobus U, Weber H. 1999. Seed maturation: genetic programmes and control signals. *Curr. Opin. Plant Biol. Sect. Growth Dev.* 2:33–38
162. Wobus U, Weber H. 1999. Sugars as signal molecules in plant seed development. *Biol. Chem.* 380:937–44
163. Woods A, Munday MR, Scott J, Yang X, Carlson M, et al. 1994. Yeast SNF1 is functionally related to mammalian AMP-activated protein kinase and regulates acetyl-CoA carboxylase in vivo. *J. Biol. Chem.* 269:19509–515
164. Yazdi-Samadi B, Rinne RW, Seif RD. 1976. Components of developing soybean seeds: oil, protein, starch, organic acids and amino acids. *Agron. J.* 69:481–86
165. Zhang Y, Shewry PR, Jones H, Barcelo P, Lazzeri PA, et al. 2001. Expression of antisense SnRK1 protein kinase sequence causes abnormal pollen development and male sterility in transgenic barley. *Plant J.* 28:431–42
166. Zinselmeier C, Jeong BR, Boyer JS. 1999. Starch and the control of kernel number in maize at low water potentials. *Plant Physiol.* 121:25–36

The authors highlight the role of transient starch accumulation in maternal tissue related to low water potentials and the control of kernel numbers.

Cytokinesis in Higher Plants

Gerd Jürgens

ZMBP, Entwicklungsgenetik, Universität Tübingen, 72076 Tübingen, Germany;
email: gerd.juergens@zmbp.uni-tuebingen.de

Annu. Rev. Plant Biol.
2005. 56:281–99

doi: 10.1146/
annurev.arplant.55.031903.141636

First published online as a
Review in Advance on
February 15, 2005

1543-5008/05/0602-
0281$20.00

Key Words

membrane traffic, membrane fusion, phragmoplast, cell plate, cytoskeleton, cell wall, cellularization, *Arabidopsis*

Abstract

Cytokinesis partitions the cytoplasm between two or more nuclei. In higher plants, cytokinesis is initiated by cytoskeleton-assisted targeted delivery of membrane vesicles to the plane of cell division, followed by local membrane fusion to generate tubulo-vesicular networks. This initial phase of cytokinesis is essentially the same in diverse modes of plant cytokinesis whereas the subsequent transformation of the tubulo-vesicular networks into the partitioning membrane may be different between systems. This review focuses on membrane and cytoskeleton dynamics in cell plate formation and expansion during somatic cytokinesis.

Contents

INTRODUCTION

Cytokinesis partitions the cytoplasm of a dividing eukaryotic cell by laying down a stretch of plasma membrane between the forming daughter nuclei. Although common to all eukaryotes, cytokinesis is less conserved between higher plants and nonplant organisms than are other aspects of the cell cycle. Animal and fungal cells initiate cytokinesis at the periphery of the division plane, with the plasma membrane pulled in toward the center by a contractile actomyosin ring (5). Additional membrane material is delivered by vesicle trafficking to a site behind the tip of the ingrowing furrow. Ingrowth of the plasma membrane stops at the midbody, leaving a gap in the center. This gap is closed by vesicle trafficking and fusion, which completes the separation of the daughter cells (47). Although midbody closure resembles the initial stage of cytokinesis in higher plants, the overall process of cytokinesis is very different between higher plants and nonplant organisms.

Cell plate: a transient membrane compartment that is formed by the fusion of cytokinetic vesicles and eventually matures into the plasma membranes and cross-wall between daughter cells

Cellularization: simultaneous partitioning of a cell with more than two nuclei

Phragmoplast: a cytokinesis-specific array of organized microtubules, vesicles, and actin filaments that supports the formation and expansion of the cell plate

Higher plants display several cell type-specific modes of cytokinesis (61). The most common mode occurs in somatic cells (**Figure 1**). A plant-specific cytoskeletal array—the phragmoplast—delivers membrane vesicles to the center of the division plane. Vesicle fusion generates a novel transient compartment—the cell plate, which then grows out to fuse with the plasma membrane at the cell periphery. Endosperm cellularization is a closely related variant of somatic cytokinesis, involving "mini-phragmoplasts" and "mini-cell plates" and also sharing several genetically defined components (62, 63, 79). Female and male meiotic cells as well as microspores each undergo their own modes of cytokinesis, which are also distinct from embryo sac cellularization (61). However, cytokinesis of male meiotic cells in *Arabidopsis* was recently shown to resemble somatic cytokinesis in one important aspect—delivery of membrane vesicles to the plane of division and formation of membrane networks generated by vesicle fusion across the division plane, which contrasts with the previous notion of ingrowth from the cell wall (64). It is thus likely that all cell type-specific modes of cytokinesis share conserved features, which may reflect common underlying plant-specific mechanisms (51).

This review focuses on molecular mechanisms of cytokinesis, which have been mainly identified in somatic cells. Where appropriate, findings from other cell types are discussed. There is emphasis on membrane and cytoskeleton dynamics. For other aspects of plant cytokinesis, the reader is referred to excellent recent reviews (2, 9, 11, 15, 22, 51, 56, 61, 77, 88).

MEMBRANE DYNAMICS

Somatic cytokinesis starts with the accumulation and fusion of membrane vesicles in the center of the division plane from late anaphase on. The membrane fusion processes are spatially constrained such that a disk-shaped aggregate of fusion intermediates is produced. The intermediates are gradually transformed into a transient membrane compartment—the cell plate, which eventually gives rise to the plasma membranes that underlie the cross-wall between the daughter cells. The cell plate expands centrifugally by the fusion of later-arriving vesicles with its margin. Simultaneously, the cell plate undergoes a complex process of reorganization, which involves secretion of cell wall material into its lumen and removal of excess membrane material, resulting in a planar structure. Finally, the margin of the cell plate fuses with the parental plasma membrane at cortical division sites marked earlier by the transient preprophase band, physically separating the two daughter cells from one another (**Figure 1**).

Vesicle Trafficking and Initiation of the Cell Plate

There is evidence for transport vesicles being involved in cell plate formation in studies of both wild-type and mutant *Arabidopsis*, following the in-depth electron microscopy analysis of high-pressure frozen synchronized tobacco BY-2 cells (72). Noncoated vesicles were detected in the division planes of all three different cell types analyzed: meristematic cells, cellularizing endosperm, and male meiotic cells (63, 64, 74). In addition, embryo cells of cytokinesis-defective mutants impaired in vesicle fusion accumulate 60–80-nm vesicles in the plane of cell division, and these vesicles persist into

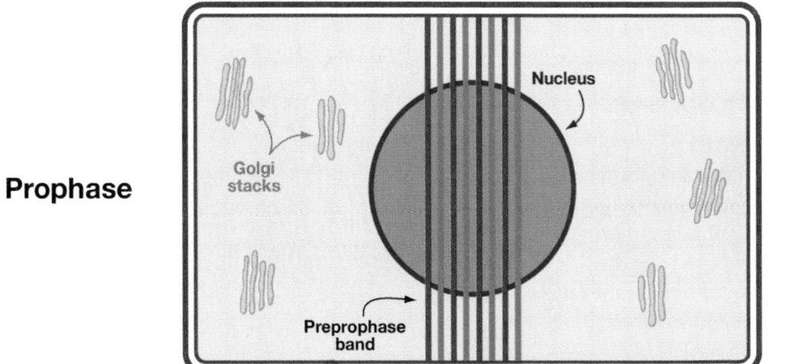

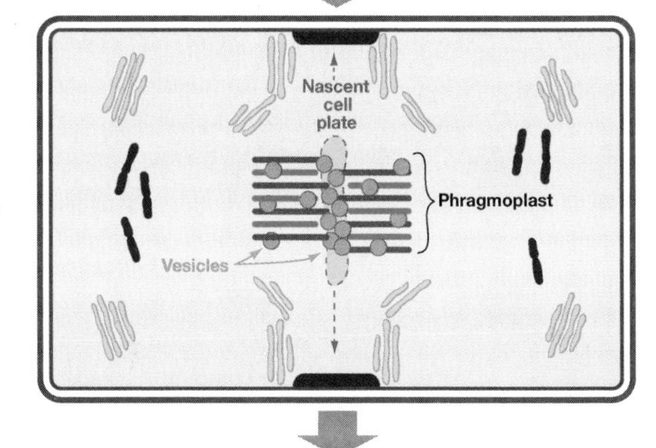

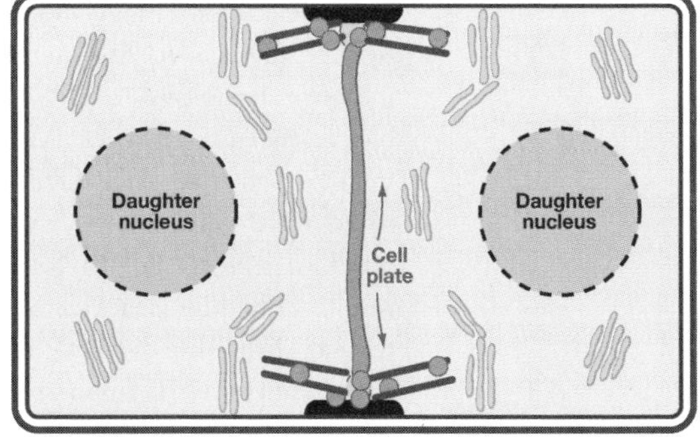

Figure 1

Cytokinesis of somatic cells. Prophase: Coaligned bundles of microtubules (*red*) and actin filaments (*green*) of the transient cortical preprophase band determine the future plane of division, marking the cortical division site. Anaphase: Spindle remnants facilitate initiation of the phragmoplast microtubules in the midzone between the two sets of daughter chromosomes. Telophase: Two antiparallel bundles, each of microtubules (*red*) and actin filaments (*green*), form the phragmoplast, their plus ends facing the nascent cell plate. Golgi stacks accumulate near the plane of division, forming a "Golgi belt." Cytokinesis: Coordinated lateral expansion of the cell plate and lateral translocation of microtubules (*red*) terminate in the fusion of the cell plate with the plasma membrane at the cortical division site.

AP: adaptor protein

SNARE: membrane-anchored protein involved in membrane fusion. A v-SNARE on the vesicle membrane (also called VAMP or R-SNARE because of a conserved arginine residue in its SNARE domain) interacts with two or three t-SNAREs on the target membrane, which are either a syntaxin (also called Qa-SNARE because of a conserved glutamine residue in its SNARE domain) and a SNAP-25 protein (also called Qb+Qc-SNARE) or a syntaxin and two t-SNARE light chains (also called Qb-SNARE and Qc-SNARE).

interphase (41, 92). Electron tomography has revealed two types of vesicles near forming cell plates: smaller dark vesicles approx. 51 nm in diameter and larger light vesicles approx. 66 nm in diameter (74). The smaller dark vesicles predominate initially, whereas the larger light vesicles appear more numerous at a later stage of cell plate formation. Both size calculations and the occurrence of hourglass-shaped putative vesicle fusion intermediates suggest that pairwise fusion of smaller dark vesicles results in larger light vesicles. Consistent with this, only smaller dark vesicles have been detected near Golgi stacks, which are thought to generate cytokinetic vesicles (74).

Golgi stacks and other organelles aggregate around the phragmoplast during telophase (**Figure 1**) (57). This spatial arrangement suggests direct delivery of Golgi-derived vesicles to the division plane via the phragmoplast. Alternatively, an intermediate endosomal compartment may be involved (see discussion in 9). A recent kinetic study of the endomembrane distribution of the endocytic tracer FM4-64 makes the latter possibility less likely as both Golgi stacks and cell plate are labeled only 30–60 min after uptake of the tracer (10). The machinery involved in the formation of cytokinetic vesicles has not been identified. However, the following components may be required, by analogy with vesicle budding in other post-Golgi trafficking pathways: an ADP-ribosylation factor (ARF)-type small GTPase, its guanine-nucleotide exchange factor (ARF-GEF) and GTPase-activating protein (ARF-GAP), AP complex coat proteins with or without clathrin, and dynamin GTPase for vesicle scission (32, 38). Cell plate formation is sensitive to the membrane-trafficking inhibitor brefeldin A (BFA) (97, 98), which blocks the activation of ARF-type GTPases by BFA-sensitive ARF-GEFs (69). Synchronized BY-2 cells treated with BFA from the onset of mitosis fail to form the cell plate, resulting in binucleate cells (98). BY-2 cells treated with BFA display characteristic Golgi abnormalities (70). In summary, several lines of evidence support the notion that the cell plate originates from Golgi-derived transport vesicles.

Membrane Fusion Machinery in Cytokinesis

Membrane vesicles arriving at the plane of cell division initially fuse with one another and later with the tubulo-vesicular network derived from earlier fusion events (**Figure 2**). By analogy with other eukaryotic fusion events, cytokinetic membrane fusion should require Rab GTPases and their effectors for vesicle tethering to target membranes as well as soluble N-ethylmaleimide-sensitive factor adaptor protein receptor (SNARE) complexes mediating membrane fusion (30). SNARE complexes consist of membrane-anchored v-SNAREs on the vesicle membrane and t-SNAREs on the target membrane, which form 4-helical bundles by association of their coiled-coil domains. In cell plate formation, neither Rab GTPases nor Rab effectors have been identified, although electron tomographic evidence suggests the occurrence of vesicle linkers in the shape of exocyst complexes (74). Exocyst complexes tether vesicles to the plasma membrane in yeast and mammalian cells (30). In contrast to Rab GTPases, SNARE proteins and some of their interactors are among the best-studied molecular components of cytokinetic vesicle fusion. The syntaxin (Qa-SNARE) KNOLLE (also known as SYP111; for nomenclature see 73) was originally identified by cytokinesis-defective mutants that accumulate unfused vesicles in the plane of cell division (41, 48). KNOLLE protein is expressed only during M phase, localizing to Golgi stacks and the cell plate (41, 90). A KNOLLE-interacting t-SNARE (Qb + Qc-SNARE), the SNAP25 homolog SNAP33, localizes to the cell plate during cytokinesis and also to the plasma membrane in interphase or postmitotic cells (23). Inactivation of SNAP33 has only a minor effect on cytokinesis, possibly because of functional overlap with two other closely related KNOLLE-interacting SNAP25 homologs, SNAP29 and SNAP30, and causes lethality for other reasons

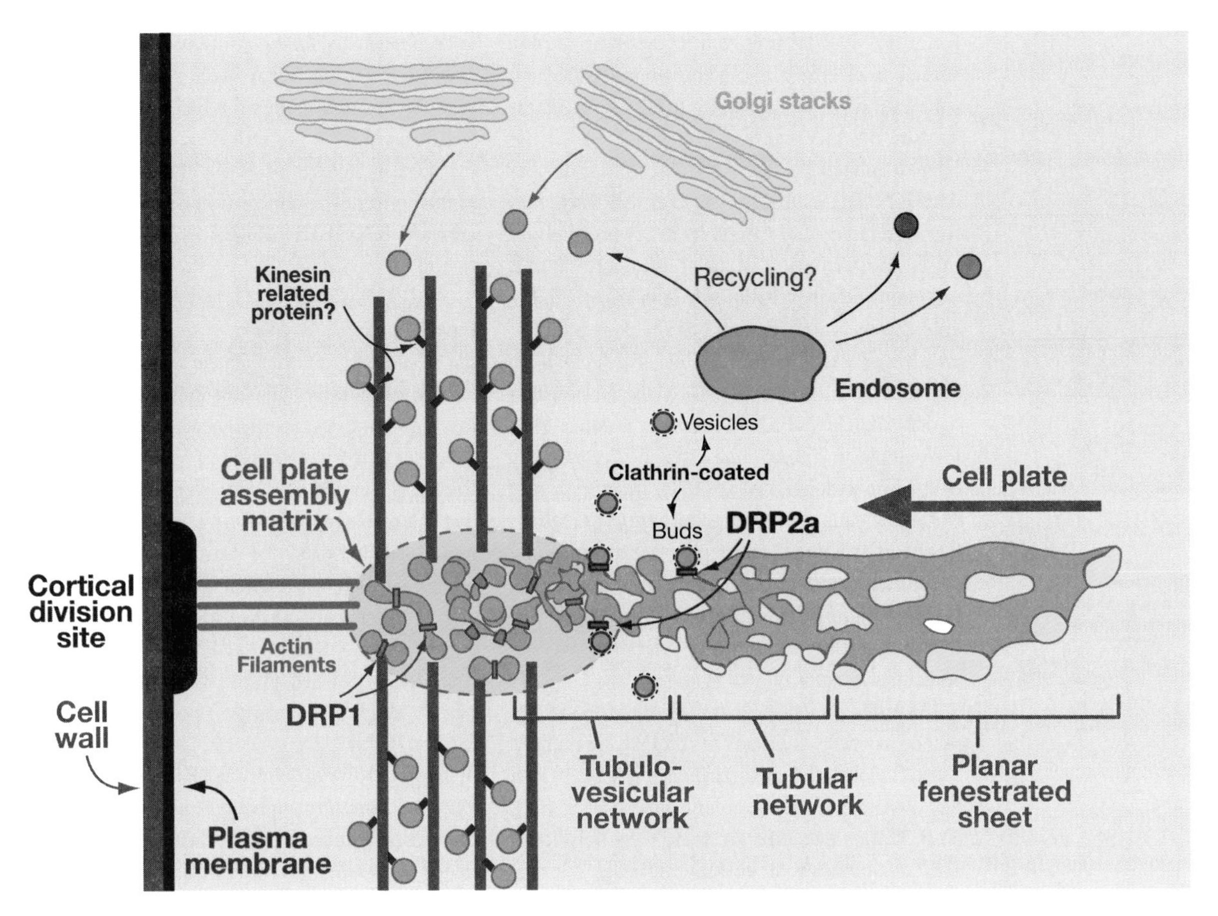

Figure 2

Membrane dynamics during cell plate development. Golgi-derived vesicles (*orange*) are delivered along phragmoplast microtubules (*red*), by a putative kinesin-related protein (*blue*), to the cell plate assembly matrix. Vesicle fusion generates fusion tubes and tubulo-vesicular networks as a result of the constricting activity of class I dynamin-related proteins (DRP1) (*magenta*). The tubulo-vesicular network is successively transformed into a tubular network and a planar fenestrated sheet. Lateral expansion of the cell plate (*large arrow*) toward the cortical division site is guided by actin filaments. Endocytosis from the tubulo-vesicular network and tubular network removes excess membrane, which is delivered to endosomes via clathrin-coated buds and vesicles. Dynamin-related protein 2a (DRP2a; *green*) is involved in the formation of clathrin-coated vesicles. The endosome sorts proteins for trafficking to various destinations (*blue, green, orange*), possibly including recycling to the margin of the cell plate.

(23). At the plasma membrane of yeast and animal cells, SNARE complexes form 4-helix bundles by interaction of the t-SNAREs syntaxin and SNAP25 with a specific v-SNARE, synaptobrevin/VAMP (R-SNARE) (30). By analogy, KNOLLE and SNAP33 might form a cytokinetic SNARE complex with an R-SNARE, which has not been identified. R-SNAREs encoded in the *Arabidopsis* genome do not resemble R-SNAREs involved in plasma membrane trafficking in nonplant eukaryotes (73). However, an unusually large R-SNARE family is related to the mammalian VAMP7 involved in post-Golgi endomembrane trafficking (73). Recently, 5 of the 14 VAMP7 family members were localized to both the plasma membrane and endosomes in *Arabidopsis* suspension cells and thus, are potential candidates

MT: microtubule

for an R-SNARE involved in cytokinesis (86). Alternatively, the missing component of the cytokinetic SNARE complex might be a plant-specific SNARE such as NPSN11, which has been localized to the cell plate and interacts with KNOLLE (101). However, NPSN11 has the signature of a Qb-SNARE, which would imply the formation of an unusual Q-SNARE only complex in cytokinesis. It is also conceivable that KNOLLE and NPSN11 are part of another SNARE complex that still lacks both a Qc-SNARE and an R-SNARE. In contrast to KNOLLE, NPSN11 persists in the newly formed plasma membrane after the disassembly of the phragmoplast MTs (101). Inactivation of NPSN11 causes no obvious phenotype, which has been attributed to its presumed functional redundancy with two closely related homologs, NPSN12 and NPSN13 (101). Thus, the composition of the cytokinetic SNARE complex(es) remains unresolved.

Another syntaxin that accumulates at the plane of cell division is SYP31, which is most closely related to the cis-Golgi syntaxin Sed5/syntaxin5 of nonplant eukaryotes (67). SYP31 does not interact with KNOLLE and also accumulates in endomembranes such as Golgi during interphase (67). SYP31 interacts with the AAA-type ATPase CDC48, which has been localized to the division plane (17), whereas KNOLLE interacts with the AAA-type ATPase NSF (67). Whether these results suggest two separate pathways for cell plate formation or simply reflect dynamics of ER fusion and/or ER-Golgi trafficking near the cell plate remains to be determined (see 74 for ER dynamics in cell plate formation).

Sec1/Munc18 related (SM) proteins interact with syntaxins or assembled SNARE complexes, which is thought to increase the specificity of SNARE action (30). The SM protein KEULE was originally identified by mutants defective in cytokinetic vesicle fusion (4, 92) and interacts with KNOLLE both genetically and biochemically, whereas KNOLLE does not interact with AtSEC1a, a close homolog of KEULE (3, 92). However, KEULE is also expressed in nonproliferating tissue and in contrast to *knolle* mutants, *keule* mutant seedlings fail to form long root hairs, suggesting another role of KEULE beyond cytokinesis (3, 78). Nonetheless, the very similar cytokinesis defects of *knolle* and *keule* mutants are consistent with the notion that KEULE may activate KNOLLE during cytokinetic vesicle fusion.

KNOLLE (SYP111) is not only a cytokinesis-specific SNARE protein but also a plant-specific member of the *Arabidopsis* SYP1 family of putative plasma-membrane syntaxins (73, 86). To determine what distinguishes KNOLLE from other syntaxins in regard to cytokinesis, several syntaxins were expressed under the cis-regulatory control of the *KNOLLE* gene and analyzed for both subcellular protein localization and their ability to functionally complement a *knolle* null mutant (53). The prevacuolar syntaxin PEP12 (SYP21) did not localize to the cell plate, whereas two members of the SYP1 family, SYP112 and SYP121 (SYR1/PEN1), accumulated at the plane of cell division. However, only SYP112 rescued the *knolle* mutant, although a *syp112* mutant had no obvious phenotype on its own nor did it enhance the *knolle* mutant phenotype. Thus, syntaxin specificity of cytokinesis is brought about by cell cycle–regulated gene expression, protein targeting, and protein activity at the cell plate. It is not known what features of KNOLLE syntaxin determine its activity during cytokinetic membrane fusion.

Delivery of Cargo to the Cell Plate

The cell plate may be viewed as an immature plasma membrane whose lumen is to be filled with cell wall material such as pectic polysaccharides and xyloglucans, which are synthesized in Golgi stacks and have to be delivered as vesicle cargo to the cell plate (100). Another major component—callose later to be replaced by cellulose—is synthesized by callose synthase at the cell plate (25). Thus, callose synthase and cell wall–modifying enzymes such as membrane-bound endo-1,4-beta-glucanase KORRIGAN (KOR) or secreted endoxyloglucan transferase (EXGT) are putative cargo proteins of cytokinetic vesicles (100, 102).

In addition to cell plate–specific cargo, a number of proteins that accumulate at the plasma membrane of, or are secreted from, the interphase cell have been localized to the cell plate (reviewed in 33). For example, the plasma membrane–localized syntaxin PEN1 (SYP121/SYR1) accumulated at the cell plate when expressed during M phase (13, 53). Conversely, when ectopically expressed in postmitotic cells, KNOLLE was also targeted to the plasma membrane (90). These observations suggest that the cell plate and the plasma membrane are interchangeable target membranes in exocytic vesicle trafficking. Exocytosis is the default pathway of membrane trafficking, which is taken in the absence of sorting signals, as supported by the following lines of evidence. GFP with a signal peptide for uptake into the ER (sp-GFP) is secreted from the cell during interphase (8) and accumulates in the cell plate of dividing tobacco BY-2 cells (100). Furthermore, the peptide ligand CLV3 fused to GFP is secreted from the cell, whereas an engineered variant carrying a C-terminal vacuolar sorting sequence accumulates in the vacuole (71). Similarly, the prevacuolar syntaxin PEP12 does not accumulate in the cell plate when expressed during M phase (53). These observations suggest that proteins trafficking to the plasma membrane or the cell plate lack sorting signals that would target them to endomembrane compartments. The proper localization of the membrane-bound endo-1,4-beta-glucanase KORRIGAN to the cell plate is proposed to involve two putative sorting motifs, a tyrosine-based motif and an acidic dileucine motif (102). However, both motifs are absent from PEN1 (SYP121/SYR1) syntaxin, which also accumulates at the cell plate when expressed during M phase (53). Thus, the significance of these sorting motifs in cytokinetic vesicle trafficking remains to be determined. In conclusion, the available evidence indicates that trafficking to the cell plate is an M phase–specific default pathway of proteins that lack sorting signals.

Membrane Dynamics During Cell Plate Expansion

The developing cell plate undergoes extensive reorganization from the initial fusion of vesicles via the formation of a lattice of membrane tubules, which is accompanied by the removal of excess membrane material to yield a planar membrane compartment (**Figure 2**). While this process is underway in the center of the division plane, newly arriving vesicles are targeted to the lateral margin of the cell plate, resulting in its expansion outward to the parental plasma membrane.

Endocytosis: internalization of membrane material from the plasma membrane, often via clathrin-coated vesicles

The formation of a lattice of tubules rather than a balloon-shaped membrane compartment has been attributed to the activity of dynamin-related proteins (DRPs) [formerly called phragmoplastins or *Arabidopsis* dynamin-like proteins (ADLs)]. Dynamin GTPases are eukaryotic mechano-enzymes that tubulate membranes and also mediate scission of clathrin-coated vesicles (66). Members of subgroup 1 of the *Arabidopsis* DRP family are associated with the developing cell plate, and the *drp1a drp1e* double mutant is embryo-lethal, displaying defects in cytokinesis (27, 28, 34–36). The tubular membranes are locally constricted by DRP1a (ADL1A), which also interacts with callose synthase and may facilitate callose deposition into the lumen of the incipient cell plate (25, 26, 63).

Removing membrane material from the maturing cell plate is estimated to reduce both surface area and volume by approximately 70% in the case of endosperm cellularization (63). Clathrin-coated buds on the cell plate and nearby clathrin-coated vesicles suggest that membrane material is removed by an endocytosis-related process (**Figure 2**) (63, 72, 74). Consistent with this notion, DRP2a (ADL6), which is related to animal classical dynamin involved in membrane scission, localizes to the cell plate, in addition to its association with the plasma membrane and Golgi stacks in interphase cells (28). Because DRP2a is required for trafficking of clathrin-coated vesicles from the trans-Golgi network to the lytic vacuole (31, 40), it may play a comparable role in the removal of excess membrane from the maturing cell plate.

Endocytosis during cell plate expansion may merely shape the cell plate. Alternatively, membrane removed from the center may be recycled, via endosomes, to the growing margin of the cell plate, which would speed up cytokinesis. At present, there is only limited evidence that recycling may occur. BFA treatment, which blocks cell plate expansion in BY-2 cells presumably by disrupting Golgi stacks (98, see above), appears to have a different effect in *Arabidopsis* root cells. BFA causes the formation of endosomal "BFA compartments," which accumulate the plasma-membrane protein PIN1, the endosomal ARF-GEF GNOM but not Golgi or trans-Golgi markers (18, 19). During cytokinesis, BFA compartments accumulate the cytokinesis-specific syntaxin KNOLLE, in addition to PIN1. In BFA-resistant GNOM transgenic seedlings, BFA compartments still accumulate KNOLLE but neither PIN1 nor GNOM, suggesting that KNOLLE trafficking from endosomes involves a different BFA-sensitive ARF-GEF (19). Nonetheless, recycling KNOLLE to the growing margin of the cell plate has not been conclusively demonstrated, in contrast to recycling PIN1 to the plasma membrane. In maize root cells, an ARF1-type GTPase has been localized to the cell plate, Golgi stacks, and the plasma membrane (14). Upon BFA treatment, ARF1 also colocalized with a Golgi marker in aggregates at the growing margin of the cell plate. Similar subcellular distributions were observed for several COPI coat proteins in both BFA-treated and -untreated cells (14). These results are difficult to interpret but may suggest the possibility of retrograde transport from Golgi stacks that accumulate near the leading edge of the expanding cell plate (57). The seemingly conflicting data on BFA effects in different plant species caution against a simplified view of membrane recycling from the expanding cell plate. Although local recycling of cell plate membrane via endosomes would provide an efficient mechanism for rapid completion of cytokinesis, additional studies are required to determine whether this is true.

Fusion with the Plasma Membrane

To seal off the two daughter cells, the cell plate has to fuse with the parental plasma membrane. This fusion event triggers further maturation of the cell plate, including breakdown of callose and its replacement by cellulose (72). According to the classic model, the cell plate expands symmetrically from the center to the periphery before fusing with the parental plasma membrane (81). However, asymmetric expansion of the cell plate ("polar cytokinesis") has been described in vacuolate shoot cells of *Arabidopsis* (16). The cell plate appears to contact, and possibly fuse with, the plasma membrane on one side early during cytokinesis and then expand along the plasma membrane to the opposite side of the cell. In BY-2 cells, polar cytokinesis has been detected with the lipophilic dye FM4-64, which stained the developing cell plate together with the continuous plasma membrane, whereas symmetrically expanding cell plates were only stained after 30–60 min of incubation (10). The asymmetric expansion of the cell plate would easily account for the cell wall stubs observed upon genetic or experimental interference with cytokinesis (16). However, cell wall stubs also occur in cytokinesis-defective mutants that accumulate unfused vesicles across the entire plane of cell division (41, 92). This suggests that in the absence of cell plate formation, cell wall stubs may originate by local membrane fusion initiated from the cortical division site of the parental plasma membrane. A comparable process occurs normally in *Arabidopsis* male meiotic cytokinesis during which no cell plate is formed: Networks of tubular membranes formed by fusion of Golgi-derived vesicles are lined up across the division plane and are successively fused with the plasma membrane from the periphery to the center (64). It is thus conceivable that in somatic cells, the fusion of the cell plate margin with the plasma membrane is affected by different machinery than are the vesicle fusion events that generate and expand the cell plate.

CYTOSKELETON DYNAMICS

Cytokinesis of somatic cells is assisted by two plant-specific cytoskeletal arrays, the preprophase band and the phragmoplast (**Figure 1**). The preprophase band forms transiently from late G2 phase to prometaphase and somehow marks the cortical division site at which the expanding cell plate fuses with the parental plasma membrane during cytokinesis. The phragmoplast originates in the interzone between the two sets of daughter chromosomes during late anaphase and undergoes lateral translocation during the expansion of the cell plate. These two cytoskeletal arrays consist of both MTs and AFs. Membrane dynamics during cytokinesis depends on MTs and AFs of the phragmoplast.

Roles of Microtubules and Actin Filaments in Phragmoplast-Assisted Cytokinesis

The phragmoplast contains MTs and AFs, which are aligned parallel to each other. Although both MTs and AFs are organized in two bundles, with their plus ends facing the plane of division, only the MTs interdigitate (81). During lateral expansion of the cell plate, MTs are translocated to its margin, whereas AFs remain present throughout the forming cell plate. A novel 190-kDa protein from tobacco BY-2 cells bundles both MTs and AFs in vitro and has been localized to the phragmoplast, suggesting a possible role in cross-linking MTs with AFs (**Figure 3*A***) (29).

The phragmoplast plays two major roles: targeted delivery of Golgi-derived membrane vesicles to the plane of cell division and lateral expansion of the cell plate toward the cortical division site. Which of these roles can be attributed to MTs and/or AFs? In interphase cells, AFs but not MTs are involved in membrane trafficking to the plasma membrane, as evidenced by both drug studies and mutant analysis (18, 82). Conversely, MTs but not AFs appear to mediate trafficking to the division plane because oryzaline inhibits delivery of KNOLLE (18). In addition, KNOLLE accumulation is dispersed rather than localized in mitotic cells of MT-deficient mutants (50, 82). Furthermore, noncoated vesicles appear to be linked, by structures resembling kinesins, to mini-phragmoplast MTs during endosperm cellularization (63). Thus, the available evidence favors a role for MTs in the targeted delivery of Golgi-derived membrane vesicles to the plane of cell division. However, no vesicle-carrying kinesin motor protein has been identified unambiguously. A possible candidate is PAKRP2, which colocalizes with KNOLLE at the cell plate and also shows a broader punctate distribution in the phragmoplast (43). However, there is no conclusive evidence that this protein is involved in transporting Golgi-derived vesicles along phragmoplast MTs.

AF: actin filament

Kinesin: a motor protein that travels along microtubules mostly to the plus end, sometimes to the minus end. Kinesins often carry cargo such as vesicles or microtubules but may perform other functions.

Expanding the cell plate toward the cortical division site is a complex process involving both MTs and AFs (24). The lateral translocation of phragmoplast MTs is required for the targeted delivery of the vesicles to the growing margin of the cell plate (see below). The role of AFs has been probed by drug treatment and by injection of profilin into dividing stamen hair cells of *Tradescantia* (52, 87). Depending on the treatment, cell plate growth is impaired or the cell plate disintegrates, suggesting that AFs play a stabilizing role. In addition, AFs link the phragmoplast with the cortical division site during cell plate expansion and may thus direct the growing margin to the division site (24). So far, a role for AFs has not been supported by genetic evidence. A dominant-negative mutation in the *ACT2* gene, which disturbs actin polymerization, affects root hair initiation and root cell elongation but does not alter cell division (60). Also, an *act2 act7* double mutant is severely stunted but develops to the adult stage (20). These results are consistent with the stunted growth phenotype of seedlings germinated on the actin-depolymerizing drug latrunculin B (6). In summary, whatever role AFs play in cytokinesis, their contribution is less pronounced than that of the phragmoplast MT.

Formation and Dynamic Stability of the Phragmoplast Microtubules

The dynamics of phragmoplast MTs has been visualized with GFP-MAP4 or GFP-TUA6 fusion proteins in live BY-2 cells (21, 85). These MTs originate alongside remaining kinetochore MTs in the midzone during late anaphase, and they become more numerous and consolidate into two short bundles with their plus ends overlapping (**Figure 1**) (81). How the transition from anaphase MTs to phragmoplast MTs occurs is not known. Mitotic cyclin B1 must be degraded during metaphase-anaphase transition for cytokinesis to occur.

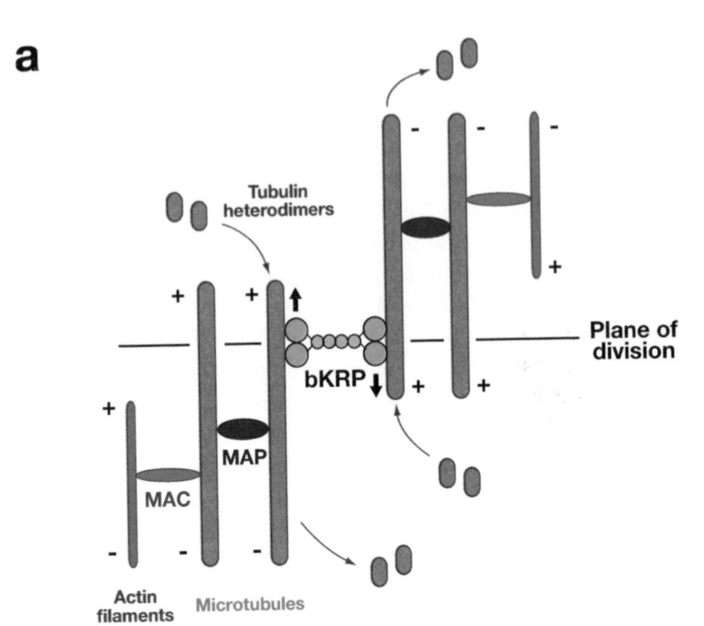

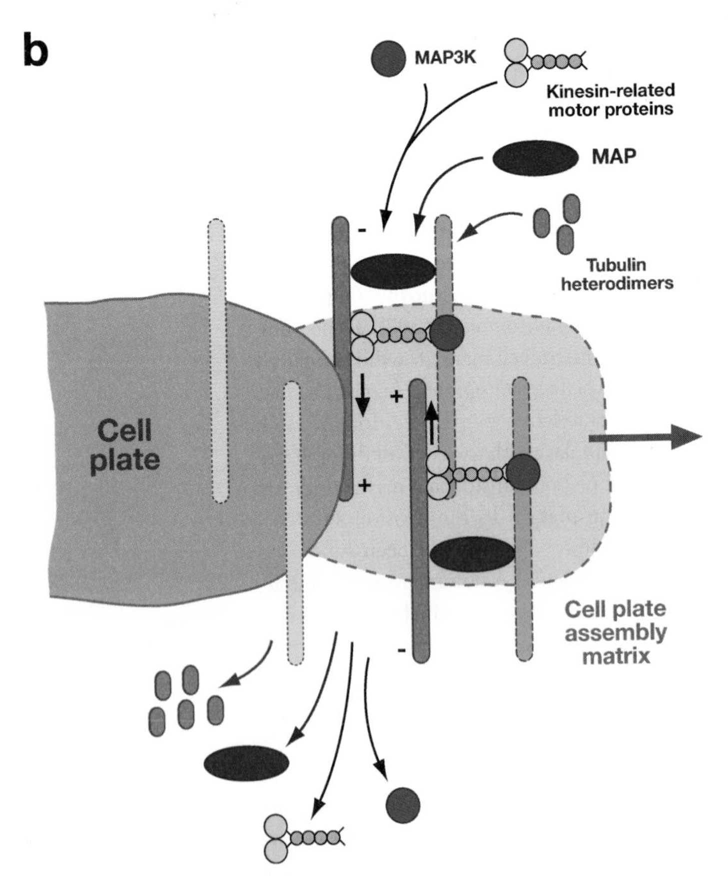

Figure 3

Cytoskeletal dynamics during cytokinesis. (*A*) Dynamic stability of phragmoplast. Microtubules form two overlapping antiparallel bundles and are cross-linked by MAPs within each bundle. Coaligned actin filaments may be cross-linked with microtubules by MAC proteins. Growth of microtubules at plus ends (+) by adding tubulin heterodimers is counteracted by plus end–directed bKRPs (*arrows*) such that the position of the phragmoplast relative to the plane of division is maintained. This simplified diagram shows generic MAPs and bKRPs (for details, see text). (*B*) Lateral translocation of phragmoplast microtubules. Microtubules underneath the developing cell plate lose their cross-linking MAPs (*blue shading*) and are depolymerized (*light shading*) into tubulin heterodimers. Microtubules (*red*) terminating in the cell plate assembly matrix at the margin of the expanding cell plate are stable and cross-link via MAPs with new microtubules (*dark shading*) that are polymerized on the outer face of the microtubule ring. The direction of lateral translocation is indicated by a large arrow. Plus (+) end–directed kinesin-related motor proteins (*arrows*) deliver active MAP3K (*magenta*) to the cell plate margin. MAP3K signaling results in microtubule depolymerization, which in turn inactivates the MAP3K. For details, see text. bKRP, bipolar kinesin-related protein; MAC, putative cross-linker between microtubules and actin filaments; MAP, microtubule-associated protein; MAP3K, mitogen-activated protein kinase kinase kinase.

Nondegradable cyclin B1 interferes with anaphase spindle disassembly and phragmoplast formation, although KNOLLE still accumulates at the midzone (93). Once phragmoplast MTs have formed their organization is dynamically maintained by MT-associated proteins (MAPs) and kinesin-related motor proteins (**Figure 3*A***) (46). For example, MAP65-1 cross-bridges MTs and is normally associated with interzone anaphase MTs and plus ends of phragmoplast MTs (75, 76). However, if cyclin B1 is not degraded, MAP65-1 does not accumulate at the midline, which may account for the failure to form a stable phragmoplast (93).

MAPs have MT-bundling activity in vitro and colocalize with MT arrays in vivo (46). MAP65, the most abundant group of MAPs, is encoded by a family of nine genes in *Arabidopsis* (75, 76). MAP65-1 associates with all MT arrays during the cell cycle, and its dimerization is necessary to form MT cross-bridges (76). Another member of the MAP65 family, PLEIADE (PLE/MAP65-3), specifically localizes to interzone spindles in anaphase and to the overlapping plus ends of phragmoplast MTs but not to other MT arrays (55). In addition, *ple* mutants display cytokinesis defects, presumably due to compromised organization of phragmoplast MTs (54, 55). In the absence of PLE activity, the width of the phragmoplast increases, suggesting that PLE is essential for the integrity of the overlapped MTs. Another MAP with an essential role in cytokinesis is GEM1/MOR1, evidenced by the aberrant division of the microspore in *gem1* mutants (84). However, GEM1/MOR1 also stabilizes interphase cortical MT arrays, as indicated by the mutant phenotype of the temperature-sensitive allele *mor1* (94). GEM1/MOR1 localizes to all MT arrays and accumulates toward the plus ends of phragmoplast MTs (84). Its tobacco homolog has been isolated from telophase cells and cross-bridges MTs (99). Thus, GEM1/MOR1 may stabilize the growing plus ends of phragmoplast MTs (**Figure 3*A***). In summary, several MAPs stabilize phragmoplast MTs but it is not known at present to what extent they are functionally distinct.

The *Arabidopsis* genome encodes some 60 kinesin-related motor proteins (KRPs) (68). Several KRPs localize to the phragmoplast and are implicated in the organization of phragmoplast MTs (**Figure 3*A***) (reviewed in 44, 46). Bipolar KRPs such as TKRP125 are plus end–directed motors that slide overlapping MTs against each other and may thus compensate for MT growth at their overlapping plus ends (1). Whereas TKRP25 localizes along the MTs, related DcKRP120-2 mainly accumulates in the zone of overlap (7). PAKRP1 and its close homolog PAKRP1L represent a different group of phragmoplast-associated KRPs (42, 65). These proteins specifically localize to the interzonal MTs at anaphase and to the overlapping plus ends throughout phragmoplast development. Both may form dimers and thus maintain the organization of overlapping antiparallel bundles. However, their precise roles need to be clarified. A third group of KRPs, represented by ATK1/KatA, has C-terminal motor domains and may counteract the action of bipolar KRPs. ATK1/KatA is a minus end-directed KRP that specifically localizes to the midzone of the mitotic apparatus and the phragmoplast (45, 49). Disrupting the *ATK1* gene alters the spindle organization of male meiotic cells but does not affect mitotic spindle or phragmoplast. This finding has been attributed to functional overlap between members of the *ATK* family (12). Another minus–end–directed motor protein is the kinesin calmodulin binding protein (KCBP). Its C-terminal motor domain and an adjacent Ca calmodulin–binding domain bundle MTs in vitro in a calcium-dependent manner (37). Structural analysis of KCBP also predicts that Ca-calmodulin binding blocks motor motility and may initiate dissociation of KCBP from MTs (89). Injecting anti-KCBP antibody into anaphase cells activates KCBP and disrupts the phragmoplast, which suggests that KCBP is normally downregulated (91). This regulation may involve the calmodulin-binding domain of KCBP because limiting the rise of calcium concentration that normally occurs during cytokinesis has a similar disruptive effect (24).

MAP: a microtubule-associated protein. Some of these form cross-bridges between microtubules and thereby stabilize microtubule arrays.

CPAM: cell plate assembly matrix

In summary, both structural MAPs and KRPs play important roles in organizing the phragmoplast MTs at the plane of cell division. Their activity results in dynamic stability of the MT array perpendicular to the division plane during both the formation and the lateral translocation of the phragmoplast.

Lateral Translocation and Disassembly of the Phragmoplast Microtubules

As the cell plate develops in the center of the division plane, the associated phragmoplast MTs start to depolymerize, except those at the margin of the MT bundles. New MTs polymerize adjacent to the remaining ones, delivering Golgi-derived vesicles to the margin of the developing cell plate. Inner MTs then depolymerize and new MTs are added to the outer face of the ring-shaped array (**Figure 3*B***). This cycle of phragmoplast translocation and cell plate expansion is repeated until the cell plate reaches the cortical division site marked earlier by the preprophase band (21, 81). In polar cytokinesis, the lateral translocation of the phragmoplast MTs occurs asymmetrically as does the expansion of the cell plate (85). What is the driving force behind the lateral progression of cytokinesis?

Lateral translocation of the phragmoplast is inhibited by the MT-stabilizing drug taxol, suggesting that polymerization of MTs at the margin requires depolymerization of MTs in the center (96). Taxol also inhibits cell plate expansion, presumably because no vesicles are delivered to the CPAM near its margin (**Figure 3*B***). More recently, two orthologous plant-specific kinesin-related proteins, HINKEL (HIK) in *Arabidopsis* and NACK1 in BY-2 cells, were shown to play a role in phragmoplast dynamics during cell plate expansion (59, 83). The *HIK* gene is expressed in a cell cycle–regulated manner, and *hik* mutant embryos display characteristic cytokinesis defects. Phragmoplast MTs are stabilized beneath the expanding cell plate, as reported for taxol-treated cells, and the lateral translocation appears to continue (83). NACK1 colocalizes with the plus ends of phragmoplast MTs during their lateral translocation (59). A dominant-negative form of NACK1 stabilizes the MTs beneath the cell plate but blocks lateral translocation of phragmoplast MTs and expansion of the cell plate, thus resembling the effect of taxol treatment (59). Regardless of the differences between the two systems, both studies indicate a role for HIK/NACK1 in the depolymerization of MTs beneath the cell plate. Further analysis shows that NACK1 activates and targets the MAP kinase kinase kinase (MAP3K) NPK1 to the plus ends of phragmoplast MTs, which is required for cytokinesis (59). Overexpression of a kinase-negative variant of NPK1 causes essentially the same defects as the dominant-negative form of NACK1 (58). It is interesting to note that both NPK1 and its target MAP2K NQK1 are inactivated by MT depolymerization (80). It is thus conceivable that NPK1 signaling is controlled by a negative feedback loop, leading to MT depolymerization, which in turn results in NPK1 inactivation. NPK1-mediated MT depolymerization does not occur in the CPAM but only underneath the cell plate and may thus be linked to membrane fusion processes in the developing cell plate. This would ensure that membrane vesicles are only delivered to the growing margin of the cell plate and that the phragmoplast is disassembled when the fully expanded cell plate fuses with the plasma membrane.

The closest homolog of HIK/NACK1 is TETRASPORE (TES)/NACK2, which is required for male meiotic cytokinesis in *Arabidopsis* (95). In *tes* mutants, the radial MT arrays that mediate the formation of the partitioning membrane are disorganized, suggesting that the two homologous kinesin-related proteins perform comparable roles in two different types of cytokinesis. There is also indirect evidence that NPK1 signaling plays a role in male meiotic cytokinesis as well (39, 80). How TES/NACK2 and NPK1 signaling might affect disassembly of the radial MT arrays remains to be determined. Note that these arrays form simultaneously across the plane of division and the membrane fusion process starts at the plasma membrane and progresses to the center (64).

In summary, the directional progression of cytokinesis can in part be explained by the interaction of the NACK1-NPK1 signaling pathway with the fusion processes occurring in the adjacent cell plate, which accounts for the depolymerization of phragmoplast MTs as well as for the disassembly of the phragmoplast at the end of cytokinesis. However, it is not known at present how new MTs polymerize on the outer face of the ring-shaped MT array and thus deliver membrane vesicles required for the further expansion of the cell plate.

CONCLUDING REMARKS

Higher-plant cytokinesis is a highly orchestrated, cytoskeleton-assisted process of membrane targeting and fusion. The recent studies of diverse modes of cytokinesis, once considered mechanistically different, have revealed a remarkable similarity of the initial phase: in all three systems studied, transport vesicles accumulate in the division plane and fuse with one another to give vesiculo-tubular networks. These results suggest common mechanisms in diverse modes of cytokinesis. However, our knowledge of the cytokinetic trafficking pathway and the fusion machinery is still fragmentary. Similarly, endocytosis from the expanding cell plate needs further study. We particularly need to determine whether there is a recycling pathway from endosomes to the growing margin of the cell plate, which would speed up the expansion of the cell plate. Regarding cytoskeleton dynamics in cytokinesis, the most important finding is the discovery of a kinesin-MAPkinase pathway that, in conjunction with fusion processes in the cell plate, mediates microtubule depolymerization. By contrast, it is not known how new microtubules polymerize at the outer face of the array, which delivers vesicles for cell plate expansion. Finally, the relative roles of microtubules and AFs have not been defined precisely. Addressing these and other cytokinesis-related problems will remain a challenge for the future.

SUMMARY POINTS

1. Higher plant cytokinesis involves a coordinated interplay of membrane and cytoskeleton dynamics.
2. Diverse modes of cytokinesis share common features during the early phase: targeted delivery of membrane vesicles, which fuse with one another to form vesiculo-tubular networks.
3. Membrane fusion during cell plate formation is mediated by a SNARE complex that contains a cytokinesis-specific syntaxin.
4. Vesiculo-tubular networks are locally constricted by dynamin-related proteins, which results in a lattice of membrane tubules.
5. Further flattening of the maturing cell plate is achieved by endocytosis of excess membrane material.
6. Cell plate expansion requires depolymerization of phragmoplast microtubules, which is mediated by a kinesin-MAPkinase pathway in conjunction with membrane fusion processes in the cell plate.

LITERATURE CITED

1. Asada T, Kuriyama R, Shibaoka H. 1997. TKRP125, a kinesin-related protein involved in the centrosome-independent organization of the cytokinetic apparatus in tobacco BY-2 cells. *J. Cell Sci.* 110:179–89
2. Assaad FF. 2001. Plant cytokinesis. Exploring the links. *Plant Physiol.* 126:509–16

3. Assaad FF, Huet Y, Mayer U, Jürgens G. 2001. The cytokinesis gene *KEULE* encodes a Sec1 protein that binds the syntaxin KNOLLE. *J. Cell Biol.* 152:531–43
4. Assaad FF, Mayer U, Wanner G, Jürgens G. 1996. The *KEULE* gene is involved in cytokinesis in *Arabidopsis*. *Mol. Gen. Genet.* 253:267–77
5. Balasubramanian MK, Bi E, Glotzer M. 2004. Comparative analysis of cytokinesis in budding yeast, fission yeast and animal cells. *Curr. Biol.* 14:R806–18
6. Baluska F, Jasik J, Edelmann HG, Salajova T, Volkmann D. 2001. Latrunculin B-induced plant dwarfism: Plant cell elongation is F-actin-dependent. *Dev. Biol.* 231:113–24
7. Barroso C, Chan J, Allan V, Doonan J, Hussey P, Lloyd C. 2000. Two kinesin-related proteins associated with the cold-stable cytoskeleton of carrot cells: characterization of a novel kinesin, DcKRP120-2. *Plant J.* 24:859–68
8. Batoko H, Zheng HQ, Hawes C, Moore I. 2000. A rab1 GTPase is required for transport between the endoplasmic reticulum and golgi apparatus and for normal golgi movement in plants. *Plant Cell* 12:2201–18
9. Bednarek SY, Falbel TG. 2002. Membrane trafficking during plant cytokinesis. *Traffic* 3:621–29
10. Bolte S, Talbot C, Boutte Y, Catrice O, Read ND, Satiat-Jeunemaitre B. 2004. FM-dyes as experimental probes for dissecting vesicle trafficking in living plant cells. *J. Microsc.* 214:159–73
11. Brown RC, Lemmon BE. 2001. The cytoskeleton and spatial control of cytokinesis in the plant life cycle. *Protoplasma* 215:35–49
12. Chen C, Marcus A, Li W, Hu Y, Calzada JP, et al. 2002. The *Arabidopsis ATK1* gene is required for spindle morphogenesis in male meiosis. *Development* 129:2401–9
13. Collins NC, Thordal-Christensen H, Lipka V, Bau S, Kombrink E, et al. 2003. SNARE-protein-mediated disease resistance at the plant cell wall. *Nature* 425:973–77
14. Couchy I, Bolte S, Crosnier MT, Brown S, Satiat-Jeunemaitre B. 2003. Identification and localization of a beta-COP-like protein involved in the morphodynamics of the plant Golgi apparatus. *J. Exp. Bot.* 54:2053–63
15. Criqui MC, Genschik P. 2002. Mitosis in plants: how far we have come at the molecular level? *Curr. Opin. Plant Biol.* 5:487–93
16. Cutler SR, Ehrhardt DW. 2002. Polarized cytokinesis in vacuolate cells of *Arabidopsis*. *Proc. Natl. Acad. Sci. USA* 99:2812–17
17. Feiler HS, Desprez T, Santoni V, Kronenberger J, Caboche M, Traas J. 1995. The higher plant *Arabidopsis thaliana* encodes a functional CDC48 homologue which is highly expressed in dividing and expanding cells. *EMBO J.* 14:5626–37
18. Geldner N, Friml J, Stierhof Y-D, Jürgens G, Palme K. 2001. Auxin-transport inhibitors block PIN1 cycling and vesicle trafficking. *Nature* 413:425–28
19. Geldner N, Anders N, Wolters H, Keicher J, Kornberger W, et al. 2003. The *Arabidopsis* GNOM ARF-GEF mediates endosomal recycling, auxin transport, and auxin-dependent plant growth. *Cell* 112:219–30
20. Gilliland LU, Kandasamy MK, Pawloski LC, Meagher RB. 2002. Both vegetative and reproductive actin isovariants complement the stunted root hair phenotype of the *Arabidopsis act2-1* mutation. *Plant Physiol.* 130:2199–209
21. Granger CL, Cyr RJ. 2000. Microtubule reorganization in tobacco BY-2 cells stably expressing GFP-MBD. *Planta* 210:502–9
22. Heese M, Mayer U, Jürgens G. 1998. Cytokinesis in flowering plants: cellular process and developmental integration. *Curr. Opin. Plant Biol.* 1:486–91

23. Heese M, Gansel X, Sticher L, Wick P, Grebe M, et al. 2001. Functional characterization of the KNOLLE-interacting t-SNARE AtSNAP33 and its role in plant cytokinesis. *J. Cell Biol.* 155:239–49
24. Hepler PK, Valster A, Molchan T, Vos JW. 2002. Roles for kinesin and myosin during cytokinesis. *Philos. Trans. R. Soc. Lond. B* 357:761–66
25. Hong Z, Delauney AJ, Verma DP. 2001. A cell plate-specific callose synthase and its interaction with phragmoplastin. *Plant Cell* 13:755–68
26. Hong Z, Zhang Z, Olson JM, Verma DP. 2001. A novel UDP-glucose transferase is part of the callose synthase complex and interacts with phragmoplastin at the forming cell plate. *Plant Cell* 13:769–79
27. Hong Z, Bednarek SY, Blumwald E, Hwang I, Jürgens G, et al. 2003. A unified nomenclature for *Arabidopsis* dynamin-related large GTPases based on homology and possible functions. *Plant Mol. Biol.* 53:261–65
28. Hong Z, Geisler-Lee CJ, Zhang Z, Verma DP. 2003. Phragmoplastin dynamics: multiple forms, microtubule association and their roles in cell plate formation in plants. *Plant Mol. Biol.* 53:297–312
29. Igarashi H, Orii H, Mori H, Shimmen T, Sonobe S. 2000. Isolation of a novel 190 kDa protein from tobacco BY-2 cells: possible involvement in the interaction between actin filaments and microtubules. *Plant Cell Physiol.* 41:920–31
30. Jahn R, Lang T, Südhof TC. 2003. Membrane fusion. *Cell* 112:519–33
31. Jin JB, Kim YA, Kim SJ, Lee SH, Kim DH, et al. 2001. A new dynamin-like protein, ADL6, is involved in trafficking from the trans-Golgi network to the central vacuole in *Arabidopsis*. *Plant Cell* 13:1511–26
32. Jürgens G, Geldner N. 2002. Protein secretion in plants: from the trans-Golgi network to the outer space. *Traffic* 3:605–13
33. Jürgens G, Pacher T. 2003. Cytokinesis: membrane trafficking by default? *Annu. Plant Rev.* 9:239–54
34. Kang BH, Busse JS, Dickey C, Rancour DM, Bednarek SY. 2001. The *Arabidopsis* cell plate-associated dynamin-like protein, ADL1Ap, is required for multiple stages of plant growth and development. *Plant Physiol.* 126:47–68
35. Kang BH, Busse JS, Bednarek SY. 2003. Members of the *Arabidopsis* dynamin-like gene family, ADL1, are essential for plant cytokinesis and polarized cell growth. *Plant Cell* 15:899–913
36. Kang BH, Rancour DM, Bednarek SY. 2003. The dynamin-like protein ADL1C is essential for plasma membrane maintenance during pollen maturation. *Plant J.* 35:1–15
37. Kao YL, Deavours BE, Phelps KK, Walker RA, Reddy AS. 2000. Bundling of microtubules by motor and tail domains of a kinesin-like calmodulin-binding protein from *Arabidopsis*: regulation by Ca^{2+}/calmodulin. *Biochem. Biophys. Res. Commun.* 267:201–7
38. Kirchhausen T. 2000. Three ways to make a vesicle. *Nat. Rev. Mol. Cell Biol.* 1:187–98
39. Krysan PJ, Jester PJ, Gottwald JR, Sussman MR. 2002. An *Arabidopsis* mitogen-activated protein kinase kinase kinase gene family encodes essential positive regulators of cytokinesis. *Plant Cell* 14:1109–20
40. Lam BC, Sage TL, Bianchi F, Blumwald E. 2002. Regulation of ADL6 activity by its associated molecular network. *Plant J.* 31:565–76
41. Lauber MH, Waizenegger I, Steinmann T, Schwarz H, Mayer U, et al. 1997. The *Arabidopsis* KNOLLE protein is a cytokinesis-specific syntaxin. *J. Cell Biol.* 139:1485–93
42. Lee YR, Liu B. 2000. Identification of a phragmoplast-associated kinesin-related protein in higher plants. *Curr. Biol.* 10:797–800

43. Lee YR, Giang HM, Liu B. 2001. A novel plant kinesin-related protein specifically associates with the phragmoplast organelles. *Plant Cell* 13:2427–39
44. Liu B, Lee YRJ. 2001. Kinesin-related proteins in plant cytokinesis. *J. Plant Growth Regul.* 20:141–50
45. Liu B, Cyr RJ, Palevitz BA. 1996. A kinesin-like protein, KatAp, in the cells of *Arabidopsis* and other plants. *Plant Cell* 8:119–32
46. Lloyd C, Hussey P. 2001. Microtubule-associated proteins in plants—why we need a MAP. *Nat. Rev. Mol. Cell Biol.* 2:40–47
47. Low SH, Li X, Miura M, Kudo N, Quiñones B, Weimbs T. 2003. Syntaxin 2 and endobrevin are required for the terminal step of cytokinesis in mammalian cells. *Dev. Cell* 4:753–59
48. Lukowitz W, Mayer U, Jürgens G. 1996. Cytokinesis in the *Arabidopsis* embryo involves the syntaxin-related *KNOLLE* gene product. *Cell* 84:61–71
49. Marcus AI, Ambrose JC, Blickley L, Hancock WO, Cyr RJ. 2002. *Arabidopsis thaliana* protein, ATK1, is a minus-end directed kinesin that exhibits non-processive movement. *Cell Motil. Cytoskel.* 52:144–50
50. Mayer U, Herzog U, Berger F, Inzé D, Jürgens G. 1999. Mutations in the *pilz* group genes disrupt the microtubule cytoskeleton and uncouple cell cycle progression from cell division in Arabidopsis embryo and endosperm. *Eur. J. Cell Biol.* 78:100–8
51. Mayer U, Jürgens G. 2004. Cytokinesis: lines of division taking shape. *Curr. Opin. Plant Biol.* 7:599–604
52. Molchan TM, Valster AH, Hepler PK. 2002. Actomyosin promotes cell plate alignment and late lateral expansion in *Tradescantia* stamen hair cells. *Planta* 214:683–93
53. **Müller I, Wagner W, Völker A, Schellmann S, Nacry P, et al. 2003. Syntaxin specificity of cytokinesis in *Arabidopsis*. *Nat. Cell Biol.* 5:531–34**

This study demonstrates that syntaxin function in cytokinesis requires strong expression during mitosis, targeting to the plane of division and KNOLLE-like protein function.

54. Müller S, Fuchs E, Ovecka M, Wysocka-Diller J, Benfey PN, Hauser MT. 2002. Two new loci, *PLEIADE* and *HYADE*, implicate organ-specific regulation of cytokinesis in *Arabidopsis*. *Plant Physiol.* 130:312–24
55. Müller S, Smertenko A, Wagner V, Heinrich M, Hussey PJ, Hauser MT. 2004. The plant microtubule-associated protein AtMAP65-3/PLE is essential for cytokinetic phragmoplast function. *Curr. Biol.* 14:412–17
56. Nacry P, Mayer U, Jürgens G. 2000. Genetic dissection of cytokinesis. *Plant Mol. Biol.* 43:719–33
57. Nebenführ A, Frohlick JA, Staehelin LA. 2000. Redistribution of Golgi stacks and other organelles during mitosis and cytokinesis in plant cells. *Plant Physiol.* 124:135–51
58. **Nishihama R, Ishikawa M, Araki S, Soyano T, Asada T, Machida Y. 2001. The NPK1 mitogen-activated protein kinase kinase kinase is a regulator of cell-plate formation in plant cytokinesis. *Genes Dev.* 15:352–63**

This study provides evidence that a MAP kinase signaling pathway mediates depolymerization of phragmoplast microtubules during cell plate expansion.

59. **Nishihama R, Soyano T, Ishikawa M, Araki S, Tanaka H, et al. 2002. Expansion of the cell plate in plant cytokinesis requires a kinesin-like protein/MAPKKK complex. *Cell* 109:87–99**

This study demonstrates that the plus end-directed kinesin-related protein NACK1 activates the NPK1 kinase and that NACK1-mediated transport of NPK1 is required for microtubule depolymerization and cell plate expansion.

60. Nishimura T, Yokota E, Wada T, Shimmen T, Okada K. 2003. An *Arabidopsis ACT2* dominant-negative mutation, which disturbs F-actin polymerization, reveals its distinctive function in root development. *Plant Cell Physiol.* 44:1131–40
61. Otegui M, Staehelin LA. 2000. Cytokinesis in flowering plants: more than one way to divide a cell. *Curr. Opin. Plant Biol.* 3:493–502
62. Otegui M, Staehelin LA. 2000. Syncytial-type cell plates: a novel kind of cell plate involved in endosperm cellularization of *Arabidopsis*. *Plant Cell* 12:933–47

63. **Otegui MS, Mastronarde DN, Kang BH, Bednarek SY, Staehelin LA. 2001. Three-dimensional analysis of syncytial-type cell plates during endosperm cellularization visualized by high resolution electron tomography. *Plant Cell* 13:2033–51**
64. **Otegui MS, Staehelin LA. 2004. Electron tomographic analysis of post-meiotic cytokinesis during pollen development in *Arabidopsis thaliana*. *Planta* 218:501–15**
65. Pan R, Lee YR, Liu B. 2004. Localization of two homologous *Arabidopsis* kinesin-related proteins in the phragmoplast. *Planta* 220:156–64
66. Praefcke GJ, McMahon HT. 2004. The dynamin superfamily: universal membrane tubulation and fission molecules? *Nat. Rev. Mol. Cell Biol.* 5:133–47
67. Rancour DM, Dickey CE, Park S, Bednarek SY. 2002. Characterization of AtCDC48. Evidence for multiple membrane fusion mechanisms at the plane of cell division in plants. *Plant Physiol.* 130:1241–53
68. Reddy AS, Day IS. 2001. Kinesins in the *Arabidopsis* genome: a comparative analysis among eukaryotes. *BMC Genomics* 2:2
69. Renault L, Guibert B, Cherfils J. 2003. Structural snapshots of the mechanism and inhibition of a guanine nucleotide exchange factor. *Nature* 426:525–30
70. Ritzenthaler C, Nebenführ A, Movafeghi A, Stussi-Garaud C, Behnia L, et al. 2002. Reevaluation of the effects of brefeldin A on plant cells using tobacco Bright Yellow 2 cells expressing Golgi-targeted green fluorescent protein and COPI antisera. *Plant Cell* 14:237–61
71. Rojo E, Sharma VK, Kovaleva V, Raikhel NV, Fletcher JC. 2002. CLV3 is localized to the extracellular space, where it activates the Arabidopsis CLAVATA stem cell signaling pathway. *Plant Cell* 14:969–77
72. Samuels AL, Giddings TH, Staehelin LA. 1995. Cytokinesis in tobacco BY-2 and root tip cells: a new model of cell plate formation in higher plants. *J. Cell Biol.* 130:1345–57
73. Sanderfoot AA, Assaad FF, Raikhel NV. 2000. The *Arabidopsis* genome. An abundance of soluble N-ethylmaleimide-sensitive factor adaptor protein receptors. *Plant Physiol.* 124:1558–69
74. **Segui-Simarro JM, Austin JR 2nd, White EA, Staehelin LA. 2004. Electron tomographic analysis of somatic cell plate formation in meristematic cells of Arabidopsis preserved by high-pressure freezing. *Plant Cell* 16:836–56**
75. Smertenko A, Saleh N, Igarashi H, Mori H, Hauser-Hahn I, et al. 2000. A new class of microtubule-associated proteins in plants. *Nat. Cell Biol.* 2:750–53
76. Smertenko AP, Chang HY, Wagner V, Kaloriti D, Fenyk S, et al. 2004. The *Arabidopsis* microtubule-associated protein AtMAP65-1: molecular analysis of its microtubule bundling activity. *Plant Cell* 16:2035–47
77. Smith LG. 2001. Plant cell division: building walls in the right places. *Nat. Rev. Mol. Cell Biol.* 2:33–39
78. Söllner R, Glässer G, Wanner G, Somerville CR, Jürgens G, Assaad FF. 2002. Cytokinesis-defective mutants of *Arabidopsis*. *Plant Physiol.* 129:678–90
79. Sorensen MB, Mayer U, Lukowitz W, Robert H, Chambrier P, et al. 2002. Cellularisation in the endosperm of *Arabidopsis thaliana* is coupled to mitosis and shares multiple components with cytokinesis. *Development* 129:5567–76
80. **Soyano T, Nishihama R, Morikiyo K, Ishikawa M, Machida Y. 2003. NQK1/NtMEK1 is a MAPKK that acts in the NPK1 MAPKKK-mediated MAPK cascade and is required for plant cytokinesis. *Genes Dev.* 17:1055–67**
81. Staehelin LA, Hepler PK. 1996. Cytokinesis in higher plants. *Cell* 84:821–24

This study demonstrates that cellularization of the endosperm is essentially a variant of phragmoplast-assisted cytokinesis of somatic cells.

This study demonstrates that in male meiotic cytokinesis, targeted delivery of vesicles and formation of membrane-tubule networks across the plane of division provide the material from which the partitioning membrane is made.

This study provides an in-depth analysis of cell plate formation in somatic-cell cytokinesis.

This study not only identifies a target for NPK1 signaling in cytokinesis but also demonstrates that the members of the MAPK cascade are inactivated by microtubule depolymerization.

82. Steinborn K, Maulbetsch C, Priester B, Trautmann S, Pacher T, et al. 2002. The *Arabidopsis PILZ* group genes encode tubulin-folding cofactor orthologs required for cell division but not cell growth. *Genes Dev.* 16:959–71
83. Strompen G, El Kasmi F, Richter S, Lukowitz W, Assaad FF, et al. 2002. The *Arabidopsis HINKEL* gene encodes a kinesin-related protein involved in cytokinesis and is expressed in a cell cycle-dependent manner. *Curr. Biol.* 12:153–58
84. Twell D, Park SK, Hawkins TJ, Schubert D, Schmidt R, et al. 2002. MOR1/GEM1 has an essential role in the plant-specific cytokinetic phragmoplast. *Nat. Cell Biol.* 4:711–14
85. Ueda K, Sakaguchi S, Kumagai F, Hasezawa S, Quader H, Kristen U. 2003. Development and disintegration of phragmoplasts in living cultured cells of a GFP::TUA6 transgenic *Arabidopsis thaliana* plant. *Protoplasma* 220:111–18
86. Uemura T, Ueda T, Ohniwa RL, Nakano A, Takeyasu K, Sato MH. 2004. Systematic analysis of SNARE molecules in *Arabidopsis*: dissection of the post-Golgi network in plant cells. *Cell Struct. Funct.* 29:49–65
87. Valster AH, Pierson ES, Valenta R, Hepler PK, Emons AMC. 1997. Probing the plant actin cytoskeleton during cytokinesis and interphase by profilin microinjection. *Plant Cell* 9:1815–24
88. Verma DPS. 2001. Cytokinesis and building of the cell plate in plants. *Annu. Rev. Plant Physiol. Plant Mol. Biol.* 52:751–84
89. Vinogradova MV, Reddy VS, Reddy AS, Sablin EP, Fletterick RJ. 2004. Crystal structure of kinesin regulated by Ca^{2+}-calmodulin. *J. Biol. Chem.* 279:23504–9
90. Völker A, Stierhof YD, Jürgens G. 2001. Cell cycle-independent expression of the *Arabidopsis* cytokinesis-specific syntaxin KNOLLE results in mistargeting to the plasma membrane and is not sufficient for cytokinesis. *J. Cell Sci.* 114:3001–12
91. Vos JW, Safadi F, Reddy AS, Hepler PK. 2000. The kinesin-like calmodulin binding protein is differentially involved in cell division. *Plant Cell* 12:979–90
92. Waizenegger I, Lukowitz W, Assaad F, Schwarz H, Jürgens G, Mayer U. 2000. The *Arabidopsis KNOLLE* and *KEULE* genes interact to promote vesicle fusion during cytokinesis. *Curr. Biol.* 10:1371–74
93. Weingartner M, Criqui MC, Meszaros T, Binarova P, Schmit AC, et al. 2004. Expression of a nondegradable cyclin B1 affects plant development and leads to endomitosis by inhibiting the formation of a phragmoplast. *Plant Cell* 16:643–57
94. Whittington AT, Vugrek O, Wei KJ, Hasenbein NG, Sugimoto K, et al. 2001. MOR1 is essential for organizing cortical microtubules in plants. *Nature* 411:610–13
95. **Yang CY, Spielman M, Coles JP, Li Y, Ghelani S, et al. 2003. *TETRASPORE* encodes a kinesin required for male meiotic cytokinesis in *Arabidopsis*. *Plant J.* 34:229–40**

This study identifies a close homologue of HINKEL/NACK1, which plays a comparable role of regulating microtubule dynamics in a different cell type.

96. Yasuhara H, Sonobe S, Shibaoka H. 1993. Effects of taxol on the development of the cell plate and of the phragmoplast in tobacco BY-2 cells. *Plant Cell Physiol.* 34:21–29
97. Yasuhara H, Sonobe S, Shibaoka H. 1995. Effects of brefeldin A on the formation of the cell plate in tobacco BY-2 cells. *Eur. J. Cell Biol.* 66:274–81
98. Yasuhara H, Shibaoka H. 2000. Inhibition of cell-plate formation by brefeldin A inhibited the depolymerization of microtubules in the central region of the phragmoplast. *Plant Cell Physiol.* 41:300–10
99. Yasuhara H, Muraoka M, Shogaki H, Mori H, Sonobe S. 2002. TMBP200, a microtubule bundling polypeptide isolated from telophase tobacco BY-2 cells is a MOR1 homologue. *Plant Cell Physiol.* 43:595–603

100. Yokoyama R, Nishitani K. 2001. Endoxyloglucan transferase is localized both in the cell plate and in the secretory pathway destined for the apoplast in tobacco cells. *Plant Cell Physiol.* 42:292–300
101. Zheng H, Bednarek SY, Sanderfoot AA, Alonso J, Ecker JR, Raikhel NV. 2002. NPSN11 is a cell plate-associated SNARE protein that interacts with the syntaxin KNOLLE. *Plant Physiol.* 129:530–39
102. Zuo J, Niu QW, Nishizawa N, Wu Y, Kost B, Chua NH. 2000. KORRIGAN, an *Arabidopsis* endo-1,4-beta-glucanase, localizes to the cell plate by polarized targeting and is essential for cytokinesis. *Plant Cell* 12:1137–52

Evolution of Flavors and Scents

David R. Gang

Department of Plant Sciences and BIO5 Institute, University of Arizona, Tucson, Arizona 85721-0036; email: gang@ag.arizona.edu

Annu. Rev. Plant Biol. 2005. 56:301–25

doi: 10.1146/annurev.arplant.56.032604.144128

First published online as a Review in Advance on January 17, 2005

1543-5008/05/0602-0301$20.00

Key Words

specialized metabolism, volatile emission, repeated evolution, *O*-methyltransferases, acyltransferases, terpene synthases

Abstract

The world is filled with flavors and scents, which are the result of volatile compounds produced and emitted by plants. These specialized metabolites are the products of specific metabolic pathways. The terpenoid, fatty acid, and phenylpropanoid pathways contribute greatly to production of volatile compounds. Mechanisms that lead to evolution of volatile production in plants include gene duplication and divergence, convergent evolution, repeated evolution, and alteration of gene expression, caused by a number of factors, followed by change in enzyme specificity. Many examples of these processes are now available for three important gene families involved in production of volatile metabolites: the small molecule *O*-methyltransferases, the acyltransferases, and the terpene synthases. Examples of these processes in these gene families are found in roses, *Clarkia breweri*, and sweet basil, among others. Finally, evolution of volatile emission will be an exciting field of study for the foreseeable future.

Contents

INTRODUCTION

We live in a world full of fragrances and flavors. Scents profoundly influence much of what humans and animals do. Many chemicals that are emitted by flowers, such as geranylacetate, phenylethanol, and methyleugenol, are pleasant to humans and to the insects such as bees and moths that use these molecules as cues to find food. Other compounds, such as cadaverine and putrescine, are repugnant to humans and are usually associated with the presence of rotting flesh. Flavors and scents are directly tied to food preference and palatability and to mate choice (41). A significant body of knowledge has grown to describe the flavor, scent, and aroma compounds produced by plants. Much of early organic chemistry research and significant efforts by the food and beverage industries have been directed toward understanding the nature of these properties of plants. The perfume and food industries know the importance of these compounds, and the artificial and natural flavor and fragrance business has been a multibillion dollar business for some time (89).

The purpose of this article is to review recent advancements in our understanding of the mechanisms responsible for the evolution of volatile compounds responsible for flavors and scents in higher plants. This review discusses the evolution of several gene families that are responsible for controlling production, in one way or another, of many compounds that are important for the flavor and scents that are widely associated with plants. Because of the broad scope of this topic, this discussion focuses on what is known about the production of scent and flavor compounds in a few selected species and what is known or can be inferred about the evolution of these gene families in the plant kingdom.

THE NATURE OF FLAVORS AND SCENTS

The unique scents, aromas, flavors, fragrances, and smells that we attribute to plants are due to the presence of one or, in most cases, several volatile compounds that are produced and emitted. Some of these compounds are emitted directly into the air, such as those from flower parts. Others are present as the free compounds inside the plant tissue that are released when cuticles, cell walls, or membranes are broken. Still others exist in vivo as conjugates, which release the volatile components when enzyme activities act on them during processes such as maceration. However, what is important for the perception of flavor or scent, aside from factors such as acidity or sugar content, is that a compound is volatilized so that it may move freely through the air and then to specific receptors in the olfactory system of the receiving organism. Thus, a discussion on the evolution of flavors and scents in plants is in effect a discussion of the evolution of plant volatile compounds.

The compounds that are responsible for the scent and flavor properties of plants are small molecules, most with low boiling points and high vapor pressures at ambient temperature. Aside from these similarities, however, groups of plant-produced flavor and scent compounds have little in common with each other. They are derived from many different biosynthetic pathways (such as the isoprenoid/terpenoid, phenylpropanoid, acetate, fatty acid, and alkaloid pathways, among others) and contain many different chemically functional groups (such as esters, ethers, aldehydes, alcohols, alkenes, amino groups, etc.). These come not only from flowers, but also from all other parts of the plant such as fruits, leaves, stems, and roots (13, 83). Although arguably the most diverse and complex group of volatile compound producers, the angiosperms do not have a corner on the plant kingdom's aroma market. Other plants produced volatiles long before the angiosperms came on the scene. Pine, spruce, and other conifers are aromatic and are well known for their characteristic aromas. Many of the compounds that we find familiar in scents and flavors, such as limonene and linalool, are common across the plant kingdom. Others, such as the glucosinolates found in the mustards (46), are restricted to specific groups, families, and even species.

FORCES DRIVING THE EVOLUTION OF VOLATILE COMPOUND PRODUCTION

One question that has intrigued researchers for decades is: Why do plants produce volatile compounds? (47) Several hypotheses have been put forth over the years in an effort to answer this question. One early proposed explanation for the production of these compounds was the supposition that these metabolites were of secondary importance to plant function, that they were really metabolic waste products. The same "function" has been attributed to many groups of plant metabolites, even to some compounds that we now know are critically important for plant growth, development, and survival (32, 70). Plant metabolites are not produced as superfluous overflows of runaway metabolic pathways, nor as byproducts of food digestion. Evidence suggests that an animal-centric view of the plant kingdom is simply not valid. Indeed, plant metabolites are almost always produced via specific metabolic pathways that are comprised of specific enzymes that often act on specific substrates to produce specific metabolites (27, 70).

Specialized metabolite: A metabolite, usually a small molecule, which is not a building block of proteins, lipids, or sugars, but that plays another role in the organism. Usually, but not always, its occurrence is restricted.

The hypothesis that has gained support in the past two decades is that these compounds have specialized and important functions in plants (27, 70). Although not required for "primary" metabolic processes such as protein synthesis or direct sugar utilization, many of these metabolites are nevertheless required for survival of the species (e.g., an individual bee pollinated plant can live without attracting bees, although it may not then reproduce) and therefore are not "secondary" in importance. However, many of such compounds are limited in their distribution across the plant kingdom, often being restricted to certain clades. Thus, the term "specialized" metabolite (as opposed to "secondary" metabolite) has gained favor in recent years and more aptly describes the vast majority of metabolites that plants produce. This is especially true of volatile specialized metabolites because the biosynthesis of such compounds is often restricted to specific tissue types and to specific stages of development. So, what vital functions do these volatile metabolites serve in the plant?

One function that has been well documented is that specialized volatiles act as attractants to pollinators and seed dispersers (47, 71, 86), perhaps even being involved in driving coevolution of both the pollinators and the pollinated (75). For many species that are unable to self-pollinate, attraction of pollinators is absolutely required for reproduction. Many volatile metabolites, such as the phenylpropene methyleugenol, are produced

Terpene synthase: A member of the large family of enzymes that catalyze the formation of terpenoid (isoprenoid) specialized metabolites from geranyl diphosphate, farnesyl diphosphate, or geranylgeranyl diphosphate via a carbocation intermediate.

Repeated evolution: This occurs when a new and identical (or very similar) genetic function arises independently in the same gene family from two or more orthologous or paralogous genes that did not share the same function.

and emitted by flowers to attract pollinating moths. Methyleugenol is emitted by floral organs, such as the stigma and petals of *Clarkia breweri* (73, 92), when the flower is receptive to pollination. Other volatiles, especially esters, are released as fruit ripens, signaling to potential seed dispersers that the fruit is ready to eat.

A second function of plant volatiles is to attract the natural enemies of herbivores. Many examples of this interaction between plant, herbivore, and parasite have been described. For example, the predatory mite, *Cotesia rubecula* is able to find its host, the spider mite *Pieris rapae*, because of chemical signals released by the host plant *Brassica oleracea* when it is damaged (47, 54, 90). Thus, plant volatiles play a crucial role in tritrophic interactions between plants and insects.

A third important function of plant volatile metabolites is that they act as direct defense compounds (23, 24, 97, 98). This function has received less attention in recent years than the others. More emphasis has been placed on the role of nonvolatile compounds in plant defense, such as the pterocarpan phytoalexins and the lignans (34), among others. However, many volatiles have potent activities against microorganisms and herbivores. An excellent example of this is eugenol, which is an antibacterial compound (55) and antifungal agent (1, 55). The production of aflatoxins by many molds is inhibited by eugenol (10). Eugenol has a marked insect antiherbivory effect (66, 80). For example, it deters feeding by and is highly toxic toward many species of beetles (43, 66). Thus, eugenol plays an important role in the general defense of plants that produce it. This is true for a large number of volatile specialized metabolites from plants. It has been suggested that plants began to emit volatiles from the reproductive tissues first as a defense mechanism, to protect these vital tissues from herbivores and pathogens (24, 47). Pollinators then presumably learned to recognize these compounds as signals for the presence of flowers. Thus began the coevolution of flowers with their pollinators.

ORIGIN OF NEW ENZYME ACTIVITIES

A few years ago, Pichersky & Gang (70) discussed the genetics and biochemistry of plant metabolites from an evolutionary perspective. At that time, the genome sequence of *Arabidopsis* was not complete, and little sequence data from rice was available. Nevertheless, several large gene families that are involved in the production of plant specialized metabolites had been characterized in detail (7, 8, 50, 53, 77, 81, 91). Clear patterns in the evolution of these gene families could be discerned (70). With the *Arabidopsis* genome sequence now complete and the rice genome sequence almost done, these patterns have become much more obvious. As described below, these patterns aptly describe many of the mechanisms that drive the evolution of volatile compounds in plants.

One way that new metabolites are produced is for new enzymatic functions to evolve. This can happen in one of two ways. First, gene duplication followed by divergence retains the original enzymatic function in the plant, while a new function evolves in the enzyme encoded by the duplicate gene (72). This appears to have been the most common means for plants to produce new enzymatic functions in specialized metabolism, as illustrated below for small molecule *O*-methyltransferases and terpene synthases. Furthermore, convergent evolution, where new functions arise independently multiple times, occurs in this process of gene duplication and divergence. A special case of convergent evolution, termed repeated evolution (70), where new genetic functions arise independently from orthologous or paralogous genes, appears to have been a force behind the radiation of the terpenoids and *O*-methylation of several groups of molecules, including the phenylpropene eugenol (see below).

A second way for new enzymatic functions to originate involves the evolution of an existing structural gene (not duplicated), which results in a new enzymatic function, but also leads to concomitant loss of the prior function. This appears to be how production of the glucosinolates evolved (15, 46, 64), with simultaneous loss of cyanogenic glycoside production. Regardless of which route produces the new enzyme, the new function is catalyzed by an enzyme that is distinct from its progenitor (70). For this to be the case, the enzyme can either catalyze a chemically similar reaction on a different substrate or it can utilize the same substrate, but carry out a different chemical reaction (70).

New metabolites may also arise due to loss of enzymatic activity. This can occur because of loss of gene expression. Examples of this would be when a gene is lost due to hybrid formation and accompanying chromosomal rearrangements or when it is damaged by point mutations or larger insertions/deletions. Alternatively, transposons or other factors may disrupt the coding region of the gene or silence gene expression (56); or the loss of gene expression may be an epigenetic effect, caused by changes in chromatin structure (5, 11, 12, 82). Changes in regulatory protein expression may also lead to changes in gene expression (22). In all of these cases, intermediate metabolites in the affected pathway may build up to levels that were not previously present, enabling other enzymes to act on these metabolites. This may even occur with enzymes that normally would not utilize these metabolites as substrates because of low affinity (i.e., high apparent K_M). However, due to increased concentration caused by the pathway blockage, these metabolites become suitable substrates for these enzymes. If these new metabolites confer a selective advantage for the plant, their production is retained and increased. This increase will be brought about in one of two ways (70): either by increased protein synthesis (67), to compensate for low efficiency of the enzyme, or by mutation of the enzyme to produce a more efficient catalyst. In specialized metabolism, loss of an enzyme activity is thought to be a driving force in the production of new metabolites, and is evidenced by the relatively high K_M values for many substrate/enzyme pairs involved in producing such compounds (70). This process contributed to the evolution of small molecule *O*-methyltransferases in sweet basil, rose, and strawberry, as described below.

Changes in regulatory protein expression may not necessarily lead to complete loss of enzyme activity from the plant, but may, for example, cause the enzyme to be produced in a different cell, tissue, or organ type. As a result, the original substrate may not be available. Alternatively, changes in upstream metabolic pathways may lead to loss of original substrate. Without the normal substrate present, the enzyme may utilize a different metabolite as substrate, albeit at low efficiency. Should the resulting product yield a selective advantage, then the scenario described above is applicable, with an increase in protein production and/or mutation of protein yielding a more efficient enzyme, and the production of the new metabolite is increased. This is true for the acyltransferases BEBT and CHAT, from *Clarkia* and *Arabidopsis*, respectively, which are involved in forming different esters, benzyl benzoate and *cis*-3-hexen-1-yl acetate (21), but which share significant homology.

Acyltransferase: An enzyme that catalyzes the formation of esters and amides via transfer from the co-enzyme A derivate of acid moieties to a receptive alcohol.

EVOLUTION OF SCENTS AND FLAVORS IN ROSACEAE

When one discusses any aspect of floral scents, from their character, to their biosynthesis, to their evolution, it is practically impossible not to think of roses, at least for comparison purposes. Roses (*Rosa* sp.) are among the most widely recognized and widely loved flowers, and are the most important crop in the floriculture industry (with an annual value of ~$10 billion) (45). Not only are roses admired for the beauty of their flowers, they also are recognized for their wonderful fragrances and flavors. Shakespeare epitomized the love of rose scent in Juliet's famous phrase: "a rose/ By any other name would smell as sweet" (78). The scent of roses consists of a fairly complex mixture of between 300 and 400 compounds (29). These compounds are usually classified into several groups, based on major chemical functional groups. These include hydrocarbons, alcohols, esters, aldehydes, aromatic ethers, and others (79). These compounds are derived for the most part from three major pathways, the phenylpropanoid, terpenoid, and fatty acid pathways (45). Investigation of these pathways in roses has led to interesting conclusions regarding the evolution of the biosynthesis of these compounds.

SMOMT: small molecule *O*-methyltransferase. An enzyme belonging to the Type I family of methyltransferases that catalyzes the *O*-methylation of small organic, mostly aromatic plant specialized metabolites, resulting in the formation of methoxy functional groups.

Although Juliet's statement may suggest otherwise, not all roses do "smell as sweet." In fact, there are two major lineages of modern roses, traceable back to wild Chinese (*R. chinensis* and *R. gigantea*) and wild European (*R. moschata*, *R. gallica*, and *R. phoenicia*) rose species. These two groups of species and their hybrid progeny differ greatly in their scent composition (74). Tea roses, derived from crosses of Chinese species with *R. moschata*, produce an array of methoxylated phenolics, such as trimethoxybenzene (TMB), dimethoxytoluene (DMT), methyleugenol, and methylisoeugenol, and a variety of alcohols and esters, such as 2-phenylethyl alcohol, citronellol, geraniol, 2-phenylethylacetate, and geranylacetate, as well as mono- and sesequiterpenes (predominantly germacrene D). Although most modern hybrid roses have lost most of their scent, Fragrant Cloud is an example of a rose variety rich in these compounds (see **Figure 1**) (45, 59). Damask roses (*R. damascena*, the result of crosses between *R. gallica* and *R. phoenicia*), in contrast, produce a scent characterized by rose oxide and that is rich in alcohols (74, 94, 95), such as 2-phenylethanol, geraniol, and nerol. These roses are utilized the most in the fragrance and flavor industries, as sources of compounds for perfumes, and as the source of rose water and rose essential oils (45). No genes that are involved in the production of scent volatiles have been identified from Damask roses. In contrast, the biosynthesis of three major groups of compounds, methoxylated phenolics, monoterpenoid esters, and sesquiterpenes, has been studied in some detail in tea and hybrid roses.

Evolution of *O*-Methyltransferases

The methoxylated phenolics in hybrid and Chinese tea roses are produced via the action of several enzymes that belong to the large family of *O*-methyltransferases (OMTs). The reactions catalyzed by these enzymes are depicted in **Figure 2**. Members of this OMT family form methylethers (i.e., methoxyl groups) on small molecules, both nonvolatile and volatile. Consequently, this family is called the small molecule OMT (SMOMT) family (69); it is also known as the "Type I" methyltransferase family (65). **Figure 3** contains a phylogenetic tree of known SMOMTs. There are 14 members of this family in *Arabidopsis* and 13 members so far in rice. At least six separate enzymes belonging to this family are involved in producing the methoxylated phenolics found in roses: the orcinol OMTs OOMT1 (AAM23004.1) and OOMT2 (AAM23005.1) from *R. hybrida* (and *R. chinensis*); RcOMT1 (BAC78826.1), RcOMT2 (BAC78827.1), and RcOMT3 (BAC78828.1) from *R. chinensis*; and POMT (BAD18975.1) from *R. chinensis*.

The first enzyme in the pathway from TMB is POMT (99), which is unique among the enzymes in this group in that it uses phloroglucinol as a substrate (see **Figure 2**). The other intermediates to TMB (or to DMT) do not serve as substrates for this enzyme. In contrast, OOMT1 and OOMT2 from *R. hybrida* (59) have slightly broader substrate specificities and can perform the last two *O*-methylation steps to form TMB and DMT, although OOMT1 prefers the dihydroxy compounds whereas OOMT2 prefers the monohydroxy compounds as substrates. These enzymes do not use other metabolites, including phloroglucinol, as substrates. For example, caffeic acid is a very poor substrate for these enzymes (59, 74). The identical genes were isolated from *R. chinensis* (74, 59). Thus, the ability to form these compounds can be directly attributed to the inheritance of these genes from the *R. chinensis* progenitor. OOMT1 and OOMT2 illustrate the processes of gene duplication followed by divergence, with OOMT1

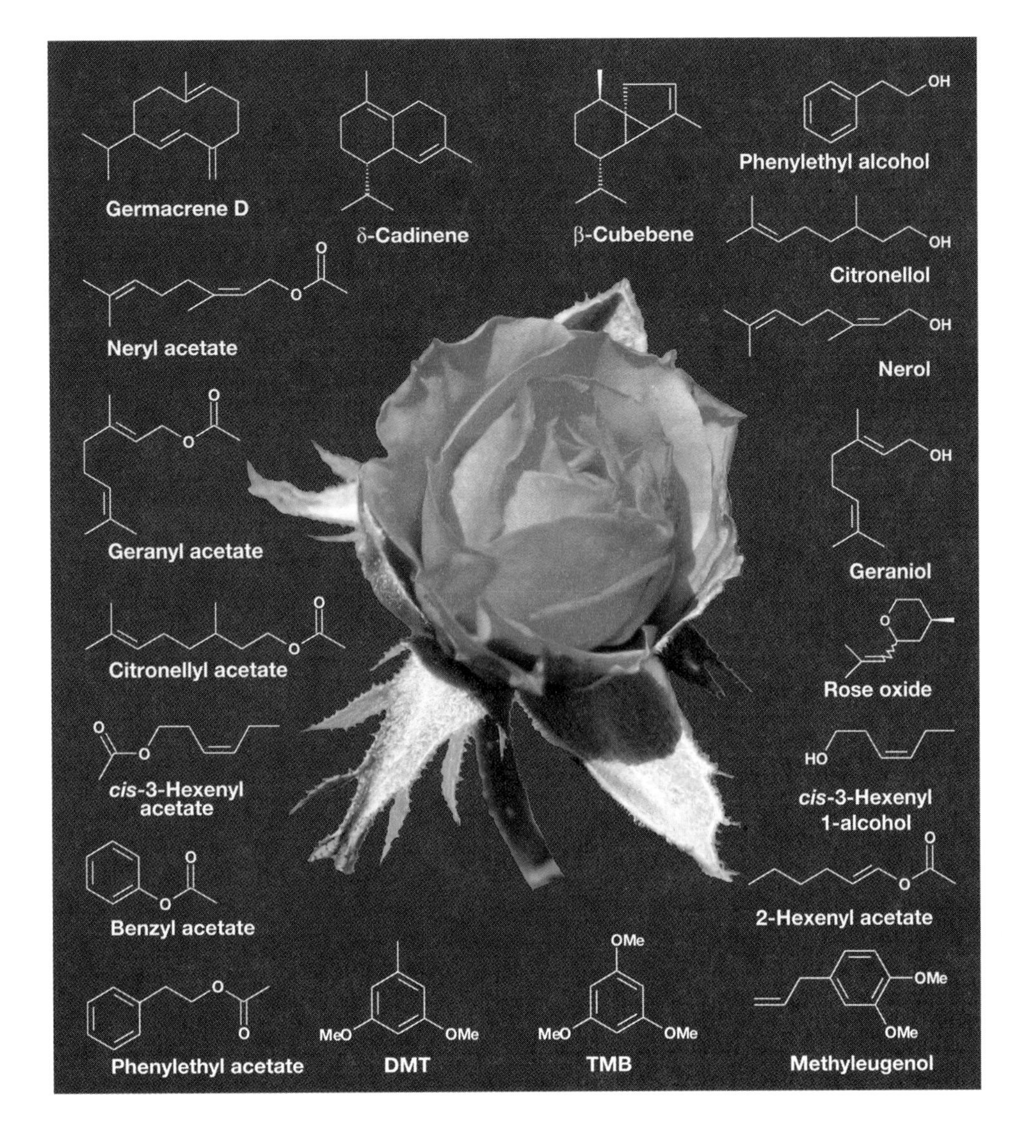

Figure 1

Roses produce a large array of scent compounds. All of the compounds shown here (except for rose oxide, geraniol, *cis*-3-hexenyl-1-alcohol, citronellol, and nerol, which are found at high levels in Damask roses) are major components of the scent produced by Chinese tea roses and hybrid roses. One such hybrid rose, Fragrant Cloud, is shown here at stage 4 of development (59), when scent production is at a maximum. Most of these compounds are commonly found in other scents and flavors.

now specializing on the earlier, and OOMT2 on the latter, step in the pathways to TMB and DMT.

Even more interesting was the discovery of a third enzyme, RcOMT2, which is able to catalyze the same reactions as OOMT1 and OOMT2 (100). Being much more efficient with the dihydroxy intermediates, RcOMT2 resembles OOMT1. However, unlike OOMT1, which does not utilize caffeic acid as a substrate, RcOMT2 can *O*-methylate this compound at the 3 position of the aromatic ring with high efficiency (100). In fact, RcOMT2 can use a broad array of metabolites as substrates. Based on substrate preference, it might appear that RcOMT2 is the progenitor of OOMT1. However, many COMTs (caffeic acid OMTs, involved in lignification and other processes) possess this same broad substrate allowance, and RcOMT2 appears to be a COMT that is diverging in catalytic ability. Furthermore, from **Figure 3** it is clear that RcOMT2 and OOMT1 are not that closely related. Another COMT (CAD29457.1) in *R. chinensis* was reported to GenBank, but its enzymatic function has not been demonstrated. RcOMT2

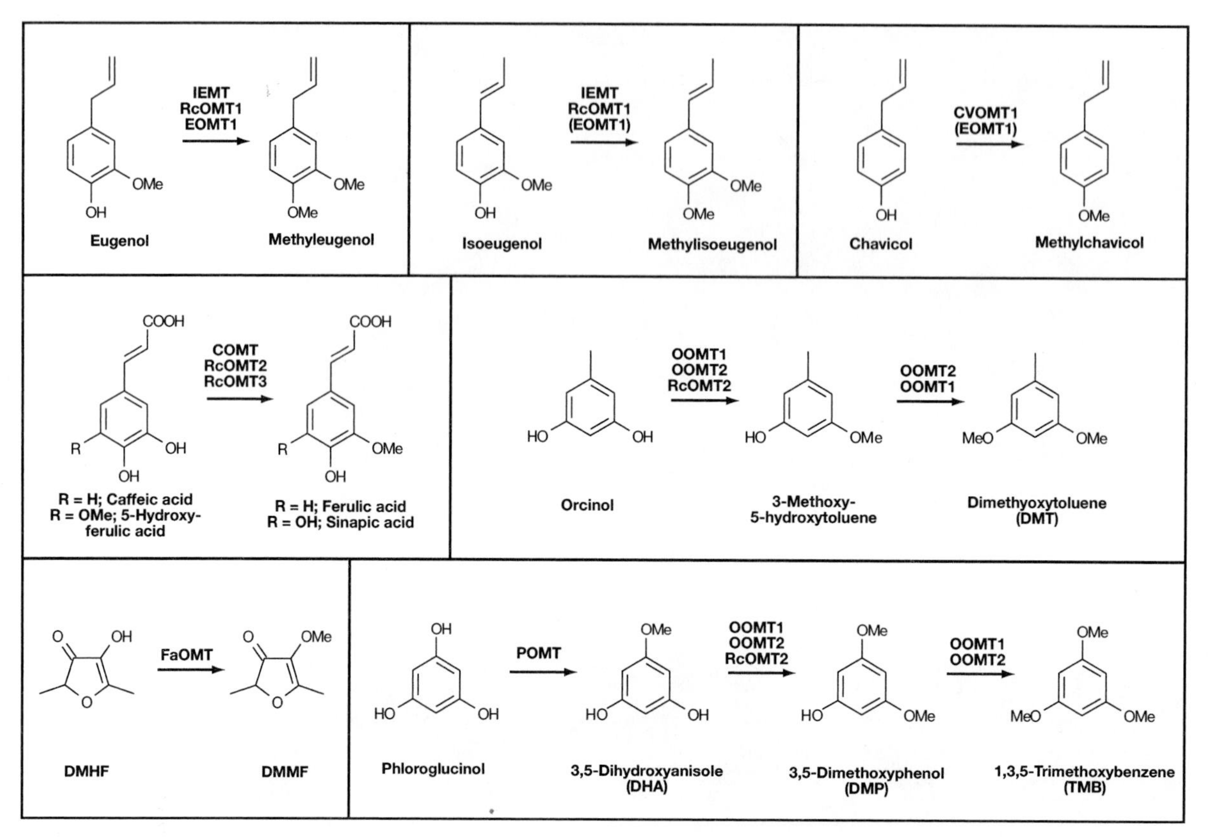

Figure 2

O-methylation reactions catalyzed by SMOMTs involved in scent and flavor production. Abbreviations over the reaction arrows indicate enzymes (described in the text) that are involved in catalyzing these reactions in vivo. Parentheses indicate poor but potential activity with this substrate.

appears to have originated from duplication of this other COMT gene. This appears to be a case of repeated evolution, where OOMT1 and RcOMT2 possess similar functions, but had divergent progenitors within the same gene family. A third COMT, RcOMT3, is also present in this species. However, it possesses a very narrow substrate specificity, being able to use only a small subset of compounds as substrates that are closely related to caffeic acid (100), and requires a methyl or larger aliphatic side chain in the position para to one of the hydroxyls in the ortho diol. And, it is not directly involved in production of volatile compounds.

In addition to TMB and DMT, other methoxylated phenols such as methylisoeugenol and methyleugenol are formed by tea roses. RcOMT1 can catalyze the *para O*-methylation of eugenol and isoeugenol. In this respect, it is similar to IEMT from *Clarkia breweri* and EOMT1 from sweet basil (see below). Unlike RcOMT2, it does not display broad substrate allowance. The same is true for IEMT (92, 93) and EOMT1 (35). However, RcOMT1's relative ratio of activity toward eugenol:isoeugenol (100:75) is different from that observed in IEMT (71:100) and EOMT1 (100:26). This discrepancy in substrate preference is easily explained by the widely separated position of these three enzymes on the phylogenetic tree of the SMOMT family. IEMT clearly evolved recently from *C. breweri* COMT (92, 93) and lies deep within the COMT cluster (red in **Figure 3**). EOMT1 resides within a cluster that contains OOMT1 and OOMT2, although it does not utilize the same substrates. This suggests divergent evolution from a common ancestor for OOMT1 and OOMT2 compared to EOMT1. But RcOMT1 lies outside both of these groups and is more closely related to scoulerine 9-O-methyltransferase, involved in berberine alkaloid biosynthesis (84). Nevertheless, RcOMT1 performs a similar function to EOMT1 and IEMT, demonstrating again the process of repeated evolution. This also supports the contention that protein sequence homology is not a good predictor of true function (27, 70).

One of the most important flavor compounds in strawberries (*Fragaria × ananassa*, Rosaceae) is 2,5-dimethyl-4-methoxy-3(2H)-furanone (DMMF). This compound imparts a subtle caramel flavor that is detectable by the human nose at remarkably low concentrations in the air (9). Nevertheless, several strawberry varieties produce significant amounts of this compound, in addition to a number of esters (see below). The enzyme responsible for catalyzing the *O*-methylation of 2,5-dimethyl-4-hydroxy-3(2H)-furanone (Furaneol®, DMHF) to form DMMF was identified (96) in strawberry and is called FaOMT (AAF28353.1). It is found in the top red cluster in **Figure 3**, deep within the COMT cluster of SMOMTs. Based on its catalytic properties, it is a typical COMT, with broad substrate specificity, and is most active with protocatechuic aldehyde and caffeic acid as substrates (96). However, it can convert DMHF to DMMF, which could either be a new property compared to other COMT enzymes in other plants, or it could be that this substrate has never before been tested with other COMT enzymes. Based on the reported activity toward DMHF, this is a poor substrate. However, this enzyme was expressed at very high level in developing fruit, and the flavor threshold for the product is so low that this may be sufficient to produce enough DMMF to flavor strawberries. This is an example of divergence in function, coupled with an increase in protein expression to facilitate volatile production.

Evolution of Acyltransferases

Based on comparison of floral volatile profiles, the formation of esters is one of the major features that distinguishes tea and hybrid roses from damask roses. These esters are formed from the corresponding alcohols via the action of enzymes that belong to the BAHD family of acyltransferases (79). This family was named for the first four characterized enzymes in this gene family (BEAT, AHCTs, HCBT, and DAT) (81). See St. Pierre & De Luca (81) for a pertinent review discussing more features of this gene family. These enzymes utilize CoA esters of several small molecule acids, particularly acetate, benzoate, and hydroxycinnamoates, as acyl donors (81). **Figure 4*A*** shows examples of reactions catalyzed by members of this gene family involved in scent and flavor production.

Recent years have revealed many more members to this family. There are 88 members of this family in *Arabidopsis*, and 45 known so far from rice. None of these enzymes from rice, and only a few from *Arabidopsis*, have been characterized for biochemical function in the plant. One of these (CHAT, AAF01587.1) uses acetyl-CoA as acyl donor and *cis*-3-hexen-1-ol as acyl acceptor (21). Another *Arabidopsis* acyltransferase characterized so far (HCT, BAB10316.1) is involved in producing precursors for lignification and prefers *p*-coumaroyl-CoA and shikimic acid as substrates (see discussion of sweet basil CST below). Because of the addition of these sequences to the databases, stronger conclusions can now be drawn about potential function of individual enzymes, based on sequence homology. For example, the red cluster in **Figure 5** contains enzymes that generally prefer hydroxycinnamoyl-CoAs as acyl donor. In contrast, the enzymes in the green clusters generally prefer acetyl-CoA and those in the blue clusters benzoyl-CoA as acyl donors. However, although it most closely resembles benzoyl-CoA-dependent acyltransferases, CHAT cannot use benzoyl-CoA as a substrate (21), suggesting again that divergence from common ancestors is an ongoing process and that assigning "putative" function to a

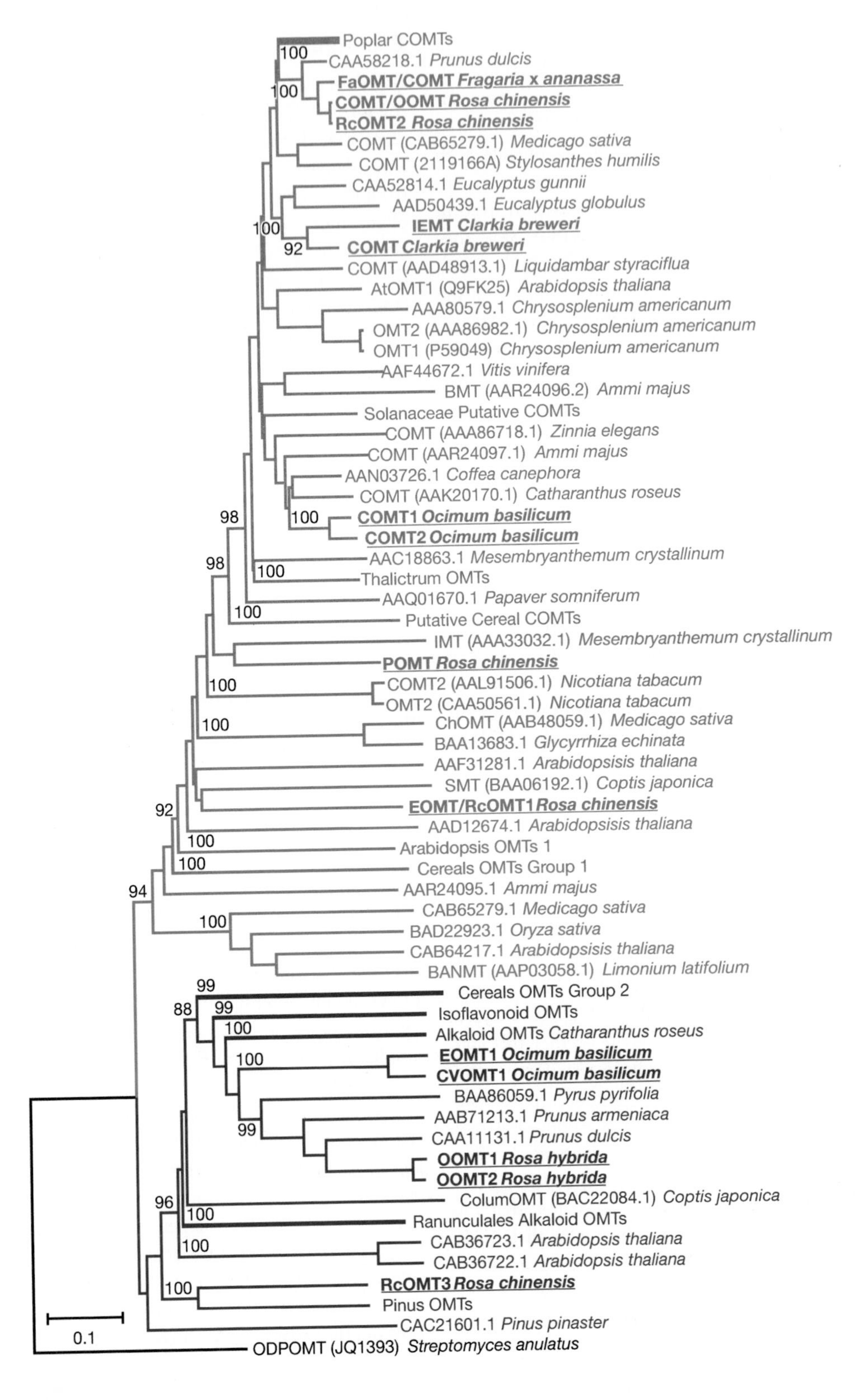

Poplar COMTs
CAA58218.1 Prunus dulcis
FaOMT/COMT Fragaria x ananassa
COMT/OOMT Rosa chinensis
RcOMT2 Rosa chinensis
COMT (CAB65279.1) Medicago sativa
COMT (2119166A) Stylosanthes humilis
CAA52814.1 Eucalyptus gunnii
AAD50439.1 Eucalyptus globulus
IEMT Clarkia breweri
COMT Clarkia breweri
COMT (AAD48913.1) Liquidambar styraciflua
AtOMT1 (Q9FK25) Arabidopsis thaliana
AAA80579.1 Chrysosplenium americanum
OMT2 (AAA86982.1) Chrysosplenium americanum
OMT1 (P59049) Chrysosplenium americanum
AAF44672.1 Vitis vinifera
BMT (AAR24096.2) Ammi majus
Solanaceae Putative COMTs
COMT (AAA86718.1) Zinnia elegans
COMT (AAR24097.1) Ammi majus
AAN03726.1 Coffea canephora
COMT (AAK20170.1) Catharanthus roseus
COMT1 Ocimum basilicum
COMT2 Ocimum basilicum
AAC18863.1 Mesembryanthemum crystallinum
Thalictrum OMTs
AAQ01670.1 Papaver somniferum
Putative Cereal COMTs
IMT (AAA33032.1) Mesembryanthemum crystallinum
POMT Rosa chinensis
COMT2 (AAL91506.1) Nicotiana tabacum
OMT2 (CAA50561.1) Nicotiana tabacum
ChOMT (AAB48059.1) Medicago sativa
BAA13683.1 Glycyrrhiza echinata
AAF31281.1 Arabidopsis thaliana
SMT (BAA06192.1) Coptis japonica
EOMT/RcOMT1 Rosa chinensis
AAD12674.1 Arabidopsis thaliana
Arabidopsis OMTs 1
Cereals OMTs Group 1
AAR24095.1 Ammi majus
CAB65279.1 Medicago sativa
BAD22923.1 Oryza sativa
CAB64217.1 Arabidopsis thaliana
BANMT (AAP03058.1) Limonium latifolium
Cereals OMTs Group 2
Isoflavonoid OMTs
Alkaloid OMTs Catharanthus roseus
EOMT1 Ocimum basilicum
CVOMT1 Ocimum basilicum
BAA86059.1 Pyrus pyrifolia
AAB71213.1 Prunus armeniaca
CAA11131.1 Prunus dulcis
OOMT1 Rosa hybrida
OOMT2 Rosa hybrida
ColumOMT (BAC22084.1) Coptis japonica
Ranunculales Alkaloid OMTs
CAB36723.1 Arabidopsis thaliana
CAB36722.1 Arabidopsis thaliana
RcOMT3 Rosa chinensis
Pinus OMTs
CAC21601.1 Pinus pinaster
ODPOMT (JQ1393) Streptomyces anulatus
0.1
100
100
100
92
98
98
100
100
100
100
100
100
92
100
100
94
100
99
88
99
100
100
99
96
100
100
100

sequence can be a very misleading practice and should only be used as an initial guess in efforts to further characterize the role of the gene in the biology of the plant.

The major esters emitted from Fragrant Cloud flowers are 2-phenylethyl acetate and *cis*-3-hexenyl acetate, although geranyl acetate, citronellyl acetate, hexyl acetate, and 2-hexenyl acetate are also abundant (79). Neryl acetate and benzyl acetate are emitted at low levels. The BAHD family of acyltransferases is the obvious gene family involved in forming these compounds. The only gene identified in rose so far from this family is RhAAT1 (*R. hybrida* alcohol acyltransferase 1). Neither 2-phenylethyl alcohol nor *cis*-3-hexen-1-ol are efficient substrates for recombinant RhAAT1 (79). Instead, geraniol is the best substrate, followed by nerol and 1-octanol. However, assays with protein extracts from the petals indicate that other acyltranferase activities that utilize 2-phenylethyl alcohol, *cis*-3-hexen-1-ol, and 1-hexanol as substrates are more active in vivo than the activity catalyzed by RhAAT1. Thus, other members of this gene family must be involved in forming 2-phenylethyl acetate, *cis*-3-hexenyl acetate, and other esters in Fragrant Cloud flowers.

Of the known BAHD acyltransferases, RhAAT1 is most closely related to an acyltransferase from strawberry, SAAT (AAG13130.1), also in the Rosaceae. In contrast to the more limited substrate specificity of RhAAT1, the strawberry enzyme can utilize a wide variety of acyl acceptors, from methanol up to 1-decanol (2). The best activity is observed with longer chain alcohols, but no activity is seen with the monoterpenoid alcohol linalool. Unfortunately, other monoterpenoid alcohols, such as geraniol, nerol, and citronellol were not evaluated as potential substrates for SAAT (2). Nevertheless, SAAT has a much wider substrate allowance than does RhAAT1. Furthermore, SAAT is only expressed during the final stages of fruit ripening (2), in contrast to RhAAT1, which is expressed only prior to fertilization

Figure 3

Phylogenetic tree of the small molecule OMT (SMOMT) gene family. Genes discussed in the text (where accession numbers are given) are indicated by bold and underline. The red cluster includes COMTs and closely related enzymes, which typically have broad substrate allowances. The blue cluster includes enzymes such as EOMT1 and RcOMT3, which typically have more restricted substrate specificities. The branches in green represent several diverse clusters of enzymes that are distinct from the other two major clusters. The outlier is *O*-demethylpuromycin *O*-methyltransferase, from *Streptomyces anulatus*. The amino acid sequences were aligned and the Neighbor-Joining tree was created and bootstrapped using ClustalX v. 1.83 (87). The tree was visualized using TreeExplorer (85). Bootstrap values (indicated on tree) were visualized using NJplot (68). Several branches were condensed to make the tree legible in this format. The full tree, containing all SMOMTs deposited in the public gene databases, is available in **Figure 1** of the Supplemental Material. (Follow the Supplemental Material link from the Annual Reviews home page at http://www.annualreviews.org.) The condensed branches contain the following accessions: Poplar COMTs (BAA08558.1, AAF63200.1, CAA44006.1, AAF60951.1, BAA08559.1, and AAB68049.1); Solanaceae Putative COMTs (CAA52461.1, AAC17455.1, and AAC78475.1); Thalictrum OMTs (AAD29843.1, AAD29845.1, AAD29841.1, and AAD29844.1); Putative Cereal COMTs (AAD10254.1, BAC54275.1, AAQ24337.1, AAQ24339.1, AAQ24338.1, AAO43609.1, CAA13175.1, AAQ67347.1, BAC99512.1, AAP23942.1, AAD10255.1, AAK68910.1, CAE51883.1, CAE51884.1, AAK68907.1, AAD10253.1, and AAC18623.1); *Arabidopsis* OMTs 1 (AAG51679.1, AAG51676.1, AAG51616.1, BAB09553.1, AAF04440.1, AAF80651.1, AAF80649.1, and AAF80648.1); Cereals OMTs Group 1 (AAC18643.1, CAD39486.2, CAD39487.2, and CAD39344.2); Cereals OMTs Group 2 (AAL31649.1, AAL31646.1, AAA18532.1, AAD10485.1, CAE03691.2, BAB89545.1, BAC07028.1, BAC07035.1, BAD01311.1, BAC99560.1, CAA54616.1, AAC12715.1, and AAO23335.1); Isoflavonoid OMTs (BAC58011.1, BAC58013.1, AAC49856.1, BAC58012.1, AAC49926.1, and AAB88294.1); Alkaloid OMTs *Catharanthus roseus* (AAR02417.1, AAR02419.1, AAR02421.1, AAM97497.1, and AAM97498.1); Ranunculales Alkaloid OMTs (AAP45314.1, AAP45313.1, BAB08005.1, BAB08004.1, and AAP45315.1); Pinus OMTs (AAC49708.1, AAD24001.1, and AAB09044.1).

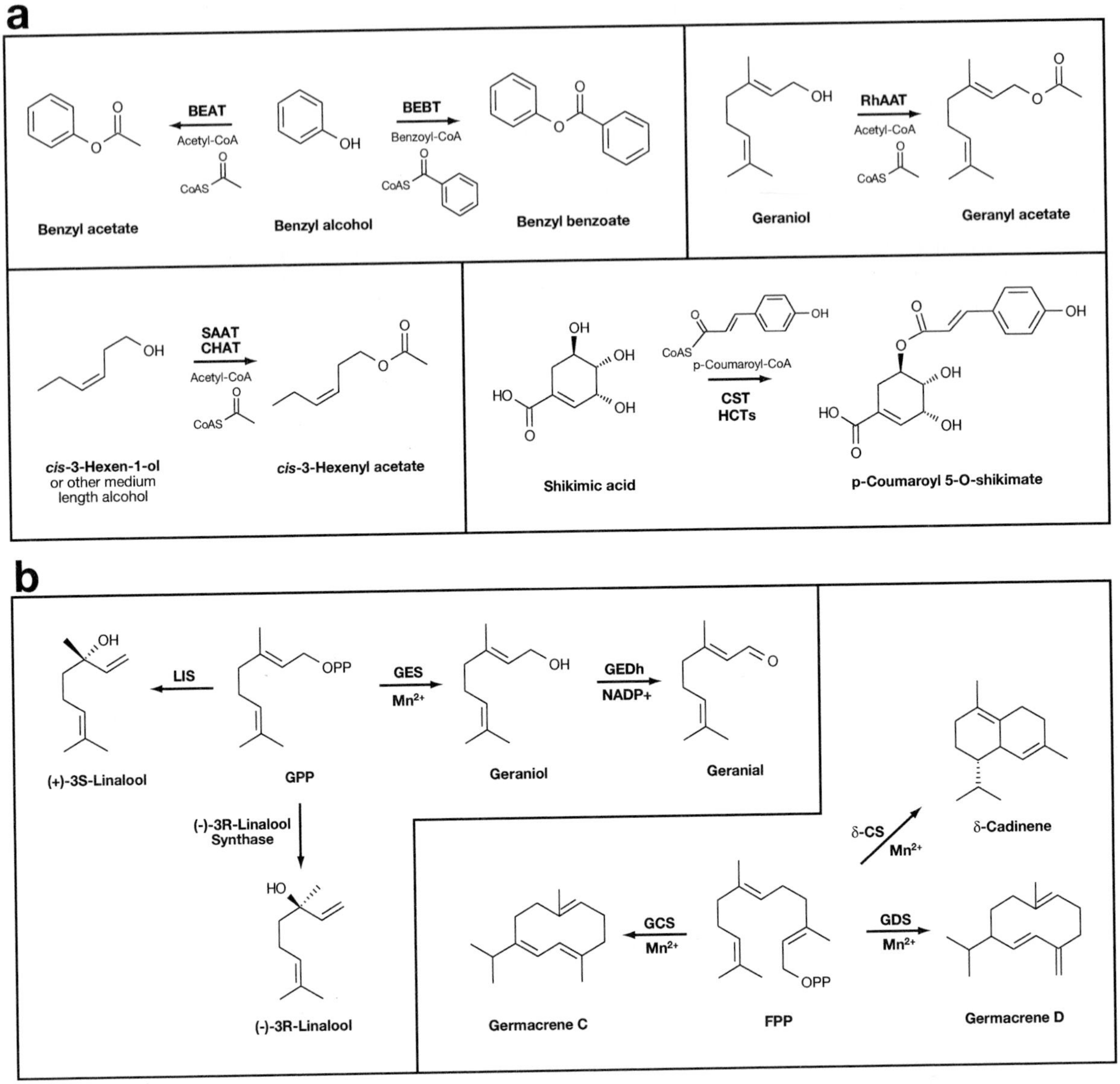

Figure 4

(*A*) Examples of acylations catalyzed by BAHD acyltransferases involved in scent and flavor production. (*B*) Examples of reactions involved in production of terpenoids that are important for scent and flavor. Abbreviations over the reaction arrows indicate enzymes (described in the text) that are involved in catalyzing these reactions in vivo.

(79). Based on sequence similarity, we can conclude that these two genes descended from the same ancestor. Thus, a transcriptional change in gene expression (different stages of development and tissues) has led to differences in function for these two enzymes. Not only are they involved in different biological processes (pollinator attractant versus seed disperser attractant), they now have different substrate specificities.

Evolution of Terpene Synthases

The terpenoids are the most diverse class of plant specialized metabolites, with the sesquiterpenoids being the most diverse group of these. Many flavor and aroma compounds are sesquiterpenoids (e.g., germacrene D, δ-cadinene, and β-cubebene), as are many phytoalexins and phytoanticipins, such as compounds derived from 5-epi-aristolochene (28), (+)-δ-cadinene (16, 17), and vetispiradiene (4). These compounds are produced in the cytosol from farnesyl diphosphate via the action of enzymes known as sesquiterpene synthases, which belong to the greater terpene synthase (TPS) class of enzymes. The terpene synthases cluster generally according to the substrate that they utilize. Thus in **Figure 6**, the top red cluster [the TPS-a group according to nomenclature by Bohlmann et al. (3, 8)] consists of angiosperm enzymes that utilize FPP as a substrate to yield, generally, sesquiterpenoids as products. The green cluster (TPS-b, angiosperms) utilizes GPP (geranyldiphosphate) to produce monoterpenes. Enzymes belonging to the angiosperm TPS-c and TPS-e classes utilize GGPP (geranylgeranyldiphosphate) to produce diterpenes. Also producing diterpenes are members of the gymnosperm TPS-d1 class, which are more similar to the other terpene synthases found in the gymnosperms, even though these produce sesquiterpenes (TPS-d2) or monoterpenes (TPS-d3). Other pertinent reviews were recently published on the evolution of TPS genes (58, 88).

Sequence similarity among TPS genes does not necessarily translate to similarity among their products (7, 8). This is definitely the case for *R. hybrida* germacrene D synthase (RhGDS), isolated from Fragrant Cloud flowers (45). Although this enzyme has greater sequence similarity to cotton (*Gossypium*) (+)-δ-cadinene synthases (**Figure 6**, bootstrap values show this clearly), it catalyzes the synthesis of a sesquiterpene (germacrene D) whose structure is more closely related to the product of tomato germacrene C synthase (18), see **Figure 4*B***. Recombinant RhGDS is specific in the volatile metabolites that it produces. It does not make δ-cadinene from FPP. Fragrant Cloud flowers, however, produce a variety of sesquiterpenoid and monoterpenoid volatile compounds, including δ-cadinene. It will be interesting to see where the genes responsible for forming compounds such as δ-cadinene, nerol, citronellol, and geraniol in rose fit into the TPS tree. Based on the trends seen so far with other TPS enzymes, this is hard to predict from the outset. Only detailed analysis of isolated genes and their encoded proteins will enable the identification of these functions. This trend will become even clearer as we discuss the formation of monoterpenoids in *Clarkia* and sweet basil below.

EVOLUTION OF *CLARKIA BREWERI* SCENT

Clarkia breweri was the first and is one of the best-studied flowers in terms of floral scent evolution. It is a small annual native of California, and it is the only species in the Onagraceae family that emits detectable (by humans or by analytical instruments) floral scent (73). The closely related *C. concinna* is virtually scentless. The major scent components that have been the subject of biochemical and evolutionary investigations in *C. breweri* are methylisoeugenol, methyleugenol, benzylacetate, and linalool which are emitted from petal tissue (26).

Evolution of *O*-Methyltransferases

Methylisoeugenol and methyleugenol are produced by *C. breweri* petals via the action of IEMT (introduced above). The evolution of this enzyme is an interesting story, and research involving this enzyme led to some important conclusions about the creation of new enzymes involved in plant specialized metabolism. IEMT evolved fairly recently via gene duplication of COMT. Wang & Pichersky (92, 93) isolated genes for IEMT (AAC01533.1) and COMT (AAB71141.1) from *C. breweri*, and found that although the proteins were 83% identical at the amino acid sequence level, only seven amino acids played a significant role in substrate specificity. Native (and recombinant) IEMT utilizes isoeugenol and eugenol (which contain no oxygen functionalities at the end of the propenyl side chain of the molecules) efficiently as substrates, but does not methylate caffeic acid or 5-hydroxyferulic acid. The substrate specificity of COMT is the opposite of IEMT, preferring caffeic acid or 5-hydroxyferulic acid. Site-directed mutagenesis demonstrated that swapping these seven amino acids was all that was required to turn IEMT into a (functional) COMT, and vice versa. The results of these investigations provided strong evidence that substrate preference of *O*-methyltransferases (and perhaps other genes involved in plant specialized metabolism) could be controlled by only a few amino acid residues. Furthermore, these results suggested that new OMTs with different substrate specificities could evolve from existing

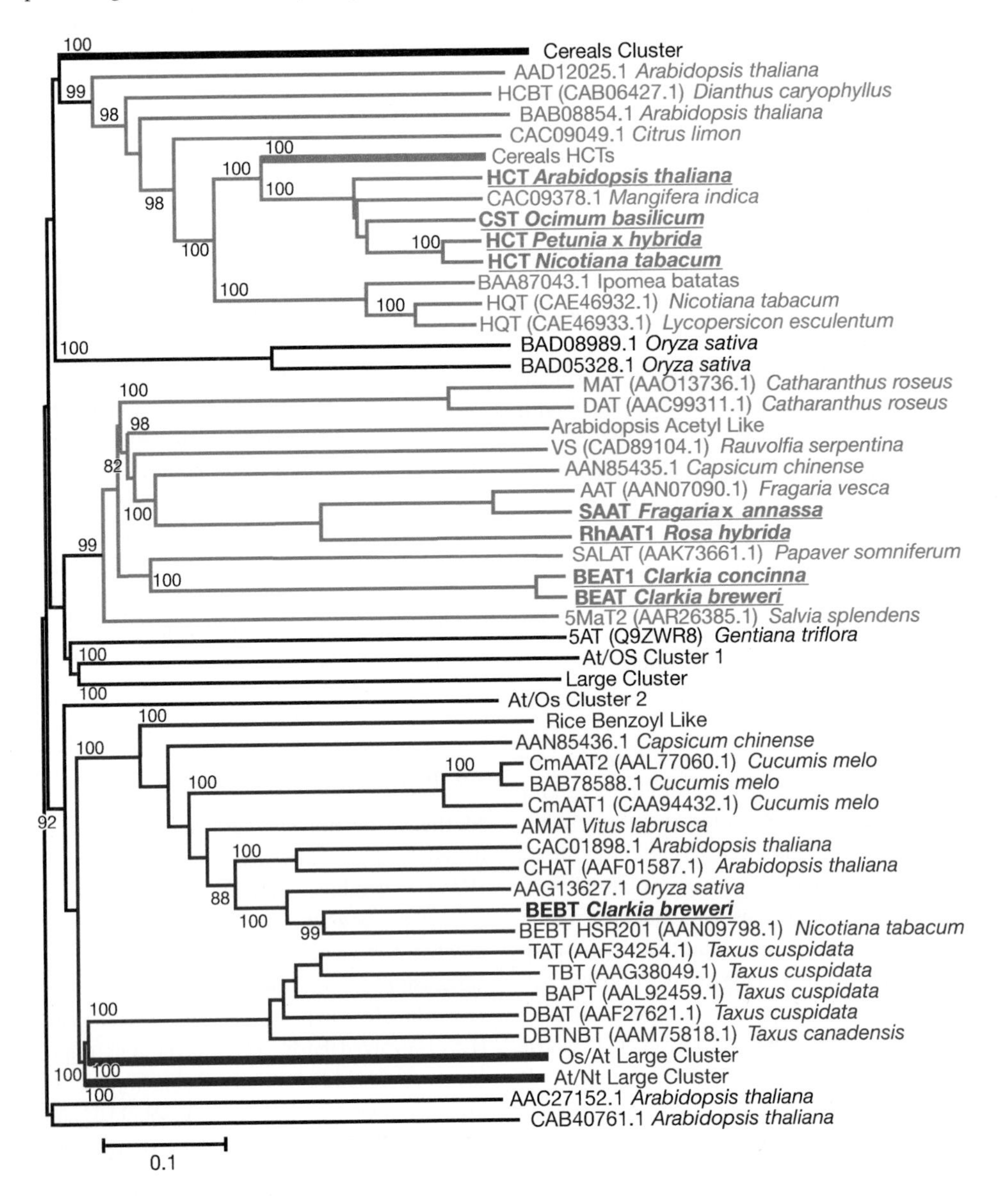

OMTs by mutation of very few amino acids. We return to this later with the discussion of sweet basil OMTs, below. The evolution of *C. breweri* IEMT (as compared to *C. breweri* COMT) is thus an ideal example of gene duplication followed by divergence. Furthermore, as outlined above, when considered with RcOMT1 and EOMT1, IEMT is a perfect example of repeated evolution.

Evolution of Acyltransferases

C. breweri also produces esters in its floral scent, the major volatile esters being benzyl acetate and benzyl benzoate. These compounds are produced via the action of two members of the BAHD family of acyltransferases (81), acetyl-CoA:benzylalcohol *O*-acetyltransferase (BEAT) (26, 63) and benzoyl-CoA:benzyl alcohol benzoyl transferase (BEBT) (21). These genes, although they utilize similar acyl acceptors, are distinct from each other, belonging to two very divergent clusters in the BAHD family tree (see **Figure 5**). The ability of both of these enzymes to utilize benzyl alcohol as the most efficient acyl acceptor substrate is another example of convergent evolution.

Evolution of Terpene Synthases

In addition to methyleugenol, methylisoeugenol, and benzylacetate, *C. breweri* flowers produce large amounts of the monoterpenes linalool and linalool oxide (73). These compounds are produced from GPP by linalool synthases (LIS). Two isoforms of LIS have been isolated from *C. breweri* (AAC49395.1 and AAD19840.1) and one from *C. concinna* (AAD19839.1). These are unusual terpene synthases, derived from a recombination event (20) between a diterpene synthase, similar to *ent*-kaurene synthases, and a monoterpene synthase. They produce (+)-3*S*-linalool as product. Other linalool synthases have been identified in *Arabidopsis* (At1g61680.1), *Mentha aquatica* (AAL99381.1), and *Artemisia annua* (AAF13356.1), and reported to GenBank for *Perilla frutescens* var. *crispa* (AAL38029.1), although the actual function of the latter has not been demonstrated. These enzymes are found in three distinct groups among the TPS-b group of monoterpene synthases (see **Figure 6**). The enzymes from *M. aquatica* and *A. annua* produce (−)-*R*-linalool (19, 52), whereas the *Arabidopsis* enzyme produces (+)-3*S*-linalool

Figure 5

Phylogenetic tree of BAHD acyltransferase gene family. The red cluster includes enzymes that generally prefer hydroxycinnamoyl-CoAs as acyl donor substrate. The green cluster includes enzymes that generally prefer acetyl-CoA. The blue cluster includes enzymes that generally prefer benzoyl-CoA. For details on generation of this figure, please see the **Figure 3** legend. Several branches were condensed to make the tree readable in this format. The full tree, containing all BAHD acyltransferases deposited in the public gene databases, is available in the **Figure 2** of the Supplemental Material. (Follow the Supplemental Material link from the Annual Reviews home page at http://www.annualreviews.org.) The condensed branches contain the following accessions: Cereals Cluster (AAO73071.1, AAO73072.1, CAE03579.1, CAE03578.1, AAP68378.1, and BAC79154.1); Cereals HCTs (BAC78635.1, BAC78633.1, BAC78634.1, BAD19683.1, BAC78636.1, and CAE01632.2); Arabidopsis Acetyl Like (BAB02229.1, CAB10318.1, BAB09047.1, AAF97955.1, BAB08720.1, CAB10319.1, AAO64765.1, BAB01067.1, and AAF97979.1); At/Os Cluster 1 (AAF31274.1, AAC17079.1, BAB09608.1, CAE03656.2, and AAG51274.1); Large Cluster (BAA93453.1, AAL67994.1, BAB10067.1, BAD09615.1, CAB62309.1, CAB62307.1, BAB11280.1, CAB62306.1, BAB10949.1, BAB10950.1, BAD09498.1, BAD09507.1, AAM69843.1, BAD09504.1, BAC22219.1, BAC21404.1, BAB09184.1, CAB62597.1, CAB62599.1, and CAB62598.1); At/Os Cluster 1 (BAB10449.1, CAB62361.1, AAK59460.1, BAB90659.1, T00918, AAD25938.1, AAD45999.1, BAB16338.1, and BAB84471.1); Rice Benzoyl Like (AAG12486.2, AAL75750.1, AAG12484.2, AAG12479.2, AAG12478.2, AAM97746.1, BAC65990.1, AAL01166.1, AAL01169.1, CAE04720.1, BAC65365.1, AAK70914.1, BAC22537.1, and BAB89606.1); Os/At Large Cluster (BAB17110.1, BAB17109.1, BAB03362.1, BAB07968.1, CAB71876.1, CAE02433.2, CAD39633.2, CAE02223.1, AAL83353.1, AAL83355.1, AAG12489.2, AAM08506.1, and AAM08507.1); At/Nt Large Cluster (BAB16426.1, AAF24555.2, AAC23766.1, NP_18,0087.1, CAB61963.1, and CAB87264.1).

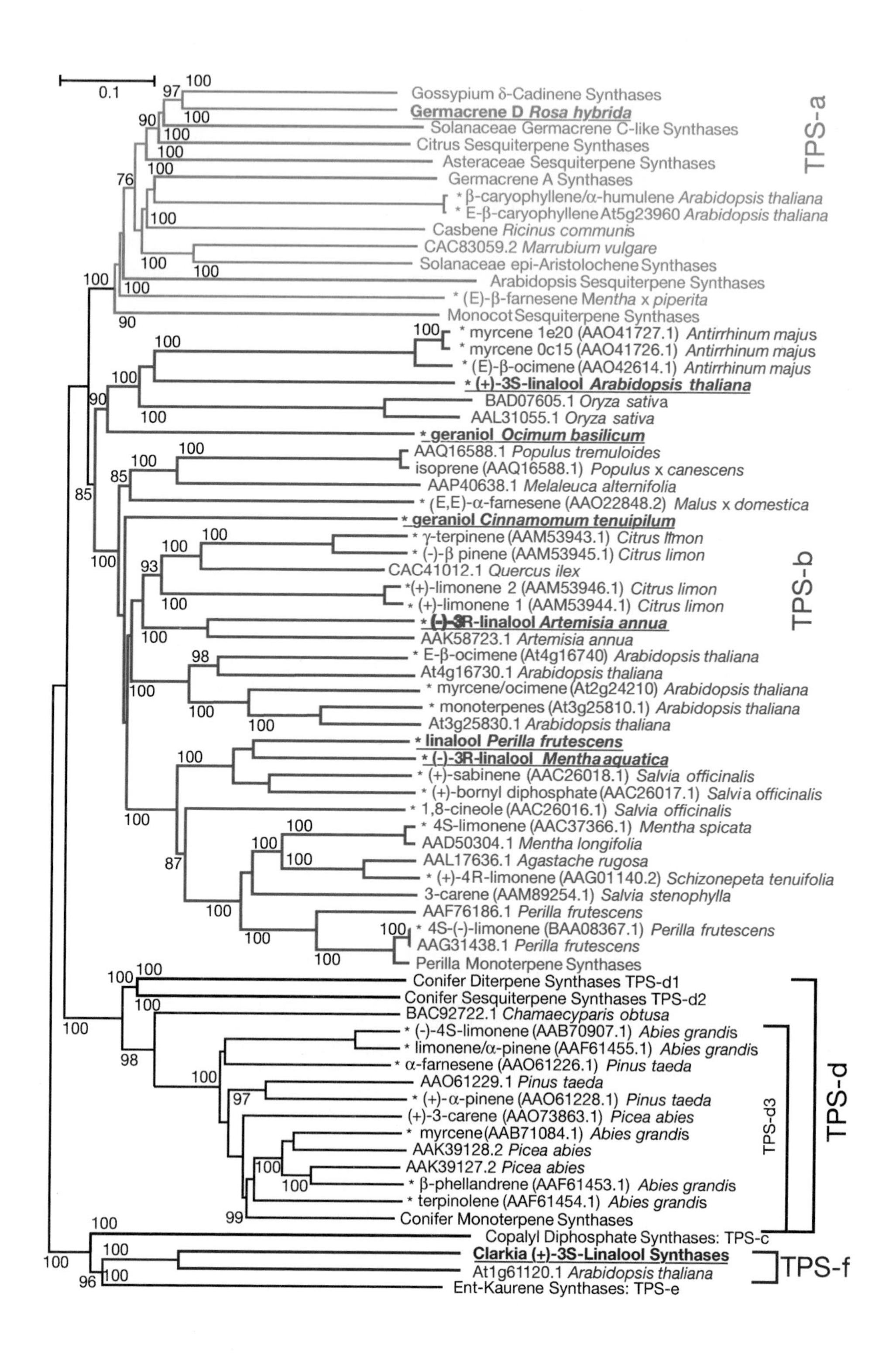
0.1
Gossypium δ-Cadinene Synthases
Germacrene D Rosa hybrida
Solanaceae Germacrene C-like Synthases
Citrus Sesquiterpene Synthases
Asteraceae Sesquiterpene Synthases
Germacrene A Synthases
* β-caryophyllene/α-humulene Arabidopsis thaliana
* E-β-caryophyllene At5g23960 Arabidopsis thaliana
Casbene Ricinus communis
CAC83059.2 Marrubium vulgare
Solanaceae epi-Aristolochene Synthases
Arabidopsis Sesquiterpene Synthases
* (E)-β-farnesene Mentha x piperita
Monocot Sesquiterpene Synthases
TPS-a
* myrcene 1e20 (AAO41727.1) Antirrhinum majus
* myrcene 0c15 (AAO41726.1) Antirrhinum majus
* (E)-β-ocimene (AAO42614.1) Antirrhinum majus
* (+)-3S-linalool Arabidopsis thaliana
BAD07605.1 Oryza sativa
AAL31055.1 Oryza sativa
* geraniol Ocimum basilicum
AAQ16588.1 Populus tremuloides
isoprene (AAQ16588.1) Populus x canescens
AAP40638.1 Melaleuca alternifolia
* (E,E)-α-farnesene (AAO22848.2) Malus x domestica
* geraniol Cinnamomum tenuipilum
* γ-terpinene (AAM53943.1) Citrus limon
* (-)-β pinene (AAM53945.1) Citrus limon
CAC41012.1 Quercus ilex
*(+)-limonene 2 (AAM53946.1) Citrus limon
* (+)-limonene 1 (AAM53944.1) Citrus limon
* (-)-3R-linalool Artemisia annua
AAK58723.1 Artemisia annua
* E-β-ocimene (At4g16740) Arabidopsis thaliana
At4g16730.1 Arabidopsis thaliana
* myrcene/ocimene (At2g24210) Arabidopsis thaliana
* monoterpenes (At3g25810.1) Arabidopsis thaliana
At3g25830.1 Arabidopsis thaliana
* linalool Perilla frutescens
* (-)-3R-linalool Mentha aquatica
* (+)-sabinene (AAC26018.1) Salvia officinalis
* (+)-bornyl diphosphate (AAC26017.1) Salvia officinalis
* 1,8-cineole (AAC26016.1) Salvia officinalis
* 4S-limonene (AAC37366.1) Mentha spicata
AAD50304.1 Mentha longifolia
AAL17636.1 Agastache rugosa
* (+)-4R-limonene (AAG01140.2) Schizonepeta tenuifolia
3-carene (AAM89254.1) Salvia stenophylla
AAF76186.1 Perilla frutescens
* 4S-(-)-limonene (BAA08367.1) Perilla frutescens
AAG31438.1 Perilla frutescens
Perilla Monoterpene Synthases
TPS-b
Conifer Diterpene Synthases TPS-d1
Conifer Sesquiterpene Synthases TPS-d2
BAC92722.1 Chamaecyparis obtusa
* (-)-4S-limonene (AAB70907.1) Abies grandis
* limonene/α-pinene (AAF61455.1) Abies grandis
* α-farnesene (AAO61226.1) Pinus taeda
AAO61229.1 Pinus taeda
* (+)-α-pinene (AAO61228.1) Pinus taeda
(+)-3-carene (AAO73863.1) Picea abies
* myrcene(AAB71084.1) Abies grandis
AAK39128.2 Picea abies
AAK39127.2 Picea abies
* β-phellandrene (AAF61453.1) Abies grandis
* terpinolene (AAF61454.1) Abies grandis
Conifer Monoterpene Synthases
TPS-d3
TPS-d
Copalyl Diphosphate Synthases: TPS-c
Clarkia (+)-3S-Linalool Synthases
At1g61120.1 Arabidopsis thaliana
TPS-f
Ent-Kaurene Synthases: TPS-e

(14), as do the enzymes from *Clarkia*. Remember that 3*R*-linalool and 3*S*-linalool are not the same molecule, they are enantiomers. Therefore, these enzymes should be viewed as catalyzing different reactions. Nevertheless, the *Arabidopsis* enzyme does catalyze the same reaction as *Clarkia* LIS. Thus, we again see repeated evolution. We are also reminded of the importance of rigorously identifying the reaction substrates and products of a given enzyme to reduce confusion regarding its actual function in plants. Other interesting aspects of LIS evolution, such as conserving introns, domain swapping, and clues about the recombination event that led to the production of *Clarkia* LIS, were previously discussed (20, 25).

EVOLUTION OF SWEET BASIL FLAVOR

Sweet basil is one of the most versatile herbs and contains a wide array of terpenoid and phenylpropanoid compounds in abundance, particularly the monoterpenes citral, linalool, 1,8-cineole, geraniol, ocimene, and terpineol, and the phenylpropanoids eugenol, methyleugenol, methylchavicol, and methylcinnamate (42, 44, 62). Short- to medium-chain alkanes, alcohols, and aldehydes, such as nonane, n-heptanol, and undecanal, are also present at varying levels in different basil varieties (37). But these compounds have not been the subject of significant investigations. Many varieties and chemotypic lines of basil have been identified (42, 44,

Figure 6

Phylogenetic tree of the terpene synthase gene family. The red cluster (TPS-a) includes enzymes that generally prefer farnesyl diphosphate as a substrate, i.e., sesquiterpene synthases. The green cluster (TPS-b) includes enzymes that generally prefer geranyl diphosphate as a substrate, i.e., monoterpene synthases. The blue cluster (TPS-f) includes (+)-3*S*-linalool synthases (LIS) from *Clarkia*, and an enzyme from *Arabidopsis* with unknown function but homology to LIS. The black clusters include several diterpenes synthases from angiosperms (TPS-c and TPS-e) and all terpene synthases from gymnosperms (TPS-d). Enzymes preceded by an asterix produce components of flavors or floral scents. For details on generation of this figure, please see the **Figure 3** legend. Several branches were condensed to make the tree readable in this format. The full tree, containing all mono- and sesquiterpene synthases and most di-terpene synthases deposited in the public gene databases, is available in **Figure 3** of the Supplemental Material. (Follow the Supplemental Material link from the Annual Reviews home page at http://www.annualreviews.org.) The condensed branches contain the following accessions: Gossypium δ-Cadinene Synthases (CAA76223.1, AAD51718.1, AAA93064.1, AAF74977.1, AAC12784.1, AAA93065.1, CAD90835.1, and AAB41259.1, CAA77191.1); Solanaceae Germacrene C-like Synthases (AAK95517.1, AAL25826.1, AAG41892.1, AAG41891.1, AAG41890.1, and AAC39431.1); Citrus Sesquiterpene Synthases (AAK54279.1, AAG01339.1, AAQ04608.1, and AAM00426.1); Asteraceae Sesquiterpene Synthases (AAG24640.2, CAC12731.1, CAB94691.1, CAC36896.1, CAE47440.1, CAE47439.1, AAL79181.1, CAC12732.1, CAB56499.1, and AAF80333.1); Germacrene A Synthases (AAM21659.1, AAM11626.1, AAM11627.1, and AAM21658.1); Solanaceae epi-Aristolochene Synthases (AAA19216.1, AAP05761.1, AAO85555.1, AAP05760.1, AAP05762.1, AAP79448.1, AAF21053.1, AAC61260.1, CAA06614.1, AAK15642.1, AAD02270.1, AAD02223.1, AAK15641.1, BAA82092.1, BAA82141.1, AAG09949.1, AAD02268.1, BAA82108.1, and AAD02269.1); Arabidopsis Sesquiterpene Synthases (At1g70080.1, At5g48110.1, At3g14490.1, At3g32030.1, At3g29410.1, At1g33750.1, At1g31950.1, At3g14520.1, At3g14540.1, At5g44630.1, At4g13280.1, At4g15870.1, At2g23230.1, At1g48800.1, At3g29110.1, At3g29190.1, At1g66020.1, and At4g20200.1); Monocot Sesquiterpene Synthases (AAC31570.2, AAR01759.1, BAC20102.1, AAG37841.1, AAL59230.1, BAB63870.1, BAC99543.1, and BAC99549.1); Perilla Monoterpene Synthases (AAG31437.1, AAF65545.1, AAG31435.1, and AAK06663.1); Conifer Diterpene Synthases TPS-d1 (AAK83566.1, AAG02257.1, and AAC24192.1); Conifer Sesquiterpene Synthases (AAL09965.1, AAB05407.1, AAK83561.1, AAC05727.1, AAK39129.2, and AAC05728.1); Conifer Monoterpene Synthases (AAB70707.1, AAO61227.1, AAO61225.1, AAB71085.1, and AAP72020.1); Copalyl Diphosphate Synthases (AAD04292.1, AAD04293.1, T06783, BAB12440.1, AAB87091.1, BAA84918.1, BAB03594.1, NP˙19,2187.1, and BAA95612.1); and Ent-Kaurene Synthases (BAD17672.1, BAD17270.1, AAD34294.1, BAB12441.1, NP_17,8064.1, BAB19275.1, and T09672).

62). Basil flavor compounds are produced and stored in specialized structures on the surface of the leaf, known as peltate glandular trichomes, which can be removed intact from the leaf and studied in isolation (36, 37), as is the case for other members of the mint family (38, 39, 61). Because of this, sweet basil is an ideal system to study the evolution of flavor compounds in plants. Most research on sweet basil flavor compound production, and on the evolution of the genes responsible for production of these compounds, has centered on the phenylpropanoid and terpenoid pathway-derived compounds. As described below, we see in basil examples of changes in active site structure in enzymes and of changes in gene regulation that affect evolution of volatile metabolite production.

Evolution of *O*-Methyltransferases

In addition to eugenol, two of the most important flavor compounds in basil are methyleugenol and methylchavicol. These compounds are derived from eugenol and chavicol, respectively, via the action of specific SMOMTs: eugenol *O*-methyltransferase (EOMT1) and chavicol *O*-methyltransferase (CVOMT1), isolated from the basil variety EMX-1 (35). These enzymes are distinct from COMT. Basil possesses at least two COMT genes, COMT1 (AAD38189.1) and COMT2 (AAD38190), which are localized to a different part of the SMOMT gene tree, and which have completely different activities, similar to COMT from *Clarkia*. Besides the discussion above of EOMT1's relationship to RcOMT1 and IEMT, we can gain more insight into the evolution of OMTs by evaluating this enzyme further, particularly in the context of its relationship to CVOMT1. EOMT1 greatly prefers eugenol over other potential substrates, with chavicol and isoeugenol (the next best substrates) being fourfold less effective as substrates. CVOMT1 is even more discriminatory, with eugenol, the next best substrate, being more than 20-fold less effective as a substrate than chavicol. EOMT1 and CVOMT1 are very similar to each other, sharing 90% sequence identity. However, the difference in activity between these two enzymes resides in the substitution of a single nucleotide. This results in substitution of a single amino acid residue in the active site, a serine in EOMT1, being converted to a phenylalanine in CVOMT1. Because this difference represents a C to T transition (comparing EOMT1 to CVOMT1, which is the most common type of mutation in DNA), it was concluded that CVOMT1 evolved from EOMT1 (35). Site-directed mutagenesis that swapped only this one amino acid in these two enzymes led to enzymes with the opposing substrate preferences. Thus, the CVOMT1 F260S mutant behaves as EOMT1, and the EOMT1 S261F mutant had the same limited substrate preference as CVOMT1, with only chavicol serving as a reasonable substrate. This comparison of EOMT1 to CVOMT1 illustrates gene duplication followed by divergence, in addition to the repeated evolution discussed above for the comparison of EOMT1 to RcOMT1 and IEMT.

Evolution of Acyltransferases

Formation of *p*-coumaroyl shikimate is required for production of methoxylated phenylpropanoids (e.g., eugenol and methyleugenol) in basil glandular trichomes (33), and in other plants as well, such as *Arabidopsis* and tobacco (30, 31, 48, 49, 76). The formation of this intermediate is catalyzed by an enzyme belonging to the BAHD family of acyltransferases, called *p*-coumaroyl-CoA:shikimic acid *p*-coumaroyl transferase (or CST for short) in sweet basil (33). The enzyme in *Arabidopsis*, tobacco, and other plants is called HCT because of the broader substrate specificity seen in this enzyme in these other species. In tobacco, for example, HCT (CAD47830.1) can utilize both *p*-coumaroyl-CoA and caffeoyl-CoA as reasonable acyl donors, although the former is preferred (49). The same is true for the HCT enzymes from *Arabidopsis* (BAB10316.1) and petunia (S. Hunter, J. Kapteyn, N. Dudareva & D.R. Gang, unpublished). Quinate, in addition to shikimate, can also act as acyl acceptor for these enzymes. However, the basil CST enzyme has a much stronger preference for *p*-coumaroyl-CoA as acyl donor, and it has a strict requirement of shikimate as acyl acceptor. No detectable activity can be seen with other potential substrates (33). Basil varieties that produce eugenol and methyleugenol actively express CST in the peltate glandular trichomes, the site of synthesis of the basil flavor compounds (37). However, basil varieties that produce high levels of methylchavicol but no methyleugenol or eugenol do not have detectable CST activity (or mRNA expression) in the glandular trichomes, although this activity is present in whole leaf tissue, stems, and roots (S. Hunter, J. Kapteyn, & D. R Gang, unpublished). Thus, differential production of specific metabolites in these basil varieties (eugenol/methyleugenol versus methylchavicol) is due to differential expression of the branch point enzyme CST in the pathway. This is an example of the evolution of gene expression (70), caused, presumably, by a change in a regulatory protein, such as a transcription factor.

In contrast to many of the other enzymes discussed in this paper, CST does not reside at or near the end of the biosynthetic pathway to the volatile metabolite in question. Instead, it resides quite a distance upstream. For this change to have an impact other than complete loss of production of volatiles from the pathway, enzymes in the pathway downstream from the mutation must be able to utilize the newly accumulating intermediate (*p*-coumaroyl-CoA, or derivative thereof) with reasonable efficiency. Thus, the "eugenol synthase" enzyme must have the ability to utilize such an intermediate when the feruloylated substrate is not available. As discussed above, EOMT1 possesses the ability to *O*-methylate chavicol. An increase in chavicol concentration, which would be toxic to the cell, would increase selection on this enzyme for production of a more efficient catalyst or for an increase in enzyme concentration. In fact, CVOMT transcripts are among the most abundant in EST databases from basil (35, 37). This may explain the provenance of CVOMT1.

Evolution of Terpene Synthases

In addition to the phenylpropanoid-derived aroma molecules, basil contains significant levels of mono- and sesquiterpenoids, which contribute greatly to the flavor of some varieties. The best known of these are the lemon basils, particularly the popular Sweet Dani cultivar (62). This variety contains high levels of citral, a mixture of neral and geranial, which is responsible for the strong lemony scent and flavor. Citral is produced from geraniol, as indicated in **Figure 4*B***, via the action of an alcohol dehydrogenase (Y. Iijima & E. Pichersky, personal communication). This requires the formation of geraniol. Two potential routes to geraniol from geranyl diphosphate (GPP) are feasible. The first involves the action of a pyrophosphatase, which hydrolyzes GPP. The second route involves a TPS, of the monoterpene synthase type. The latter turned out to be the case (51). Basil geraniol synthase (AAR11765.1) resides in the top cluster of the TPS-b group in **Figure 6**. A second geraniol synthase (CAD29734) from *Cinnamomum tenuipilum* has been reported in GenBank. Because these two genes belong to distinct subgroups of the TPS-b cluster, if this second enzyme catalyzes the same reaction, then we again have an example of repeated evolution.

EVOLUTION OF VOLATILE COMPOUND EMISSION

The mechanisms that lead to metabolite volatilization/emission in plants are still not well understood. It is possible that many volatile compounds, due to their hydrophobic nature, diffuse readily out of plant tissues (hydrophilic places) and into the air. However, most plant tissues are covered with impermeable cuticles and epiculticular waxes. This is even the case for tissues such as floral parts that are well known for their role in emitting volatile compounds (40). Furthermore, many plant specialized metabolites, such as those that accumulate under the cuticular membrane of peltate glandular trichomes in the mint family (37–39, 61), are hydrophobic and volatile. However, they often remain in these glandular oil sacs until the cuticle is burst by abrasion (37). Passive diffusion is not sufficient to explain the changes in emission profiles that have been observed in many flowers, where several compounds that are sufficiently volatile to be released into the atmosphere, such as benzyl alcohol, benzylbenzoate, phenylethanol, phenylethyl benzoate, isoeugenol, and eugenol, accumulate to significant levels in the floral tissue, but are not observed in the emitted floral scent (6). Obviously, something either retains these molecules within the tissue or their emission requires the action of some component to facilitate their emission (presumably a transporter protein or something similar). The former is not likely because similar molecules (methylbenzoate, benzaldehyde, and phenylacetaldehyde) are emitted as the main constituents of petunia floral scent (6). Thus, the latter is the case. Several proteins, such as lipid transfer proteins and ABC transporters (57, 60), have been hypothesized to be involved in the secretion/emission of hydrophobic molecules from plant tissues. However, the exact role that these proteins play in regulating emission has not been determined, and other proteins/mechanisms may also be involved. Thus, another factor that contributes to the evolution of floral scents is the evolution of the ability to emit a given metabolite from the flower tissue into the surrounding environment. This should be an exciting field of study in the foreseeable future.

ACKNOWLEDGMENTS

I wish to thank Eran Pichersky and Natalia Dudareva for many discussions on this topic. The National Science Foundation Plant Genome Program and Metabolic Biochemistry Program and the United States Department of Agriculture Plant Biochemistry Program supported research in this laboratory.

LITERATURE CITED

1. Adams S, Weidenborner M. 1996. Mycelial deformations of *Cladosporium herbarum* due to the application of eugenol or carvacrol. *J. Essent. Oil. Res.* 8:535–40
2. Aharoni A, Keizer LCP, Bouwmeester HJ, Sun ZK, Alvarez-Huerta M, et al. 2000. Identification of the SAAT gene involved in strawberry flavor biogenesis by use of DNA microarrays. *Plant Cell* 12:647–61
3. Aubourg S, Lecharny A, Bohlmann J. 2002. Genomic analysis of the terpenoid synthase (AtTPS) gene family of Arabidopsis thaliana. *Mol. Gen. Genomics* 267:730–45
4. Back K, Chappell J. 1995. Cloning and bacterial expression of a sesquiterpene cyclase from *Hyoscyamus muticus* and its molecular comparison to related terpene cyclases. *J. Biol. Chem.* 270:7375–81
5. Bender J. 2004. DNA methylation and epigenetics. *Ann. Rev. Plant Biol.* 55:41–68
6. Boatright J, Negre F, Chen X, Kish CM, Wood B, et al. 2004. Understanding in vivo benzenoid metabolism in petunia petal tissue. *Plant Physiol.* 135:1993–2011
7. Bohlmann J, Martin D, Oldham NJ, Gershenzon J. 2000. Terpenoid secondary metabolism in *Arabidopsis thaliana*: cDNA cloning, characterization, and functional expression of a myrcene/(*E*)-β-ocimene synthase. *Arch. Biochem. Biophys.* 375:261–69
8. Bohlmann J, Meyer-Gauen G, Croteau R. 1998. Plant terpenoid synthases: molecular biology and phylogenetic analysis. *Proc. Nat. Acad. Sci. USA* 95:4126–33
9. Bood KG, Zabetakis I. 2002. The biosynthesis of strawberry flavor (II): biosynthetic and molecular biology studies. *J. Food Sci.* 67:2–8

10. Bullerman LB, Lieu FY, Seier SA. 1977. Inhibition of growth and aflatoxin production by cinnamon and clove oils. *J. Food Sci.* 42:1107–9, 16
11. Chandler VL, Stam M. 2004. Chromatin conversations: mechanisms and implications of paramutation. *Nat. Rev. Genet.* 5:532–44
12. Chandler VL, Stam M, Sidorenko LV. 2002. Long-distance cis and trans interactions mediate paramutation. *Adv. Med. Incorp. Molec. Genet. Med.* 46:215–42
13. Chen F, Ro D-K, Petri J, Gershenzon J, Bohlmann J, et al. 2004. Characterization of a root-specific Arabidopsis terpene synthase responsible for the formation of the volatile monoterpene 1,8-cineole. *Plant Physiol.* 135:1956–66
14. Chen F, Tholl D, D'Auria JC, Farooq A, Pichersky E, Gershenzon J. 2003. Biosynthesis and emission of terpenoid volatiles from Arabidopsis flowers. *Plant Cell* 15:481–94
15. Chen SX, Glawischnig E, Jorgensen K, Naur P, Jorgensen B, et al. 2003. CYP79F1 and CYP79F2 have distinct functions in the biosynthesis of aliphatic glucosinolates in Arabidopsis. *Plant J.* 33:923–37
16. Chen XY, Chen Y, Heinstein P, Davisson VJ. 1995. Cloning, expression, and characterization of (+)-delta-cadinene synthase: a catalyst for cotton phytoalexin biosynthesis. *Arch. Biochem. Biophys.* 324:255–66
17. Chen XY, Wang MS, Chen Y, Davisson VJ, Heinstein P. 1996. Cloning and heterologous expression of a second (+)-delta-cadinene synthase from *Gossypium arboreum*. *J. Nat. Prod.* 59:944–51
18. Colby S, Crock J, Dowdle-Rizzo B, Lemaux P, Croteau R. 1998. Germacrene C synthase from *Lycopersicon esculentum* cv. VFNT cherry tomato: cDNA isolation, characterization, and bacterial expression of the multiple product sesquiterpene cyclase. *Proc. Nat. Acad. Sci. USA* 95:2216–21
19. Crowell AL, Williams DC, Davis EM, Wildung MR, Croteau R. 2002. Molecular cloning and characterization of a new linalool synthase. *Arch. Biochem. Biophys.* 405:112–21
20. Cseke L, Dudareva N, Pichersky E. 1998. Structure and evolution of linalool synthase. *Mol. Biol. Evol.* 15:1491–98
21. D'Auria JC, Chen F, Pichersky E. 2002. Characterization of an acyltransferase capable of synthesizing benzylbenzoate and other volatile esters in flowers and damaged leaves of *Clarkia breweri*. *Plant Physiol.* 130:466–76
22. Davies KM, Schwinn KE. 2003. Transcriptional regulation of secondary metabolism. *Funct. Plant Biol.* 30:913–25
23. Dobson HEM, Bergstrom G. 2000. The ecology and evolution of pollen odors. *Plant Syst. Evol.* 222:63–87
24. Dobson HEM, Groth I, Bergstrom G. 1996. Pollen advertisement: chemical contrasts between whole-flower and pollen odors. *Am. J. Bot.* 83:877–85
25. Dudareva N, Cseke L, Blanc VM, Pichersky E. 1996. Evolution of floral scent in *Clarkia*: novel patterns of S-linalool synthase gene expression in the *C. breweri* flower. *Plant Cell* 8:1137–48
26. Dudareva N, D'Auria JC, Nam KH, Raguso RA, Pichersky E. 1998. Acetyl-CoA:benzylalcohol acetyltransferase—an enzyme involved in floral scent production in *Clarkia breweri*. *Plant J.* 14:297–304
27. Dudareva N, Pichersky E, Gershenzon J. 2004. Biochemistry of plant volatiles. *Plant Physiol.* 135:1893–902
28. Facchini PJ, Chappell J. 1992. Gene family for an elicitor-induced sesquiterpene cyclase in tobacco. *Proc. Nat. Acad. Sci. USA* 89:11088–92

29. Flament I, Debonneville C, Furrer A. 1993. Volatile constituents of roses. In *Bioactive Volatile Compounds from Plants*, ed. R Teranishi, R Buttery, H Sugisawa, pp. 269–81. Washington, DC: Am. Chem. Soc.
30. Franke R, Hemm MR, Denault JW, Ruegger MO, Humphreys JM, Chapple C. 2002. Changes in secondary metabolism and deposition of an unusual lignin in the ref8 mutant of Arabidopsis. *Plant J.* 30:47–59
31. Franke R, Humphreys JM, Hemm MR, Denault JW, Ruegger MO, et al. 2002. The Arabidopsis REF8 gene encodes the 3-hydroxylase of phenylpropanoid metabolism. *Plant J.* 30:33–45
32. Fujita M, Gang DR, Davin LB, Lewis NG. 1999. Recombinant pinoresinol-lariciresinol reductases from western red cedar (*Thuja plicata*) catalyze opposite enantiospecific conversions. *J. Biol. Chem.* 274:618–27
33. Gang DR, Beuerle T, Ullmann P, Werck-Reichhart D, Pichersky E. 2002. Differential production of meta hydroxylated phenylpropanoids in sweet basil peltate glandular trichomes and leaves is controlled by the activities of specific acyltransferases and hydroxylases. *Plant Physiol.* 130:1536–44
34. Gang DR, Dinkova-Kostova AT, Davin LB, Lewis NG. 1997. Phylogenetic links in plant defense systems: lignans, isoflavonoids and their reductases. *ACS Symp. Ser.* 658:58–89
35. Gang DR, Lavid N, Zubieta C, Chen F, Beuerle T, et al. 2002. Characterization of phenylpropene *O*-methyltransferases from sweet basil: facile change of substrate specificity and convergent evolution within a plant *O*-methyltransferase family. *Plant Cell* 14:505–19
36. Gang DR, Simon J, Lewinsohn E, Pichersky E. 2002. Peltate glandular trichomes of *Ocimum basilicum* L. (Sweet Basil) contain high levels of enzymes involved in the biosynthesis of phenylpropenes. *J. Herbs Spices Med. Plants* 9:189–95
37. Gang DR, Wang J, Dudareva N, Nam KH, Simon JE, et al. 2001. An investigation of the storage and biosynthesis of phenylpropenes in sweet basil. *Plant Physiol.* 125:539–55
38. Gershenzon J, Maffei M, Croteau R. 1989. Biochemical and histochemical localization of monoterpene biosynthesis in the glandular trichomes of spearmint (*Mentha spicata*). *Plant Physiol.* 89:1351–57
39. Gershenzon J, McCaskill D, Rajaonarivony JIM, Mihaliak C, Karp F, Croteau R. 1992. Isolation of secretory cells from plant glandular trichomes and their use in biosynthetic studies of monoterpenes and other gland products. *Anal. Biochem.* 200:130–38
40. Goodwin SM, Kolosova N, Kish CM, Wood KV, Dudareva N, Jenks MA. 2003. Cuticle characteristics and volatile emissions of petals in *Antirrhinum majus*. *Physiol. Plant.* 117:435–43
41. Grammer K, Fink B, Moller AP, Thornhill R. 2003. Darwinian aesthetics: sexual selection and the biology of beauty. *Biol. Rev.* 78:385–407
42. Grayer R, Kite G, Goldstone F, Bryan S, Paton S, Putievsky E. 1996. Infraspecific taxonomy and essential oil chemotypes in sweet basil, *Ocimum basilicum*. *Phytochemistry* 43:1033–39
43. Grossman J. 1993. Botanical pesticides in Africa. *IPM Pract.* 15:1–9
44. Gupta S. 1994. Genetic analysis of some chemotypes in *Ocimum basilicum* var. *glabratum*. *Plant Breed.* 112:135–40
45. Guterman I, Shalit M, Menda N, Piestun D, Dafny-Yelin M, et al. 2002. Rose scent: genomics approach to discovering novel floral fragrance-related genes. *Plant Cell* 14:2325–38
46. Halkier BA, Hansen CH, Mikkelsen MD, Naur P, Wittstock U. 2002. The role of cytochromes P450 in biosynthesis and evolution of glucosinolates. *Recent Adv. Phytochem.* 36:223–48
47. Harrewijn P, Minks AK, Mollema C. 1995. Evolution of plant volatile production in insect-plant relationships. *Chemoecology* 5/6:55–73

48. Hoffmann L, Besseau S, Geoffroy P, Ritzenthaler C, Meyer D, et al. 2004. Silencing of hydroxycinnamoyl-coenzyme A shikimate/quinate hydroxycinnamoyltransferase affects phenylpropanoid biosynthesis. *Plant Cell* 16:1446–65
49. Hoffmann L, Maury S, Martz F, Geoffroy P, Legrand M. 2003. Purification, cloning, and properties of an acyltransferase controlling shikimate and quinate ester intermediates in phenylpropanoid metabolism. *J. Biol. Chem.* 278:95–103
50. Ibrahim RK, Muzac I. 2000. The methyltransferase gene superfamily: a tree with multiple branches. *Recent Adv. Phytochem.* 34:349–84
51. Iijima Y, Gang DR, Fridman E, Lewinsohn E, Pichersky E. 2004. Characterization of geraniol synthase from the peltate glands of sweet basil. *Plant Physiol.* 134:370–79
52. Jia JW, Crock J, Lu S, Croteau R, Chen XY. 1999. (3*R*)-linalool synthase from *Artemisia annua* L.: cDNA isolation, characterization, and wound induction. *Arch. Biochem. Biophys.* 372:143–49
53. Kahn RA, Durst F. 2000. Function and evolution of plant cytochrome P450. *Recent Adv. Phytochem.* 34:151–89
54. Kaori S, Maeda T, Arimura G, Ozawa R, Shimoda T, Takabayashi J. 2002. Functions of plant infochemicals in tritrophic interactions between plants, herbivores and carnivorous natural enemies. *Jap. J. Appl. Entomol. Zool.* 46:117–33
55. Karapinar M, Aktug S. 1987. Inhibition of foodborne pathogens by thymol, eugenol, menthol and anethole. *Int. J. Food Micro.* 4:161–66
56. Kerschen A, Napoli CA, Jorgensen RA, Muller AE. 2004. Effectiveness of RNA interference in transgenic plants. *FEBS Lett.* 566:223–28
57. Kunst L, Samuels AL. 2003. Biosynthesis and secretion of plant cuticular wax. *Prog. Lipid Res.* 42:51–80
58. Lange BM, Ghassemian M. 2003. Genome organization in Arabidopsis thaliana: a survey for genes involved in isoprenoid and chlorophyll metabolism. *Plant Mol. Biol.* 51:925–48
59. Lavid N, Wang J, Shalit M, Guterman I, Bar E, et al. 2002. O-methyltransferases involved in the biosynthesis of volatile phenolic derivatives in rose petals. *Plant Physiol.* 129:1899–907
60. Martinoia E, Klein M, Geisler M, Bovet L, Forestier C, et al. 2002. Multifunctionality of plant ABC transporters—more than just detoxifiers. *Planta* 214:345–55
61. McCaskill D, Gershenzon J, Croteau R. 1992. Morphology and monoterpene biosynthetic capabilities of secretory cell clusters isolated from glandular trichomes of peppermint (*Mentha piperita* L.). *Planta* 187:445–54
62. Morales MR, Simon JE. 1997. 'Sweet Dani': a new culinary and ornamental lemon basil. *Hortscience* 32:148–49
63. Nam KH, Dudareva N, Pichersky E. 1999. Characterization of benzylalcohol acetyltransferases in scented and non-scented Clarkia species. *Plant Cell Physiol.* 40:916–23
64. Naur P, Petersen BL, Mikkelsen MD, Bak S, Rasmussen H, et al. 2003. CYP83A1 and CYP83B1, two nonredundant cytochrome P450 enzymes metabolizing oximes in the biosynthesis of glucosinolates in Arabidopsis. *Plant Physiol.* 133:63–72
65. Noel JD, Dixon RA, Pichersky E, Zubieta C, Ferrer J. 2003. Structural, functional, and evolutionary basis for methylation of plant small molecules. In *Recent Advances in Phytochemistry (Integrative Phytochemistry: From Ethnobotany to Molecular Ecology)*, ed. J Romeo, pp. 37–58. Oxford: Elsevier
66. Obeng-Ofori D, Reichmuth C. 1997. Bioactivity of eugenol, a major component of essential oil of *Ocimum suave* (Wild.) against four species of stored-product Coleoptera. *Int. J. Pest. Manag.* 43:89–94

67. Ober D, Hartmann T. 1999. Homospermidine synthase, the first pathway-specific enzyme of pyrrolizidine alkaloid biosynthesis, evolved from deoxyhypusine synthase. *Proc. Nat. Acad. Sci. USA* 96:14777–82
68. Perrière G, Gouy M. 1996. WWW-Query: an on-line retrieval system for biological sequence banks. *Biochimie* 78:364–69
69. Pichersky E, Dudareva N, Wang J, Cseke L, Lewinsohn E. 1999. Biosynthesis of scent and flavor compounds. *Curr. Plant Sci. Biotech. Agric.* 36:601–4
70. Pichersky E, Gang DR. 2000. Genetics and biochemistry of secondary metabolites in plants: an evolutionary perspective. *Trends Plant Sci.* 5:439–45
71. Piechulla B, Pott MB. 2003. Plant scents—mediators of inter- and intraorganismic communication. *Planta* 217:687–89
72. Qi X, Bakht S, Leggett M, Maxwell C, Melton R, Osbourn A. 2004. A gene cluster for secondary metabolism in oat: implications for the evolution of metabolic diversity in plants. *Proc. Nat. Acad. Sci. USA* 101:8233–38
73. Raguso RA, Pichersky E. 1995. Floral volatiles from *Clarkia breweri* and *C. concinna* (Onagraceae): recent evolution of floral scent and moth pollination. *Plant Syst. Evol.* 194:55–67
74. Scalliet G, Journot N, Jullien F, Baudino S, Magnard JL, et al. 2002. Biosynthesis of the major scent components 3,5-dimethoxytoluene and 1,3,5-trimethoxybenzene by novel rose *O*-methyltransferases. *FEBS Lett.* 523:113–18
75. Schiestl FP, Ayasse M. 2002. Do changes in floral odor cause speciation in sexually deceptive orchids? *Plant Syst. Evol.* 234:111–19
76. Schoch G, Goepfert S, Morant M, Hehn A, Meyer D, et al. 2001. CYP98A3 from *Arabidopsis thaliana* is a 3′-hydroxylase of phenolic esters, a missing link in the phenylpropanoid pathway. *J. Biol. Chem.* 276:36566–74
77. Schuler MA. 1996. Plant cytochrome P450 monooxygenases. *Crit. Rev. Plant Sci.* 15:235–84
78. Evans GB, ed. 1974. The Tragedy of Romeo and Juliet, Act II, Scene II, Lines 43–44. In *The Riverside Shakespeare*, pp. 1058–99. Boston: Houghton Mifflin
79. Shalit M, Guterman I, Volpin H, Bar E, Tamari T, et al. 2003. Volatile ester formation in roses. Identification of an acetyl-coenzyme A: geraniol/citronellol acetyltransferase in developing rose petals. *Plant Physiol.* 131:1868–76
80. Sisk C, Shorey H, Gerber R, Gaston L. 1996. Semiochemicals that disrupt foraging by the Argentine ant (Hymenoptera: Formicidae): laboratory bioassays. *J. Econ. Entom.* 89:381–85
81. St-Pierre B, De Luca V. 2000. Evolution of acyltransferase genes: origin and diversification of the BAHD superfamily of acyltransferases involved in secondary metabolism. *Recent Adv. Phytochem.* 34:285–315
82. Stam M, Belele C, Dorweiler JE, Chandler VL. 2002. Differential chromatin structure within a tandem array 100 kb upstream of the maize b1 locus is associated with paramutation. *Genes Dev.* 16:1906–18
83. Steeghs M, Bais HP, Gouw JD, Goldan P, Kuster W, et al. 2004. Proton-transfer-reaction mass spectrometry as a new tool for real time analysis of root-secreted volatile organic compounds in Arabidopsis. *Plant Physiol.* 135:47–58
84. Takeshita N, Fujiwara H, Mimura H, Fitchen JH, Yamada Y, Sato F. 1995. Molecular-cloning and characterization of *S*-adenosyl-L-methionine-scoulerine-9-*O*-methyltransferase from cultured cells of *Coptis japonica*. *Plant Cell Physiol.* 36:29–36
85. Tamura K. 1999. TreeExplorer. TMU Evolutionary Genetics Laboratory. (http://evolgen.biol.metro-u.ac.jp/TE/TE_man.html)
86. Theis N, Lerdau M. 2003. The evolution of function in plant secondary metabolites. *Int. J. Plant Sci.* 164:S93–S102

87. Thompson JD, Gibson TJ, Plewniak F, Jeanmougin F, Higgins DG. 1997. The ClustalX windows interface: flexible strategies for multiple sequence alignment aided by quality analysis tools. *Nuclic Acids Res.* 25:4876–82
88. Trapp SC, Croteau RB. 2001. Genomic organization of plant terpene synthases and molecular evolutionary implications. *Genetics* 158:811–32
89. Verlet N. 1993. Commercial aspects. In *Volatile Crops: Their Biology, Biochemistry and Production*, ed. RKM Hay, PG Waterman, pp. 137–74. Essex, England: Longman Scientific & Technical
90. Vinson SB. 1999. Parasitoid manipulation as a plant defense strategy. *Annals Entomol. Soc. Am.* 92:812–28
91. Vogt T. 2000. Glycosyltransferases involved in plant secondary metabolism. *Recent Adv. Phytochem.* 34:317–47
92. Wang J, Pichersky E. 1998. Characterization of *S*-adenosyl-L-methionine:(iso)eugenol *O*-methyltransferase involved in floral scent production in *Clarkia breweri*. *Arch. Biochem. Biophys.* 349:153–60
93. Wang J, Pichersky E. 1999. Identification of specific residues involved in substrate discrimination in two plant *O*-methyltransferases. *Arch. Biochem. Biophys.* 368:172–80
94. Watanabe N, Washio H, Straubinger M, Knapp H, Winterhalter P. 1998. Occurrence of a glucosidic progenitor of rose oxide in rose flowers, *Rosa damascena* Mill. *Nat. Prod. Lett.* 12:5–10
95. Watanabe S, Hayashi K, Yagi K, Asai T, MacTavish H, et al. 2002. Biogenesis of 2-phenylethanol in rose flowers: incorporation of H-2(8) L-phenylalanine into 2-phenylethanol and its beta-D-glucopyranoside during the flower opening of *Rosa* 'Hoh-Jun' and *Rosa damascena* Mill. *Biosci. Biotech. Biochem.* 66:943–47
96. Wein M, Lavid N, Lunkenbein S, Lewinsohn E, Schwab W, Kaldenhoff R. 2002. Isolation, cloning and expression of a multifunctional *O*-methyltransferase capable of forming 2,5-dimethyl-4-methoxy-3(2H)-furanone, one of the key aroma compounds in strawberry fruits. *Plant J.* 31:755–65
97. Wink M. 2003. Evolution of secondary metabolites from an ecological and molecular phylogenetic perspective. *Phytochemistry* 64:3–19
98. Wink M, Mohamed GIA. 2003. Evolution of chemical defense traits in the Leguminosae: mapping of distribution patterns of secondary metabolites on a molecular phylogeny inferred from nucleotide sequences of the rbcL gene. *Biochem. Syst. Ecol.* 31:897–917
99. Wu SQ, Watanabe N, Mita S, Dohra H, Ueda Y, et al. 2004. The key role of phloroglucinol *O*-methyltransferase in the biosynthesis of *Rosa chinensis* volatile 1,3,5-trimethoxybenzene. *Plant Physiol.* 135:95–102
100. Wu SQ, Watanabe N, Mita S, Ueda Y, Shibuya M, Ebizuka Y. 2003. Two *O*-methyltransferases isolated from flower petals of *Rosa chinensis* var. *spontanea* involved in scent biosynthesis. *J. Biosci. Bioeng.* 96:119–28

Biology of Chromatin Dynamics

Tzung-Fu Hsieh and Robert L. Fischer

Department of Plant and Microbial Biology, University of California, Berkeley, California 94720-3102; email: tzungfu@berkeley.edu, rfischer@berkeley.edu

Annu. Rev. Plant Biol.
2005. 56:327–51

doi: 10.1146/
annurev.arplant.56.032604.144118

First published online as a Review in Advance on January 17, 2005

1543-5008/05/0602-0327$20.00

Key Words

chromatin remodeling, epigenetic inheritance, histone modification, DNA methylation, RNA interference

Abstract

During the development of a multicellular organism, cell differentiation involves activation and repression of transcription programs that must be stably maintained during subsequent cell divisions. Chromatin remodeling plays a crucial role in regulating chromatin states that conserve transcription programs and provide a mechanism for chromatin states to be maintained as cells proliferate, a process referred to as epigenetic inheritance. A large number of factors and protein complexes are now known to be involved in regulating the dynamic states of chromatin structure. Their biological functions and molecular mechanisms are beginning to be revealed.

Contents

Homeotic gene: regulatory genes that specify identity of body or tissue types

PcG and trxG: proteins originally discovered in *Drosophila* that form complexes and modify chromatin structure to repress (PcG) or activate (trxG) gene transcription

INTRODUCTION

Proper development of a multicellular organism depends on the establishment and maintenance of differential transcription programs. Because most somatic cells in a multicellular organism are genetically identical, the diversity of cell identities must be determined by which sets of genes they express. For example, spatial and temporal expression of homeotic genes during early animal development determines the formation of the body plan. Once established, their transcriptional states are faithfully maintained through mitosis by Trithorax Group (trxG) and Polycomb Group (PcG) protein complexes (44, 55). Similarly, plant trxG and PcG complexes also function to maintain expression states of key regulatory genes and play important roles in phase transition throughout development (5, 78). The functions of trxG and PcG, and other modifications in chromatin structure described below, represent a biochemical mechanism that influences gene activities without altering their primary DNA sequences. Such epigenetic control of gene transcription plays a crucial role in cell differentiation and development, and provides an additional layer of regulation over primary DNA sequences. It is now evident that chromatin structure is intimately linked to the activity of underlying genes, and that epigenetic control of transcription is mediated through different states of chromatin structure.

The genomic DNA of a eukaryotic cell is compacted and organized inside the nucleus as chromatin. The building block of chromatin is the nucleosome. The nucleosome contains DNA wrapped around the histone octamer, which is composed of two copies each of histones H2A, H2B, H3, and H4 (71). Hundreds of thousands of nucleosomes are organized on a continuous DNA strand, which gives chromatin the appearance of a "beads-on-a-string" fiber under an electron microscope. Higher-order chromatin structures, such as the 30-nm chromatin fibers seen in interphase chromatin, are formed by interacting nucleosomes. The organization of DNA into chromatin poses immediate challenges for processes that require DNA as templates. Enzymatic reactions such as replication, transcription, recombination, and DNA repair all involve direct binding of protein factors to their target DNA elements, a process that requires the tightly packed chromatin to be partially unraveled. It is now widely appreciated that altering chromatin structure is an important regulatory mechanism. The term "chromatin remodeling" is used to describe a wide range of biochemical processes that lead to an altered or reconfigured chromatin structure and change its accessibility to a variety of enzymatic or chemical probes (1). In a broader sense, chromatin remodeling can be regarded as modifications that result in changes in chromatin functionality.

Chromatin is historically divided into two distinct domains, heterochromatin and euchromatin. Heterochromatin is defined as chromosome regions that remain densely stained and highly condensed throughout the cell cycle, whereas euchromatin is decondensed during interphase (72). Heterochromatin is often associated with telomeres and pericentric regions of chromosomes, and is rich in repetitive sequences and low in gene density. These regions are enriched for transposable elements and tandemly repeated DNA. The condensed feature of heterochromatin is reflected in its regularly packaged nucleosomal arrays and resistance to nuclease digestion. In contrast, euchromatic chromosome regions are rich in genes and irregularly spaced nucleosome arrays, decorated with nucleosome-free, nuclease hypersensitive sites characteristic of active genes (64, 160). Although constitutive heterochromatin refers to the condensed and permanently inactive chromosomal regions associated with telomeres and pericentromeres, facultative heterochromatin exists in different locations within euchromatic regions and may be transcriptionally silenced in one tissue type, cell lineage, or developmental stage while active in others (64). Thus, heterochromatinization of active euchromatic genes provides an efficient way to lock in a clonally stable transcriptional state once a new developmental fate is established (cellular memory).

There are several ways that chromatin structure can be modulated. First, the positioning of nucleosomes on DNA can be disrupted and reconfigured by ATP-dependent remodeling complexes, and can temporarily open or close chromatin for specific enzymatic reactions. Second, posttranslational modifications of histone proteins by histone acetylases, deacetylases, and methyltransferases can generate localized distinct chromosomal domains by recruiting diverse chromatin-binding protein complexes. Third, the composition of nucleosomes can be modified by replacing major histones with variants (2, 146, 153). Finally, methylation of cytosine by DNA methyltransferases can provide a stable and heritable epigenetic mark and modulate chromatin structure by recruiting protein complexes that bind to methylated DNA. Interplay between all of these mechanisms also exists, and might act in concert to define a specific chromatin state. Here we review a number of mechanisms known to affect chromatin structures, and discuss specific examples in which the influence of chromatin dynamics on plant development is understood.

Chromosomal domain: region of chromosome with a common chromatin structure

FACTORS THAT INFLUENCE CHROMATIN STRUCTURE

ATP-Dependent Remodeling Complex

Integral to the dynamic nature of chromatin are ATP-dependent remodeling complexes. They facilitate the movement of histones relative to DNA, leading to a condensing or decondensing of chromatin. The SWI/SNF ATP-dependent remodeling complex was first discovered in *Saccharomyces cervisiae* (180a). Subsequently, the SWI/SNF complexes were shown to be required for obtaining a transcriptionally active or repressed chromatin state as assayed by increased or decreased nuclease sensitive at its target promoter (58, 184). In vitro biochemical studies demonstrate that SWI/SNF complexes cause ATP-dependent disruption of nucleosome structure, and facilitate binding of transcription factors to their target sites on nucleosomal templates (92). The SWI/SNF family is subdivided into several classes (SWI2/SNF2, ISWI, and CHD) based on the presence of other protein motifs in addition to the ATPase domain (101, 112). Current data from studies in animals suggest that the biochemical mechanism each subfamily uses is distinct and may account for the unique biological functions of each complex. For example, some SWI2/SNF2 subfamily members change the winding of DNA around the histone core, making DNA more accessible to nuclease and restriction enzyme digestion. In contrast, the ISWI members relocate nucleosomes by sliding the histone octamers along the DNA template (101). Whereas the SWI2/SNF2 subfamily functions mainly in transcription activation or repression (114), the ISWI subfamily has additional roles in maintaining stable higher-order chromatin structure and chromatin assembly (46, 80).

The *Arabidopsis* genome encodes more than 40 SWI2/SNF2-like proteins (23, 143). Among them, only AtBRM (48), SPLAYED (SYD) (176), PICKLE (PKL) (126, 127), DECREASE IN DNA METHYLATION 1 (DDM1) (85), MORPHEUS MOLECULE (MOM) (6), DRD1 (88), and PHOTOPERIOD INDEPENDENT EARLY FLOWERING1 (PIE1) (125) have been functionally characterized (see **Table 1** for summary). Recombinant DDM1 has in vitro chromatin remodeling activity (29).

In addition to the ATPase motor subunit, metazoan SWI/SNF complexes contain 8 to 11 proteins with SNF5 and SWI3 subunits as their core set of conserved components (92). Both in vitro and in vivo studies indicate that SNF5 and SWI3 subunits stimulate the ATPase motor protein activity and are required for remodeling the target chromatin, respectively (61, 138). Several plant SWI3 and SNF5 homologs have been isolated, and some of their functions in plant development have been revealed (**Table 1**).

Table 1 Arabidopsis homologs of the SWI/SNF complex components with known functions

SWI/SNF component	Subfamily/domain characteristics	Functions	Reference
AtBRM	SWI2/SNF2-like ATPase Brahma homolog	Control shoot development and flowering	(48)
SYD	SWI2/SNF2-like ATPase Partial bromodomain	Regulate homeotic gene expression Maintain floral meristem Regulate floral transition and reproduction	(176)
MOM1	SWI2/SNF2-like ATPase Missing half of helicase domain	Maintain transgene silencing independent of DNA methylation	(6)
DDM1	Close to ISWI-type ATPase Missing SANT domain	Maintain DNA methylation and H3K9 methylation Maintain gene silencing and transgene silencing Maintain transposon/ retrotransposon silencing Maintain nucleolar integrity during interphase Chromatin remodeling activity in vitro	(29, 60, 77, 84, 152)
PKL/GYM	CHD/Mi-2-type ATPase	Repress embryonic programs after germination Regulate carpel development	(47, 126, 127)
PIE	ISWI-type ATPase	Control flowering time Regulate FLC expression	(125)
DRD1	Novel, close to Drosophila DRD54	Regulate transgene silencing via RNA-induced non-CG methylation	(88)
BSH	SNF5 homolog, subunit of complex	Antisense leads to bushy plants with sterile flowers	(30)
CHB2/ AtSWI3	SWI3 homolog	Antisense leads to abnormal seedling and leaf phenotypes, dwarfism, delayed flowering and loss of apical dominance	(148, 191)

Histone Modification

Core histone proteins are structurally conserved through evolution and contain flexible N-terminal tails protruding outward from the nucleosomes. The tails of the histones are subject to numerous posttranslational, covalent modification, including acetylation, methylation, phosphorylation, ubiquitination, ribosylation, glycosylation, and sumoylation (86, 170). Posttranslational modification of histones modulates chromatin structure, influences gene activity, and plays an important role in mediating chromatin inheritance (151). The histone code theory proposes that distinct histone modifications act sequentially or in combination to recruit binding proteins for specific tasks (159). It also suggests that regions of the chromosome are functionally different from each other owing to the complexity of differential histone modifications, resulting in the recruitment of diverse chromatin-binding complexes and distinct functional readout. For example, HETEROCHROMATIN PROTEIN 1 (HP1) in animals (SWI6 in fission yeast) recognizes and binds to methylated H3K9 and is an integral part of heterochromatin (65). Some aspects of the histone code have universal biological consequences, and their regulatory roles in gene expression control, replication, recombination, and repair are subjects of intense study (79).

Histone acetylation and deacetylation. Acetylated histones localize to regions of chromatin with high DNAse I sensitivity, which correlates with transcriptional competence (100). Acetylation results in charge neutralization of histones and weakens histone/DNA contacts, facilitates the binding of bromodomain proteins, and promotes transcription (56, 145). Lysine residues at the N-terminal tails of histone proteins are the predominant sites for acetylation (e.g., K9, 14, 18, 23 of H3; K5, 8, 12, 16 of H4) (170), and the regulatory role of the modification was proposed some 40 years ago (3). It was not until the isolation of HATs as transcriptional coactivators that the role of histone acetylation in transcriptional activation became clear (28). Conversely, many corepressors have HDAC activity (167). Histone acetyltransferases (HATs) and histone deacetylases (HDACs) associate with large, multisubunit complexes (128, 145). Studies in yeast show that both HATs and HDACs can be recruited to target sites through interaction with DNA-binding proteins (87, 99). In addition to their roles in transcriptional regulation, HATs and HDACs are also involved in DNA replication, DNA repair, maintaining the silenced heterochromatic states at the mating-type loci and subtelomeric regions, and preventing the spread of silenced chromatin to the neighboring euchromatic regions (100).

Histone code: combined covalent modifications of histones that recruit proteins or complexes that control chromatin structure

COR: cold-regulated

In plants, histone hyperacetylation is also correlated with gene activity. Hyperacetylation of H3 and H4 in the pea plastocyanin gene (*PetE*) promoter region correlates with increased *PetE* transcription (42, 43). The *Arabidopsis* genome encodes at least 12 HATs and 17 HDACs (135). For example, the *Arabidopsis* GCN5 has HAT activity in vitro, can interact with the *Arabidopsis* ADA2 protein (homolog of the yeast ADA protein, a component of the yeast GCN5/HAT complex), and might function as a subunit in a larger complex to regulate COR gene expression (158, 173). GCN5 is also involved in regulating floral meristem activity (17). Three distinct families of HDACs have been found in plants (135). In addition to the conserved RPD3/HDA family and the sirtuin family related to yeast SIR2, plant genomes encode a third, plant-specific family of HDACs called HD2 (98, 103, 111). Biological and developmental functions of these enzymes are beginning to be revealed (**Table 2**) (15, 109). For example, HDA6 plays roles in transgene silencing (122), RNA-induced transgene methylation (7), and maintenance of chromatin structure in the rDNA repeats (139).

Histone methylation. The modification of histone by methylation has been known for decades, but its role in epigenetic regulation was only revealed recently. The identification of the mammalian SET (SU(VAR) 3-9, E(Z) and

Table 2 **Plant histone acetylases and deacetylases with known functions**

Name	Organism	Family	Characteristic or Function	Reference
AtGCN5	*Arabidopsis*	GNAT	HAT activity in vitro Control induction of COR gene transcription Regulate floral meristem	(17, 158, 173)
PCAT2	*Arabidopsis*	P300/CBP	HAT activity in vitro	(21, 22)
ZmGCN5	Maize	GNAT	Regulate histone mRNA transcription Regulate HD1B-I HDAC gene expression	(18)
AtHD1/RPD3A	*Arabidopsis*	RPD3	Putative global transcriptional regulator	(168, 169, 182)
HDA6/RPD3B	*Arabidopsis*	RPD3	Transgene silencing RNA-induced de novo transgene methylation Chromatin structure in the rDNA repeats	(7, 57, 122, 139)
HD2A	*Arabidopsis*	HD2	Maintain nucleolar dominance in genetic hybrid Overexpression leads to pleiotropic developmental abnormalities and seed abortion	(102, 183, 190)
HD2	Maize	HD2	Acidic nucleolar phosphoprotein might regulate ribosomal chromatin structure and function	(111)

Imprinting: when expression of alleles depends on their parental origin

Trithorax) domain protein SU(VAR) 3-9, and its homolog in yeast as H3K9-specific methyltransferases, was a landmark discovery in chromatin research (124, 140). These studies defined that the SET domains are responsible for the methyltransferase activities and established links between H3K9 methylation and its silencing effects in regulating heterochromatin structure. Histone methylation can occur at lysines 4, 9, 27, and 36 of H3 and lysine 20 of H4 (109, 166). H3 can also be methylated at K79 inside the globular domain (49). To date, methylation of H3K4, K36, and K79 correlate with active transcription, whereas methylation of H3K9, K27, and H4K20 are imprints for silenced chromatin (54). In addition, lysine residues can be mono-, di-, or trimethylated in vivo, and each distinct methylation might have unique biological relevance. For example, in *S. cerevisiae*, dimethylated H3K4 occurs at both active and inactive euchromatic genes whereas trimethylated H3K4 is present exclusively at active genes (147). In *Neurospora*, trimethylated H3K9 is the mark for DNA methylation (164). The recent development of antibodies specific to each state of methylation, along with novel mass spectrometry approaches, has demonstrated that in mammals H3K27 monoacetylation and H3K9 trimethylation are selectively enriched in pericentric heterochromatin. In *suv39h* double null cells, this methylation pattern is converted to trimethylated H3K27 and monoacetylated H3K9 (144). Thus, levels of modifications on a single residue deepen the complexity of the histone code owing to the combinatorial possibilities of different histone modifications.

In *Arabidopsis*, heterochromatin is associated with a high level of H3K9 dimethylation whereas euchromatin is rich in dimethylated H3K4 (83). A total number of 32 and 22 SET domain proteins encoded in the *Arabidopsis* and Maize genomes, respectively, have been identified, although most have not been functionally characterized (4, 11, 155). KRYPTONITE (KYP) is the first plant histone methyltransferase to be discovered, and it methylates H3K9 in vitro. *kyp* mutants show loss of cytosine methylation in a CNG trinucleotide sequence context and reactivation of epigenetically silenced *SUPERMAN* (*SUP*) and *PAI* loci as well as retrotransposon sequences (82, 113). ATX1, an *Arabidopsis* homolog of the *Drosophila* trithorax protein methylates H3 at K4 position in vitro (5).

Polycomb Group and Trithorax Group

During the development of multicellular organisms, epigenetic regulation of transcription is crucial for preserving lineage-specific cell identity. The PcG and trxG protein complexes are important for maintaining the cellular memory of key expression programs. In *Drosophila*, mammals, and plants, the PcG and trxG complexes maintain transcription states of homeotic genes during development. In general, PcG complexes function to maintain the silenced state of transcription by compacting chromatin whereas trxG complexes maintain the active state of transcription by opening chromatin structure (27). Recent progress in understanding the biochemical mechanisms of PcG and trxG comes from the purification and characterization of PcG and trxG protein complexes. The *Drosophila* ESC/E(Z) complex and its mammalian counterparts function as a histone methyltransferase that methylates K27 of histone H3, and it is the SET domain within the complex that is responsible for the histone methyltransferase activity (32). Methylated H3K27 further recruits PRC1 complexes (Polycomb Repressive Complex 1, a distinct animal PcG complex not present in plants) for stable, long-term repression through interaction with the chromodomain of the POLYCOMB subunit.

Studies of the trxG complex reveal that the *Drosophila* ASH1, and its mammalian homolog ALL-1/MLL, also function as histone methyltransferases. ASH1 methylates K4 and K9 on histone H3, and K20 on histone H4, in a trivalent manner (12, 31), whereas ALL-1/MLL methylates H3K4, resulting in gene activation (119, 123).

Several PcG proteins have been identified in plants that control many aspects of plant development (78, 142). For example, the *Arabidopsis* MEA/FIE/FIS2/MSI1 complex regulates gametophytic to sporophytic transition by preventing initiation of the endosperm development program in the central cell prior to fertilization (66, 78, 96, 142). Maternally inherited mutations in any component of this complex result in autonomous endosperm development in the absence of fertilization, and in seed abortion after fertilization. The parent-of-origin effect on seed development is likely the result of epigenetic functions exerted by this PcG complex in the central cell. A target gene of the MEA/FIE/FIS2/MSI1 complex, *PHERES1* (*PHE1*), has been identified (97). *PHE1* (encodes a type I MADS domain protein) is expressed transiently after fertilization and likely plays a role in seed development. The MEA/FIE/FIS2/MSI1 complex is required to maintain repression of *PHE1*.

Another SET domain PcG protein in *Arabidopsis*, CURLY LEAF (CLF), is involved in regulating the transition from vegetative development to reproduction (63). Mutations in *CLF* cause ectopic expression of the MADS box homeotic gene *AGAMOUS* (*AG*) in leaves, inflorescence stems, and flowers, indicating that CLF is required to keep *AG* repressed in these tissues. Thus, the CLF complex presumably functions to maintain chromatin of the *AG* locus at a repressed state by histone H3K27 methylation. CLF likely functions by interacting with the FIE and EMF2 Polycomb proteins (93, 188, 39a). The FIE and CLF Polycomb proteins interact in vitro and regulate homeobox gene expression during sporophyte development (89).

There are at least five trxG-like proteins present in the *Arabidopsis* genome (4). Among them, *ATX-1* and *ATX-2* have been studied in

MBD: methylcytosine-binding domain

greater detail. Both proteins contain a PWWP motif (a conserved proline-tryptophan-tryptophan-proline sequence within a module of about 100 amino acids that is found in many chromatin associated proteins) that is not found in other members of the trxG family, suggesting that they might have plant-specific functions. *ATX1* is required for normal flower development and floral organ identity determination. In *atx1* mutants, expression of several floral homeotic genes is reduced, indicating that *ATX1* positively regulates their expression. In addition, the SET domain and flanking region of ATX1 displays H3K4 methyltransferase activity in vitro. Thus, ATX1 shows conserved functional features of its animal counterparts and will likely counteract PcG activity in the epigenetic regulation of floral homeotic gene expression (5).

DNA Methylation

DNA methylation is one of the most abundant epigenetic modifications in higher plants and animals. The reaction involves covalent addition of a methyl group to the 5-position of cytosine by a family of enzymes called DNA methyltransferases. DNA methylation in the symmetric dinucleotide CG sequence context is an evolutionarily conserved modification found in vertebrates, plants, and some fungi (19, 51). Patterns of symmetric CG methylation are maintained after DNA replication, where each parental strand retains half of the methylation imprints. Maintenance DNA methyltransferases, like DNMT1 in mammals (17a) and MET1 in plants (50a), copy the methylation marks onto the daughter DNA strand. Thus, DNA methylation provides an efficient mechanism to store epigenetic information that can be stably inherited through cell divisions.

Plants differ from animals in that significant levels of cytosine methylation at symmetric CNG and non-symmetric CNN sequences (N = A, C, or T) exist (13, 166). Chromomethylases (CMTs) are plant-specific DNA methyltransferases for CNG methylation (74, 136). In *Arabidopsis*, CMT3 is the major enzyme for CNG methylation. Mutations in *CMT3* do not result in obvious phenotypic abnormalities that are commonly seen in the *met1* background, indicating that CG methylation is likely a more prevalent DNA methylation sequence context at most target sites, whereas non-CG methylation occurs in a locus-specific manner (8, 34, 106). Two *Arabidopsis* domain rearranged methyltransferase (DRM) class, DRM1 and DRM2, have been identified as de novo methyltransferases (33–35, 35a). DRMs resemble the mammalian DNMT3 members, which are involved in establishing new DNA methylation imprints during early embryogenesis, but with different organization of catalytic sequence motifs (129). DRM1 and DRM2 are responsible for the establishment, but not the maintenance, of silencing at *FWA* and *SUPERMAN* loci, and *drm1/drm2* double mutants are impaired in de novo methylation at non-CG sites in *FWA* and *SUP* loci (35). DRMs are also required for the initial establishment of RNA-directed DNA methylation (RdDM) at all sequence contexts (33). Like *cmt3*, *drm1/drm2* double mutants do not display overt phenotypes, but exhibit locus-specific loss of gene silencing (35).

DNA methylation is often associated with gene silencing and is prominently present in heterochromatin and transposons. It is also generally regarded as a cellular mechanism to defend against invading retroviruses or transposable elements (19, 115). In animals, the effect of DNA methylation on chromatin structure is linked to histone deacetylation and gene silencing through proteins that contain MBDs, which bind to methylated DNA and recruit HDACs (20). Homologs of MBD proteins have been found in *Arabidopsis* and maize, suggesting that interplay between DNA methylation and histone deacetylation might also exist in plants (14, 189). In addition, DNA methylation and histone methylation are intimately linked in plants, animals, and fungi. In *Neurospora*, trimethylation of H3K9 is required for DNA methylation (165). In *Arabidopsis*, mutations in *KYP* gene cause reductions in CNG methylation (81) and fail to maintain DNA methylation of an endogenous *PAI2* gene (113). The requirement of DDM1 for the maintenance of genome-wide DNA methylation further strengthens the relationship between DNA methylation and chromatin (84, 85). In *Arabidopsis*, DNA methylation plays a central role in gene silencing, imprinting, transposon regulation, and transgene silencing (13, 115).

Histone Variants

Epigenetic inheritance of gene expression can be achieved by a combination of DNA methylation, histone modification, and binding of non-nucleosomal proteins such as PcG and trxG complexes, or HP1. According to the histone code hypothesis, the combined histone modifications are a nucleosomal blueprint for maintaining the chromatin state during subsequent cell divisions. Less well understood, however, is how the code is perpetuated when the cell divides. Unlike DNA methylation, where maintenance DNA methyltransferases faithfully restore symmetric methylation patterns based on the methylated parental template strand, DNA and histone octamers need to be physically separated and rejoined after replication. Embedded in the process is the random mixing of old and newly synthesized histones that need to be reassembled into histone octamers and wrapped by each of the daughter chromosomes. Thus, how to maintain the integrity of the code presents a great challenge to our understanding of the mechanism of inheriting chromatin structure. With no enzyme that can remove methyl groups from lysine residues identified so far, how is silent chromatin that was once marked with H3K9 methylation reactivated?

Recent findings that active chromatin is enriched for certain histone variants, and the discovery of nucleosome assembly complexes responsible for their deposition, have shed light on the problems described above. Variants of histones H2A and H3 have been known for decades and are found in all eukaryotes. During the cell cycle, the expression of histone genes is tightly regulated in S-phase to meet the demand of packaging newly replicated DNA (131). However, other situations occur outside the S phase, such as the repair of DNA damage, where limited supply of histone proteins might endanger the integrity of the chromosome. Therefore, most organisms contain alternative copies of variant histone genes constitutively expressed at low levels throughout cell cycle. These variant histones serve as replacements and are deposited through the RI pathway. By contrast, the bulk histones assembled at the replication forks are deposited through the RC process. Recently, it was discovered that histone variant H3.3 is deposited to active chromatin throughout the cell cycle by the RI nucleosome assembly, suggesting that nucleosomes are replaced during transcription (2). McKittrick et al. (117) showed that the amount of cellular H3.3 is sufficient to mark all the active gene loci, and H3.3 variants are enriched in modifications associated with transcription activity (e.g., di- and tri-methylated K4) and deficient in dimethylated K9, which is associated with heterochromatin. These results indicate that H3.3 marks the active chromatin, and displacement of H3 with H3.3 provides a dynamic mechanism for rapid activation of chromatin that had been marked inactive, for example, by K9 methylation (75). The mechanism of H3.3 displacement for gene activity might be conserved in plants as differential levels of acetylation and methylation of H3 and its variant have been observed in alfalfa (178, 179).

A parallel study identified two nucleosome assembly complexes in which the CAF-1 (CHROMATIN ASSEMBLY FACTOR 1) complex is associated with H3-H4 dimers, whereas the HIRA (HISTONE REGULATION) complex is specific for the H3.3-H4 dimer (163). The discovery is in contrast to the established view of how the histone octamer of the parental DNA strand is disassembled (into one H3-H4 tetramer and two H2A-H2B dimers) during replication (181). Instead, it raises the exciting possibility that H3-H4 are deposited by the CAF-1 complex as dimers (rather than tetramers) onto two daughter DNA strands during replication, and thus each nucleosome retains half of the parental

Epigenetic inheritance: inheritance of a gene activity state that is not specified by DNA sequence

RI: replication independent

RC: replication coupled

PTGS: posttranscriptional gene silencing

RNA interference (RNAi): a cellular defense mechanism in which the introduction of dsRNA inhibits translation, or causes degradation, of complementary mRNA sequences

RISC: RNA-induced silencing complex

TGS: transcriptional gene silencing

LTR: long terminal repeat

epigenetic marks (163). This would suggest that CAF-1 might play an important role in the epigenetic regulation of chromatin structure. Supporting this hypothesis, the yeast CAF-1 is required for stable inheritance of a transcriptionally silenced chromatin state in telomeres (90, 120). However, it is important to caution that so far there is no direct biochemical data to support that nucleosomes are replicated semiconservatively (75).

RNA-Induced Gene Silencing

RNA-induced gene silencing, first discovered in plants more than 15 years ago, involves processing aberrant double-stranded RNA species (dsRNAs) into short RNAs of 21–24 nucleotides in length and triggers two types of gene silencing. PTGS occurs primarily in the cytoplasm and through a mechanism similar to quelling in *Neurospora* and RNAi in animals (116). In animals, introducing dsRNAs by microinjection effectively silences endogenous genes with complementary sequences to the dsRNAs (53). Double-stranded RNAs are processed by an RNaseIII-like enzyme DICER, generating siRNAs 21–26 nucleotides in length (16). Similar to small interfering RNAs (siRNAs), endogenous RNA species called micro RNAs (miRNAs) function to suppress endogenous genes important for regulating developmental programs in animals and plants (36). In animals, both siRNAs and miRNAs silence their target genes at the posttranscriptional level. An endonuclease-containing RISC complex associates with siRNAs and targets their complementary mRNAs for degradation (70). In contrast, miRNAs primarily act by base pairing with the 3′-UTR of mRNAs and inhibiting their translation (130). In plants, some miRNAs direct mRNA degradation (108, 134), whereas others work to block translation (40).

TGS, on the other hand, occurs in the nucleus and blocks the transcription of target genes. In plants, dsRNAs can target their homologous chromosomal sequences for methylation, a process called RdDM, which results in gene silencing and involves siRNAs (69, 118, 150, 177). In *Drosophila*, TGS also involves the RNAi pathway (133). Recently, siRNAs were shown to direct homologous sequences for DNA and H3K9 methylation and to result in gene silencing in human cells (91, 121). Thus, siRNA-induced DNA and gene silencing appears to be a conserved mechanism during evolution.

The role of RNAi in chromatin modification has been demonstrated by several studies in *Schizosaccharomyces pombe*, where RNAi is required for heterochromatin formation. Small RNAs corresponding to centromeric repeats and the presence of a functional RNAi machinery are essential for proper histone H3K9 methylation in centromeric repeats and centromere function (67, 68, 174, 175). Likewise, centromeric repeats in the mating-type locus in yeast are also targeted for RNAi-mediated heterochromatin formation (68). Furthermore, RNAi is responsible for silencing the LTRs of retrotransposons through heterochromatin formation. Silencing that occurs on the LTRs can spread to neighboring genes and creates heritable repression of those genes throughout development (149). Thus, RNAi-mediated chromatin modification is a general mechanism for gene silencing and transcription regulation in *S. pombe*. Heterochromatin silencing and HP1 localization also depend on the RNAi machinery in *Drosophila* (132).

In *Arabidopsis*, mutations in *AGO4* (a component of RNAi) decrease CNG methylation and histone H3K9 methylation, causing derepression of the silent *SUPERMAN* (*SUP*) epialleles. In addition, *ago4-1* also eliminates cytosine and H3K9 methylation of a constitutive heterochromatin locus, *AtSN1*, due to failure of siRNA accumulation (192). Furthermore, loss of endogenous siRNAs in dicer-like 3 (*dcl3*) mutants results in loss of heterochromatin marks (e.g., cytosine and H3K9 methylation) and increased transcription at some loci. These observations strongly suggest that a parallel RNAi-mediated chromatin modification pathway might exist in *Arabidopsis* with *DCL3*,

AGO4, and *RDR2* as important components (186).

BIOLOGY OF PLANT CHROMATIN DYNAMICS

The life cycle of a higher plant is distinct and is organized by successive developmental phase changes. Formation of new organs and cell types, produced throughout the life cycle, frequently follows each developmental switch. Chromatin remodeling plays a critical role in regulating transcriptional states throughout plant development and provides an epigenetic mechanism for inheriting expression states. In this section, we discuss a few specific examples where the developmental consequences of chromatin regulation in plants are best understood. The role of RNAi in the regulation of chromatin structure is increasingly being appreciated.

Vernalization

In *Arabidopsis*, naturally occurring flowering time variations among different accessions are caused by allelic variations in two loci, *FRIGIDA* (*FRI*) and *FLOWERING LOCUS C* (*FLC*). In winter-annual accessions, *FRI* acts as a positive regulator of *FLC*, a strong suppressor of flowering, and vernalization is required for rapid flowering (73, 160a, 161). Vernalization is a cold treatment that promotes the vegetative to reproductive transition of the shoot meristem, and it is an excellent example of how the environment influences plant development by modifying chromatin structure. To ensure that flowering occurs after the winter season, winter-annual plants evolved a strategy to determine when they will flower by measuring how long they have been exposed to low temperature. By this mechanism winter-annual plants remember and respond to vernalization. At the molecular level, vernalization counteracts the effect of *FRI* by repressing *FLC* expression, and this repression of *FLC* needs to be mitotically maintained long after the cold treatment in order for the plants to flower at the proper time. Results from genetic screens and molecular analyses highlight the importance of chromatin modifications in the control of *FLC* expression (73, 160a, 161).

Vernalization: prolonged period of cold treatment required for winter-annual plants to flower the following spring

VERNALIZATION INSENSITIVITY3 (*VIN3*) encodes a plant homeodomain zinc (PHD) finger protein and is responsible for the initial downregulation of *FLC* expression (162). *VIN3* expression is only induced after prolonged exposure to cold, providing a molecular mechanism to ensure that plants only respond to a complete winter season instead of temperature fluctuations in autumn. *VERNALIZATION1* (*VRN1*) and *VRN2* encode a Myb-related DNA-binding protein (105) and zinc finger PcG protein (59), respectively, and are required to mitotically maintain the transcriptionally repressed state of *FLC* that was established by *VIN3*. Establishing the transcriptionally repressed state of *FLC* correlates with hypoacetylation of histones H3 lysine 9 and lysine 14 at the *FLC* locus. This is closely followed by an increase in H3K9 and H3K27 methylation, hallmarks of transcriptional repression by chromatin (10). In *vin3* mutants, none of these histone modifications occur during vernalization, indicating that *VIN3* is crucial in establishing the vernalization-induced chromatin remodeling at the *FLC* locus. This histone hypoacetylation at the *FLC* locus can be detected in *vrn1* and *vrn2* plants, but the modifications are not stably maintained once the plants return to warm temperatures. Furthermore, H3K27 methylation is lost in *vrn2* plants, and methylation at both H3K9 and H3K27 is lost in *vrn1* (10, 162). Thus, VRN1 and VRN2 are required to maintain the repressed *FLC* state. In summary, vernalization acts through the chromatin remodeling proteins (VIN3, VRN1, VRN2) to suppress *FLC* expression and maintain *FLC* in a repressed state for plants to flower in the spring.

The chromatin modifying effect of vernalization is not limited to the *FLC* gene. A recent study identified a small cluster of genes flanking *FLC* whose expression is coordinately regulated by cold treatment (52). The cluster includes the gene *UPSTREAM OF FLC* (*UFC*, located

Paramutation: an interaction between homologous alleles in which one allele (paramutagenic) can influence the expression state of the other (paramutable) allele

4.7-kb upstream of *FLC*), *FLC*, and At5g10130 (located 6.9-kb downstream of *FLC*). The expression of these three genes is also coregulated by different genetic modifiers of *FLC*. Noncold regulated transgenes inserted into the cluster respond to cold treatment. When *FLC* is inserted into ectopic locations in the genome, the *FLC* transgene maintains its response to vernalization and imposes a similar response on a flanking gene. Together, these observations strongly suggest that chromatin in this cluster constitutes a chromosomal domain, and *FLC* contains sequences that can modulate chromatin structure beyond itself. Because many of the genetic modifiers of *FLC* expression are chromatin modifiers, the chromatin flanking the *FLC* locus might be modified too, thereby making the promoters of flanking genes inaccessible to transcription factors. This is the first example of a plant chromosomal domain. In animals, coregulated gene clusters have been reported in many cases (24, 104, 110). It has been proposed that chromatin structure may subdivide genomes into chromosomal domains within which a cluster of genes is coregulated (180).

Paramutation

Paramutation is an allelic interaction in which one allele, called paramutagenic, can cause a heritable change in the expression of a homologous, paramutable allele. Originally reported more than 50 years ago (25), paramutation has been intensively studied in maize. To date, paramutation has been described in four maize genes (*r1*, *b1*, *pl1*, and *p1*) that all encode transcription factors that control the biosynthesis of flavonoid pigments (38). Recent genetic and molecular studies have started to reveal the underlining molecular mechanism of paramutation (39). It is now clear that paramutation involves communication between homologous loci *in trans*, and the communication results in heritable chromatin changes and alteration of DNA methylation. One of the best-studied paramutation systems is the maize *booster1* (*b1*) gene. The paramutable *B-I* allele has a high level of transcription and a dark pigmentation phenotype, whereas the paramutagenic *B′* is weakly transcribed with light pigmentation. In plants heterozygous for *B-I* and *B′*, the transcription of *B-I*'s allele is always downregulated to the *B′* transcription level, causing a dramatic decrease in pigmentation. The change of *B-I* to *B′* is heritable and the new *B′* is paramutagenic. A 6-kb enhancer region located ~100-kb upstream of the *b1* coding region is required for paramutation (157). The enhancer regions of both *B-I* and *B′* alleles have seven tandem repeats of 853-bp sequence, whereas other neutral (nonparamutation) *b* alleles have only one copy of the repeat. The 6-kb region is identical in *B-I* and *B′*, indicating that the mechanism of paramutation is epigenetic in nature. The active and repressed states of transcription associated with *B-I* and *B′* alleles are reflected in their respective chromatin structures. The repeats of the active *B-I* allele have an open chromatin structure as assessed by nuclease sensitivity assay, whereas repeats of the repressed *B′* allele assume a more closed chromatin state. At the DNA level, the *B-I* repeats are hypermethylated, whereas the *B′* repeats are hypomethylated (156). The mechanistic base of such differential methylation remains to be elucidated. The possibility that hypomethylation of the repeats in *B′* allele might allow transcription of noncoding RNA and generate siRNA to guide formation of silent chromatin has been proposed (39, 45).

Imprinting

Imprinting is an epigenetic regulation of gene expression in which one of the two parental copies is expressed while the other is silenced. In mammals, imprinted genes are usually found in chromosomal clusters and can be associated with differential DNA methylation at *cis*-acting imprinting control regions (141). The imprinting of many genes is disrupted in embryos null for *DNMT1* (50). Thus, DNA methylation is an important mechanism that differentially marks the parental alleles during gametogenesis (9). Because DNA methylation and histone methylation are intimately linked, it is reasonable to propose that chromatin structure ultimately determines the "on" or "off" state of the imprinted genes. Correlation of H3K9 and H3K4 methylation found in the silent and the active alleles of the Prader-Willi syndrome imprinting center supports the hypothesis (187). In addition, a recent study in mammals also suggests the influence of chromatin structure in regulating the imprinting of two loci, *Igf2r* (insulin-like growth factor 2 receptor, maternally expressed) and *Rasgrf1* (RAS protein-specific guanine nucleotide-releasing factor 1, paternally expressed). Herman and colleagues demonstrated that adding Region 2 of *Igf2r* in a chimera *Rasgrf1* allele is sufficient to silence the maternal chimeric allele in cis and cause derepression of endogenous paternal *Rasgf1* allele *in trans* in some progeny (76). Trans silencing of the paternal *Rasgf1* allele by the chimera allele resembles paramutation and is likely caused by changes in chromatin structures and DNA methylation patterns (45).

Recent studies link DNA methylation to the imprinting of two *Arabidopsis* genes, *MEA* and *FWA*. Both *MEA* and *FWA* are imprinted in the endosperm; only the maternal alleles are active, whereas the paternal alleles are silenced (94, 95, 172). The expression of the maternal *MEA* and *FWA* alleles in the endosperm is controlled by the *DEMETER* (*DME*) gene, suggesting an important role for DME in regulating gene imprinting in *Arabidopsis* (41, 94). *DME* encodes a large novel protein with a DNA glycosylase domain. Proteins with glycosylase domains are usually involved in base excision DNA repair. DME acts directly at the *MEA* promoter, as ectopic *DME* expression results in nicks in the *MEA* promoter (41). Nicks indicate the occurrence of a base excision process. Mutations in *DME* are suppressed by mutations in *MET1* (185). This suppression acts at or upstream of the *MEA* locus specifically in the female gametyphyte, providing a link between DNA methylation and the regulation of *MEA* imprinting by DME. Three regions of the *MEA* promoter are subject to DNA methylation, and this methylation is reduced in *met1* mutant seeds. However, it is unknown whether *MEA* promoter methylation regulates *MEA* expression directly. It is also not known how DME overcomes the *MET1*-mediated suppression of *MEA*. One possibility is that *DME* may activate *MEA* by removing methylated cytosine residues, as suggested by a *DME*-like gene, *REPRESSOR OF SILENCING 1* (*ROS1*), which can excise 5-methylcytosine in vitro (62).

The regulation of *FWA* expression by *MET1* has been known for some time (154), although only recently was *FWA* identified as an imprinted gene (94). *FWA* is not expressed in wild-type adult tissues, and silencing of *FWA* after embryogenesis is associated with hypermethylation of *FWA* promoter repeats. In wild-type *Arabidopsis*, *FWA* is only expressed in endosperm tissue from the maternal allele. The promoter repeats are hypomethylated in endosperm, but are hypermethylated in other tissues where *FWA* is not expressed (i.e., leaf, embryo, and seed coat). Crossing a wild-type female to a *met1*male results in derepression of paternal *FWA* expression in the endosperm and embryo. Thus, paternal *MET1* is required for *FWA* paternal allele silencing in the embryo and endosperm (94).

RNAi-Dependent Histone and DNA Methylation

A detailed study on the role of RNAi in regulating plant heterochromatin was just reported. Lippman et al. (107) applied a novel genomic approach to systematically profile histone and DNA methylation patterns in a heterochromatin knob *hk4s* in *Arabidopsis*. As expected, *hk4s* has high levels of dimethylated H3K9 and DNA methylation (hallmarks of heterochromatin), and is relatively depleted of H3K4 methylation (mark for active chromatin). The distribution of H3K9 methylation correlates significantly with the location of transposable elements, indicating that heterochromatin formation is determined by the presence of transposable elements. DDM1 is crucial in

RITS: RNA-induced initiation of transcriptional silencing

controlling heterochromatin silencing because in *ddm1* mutants DNA methylation and H3K9 methylation of *hk4s* are lost. Instead, DNA and H3K9 methylation is replaced by uniform H3K4 methylation with concurrent reactivation of some transposable elements. Hypothesizing that siRNA might target transposable elements for H3K9 methylation, the authors identified that the main source of siRNAs complementary to the *hk4s* region is derived from transposable elements. Therefore, transcription of transposable elements targets them for histone and DNA methylation and heterochromatin formation. This is consistent with the siRNA guided histone methylation mechanism found in yeast (68, 174, 175). Furthermore, transposable elements with corresponding siRNAs are the ones typically reactivated in the *ddm1* background, suggesting that DDM1 might be guided by the siRNAs (107).

In addition to regulating repeats and transposable elements, RNAi-dependent histone and DNA methylation can also silence euchromatic genes located within close vicinity of repeats or transposons. For example, the *FWA* promoter and 5′ untranslated region contain two sets of tandem repeats, encompassed within a *SINE3* retrotransposon, which generate corresponding siRNAs and target *FWA* for histone and DNA methylation under DDM1 control (107, 154). The requirement of siRNA in silencing *FWA* was also demonstrated independently by the studies of *Arabidopsis* mutants defective in the RNAi machinery (37). Thus, inserting a retrotransposon can induce a silencing effect, impose epigenetic repression of a neighboring gene, and might contribute to the imprinting of *FWA*. Misregulation of *FWA* results in promoter demethylation and ectopic expression of *FWA*, which leads to later flowering phenotype (154).

The precise mechanism of how siRNAs find their homologous sequences for silencing is still unknown. A RITS complex in fission yeast appears to target siRNAs to their homologous loci (171). One of the RITS components is a chromodomain protein CHP1 previously known to bind centromeres, providing a direct link to the chromatin (137). This finding presents an exciting new lead toward our understanding of this fascinating epigenetic phenomenon.

CONCLUDING REMARKS

Chromatin remodeling factors have been identified as genetic modifiers of developmental mutations in plants (**Table 1** and **2**). Mutations in many chromatin remodeling factors result in lethality in metazoans, but are viable in plants. This might reflect the profound differences between the developmental processes of animals and plants. It also underscores the advantage of using plants as a model system to study the mechanism of chromatin-mediated developmental processes. In recent years we have witnessed rapid progress in the field of plant chromatin biology. In addition, plant research has contributed significantly to general chromatin biology on many fronts. One such research area is RNAi in the regulation of genome function. RNA-dependent DNA and histone methylation provide a mechanism for sequence-specific silencing, and will likely be important for silencing transposons, repetitive elements, and euchromatic genes located near transposons. The phenomenon described for *FWA* resembles the silencing of reterotransposon LTRs in fission yeast, which results in the formation and spread of heterochromatin to neighboring genes. Thus, transposons and repeats might fortuitously provide regulatory sequences for adjacent genes and influence developmental outcomes by subjecting them to epigenetic silencing (149). It has long been postulated that related repetitive elements might regulate differential expression of gene networks (26). The emerging paradigm places RNAi as a central player in establishing epigenetic silencing of homologous genes by means of heterochromatin formation. Future work is needed to determine whether RNAi ultimately triggers chromatin changes that establish and maintain paramutation and locus-specific gene imprinting in plants.

NOTE ADDED IN PROOF

After completion of this review, a mammalian transcriptional corepressor (*LSD1*) with an amine oxidase domain was reported to have histone H3K4-specific demethylase activity (195). The Arabidopsis *FLOWERING LOCUS D* is a *LSD1* homolog that is required for repressing *FLC* expression in the autonomous pathway (193). In addition, the Arabidopsis *PHERES1* (*PHE1*) gene has been shown to be imprinted (paternally expressed, maternally silenced), and the PcG protein MEA is required for the repression of the maternal *PHE1* allele (194).

SUMMARY POINTS

1. Gene transcription is regulated at the level of chromatin structure. Condensed chromatin tends to repress transcription whereas decondensed chromatin favors activation of transcription.
2. Chromatin structure is stably transmitted as cells proliferate. Chromatin structure is also dynamic and modulated during development.
3. Chromatin structure and gene transcription are influenced by ATP-dependent chromatin remodeling complexes, histone modifications, PcG and trxG protein complexes, DNA methylation, variant histone proteins, and RNAi.
4. Vernalization affects flowering time by modifying the structure of chromatin encompassing the FLC gene.
5. Paramutation involves communication between homologous alleles, resulting in heritable chromatin changes and altered DNA methylation.
6. A DNA glycosylase and a DNA methyltransferase control gene imprinting in the endosperm.
7. RNAi-dependent histone and DNA methylation are important regulators of chromatin dynamics and gene silencing.

ACKNOWLEDGMENTS

We thank Mary Gehring and Jon Penterman for their critical comments and discussion throughout the writing of this review. We apologize for not citing many important manuscripts due to space limitations.

LITERATURE CITED

1. Aalfs JD, Kingston RE. 2000. What does 'chromatin remodeling' mean? *Trends Biochem. Sci.* 25:548–55
2. **Ahmad K, Henikoff S. 2002. The histone variant H3.3 marks active chromatin by replication-independent nucleosome assembly. *Mol. Cell* 9:1191–200**
3. Allfrey VG, Faulkner R, Mirsky AE. 1964. Acetylation and methylation of histones and their possible role in the regulation of RNA synthesis. *Proc. Natl. Acad. Sci. USA* 51:786–94
4. Alvarez-Venegas R, Avramova Z. 2002. SET-domain proteins of the Su(var)3–9, E(z) and trithorax families. *Gene* 285:25–37

The discovery of histone variant H3.3 suggests a mechanism to reverse and reactivate genes that were silenced by H3K9 methylation.

5. Alvarez-Venegas R, Pien S, Sadder M, Witmer X, Grossniklaus U, Avramova Z. 2003. ATX-1, an Arabidopsis homolog of trithorax, activates flower homeotic genes. *Curr. Biol.* 13:627–37
6. Amedeo P, Habu Y, Afsar K, Scheid OM, Paszkowski J. 2000. Disruption of the plant gene MOM releases transcriptional silencing of methylated genes. *Nature* 405:203–6
7. Aufsatz W, Mette MF, van der Winden J, Matzke M, Matzke AJ. 2002. HDA6, a putative histone deacetylase needed to enhance DNA methylation induced by double-stranded RNA. *EMBO J.* 21:6832–41
8. Bartee L, Malagnac F, Bender J. 2001. Arabidopsis cmt3 chromomethylase mutations block non-CG methylation and silencing of an endogenous gene. *Genes Dev.* 15:1753–58
9. Bartolomei MS, Tilghman SM. 1997. Genomic imprinting in mammals. *Annu. Rev. Genet.* 31:493–525
10. Bastow R, Mylne JS, Lister C, Lippman Z, Martienssen RA, Dean C. 2004. Vernalization requires epigenetic silencing of FLC by histone methylation. *Nature* 427:164–67
11. Baumbusch LO, Thorstensen T, Krauss V, Fischer A, Naumann K, et al. 2001. The Arabidopsis thaliana genome contains at least 29 active genes encoding SET domain proteins that can be assigned to four evolutionarily conserved classes. *Nucleic Acids Res.* 29:4319–33
12. Beisel C, Imhof A, Greene J, Kremmer E, Sauer F. 2002. Histone methylation by the Drosophila epigenetic transcriptional regulator Ash1. *Nature* 419:857–62
13. Bender J. 2004. DNA methylation and epigenetics. *Annu. Rev. Plant. Biol.* 55:41–68
14. Berg A, Meza TJ, Mahic M, Thorstensen T, Kristiansen K, Aalen RB. 2003. Ten members of the Arabidopsis gene family encoding methyl-CpG-binding domain proteins are transcriptionally active and at least one, AtMBD11, is crucial for normal development. *Nucleic Acids Res.* 31:5291–304
15. Berger F, Gaudin V. 2003. Chromatin dynamics and Arabidopsis development. *Chromosome Res.* 11:277–304
16. Bernstein E, Caudy AA, Hammond SM, Hannon GJ. 2001. Role for a bidentate ribonuclease in the initiation step of RNA interference. *Nature* 409:363–66
17. Bertrand C, Bergounioux C, Domenichini S, Delarue M, Zhou DX. 2003. Arabidopsis histone acetyltransferase AtGCN5 regulates the floral meristem activity through the WUSCHEL/AGAMOUS pathway. *J. Biol. Chem.* 278:28246–51
17a. Bestor TH. 2000. The DNA methyltransferases of mammals. *Hum. Mol. Genet.* 9:2395–402
18. Bhat RA, Riehl M, Santandrea G, Velasco R, Slocombe S, et al. 2003. Alteration of GCN5 levels in maize reveals dynamic responses to manipulating histone acetylation. *Plant J.* 33:455–69
19. Bird A. 2002. DNA methylation patterns and epigenetic memory. *Genes Dev.* 16:6–21
20. Bird AP, Wolffe AP. 1999. Methylation-induced repression–belts, braces, and chromatin. *Cell* 99:451–54
21. Bordoli L, Husser S, Luthi U, Netsch M, Osmani H, Eckner R. 2001. Functional analysis of the p300 acetyltransferase domain: the PHD finger of p300 but not of CBP is dispensable for enzymatic activity. *Nucleic Acids Res.* 29:4462–71
22. Bordoli L, Netsch M, Luthi U, Lutz W, Eckner R. 2001. Plant orthologs of p300/CBP: conservation of a core domain in metazoan p300/CBP acetyltransferase-related proteins. *Nucleic Acids Res.* 29:589–97
23. Bork P, Koonin EV. 1993. An expanding family of helicases within the 'DEAD/H' superfamily. *Nucleic Acids Res.* 21:751–52
24. Boutanaev AM, Kalmykova AI, Shevelyov YY, Nurminsky DI. 2002. Large clusters of co-expressed genes in the Drosophila genome. *Nature* 420:666–69

25. Brink RA. 1956. A genetic change associated with the R locus in maize which is directed and potentially reversible. *Genetics* 41:872–89
26. Britten RJ, Davidson EH. 1969. Gene regulation for higher cells: a theory. *Science* 165:349–57
27. Brock HW, van Lohuizen M. 2001. The Polycomb group–no longer an exclusive club? *Curr. Opin. Genet. Dev.* 11:175–81
28. Brownell JE, Zhou J, Ranalli T, Kobayashi R, Edmondson DG, et al. 1996. Tetrahymena histone acetyltransferase A: a homolog to yeast Gcn5p linking histone acetylation to gene activation. *Cell* 84:843–51
29. Brzeski J, Jerzmanowski A. 2003. Deficient in DNA methylation 1 (DDM1) defines a novel family of chromatin-remodeling factors. *J. Biol. Chem.* 278:823–28
30. Brzeski J, Podstolski W, Olczak K, Jerzmanowski A. 1999. Identification and analysis of the Arabidopsis thaliana BSH gene, a member of the SNF5 gene family. *Nucleic Acids Res.* 27:2393–99
31. Byrd KN, Shearn A. 2003. ASH1, a Drosophila trithorax group protein, is required for methylation of lysine 4 residues on histone H3. *Proc. Natl. Acad. Sci. USA* 100:11535–40
32. Cao R, Zhang Y. 2004. The functions of E(Z)/EZH2-mediated methylation of lysine 27 in histone H3. *Curr. Opin. Genet. Dev.* 14:155–64
33. Cao X, Aufsatz W, Zilberman D, Mette MF, Huang MS, et al. 2003. Role of the DRM and CMT3 methyltransferases in RNA-directed DNA methylation. *Curr. Biol.* 13:2212–17
34. Cao X, Jacobsen SE. 2002. Locus-specific control of asymmetric and CpNpG methylation by the DRM and CMT3 methyltransferase genes. *Proc. Natl. Acad. Sci. USA* 99(Suppl. 4): 16491–98
35. Cao X, Jacobsen SE. 2002. Role of the Arabidopsis DRM methyltransferases in de novo DNA methylation and gene silencing. *Curr. Biol.* 12:1138–44

35a. Cao X, Springer NM, Muszynski MG, Phillips RL, Kaeppler S, Jacobsen SE. 2000. Conserved plant genes with similarity to mammalian *de novo* DNA methyltransferases. *Proc. Natl. Acad. Sci. USA* 97:4979–84

36. Carrington JC, Ambros V. 2003. Role of microRNAs in plant and animal development. *Science* 301:336–38
37. Chan SW, Zilberman D, Xie Z, Johansen LK, Carrington JC, Jacobsen SE. 2004. RNA silencing genes control de novo DNA methylation. *Science* 303:1336
38. Chandler VL, Eggleston WB, Dorweiler JE. 2000. Paramutation in maize. *Plant Mol. Biol.* 43:121–45
39. Chandler VL, Stam M. 2004. Chromatin conversations: mechanisms and implications of paramutation. *Nat. Rev. Genet.* 5:532–44

39a. Chanvivattana Y, Bishopp A, Schubert D, Stock C, Moon YH. 2004. Interaction of Polycomb-group proteins controlling flowering in Arabidopsis. *Development* 131:5263–76

40. Chen X. 2004. A microRNA as a translational repressor of APETALA2 in Arabidopsis flower development. *Science* 303:2022–25
41. **Choi Y, Gehring M, Johnson L, Hannon M, Harada JJ, et al. 2002. DEMETER, a DNA glycosylase domain protein, is required for endosperm gene imprinting and seed viability in Arabidopsis. *Cell* 110:33–42**
42. Chua YL, Brown AP, Gray JC. 2001. Targeted histone acetylation and altered nuclease accessibility over short regions of the pea plastocyanin gene. *Plant Cell* 13:599–612
43. Chua YL, Watson LA, Gray JC. 2003. The transcriptional enhancer of the pea plastocyanin gene associates with the nuclear matrix and regulates gene expression through histone acetylation. *Plant Cell* 15:1468–79

The authors identified a novel DNA glycosylase protein that regulates gene transcription. DME is required for imprinting in the Arabidopsis endosperm.

44. Cunliffe VT. 2003. Memory by modification: the influence of chromatin structure on gene expression during vertebrate development. *Gene* 305:141–50

45. Della Vedova CB, Cone KC. 2004. Paramutation: the chromatin connection. *Plant Cell* 16:1358–64

46. Deuring R, Fanti L, Armstrong JA, Sarte M, Papoulas O, et al. 2000. The ISWI chromatin-remodeling protein is required for gene expression and the maintenance of higher order chromatin structure in vivo. *Mol. Cell* 5:355–65

47. Eshed Y, Baum SF, Bowman JL. 1999. Distinct mechanisms promote polarity establishment in carpels of Arabidopsis. *Cell* 99:199–209

48. Farrona S, Hurtado L, Bowman JL, Reyes JC. 2004. The Arabidopsis thaliana SNF2 homolog AtBRM controls shoot development and flowering. *Development* 131:4965–75

49. Feng Q, Wang H, Ng HH, Erdjument-Bromage H, Tempst P, et al. 2002. Methylation of H3-lysine 79 is mediated by a new family of HMTases without a SET domain. *Curr. Biol.* 12:1052–58

50. Ferguson-Smith AC, Surani MA. 2001. Imprinting and the epigenetic asymmetry between parental genomes. *Science* 293:1086–89

50a. Finnegan EJ, Dennis ES. 1993. Isolation and identification by sequence homology of a putative cytosine methyltransferase from *Arabidopsis thaliana*. *Nucleic Acids Res.* 21:2383–88

51. Finnegan EJ, Peacock WJ, Dennis ES. 2000. DNA methylation, a key regulator of plant development and other processes. *Curr. Opin. Genet. Dev.* 10:217–23

52. **Finnegan EJ, Sheldon CC, Jardinaud F, Peacock WJ, Dennis ES. 2004. A cluster of Arabidopsis genes with a coordinate response to an environmental stimulus. *Curr. Biol.* 14:911–16**

This paper revealed that genes clustered around FLC are coregulated. The study also identified sequences within the FLC locus that control chromatin structure. This is the first chromosomal domain reported in plants.

53. Fire A, Xu S, Montgomery MK, Kostas SA, Driver SE, Mello CC. 1998. Potent and specific genetic interference by double-stranded RNA in Caenorhabditis elegans. *Nature* 391:806–11

54. Fischle W, Wang Y, Allis CD. 2003. Histone and chromatin cross-talk. *Curr. Opin. Cell. Biol.* 15:172–83

55. Francis NJ, Kingston RE. 2001. Mechanisms of transcriptional memory. *Nat. Rev. Mol. Cell. Biol.* 2:409–21

56. Fransz PF, de Jong JH. 2002. Chromatin dynamics in plants. *Curr. Opin. Plant Biol.* 5:560–67

57. Furner IJ, Sheikh MA, Collett CE. 1998. Gene silencing and homology-dependent gene silencing in Arabidopsis: genetic modifiers and DNA methylation. *Genetics* 149:651–62

58. Gavin IM, Simpson RT. 1997. Interplay of yeast global transcriptional regulators Ssn6p-Tup1p and Swi-Snf and their effect on chromatin structure. *EMBO J.* 16:6263–71

59. Gendall AR, Levy YY, Wilson A, Dean C. 2001. The VERNALIZATION 2 gene mediates the epigenetic regulation of vernalization in Arabidopsis. *Cell* 107:525–35

60. Gendrel AV, Lippman Z, Yordan C, Colot V, Martienssen RA. 2002. Dependence of heterochromatic histone H3 methylation patterns on the Arabidopsis gene DDM1. *Science* 297:1871–73

61. Geng F, Cao Y, Laurent BC. 2001. Essential roles of Snf5p in Snf-Swi chromatin remodeling in vivo. *Mol. Cell. Biol.* 21:4311–20

62. **Gong Z, Morales-Ruiz T, Ariza RR, Roldan-Arjona T, David L, Zhu JK. 2002. ROS1, a repressor of transcriptional gene silencing in Arabidopsis, encodes a DNA glycosylase/lyase. *Cell* 111:803–14**

The authors identified a DNA glycosylase protein that suppresses gene silencing in vivo and excises 5-methylcytosine residues from DNA in vitro.

63. Goodrich J, Puangsomlee P, Martin M, Long D, Meyerowitz EM, Coupland G. 1997. A Polycomb-group gene regulates homeotic gene expression in Arabidopsis. *Nature* 386:44–51
64. Grewal SI, Elgin SC. 2002. Heterochromatin: new possibilities for the inheritance of structure. *Curr. Opin. Genet. Dev.* 12:178–87
65. Grewal SI, Moazed D. 2003. Heterochromatin and epigenetic control of gene expression. *Science* 301:798–802
66. Guitton AE, Page DR, Chambrier P, Lionnet C, Faure JE, et al. 2004. Identification of new members of Fertilisation Independent Seed Polycomb Group pathway involved in the control of seed development in Arabidopsis thaliana. *Development* 131:2971–81
67. Hall IM, Noma K, Grewal SI. 2003. RNA interference machinery regulates chromosome dynamics during mitosis and meiosis in fission yeast. *Proc. Natl. Acad. Sci. USA* 100:193–98
68. Hall IM, Shankaranarayana GD, Noma K, Ayoub N, Cohen A, Grewal SI. 2002. Establishment and maintenance of a heterochromatin domain. *Science* 297:2232–37
69. Hamilton A, Voinnet O, Chappell L, Baulcombe D. 2002. Two classes of short interfering RNA in RNA silencing. *EMBO J.* 21:4671–79
70. Hammond SM, Bernstein E, Beach D, Hannon GJ. 2000. An RNA-directed nuclease mediates post-transcriptional gene silencing in Drosophila cells. *Nature* 404:293–96
71. Hayes JJ, Hansen JC. 2001. Nucleosomes and the chromatin fiber. *Curr. Opin. Genet. Dev.* 11:124–29
72. Heitz E. 1928. Das heterochromatin der Moose. *Jehrb. Wiss. Botanik* 69:762–818
73. Henderson IR, Shindo C, Dean C. 2003. The need for winter in the switch to flowering. *Annu. Rev. Genet.* 37:371–92
74. Henikoff S, Comai L. 1998. A DNA methyltransferase homolog with a chromodomain exists in multiple polymorphic forms in Arabidopsis. *Genetics* 149:307–18
75. Henikoff S, Furuyama T, Ahmad K. 2004. Histone variants, nucleosome assembly and epigenetic inheritance. *Trends Genet.* 20:320–26
76. Herman H, Lu M, Anggraini M, Sikora A, Chang Y, et al. 2003. Trans allele methylation and paramutation-like effects in mice. *Nat. Genet.* 34:199–202
77. Hirochika H, Okamoto H, Kakutani T. 2000. Silencing of retrotransposons in arabidopsis and reactivation by the ddm1 mutation. *Plant Cell* 12:357–69
78. Hsieh TF, Hakim O, Ohad N, Fischer RL. 2003. From flour to flower: how Polycomb group proteins influence multiple aspects of plant development. *Trends Plant Sci.* 8:439–45
79. Iizuka M, Smith MM. 2003. Functional consequences of histone modifications. *Curr. Opin. Genet. Dev.* 13:154–60
80. Ito T, Bulger M, Pazin MJ, Kobayashi R, Kadonaga JT. 1997. ACF, an ISWI-containing and ATP-utilizing chromatin assembly and remodeling factor. *Cell* 90:145–55
81. Jackson JP, Lindroth AM, Cao X, Jacobsen S. 2002. Control of CpNpG DNA methylation by the KRYPTONITE histone H3 methyltransferase. *Nature* 416:556–60
82. Jackson JP, Lindroth AM, Cao X, Jacobsen SE. 2002. Control of CpNpG DNA methylation by the KRYPTONITE histone H3 methyltransferase. *Nature* 416:556–60
83. Jasencakova Z, Soppe WJ, Meister A, Gernand D, Turner BM, Schubert I. 2003. Histone modifications in Arabidopsis—high methylation of H3 lysine 9 is dispensable for constitutive heterochromatin. *Plant J.* 33:471–80
84. Jeddeloh JA, Bender J, Richards EJ. 1998. The DNA methylation locus DDM1 is required for maintenance of gene silencing in Arabidopsis. *Genes Dev.* 12:1714–25
85. Jeddeloh JA, Stokes TL, Richards EJ. 1999. Maintenance of genomic methylation requires a SWI2/SNF2-like protein. *Nat. Genet.* 22:94–97

86. Jenuwein T, Allis CD. 2001. Translating the histone code. *Science* 293:1074–80
87. Kadosh D, Struhl K. 1998. Targeted recruitment of the Sin3-Rpd3 histone deacetylase complex generates a highly localized domain of repressed chromatin in vivo. *Mol. Cell. Biol.* 18:5121–27
88. Kanno T, Mette MF, Kreil DP, Aufsatz W, Matzke M, Matzke AJ. 2004. Involvement of putative SNF2 chromatin remodeling protein DRD1 in RNA-directed DNA methylation. *Curr. Biol.* 14:801–5
89. Katz A, Oliva M, Mosquna A, Hakim O, Ohad N. 2004. FIE and CURLY LEAF polycomb proteins interact in the regulation of homeobox gene expression during sporophyte development. *Plant J.* 37:707–19
90. Kaufman PD, Kobayashi R, Stillman B. 1997. Ultraviolet radiation sensitivity and reduction of telomeric silencing in Saccharomyces cerevisiae cells lacking chromatin assembly factor-I. *Genes Dev.* 11:345–57
91. Kawasaki H, Taira K. 2004. Induction of DNA methylation and gene silencing by short interfering RNAs in human cells. *Nature* 431:211–17
92. Kingston RE, Narlikar GJ. 1999. ATP-dependent remodeling and acetylation as regulators of chromatin fluidity. *Genes Dev.* 13:2339–52
93. Kinoshita T, Harada JJ, Goldberg RB, Fischer RL. 2001. Polycomb repression of flowering during early plant development. *Proc. Natl. Acad. Sci. USA* 98:14156–61
94. Kinoshita T, Miura A, Choi Y, Kinoshita Y, Cao X, et al. 2004. One-way control of FWA imprinting in Arabidopsis endosperm by DNA methylation. *Science* 303:521–23
95. Kinoshita T, Yadegari R, Harada JJ, Goldberg RB, Fischer RL. 1999. Imprinting of the MEDEA polycomb gene in the Arabidopsis endosperm. *Plant Cell* 11:1945–52
96. Kohler C, Hennig L, Bouveret R, Gheyselinck J, Grossniklaus U, Gruissem W. 2003. Arabidopsis MSI1 is a component of the MEA/FIE Polycomb group complex and required for seed development. *EMBO J.* 22:4804–14
97. Kohler C, Hennig L, Spillane C, Pien S, Gruissem W, Grossniklaus U. 2003. The Polycomb-group protein MEDEA regulates seed development by controlling expression of the MADS-box gene PHERES1. *Genes Dev.* 17:1540–53
98. Kolle D, Brosch G, Lechner T, Pipal A, Helliger W, et al. 1999. Different types of maize histone deacetylases are distinguished by a highly complex substrate and site specificity. *Biochemistry* 38:6769–73
99. Kuo MH, vom Baur E, Struhl K, Allis CD. 2000. Gcn4 activator targets Gcn5 histone acetyltransferase to specific promoters independently of transcription. *Mol. Cell* 6:1309–20
100. Kurdistani SK, Grunstein M. 2003. Histone acetylation and deacetylation in yeast. *Nat. Rev. Mol. Cell. Biol.* 4:276–84
101. Langst G, Becker PB. 2004. Nucleosome remodeling: one mechanism, many phenomena? *Biochim. Biophys. Acta* 1677:58–63
102. Lawrence RJ, Earley K, Pontes O, Silva M, Chen ZJ, et al. 2004. A concerted DNA methylation/histone methylation switch regulates rRNA gene dosage control and nucleolar dominance. *Mol. Cell* 13:599–609
103. Lechner T, Lusser A, Pipal A, Brosch G, Loidl A, et al. 2000. RPD3-type histone deacetylases in maize embryos. *Biochemistry* 39:1683–92
104. Lercher MJ, Urrutia AO, Hurst LD. 2002. Clustering of housekeeping genes provides a unified model of gene order in the human genome. *Nat. Genet.* 31:180–83
105. Levy YY, Mesnage S, Mylne JS, Gendall AR, Dean C. 2002. Multiple roles of Arabidopsis VRN1 in vernalization and flowering time control. *Science* 297:243–46

106. Lindroth AM, Cao X, Jackson JP, Zilberman D, McCallum CM, et al. 2001. Requirement of CHROMOMETHYLASE3 for maintenance of CpXpG methylation. *Science* 292:2077–80
107. Lippman ZL, Gendrel AV, Black M, Vaughn MW, Dedhia N, et al. 2004. Role of transposable elements in heterochromatin and epigenetic control. *Nature* 430:471–76
108. Llave C, Xie Z, Kasschau KD, Carrington JC. 2002. Cleavage of Scarecrow-like mRNA targets directed by a class of Arabidopsis miRNA. *Science* 297:2053–56
109. Loidl P. 2004. A plant dialect of the histone language. *Trends Plant Sci.* 9:84–90
110. Lunyak VV, Burgess R, Prefontaine GG, Nelson C, Sze SH, et al. 2002. Corepressor-dependent silencing of chromosomal regions encoding neuronal genes. *Science* 298:1747–52
111. Lusser A, Brosch G, Loidl A, Haas H, Loidl P. 1997. Identification of maize histone deacetylase HD2 as an acidic nucleolar phosphoprotein. *Science* 277:88–91
112. Lusser A, Kadonaga JT. 2003. Chromatin remodeling by ATP-dependent molecular machines. *Bioessays* 25:1192–200
113. Malagnac F, Bartee L, Bender J. 2002. An Arabidopsis SET domain protein required for maintenance but not establishment of DNA methylation. *EMBO J.* 21:6842–52
114. Martens JA, Winston F. 2003. Recent advances in understanding chromatin remodeling by Swi/Snf complexes. *Curr. Opin. Genet. Dev.* 13:136–42
115. Martienssen RA, Colot V. 2001. DNA methylation and epigenetic inheritance in plants and filamentous fungi. *Science* 293:1070–74
116. Matzke M, Matzke AJ, Kooter JM. 2001. RNA: guiding gene silencing. *Science* 293:1080–83
117. McKittrick E, Gafken PR, Ahmad K, Henikoff S. 2004. Histone H3.3 is enriched in covalent modifications associated with active chromatin. *Proc. Natl. Acad. Sci. USA* 101:1525–30
118. Mette MF, Aufsatz W, van der Winden J, Matzke MA, Matzke AJ. 2000. Transcriptional silencing and promoter methylation triggered by double-stranded RNA. *EMBO J.* 19:5194–201
119. Milne TA, Briggs SD, Brock HW, Martin ME, Gibbs D, et al. 2002. MLL targets SET domain methyltransferase activity to Hox gene promoters. *Mol. Cell* 10:1107–17
120. Monson EK, de Bruin D, Zakian VA. 1997. The yeast Cac1 protein is required for the stable inheritance of transcriptionally repressed chromatin at telomeres. *Proc. Natl. Acad. Sci. USA* 94:13081–86
121. Morris KV, Chan SW, Jacobsen SE, Looney DJ. 2004. Small interfering RNA-induced transcriptional gene silencing in human cells. *Science* 305:1289–92
122. Murfett J, Wang XJ, Hagen G, Guilfoyle TJ. 2001. Identification of Arabidopsis histone deacetylase HDA6 mutants that affect transgene expression. *Plant Cell* 13:1047–61
123. Nakamura T, Mori T, Tada S, Krajewski W, Rozovskaia T, et al. 2002. ALL-1 is a histone methyltransferase that assembles a supercomplex of proteins involved in transcriptional regulation. *Mol. Cell* 10:1119–28
124. Nakayama J, Rice JC, Strahl BD, Allis CD, Grewal SI. 2001. Role of histone H3 lysine 9 methylation in epigenetic control of heterochromatin assembly. *Science* 292:110–13
125. Noh YS, Amasino RM. 2003. PIE1, an ISWI family gene, is required for FLC activation and floral repression in Arabidopsis. *Plant Cell* 15:1671–82
126. Ogas J, Cheng JC, Sung ZR, Somerville C. 1997. Cellular differentiation regulated by gibberellin in the Arabidopsis thaliana pickle mutant. *Science* 277:91–94
127. Ogas J, Kaufmann S, Henderson J, Somerville C. 1999. PICKLE is a CHD3 chromatin-remodeling factor that regulates the transition from embryonic to vegetative development in Arabidopsis. *Proc. Natl. Acad. Sci. USA* 96:13839–44

The study showed that H3K9 methylation is closely associated with transposable elements, indicating that transposable elements are a major determinant of heterochromatin formation in Arabidopsis.

This study provides evidence that there is sufficient H3.3 histone to mark all the active chromatin in the nucleus. They show that purified H3.3 is enriched for modifications associated with active chromatin while depleted for repressive modifications like H3K9 methylation.

128. Ogryzko VV. 2001. Mammalian histone acetyltransferases and their complexes. *Cell. Mol. Life Sci.* 58:683–92
129. Okano M, Bell DW, Haber DA, Li E. 1999. DNA methyltransferases Dnmt3a and Dnmt3b are essential for de novo methylation and mammalian development. *Cell* 99:247–57
130. Olsen PH, Ambros V. 1999. The lin-4 regulatory RNA controls developmental timing in Caenorhabditis elegans by blocking LIN-14 protein synthesis after the initiation of translation. *Dev. Biol.* 216:671–80
131. Osley MA. 1991. The regulation of histone synthesis in the cell cycle. *Annu. Rev. Biochem.* 60:827–61
132. Pal-Bhadra M, Leibovitch BA, Gandhi SG, Rao M, Bhadra U, et al. 2004. Heterochromatic silencing and HP1 localization in Drosophila are dependent on the RNAi machinery. *Science* 303:669–72
133. Pal-Bhadra M, Bhadra U, Birchler JA. 2002. RNAi related mechanisms affect both transcriptional and posttranscriptional transgene silencing in Drosophila. *Mol. Cell* 9:315–27
134. Palatnik JF, Allen E, Wu X, Schommer C, Schwab R, et al. 2003. Control of leaf morphogenesis by microRNAs. *Nature* 425:257–63
135. Pandey R, Muller A, Napoli CA, Selinger DA, Pikaard CS, et al. 2002. Analysis of histone acetyltransferase and histone deacetylase families of Arabidopsis thaliana suggests functional diversification of chromatin modification among multicellular eukaryotes. *Nucleic Acids Res.* 30:5036–55
136. Papa CM, Springer NM, Muszynski MG, Meeley R, Kaeppler SM. 2001. Maize chromomethylase Zea methyltransferase2 is required for CpNpG methylation. *Plant Cell* 13:1919–28
137. Partridge JF, Scott KS, Bannister AJ, Kouzarides T, Allshire RC. 2002. cis-acting DNA from fission yeast centromeres mediates histone H3 methylation and recruitment of silencing factors and cohesin to an ectopic site. *Curr. Biol.* 12:1652–60
138. Phelan ML, Sif S, Narlikar GJ, Kingston RE. 1999. Reconstitution of a core chromatin remodeling complex from SWI/SNF subunits. *Mol. Cell* 3:247–53
139. Probst AV, Fagard M, Proux F, Mourrain P, Boutet S, et al. 2004. Arabidopsis histone deacetylase HDA6 is required for maintenance of transcriptional gene silencing and determines nuclear organization of rDNA repeats. *Plant Cell* 16:1021–34
140. Rea S, Eisenhaber F, O'Carroll D, Strahl BD, Sun ZW, et al. 2000. Regulation of chromatin structure by site-specific histone H3 methyltransferases. *Nature* 406:593–59
141. Reik W, Walter J. 2001. Genomic imprinting: parental influence on the genome. *Nat. Rev. Genet.* 2:21–32
142. Reyes JC, Grossniklaus U. 2003. Diverse functions of Polycomb group proteins during plant development. *Semin. Cell. Dev. Biol.* 14:77–84
143. Reyes JC, Hennig L, Gruissem W. 2002. Chromatin-remodeling and memory factors. New regulators of plant development. *Plant Physiol.* 130:1090–101
144. Rice JC, Briggs SD, Ueberheide B, Barber CM, Shabanowitz J, et al. 2003. Histone methyltransferases direct different degrees of methylation to define distinct chromatin domains. *Mol. Cell* 12:1591–98
145. Roth SY, Denu JM, Allis CD. 2001. Histone acetyltransferases. *Annu. Rev. Biochem.* 70:81–120
146. Santisteban MS, Kalashnikova T, Smith MM. 2000. Histone H2A.Z regulates transcription and is partially redundant with nucleosome remodeling complexes. *Cell* 103:411–22
147. Santos-Rosa H, Schneider R, Bannister AJ, Sherriff J, Bernstein BE, et al. 2002. Active genes are tri-methylated at K4 of histone H3. *Nature* 419:407–11

148. Sarnowski TJ, Swiezewski S, Pawlikowska K, Kaczanowski S, Jerzmanowski A. 2002. AtSWI3B, an Arabidopsis homolog of SWI3, a core subunit of yeast Swi/Snf chromatin remodeling complex, interacts with FCA, a regulator of flowering time. *Nucleic Acids Res.* 30:3412–21
149. Schramke V, Allshire R. 2003. Hairpin RNAs and retrotransposon LTRs effect RNAi and chromatin-based gene silencing. *Science* 301:1069–74
150. Sijen T, Vijn I, Rebocho A, van Blokland R, Roelofs D, et al. 2001. Transcriptional and posttranscriptional gene silencing are mechanistically related. *Curr. Biol.* 11:436–40
151. Sims RJ 3rd, Nishioka K, Reinberg D. 2003. Histone lysine methylation: a signature for chromatin function. *Trends Genet.* 19:629–39
152. Singer T, Yordan C, Martienssen RA. 2001. Robertson's mutator transposons in A. thaliana are regulated by the chromatin-remodeling gene decrease in DNA methylation (DDM1). *Genes Dev.* 15:591–602
153. Smith MM. 2002. Histone variants and nucleosome deposition pathways. *Mol. Cell* 9:1158–60
154. Soppe WJ, Jacobsen SE, Alonso-Blanco C, Jackson KJ, Kakutani T, et al. 2000. The late flowering phenotype of fwa mutants is caused by gain-of-function epigenetic alleles of a homeodomain gene. *Mol. Cell* 6:791–802
155. Springer NM, Napoli CA, Selinger DA, Pandey R, Cone KC, et al. 2003. Comparative analysis of SET domain proteins in maize and Arabidopsis reveals multiple duplications preceding the divergence of monocots and dicots. *Plant Physiol.* 132:907–25
156. Stam M, Belele C, Dorweiler JE, Chandler VL. 2002. Differential chromatin structure within a tandem array 100 kb upstream of the maize b1 locus is associated with paramutation. *Genes Dev.* 16:1906–18
157. Stam M, Belele C, Ramakrishna W, Dorweiler JE, Bennetzen JL, Chandler VL. 2002. The regulatory regions required for B′ paramutation and expression are located far upstream of the maize b1 transcribed sequences. *Genetics* 162:917–30
158. Stockinger EJ, Mao Y, Regier MK, Triezenberg SJ, Thomashow MF. 2001. Transcriptional adaptor and histone acetyltransferase proteins in Arabidopsis and their interactions with CBF1, a transcriptional activator involved in cold-regulated gene expression. *Nucleic Acids Res.* 29:1524–33
159. Strahl BD, Allis CD. 2000. The language of covalent histone modifications. *Nature* 403:41–45
160. Sun FL, Cuaycong MH, Elgin SC. 2001. Long-range nucleosome ordering is associated with gene silencing in Drosophila melanogaster pericentric heterochromatin. *Mol. Cell. Biol.* 21:2867–79
160a. Sung S, Amasino RM. 2005. Remembering winter: toward a molecular understanding of vernalization. *Annu. Rev. Plant. Biol.* 56:491–508
161. Sung S, Amasino RM. 2004. Vernalization and epigenetics: how plants remember winter. *Curr. Opin. Plant Biol.* 7:4–10
162. Sung S, Amasino RM. 2004. Vernalization in Arabidopsis thaliana is mediated by the PHD finger protein VIN3. *Nature* 427:159–64
163. Tagami H, Ray-Gallet D, Almouzni G, Nakatani Y. 2004. Histone H3.1 and H3.3 complexes mediate nucleosome assembly pathways dependent or independent of DNA synthesis. *Cell* 116:51–61
164. Tamaru H, Selker EU. 2001. A histone H3 methyltransferase controls DNA methylation in *Neurospora crassa*. *Nature* 414:277–83

This and the following paper (157) identify a promoter enhancer region important for paramutation. The studies also demonstrate that chromatin structure and DNA methylation patterns correlate with paramutagenetic and paramutable alleles, providing evidence that differential chromatin structures are involved in paramutation.

The authors show how flowering time is influenced by vernalization at the chromatin level. They describe a series of vernalization-induced histone modifications leading to stable repression of FLC expression.

This study finds that nucleosome assembly complexes are associated with H3-H4 dimers rather than tetramers, raising the possibility that epigenetic marks on the nucleosomes can be replicated and inherited in a semiconservative fashion.

165. Tamaru H, Zhang X, McMillen D, Singh PB, Nakayama J, et al. 2003. Trimethylated lysine 9 of histone H3 is a mark for DNA methylation in Neurospora crassa. *Nat. Genet.* 34:75–79
166. Tariq M, Paszkowski J. 2004. DNA and histone methylation in plants. *Trends Genet.* 20:244–51
167. Taunton J, Hassig CA, Schreiber SL. 1996. A mammalian histone deacetylase related to the yeast transcriptional regulator Rpd3p. *Science* 272:408–11
168. Tian L, Fong MP, Wang JJ, Wei NE, Jiang H, et al. 2004. Reversible histone acetylation and deacetylation mediate genome-wide, promoter-dependent, and locus-specific changes in gene expression during plant development. *Genetics* In press
169. Tian L, Wang J, Fong MP, Chen M, Cao H, et al. 2003. Genetic control of developmental changes induced by disruption of Arabidopsis histone deacetylase 1 (AtHD1) expression. *Genetics* 165:399–409
170. Turner BM. 2002. Cellular memory and the histone code. *Cell* 111:285–91
171. Verdel A, Jia S, Gerber S, Sugiyama T, Gygi S, et al. 2004. RNAi-mediated targeting of heterochromatin by the RITS complex. *Science* 303:672–76
172. Vielle-Calzada JP, Thomas J, Spillane C, Coluccio A, Hoeppner MA, Grossniklaus U. 1999. Maintenance of genomic imprinting at the Arabidopsis medea locus requires zygotic DDM1 activity. *Genes Dev.* 13:2971–82
173. Vlachonasios KE, Thomashow MF, Triezenberg SJ. 2003. Disruption mutations of ADA2b and GCN5 transcriptional adaptor genes dramatically affect Arabidopsis growth, development, and gene expression. *Plant Cell* 15:626–38
174. Volpe T, Schramke V, Hamilton GL, White SA, Teng G, et al. 2003. RNA interference is required for normal centromere function in fission yeast. *Chromosome Res.* 11:137–46
175. Volpe TA, Kidner C, Hall IM, Teng G, Grewal SI, Martienssen RA. 2002. Regulation of heterochromatic silencing and histone H3 lysine-9 methylation by RNAi. *Science* 297:1833–37
176. Wagner D, Meyerowitz EM. 2002. SPLAYED, a novel SWI/SNF ATPase homolog, controls reproductive development in Arabidopsis. *Curr. Biol.* 12:85–94
177. Wassenegger M, Heimes S, Riedel L, Sanger HL. 1994. RNA-directed de novo methylation of genomic sequences in plants. *Cell* 76:567–76
178. Waterborg JH. 1993. Dynamic methylation of alfalfa histone H3. *J. Biol. Chem.* 268:4918–21
179. Waterborg JH. 1993. Histone synthesis and turnover in alfalfa. Fast loss of highly acetylated replacement histone variant H3.2. *J. Biol. Chem.* 268:4912–17
180. Weitzman JB. 2002. Transcriptional territories in the genome. *J. Biol.* 1:2
180a. Winston F, Carlson M. 1992. Yeast SNF/SWI transcriptional activators and the SPT/SIN chromatin connection. *Trends Genet.* 8:387–91
181. Wolffe AP. 1998. *Chromatin: Structure and Function*. London: Academic
182. Wu K, Malik K, Tian L, Brown D, Miki B. 2000. Functional analysis of a RPD3 histone deacetylase homologue in Arabidopsis thaliana. *Plant Mol. Biol.* 44:167–76
183. Wu K, Tian L, Malik K, Brown D, Miki B. 2000. Functional analysis of HD2 histone deacetylase homologues in Arabidopsis thaliana. *Plant J.* 22:19–27
184. Wu L, Winston F. 1997. Evidence that Snf-Swi controls chromatin structure over both the TATA and UAS regions of the SUC2 promoter in Saccharomyces cerevisiae. *Nucleic Acids Res.* 25:4230–34
185. Xiao W, Gehring M, Choi Y, Margossian L, Pu H, et al. 2003. Imprinting of the MEA Polycomb gene is controlled by antagonism between MET1 methyltransferase and DME glycosylase. *Dev. Cell* 5:891–901

186. Xie Z, Johansen LK, Gustafson AM, Kasschau1 KD, Lellis AD, et al. 2004. Genetic and functional diversification of small RNA pathways in plants. *PLoS Biol.* 2:E104
187. Xin Z, Allis CD, Wagstaff J. 2001. Parent-specific complementary patterns of histone H3 lysine 9 and H3 lysine 4 methylation at the Prader-Willi syndrome imprinting center. *Am. J. Hum. Genet.* 69:1389–94
188. Yoshida N, Yanai Y, Chen L, Kato Y, Hiratsuka J, et al. 2001. EMBRYONIC FLOWER2, a novel polycomb group protein homolog, mediates shoot development and flowering in Arabidopsis. *Plant Cell* 13:2471–81
189. Zemach A, Grafi G. 2003. Characterization of Arabidopsis thaliana methyl-CpG-binding domain (MBD) proteins. *Plant J.* 34:565–72
190. Zhou C, Labbe H, Sridha S, Wang L, Tian L, et al. 2004. Expression and function of HD2-type histone deacetylases in Arabidopsis development. *Plant J.* 38:715–24
191. Zhou C, Miki B, Wu K. 2003. CHB2, a member of the SWI3 gene family, is a global regulator in Arabidopsis. *Plant Mol. Biol.* 52:1125–34
192. Zilberman D, Cao X, Jacobsen SE. 2003. ARGONAUTE4 control of locus-specific siRNA accumulation and DNA and histone methylation. *Science* 299:716–19
193. He Y, Michaels SD, Amasino RM. 2003. Regulation of flowering time by histone acetylation in Arabidopsis. *Science* 302:1751–54
194. Kohler C, Page DR, Gagliardini V, Grossniklaus U. 2005. The *Arabidopsis thaliana* MEDEA Polycomb group protein controls expression of PHERES1 by parental imprinting. *Nat. Genet.* 37:28–30
195. Shi Y, Lan F, Matson C, Mulligan P, Whetstine JR, et al. 2004. Histone demethylation mediated by the nuclear amine oxidase homolog LSD1. *Cell* 119:941–53

Shoot Branching

Paula McSteen[1] and Ottoline Leyser[2]

[1]Department of Biology, Pennsylvania State University, University Park, Pennsylvania 16802; email: pcm11@psu.edu

[2]Department of Biology, University of York, Heslington, York, YO10 5YW, United Kingdom; email: hmol1@york.ac.uk

Annu. Rev. Plant Biol. 2005. 56:353–74

doi: 10.1146/annurev.arplant.56.032604.144122

First published online as a Review in Advance on January 17, 2005

1543-5008/05/0602-0353$20.00

Key Words

axillary meristem, branch, auxin, spikelet, inflorescence

Abstract

All plant shoots can be described as a series of developmental modules termed phytomers, which are produced from shoot apical meristems. A phytomer generally consists of a leaf, a stem segment, and a secondary shoot meristem. The fate and activity adopted by these secondary, axillary shoot meristems is the major source of evolutionary and environmental diversity in shoot system architecture. Axillary meristem fate and activity are regulated by the interplay of genetic programs with the environment. Recent results show that these inputs are channeled through interacting hormonal and transcription factor regulatory networks. Comparison of the factors involved in regulating the function of diverse axillary meristem types both within and between species is gradually revealing a pattern in which a common basic program has been modified to produce a range of axillary meristem types.

Contents

INTRODUCTION

SAM: shoot apical meristem

AM: axillary meristem

The basic body plan of higher plants is established during embryogenesis, with the development of an apical-basal axis defined by a root apical meristem at one end and a SAM at the other. However, this simple bipolar organization is only the starting point for an impressive diversity of postembryonic body plans. The complexity of plant form depends on the ability to establish new axes of growth, through the production of secondary AMs, which can recapitulate or deviate from the developmental program of the primary meristems. Despite the diverse fates possible for secondary meristems, the entire plant body consists of a series of similar modules produced by the meristems. In the shoot, each module, or phytomer, consists of a stem segment, a node bearing one or more leaves or leaf-like structures, and one or more AMs in each leaf axil (94). The astonishing diversity of shoot form, both within and between species, is almost entirely a consequence of variation in the number of phytomers produced, the timing of their production, and the developmental fate that they adopt. Recent years have seen significant progress in understanding AM fate and the main conclusions that have emerged are reviewed here. For ease of reference, we first discuss dicots, followed by monocots, when we highlight similarities and differences. To avoid confusion, the entire review is written using *Arabidopsis* nomenclature.

BRANCHING IN DICOTS

Phytomer Diversity

SAMs can grow indeterminately, producing an endless succession of phytomers, the fate of which varies along the shoot axis. Alternatively, they can undergo a determinate developmental program, producing a fixed number of phytomers, usually terminating in the production of reproductive structures.

For example, in *Arabidopsis* (**Figure 1*a***), the primary SAM maintains an indeterminate growth pattern, first producing rosette phytomers consisting of a large leaf, a short stem segment, and a morphologically undetectable AM. At floral transition, several phytomers are produced consisting of a smaller leaf, a greatly elongated stem segment, and a large indeterminate AM. The series of phytomers that follows consists of a cryptic leaf, an elongated stem segment, and a determinate AM that forms a flower consisting of successive whorls (phytomers) of sepals, petals, stamens, and carpels. The flower is a compressed shoot with short internodes and suppressed AMs. Unlike some species, such as pea, only a single AM arises in each leaf axil, but additional meristems, termed accessory meristems, can arise later, adjacent to the original AM (34).

A superficially different growth habit is observed in tomato (**Figure 1*b***). In fact, the pattern of progression through different phytomer types is in many ways similar to *Arabidopsis* (84). The vegetative phytomers are produced in the same way, except with greater stem elongation than in *Arabidopsis*. Upon floral transition, the primary apex becomes determinate, initiating a fixed number of leaves before terminating in a flower. The AMs in these leaves immediately terminate in a flower, producing the familiar

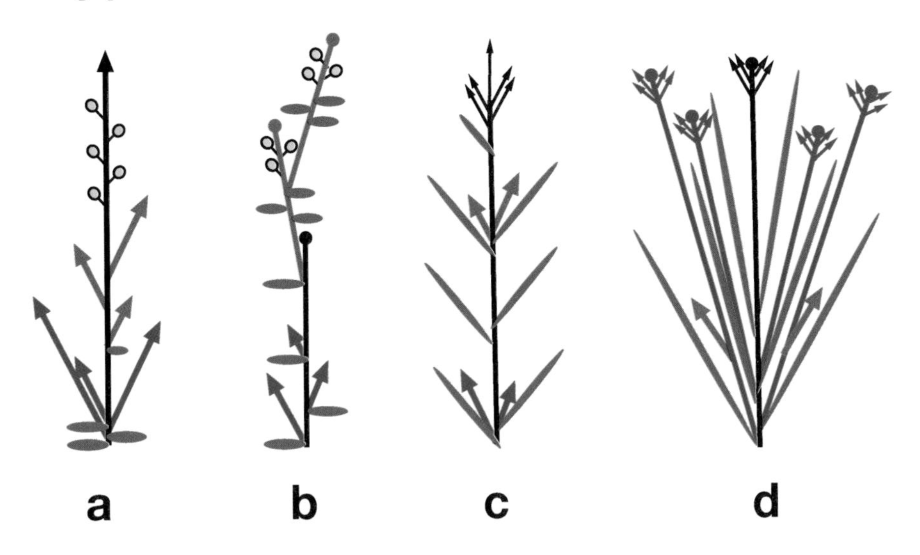

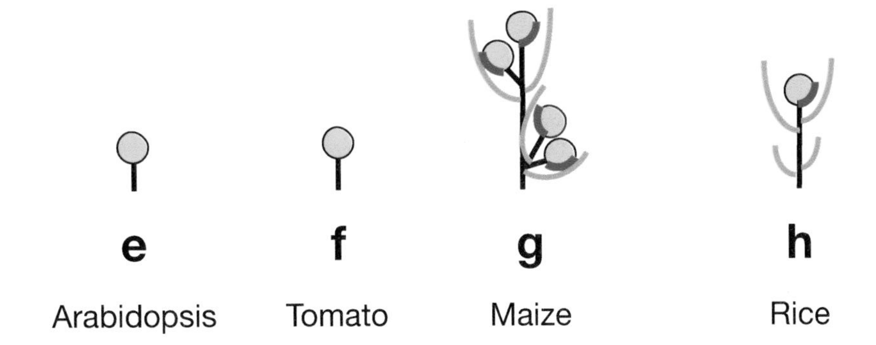

Figure 1

Axillary branching in dicots and monocots. The schematic indicates the activity of axillary meristems during development. Arrows represent indeterminate meristems and circles represent determinate meristems. The figure shows body plans of (*a*) *Arabidopsis*, (*b*) tomato, (*c*) maize, and (*d*) rice. The black lines indicate the main axis. The green lines indicate leaves. The blue arrows indicate the most basal-arising branches. The red arrows indicate axillary meristems arising later during development. The yellow circles represent flowers. Maize and rice also produce lateral branches in the inflorescence indicated by the black or blue lines with smaller arrowheads. The units of the inflorescence are shown in (*e*) *Arabidopsis*, (*f*) tomato, (*g*) maize, and (*d*) rice. The yellow circles represent flowers. Maize and rice flowers are enclosed by small leaf-like organs called glumes (*orange lines*). The purple lines indicate the position of the lemma, which is the bract subtending the flower. The rice spikelet consists of one flower and four glumes. Maize produces spikelets as pairs each consisting of two flowers and two glumes.

trusses of tomatoes. Meanwhile, the AM of the last vegetative leaf, called the sympodial meristem, takes over as the SAM. This sympodial meristem is as big as the now-floral primary SAM at its initiation and hence, morphologically, the primary apical meristem bifurcates at floral transition, but the axillary origin of the sympodial meristem is clear from its subtending leaf. The sympodial meristem then goes through a short vegetative phase, producing phytomers consisting of a leaf, a stem segment, and an indeterminate AM before becoming floral and repeating the sympodial branching pattern. Thus, variation in stem length, leaf size, AM size, and AM determinacy can account for the apparent differences between tomato (sympodial) and *Arabidopsis* (monopodial) architecture. As described below, similar arguments can be applied to the even more divergent growth habits of monocots (**Figure 1**).

The essential equivalence of all these phytomers is demonstrated by using molecular probes that can penetrate morphological disguises, and by the observation that apparent homeotic transformations between phytomer types can result from mutation in single genes. For example, the cryptic leaf of the flower bearing nodes of *Arabidopsis*, although invisible morphologically, can be detected as a result of its expression of the *AINTEGUMENTA* (*ANT*) leaf identity marker gene (58). The equivalence of determinate versus indeterminate meristems is supported by the phenotype of mutations in the *TERMINAL FLOWER1* (*TFL1*) gene, in which indeterminate meristems are converted to determinate ones (87). Note that in this context there is a major difficulty in interpreting such "homeotic" mutant phenotypes. In a wild-type plant, the developmental fate of phytomers changes both with position along the shoot axis and with time and thus it is not possible and/or meaningful to distinguish between homeotic mutants, where one structure replaces another, and heterochronic mutants, where the timing of the transition between different phases of growth is altered. In the *Arabidopsis tfl1* loss-of-function mutant, fewer of each type of phytomer is produced (85); when *TFL1* is overexpressed from the CaMV 35S promoter, more of each phytomer type is produced (79). The *TFL1* gene functions in a general way, both transcriptionally and post-transcriptionally, to delay expression of floral meristem identity genes such as *LEAFY* (*LFY*) and *APETALA1* (*AP1*). This effect delays the progression through different phytomer types and maintains apical meristem indeterminacy, which is an extreme case of its role in excluding *LFY* and *AP1* activity indefinitely from the center of indeterminate meristems (78).

Axillary Meristem Initiation in Dicots

The developmental origin of dicot axillary meristems. Although some diversity in form can be achieved simply by variation in the leaf and stem components of the phytomer, the major source of body plan variation derives from variation in the behavior of AMs. The ontogeny of AMs is the subject of a long-running debate. There are two opposing explanations for their origins. The first is that AMs initiate de novo in leaf axils (89), and the alternative is that they derive from groups of meristem cells that detach from the SAM at the time of leaf initiation and never lose their meristematic identity (32, 93). For the large AMs formed in the later phases of growth described above for *Arabidopsis* and tomato, there is little doubt that the cells of the meristem are derived directly from the primary SAM. However, for the early leaves, there is no morphologically distinguishable AM at the time of leaf initiation. Here, morphologically at least, the AM arises de novo. Furthermore, even in the more apical nodes in *Arabidopsis*, AMs can arise distal to the first AM, further up the petiole (34). These meristems also appear morphologically to arise de novo from the leaf.

Given the arguments above, that all phytomers are equivalent, the idea that different AMs have such different ontogenies seems unlikely. Attempts to resolve the issue using molecular markers have proven inconclusive. For example, the *SHOOTMERISTEMLESS* (*STM*) gene of *Arabidopsis* is expressed in the SAM and its function is required for SAM activity (59). Furthermore, *STM* expression is downregulated in initiating leaves, and taken together these observations make *STM* expression a good marker for meristematic identity (58, 59). There have been several attempts to determine whether the axils of vegetative nodes of *Arabidopsis* that lack morphologically defined meristems maintain *STM* expression (34, 35, 58). These results identify a clear focusing and intensification of *STM* expression immediately preceding the appearance of a morphological AM. The correlation between these events and AM formation is strengthened further by the observation that they fail to occur in plants mutant at the *REVOLUTA* (*REV*) locus, which also fail to develop AMs (74). *REV* transcription is also upregulated in leaf axils, preceding *STM* upregulation. However, despite the existence of these markers for AM differentiation, it is not possible to assess whether there is a phase prior to upregulation when *REV* and *STM* expression are completely absent, suggesting a complete loss of meristematic identity and hence de novo meristem initiation. Note that *REV* expression is also required during the initiation of the determinate floral AMs that arise in the cryptic leaves initiated by the primary SAM after floral transition (74, 96). This observation reenforces the idea that all these AMs are fundamentally similar.

Fortunately, an interesting resolution to the detached versus de novo AM debate is emerging from the study of the *LATERAL SUPPRESSOR* (*LS*) gene of tomato (86) and its *Arabidopsis* ortholog *LAS* (35). Tomato plants carrying loss-of-function mutations at the *LATERAL SUPPRESSOR* (*LS*) locus have reduced branching from vegetative nodes due to failure in the development of AMs specifically at these nodes. Similarly, mutations in the *Arabidopsis* ortholog *LAS* confer reduced branching due to failure in the development of AMs in the rosette phytomers. *LAS* encodes a transcription factor of the GRAS family. *LAS* transcripts are localized to the axils of all leaf primordia derived from the primary SAM and begin to accumulate early during leaf initiation on the SAM just as the primordium begins to form a bulge (35). Expression is in the *STM*-positive zone adjacent to the *STM*-negative region that defines the primordium. This leaf axil expression persists until the time of *STM* focusing and upregulation, which precede AM organization. In the *las* mutant, neither *REV* nor *STM* upregulation occurs in the rosette leaf axils, and AMs fail to develop in the rosette nodes. Accessory AMs also fail to develop, and thus meristems such as floral meristems that arise rapidly after lateral organ initiation on the primary SAM are unaffected, whereas those that arise after a significant delay are lost. This led to the hypothesis that *LS/LAS* functions to protect the leaf axil from full differentiation, maintaining its meristematic potential.

This attractive hypothesis is essentially a compromise between the detached and de novo ideas and maintains the same basic mechanism for developing all axil types. A group of cells with enhanced meristematic potential is maintained, but they lose the full suite of meristem characteristics, such as expression of the *CLV3* and *WUS* genes (74). This hypothesis links meristematic potential with adaxial position during leaf initiation and is strengthened by the observation that leaf polarity mutants that have adaxialized leaves produce AMs in a ring encircling the base of their petioles (63). Although it is still unknown what restricts meristematic potential in the leaf proximo-distal axis, limiting it to the petiole base, it is becoming clear that AM initiation starts at the time of leaf initiation with the definition of an adaxial zone of high meristematic potential, which can either rapidly develop into a fully fledged AM, or, if *LAS* is present, can remain in an undifferentiated state.

Auxin and axillary meristem initiation. Because AM formation is part of the general lateral organ initiation program at the SAM, it is not surprising that auxin is involved. A marriage of classical micromanipulation techniques and probes that predict auxin transport pathways provides convincing evidence that dynamic auxin fluxes pattern organ initiation at the shoot apex (80, 81). Mutations, such as *pinformed1* (*pin1*), *pinoid1* (*pid1*), and *monopteros* (*mp*), which disrupt auxin transport or signaling and organ formation also disrupt AM formation, showing that auxin or auxin transport is required for AM formation, at least during floral growth (5, 73, 77, 81, 99).

However, there is also evidence for a role for auxin in subsequent AM development. The *auxin resistant1* (*axr1*) auxin-resistant mutant of *Arabidopsis* has increased shoot branching, principally because of reduced AM dormancy, following initiation (see below), but the mutant also provides evidence for a role of auxin earlier in AM development. Although *las* is largely epistatic to *axr1* in the *las, axr1* double mutant, as would be expected if auxin acts in AM activation, *las* is not completely epistatic (35). When the number of rosette branches produced following decapitation is used as an assay for the number of rosette-borne AMs, the *las* mutant has fewer than wild type (0–2 compared to 5–13), but the *axr1, las* double mutant produces up to 5 branches. This is far fewer than the *axr1* single mutant (9–23) but clearly does not represent complete epistasis. This suggests either that the auxin-resistant mutant may organize its AMs earlier than wild type, reducing the requirement for *LAS* function, or that reduced auxin response in the axils delays terminal differentiation of the axils, thus reducing the requirement for *LAS* function. This latter explanation is more likely because the timing of AM initiation in *axr1* mutants appears to be wild type (91).

Axillary Meristem Outgrowth in Dicots

The fate of dicot axillary meristems. As described above, AMs can adopt a range of fates from indeterminate vegetative meristems to determinate flowers. Their fate is closely linked to that of the primary SAM. This is clear from the gross morphological differences in the size of the AM at its inception, depending on the floral status of the primary SAM. As mentioned above, this influence is mediated by the expression of meristem identity genes such as *LFY* and *AP1*. It is somewhat difficult to distinguish between the influence of the primary SAM and direct environmental regulation of the AMs because usually both the primary and AMs experience the same environment. However, experimental manipulation makes it possible to dissect these factors and demonstrate significant input from the primary SAM. For example, in *Arabidopsis*, plants that are shifted from nonflorally inductive conditions to florally inductive conditions switch to the production of phytomers where the AMs adopt a floral fate. If such plants are switched back to nonflorally inductive conditions, flowers continue to be produced, despite the fact that the new AMs have never been exposed to florally inductive conditions (14). This influence of the primary apex on AM fate is strengthened by the observation that in lines where the irreversibility of floral induction is compromised, such as in the Sy-O ecotype, as a result of a dominant allele at the *AERIAL ROSETTE1* locus, AMs follow the state of the primary apex and develop vegetatively (76).

Despite the obvious influence of the primary apex, it is important to note that AM fate is not strictly coupled to primary meristem fate. For example, the indeterminate AMs that arise in the cauline and rosette leaf axils of *Arabidopsis* undergo floral transition after the floral transition of the primary shoot apex, as indicated by the fact that they produce leaves at a time when the primary apex is producing flowers.

Dormancy in dicot axillary meristems. An important component in the plasticity of plant architectures derives from the ability to modulate the activity of AMs after their initiation. After initiation, and the formation of a few unexpanded phytomers, AMs can arrest their growth and produce a dormant bud. The bud may later reactivate, growing out to form a branch, or it may remain dormant for the entire life cycle of the plant (88).

Several different kinds of axillary bud dormancy have been described (39). Ecodormancy is imposed by external environmental factors and is experimentally characterized by the ability to break dormancy by changing the environmental conditions. Endodormancy is imposed by factors within the bud and cannot be broken by changing the environment. These categories are useful in analyzing long-term dormancy such as overwintering, where typically a phase of ecodormancy in autumn, during which a return to warm weather and longer days releases dormancy, is followed by a phase of endodormancy, during which dormancy is maintained regardless of environmental fluctuation and cannot be broken without a prolonged period of cold treatment (vernalization). In annual plants, to which the rest of the discussion is restricted, endodormancy is largely limited to the seed, and most bud dormancy is either environmentally imposed or imposed by factors outside the bud but internal to the rest of the plant (paradormancy). Because these factors are interconnected, it is both difficult and not useful to distinguish between them; instead it is better to consider the activity of the bud as regulated by a complex interplay between external environmental factors, physiological status, developmental status, and, of course, genotype.

Developmental regulation of bud activity. In many species, the pattern of AM activation depends on distance from, and developmental status of, the primary shoot apex. For example, in *Arabidopsis*, before the floral transition, buds are activated in an acropetal, bottom-up gradient, with those in the cauline nodes firing first (**Figure 2**). After floral transition, the gradient flips and the most apical buds activate first, producing a top-down basipetal gradient of bud outgrowth (**Figure 2**).

The influence of the primary apex is evident in the classical, and near-universally observed, phenomenon of apical dominance, in which removing the primary SAM results in the activation of AMs below it, as in the decapitation assay described above (22). In 1933, Thimann & Skoog (97) proposed that this influence of the apex was mediated by auxin because substituting the decapitated apex with auxin prevented bud activation. Over the years, a large body of evidence has accumulated to support this hypothesis (21, 54). The primary shoot apex is an excellent source of auxin, probably synthesized in young leaves, which is actively pumped down the plant in the polar transport stream (57). An apical source of auxin as well as active polar auxin transport, are required for apical dominance (54). Furthermore, auxin response mutants, such as the *axr1* auxin-resistant mutant, have disrupted apical dominance and bud outgrowth resistant to the apical application of auxin. For *axr1*, as described above, decapitation results in the release of more axillary buds than in the wild type. In an excised node assay where in wild-type, apical, but not basal, supply of auxin to the primary axis inhibits the activity of the bud, *axr1* buds are resistant to apical auxin (16, 91). The observation that these mutants also have altered branching levels in intact plants indicates that the influence of auxin is not limited to decapitated plants or excised nodes, but rather it plays a widely relevant role in shoot branching control. Importantly, both the acropetal and basipetal gradients of bud outgrowth are flatter and extend further in the *axr1* mutant, indicating that despite their different polarities, both gradients are regulated by auxin (91). Thus, it is possible to unify the pre- and postfloral transition gradients with the ideas of auxin-mediated apical dominance. In this model, it is proposed that the prefloral-transition gradient results from diminishing auxin-mediated apical influence with increasing distance from the primary apex. This would result in release of buds in an acropetal gradient as they separate from the primary apex by adding new phytomers. It is widely assumed that this effect results from reduced auxin concentration in the stem with increasing distance from the apex but, astonishingly, this has never been substantiated and it is equally possible that there is a reduction in auxin sensitivity or changes in auxin transport. The flip in the orientation of the gradient of bud outgrowth at floral transition can be explained if the floral apex produces or exports less auxin than the vegetative apex. Then at floral transition the level of auxin in the stem would drop in a basipetal progression such that the most apical bud would

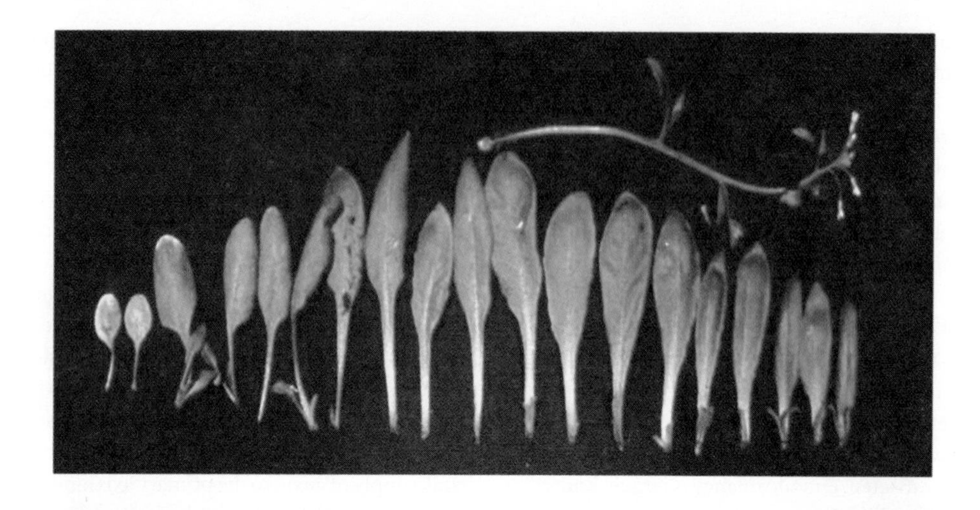

Figure 2

Position along the shoot axis affects bud activity. *Arabidopsis* shows two gradients of indeterminate axillary meristem outgrowth: an acropetal pattern characteristic of vegetative development and a basipetal pattern that follows the floral transition of the primary apical meristem.

be released first, resulting in an acropetal pattern of bud outgrowth. All axillary shoots go through a vegetative phase before making the floral transition. During these vegetative stages, the axillary branches are predicted to export significant amounts of auxin and this would slow the progression of the acropetal gradient of axillary meristem activation as auxin is transported into the primary axis from successive active axillary branches during their prefloral phase.

The mode of auxin action. Although there is ample evidence for a substantive role of auxin in the regulation of bud outgrowth, it is still unclear how it acts. Radiolabelled auxin, applied apically, can inhibit bud outgrowth but does not enter the bud (12, 38, 68). To enter it would have to move acropetally from the stem into the petiole and AM, and this does not happen. Furthermore, directly applying auxin to buds does not inhibit their growth, and endogenous auxin levels tend to be higher in active than in dormant buds (38). Expression of the *AXR1* gene in an *axr1* mutant background from tissue-specific promoters suggests that auxin sensitivity in the xylem-associated tissues of the stem is sufficient for auxin-mediated bud inhibition (12). These tissues express the *PIN1* protein that is required for polar auxin transport, and thus auxin moving in the polar transport stream may be sufficient for bud inhibition (31). This raises the interesting question of how auxin moving down the primary stem influences the activity of buds in the axils of laterally positioned leaves.

Second messengers for auxin action. There are essentially two hypotheses to explain the remote action of auxin. The first is that auxin, moving basipetally in the stem, influences the supply of an acropetally moving second messenger that enters the bud and directly affects its activity. There are two good candidates for this second messenger. The first is cytokinin, which is synthesized both in the roots and locally in the stem and is transported acropetally in the transpiration stream in the xylem (29, 72). Cytokinin promotes the activity of buds, and, unlike auxin, influences activity when applied directly to the bud (21). Auxin inhibits cytokinin synthesis and reduces the amount of cytokinin exported from the roots in the xylem (3, 29, 72). Thus, one mode for auxin-mediated bud inhibition is likely reducing cytokinin supply to the bud.

The second candidate for the second messenger for auxin is at least one novel and as yet unidentified signal or signals, the synthesis of which depends on the *MORE AXILLARY GROWTH (MAX)1, 3*, and *4* genes of *Arabidopsis*, the *RAMOSUS (RMS)1* and *5* genes of pea, and the *DECREASED APICAL DOMINANCE 1 (DAD1)* gene of petunia. Loss of function of all these genes results in increased shoot branching, affecting both the acropetal and basipetal gradients of bud outgrowth (6, 101). Bud activity, in all the mutants where it has been tested, is auxin resistant, demonstrating that the genes are required for auxin-mediated bud inhibition (6, 90). The evidence that these genes are required for the production of at least one long-range, acropetally moving signal comes from the observation that the branching habits of their mutant shoot systems can be restored to wild type by grafting them to wild-type root stocks (9, 70, 90, 98). Wild-type shoots retain their wild-type branching habits when grafted to mutant roots, indicating that the signal(s) can be made in the shoot as well as in the roots. Furthermore, small interstock grafts of wild-type tissue sandwiched between mutant roots and shoots are sufficient to restore wild-type branching to the mutant shoot systems, indicating that either the signal(s) is highly active or a small tissue segment can synthesize a large amount of it (13, 30).

The evidence that this signal (or signals) is a novel hormone comes from two sources. First, measurements of hormone levels in the mutants are inconsistent with any previously identified hormones being the cause of the branching defects. For example, in the pea *rms1* and *rms5* mutants, xylem sap cytokinin is much lower than wild type, and auxin levels and transport rates are either wild type or higher than wild type (9, 70). These are completely opposite of changes that might account for the increased branching. Second, the *MAX3* and *MAX4* genes were cloned and encode divergent members of the carotenoid cleavage dioxygenase family that have not been implicated in the synthesis of known hormones (13, 90). The *RMS1* gene is orthologous to *MAX4*, establishing similarities in the branch-regulating networks of pea and *Arabidopsis* (90). There are also significant differences because in pea auxin upregulates transcription of *RMS1* at the node, consistent with a role for auxin in regulating supply of the novel signal to the bud, but in *Arabidopsis*, auxin has no effect on the transcription of *MAX4* at the node, but upregulates its expression in roots that are known not to be required for normal branching habit (90).

Taken together, these data are consistent with the RMS/DAD/MAX signal(s) acting to relay the auxin message into the bud, but the evidence is not yet conclusive. In particular, it is not yet known where this novel signal(s) acts. There is no evidence that it acts directly in the bud, which would be a requirement for a true second messenger. Help should come from analyzing the requirements for expression of the genes required for the perception of this signal or signals. The *max2* mutant of *Arabidopsis* and the *rms3* and *rms4* mutants of pea are phenotypically similar to the *max* and *rms* mutants described here, except that they are not graft rescuable and hence likely act locally in signal perception (7, 92). This idea is consistent with the observation that *MAX2* encodes an F-box protein (92). F-box proteins have been implicated in the signal transduction pathways of many hormones and act by targeting specific regulatory proteins for destruction (37).

Auxin transport autoinhibition. Nud and coworkers (4) proposed an interesting additional mechanism by which auxin in the polar transport stream of the primary stem may inhibit bud activity. The mechanism is based on the observed strong correlation between the ability of a shoot to transport auxin with its active growth and its ability to inhibit the growth of other buds (69). This led to the proposal that active transport down the primary stem inhibits auxin export from lateral buds into the primary stem polar transport stream, and this export is required for bud activity. Thus, removing the primary shoot apex, or blocking

ATA: auxin transport autoinhibition

Tiller: branch from basal node

polar auxin transport in the primary stem, allows auxin exported from lateral buds to have access to the polar transport stream in the primary stem. The buds can then establish auxin export and, according to the model, this would activate their outgrowth. This proposed mechanism is termed ATA. It is unclear how it would work at a molecular level, and current evidence to support it is largely correlative. In its simplest form it requires the polar transport stream in the primary stem to discriminate between auxin from lateral versus apical sources, or to inhibit transport specifically from a lateral direction. Circumstantial evidence suggests that loss of polar auxin transport in the dominated shoot is caused by loss of polar localization of auxin efflux carriers (69).

Interestingly, the *rms* mutants have somewhat elevated auxin transport rates in the primary stem, but increased shoot branching (8). This runs counter to the basic ATA model, but one explanation is that the *RMS* signal(s) is required for ATA and constitutes part of the mechanism, whereby polar transport in the primary apex inhibits transport from lateral buds. Then, in the *rms* mutants, auxin transport would run unchecked and buds would activate. In a similar vein, the *tt4(2YY6)* allele confers increased auxin transport and increased shoot branching, presumably because of its inability to synthesize auxin-transport-inhibiting flavonoids (15).

Environmental regulation of bud activity. As described above, there are at least three interacting long-range signals that can regulate bud activity—auxin, cytokinin, and one or more RMS/DAD/MAX-dependent signals. Modulation of any or all of these signals by the environment could couple bud activity to environmental conditions. The auxin supply can generally be considered a reporter of the health of the primary apex and this will be influenced by a wealth of environmental factors, perhaps most obviously damage through herbivory. Similarly, xylem sap cytokinin levels at least partly report root health and, furthermore, their levels rise with increasing soil nutrient supply, which also increases shoot branching (66). Thus, most evidence points to a model in which diverse inputs are filtered through the network of interacting hormonal signals that regulate bud growth.

The combinatorial effects of, and interactions between, this hormonal regulatory network with the regulatory network that specifies phytomer and AM fate controls shoot system architecture and enables environmentally sensitive execution of the genetic programs required for life-cycle progression.

BRANCHING IN MONOCOTS

Branching during vegetative development in monocots shares similarities with dicot branching. There are the same two main patterns of branching on the primary axis (**Figure 1**). During vegetative growth, branches called tillers arise from the basal nodes of the plant and grow out in an acropetal sequence (**Figure 1*c*,*d***). The basal phytomers that bear tillers are characterized by short internodes, leaves with juvenile characteristics, and activation of AM outgrowth early during vegetative development. Tillers are responsible for the bushy architecture and yield of grasses such as rice, wheat, and barley. Some grasses, such as maize, have been selected during domestication to produce few tillers to concentrate resources in the main shoot. The second type of axillary branch differs from tillers in position and timing. These branches are usually called axillary or secondary branches rather than tillers and they grow out later during development, further up the stem or as branches arising from the tillers. The phytomers bearing these branches are characterized by longer internodes, leaves with adult characteristics, a delayed AM activity, and a basipetal sequence of branch outgrowth. These AMs also play an important role, for example in maize, where the ear shoots form from axillary branches located a few nodes below the tassel (**Figure 1*c***).

Inflorescence development in the grasses differs from dicots in having additional orders of branching (**Figure 1**). Both maize and rice have long lateral branches at the base of the inflorescence (**Figure 1*c,d***). Grasses bear their flowers in structures called spikelets, which are equivalent to short branches. Maize and grasses in the Andropogonae initiate spikelets in pairs. Rice and other cereals initiate spikelets singly. In maize, the spikelet consists of two florets encased in two reduced leaf-like structures called glumes (**Figure 1*g***). In rice, the spikelet consists of one floret encased in two sets of glumes, one of which is reduced (**Figure 1*h***). The number of branches, spikelets, and florets is specific for each grass species and varies between species. Therefore, regulation of branch, spikelet, and floret number is a key feature in the morphological evolution of the grasses (47).

Many genes involved in AM initiation affect both vegetative and inflorescence branching, whereas genes controlling AM determinacy/identity are usually specific to either the vegetative or reproductive phase and therefore are discussed separately.

Axillary Meristem Initiation During Vegetative Development

A recent example of conservation in dicot and monocot AM initiation came from cloning *MONOCULM1* (*MOC1*) in rice. *MOC1* encodes a transcription factor of the GRAS family orthologous to *LAS* of *Arabidopsis* and *LS* of tomato (55). *moc1* mutants have no tillers due to a failure to initiate AMs. The similar phenotype of the *ls* and *moc1* mutants implies that basal branches in dicots and tillers in monocots are equivalent and that *LS/LAS/MOC1* plays a conserved role in maintaining the potential for AM initiation in the leaf axils of both monocots and dicots. Increase in *MOC1* expression increases tiller number, implying that *MOC1* promotes AM outgrowth and initiation. In contrast to the dicot systems studied so far, the *moc1* mutation has defects during inflorescence branching, leading to a drastic reduction in the number of branches and spikelets. However, a closer look at the literature shows that *ls* mutants in tomato have half the number of flowers as normal (102). A useful test for the "maintenance of meristematic potential" theory for *LAS* action would come from analyzing the correlation between the effect of loss of *LAS* function with the length of time between AM initiation and the time of axil formation in the primary apex in these various systems.

A different kind of defect in AM formation is illustrated by the *uniculm2* (*cul2*) mutant of barley, which also has no tillers (2). AMs are initiated in *cul2* as a meristem-like bulge visible in the axils of embryonic leaves. However, by the next plastrochron the cells in the bulge become enlarged and vacuolated, implying that *cul2* mutants can initiate but not maintain AMs. In support of this early defect, absence of tillers in *cul2* is epistatic to nine other mutants with either increased or decreased numbers of tillers (2). Similarly, *cul2* mutants also affect spikelet development, especially at the distal tip of the inflorescence, implying that similar genetic mechanisms control tiller and spikelet initiation. In addition, a number of double-mutant combinations had severe effects on the production of spikelets in the inflorescence, implying that there is also redundant control of AM branching in barley (2).

Glume: reduced leaf-like organ that encloses the florets

Spikelet: short branch bearing the florets

Axillary Meristem Outgrowth During Vegetative Development

The evolution of maize from its wild ancestor teosinte has involved suppression of tillers such that most modern varieties of maize tiller very little. In the maize mutant, *teosinte branched1* (*tb1*), tillers and secondary and tertiary branches grow out, leading to a bushy architecture (40). *TB1* encodes a bHLH transcription factor of the TCP family (25, 27). It has been proposed that changes in the expression of *TB1* caused the suppression of tillering that occurred during the evolution of maize from teosinte (27, 100). The expression pattern of the *TB1* gene is consistent with this proposal (27, 40). In maize, *TB1* is expressed in AMs as soon as they become visible (40). This is in contrast to teosinte, where *TB1* is not expressed in axillary buds, which grow out. *TB1* expression is also correlated with growth suppression in other tissues in maize and teosinte. In teosinte, *TB1* is expressed in

QTL: quantitative trait locus

ear pedicellate spikelets, which are suppressed in teosinte. In maize, *TB1* is also proposed to play a role in axillary branches after they initiate. *TB1* is expressed in stamen primordia in maize ears, which undergo growth suppression during sex determination, and is also expressed in ear shoots, where it may play a role in shortening the shank and reducing husk leaves. The rice ortholog of *TB1* was recently reported to be involved in growth suppression after AM initiation (95). Overexpression of *OsTB1* under the control of the actin promoter led to a suppression of tillering in rice (95). Moreover, the *fine culm1* (*fc1*) mutation has increased tillering due to a loss-of-function mutation in *OsTB1* (95).

There is currently no evidence that TCP genes play a role in regulating outgrowth of dicot axillary branches. However, expression of *CYC* and *DICH* in the upper region of *Antirrhinum* floral meristems is proposed to cause growth suppression (60, 61). It is also unclear whether *TB1* mediates hormonal regulation of shoot branching in monocots. In particular, it will be of interest to determine if *TB1* acts together with auxin, the second messenger for auxin, or cytokinin, as described above for the role of these hormones in dicot apical dominance.

The existence of additional genes controlling tiller number is suggested by multiple QTL studies in many grasses. One recent example from Setaria is discussed. *Setaria italica*, or Foxtail millet, has few tillers and is proposed to have been derived from *Setaria viridis*, or green millet, which is highly tillered (28). *S. italia* was crossed with *S. viridis* and QTLs for tiller number were identified in the F2. Although a minor QTL for *TB1* was identified, most of the major QTL mapped elsewhere in the genome, implicating additional genes in the control of tillering in grasses. None of the most significant QTLs affected both tillering and later arising axillary branches, indicating that the control of branching in different parts of the plant is genetically separable. Whether these QTL identify genes regulating AM initiation or outgrowth remains to be determined.

Axillary Meristem Initiation During Inflorescence Development

To generate the highly branched grass inflorescence, multiple types of AM are produced by the apical inflorescence meristem (11, 17, 42, 65). The first or primary AMs produced by the inflorescence meristem are the branch meristems. In maize, there are two types of branch meristems: branch meristems that form at the base of the tassel that give rise to long branches and branch meristems (also called spikelet pair meristems) that arise on the main spike and the lateral branches, which give rise to two spikelet meristems. Spikelet meristems then give rise to two floral meristems. Therefore, maize makes four types of AMs (branch, spikelet pair, spikelet, and floral), whereas dicots such as *Arabidopsis* produce two types (branch and floral) (65).

The phytomer concept can also be used to analyze branching in the maize inflorescence. At all nodes, internodes are short and leaves subtending the different types of AM are suppressed or absent, so development is controlled mainly by AM activity. The branch meristem is an indeterminate AM, whereas the spikelet pair, spikelet, and floral meristems are determinate AMs that produce a fixed number of products. The branch and spikelet pair meristems arise in the axils of cryptic bracts similar to the cryptic bracts that subtend floral meristems in *Arabidopsis*. The spikelet pair meristem provides a conundrum because it appears to split to form two spikelet meristems and it is unclear if this is a case of lateral branching or meristem conversion (discussed further in side bar: Alternate Branching Models). However, the spikelet can be described as a compressed shoot. Each spikelet meristem initiates an inner and an outer reduced leaf-like glume (**Figure 1g**). The glumes are described as sterile organs with no AMs in their axils. The spikelet meristem then initiates a reduced leaf-like organ called the lemma internal to each glume and a floral meristem forms in the axil of each lemma. Therefore, the lemma is the "bract" in whose axil the floral meristem forms (**Figure 1g**). The floral meristem initiates a palea (equivalent to dicot sepals), followed by the three lodicules (homologous to dicot petals), three stamens, and a tricarpellate gynoecium. In the tassel, the gynoecium aborts, resulting in the production of male flowers, and in the ear, the stamens and the lower floret are suppressed, resulting in the production of single female florets (11, 17, 41, 65).

Because branches, spikelets, and florets arise from AMs, maize mutants with fewer branches, spikelets, and florets identify genes required for AM initiation and/or maintenance. These mutants form a large class in maize called *barren inflorescence* or *bif*. These mutants have similarities to *Arabidopsis* mutants that fail to make floral meristems such as *pid*, *pin1*, and *mp*, which are implicated in auxin signaling or transport (5, 73, 77, 81). *bif2* mutants produce very few branches and spikelets (64). RNA in situ hybridization with *KNOTTED1*, which is homologous to *STM*, confirms that the defect is due to failure to initiate AMs. When spikelets are produced, they occur singly instead of in pairs, implying that when branch meristems initiate they have defects in determinacy or maintenance. Occasionally, single florets form but more often florets with reduced numbers of organs form (64). The defects in floret initiation point to a defect in spikelet meristem function, whereas the production of fewer floral organs in the floret points to a defect in floral meristem maintenance (64). Therefore, *bif2* mutants affect initiation and maintenance of all the AMs produced by the inflorescence. In addition, *bif2* mutants often fail to produce ear shoots, which indicates that *bif2* is also required for initiation and maintenance of the ear shoot AM. At least three other mutants [*barren stalk1* (*ba1*) (23, 82), *suppressor of sessile spikelets1* (*Sos1*) (26), and *barren inflorescence1* (*Bif1*) (23)] have the same suite of defects as *bif2*, although there are subtle differences. The fact that all of these mutants have similar phenotypes, including fewer branches, single spikelets, single florets, reduced numbers of floral organs, and fewer ear shoots, implicates a common genetic program controlling AM initiation in all AMs.

Tillering is generally suppressed in maize lines, so determining whether a mutant affects vegetative axillary branching requires backcrossing the mutant into a highly tillered line or a line carrying the *tb1* mutation. *ba1*:*tb1* double mutants fail to initiate tillers, showing that *BA1* is also required for AM initiation during vegetative development (82). A similar result was obtained for the *bif2*:*tb1* double mutant, showing that *BIF2* is also required for tiller initiation (P. McSteen & S. Hake, unpublished results). Using a comparable approach, *lax midrib1-O* was shown to have reduced tiller number by backcrossing it to several highly tillered genetic backgrounds (83).

LAX1 of rice was the first gene controlling AM initiation in the inflorescence cloned from the grasses (48). Surprisingly, no gene

ALTERNATE BRANCHING MODELS

There are two alternate models for branching of the spikelet meristem during floret initiation. The "conversion model" was proposed by Erin Irish (42) with the analysis of the *tasselseed4* (*ts4*) and (*ts6*) mutants. Irish proposed that the spikelet meristem initiates a floral meristem and is then converted to a floral meristem. According to this model, *ts6* mutants, which make extra florets, are interpreted as having a delay in conversion (42). Support for the alternative "lateral branching model" came with the analysis of the *indeterminate spikelet1* (*ids1*) mutant (18). In the lateral branching model, the spikelet meristem produces two floral meristems laterally leaving a residual spikelet meristem between the two florets. Support for this model comes from the expression of *IDS1*, which, in addition to being expressed elsewhere, is expressed in a region between the two spikelets, which could represent a residual spikelet meristem. According to this model, *IDS1* is required to suppress indeterminate branching of the spikelet meristem. The two models were partially reconciled with a revised model by Irish (43) that suggests that a residual spikelet meristem remains after conversion. A recent paper analyzing *ids1* interactions with *reversed germ orientation1* (*rgo1*) reinterprets the *ids1* mutant phenotype using the conversion model and proposes that *RGO1* and *IDS1* are required for converting spikelet to floral meristem identity (46). The proposal that combinatorial interactions of transcription factors determine meristem identity may lend support to the conversion model. Both models can also be used to explain the branching of the spikelet pair meristem to make two spikelet meristems.

orthologous to *LAX1* had previously been identified in *Arabidopsis*. *lax1* mutants (the name refers to the sparse appearance of the inflorescence) have reduced numbers of long branches and reduced numbers of spikelets. Spikelets that form are defective with missing or sterile organs. RNA in situ hybridization with *OSH1*, which is homologous to *STM*, shows that the defect is due to failure to initiate AMs (49). A similar mutant, *small panicle1* (*spa1*), has reduced numbers of long branches and reduced numbers of spikelets, especially at lower nodes (48). *lax1:spa1* double mutants have an enhanced phenotype with no branches or spikelets in the tassel and no tillers, indicating that *LAX1* and *SPA1* play redundant roles in AM initiation in both reproductive and vegetative tissues (48). These studies emphasize the importance of constructing double mutants to identify all the functions of a gene. *LAX1* encodes a bHLH transcription factor expressed early in AM ontogeny at the junction with the subtending cryptic bract leaf. Surprisingly, overexpression of *LAX1* under the control of the actin promoter had little effect on inflorescence or vegetative branching. However, there were other defects in the plant, including dwarfing and leaf defects, implying that there were perturbations in hormone function (48). Therefore, *LAX1* may provide a link to the control of AM initiation by hormones.

Axillary Meristem Outgrowth During Inflorescence Development

The production of multiple AM types in grass inflorescences (branch, spikelet pair, spikelet, floret) raises the question of the homology between these meristems and the AMs in dicots (branch, flower). Although it seems to be universally possible to describe all structures in terms of the phytomer concept, assignation of homology in meristem fate between species is challenging in this instance. Are the floral meristems of maize homologous to dicot floral meristems? They seem to have similarities, as genes involved in floral organ identity are conserved in function and expression (1, 67). Branch meristems may be equivalent to secondary inflorescences because they reiterate the development of the main stem. This leaves us with the question of the spikelet pair and the spikelet meristem. Do these meristem types have anything in common with flowers and inflorescences? As they express homologs of floral meristem identity genes and inflorescence meristem identity genes (discussed further below), they could have an identity intermediate between inflorescences and flowers. So, can all of grass inflorescence development be explained by reference to what is known about inflorescence and floral meristem identity in dicots? Apparently not, as grasses have coopted additional transcription factors to specify meristem fate in these additional meristem types. There is increasing evidence that the different meristem types are specified by the combinatorial interactions of meristem identity genes, such that loss of meristem identity genes causes the meristem to revert to an earlier meristem fate. These results again illustrate the utility of the phytomer model in describing AM fate. As phytomers continually reiterate, perhaps it is not surprising that AMs can revert to an earlier meristem fate when meristem fate/identity is compromised.

Floral and inflorescence meristem identity genes. Because of the mutually inhibitory interactions between the inflorescence meristem identity gene, *TFL1*, and floral meristem identity genes, *LFY* and *AP1*, mutations in *LFY* and *AP1* in *Arabidopsis* have increased branching due to ectopic expression of *TFL1* and the production of indeterminate AMs in place of flowers (56, 78). Homologs of *TFL1* play a similar role in the grasses as transgenic overexpression causes increased production of indeterminate branches similar to the effects of *TFL1* overexpression in *Arabidopsis* (45, 71, 79). Homologs of *AP1* are expressed in floral meristems and floral organs in the grasses (51, 75). However, recent results show that homologs of *LFY* are only partially conserved in the grasses.

Maize is one of the few plant species with two copies of *LFY*, *ZFL1* on chromosome

10 and *ZFL2* on chromosome 2 (10). RNA in situ hybridization shows that both *ZFL* genes are strongly expressed in spikelet pair meristems, spikelet meristems, floral meristems, and all floral organs, compared to *Arabidopsis*, in which *LFY* is most strongly expressed in floral meristems and floral organs. This is of interest as it implies that spikelet pair meristems and spikelet meristems have floral identity. Plants that are double mutant for *zfl1;zfl2* have pleiotropic effects, some of which are consistent with the known functions of *lfy* in dicots. *zfl1;zfl2* double-mutant flowers have defects reminiscent of BC double mutants, indicating that *LFY* activates similar genes in grasses as it does in dicots. However, only occasionally are florets converted to inflorescence branches similar to *LFY* defects in floral meristem identity. The double mutants have a most unusual inflorescence structure, including the production of fewer long branches in the tassel. This is the opposite effect expected of mutations in a *LFY* homolog. The plants also have delayed flowering and leaves in the tassel. Therefore, the branching defect in the tassel could be interpreted as a defect in the vegetative to reproductive transition, which would be similar to the role of *LFY* in dicots.

On the other hand, the rice *LFY* gene (*RFL*) has diverged in both expression pattern and some aspects of function compared to *Arabidopsis* and maize (20, 52). *RFL* is expressed in the inflorescence of rice but is specifically turned off when the branch primordia initiate. After the branches grow, *RFL* is visible in the branches but is absent from initiating spikelet primordia (52). It is not expressed in floral meristems, as seen in dicots, or in branch and spikelet meristems, as in maize. In fact, *RFL* expression in rice is the inverse of the *LFY* expression pattern seen in maize and dicots. There are also some differences in the protein-coding region (20, 52). Although *RFL* driven by the *LFY* promoter could partially complement the *lfy* mutation there were additional morphological defects in the leaves and plant development (20). Although overexpression of the *Arabidopsis LFY* gene in rice causes early flowering (36), overexpression of *RFL* in *Arabidopsis* causes defects in leaf and plant development (52). Therefore, *LFY* expression and function have diverged in the grasses (33). *SEP* homologs are also widely diverged in their expression in the grasses (62). Genes that are diverged in their expression between monocots and dicots and within the grasses are candidates for genes controlling morphological evolution, emphasizing the importance of the comparative approach.

Novel meristem identity genes. Mutant analysis has identified genes required for meristem identity and/or determinacy of each AM type in the maize inflorescence (65). Recent results have isolated genes required for two of these stages, spikelet meristem identity and the conversion from spikelet to floral meristem identity.

Spikelet meristem determinacy/identity. Spikelet meristems are determinate, producing two florets in maize (**Figure 1g**). Several mutants have been identified that affect the number of florets produced by the spikelet meristem, and the genes involved have been interpreted as required for spikelet meristem determinacy. Double-mutant combinations reveal that these genes also play a role in spikelet meristem identity and in the identity of other meristem types. As these double mutants cause the reversion of meristem identity to an earlier meristem identity these results illustrate the utility of the phytomer model in demonstrating the equivalence of AMs.

INDETERMINATE SPIKELET1 (*IDS1*) encodes an AP2-like transcription factor (18). Loss-of-function mutants of *IDS1* produced 3 to 10 florets per spikelet instead of 2. Based on this analysis, it was proposed that *IDS1* plays a role in spikelet meristem determinacy (18). The *reversed germ orientation1* (*rgo1*) mutant has 3 florets instead of 2 (46). Surprisingly, *ids1;rgo1* double mutants have a synergistic interaction in which the spikelet meristem reverts to spikelet pair meristem identity after initiating a number of glumes (46). Therefore, *IDS1* and *RGO1* are required to maintain spikelet meristem identity

SEM: scanning electron microscopy

as well as determinacy. *ids1:rgo1* double mutants also affect identity and determinacy of the spikelet pair meristem, which is not affected in either single mutant, indicating that *IDS1* and *RGO1* also play a redundant role in the spikelet pair meristem. These results provide evidence for a model in which combinatorial interactions specify meristem identity. The *rgo1* and *ids1* mutants show nonallelic noncomplementation in the F1, providing further support to this model (46).

Indeterminate floral apex1 (*ifa1*) mutants affect determinacy in all the determinate meristems of the inflorescence including the spikelet meristem, which produces extra florets (53). In *ids1;ifa1* double mutants, spikelet meristems in the ear convert to branch meristem identity, whereas spikelet meristems in the tassel convert to spikelet pair meristem identity. This implies that *IDS1* and *IFA1* are required for spikelet meristem identity as well as determinacy and that in the absence of both genes the meristem converts to an earlier meristem type (53). These results provide further support for a combinatorial model of determination of meristem identity.

These models have important implications for how mutants are interpreted. The phytomer model can support either model because lateral branching and meristem conversion can occur, as in the case of sympodial branching in tomato. SEM does not provide a clear picture of what is going on, as it is impossible to know when a spikelet meristem becomes a floral meristem or when a spikelet pair meristem becomes a spikelet meristem. The issue may be resolved with the cloning of the many meristem identity genes in maize, which should provide molecular markers for each meristem type (65).

The transition from spikelet to floral meristem identity. The recent cloning of *BRANCHED SILKLESS1* (*BD1*) of maize identified the first gene required for the transition from spikelet to floral meristem identity in the grasses (19). In *bd1* mutant ears, spikelet meristems produce branched structures that never progress to making flowers. The lack of floral meristem identity was shown by the absence of expression of floral MADS box genes *ZAG1* and *ZMM2* (24). *BD1* encodes an ERF-type transcription factor expressed in the spikelet meristem in the axil of the glume (19). It was proposed that *BD1* specifies spikelet meristem identity by acting as a transcriptional repressor of branch meristem identity (19).

The ortholog of *BD1* in rice is *FRIZZY PANICLE* (*FZP*) (50, 103). SEM of developing *fzp* inflorescences clearly shows that the spikelet meristems initiate rudimentary glumes. These glumes, which are normally sterile, then produce AMs in their axils. This process reiterates producing a branch. This phenotype is also seen in *bd1* tassels and in a weak allele of *bd1* (19, 24). Therefore, *FZP* acts to inhibit AM initiation in the axil of the glume. This is supported by the expression of *BD1/FZP* in the spikelet meristem in the axil of the glume. This result confirms the classical interpretation that glumes are leaf-like organs with suppressed AMs. This phenotype is reminiscent of the *ap1* mutant phenotype in which AMs form in the axils of the sepals (44). Furthermore, the activation of these AMs is correlated with an inability to progress to floral meristem identity. Because *fzp* mutants initiate glumes, it was proposed that *fzp* mutants had already switched to spikelet meristem identity and that the function of *FZP* was to promote the transition from spikelet to floral meristem identity (50). This interpretation could equally apply to *bd1*. Komatsu et al. (50) showed that *FZP* is a transcriptional activator rather than repressor in transient expression studies. Therefore, *FZP* and *BD1* may act to activate floral meristem identity rather than repress branch identity.

CONCLUSION

The phytomer concept provides a useful framework to analyze shoot structure in diverse species and to characterize mutant phenotypes. Such analyses have provided evidence for a commonality between all AMs. However, these studies have also demonstrated that there is clear divergence between meristem types both within and between species, with the co-option of new genes to regulate AM fate and activity. Because of redundancy caused by gene duplications and by pathways with overlapping phenotypic outcomes, double mutants are often required to understand gene function fully. Such double-mutant analysis has also been instrumental in establishing combinatorial effects in the specification of meristem fate.

There are many emerging overlaps between monocots and dicots in the transcription factor regulatory networks that control AM fate, but unsurprisingly, given the additional AM types found in monocots, there are several monocot-specific factors. It is less easy to compare hormonal regulatory networks in monocots and dicots with the data currently available. This is partly because comparable data from classical decapitation and grafting experiments are difficult to obtain in monocots, particularly during vegetative growth when the apical meristem is at the base of the plant. It is widely accepted that apical dominance in dicots is an adaptation to herbivory. In dicots, leaves are carried up to the light by stem elongation, which necessarily carries the SAM up into a vulnerable position. AMs can be considered partly as spares, for use if the primary meristem is lost. In monocots, the leaves are carried upward by their own sheathes and elongated blades, AMs are active (in grasses that tiller), and the apical meristem is protected by the leaf sheaths and unlikely to suffer damage from grazing animals (or lawn mowers). It is interesting to speculate that AM activation in monocots might be less dominated by the primary apical meristem and more linked to leaf health or other environmental factors such as shading or nutrition. A comparative analysis in monocots and dicots will continue to provide valuable information on the diversity of plant form due to AM activity.

LITERATURE CITED

1. Ambrose BA, Lerner DR, Ciceri P, Padilla CM, Yanofsky MF, Schmidt RJ. 2000. Molecular and genetic analyses of the *silky1* gene reveal conservation in floral organ specification between eudicots and monocots. *Mol. Cell* 5:569–79
2. Babb S, Muehlbauer GJ. 2003. Genetic and morphological characterization of the barley *uniculm2* (*cul2*) mutant. *Theor. Appl. Genet.* 106:846–57
3. Bangerth F. 1994. Response of cytokinin concentration in the xylem exudate of bean (*phaseolus-vulgaris* l) plants to decapitation and auxin treatment, and relationship to apical dominance. *Planta* 194:439–42
4. Bangerth F, Li CJ, Gruber J. 2000. Mutual interaction of auxin and cytokinins in regulating correlative dominance. *Plant Growth Reg.* 32:205–17
5. Bennett SRM, Alvarez J, Bossinger G, Smyth DR. 1995. Morphogenesis in *pinoid* mutants of *Arabidopsis thaliana*. *Plant J.* 8:505–20
6. Beveridge CA. 2000. Long-distance signalling and a mutational analysis of branching in pea. *Plant Growth Reg.* 32:193–203
7. Beveridge CA, Ross JJ, Murfet IC. 1996. Branching in pea—action of genes *rms3* and *rms4*. *Plant Physiol.* 110:859–65
8. Beveridge CA, Symons GM, Turnbull CGN. 2000. Auxin inhibition of decapitation-induced branching is dependent on graft-transmissible signals regulated by genes rms1 and rms2. *Plant Physiol.* 123:689–97
9. Beveridge CA, Symons GM, Murfet IC, Ross JJ, Rameau C. 1997. The *rms1* mutant of pea has elevated indole-3-acetic acid levels and reduced root-sap zeatin riboside content but increased branching controlled by graft-transmissible signal(s). *Plant Physiol.* 115:1251–58
10. Bomblies K, Wang RL, Ambrose BA, Schmidt RJ, Meeley RB, Doebley J. 2003. Duplicate *floricaula/leafy* homologs *zfl1* and *zfl2* control inflorescence architecture and flower patterning in maize. *Development* 130:2385–95

11. Bonnett OT. 1948. Ear and tassel development in maize. *Ann. Missouri Bot. Garden* 35:269–87
12. Booker J, Chatfield S, Leyser O. 2003. Auxin acts in xylem-associated or medullary cells to mediate apical dominance. *Plant Cell* 15:495–507
13. Booker J, Auldridge M, Wills S, McCarty D, Klee H, Leyser O. 2004. *Max3/ccd7* is a carotenoid cleavage dioxygenase required for the synthesis of a novel plant signalling molecule. *Curr. Biol.* 14:1232–38
14. Bradley D, Ratcliffe O, Vincent C, Carpenter R, Coen E. 1997. Inflorescence commitment and architecture in *Arabidopsis*. *Science* 275:80–83
15. Brown DE, Rashotte AM, Murphy AS, Normanly J, Tague BW, et al. 2001. Flavonoids act as negative regulators of auxin transport in vivo in *Arabidopsis*. *Plant Physiol.* 126:524–35
16. Chatfield SP, Stirnberg P, Forde BG, Leyser O. 2000. The hormonal regulation of axillary bud growth in *Arabidopsis*. *Plant J.* 24:159–69
17. Cheng PC, Greyson RI, Walden DB. 1983. Organ initiation and the development of unisexual flowers in the tassel and ear of *zea mays*. *Am. J. Bot.* 70:450–62
18. Chuck G, Meeley RB, Hake S. 1998. The control of maize spikelet meristem fate by the *apetala2*-like gene *indeterminate spikelet1*. *Gene Dev.* 12:1145–54
19. Chuck G, Muszynski M, Kellogg E, Hake S, Schmidt RJ. 2002. The control of spikelet meristem identity by the *branched silkless1* gene in maize. *Science* 298:1238–41
20. Chujo A, Zhang Z, Kishino H, Shimamoto K, Kyozuka J. 2003. Partial conservation of *lfy* function between rice and *Arabidopsis*. *Plant Cell Physiol.* 44:1311–19
21. Cline MG. 1991. Apical dominance. *Bot. Rev.* 57:318–58
22. Cline MG. 1996. Exogenous auxin effects on lateral bud outgrowth in decapitated shoots. *Ann. Bot. London* 78:255–66
23. Coe EH, Neuffer MG, Hoisington DA. 1988. The genetics of corn. In *Corn and Corn Improvement*, ed. GF Sprague, JW Dudley, pp. 81–258 . Madison, WI: ASA-CSSA-SSSA
24. Colombo L, Marziani G, Masiero S, Wittich PE, Schmidt RJ, et al. 1998. *Branched silkless* mediates the transition from spikelet to floral meristem during *zea mays* ear development. *Plant J.* 16:355–63
25. Cubas P, Lauter N, Doebley J, Coen E. 1999. The tcp domain: a motif found in proteins regulating plant growth and development. *Plant J.* 18:215–22
26. Doebley J, Stec A, Kent B. 1995. *Suppressor of sessile spikelets1* (*sos1*)—a dominant mutant affecting inflorescence development in maize. *Am. J. Bot.* 82:571–77
27. Doebley J, Stec A, Hubbard L. 1997. The evolution of apical dominance in maize. *Nature* 386:485–88
28. Doust AN, Devos KM, Gadberry MD, Gale MD, Kellogg EA. 2004. Genetic control of branching in foxtail millet. *Proc. Natl. Acad. Sci. USA* 101:9045–50
29. Eklof S, Astot C, Blackwell J, Moritz T, Olsson O, Sandberg G. 1997. Auxin-cytokinin interactions in wild-type and transgenic tobacco. *Plant Cell Physiol.* 38:225–35
30. Foo E, Turnbull CGN, Beveridge CA. 2001. Long-distance signaling and the control of branching in the *rms1* mutant of pea. *Plant Physiol.* 126:203–9
31. Galweiler L, Guan CH, Muller A, Wisman E, Mendgen K, et al. 1998. Regulation of polar auxin transport by atpin1 in *Arabidopsis* vascular tissue. *Science* 282:2226–30
32. Garrison R. 1955. Studies in the development of axillary buds. *Am. J. Bot.* 42:257–66
33. Gocal GFW, King RW, Blundell CA, Schwartz OM, Andersen CH, Weigel D. 2001. Evolution of floral meristem identity genes. Analysis of lolium temulentum genes related to *apetala1* and *leafy* of Arabidopsis. *Plant Physiol.* 125:1788–801
34. Grbic V, Bleecker AB. 2000. Axillary meristem development in Arabidopsis thaliana. *Plant J.* 21:215–23

35. Greb T, Clarenz O, Schafer E, Muller D, Herrero R, et al. 2003. Molecular analysis of the *lateral suppressor* gene in *Arabidopsis* reveals a conserved control mechanism for axillary meristem formation. *Gene Dev.* 17:1175–87
36. He ZH, Zhu Q, Dabi T, Li DB, Weigel D, Lamb C. 2000. Transformation of rice with the Arabidopsis floral regulator *leafy* causes early heading. *Transgenic Res.* 9:223–27
37. Hellmann H, Estelle M. 2002. Plant development: regulation by protein degradation. *Science* 297:793–97
38. Hillman JR, Math VB, Medlow GC. 1977. Apical dominance and levels of indole acetic acid in *Phaseolus* lateral buds. *Planta* 137:191–93
39. Horvath DP, Anderson JV, Chao WS, Foley ME. 2003. Knowing when to grow: signals regulating bud dormancy. *Trends Plant Sci.* 8:534–40
40. Hubbard L, McSteen P, Doebley J, Hake S. 2002. Expression patterns and mutant phenotype of *teosinte branched1* correlate with growth suppression in maize and teosinte. *Genetics* 162:1927–35
41. Irish EE. 1996. Regulation of sex determination in maize. *Bioessays* 18:363–69
42. Irish EE. 1997. Class ii *tassel seed* mutations provide evidence for multiple types of inflorescence meristems in maize (*poaceae*). *Am. J. Bot.* 84:1502–15
43. Irish EE. 1998. Grass spikelets: a thorny problem. *Bioessays* 20:789–93
44. Irish VF, Sussex IM. 1990. Function of the *apetala1* gene during Arabidopsis flower development. *Plant Cell* 2:741–53
45. Jensen CS, Salchert K, Nielsen KK. 2001. A terminal flower1-like gene from perennial ryegrass involved in floral transition and axillary meristem identity. *Plant Physiol.* 125:1517–28
46. Kaplinsky NJ, Freeling M. 2003. Combinatorial control of meristem identity in maize inflorescences. *Development* 130:1149–58
47. Kellogg EA. 2001. Evolutionary history of the grasses. *Plant Physiol.* 125:1198–205
48. Komatsu K, Maekawa M, Ujiie S, Satake Y, Furutani I, et al. 2003. *Lax* and *spa*: major regulators of shoot branching in rice. *Proc. Natl. Acad. Sci. USA* 100:11765–70
49. Komatsu M, Maekawa M, Shimamoto K, Kyozuka J. 2001. The *lax1* and *frizzy panicle2* genes determine the inflorescence architecture of rice by controlling rachis-branch and spikelet development. *Dev. Biol.* 231:364–73
50. Komatsu M, Chujo A, Nagato Y, Shimamoto K, Kyozuka J. 2003. *Frizzy panicle* is required to prevent the formation of axillary meristems and to establish floral meristem identity in rice spikelets. *Development* 130:3841–50
51. Kyozuka J, Kobayashi T, Morita M, Shimamoto K. 2000. Spatially and temporally regulated expression of rice mads box genes with similarity to Arabidopsis class a, b and c genes. *Plant Cell Physiol.* 41:710–18
52. Kyozuka J, Konishi S, Nemoto K, Izawa T, Shimamoto K. 1998. Down-regulation of *rfl*, the *flo/lfy* homolog of rice, accompanied with panicle branch initiation. *Proc. Natl. Acad. Sci. USA* 95:1979–82
53. Laudencia-Chingcuanco D, Hake S. 2002. The *indeterminate floral apex1* gene regulates meristem determinacy and identity in the maize inflorescence. *Development* 129:2629–38
54. Leyser O. 2003. Regulation of shoot branching by auxin. *Trends Plant Sci.* 8:541–45
55. Li XY, Qian Q, Fu ZM, Wang YH, Xiong GS, et al. 2003. Control of tillering in rice. *Nature* 422:618–21
56. Liljegren SJ, Gustafson-Brown C, Pinyopich A, Ditta GS, Yanofsky MF. 1999. Interactions among *apetala1*, *leafy*, and *terminal flower1* specify meristem fate. *Plant Cell* 11:1007–18
57. Ljung K, Bhalerao RP, Sandberg G. 2001. Sites and homeostatic control of auxin biosynthesis in Arabidopsis during vegetative growth. *Plant J.* 28:465–74

58. Long J, Barton MK. 2000. Initiation of axillary and floral meristems in *Arabidopsis*. *Dev. Biol.* 218:341–53
59. Long JA, Moan EI, Medford JI, Barton MK. 1996. A member of the *knotted* class of homeodomain proteins encoded by the *stm* gene of *Arabidopsis*. *Nature* 379:66–69
60. Luo D, Carpenter R, Vincent C, Copsey L, Coen E. 1996. Origin of floral asymmetry in antirrhinum. *Nature* 383:794–99
61. Luo D, Carpenter R, Copsey L, Vincent C, Clark J, Coen E. 1999. Control of organ asymmetry in flowers of antirrhinum. *Cell* 99:367–76
62. Malcomber ST, Kellogg EA. 2004. Heterogeneous expression patterns and separate roles of the *sepallata* gene *leafy hull sterile1* in grasses. *Plant Cell* 16:1692–706
63. McConnell JR, Barton MK. 1998. Leaf polarity and meristem formation in Arabidopsis. *Development* 125:2935–42
64. McSteen P, Hake S. 2001. *Barren inflorescence2* regulates axillary meristem development in the maize inflorescence. *Development* 128:2881–91
65. McSteen P, Laudencia-Chingcuanco D, Colasanti J. 2000. A floret by any other name: control of meristem identity in maize. *Trends Plant Sci.* 5:61–66
66. Mediene S, Pages L, Jordan MO, Le Bot J, Adamowicz S. 2002. Influence of nitrogen availability on shoot development in young peach trees [prunus persica (l.) batsch]. *Trees-Struct. Funct.* 16:547–54
67. Mena M, Ambrose BA, Meeley RB, Briggs SP, Yanofsky MF, Schmidt RJ. 1996. Diversification of c-function activity in maize flower development. *Science* 274:1537–40
68. Morris DA. 1977. Transport of exogenous auxin in two-branched dwarf pea seedlings (*pisum sativum l.*). *Planta* 136:91–96
69. Morris DA, Johnson CF. 1990. The role of auxin efflux carriers in the reversible loss of polar auxin transport in the pea (*pisum sativum l.*) stem. *Planta* 181:117–24
70. Morris SE, Turnbull CGN, Murfet IC, Beveridge CA. 2001. Mutational analysis of branching in pea. Evidence that *rms1* and *rms5* regulate the same novel signal. *Plant Physiol.* 126:1205–13
71. Nakagawa M, Shimamoto K, Kyozuka J. 2002. Overexpression of *rcn1* and *rcn2*, rice *terminal flower 1/centroradialis* homologs, confers delay of phase transition and altered panicle morphology in rice. *Plant J.* 29:743–50
72. Nordstrom A, Tarkowski P, Tarkowska D, Norbaek R, Astot C, et al. 2004. Auxin regulation of cytokinin biosynthesis in Arabidopsis thaliana: a factor of potential importance for auxin-cytokinin-regulated development. *Proc. Natl. Acad. Sci. USA* 101:8039–44
73. Okada K, Ueda J, Komaki MK, Bell CJ, Shimura Y. 1991. Requirement of the auxin polar transport system in early stages of *Arabidopsis* floral bud formation. *Plant Cell* 3:677–84
74. Otsuga D, DeGuzman B, Prigge MJ, Drews GN, Clark SE. 2001. *Revoluta* regulates meristem initiation at lateral positions. *Plant J.* 25:223–36
75. Pelucchi N, Fornara F, Favalli C, Masiero S, Lago C, et al. 2002. Comparative analysis of rice mads-box genes expressed during flower development. *Sex Plant Reprod.* 15:113–22
76. Poduska B, Humphrey T, Redweik A, Grbic V. 2003. The synergistic activation of *flowering locus c* by *frigida* and a new flowering gene *aerial rosette 1* underlies a novel morphology in Arabidopsis. *Genetics* 163:1457–65
77. Przemeck GKH, Mattsson J, Hardtke CS, Sung ZR, Berleth T. 1996. Studies on the role of the Arabidopsis gene monopteros in vascular development and plant cell axialization. *Planta* 200:229–37

78. Ratcliffe OJ, Bradley DJ, Coen ES. 1999. Separation of shoot and floral identity in arabidopsis. *Development* 126:1109–20
79. Ratcliffe OJ, Amaya I, Vincent CA, Rothstein S, Carpenter R, et al. 1998. A common mechanism controls the life cycle and architecture of plants. *Development* 125:1609–15
80. Reinhardt D, Mandel T, Kuhlemeier C. 2000. Auxin regulates the initiation and radial position of plant lateral organs. *Plant Cell* 12:507–18
81. Reinhardt D, Pesce ER, Stieger P, Mandel T, Baltensperger K, et al. 2003. Regulation of phyllotaxis by polar auxin transport. *Nature* 426:255–60
82. Ritter MK, Padilla CM, Schmidt RJ. 2002. The maize mutant *barren stalk1* is defective in axillary meristem development. *Am. J. Bot.* 89:203–10
83. Schichnes DE, Freeling M. 1998. *Lax midrib1-o* heterochronic mutant of maize. *Am. J. Bot.* 85:481–91
84. Schmitz G, Theres K. 1999. Genetic control of branching in Arabidopsis and tomato. *Curr. Opin. Plant Biol.* 2:51–55
85. Schultz EA, Haughn GW. 1993. Genetic-analysis of the floral initiation process (flip) in *Arabidopsis*. *Development* 119:745–65
86. Schumacher K, Schmitt T, Rossberg M, Schmitz C, Theres K. 1999. The *lateral suppressor* (*ls*) gene of tomato encodes a new member of the vhiid protein family. *Proc. Natl. Acad. Sci. USA* 96:290–95
87. Shannon S, Meekswagner DR. 1991. A mutation in the Arabidopsis *tfl1* gene affects inflorescence meristem development. *Plant Cell* 3:877–92
88. Shimizu-sato S, Mori H. 2001. Control in outgrowth and dormancy in axillary buds. *Plant Physiol.* 127:1405–13
89. Snow M, Snow R. 1942. The determination of axillary buds. *New Phytol.* 41:13–22
90. Sorefan K, Booker J, Haurogne K, Goussot M, Bainbridge K, et al. 2003. *Max4* and *rms1* are orthologous dioxygenase-like genes that regulate shoot branching in Arabidopsis and pea. *Gene Dev.* 17:1469–74
91. Stirnberg P, Chatfield SP, Leyser HMO. 1999. *Axr1* acts after lateral bud formation to inhibit lateral bud growth in Arabidopsis. *Plant Physiol.* 121:839–47
92. Stirnberg P, van de Sande K, Leyser HMO. 2002. *Max1* and *max2* control shoot lateral branching in Arabidopsis. *Development* 129:1131–41
93. Sussex IM. 1955. Morphogenesis in *solanum tuberosum* l.: experimental investigation of leaf dorsiventrality and orientation in the juvenile shoot. *Phytomorphology* 5:286–300
94. Sussex IM. 1989. Developmental programming of the shoot meristem. *Cell* 56:225–29
95. Takeda T, Suwa Y, Suzuki M, Kitano H, Ueguchi-Tanaka M, et al. 2003. The *ostb1* gene negatively regulates lateral branching in rice. *Plant J.* 33:513–20
96. Talbert PB, Adler HT, Parks DW, Comai L. 1995. The *revoluta* gene is necessary for apical meristem development and for limiting cell divisions in the leaves and stems of Arabidopsis-thaliana. *Development* 121:2723–35
97. Thimann K, Skoog F. 1933. Studies on the growth hormone of plants iii: the inhibitory action of the growth substance on bud development. *Proc. Natl. Acad. Sci. USA* 19:714–16
98. Turnbull CGN, Booker JP, Leyser HMO. 2002. Micrografting techniques for testing long-distance signalling in Arabidopsis. *Plant J.* 32:255–62
99. Vernoux T, Kronenberger J, Grandjean O, Laufs P, Traas J. 2000. *Pin-formed 1* regulates cell fate at the periphery of the shoot apical meristem. *Development* 127:5157–65
100. Wang RL, Stec A, Hey J, Lukens L, Doebley J. 1999. The limits of selection during maize domestication. *Nature* 398:236–39

101. Ward SP, Leyser O. 2004. Shoot branching. *Curr. Opin. Plant Biol.* 7:73–78
102. Williams W. 1960. The effect of selection on the manifold expression of the "suppressed lateral" gene in tomato. *Heredity* 14:285–96
103. Zhu Q-H, Hoque MS, Dennis ES, Upadhyaya NM. 2003. Ds tagging of *branched floretless 1* (*bfl1*) that mediates the transition from spikelet to floret meristem in rice (oryza sativa l). *BMC Plant Biol.* 3:6

Protein Splicing Elements and Plants: From Transgene Containment to Protein Purification

Thomas C. Evans, Jr., Ming-Qun Xu, and Sriharsa Pradhan

New England Biolabs, Inc., Beverly, Massachusetts 01915; email: evans@neb.com

Annu. Rev. Plant Biol.
2005. 56:375–92

doi: 10.1146/
annurev.arplant.56.032604.144242

First published online as a
Review in Advance on
January 17, 2005

1543-5008/05/0602-
0375$20.00

Key Words

gene containment, intein, protein expression, protein splicing

Abstract

Protein splicing elements, termed inteins, have been discovered in all the domains of life. Basic research on inteins has led to a greater understanding of how they mediate the protein splicing process. Because inteins are natural protein engineering elements they have been harnessed for use in a number of applications, including protein purification, protein semisynthesis, and in vivo and in vitro protein modifications. This review focuses on the use of inteins in plants. A split-gene technique utilizes inteins to reconstitute the activity of a transgene product with the goal of limiting the spread of transgenes from a genetically modified plant to a weedy relative. Furthermore, merging the intein tag for protein purification with the large protein yields possible with plants has the potential to produce pharmaceutically important proteins. Finally, relevant techniques that may be used in plants in the future are discussed.

Contents

INTRODUCTION

The discovery of proteins that excise themselves, termed inteins, from a protein precursor and the elucidation of the splicing mechanism, have led to a number of intein-based applications. These applications include protein purification, protein semisynthesis, in vivo protein modification, and a proposed method of transgene containment. The first and last methods have obvious relevance to this review because they have either been used in plants or pertain to a plant-related concern. However, this review also discusses other aspects of intein function and application that may not have yet been used in a plant system. We present these methodologies and background because inteins may prove as useful in plants as they have been in bacterial, yeast, insect, and mammalian cells.

Intein: a protein splicing element

Protein splicing: the process in which an intein excises itself from a precursor molecule with the concomitant ligation of the flanking protein sequences

Extein: the amino acid sequence flanking an intein

PROTEIN SPLICING MECHANISM

Protein splicing elements, termed inteins (44), were first reported in 1990 by two groups (23, 26). The vast majority of experimentally investigated inteins utilize a common splicing mechanism (10, 73). Protein splicing is dependent on the chemical nature of the splice junction amino acid residues. Most inteins possess a cysteine or serine at the N terminus (**Figure 1**). At the C terminus and just downstream of the C terminus there is typically an asparagine followed by a cysteine, serine, or threonine.

An N-S acyl shift, utilizing the cysteine side chain, or a N-O acyl shift, utilizing a serine side chain, henceforth grouped and represented as a N-O(S) acyl shift, is the first step in the common splicing mechanism (**Figure 2**). This results in an ester or thioester, represented herein as (thio)ester, depending on whether a serine or cysteine, respectively, is involved. The (thio)ester is formed between the N terminus of the intein and the polypeptide fused to the intein N terminus (termed the N-extein) (**Figure 1**) in place of the normal peptide bond. The (thio)ester interrupting the peptide backbone is attacked by a downstream nucleophile, the side chain of a cysteine, serine, or threonine, present C terminal to the last intein residue. An intermediate with two N termini is formed, termed the branched intermediate. The intein now has a free N terminus and scission of the C-terminal peptide bond will release it. This occurs by nucleophilic attack of the asparagine side chain nitrogen on the carbonyl carbon of the peptide bond joining the intein and the downstream polypeptide (C-extein). The amino acid sequences originally flanking the intein, the N- and C-exteins, are covalently attached, but through a non-native (thio)ester bond. A spontaneous O(S)-N acyl rearrangement results in a native peptide bond joining the N- and C-exteins; there is no "scar" indicating that the intein was ever present.

Although the protein splicing mechanism described above appears to be the most common, there is a variation in a subfamily of inteins possessing an N-terminal alanine. These inteins bypass the first N-O(S) acyl shift and the downstream nucleophile directly attacks the carbonyl carbon of the peptide bond upstream of the intein (54). Furthermore, the recently described bacterial intein-like proteins may utilize a different splicing mechanism (2).

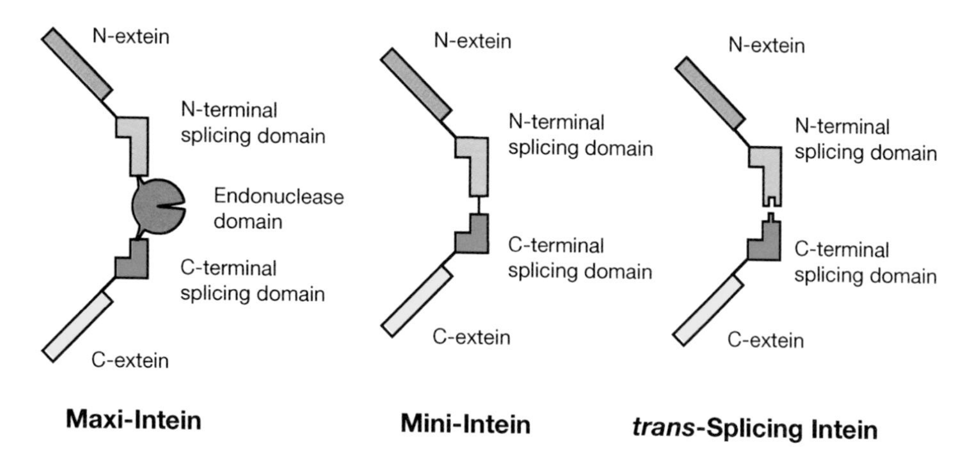

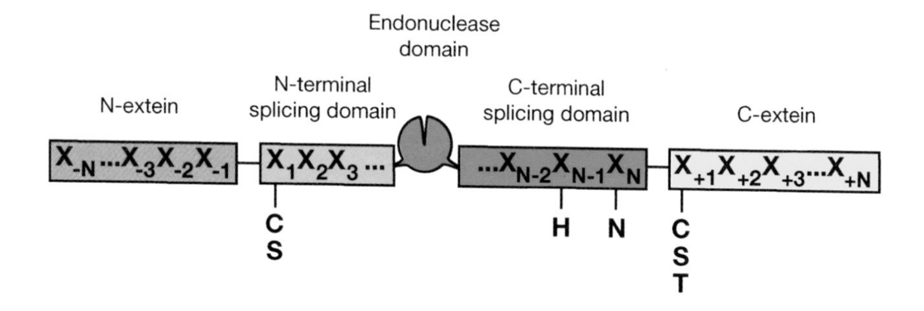

Figure 1

Intein types and nomenclature. (*A*) Maxi-inteins contain a homing endonuclease necessary for intein homing whereas mini-inteins lack this domain. The *trans*-splicing inteins do not have the N- and C-terminal splicing domains covalently linked. (*B*) The amino acid residues of the intein are numbered sequentially. The N-extein residues are numbered beginning with the first residue adjacent to the intein, the −1 residue. The C-extein is numbered sequentially starting with the first amino acid downstream of the intein, the +1 residue. Conserved amino acid residues are represented as their single-letter codes below the appropriate splicing domain position.

INTEIN TYPES

Inteins can be grouped in a number of ways. There are so-called maxi-inteins, mini-inteins, *cis*-splicing, *trans*-splicing, and alanine inteins. There are also the bacterial intein-like proteins (2), which are not covered in this review. Of the known inteins, most are maxi-inteins [see InBase (43)]. The maxi-inteins range in size from 244 to 1650 amino acids with most around 500 amino acid residues. A maxi-intein contains an endonuclease domain between its N- and C-terminal splicing domains (**Figure 1**). It has been speculated that the presence of the endonuclease domain facilitates intein spread by a homing mechanism (20). The counterpart to the maxi-inteins are the mini-inteins that lack the homing endonuclease motif. Mini-inteins range in size from 128 to 1309 amino acid residues, with most around 160 amino acids.

In addition to grouping inteins according to the presence or absence of an endonuclease domain, they are also classified by whether or not the N- and C-terminal splicing domains are covalently attached. The majority of inteins are expressed as continuous polypeptides with the N- and C-terminal splicing domains present in the same precursor protein. These are referred to as *cis*-splicing inteins. Protein splicing elements in which the N- and C-terminal splicing

Trans-splicing: a process in which the noncovalently attached N- and C-terminal protein splicing domains of an intein associate and undergo protein splicing

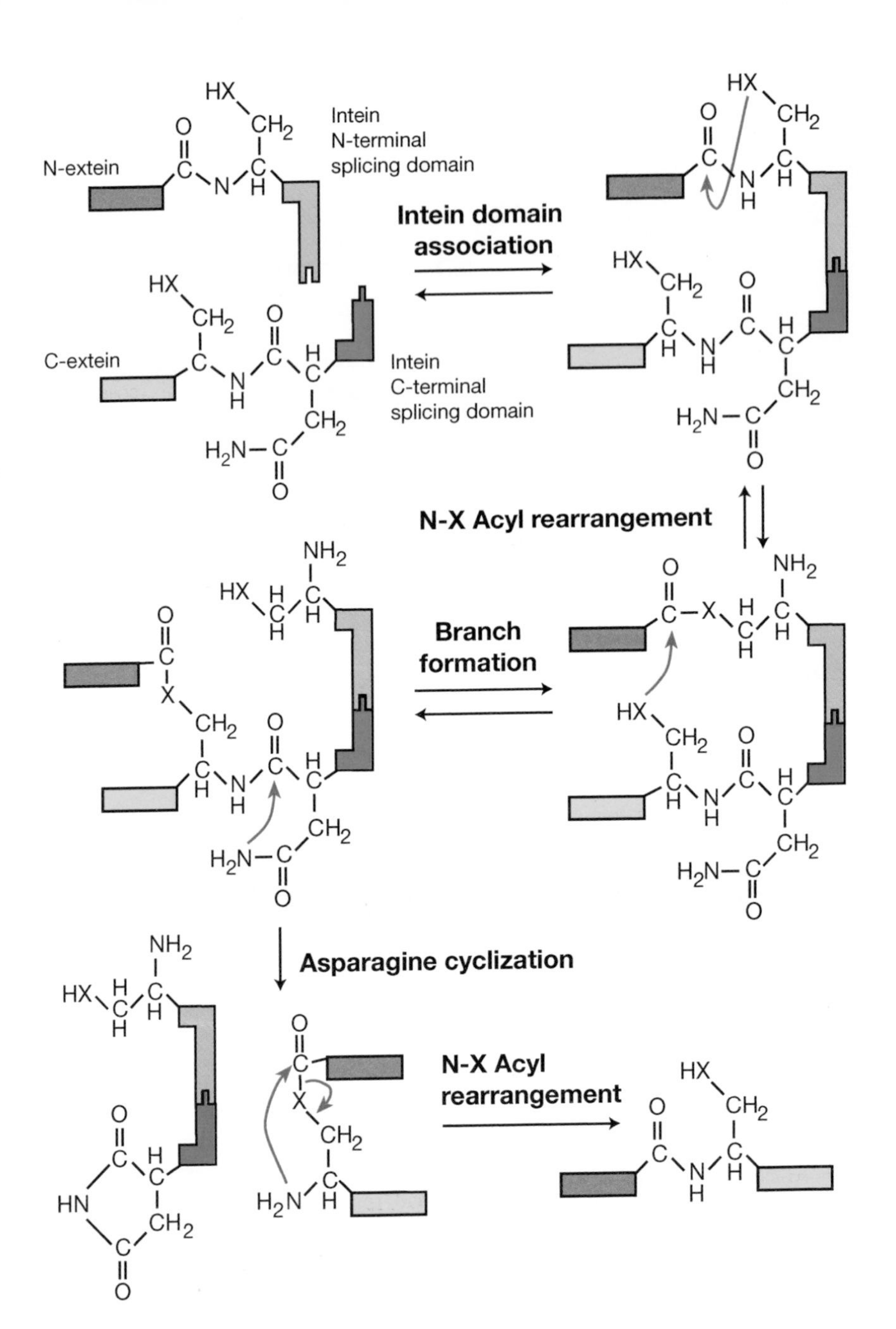

Figure 2

Protein *trans*-splicing mechanism. Protein *trans*-splicing and *cis*-splicing are essentially the same except that *trans*-splicing has an initial association step. The X represents either a sulfur or oxygen atom.

domains are not covalently attached are termed *trans*-splicing inteins (34, 52, 69). The first naturally occurring *trans*-splicing intein discovered, the Ssp DnaE intein (68), has served a double role as an experimental window into intein properties as well as a powerful tool in protein engineering.

INTEINS IN TRANSGENE CONTAINMENT

Trans-Splicing Inteins in Transgene Containment

There are a number of potential concerns with introducing transgenic plants into the environment. The magnitude of the effect that genetically modified plants have on the environment is still being assessed. However, regardless of whether the effect is small or large it would be advantageous to limit the spread of transgenes from genetically modified plants to weedy relatives. With that goal in mind inteins have been used in a novel manner to potentially limit transgene spread (6, 7, 56). The methodology builds on the idea of preventing gene spread by localizing the transgene into the chloroplast (12). The chloroplast genome is not inherited paternally in most commercially relevant plant species and therefore spread of a plastid integrated transgene should not be possible via pollen. To further decrease the possibility that a complete and active transgene is passed to the environment it can be divided into two fragments, each expressing an inactive, truncated protein. These gene fragments can be placed into different locations, such as the nuclear and plastid genomes. A plastid localization signal can target the gene product encoded in the nucleus, and translated in the cytoplasm, to the chloroplast. However, unless the proteins are capable of functional complementation they will be inactive.

Inteins make it feasible to apply the split transgene idea to a wider range of proteins (**Figure 3**). The method takes advantage of the *trans*-splicing activity of certain inteins. The published reports used the naturally occurring *trans*-splicing Ssp DnaE intein, but any intein capable of *trans*-splicing could be substituted. A gene encoding the N-terminal splicing domain of an intein fused to the N-terminal target protein fragment is created. Also generated is the complementary gene encoding the C-terminal splicing domain fused to the C-terminal target protein fragment. Separate integration of the fusion genes into the nuclear and plastid genomes followed by their expression, translocation of the nuclear-encoded gene product to the chloroplast via a signal sequence, and intein-mediated *trans*-splicing would generate a full-length transgene product. Pollen DNA from such a plant will encode only a truncated, inactive transgene product. Furthermore, should some plastid DNA escape it also encodes only a portion of the desired protein. Such a system may significantly reduce the possibility of transgene spread, although this needs further experimental verification.

Conceptually, the split transgene methodology is very straightforward, but it is not trivial to find a site to split a protein and subsequently have the truncated protein fragments form an active complex. Truncating a protein may result in a lowest energy structure having a different fold than needed for the activity of the full-length protein. However, even if the protein fragments are misfolded, after *trans*-splicing they may spontaneously rearrange to form the proper architecture because the newly fused polypeptide would have a different global energy minimum than the starting reactants. Furthermore, because the fusion occurs in living tissue, there are chaperonins to facilitate refolding of the ligated protein fragments.

In addition to folding issues, the intein *trans*-splicing reaction must be considered. All amino acid residues critical to splicing, except one, are found in the intein itself. The exception lies directly downstream of the intein and must be present at the ligation (split) site. In the case of the Ssp DnaE intein, this is a cysteine residue, although serine can be substituted with an alteration in the splicing kinetics (41, 48).

A further consideration is that inteins splice to differing extents depending on the surrounding amino acid residues. Most notably, the −1 (see **Figure 1**) amino acid residue can have a profound effect on protein splicing. To complicate matters, even if the ideal −1 and +1 amino acids are present the intein may not splice effectively. The reason is likely protein dependent, but may be due to the polypeptides flanking the intein folding into a structure that distorts the intein active site and, therefore, disrupts the splicing reaction.

Splitting Acetolactate Synthase

Two schemes were utilized to simplify the site selection process: a rational approach (56) and

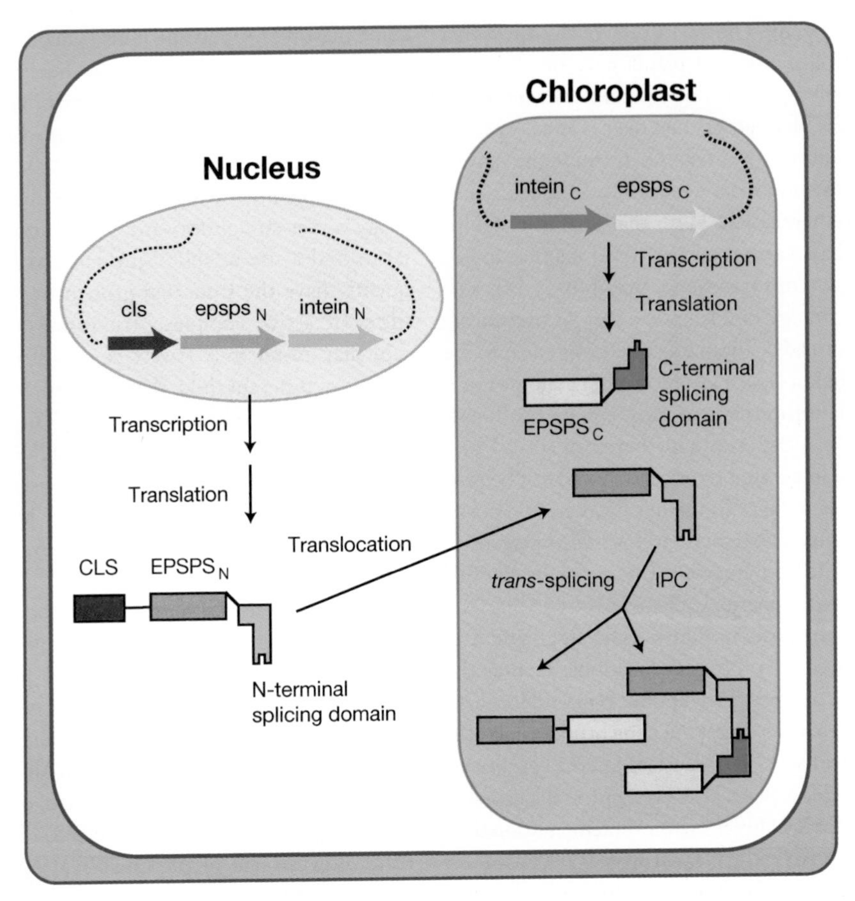

Figure 3

Transgene containment using protein *trans*-splicing. A transgene, such as that encoding the herbicide-resistant epsps gene, is split and a fragment is placed into the nucleus, and the rest is placed into the chloroplast. The nuclear gene fragment is fused with DNA encoding the N-terminal splicing domain of a *trans*-splicing intein ($intein_N$) and also encoding a chloroplast localization sequence (cls). The plastid integrated epsps fragment is fused to DNA encoding the C-terminal domain of a *trans*-splicing intein ($intein_C$). After transcription and translation the CLS-$EPSPS_N$-$Intein_N$ moves to the chloroplast where it can form an active EPSPS protein either by *trans*-splicing or intein-mediated protein complementation (IPC).

a library-based approach (6). Both studies used genes that had phenotypes in both bacteria, specifically *E. coli*, as well as in plants. Because of the short *E. coli* generation times this greatly increased the rate of screening for potential sites to split a target protein. The rational approach was applied to the ALS gene, which catalyzes a common step in the biosynthesis of branched chain amino acids. Sulfonylurea herbicides, such as sulfometuron methyl (SM), target the acetolactate synthase (ALS) enzyme in plants and bacteria and lead to cell death.

SM: sulfometuron methyl

ALS: acetolactate synthase

In general, when considering a region to split a protein and retain activity one would typically look for flexible linker regions present between folding domains. Keeping the structural units intact may result in correct domain folding in both the truncated and, subsequently, the ligated protein. Furthermore, placing the intein into a flexible linker may allow it to adopt the necessary splicing structure without interference from a domain attempting to fold.

Once a suitable flexible linker is found then it must be determined whether a naturally occurring cysteine, serine, or threonine is present. As described above, all the necessary catalytically important amino acids are contained within the intein sequence except for one, a cysteine, serine, or threonine immediately downstream of the intein. This residue persists in the spliced protein following auto-excision of the intein. Of course, the required residue could be engineered into the protein at the split site, but one would avoid this when possible, especially for a cysteine, which, if left unpaired, can affect protein folding and/or activity.

A suitable linker was identified by comparing the amino acid sequences of ALS from different organisms. The regions of high similarity are probably important for ALS function and were not considered. However, Gln327 of *E. coli* ALSII (the ALS being investigated) was near a variable region. In fact, the *E. coli* ALSII has a deletion in that region compared with the tobacco and corn ALS, indicating not only amino acid sequence flexibility but also structural flexibility.

The *E. coli* ALSII used contained an Ala26Val change (mALSII) that resulted in SM resistance. The gene encoding mALSII was split between the codons for Gln327 and Cys328. A gene was created that encoded the first 327 amino acids of mALSII fused to 7 amino acids favorable for *trans*-splicing and the Ssp DnaE intein N-terminal splicing domain. The fusion gene was cloned into an expression vector. The gene encoding the complementary fusion protein, consisting of the C-terminal Ssp DnaE intein splicing domain, 7 splicing-favorable amino acid residues, and the amino acids 328 to 548 of mALSII, was cloned into a separate plasmid. The two plasmid constructs had different selectable markers, compatible origins of replication, and isopropyl-thio-β-galactopyranoside (IPTG)-inducible protein expression.

Transformation of *E. coli* with these plasmids and induction with IPTG resulted in *trans*-splicing, as determined by Western blot analysis. There was, however, a temperature dependence to the *trans*-splicing activity. At 15–30°C spliced product was detected, whereas at 37°C there was no splicing. Temperature-sensitive splicing activity with the Ssp DnaE intein was consistent with previous observations (17).

The goal was to produce active protein, not just a spliced product. Therefore, the activity of the spliced product was assayed in vivo. The assay system took advantage of the fact that *E. coli* ALSI and ALSIII are sensitive to valine inhibition, but ALSII is not. Therefore, *E. coli* will only grow on M9 minimal media plates supplemented with valine and SM if it possesses an active mALSII protein. Plating *E. coli* co-transformed with plasmids expressing intein fusions of the N and C termini of mALSII on M9 minimal media supplemented with valine and SM resulted in colony formation when incubated at 30°C, indicative of mALSII activity. However, no growth was seen at 37°C or when either or both intein fragments were absent—situations in which *trans*-splicing is not expected to occur. Therefore, an herbicide-resistant phenotype can be generated using a split mALSII, but this phenotype appeared dependent on the *trans*-splicing activity of the intein.

Dividing 5-Enolpyruvylshikimate-3-Phosphate Synthase

The requirement for *trans*-splicing to generate active ALS contrasts with the findings using a different herbicide resistance gene, a mutant form of 5-enolpyruvylshikimate-3-phosphate synthase (EPSPS), and a library-based approach (6). Glyphosate, a popular herbicide, inhibits EPSPS and leads to death in plants and bacteria, but not humans. A Pro101Ser change in the EPSPS from *Salmonella typhimurium* makes the enzyme effectively resistant to glyphosate (55) and therefore this mutant was chosen for testing in the model system.

The library-based approach used linker scanning to randomly insert five amino acid residues into EPSPS. It was assumed that regions of EPSPS that tolerate insertions would be a good starting place for examining intein integration. There are many advantages to screening a library as compared to rational design. Very little biochemical information about the target protein is needed. The entire protein sequence, not just linker regions, can be scanned. However, an easy method of determining protein activity is needed and sites that tolerate amino acid insertions are not necessarily good sites to split a protein.

To create the library of EPSPS genes encoding proteins with randomly inserted amino acids, a transposon-based linker scanning approach was adopted (4, 6). Plasmids with a transposon, Tn7, modified to incorporate PmeI sites at the ends was mixed with plasmids containing the EPSPS gene and the necessary enzymes. The modified transposon typically inserts once per target plasmid. To avoid contamination from genes lacking a transposon, the EPSPS gene was excised from the plasmids following transposition. The transposon was large, ~1.5 kb, and the EPSPS genes containing a transposon were easily separated from those lacking one by agarose gel electrophoresis.

The EPSPS genes with a transposon were ligated into an expression vector. After PmeI digestion and subsequent DNA ligation a plasmid library was generated in which a 15-nucleotide stretch was randomly distributed throughout the EPSPS genes. The library was transformed into an *E. coli* strain that lacked an active EPSPS and therefore could not grow on minimal media. These transformed *E. coli* were plated on minimal media and colonies, indicating an active EPSPS containing a five-amino acid insert were picked. Plasmids from these colonies were sequenced to determine the exact site of nucleotide insertion.

The PmeI site used to excise the transposon was regenerated after ligation. Therefore, not only did the transposon-based system create a library of genes with 15-bp insertions, it also permitted facile genetic manipulation of any potential hits. In the published example (6), a number of sites in EPSPS tolerated five amino acid insertions. These were further investigated for their ability to support splicing activity and to determine if the spliced product was active by subcloning a *cis*-splicing intein directly into the newly inserted PmeI site. The *cis*-splicing intein was the Ssp DnaE intein because it was also used for *trans*-splicing, in which the normally noncovalently linked N- and C-terminal splicing domains were joined. Two and three extra amino acid residues were included at the N and C termini of the intein, respectively, to facilitate the splicing reaction.

The screens yielded three potential hits that were further tested by determining growth in cultures of M9 minimal media. One, with the intein inserted at amino acid 235, outgrew the others and, in fact, grew at the same rate as an *E. coli* strain expressing the *S. typhimurium* EPSPS gene lacking an insert. Furthermore, spliced EPSPS was detected in crude cell lysates. An unexpected result was discovered when the negative control was examined. The negative control consisted of an EPSPS protein with an inactive, mutant Ssp DnaE intein present at position 235. *E. coli* expressing an EPSPS containing the splicing deficient intein grew just as well in cultures of M9 minimal media as bacteria expressing an EPSPS that lacked an insert. This is probably due to the nature of the intein fold. An intein folds so that its N and C termini come into close proximity on one side of the protein (13, 28, 35, 39, 45, 47, 62). When inserted into another protein sequence the intein probably forms a separate domain-like structure protruding from the target protein. The close proximity of the intein N and C termini creates a relatively small disturbance in the target protein.

Position 235 in *S. typhimurium* EPSPS tolerated an intein insertion. Therefore, the EPSPS gene was divided at this position and fused to the genes encoding the appropriate *trans*-splicing domains of the Ssp DnaE intein. Growth of *E. coli* lacking an endogenous EPSPS gene could be rescued by coexpressing the split EPSPS-intein constructs. This was seen whether or not the intein carried a mutation that abolished splicing. The presence of both EPSPS fragments and both intein splicing domains was required, however. The intein splicing domains appeared to associate and bring together the N- and C-terminal fragments of EPSPS in a correct orientation for enzyme activity. The intein fold, as described above, undoubtedly is a crucial component in this ability. In addition to permitting growth in minimal media, the split EPSPS-intein system allowed growth on glyphosate.

Unlike the observation that came from using the intein splicing domains, later studies with dimerization domains such as GCN4 or the id1B-MyoD pair did not produce an active EPSPS protein dimer (L. Chen & T. Evans, unpublished observations). Inteins have certain properties that are advantageous for splicing activity and also make them potentially useful dimerization domains. The first is the close proximity of the intein N and C termini. Second, they bind as heterodimers, as compared to the homodimerization of GCN4, allowing more control of the system. Finally, they appear to bind very tightly and once bound do not dissociate for many hours (32).

An improvement to the library-based approach would be to insert a cassette encoding an intein into the PmeI site of the library before any screening is performed. The cassette could encode either a *cis*- or *trans*-splicing intein construct. This new library, with an intein randomly inserted into the target gene, could then be screened for activity. This saves time and resources used to isolate and sequence clones that will not be active with an intein inserted at that position.

Which is the best approach, a rational or a library-based method? As is often the case, it is probably a mixture of the two. There are too many variables, i.e., protein folding or intein context, to predict them confidently in every situation. Furthermore, there may not be sufficient biochemical or structural information available for a particular gene of interest to use a rational design. A potential difficulty with the library may be in testing all the necessary insertion sites. The most efficient method would probably be to first determine which portion of a protein would be the most desirable place to split it by rational design and then probe that protein region using a transposon-based system and intein insertion.

The utility of merging the two approaches was illustrated during experimentation with the *S. typhimurium* EPSPS. Based on structural and biochemical data the best site to split EPSPS was thought to be around amino acid 239. Upon screening the library it was determined that amino acid insertion between residues 238 and 239 as well as 240 and 241 were not tolerated, marking this region as questionable for placing an intein. However, an insertion at amino acid 235 was not only tolerated, but it was the best site evaluated.

Intein Transgenes in Plants

Two groups nearly simultaneously published articles demonstrating splicing of exogenous inteins in plants (7, 74), and one recreated the entire transgene containment model system in *Nicotiana tabacum*. A group at Dupont demonstrated protein *trans*-splicing in *Arabidopsis* cells by fusing the N-terminal Ssp DnaE intein splicing domain ($DnaE_N$) to the N-terminal half of β-glucuronidase (GUS_N) and the Ssp DnaE intein C-terminal splicing domain ($DnaE_C$) to the other half of GUS (GUS_C). Transfection of a cassette carrying both fusion fragments into *Arabidopsis* using *Agrobacterium tumefaciens* resulted in integration of the transgene and in vivo reassembly of functional GUS activity. Although protein *trans*-splicing requires optimal temperature, pH, and other conditions, the transgenic *Arabidopsis* showed GUS expression throughout the plant, suggesting a broad expression pattern. Because the *trans*-spliced GUS was catalytically functional, the authors concluded that the *trans*-splicing process also fosters correct protein folding.

In the second set of experiments the authors introduced either GUS_N-Ssp $DnaE_N$ or $DnaE_C$-GUS_C fusions into *Arabidopsis* and by genetic crossing brought both fragments to the same progeny. These plants expressed functional GUS activity as well as spliced enzyme, demonstrating that transgene fusions from different loci or chromosomes can participate in protein *trans*-splicing. Furthermore, intein-mediated complementation of GUS activity was also reported in a cross with an intact GUS_N and a defective $DnaE_N$, as observed previously in *E. coli* with EPSPS. This was the first example of protein *trans*-splicing or intein-mediated complementation in plant cell cytoplasm.

The second group demonstrated that *trans*-splicing could occur in the chloroplast of tobacco plants (7), a property vital for the use of inteins in transgene containment. Biolistic integration of a cassette containing a gene encoding green fluorescent protein fused to the N-terminal splicing domain of the Ssp DnaE intein (GFP-$DnaE_N$) and a separate gene encoding the C-terminal Ssp DnaE intein splicing domain fused to aminoglycoside-3-adenyltransferase ($DnaE_C$-aadA) into the plastid genome did not affect *trans*-plastomic plant development. These plants expressed both fusion proteins inside the chloroplast and a *trans*-spliced hybrid GFP-aadA was observed.

Transgenic tobacco plants were created to test the feasibility of intein-based transgene containment. A tobacco plant was generated that had a gene encoding the N-terminal fragment of the herbicide-resistant form of EPSPS, amino acids 1-235, fused to $DnaE_N$ in the nuclear genome, and a gene encoding $DnaE_C$ fused to the complementary EPSPS C-terminal fragment, amino acids 236–427, in the plastid genome. The DNA encoding the signal peptide sequence of the rubisco large subunit was also fused to the $DnaE_N$-$EPSPS_N$ gene. This permits translocation of the protein product to the chloroplast. Western blot analysis of crude cell lysates indicated that *trans*-splicing had occurred to produce full-length EPSPS. Furthermore, these plants were resistant to the herbicide glyphosate (**Figure 4**). These data demonstrate that a transgene can be split, localized into separate cellular compartments, and still give rise to the desired protein activity in the hope of creating more environmentally friendly transgenic plants.

A potential refinement of the split gene concept for transgene containment would be to integrate the nuclear transgene fragment into a genome that is not compatible with weedy relatives (see 22 for a discussion of genome hybridization). Cultivated oilseed rape and wheat have multiple genomes acquired from different wild relatives. Bread wheat, for example, contains three genomes designated A, B, and

Figure 4

Transgenic tobacco with increased resistance to glyphosate. A transgenic tobacco plant with the epsps gene split between the chloroplast and nucleus and fused to a *trans*-splicing intein, as described in Figure 3, was treated with glyphosate (0.2 mM). T is the transgenic plant, C is an unmodified control plant. The days post-application of glyphosate are indicated above the pictures.

D (22). A weedy relative of wheat is jointed goatgrass, which contains two genomes, C and D. The D genomes of bread wheat and jointed goatgrass can readily hybridize. Studies indicate that transgenes present in the D genome of bread wheat are passed to jointed goatgrass at substantially higher rates compared with transgenes present in the A or B genomes (22). Therefore, placing the nuclear-encoded split gene into the B genome would drastically reduce the possibility of transgene spread compared with the split gene system alone. There may also be instances in which it is more effective to place the complementary split genes on different genomes, for example A and B of bread wheat, than to place one half into the chloroplast. This, of course, would need to be considered on a case-by-case basis. In summary, genetic engineering of crop species with the *trans*-splicing technique may offer an additional level of protection for transgene containment.

INTEIN TAGS TO PURIFY PROTEINS FROM PLANTS

Inteins have been modified for use as self-cleaving affinity tags. The commercially available systems contain expression vectors, available from New England Biolabs, for producing proteins in *E. coli*. Intein tags were developed as the result of work to elucidate the protein splicing mechanism. Mutations in key residues abolished protein splicing, but steps in the splicing reaction could still proceed. Furthermore, this could be used to site-specifically cleave a peptide bond. Coupling a controllable cleavage element to an affinity tag greatly aids purification technology because the tag can be removed without using proteases.

There are a number of intein tags available. The first reported intein tag (8) used the Sce VMA intein fused to a CBD from *Bacillus circulans* (61) (**Figure 5*a***). The intein tag was fused to the C terminus of the target protein and was released by a nucleophilic reagent such as dithiothreitol (DTT). The same intein was further modified to be fused to the N terminus, instead of the C terminus, of the target protein. This permitted the isolation of proteins starting with an amino acid residue other than a methionine (9). Subsequent publications describe the use of other inteins as purification tags, including mini-inteins and constructs in which peptide bond cleavage is induced by a temperature and/or pH shift (**Figure 5*c***) (15–17, 33, 53, 65, 66, 70).

CBD: chitin binding domain

DTT: dithiothreitol

Intein tags are used most frequently to purify proteins from *E. coli*, but other hosts such as insect cells (46) and even plants (38) have also been used. The last report used an intein tag to purify SMAP-29 from *N. tabacum*. SMAP-29 is a small (3.2-kDa), cationic, broad-spectrum antimicrobial peptide that was first identified in sheep myeloid cells (3). It is active against both bacterial and fungal pathogens, including strains resistant to conventional antibiotics. Therefore, it may be useful as an antibiotic in humans.

SMAP-29 or any pharmaceutically relevant protein must be produced in large quantities. Plants represent a possible solution to this problem. The current cost of creating transgenic plants limits their use to commercially viable proteins such as pharmaceuticals, although recent publications suggest that this may change (21, 31). SMAP-29 illustrates another potential advantage of the plant system because it is bacteriocidal and fungicidal, thereby eliminating the use of some hosts for protein production.

To express intein-tagged proteins in plants the intein from the *E. coli* expression vector pCYB3 was subcloned into the *Agrobacterium tumefaciens* binary vector pBI121. This placed recombinant protein expression under control of the cauliflower mosaic virus 35S promoter. The DNA encoding the SMAP-29 peptide was cloned into the new vector along with a β-conglycinin transit peptide. The β-conglycinin transit peptide was added to target the SMAP-29-intein tag fusion protein to the apoplast compartment of the endoplasmic reticulum to facilitate its recovery in the soluble protein fraction.

Purification of the SMAP-29-intein tag fusion protein was initiated by applying the crude leaf extract to a chitin column. Unbound

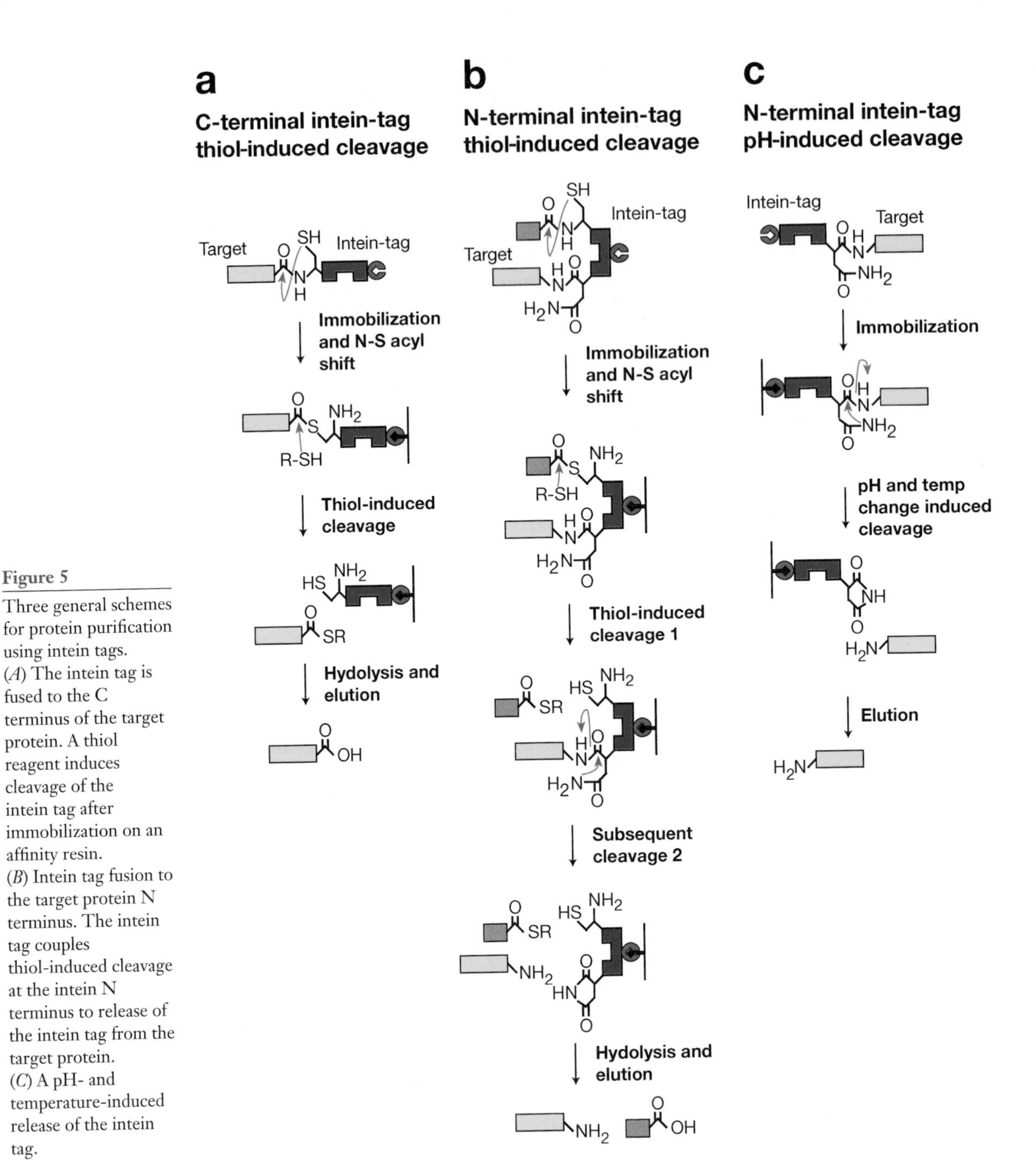

Figure 5

Three general schemes for protein purification using intein tags. (*A*) The intein tag is fused to the C terminus of the target protein. A thiol reagent induces cleavage of the intein tag after immobilization on an affinity resin. (*B*) Intein tag fusion to the target protein N terminus. The intein tag couples thiol-induced cleavage at the intein N terminus to release of the intein tag from the target protein. (*C*) A pH- and temperature-induced release of the intein tag.

protein was washed from the resin and intein tag self-cleavage was initiated by adding DTT. Interestingly, instead of the 3.2-kDa band expected for SMAP-29, a 28-kDa band was detected. This was due to the binding of an endogenous plant protein to the SMAP-29. Removing the endogenous plant protein resulted in an active SMAP-29 peptide with the expected mobility on SDS-PAGE. Therefore, as the authors point out, the intein tag behaved as expected and may be a viable option in purifying proteins from plant tissue in the future.

BIOFIBER PRODUCTION

Merging inteins and plants has the potential to contain transgene spread and facilitate protein purification, but there are other potential applications. Plants permit the relatively inexpensive production of large quantities of protein, whereas inteins have already been used to perform facile protein engineering in vitro and in vivo. This section focuses on the potential of forming unique biofibers using inteins.

Polymers are an invaluable part of modern life. However, despite the amazing plethora of synthetic polymers, science has yet to produce fibers that compare to natural silks, such as spider dragline silk (64). A common characteristic shared by silk proteins is that the bulk of their sequence is composed of amino acid repeating units. Furthermore, they can be >1,000,000 daltons in size. Cloning these large silk genes into *E. coli* has been difficult because of the DNA size and the presence of large numbers of repeated nucleotide blocks. Therefore, to complement the cloning of silk genes, perhaps inteins can be used to create artificial biofibers.

We and others experimented with inteins as tools of protein cyclization (15, 24, 25, 27, 48, 49, 63, 72). It was discovered that, in addition to cyclic proteins, there were species that appeared to be multimeric (17). The cyclization strategy utilized a chimeric protein consisting of the C-terminal Ssp DnaE intein splicing domain fused to the N terminus of maltose binding protein (MBP) and the N-terminal Ssp DnaE intein splicing domain fused to the C terminus of the same MBP (**Figure 6**).

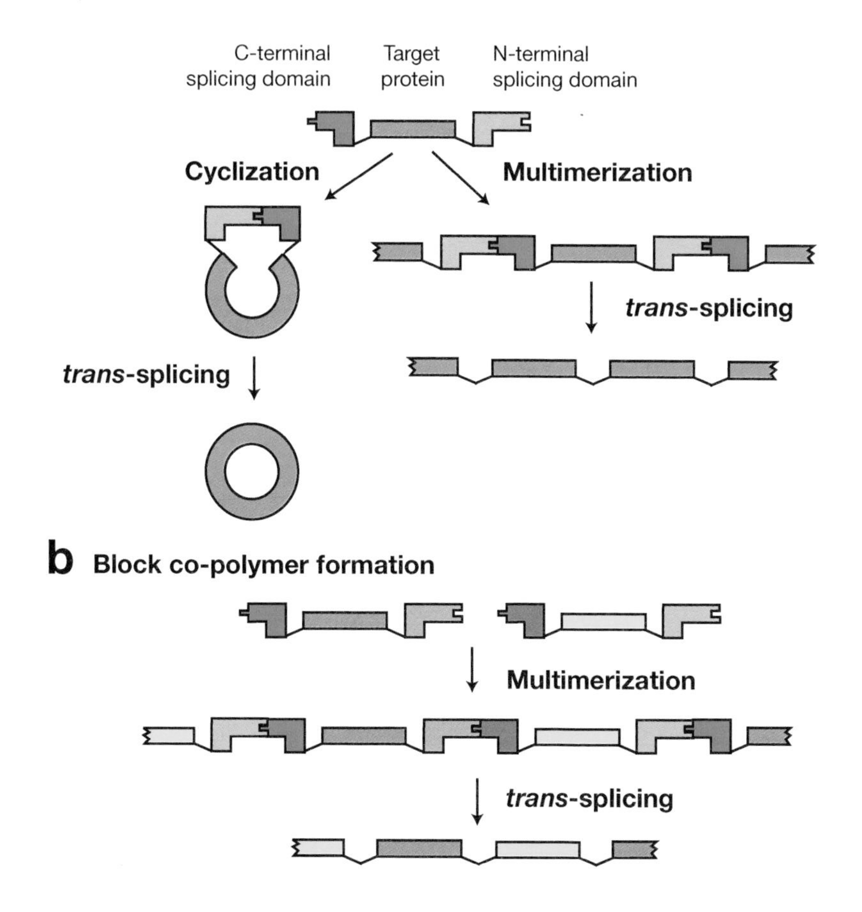

Figure 6
Formation of a biofiber via *trans*-splicing. (*A*) Fusion of protein splicing domains from the same intein to a target protein can result in cyclization or multimerization. (*B*) Using splicing domains from different inteins fused to the same target protein eliminates formation of monomeric circles and permits the precise assembly of block copolymers. Tight intein binding interactions would permit the formation of multimers with interesting properties in the absence of protein splicing.

By extension, a silk protein-like multimer may be produced if the DNA encoding MBP were replaced by a gene encoding a repeating unit found in silk. After expression, the silk protein monomer units, fused to intein splicing domains, would undergo *trans*-splicing to form large polymers. The silk protein-like species would be produced as a population of multimers with a heterogeneous size distribution. This is in contrast to the native state, in which very large genes produce silk proteins of uniform size. This may not be problematic because current synthetic polymers consist of a population of molecules with differing numbers of repeating units.

A potential difficulty with the described methodology is that monomer cyclization might be a significant, if not a major, product of the *trans*-splicing reaction. This issue can be addressed. In fact, it leads to the ability to accurately incorporate different repeating units into a biofiber (**Figure 6**). It is based on the fact that intein splicing domains cannot, to date, be exchanged. Therefore, for example, repeating unit A is fused between the C-terminal splicing domain of the Ssp DnaE intein and the N-terminal splicing domain of the Ssp DnaB intein. A second construct is made consisting of repeating unit B fused between the C-terminal splicing domain of the Ssp DnaB intein and the N-terminal splicing domain of the Ssp DnaE intein. The intein splicing domains present on each repeating unit should not be able to *trans*-splice and thereby prevent the formation of the cyclic monomer side product. If repeating unit A is the same as B then a homopolymer is produced; however, if A and B are different a block copolymer results. The specificity of the intein splicing domains ensures that the resulting block copolymer has an . . .A–B–A–B. . . arrangement. With more than 170 known inteins it should be possible to precisely create very complex polymer arrangements. Finally, because intein domains bind so tightly it should be possible to produce polymers in which the monomer units are bound together through the noncovalent interaction of the intein splicing domains. These biofibers may have unique and desirable characteristics.

CONCLUSION

Inteins are active in plants and useful in purifying proteins expressed in plants. The utility of inteins is that they permit both in vitro and in vivo protein modifications. This review covers the intein applications with the most obvious relevance to plants, but does not cover other very exciting techniques and examples that may be used in transgenic plants in the future (1, 5, 11, 14, 18, 19, 29, 30, 36, 37, 40, 42, 50, 51, 57–60, 67, 71, 75). We hope that the intein and plant fields will continue to mingle and will produce some very exciting outcomes.

SUMMARY POINTS

1. The application of a split-gene approach, coupled with intein technology, may provide superior transgene containment.
2. Intein tags permit the facile purification of proteins expressed in transgenic plants.
3. A site to split a protein for transgene containment can be determined using sequence comparison to related proteins and biochemical data.
4. Transposon-based linker scanning simplifies the identification of a site within a protein to insert an intein.
5. Novel biofibers can be created using inteins.
6. The majority of known protein splicing elements utilize a common splicing mechanism.

LITERATURE CITED

1. Alexandrov K, Heinemann I, Durek T, Sidorovitch V, Goody RS, Waldmann H. 2002. Intein-mediated synthesis of geranylgeranylated Rab7 protein in vitro. *J. Am. Chem. Soc.* 124:5648–49
2. Amitai G, Belenkiy O, Dassa B, Shainskaya A, Pietrokovski S. 2003. Distribution and function of new bacterial intein-like protein domains. *Mol. Microbiol.* 47:61–73
3. Bagella L, Scocchi M, Zanetti M. 1995. cDNA sequences of three sheep myeloid cathelicidins. *FEBS Lett.* 376:225–28
4. Biery MC, Stewart FJ, Stellwagen AE, Raleigh EA, Craig NL. 2000. A simple in vitro Tn7-based transposition system with low target site selectivity for genome and gene analysis. *Nucleic Acids Res.* 28:1067–77
5. Buskirk AR, Ong YC, Gartner ZJ, Liu DR. 2004. Directed evolution of ligand dependence: small-molecule-activated protein splicing. *Proc. Natl. Acad. Sci. USA* 101:10505–10
6. Chen L, Pradhan S, Evans TC Jr. 2001. Herbicide resistance from a divided EPSPS protein: the split Synechocystis DnaE intein as an in vivo affinity domain. *Gene* 263:39–48
7. Chin HG, Kim GD, Marin I, Mersha F, Evans TC Jr, et al. 2003. Protein trans-splicing in transgenic plant chloroplast: reconstruction of herbicide resistance from split genes. *Proc. Natl. Acad. Sci. USA* 100:4510–15
8. Chong S, Mersha FB, Comb DG, Scott ME, Landry D, et al. 1997. Single-column purification of free recombinant proteins using a self-cleavable affinity tag derived from a protein splicing element. *Gene* 192:271–81
9. Chong S, Montello GE, Zhang A, Cantor EJ, Liao W, et al. 1998. Utilizing the C-terminal cleavage activity of a protein splicing element to purify recombinant proteins in a single chromatographic step. *Nucleic Acids Res.* 26:5109–15
10. Chong S, Shao Y, Paulus H, Benner J, Perler FB, Xu MQ. 1996. Protein splicing involving the Saccharomyces cerevisiae VMA intein. The steps in the splicing pathway, side reactions leading to protein cleavage, and establishment of an in vitro splicing system. *J. Biol. Chem.* 271:22159–68
11. Cotton GJ, Ayers B, Xu R, Muir TW. 1999. Insertion of a synthetic peptide into a recombinant protein framework: a protein biosensor. *J. Am. Chem. Soc.* 121:1100–1
12. Daniell H, Ruiz ON, Dhingra A. 2004. Chloroplast genetic engineering to improve agronomic traits. *Methods Mol. Biol.* 286:111–38
13. Ding Y, Xu MQ, Ghosh I, Chen X, Ferrandon S, et al. 2003. Crystal structure of a mini-intein reveals a conserved catalytic module involved in side chain cyclization of asparagine during protein splicing. *J. Biol. Chem.* 278:39133–42
14. Evans TC Jr, Benner J, Xu MQ. 1998. Semisynthesis of cytotoxic proteins using a modified protein splicing element. *Protein Sci.* 7:2256–64
15. Evans TC Jr, Benner J, Xu MQ. 1999. The cyclization and polymerization of bacterially expressed proteins using modified self-splicing inteins. *J. Biol. Chem.* 274:18359–63
16. Evans TC Jr, Benner J, Xu MQ. 1999. The in vitro ligation of bacterially expressed proteins using an intein from Methanobacterium thermoautotrophicum. *J. Biol. Chem.* 274:3923–26
17. Evans TC Jr, Martin D, Kolly R, Panne D, Sun L, et al. 2000. Protein trans-splicing and cyclization by a naturally split intein from the dnaE gene of Synechocystis species PCC6803. *J. Biol. Chem.* 275:9091–94
18. Evans TC Jr, Xu MQ. 1999. Intein-mediated protein ligation: harnessing nature's escape artists. *Biopolymers* 51:333–42
19. Gangopadhyay JP, Jiang SQ, Paulus H. 2003. An in vitro screening system for protein splicing inhibitors based on green fluorescent protein as an indicator. *Anal. Chem.* 75:2456–62

20. Gimble FS, Thorner J. 1992. Homing of a DNA endonuclease gene by meiotic gene conversion in Saccharomyces cerevisiae. *Nature* 357:301–6
21. Gleba Y, Marillonnet S, Klimyuk V. 2004. Engineering viral expression vectors for plants: the 'full virus' and the 'deconstructed virus' strategies. *Curr. Opin. Plant Biol.* 7:182–88
22. Hedge SG, Waines JG. 2004. Hybridization and introgression between bread wheat and wild and weedy relatives in North America. *Crop Sci.* 44:1145–55
23. Hirata R, Ohsumk Y, Nakano A, Kawasaki H, Suzuki K, Anraku Y. 1990. Molecular structure of a gene, VMA1, encoding the catalytic subunit of H(+)-translocating adenosine triphosphatase from vacuolar membranes of Saccharomyces cerevisiae. *J. Biol. Chem.* 265:6726–33
24. Iwai H, Lingel A, Pluckthun A. 2001. Cyclic green fluorescent protein produced in vivo using an artificially split PI-PfuI intein from Pyrococcus furiosus. *J. Biol. Chem.* 276:16548–54
25. Iwai H, Pluckthun A. 1999. Circular beta-lactamase: stability enhancement by cyclizing the backbone. *FEBS Lett.* 459:166–72
26. Kane PM, Yamashiro CT, Wolczyk DF, Neff N, Goebl M, Stevens TH. 1990. Protein splicing converts the yeast TFP1 gene product to the 69-kD subunit of the vacuolar H(+)-adenosine triphosphatase. *Science* 250:651–57
27. Kinsella TM, Ohashi CT, Harder AG, Yam GC, Li W, et al. 2002. Retrovirally delivered random cyclic peptide libraries yield inhibitors of interleukin-4 signaling in human B cells. *J. Biol. Chem.* 277:37512–18
28. Klabunde T, Sharma S, Telenti A, Jacobs WR Jr, Sacchettini JC. 1998. Crystal structure of GyrA intein from Mycobacterium xenopi reveals structural basis of protein splicing. *Nat. Struct. Biol.* 5:31–36
29. Lovrinovic M, Seidel R, Wacker R, Schroeder H, Seitz O, et al. 2003. Synthesis of protein-nucleic acid conjugates by expressed protein ligation. *Chem. Commun.* (*Camb.*): 822–23
30. Lue RY, Chen GY, Zhu Q, Lesaicherre ML, Yao SQ. 2004. Site-specific immobilization of biotinylated proteins for protein microarray analysis. *Methods Mol. Biol.* 264:85–100
31. Marillonnet S, Giritch A, Gils M, Kandzia R, Klimyuk V, Gleba Y. 2004. In planta engineering of viral RNA replicons: efficient assembly by recombination of DNA modules delivered by Agrobacterium. *Proc. Natl. Acad. Sci. USA* 101:6852–57
32. Martin DD, Xu MQ, Evans TC Jr. 2001. Characterization of a naturally occurring trans-splicing intein from Synechocystis sp. PCC6803. *Biochemistry* 40:1393–402
33. Mathys S, Evans TC, Chute IC, Wu H, Chong S, et al. 1999. Characterization of a self-splicing mini-intein and its conversion into autocatalytic N- and C-terminal cleavage elements: facile production of protein building blocks for protein ligation. *Gene* 231:1–13
34. Mills KV, Lew BM, Jiang S, Paulus H. 1998. Protein splicing in trans by purified N- and C-terminal fragments of the Mycobacterium tuberculosis RecA intein. *Proc. Natl. Acad. Sci. USA* 95:3543–48
35. Mizutani R, Nogami S, Kawasaki M, Ohya Y, Anraku Y, Satow Y. 2002. Protein-splicing reaction via a thiazolidine intermediate: crystal structure of the VMA1-derived endonuclease bearing the N- and C-terminal propeptides. *J. Mol. Biol.* 316:919–29
36. Mootz HD, Blum ES, Tyszkiewicz AB, Muir TW. 2003. Conditional protein splicing: a new tool to control protein structure and function in vitro and in vivo. *J. Am. Chem. Soc.* 125:10561–69
37. Mootz HD, Muir TW. 2002. Protein splicing triggered by a small molecule. *J. Am. Chem. Soc.* 124:9044–45
38. Morassutti C, De Amicis F, Skerlavaj B, Zanetti M, Marchetti S. 2002. Production of a recombinant antimicrobial peptide in transgenic plants using a modified VMA intein expression system. *FEBS Lett.* 519:141–46

39. Moure CM, Gimble FS, Quiocho FA. 2002. Crystal structure of the intein homing endonuclease PI-SceI bound to its recognition sequence. *Nat. Struct. Biol.* 9:764–70
40. Muir TW, Sondhi D, Cole PA. 1998. Expressed protein ligation: a general method for protein engineering. *Proc. Natl. Acad. Sci. USA* 95:6705–10
41. Nichols NM, Evans TC Jr. 2004. Mutational analysis of protein splicing, cleavage, and self-association reactions mediated by the naturally split Ssp DnaE intein. *Biochemistry* 43:10265–76
42. Otomo T, Ito N, Kyogoku Y, Yamazaki T. 1999. NMR observation of selected segments in a larger protein: central-segment isotope labeling through intein-mediated ligation. *Biochemistry* 38:16040–44
43. Perler FB. 2002. InBase: the intein database. *Nucleic Acids Res.* 30:383–84
44. Perler FB, Davis EO, Dean GE, Gimble FS, Jack WE, et al. 1994. Protein splicing elements: inteins and exteins—a definition of terms and recommended nomenclature. *Nucleic Acids Res.* 22:1125–27
45. Poland BW, Xu MQ, Quiocho FA. 2000. Structural insights into the protein splicing mechanism of PI-SceI. *J. Biol. Chem.* 275:16408–13
46. Pradhan S, Bacolla A, Wells RD, Roberts RJ. 1999. Recombinant human DNA (cytosine-5) methyltransferase. I. Expression, purification, and comparison of de novo and maintenance methylation. *J. Biol. Chem.* 274:33002–10
47. Romanelli A, Shekhtman A, Cowburn D, Muir TW. 2004. Semisynthesis of a segmental isotopically labeled protein splicing precursor: NMR evidence for an unusual peptide bond at the N-extein-intein junction. *Proc. Natl. Acad. Sci. USA* 101:6397–402
48. Scott CP, Abel-Santos E, Jones AD, Benkovic SJ. 2001. Structural requirements for the biosynthesis of backbone cyclic peptide libraries. *Chem. Biol.* 8:801–15
49. Scott CP, Abel-Santos E, Wall M, Wahnon DC, Benkovic SJ. 1999. Production of cyclic peptides and proteins in vivo. *Proc. Natl. Acad. Sci. USA* 96:13638–43
50. Shah RV, Finkelstein JM, Verdine GL. 2002. Toward a small-molecule activated protein splicing system. *The Nucleus* 16–21
51. Sharma S, Zhang A, Wang H, Harcum SW, Chong S. 2003. Study of protein splicing and intein-mediated peptide bond cleavage under high-cell-density conditions. *Biotechnol. Prog.* 19:1085–90
52. Southworth MW, Adam E, Panne D, Byer R, Kautz R, Perler FB. 1998. Control of protein splicing by intein fragment reassembly. *Embo J.* 17:918–26
53. Southworth MW, Amaya K, Evans TC, Xu MQ, Perler FB. 1999. Purification of proteins fused to either the amino or carboxy terminus of the Mycobacterium xenopi gyrase A intein. *Biotechniques* 27:110–20
54. Southworth MW, Benner J, Perler FB. 2000. An alternative protein splicing mechanism for inteins lacking an N-terminal nucleophile. *Embo J.* 19:5019–26
55. Stalker DM, Hiatt WR, Comai L. 1985. A single amino acid substitution in the enzyme 5-enolpyruvylshikimate-3-phosphate synthase confers resistance to the herbicide glyphosate. *J. Biol. Chem.* 260:4724–28
56. Sun L, Ghosh I, Paulus H, Xu MQ. 2001. Protein trans-splicing to produce herbicide-resistant acetolactate synthase. *Appl. Environ. Microbiol.* 67:1025–29
57. Sun L, Ghosh I, Xu MQ. 2003. Generation of an affinity column for antibody purification by intein-mediated protein ligation. *J. Immunol. Methods* 282:45–52
58. Sun W, Yang J, Liu XQ. 2004. Synthetic two-piece and three-piece split inteins for protein trans-splicing. *J. Biol. Chem.* 279:35281–86

59. Sydor JR, Mariano M, Sideris S, Nock S. 2002. Establishment of intein-mediated protein ligation under denaturing conditions: C-terminal labeling of a single-chain antibody for biochip screening. *Bioconjug. Chem.* 13:707–12
60. Takeda S, Tsukiji S, Nagamune T. 2004. Site-specific conjugation of oligonucleotides to the C-terminus of recombinant protein by expressed protein ligation. *Bioorg. Med. Chem. Lett.* 14:2407–10
61. Watanabe T, Ito Y, Yamada T, Hashimoto M, Sekine S, Tanaka H. 1994. The roles of the C-terminal domain and type III domains of chitinase A1 from Bacillus circulans WL-12 in chitin degradation. *J. Bacteriol.* 176:4465–72
62. Werner E, Wende W, Pingoud A, Heinemann U. 2002. High resolution crystal structure of domain I of the Saccharomyces cerevisiae homing endonuclease PI-SceI. *Nucleic Acids Res.* 30:3962–71
63. Williams NK, Prosselkov P, Liepinsh E, Line I, Sharipo A, et al. 2002. In vivo protein cyclization promoted by a circularly permuted Synechocystis sp. PCC6803 DnaB mini-intein. *J. Biol. Chem.* 277:7790–98
64. Wong Po Foo C, Kaplan DL. 2002. Genetic engineering of fibrous proteins: spider dragline silk and collagen. *Adv. Drug Deliv. Rev.* 54:1131–43
65. Wood DW, Derbyshire V, Wu W, Chartrain M, Belfort M, Belfort G. 2000. Optimized single-step affinity purification with a self-cleaving intein applied to human acidic fibroblast growth factor. *Biotechnol. Prog.* 16:1055–63
66. Wood DW, Wu W, Belfort G, Derbyshire V, Belfort M. 1999. A genetic system yields self-cleaving inteins for bioseparations. *Nat. Biotechnol.* 17:889–92
67. Wood RJ, Pascoe DD, Brown ZK, Medlicott EM, Kriek M, et al. 2004. Optimized conjugation of a fluorescent label to proteins via intein-mediated activation and ligation. *Bioconjug. Chem.* 15:366–72
68. Wu H, Hu Z, Liu XQ. 1998. Protein trans-splicing by a split intein encoded in a split DnaE gene of Synechocystis sp. PCC6803. *Proc. Natl. Acad. Sci. USA* 95:9226–31
69. Wu H, Xu MQ, Liu XQ. 1998. Protein trans-splicing and functional mini-inteins of a cyanobacterial dnaB intein. *Biochim. Biophys. Acta* 1387:422–32
70. Wu W, Wood DW, Belfort G, Derbyshire V, Belfort M. 2002. Intein-mediated purification of cytotoxic endonuclease I-TevI by insertional inactivation and pH-controllable splicing. *Nucleic Acids Res.* 30:4864–71
71. Xu J, Sun L, Ghosh I, Xu MQ. 2004. Western blot analysis of Src kinase assays using peptide substrates ligated to a carrier protein. *Biotechniques* 36:976–78, 80–81
72. Xu MQ, Evans TC Jr. 2001. Intein-mediated ligation and cyclization of expressed proteins. *Methods* 24:257–77
73. Xu MQ, Perler FB. 1996. The mechanism of protein splicing and its modulation by mutation. *Embo J.* 15:5146–53
74. Yang J, Fox GC Jr, Henry-Smith TV. 2003. Intein-mediated assembly of a functional beta-glucuronidase in transgenic plants. *Proc. Natl. Acad. Sci. USA* 100:3513–18
75. Zeidler MP, Tan C, Bellaiche Y, Cherry S, Hader S, et al. 2004. Temperature-sensitive control of protein activity by conditionally splicing inteins. *Nat. Biotechnol.* 22:871–76

Molecular Genetic Analyses of Microsporogenesis and Microgametogenesis in Flowering Plants

Hong Ma

Department of Biology and the Huck Institutes of the Life Sciences, Pennsylvania State University, University Park, Pennsylvania 16802; email: hxm16@psu.edu

Annu. Rev. Plant Biol.
2005. 56:393–434

doi: 10.1146/
annurev.arplant.55.031903.141717

First published online as a
Review in Advance on
January 17, 2005

1543-5008/05/0602-
0393$20.00

Key Words

stamen identity, anther differentiation, tapetum, meiosis, pollen development

Abstract

In flowering plants, male reproductive development requires the formation of the stamen, including the differentiation of anther tissues. Within the anther, male meiosis produces microspores, which further develop into pollen grains, relying on both sporophytic and gametophytic gene functions. The mature pollen is released when the anther dehisces, allowing pollination to occur. Molecular studies have identified a large number of genes that are expressed during stamen and pollen development. Genetic analyses have demonstrated the function of some of these genes in specifying stamen identity, regulating anther cell division and differentiation, controlling male meiosis, supporting pollen development, and promoting anther dehiscence. These genes encode a variety of proteins, including transcriptional regulators, signal transduction proteins, regulators of protein degradation, and enzymes for the biosynthesis of hormones. Although much has been learned in recent decades, much more awaits to be discovered and understood; the future of the study of plant male reproduction remains bright and exciting with the ever-growing tool kits and rapidly expanding information and resources for gene function studies.

Contents

INTRODUCTION

In flowering plants, the angiosperms, male reproductive development begins in the sporophytic generation with the initiation and formation of the male reproductive organ stamen in the flower. The development of stamen involves the formation of an anther that has multiple specialized cell types and that houses male meiotic cells and a filament that supports the anther. Meiosis then results in the formation of microspores; therefore, microsporogenesis requires the entire process from the initiation of stamen primordium to meiosis. Following meiosis, the microspore undergoes two rounds of mitoses during pollen development, culminating in the formation of two microgametes, the sperm cells. The process of microgametogenesis occurs within the developing pollen, yet it depends on sporophytic functions provided by the surrounding anther tissues.

Microsporogenesis and microgametogenesis are essential for the propagation of flowering plants. At the same time, these developmental events involve several fundamental cellular processes, including cell division, cell differentiation, cell-cell communication, and cell death. The reproductive structures of plants provide major sources of food for humans and animals. The ability to control male reproductive development is very important for plant breeding and agriculture because male-sterile varieties are valuable resources that greatly facilitate the production of hybrids via cross-pollination. Cross-pollination may represent a major means of gene flow between populations and related species; in particular, the spread of transgenes via pollen has been a major concern regarding possible ecological impact of transgenic crop plants. Therefore, understanding stamen and pollen development has significant potential impact on basic plant development, agriculture, and the environment.

Over the last quarter century, a great deal has been uncovered about the molecular genetic control of stamen and pollen development in several model systems. Early molecular studies in tobacco, Brassica, and tomato provided estimates of the number of genes expressed in the developing stamen and identified anther- or pollen-specific genes. Genetic screens have isolated many mutants that exhibit defects in stamen/anther development, pollen development, and pollen release and delivery. More recently, specific genes required for male fertility were cloned and analyzed at the molecular level.

In this review, the normal stamen development, anther differentiation, and pollen development are summarized, using *Arabidopsis thaliana* as an example. A discussion on the specification of stamen identity, including genes in the ABC model and other genes, follows. Genetic and molecular studies on anther cell differentiation and function are also presented. Next, cell biological and molecular genetic studies of male meiosis are discussed. Pollen development and its gametophytic control are described. Due to space constraints, early studies will be briefly summarized and morphological studies in most plants will not be included. The reader is referred to a number of excellent reviews that were published over the past 15 years (23, 43, 50, 57, 62, 72, 92, 97, 124, 125, 138, 139, 144, 153, 177, 178, 202–204, 236, 238–240, 251, 254, 260).

STAMEN DEVELOPMENT

Although there is a great deal of variation in floral morphology among angiosperms, in the vast majority of species, stamens occupy positions interior to the sterile perianth organs and surround the central gynoecium. The number of stamens can vary from species to species, ranging from a few to dozens; some species produce sterile stamen-like organs called staminodes, which may attract pollinators. In *Arabidopsis*, there are six stamens, each consisting of a filament and an anther (**Figure 1**). The filament has an epidermis and a vascular bundle and the anther has four lobes that are attached to the vascular bundle by connective tissues. At maturity, the anther is about four times as wide and twice as thick as the diameter of the filament and about one tenth the length compared to the filament. The *Arabidopsis* flower development has been divided into 12 stages using morphological landmarks observed from scanning electron microscopy (210). Early stages involve the initiation and growth of the floral meristems, with sepal primordia forming at stage 3. At stage 5, six stamen primordia are initiated at the same time as the initiation of petal primordia. Cell division and differentiation occur from stages 6 to 9; the four long stamens are slightly more advanced than the two short ones and they form an anther at the top and a stalk at the base by stage 7. At stage 8, the anthers of the long stamens develop four lobes in which meiosis will take place and the pollen will develop. Male meiosis occurs at stage 9 when petal primordia are still circular and small and the filament has yet to elongate (**Figure 1*C***). This is followed by pollen development at stages 10 through 12. The anther dehisces afterward, when the flower opens.

Stamen development was also examined in detail using sections in both tobacco and *Arabidopsis* (71, 106, 189). Because the anther is much more complex than the filament, anther development has been described more thoroughly and has been divided into two phases on the basis of morphological, cellular, and molecular events. During phase 1, cell division and differentiation result in the formation of anther tissues. The anther has four lobes, each with several sporophytic (somatic) tissues that surround the MMCs, also called PMCs, or male sporocytes. The anther lobes are laterally attached to a vascular bundle by a contiguous connective tissue; the vascular bundle is continuous with the vasculature of the filament. There are two additional nonreproductive anther tissues: the circular cell cluster and stomium, both of which are important for pollen release. During phase 2, the filament elongates greatly; at the same time, the anther further enlarges and microspores develop into pollen grains, followed by the degeneration of anther tissues, anther dehiscence, and pollen release.

MMC: microspore mother cell

PMC: pollen mother cell

Gynoecium: female reproductive organs collectively in angiosperms

In *Arabidopsis*, anther development is divided into 14 stages using morphological and cellular features from light microscopy (189) (**Figure 2**). Stages 1–8 make up the aforementioned phase 1 of anther development and

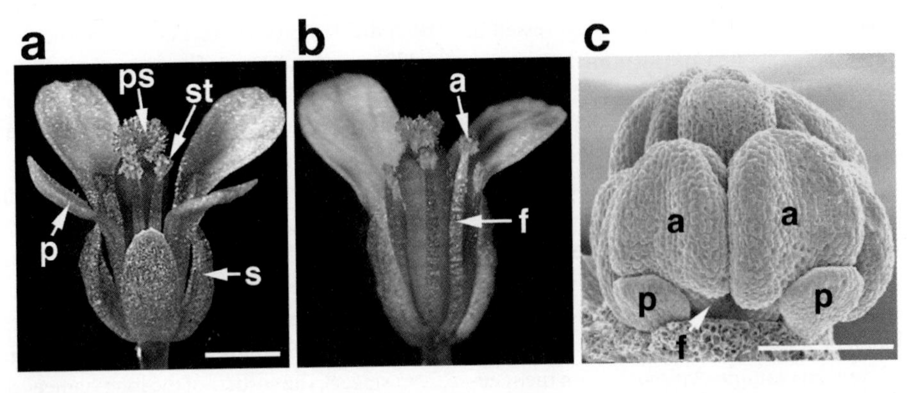

Figure 1

Arabidopsis flower. (*A*) An intact mature (stage 13) *Arabidopsis* flower with four types of organs, sepals (s), petals (p), stamens (st), and the pistil (ps). (*B*) A mature flower with one sepal and two petals removed to reveal some of the stamens, which have an anther (a) and a filament (f). A and B are of the same magnification; bar = 1.0 mm. (*C*) A scanning electron micrograph of a stage 9 flower, with its sepals removed to show the inner organs; two petal primordia (p) are round and small; two of the four long stamens can be easily seen with their anthers (a) having attained the characteristic lobed shape and the filaments (f) still very short. Bar = 100 μm. These images were provided by Y. Hu.

Anticlinal: the cell divisional plane is perpendicular to the cell layer

stages 9–14 constitute phase 2. At stage 1, divisions in the L1, L2, and L3 layers of the floral meristem result in the formation of the stamen (anther) primordium. From stages 2 to 5, anticlinal cell division in L1 expands the surface area of the anther to form the epidermis. At the same time, cells in the L3 layer divide and differentiate to form the connective and vascular tissues. Periclinal and anticlinal cell divisions are responsible for the formation of internal cell layers in the four anther lobes. Specifically, four clusters of archesporial cells are formed at stage 2 from periclinal division of L2 cells. The archesporial cells then divide at stage 3 to form both the primary parietal layer and the sporogenous cells; the primary parietal layer is just beneath the epidermal layer and surrounds the sporogenous cells. Cell division in the primary parietal layer then forms two secondary parietal layers. The outer secondary parietal layer further divides and differentiates at stage 4 into an endothecium layer and a middle layer. During this stage, cells of the inner secondary parietal layer divide and develop into the tapetum layer.

By stage 5, the anther assumes the characteristic four-lobed morphology and anther morphogenesis is complete. In each lobe there are four nonreproductive layers from the surface to the interior: the epidermis, the endothecium, the middle layer, and the tapetum. Surrounded by these somatic layers are the reproductive MMCs. Subsequently, at stage 6, the MMCs undergo meiosis; also, callose is deposited on the wall of the meiotic cells during this time. The meiotic cells then dissociate from each other and from the tapetum, resulting in a space called the locule, interior to the tapetum. Stage 7 is marked by the completion of meiosis and the formation of a tetrad of four haploid microspores. In stage 8, the microspores are released from the tetrad following the degradation of the callose wall. Microspores then develop into pollen grains during stages 9–12. During stages 10 and 11, the tapetum degenerates. At stage 12, pollen mitotic divisions occur, resulting in tricellular haploid pollen grains. At stage 13, anther dehiscence occurs, followed by the shrinkage of anther cells at stage 14.

The anther stages 1 through 4 (189) correspond approximately to stages 5 through 8 of flower development, as defined by Smyth et al. (210). Anther stages 5–8 make up the long floral stage 9. From anther stages 9 to 12 and floral stages 10 to 12, these two systems use, respectively, internal and external markers that are not easily matched. Because the size of floral organs

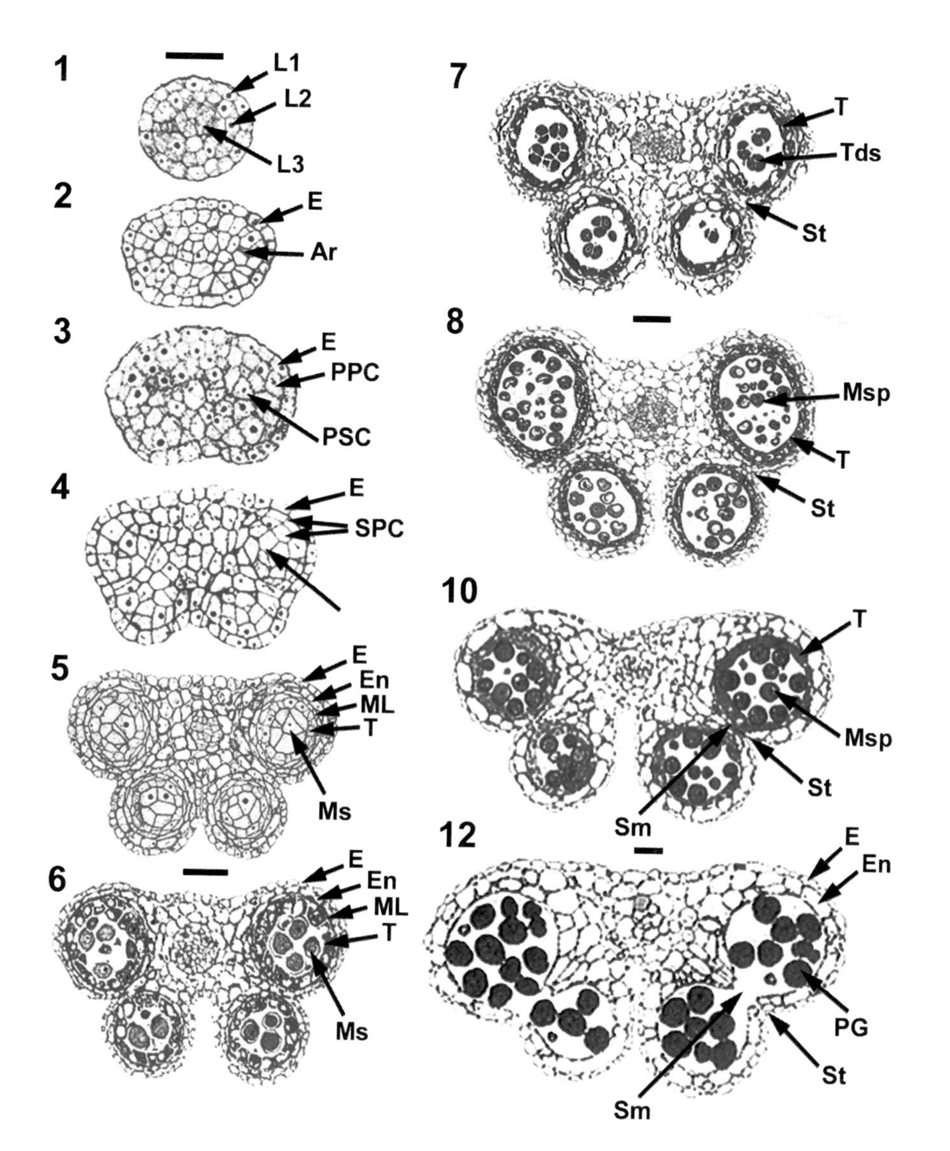

Figure 2

Anther cell differentiation. Shown are the stages of wild-type *Arabidopsis thaliana* anther development. The numbers indicate stages. Bar = 25 μm and stages 1 to 5, 6 and 7, 8 and 10, and 12 alone have same-sized bars. Ar, archesporial; E, epidermis; En, endothecium; L1, L2, L3, Layer 1, 2, 3; ML, middle layer; Ms, microsporocytes; Msp, microspore; PG, pollen grain; PPC, primary parietal cell; PSC, primary sporogenous cell; Sm, septum; SPC, secondary parietal cell; St, stomium; T, tapetum; Tds, tetrads. These images were provided by D. Zhao.

can vary under different growth conditions, it is not feasible to use only organ size to stage anther or flower. At stage 13 for both systems, anthesis occurs and pollen is released. In the remainder of this review, the 14 stages of anther development (189) are used to avoid confusion.

GENETIC CONTROL OF STAMEN IDENTITY

The morphological studies of stamen development have established a structural foundation upon which genetic and molecular analyses can be carried out. Over the past 15 years, a great deal has been learned about the molecular control of stamen identity and development, primarily from molecular genetic studies of *Arabidopsis thaliana* and *Antirrhinum majus* (23, 43, 92, 124, 125, 144, 153, 177, 178, 202, 236, 238–240, 251, 254, 260).

The Classical and Revised ABC Models

Characterizing a series of *Arabidopsis* and *Antirrhinum* homeotic mutants that exhibit

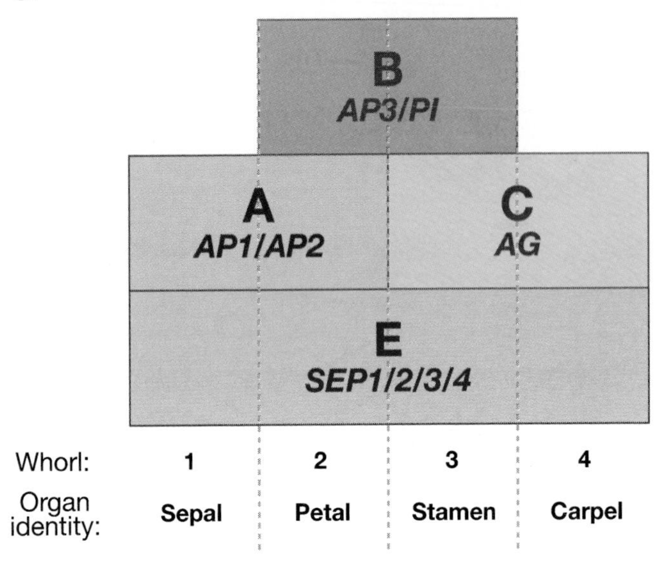

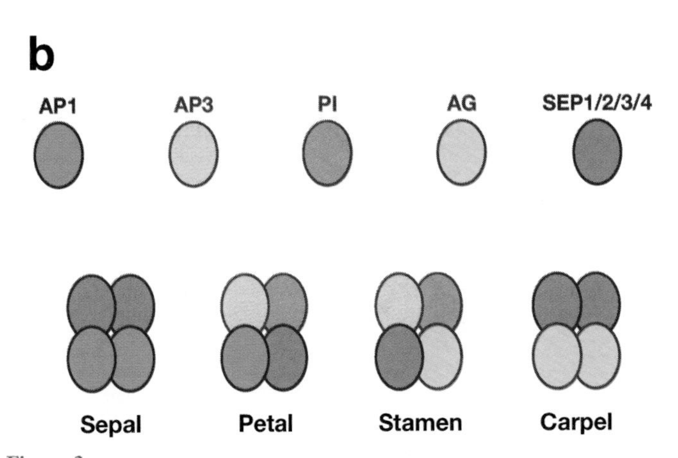

Figure 3

The ABC model. (*A*) The ABC functions are indicated as boxes with the *Arabidopsis* genes. The *SEP* genes are shown below the regions for A, B, and C functions. The normal floral organ identities corresponding to the whorls 1 through 4 are shown at the bottom. (*B*) The quartet model, with an expanded E function as shown in A. ABCE MADS-domain proteins are shown as ovals, with different shading as indicated at the top. Tetrameric complexes are shown for each organ identity. Part B of this figure is adopted from (228).

conversion of one set of organs to another led to the proposal of the original ABC model for controlling floral organ identities (43, 124, 125, 239) (**Figure 3**). Specifically, the B and C functions are required to specify the stamen identity. The A function is indirectly involved in controlling stamen identity because it antagonizes the C function, such that in the absence of A function, stamens are formed ectopically. In *Arabidopsis*, the A function requires the *APETALA1* (*AP1*) and *APETALA2* (*AP2*) genes, the B function needs the *APETALA3* (*AP3*) and *PISTILLATA* (*PI*) genes, and the *AGAMOUS* (*AG*) gene is essential for C function. Therefore, mutations in any of the *AP3*, *PI*, and *AG* genes cause defects in stamen identity (24). Molecular studies revealed that *AP3*, *PI*, and *AG* are all expressed initially in regions of the floral meristem that will form stamen primordia and continue to be expressed in stamen primordia and developing stamens (75, 93, 255). Furthermore, ectopic expression of *AP3*, *PI*, *AG* expression causes the formation of ectopic stamens in the flower (87, 109, 147).

Nearly all ABC genes, except *AP2*, belong to the MADS-box gene family, which encodes conserved proteins that share the MADS-domain that is also found in known transcription factors mini-chromosome maintenance1 (MCM1) and SRF (serum response factor), from yeast and human, respectively (124, 153, 177, 202). In fact, the MADS-box was named after the founding members: MCM1, AG, DEF (the AP3 homolog from *Antirrhinum*), and SRF. The MADS-domain is part of the DNA-binding domains of these proteins and a number of other MADS-domain proteins (85, 148). The ABC genes and many other plant MADS-box genes have a so-called MADS-box, MIKC (Intervening sequence, K-box, C region) structure, which contains another domain, named K-box for its similarity to the karatin protein of intermediate filaments, and two nonconserved regions, I and C (126).

Although the B and C genes can cause the formation of ectopic reproductive organs within the flower when misexpressed, they are not sufficient for converting leaves into floral organs. This suggests that additional flower-specific genes are required for floral organ identity. In fact, several other flower-specific MADS-box genes have been identified on the basis of their sequence similarity to *AG*, and were named *AGL* genes for *AG-Like* (126, 129, 130, 183, 191). In particular, *AGL2*, *AGL4*, and *AGL9* are expressed in the floral meristem just prior to the onset of the expression of the known B and C genes *AP3*, *PI*, and *AG* (64, 126, 130, 191). AG and AGL2 can form a heterodimer that binds to DNA in vitro (85). Yeast–two-hybrid experiments revealed that AG can interact with AGL2, AGL4, and AGL9, which are highly similar in sequence, but not some other MADS-domain proteins (59), suggesting that these three AGL proteins may participate in AG-dependent functions. Similarly, AGL9 interacts with AP1 and the PI-AP3 complex (81, 170). These results support the hypothesis that AGL9 can form a tetramer with PI, AP3, and AG.

Definitive evidence for the importance of these genes in controlling organ identity was obtained when insertional mutations were combined in a triple mutant, which produced flowers with only sepals, a phenotype expected for the loss of B and C functions (169). Because of this striking phenotype, *AGL2*, *AGL4*, and *AGL9* were then renamed as *SEPALLATA1*, *SEPALLATA2*, and *SEPALLATA3* (*SEP1*, *SEP2*, *SEP3*), respectively (169). Ectopic expression of *SEP3* together with B and C genes caused the conversion of leaves into staminoid organs (76, 81, 171). Recently, the *AGL3* gene (85a) was found to play redundant roles with *SEP1*, *SEP2*, *SEP3* in specifying sepal identity and regulating meristem identity, thus *AGL3* has been renamed *EPALLATA4* (*SEP4*) (53a). Therefore, on basis of both molecular-genetic and protein-protein interaction studies, the classic ABC model was revised as follows (**Figure 3*A***): A function and SEP1/2/3/4 specify sepal identity; A and B functions plus SEP1/2/3/4 determine petal identity; B and C functions and SEP1/2/3/4 control stamen identity; and C function plus SEP1/2/3/4 specify carpel identity. In particular, on the basis on molecular genetic studies and protein-protein and protein-DNA interactions, a quartet model was proposed that explicitly states that a pair of dimers controls each organ identity (228) (**Figure 3*B***).

Regulation of B Function Genes

In addition to the ABC genes, other genes have also been identified that affect stamen organ identity and development (**Figure 4**). Most of these genes interact genetically with the B

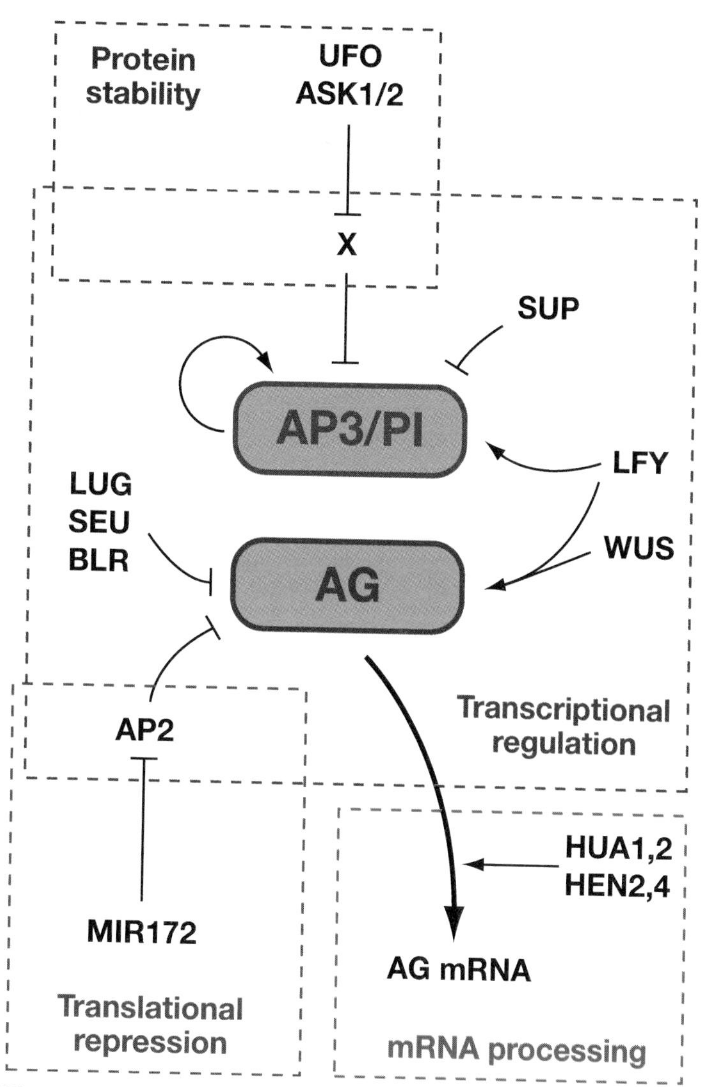

Figure 4

Genes regulating B and C function genes. The B function *AP3* and *PI* and the C function *AG* genes are shown in their respective boxes with thick solid lines. Positive regulation is represented by an arrow, whereas negative regulation is represented by a line with a short bar at the end. Four boxes of dashed lines indicate that the proteins within each box perform one type of biochemical function. Specifically, the proteins in the central large box (*blue*) are putative transcription factors, whereas the proteins in the other boxes regulate posttranscriptional processes.

function genes, *AP3* and *PI*, and/or the C function gene *AG*. Several mutations that affect stamen identity regulate B function genes by genetic and molecular studies (92, 260). One of the most important among these is the meristem identity gene *LEAFY* (*LFY*) encoding a plant-specific transcription factor (242). Flowers of severe *lfy* mutants lack stamens and weak *lfy* mutant flowers produce abnormal stamens (84, 242). In addition, mutations in the *UFO* (*UNUSUAL FLORAL ORGAN*) and *ASK1* (*ARABIDOPSIS SKP1-LIKE1*) genes cause defects in stamen identity and development (114, 244). Flowers of severe *ufo* mutants have substantially reduced stamen numbers, as well as chimeric petal-stamen organs. *ufo* flowers seem to produce filaments in place of stamens. In the *ask1-1* mutant, some flowers exhibit defects in stamen development, including a reduction in number and size, and other abnormalities, similar to the weak *ufo* and *lfy* mutants (257, 258).

How do the *LFY*, *UFO*, and *ASK1* genes regulate stamen identity and development? Several pieces of evidence support the hypothesis that *LFY*, *UFO*, and *ASK1* regulate the expression of the B function genes *AP3* and *PI*. First, severe *lfy* mutants exhibit greatly reduced levels of *AP3* and *PI* expression in early floral primordia (241, 242). Although *AP3* and *PI* expression is not severely affected in flowers of *ufo* or *ask1* single mutants, nor in the weak *lfy-5* mutant, their expression is greatly reduced in *ask1 lfy-5* or *ufo lfy-5* double mutants, which have floral phenotypes similar to the severe *lfy-6* mutant (258). Expression of *AP3* and *PI* from the 35S promoter can partially restore stamens and petals in *lfy* and *ufo* mutants (94). Also, *AP3* is expressed precociously and ectopically in plants with a 35S-*UFO* transgene (112). Finally, the ectopic expression of *UFO* and *LFY* could induce *AP3* expression in seedlings (159).

LFY is expressed throughout young floral meristems but only activates the expression of *AP3* and *PI* genes in the second and third floral whorls (152, 159). Therefore, another factor(s) must be responsible for the whorl specificity. *UFO* is not sufficient for activating *AP3* and *PI* expression because the levels of *UFO* expression are not significantly reduced in *lfy* mutants and the 35S-*UFO* transgene cannot restore petal and stamen development in *lfy* mutants (112). These observations suggest that *LFY* and *UFO* cooperate to activate *AP3* and *PI* expression. Although *UFO* is expressed outside the flower, its expression domain in the flower is similar to that of *AP3* and *PI*, suggesting that *LFY* provides floral specificity and *UFO* provides regional specificity in the flower (112, 122). *LFY* encodes a transcription factor, although there is not sufficient evidence for a direct transcriptional regulation of *AP3* by LFY. UFO is an F-box protein and interacts with ASK1, which is a homolog of SKP1 (173, 187, 249). SKP1 and F-box proteins are subunits of SKP1-Cullin/CDC53-F-box protein (SCF) ubiquitin ligase(s) that mediate protein degradation by the 26S proteasome (48, 179, 198, 209, 261). Therefore, it is possible that UFO and ASK1 are components of a SCF that facilitates the proteolysis of a protein that antagonizes LFY (258). Recent results with genes encoding other SCF subunits suggest that additional SCF complexes are likely involved in regulating stamen development (154).

In addition to the positive regulators, the maintenance of proper spatial expression of *AP3* and *PI* also requires the negative regulation by the *Arabidopsis SUPERMAN* (*SUP*, also called *FLO10*) gene. Mutations in the *SUP* gene cause the production of extra stamens at the expense of carpels (25, 199). *sup* mutant flowers exhibit an expansion of *AP3* and *PI* expression to most of the region of the floral meristem that corresponds to the fourth whorl, suggesting that SUP represses *AP3* and *PI* expression in the fourth whorl. The predicted SUP protein is a putative transcription factor with a C2H2-type zinc-finger and C-terminal repression domain (185, 186). The *SUP* gene is expressed in the third whorl and its expression depends largely on *AP3* and *PI* because early *SUP* expression is decreased and late *SUP* expression is not detected in *ap3* and *pi* mutants. These results suggest that there is a feedback loop between *SUP* and *AP3*/*PI*: *AP3* and *PI* activate *SUP* expression, which in turn represses *AP3* and *PI* expression, although it is not known whether these regulations are direct.

Regulation of the C Function Gene *AG*

The C function gene *AG* is expressed in a temporally and spatially highly specific manner, starting at the time when sepal primordia have been initiated and only in the central region of the floral meristem (55, 255). The degree of specificity is achieved through both positive and negative regulation. As is the case for the B function genes, *LFY* is an important positive regulator of *AG* expression. In strong *lfy* mutants, abnormal flowers lack petals and stamens and produce irregularly fused carpels (242), and *AG* expression is delayed and in a smaller domain than that in the wild type (241). In addition, *AG* expression is precocious in plants with constitutive expression of an activated LFY, and *AG* mRNA is detected throughout these flowers. Consequently, flowers of these plants produce carpels and stamens in place of sepals and petals (159), similar to the flowers of 35S-*AG* transgenic plants (147). Furthermore, expression of the activated LFY is sufficient to activate *AG* in vegetative tissues. LFY binds to *cis* elements in an unusually large *AG* intron. Mutations in the *cis* elements that eliminate LFY binding also cause defects in expression of a reporter gene, indicating that LFY is a direct positive regulator of *AG* (28).

As mentioned before, *LFY* is expressed throughout the floral meristem. Therefore, other factors are needed to provide positional cues for *AG* expression in whorls three and four. Molecular genetic studies have identified WUS (WUSCEL) as an activator of *AG* expression in the center of flowers (113, 123, 136). First, *WUS* is expressed earlier than *AG* in the center of the floral meristem. In *wus* mutants, meristem defects cause a great reduction in the number of flowers; a small number of flowers that are produced in *wus* mutants lack carpels and most stamens. When *WUS* expression is reduced, *AG* is expressed in a reduced domain. In contrast, ectopic *WUS* expression can cause the ectopic formation of stamens and carpels. Furthermore, WUS can bind to the same *AG cis* regulatory region as that bound by LFY, although at different sites. Both LFY- and WUS-binding sites are needed for high levels of transcription because mutations of the WUS-binding sites strongly reduce the level of expression (113, 123).

Additional factors that promote normal *AG* expression were identified by screening for second-site enhancers of the weak *ag-4* allele, which produces flowers with stamens in the third whorl. The initial studies uncovered two genes, *HUA1* and *HUA2*, mutations in which together resulted in the loss of stamen identity in the *ag-4* background, such that the *ag-4 hua1 hua2* triple mutant flowers resemble those of strong *ag* mutants (39). The single *hua1* or *hua2* mutations did not have obvious floral defects, and individually they only weakly enhanced the *ag-4* phenotypes (39). These observations suggest that *HUA1* and *HUA2* have genetically redundant functions. The *hua1 hua2* double mutant has very mild defects in stamens and carpels, and was used to screen for additional enhancers, resulting in the identification of *hua enhancer1* through *hua enhancer4* mutations (39, 40, 115, 243).

How do these genes affect reproductive organ identity? The HUA1 protein contains a CCCH zinc-finger and can bind to single-stranded RNA and DNA, but not double-stranded DNA (115). The predicted HUA2 protein contains a putative nuclear localization signal and possesses transcription activation activity (39). *HUA ENHANCER4* (*HEN4*) was recently cloned and encodes a putative RNA-binding protein with a K homology domain (42). HUA1 binds to the AG pre-RNA and HEN4 can interact with HUA1. Further experiments demonstrate that HUA1, HUA2, and HEN4 proteins facilitate the processing of the AG pre-RNA, particularly the splicing of the unusually large intron, to produce the mature mRNA (42). The *HUA ENHANCER2* (*HEN2*) gene encodes a protein with a nuclear localization signal and sequence similarity to DExH-box RNA helicases, including the yeast

Dob1p helicase (243). Analysis of the expression of ABC genes at different floral developmental stages in mutant backgrounds defective in *HEN2* suggests that *HEN2* may be involved in maintaining, but not establishing, normal levels of B and C gene expression.

Although *HUA1* and *HUA2* are important for normal *AG* expression and the control of reproductive organ identity, *HUA1* and *HUA2* are expressed widely and have functions outside the flower. Similarly, the *HUA ENHANCER1* (*HEN1*) gene is also expressed throughout the plant. In addition to the enhancement of the *hua1 hua2* double mutant floral phenotypes, the *hen1* mutation causes pleiotropic phenotypes, including small organs and altered leaf shape. *HEN1* encodes a novel nuclear protein (40) that shares sequence similarity with predicted proteins from bacterial, fungal, and animal genomes, suggesting that it has a function conserved in eukaryotes and prokaryotes. Additional studies further indicate that HEN1 is important for the normal accumulation of miRNAs, which have become increasingly recognized as important regulators of development, particularly in plants, in recent years (15). The *hen1* mutant phenotype is similar to another mutant, *caf-1*, that is also defective in miRNA accumulation (163, 176). Therefore, HEN1 may exert its effects through the regulation of levels of miRNAs. A recent report showed that a miRNA is critical for regulating *AG* expression via the translational control of AP2 [(38); see below for a discussion of the repression of *AG* by AP2].

miRNA: microRNA

The discovery and further characterization of *HUA ENHANCER3* (*HEN3*) revealed another level of control in the specification of reproductive organ identity (235). *HEN3* is important for specifying stamen and carpel identity and for terminating floral meristem activity. The *HEN3* gene encodes a predicted cyclin-dependent kinase (CDKE), a homolog of CDK8 from mammals. *HEN3* is expressed in tissues with dividing cells, such as meristems and organ primordia; the fact that *HEN3* expression is uniform suggests that its level is not regulated by the cell-cycle machinery. Mutant analysis indicates that *HEN3* is important in promoting cell differentiation and expansion.

CDKE: cyclin-dependent kinase

As mentioned before, *AG* is negatively regulated in the flower by the A function. This is due to the repression by the AP2 protein (55), which is the founding member of a plant-specific family of transcription factors; members of this family have one or more AP2 repeat(s), which is the DNA-binding domain (95, 178). In *ap2* mutants, *AG* expression expands to the periphery of the floral meristem, resulting in the production of carpels and stamens in the outer whorls of the flower. AP2 binds to a region in the large *AG* intron and directly represses *AG* expression in the outer whorls of the floral meristem (22, 208). Unlike the other ABC genes, which are expressed in the region of the floral meristem in which they function, *AP2* mRNA is detected throughout the floral meristem and all floral organ primordia (95). Why, then, does AP2 not repress *AG* expression in the center of the flower? This is because AP2 translation is inhibited in the center, by a miRNA called miRNA172 (38). An increase in the miRNA172 level causes a mutant phenotype similar to those of *ap2* mutants, and expression of mutated *AP2* mRNAs that do not pair well with miRNA172 leads to an increase in AP2 protein levels and abnormal floral phenotypes resembling those of the *ag* mutants. These results indicate that inhibiting AP2 translation by a miRNA is important for normal *AG* expression at the center.

AG expression is also negatively regulated by three other genes encoding proteins that may form a transcriptional repressor complex. These three genes, *LEUNIG* (*LUG*), *SEUSS* (*SEU*), and *BELLRINGER* (*BLR*), were identified by genetic screens, initially for enhancers of the weak *ap2-1* mutant, by looking for more severe conversion of the perianth into reproductive organs (14, 45, 68, 121, 216). In addition to causing enhanced *ap2-1* phenotypes, *lug* mutations alone also cause weak *ap2*-like phenotypes. *AG* is ectopically expressed in the peripheral regions of the floral meristem due to *lug* mutations. *LUG* encodes a nuclear protein with 7 WD40 repeats and a glutamine-rich region, resembling Tup1 and Groucho, proteins that are transcriptional repressors in yeast and *Drosophila*, respectively. Similarly, *seu* mutations enhance *ap2-1* phenotypes and allow ectopic *AG* expression (68). *seu* mutations also enhance *lug* phenotypes. Molecular studies reveal that the SEU protein contains a conserved domain similar to the dimerization domain in animal proteins that binds to the LIM-domain transcription factors. Further experiments demonstrate that LUG and SEU interact in vivo and in vitro and that LUG can repress transcription from heterologous promoters, suggesting they may directly repress *AG* expression in the outer region of the floral meristem (216). However, neither LUG nor SEU is a DNA-binding protein. This last piece of puzzle was found by screening for additional mutants that exhibit weak *ap2*-like phenotypes. Mutations in the *BELLRINGER* (*BLR*) gene cause the formation of carpels in place of sepals in flowers produced very late in inflorescence development; the mutant phenotypes were greatly enhanced at high temperatures and by *lug* or *seu* mutations (14). The BLR protein is a homeodomain protein of the BELL subfamily and can bind to AG *cis* elements that contain sites similar to homeodomain-binding sites. Such BLR-binding sites were found in putative *cis* regulatory regions of AG homologs from related species. Therefore, genetic and molecular results support the model that LUG, SEU, and BLR form a transcriptional repressor complex, with LUG serving as the corepressor and BLR being the DNA-binding protein.

Conservation and Divergence of B and C Genes

Molecular genetic studies indicate that homologous MADS-box genes controlling reproductive organ identities from *Arabidopsis thaliana* and *Antirrhinum majus* have very similar functions (26, 75, 88, 93, 211, 229, 255). Phylogenetic studies indicate that *Arabidopsis thaliana* and *Antirrhinum majus* are members of the eudicots, which form the largest monophyletic group of angiosperms. Other eudicots include many commercially important plants such as soybean, apple, cabbage, squash, tomato, lettuce, orange, oak, and many others. Monocots are another monophyletic group and include cereals such as rice, wheat, maize, and other grasses, as well as vegetables such as onions and asparagus, ornamentals including lilies and iris, and palm trees. The remaining flowering plants are not monophyletic and are referred to as the basal angiosperms. Molecular studies have identified homologs of B and C genes in other angiosperms as well as gymnosperms (227). Although most of these homologs have not been studied functionally, there is strong evidence supporting the conservation of B and C function genes in eudicots (125). For example, the putative *AG* orthologs play similar roles in many eudicot plants, such as Brassica (131), petunia (230), tobacco (103), and tomato (172), and possibly also cucumber (101), eucalypt (214), and poplar (27).

B and C function genes were also characterized in maize and rice, two members of the grass family (76, 125, 196). *ZAG1* and *ZMM2* are putative maize orthologs of *AG* and are expressed in stamens and carpels; however, *ZAG1* is expressed at a higher level in carpels than stamens and *ZMM2* is expressed more strongly in the stamen than the carpel (140, 141). The *zag1* mutant produces flowers with abnormal carpels (141), suggesting that *ZMM2* is sufficient for stamen identity but does not provide the needed function for carpel development. Therefore, *ZAG1* and *ZMM2* may have overlapping as well as distinct functions and together they may provide the C function. The *silky1* (*si1*) gene is homologous to *AP3* and is expressed in both the lodicule and stamen primordia (5). The observation that *si1* mutant flowers form carpels in place of stamens indicates that at least part of the B function is conserved in maize. Similarly, molecular genetic experiments indicate that rice homologs of B function genes *AP3* and *PI*, as well as *AG*, likely have conserved functions (100, 149).

Homologs of B function genes have also been isolated from basal eudicots, such as members of the Ranunculalese (107, 108). The expression of these genes in stamen primordia supports the hypothesis that they have similar functions in controlling stamen identity as do their counterparts in core eudicots. In addition, putative orthologs of B and C function genes have also been isolated from several gymnosperm species, such as the gnetophyte *Gnetum gnemon* and conifers (150, 222). The observation that the gymnosperms B and C function homologs are expressed in male reproductive organs suggests that these genes specify the male identity (150, 184, 222, 224, 225, 246). Therefore, B and C functions may be important for specifying male reproductive organs in angiosperms and gymnosperms.

MOLECULAR GENETIC ANALYSIS OF ANTHER CELL DIFFERENTIATION

Early molecular studies using solution hybridization experiments suggested that as many as 25 × 10^3 genes may be expressed in the stage 6 tobacco anther; among these, ~10 × 10^3 were proposed to be anther specific (98, 99). More recently, similar studies suggested that ~3.5 × 10^3 genes may be anther specific in *Arabidopsis* (71, 203). Genetic studies in *Arabidopsis* have identified a large number of mutants that are male sterile or reduced in male fertility. In particular, Sanders et al. (189, 203) described large-scale screens of T-DNA and chemically mutagenized *Arabidopsis* plants and identified nearly 900 mutants that fall into nine classes with different defects in anther development. Therefore, thousands of genes are expressed during normal anther development and genetic studies indicate that many are required for anther and/or pollen development. In addition to *Arabidopsis* mutants, many maize male-sterile mutants are known (34).

SPL/NZZ is Required for Anther Cell Division and Differentiation

As mentioned earlier, cell division and differentiation during phase 1 lead to the formation of four nonreproductive cell layers, including the tapetum, that make up the wall of the anther lobes and surround the microsporocytes. Several genes have been identified genetically that are critical for early anther development (203). One of the earliest acting genes is the *Arabidopsis SPOROCYTELESS* (*SPL*)/*NOZZLE* (*NZZ*) gene, which is important for early anther cell division and differentiation, both of which are required for sporogenesis (194, 252). In *spl/nzz* mutant anthers, the formation of archesporial cells seems normal but subsequent cells divisions are defective, resulting in the absence of sporogenous cells and nonreproductive tissues, including the tapetum (194, 252). Consequently, the *spl/nzz* mutants fail to produce pollen and are male sterile. *SPL/NZZ* is expressed in stage 3 anthers; subsequently, the *SPL/NZZ* expression continues at a high level in subepidermal cells and at a reduced level in the epidermis. In stage 5 anthers, *SPL/NZZ* expression is strong in the microsporocytes and tapetum. *SPL/NZZ* encodes a nuclear protein with some features of transcription factors, suggesting that it may regulate gene expression during early sporogenesis in the anther (252).

Recently, it was shown that the *AG* gene promotes the expression of *SPL/NZZ* and supports sporogenesis (91), indicating that part of AG's role on controlling male fertility is mediated by *SPL/NZZ*. Using an inducible system for AG activity, *SPL* was one of the genes activated in mRNA levels following the induction of AG activity, even in the presence of a protein synthesis inhibitor, suggesting that the AG protein may be a direct regulator of *SPL* expression. A sequence very similar to the AG-binding consensus was found downstream of the *SPL* transcribed region and can bind to AG in vitro. A mutation in this AG-binding site that eliminates the in vitro AG binding also leads to reduced expression of a reporter gene, further supporting the idea that AG binding is important for *SPL* expression. *SPL* expression is sufficient to induce sporogenesis in locules that formed on petaloid floral organs in the absence of *AG* function (91).

EMS1/EXS and *TPD1* Define a Signaling Pathway for Tapetum Differentiation

Subsequent to the action of *SPL/NZZ*, additional *Arabidopsis* genes have been uncovered by mutants that exhibit defects in the differentiation of anther cell layers. In particular, mutations in the *EXCESS MALE SPOROCYTES1* (*EMS1*) gene [also called *EXTRA SPOROGENOUS CELLS* (*EXS*)] cause the formation of additional male sporocytes (MMCs) along with a lack of tapetal cells (31, 259). As a result, these mutants cannot produce any viable pollen grains and are male sterile. Detailed histological analysis of the mutant and the wild-type anthers supports the idea that the number of cells prior to the formation of MMCs in the *ems1* mutant is normal, suggesting that cell division during early anther development is not severely affected by the mutation. Further studies using molecular and cytological markers support the idea that excess male sporocytes likely originated from the precursor cells of the tapetum.The total number of male sporocytes in the *ems1* mutant anther is close to the sum of tapetal cells and male sporocytes. Even though the *ems1/exs* mutant anthers lack the tapetum, meiocytes in the *ems1* mutant can complete the meiotic nuclear division, suggesting that the tapetum is not necessary for the meiotic nuclear division. Although the *ems1* male sporocytes can complete meiotic nuclear divisions, they fail to carry out cytokinesis. Instead, these cells undergo cell degeneration and thus fail to produce microspores. Therefore, *EMS1* function is required for meiotic cytokinesis, either by acting in the meiocytes, or by promoting the formation of the tapetum, which may be required for the meiotic cell division.

EMS1 is critical for the normal differentiation of anther cells, particularly the tapetum. How does *EMS1* achieve this function? Molecular cloning indicates that the *EMS1/EXS* gene encodes a putative leucine-rich repeat receptor protein kinase that likely localizes to the cell surface (259). Therefore, cell-cell communication is probably important for normal cell differentiation during anther development and the *EMS1/EXS* gene product is an important component of a signaling pathway for such cell-cell communication. The action of EMS1/EXS is hypothesized to be a trigger for a signaling pathway essential for tapetal cell differentiation.

If the EMS1/EXS protein is a receptor, what is its ligand and where is it from? One reasonable hypothesis is that the ligand originates from the cells near the precursors of the tapetum. The fact that male sporocyte is the default cell fate in the absence of the EMS1/EXS function suggests that the signal might be provided by the male sporocytes (259). A recent report revealed that a mutation in the *TPD1* (*TAPETUM DETERMINANT1*) gene causes the same phenotype as do the *ems1/exs* mutations: excess male sporocytes and lack of tapetum (250). Additional experiments using molecular markers further demonstrated that the *tpd1* mutant is remarkably similar to the *ems1* mutant. The *TPD1* gene encodes a predicted small protein with a putative signal peptide for secretion, supporting the idea that it is a ligand. Furthermore, RNA in situ hybridization indicates that it is expressed in sporogenous cells and tapetum precursors and becomes more strongly expressed in the male sporocytes than the tapetum as these two cell types differentiate. Conversely, *EMS1/EXS* is also expressed initially in the precursors of sporogenous cells and tapetum, then its expression becomes stronger in the tapetum than that in the male sporocytes (259).

A Model for EMS1/TPD1 Action

As the phenotypes of *ems1/exs* and *tpd1* mutants indicate, the precursors of male sporocytes and tapetum can differentiate into male sporocytes. Initially, *EMS1/EXS* and *TPD1* are expressed in the precursors of both cell types. The available results suggest the following model (**Figure 5**). It is possible that an unknown factor triggers the differentiation of male sporocytes at the center. In the differentiating male sporocytes, there is a decrease in the level of EMS1 and an increase in the level of TPD1, which is secreted and binds to EMS1 on the surface of cells surrounding

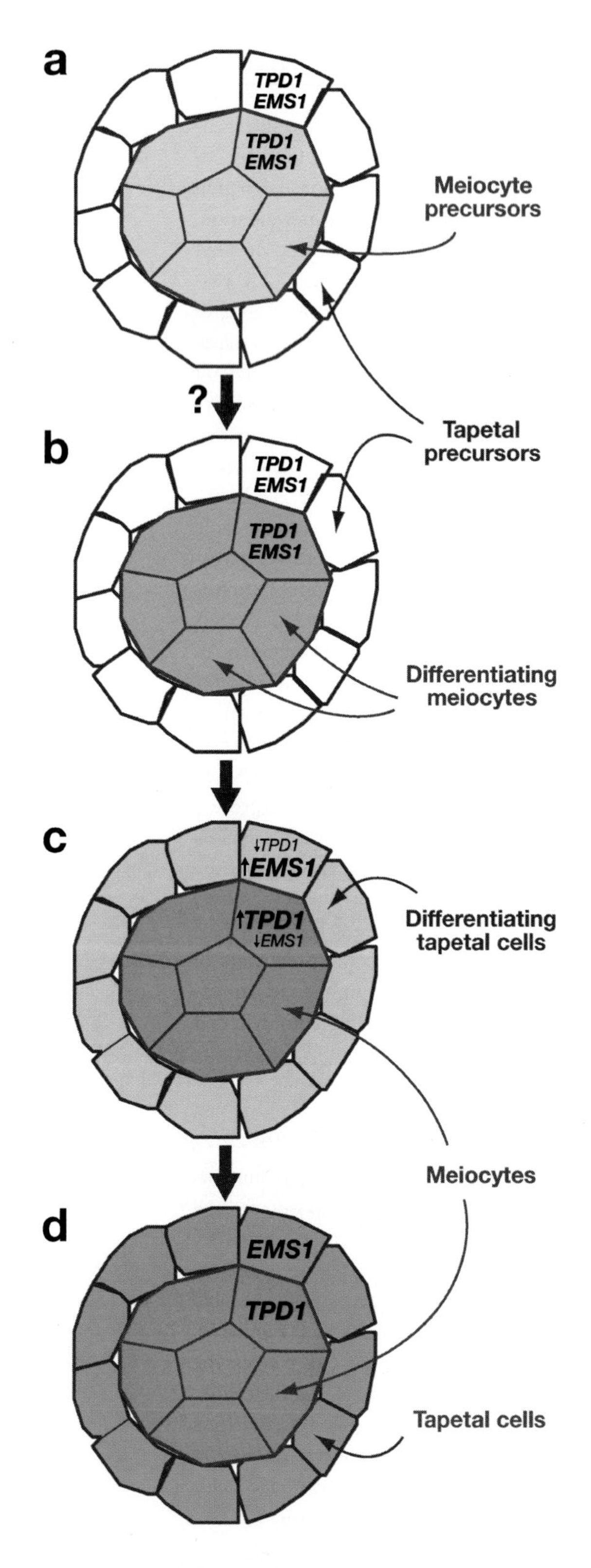

the newly differentiated male sporocytes. The interaction of TPD1 with EMS1 activates a pathway in these surrounding cells to promote tapetum differentiation and causes a decrease of TPD1. The differentiation of the tapetum occurs at the expense of the potential outer male sporocytes. The anther wall is more complex than the wall of the microsporangium in gymnosperms and ferns; therefore, it is possible that the EMS1/TPD1-dependent pathway(s) contributed to the evolution of the anther wall by directing the differentiation of outer portions of precursors for male sporocytes into the tapetum.

The general significance of genes such as *EMS1/EXS* and *TPD1* is further supported by the discovery that the rice *MULTIPLE SPOROCYTE1* (*MSP1*) gene is required for normal anther cell differentiation (155). The *msp1* mutant exhibits an increased number of male sporocytes and a lack of the tapetum; these phenotypes are remarkably similar to those of the *ems1* and *tpd1* mutants, although the *msp1* mutant also has other defects in anther wall formation and produces excess female sporocytes. The fact that *MSP1* also encodes a predicted leucine-rich repeat receptor-like protein kinase suggests that *MSP1* and *EMS1* may represent related functions in promoting tapetum differentiation from precursor cells in monocot and eudicot lineages. Because eudicots and monocots represent more than 90% of angiosperms

Figure 5

Model for *EMS1/EXS* and *TPD1* function. (*A*) Initially, TPD1 and EMS1 are expressed in the precursors of both meiocytes (the central group) and tapetal cells (the outer ring). (*B*) An unknown trigger (?) activates the differentiation of meiocytes, as indicated by shading. (*C*) In the differentiating meiocytes, the level of TPD1 increases and that of EMS1 decreases. The TPD1 protein is secreted and binds to the EMS1 receptors on the neighboring cells, causing an elevation of EMS1 levels in these cells and a drop in TPD1 levels. EMS1 then activates a pathway for tapetal differentiation. (*D*) Further reduction of EMS1 and TPD1 in the meiocytes and tapetal cells, respectively, stabilizes the differentiation of tapetal cells.

and diverged at least 120 million years ago (247), it is possible that a signaling pathway like the ones defined by *EMS1/TPD1* and *MSP1* existed in early angiosperms.

Other Genes Important for Anther Cell Differentiation and Function

One possible scenario for EMS1/TPD1 function is that, following the binding of TPD1 to EMS1, EMS1 phosphorylates one or more target proteins, which in turn regulate additional steps in a pathway. The pathway then regulates the expression of a number of genes, which collectively promote tapetum differentiation and carry out other possible functions. What might be some of the downstream genes of TPD1/EMS1? Although the direct targets of EMS1 are not yet identified, molecular genetic studies have uncovered *Arabidopsis* genes that might encode proteins that could be regulated by the EMS1-dependent pathway (79, 213, 245). One of these genes is called *ABORTED MICROSPORES* (*AMS*) (213). Although *ams* anthers have both the tapetum and male sporocytes and can complete meiosis, the tapetum and microspores degenerate soon after meiosis. The *AMS* gene encodes a bHLH-type putative transcription factor and its expression begins at a low level premeiotically and increases in postmeiotic flowers, as indicated by reverse-transcriptase polymerase chain reaction (RT-PCR) analysis (213). Another gene, *MALE STERILE1* (*MS1*), is also required for early microspore development (245). The *ms1* mutant anthers exhibit microspore degeneration following their release from the tetrad, and tapetum becomes abnormal at this time. Similar to *AMS*, *MS1* is also expressed at the time of meiosis and when microspores are released, but not in open flowers. The MS1 protein contains a PHD-finger domain that is found in some transcriptional regulators (3, 104), suggesting that MS1 may regulate transcription. Furthermore, a MYB-domain gene, *AtMYB103*, is expressed specifically in the tapetum and is required for normal tapetum morphology and pollen development (79). The *AMS*, *MS1*, and *AtMYB103* genes are expressed beginning when *EMS1* expression is at its highest levels and are important for normal tapetum morphology and microspore development, suggesting that they function subsequent to the action of EMS1 and may be indirectly regulated by the EMS1-dependent pathway. (See section on pollen development for further discussion of these genes.)

Genetic studies have also identified other mutants with abnormal anther cell differentiation. A particularly interesting *Arabidopsis* mutant is called *fat tapetum*, which seems normal in early anther development, but the mutant tapetum becomes enlarged at the time of meiosis and the middle layer persists and also enlarges in a way similar to the tapetum (189). The fact that the mutant middle layer appears to be similar to the tapetum suggests that this gene may be a critical regulator of proper differentiation of the middle layer. Two other mutants, *gne1* and *gne4*, were isolated by looking for altered expression of an anther-specific reporter gene; these mutants show similar phenotypes to those of the *fat tapetum* mutant (212).

Genetic and molecular analyses have already uncovered a number of important regulators for several aspects of anther development, including anther cell differentiation, tapetum function, and microspore development. At the same time, molecular studies and mutant analyses [e.g., Sanders et al. (189)] indicate that additional genes likely participate in controlling the differentiation of anther cell layers. Further characterization of the reported mutants and isolation of additional mutants will facilitate the identification and understanding of those undiscovered genes that are important for anther cell differentiation.

PLANT MEIOSIS AND ITS GENETIC CONTROL

In the anthers of flowering plants, meiosis produces haploid spores (50, 251). Meiosis is a conserved cell division that is essential for eukaryotic sexual reproduction and that produces haploid cells from diploid parental cells.

Following a single round of DNA replication, there are two rounds of nuclear division, meiosis I and meiosis II. Meiosis I is unique and involves the segregation of homologous chromosomes (homologs), whereas meiosis II is similar to mitosis and results in the segregation of sister chromatids. Following meiosis II, the cell undergoes cytokinesis to produce four haploid cells.

To ensure proper homolog segregation at anaphase I, it is necessary for the homologs to condense and recognize each other in prophase I, which is further subdivided into five substages: leptotene, zygotene, pachytene, diplotene, and diakinesis. The recognition between homologs is achieved through homolog pairing and stabilized by synapsis. Homolog pairing is a transient interaction between homologs that allows homology search, whereas synapsis is a more extensive and stable interaction that results in the formation of a complex proteinaceous structure called the SC (180). In addition to pairing and synapsis, homologous recombination also occurs during the prophase I. Recombination, along with sister chromatid cohesion, results in the formation of chiasmata, which provide the physical link between homologs in a bivalent.

SC: synaptonemal complex

Plant meiosis has been studied extensively for many years using cytological and genetic approaches; due to space constraints, the reader is referred to excellent reviews for information on cytological and genetic studies in a variety of plants (50, 73, 264, 265). Over the past decade, several groups have demonstrated that it is possible to investigate meiosis in *Arabidopsis* using cytological approaches (9, 12, 32, 33, 36, 77, 105, 168, 181, 182, 249). As shown in **Figure 6**, *Arabidopsis* male meiotic chromosomes begin to condense and can be observed as thin lines in leptotene (**Figure 6*A***). Homolog pairing initiates at late leptotene to early zygotene (**Figure 6*B***), and synapsis and recombination occur from zygotene through pachytene (**Figure 6*C***). Fully synapsed chromosomes are seen as thick threads in light micrographs, although synaptonemal complexes can only be observed using transmission electron microscopy. The homologs separate except at specific regions (chiasmata) at diplotene (**Figure 6*D***), and five bivalents are formed at the diakinesis (**Figure 6*E***) as highly condensed entities. The bivalents then are aligned at the division plane at metaphase I (**Figure 6*F***). Following the separation of homologs, they move to opposite poles of the spindle (not seen here) at anaphase I (**Figure 6*G***). At the end of meiosis I, five chromosomes at each pole form a cluster and chromosomes decondense at telophase I (**Figure 6*H***). At prophase II (**Figure 6*I***), chromosomes recondense and are separated by an organelle band; subsequently they are again aligned at two division planes at metaphase II (**Figure 6*J***). At anaphase II, sister chromatids separate and are segregated to form four clusters, each consisting of five newly formed chromosomes (**Figure 6*K***). These clusters then undergo decondensation and form four haploid nuclei within the male meiocytes at telophase II (**Figure 6*L***). Cytokinesis then yields four microspores in a tetrad.

Meiosis in *Arabidopsis* has been studied using both forward and reverse genetic approaches. Because meiotic defects often affect normal pollen production, novel meiotic genes and homologs of known genes in other organisms have been successfully uncovered using sterility or reduced fertility as mutant phenotypes. Analysis of mutations in homologs of known meiotic genes has also demonstrated important meiotic functions for the plant homologs.

Sister Chromatid Cohesion and Chromosome Condensation

Sister chromatid cohesion maintains the association between sister chromatids following DNA replication in mitosis and meiosis (145, 151). During meiosis before anaphase I, sister chromatid cohesion, along with recombination, is also needed for the integrity of chiasmata, or links between homologs (158). Sister chromatid cohesion requires the cohesion complex consisting of several subunits. In yeast, the *REC8* gene encodes a meiosis-specific cohesin subunit. Mutations in the *Arabidopsis SYN1/DIF1*

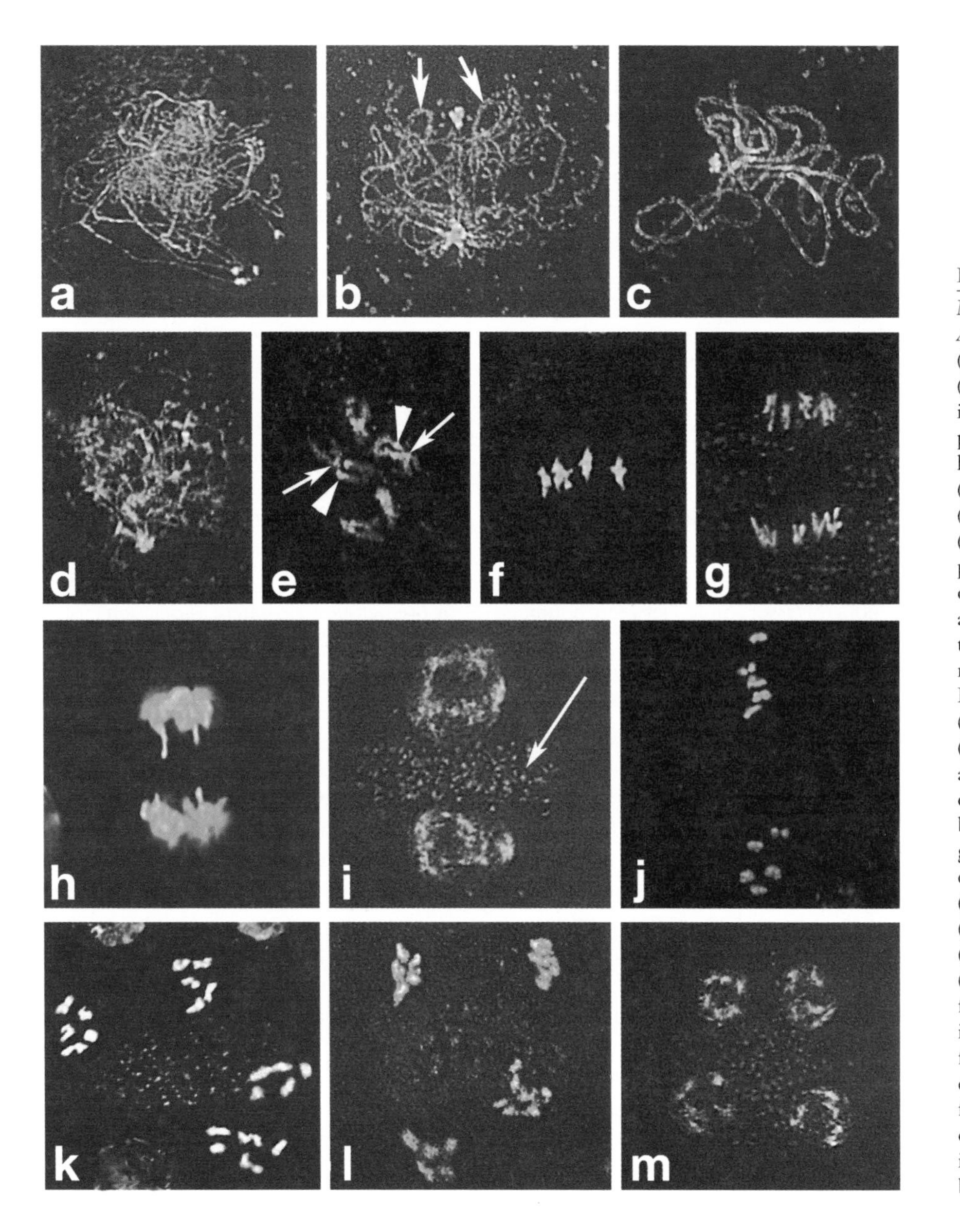

Figure 6

Male meiosis in *Arabidopsis thaliana*. (*A*) Leptotene. (*B*) Zygotene. Arrows indicate regions of pairing between homologs. (*C*) Pachytene. (*D*) Diplotene. (*E*) Diakinesis. Arrows point to the position of chiasmata and arrowheads indicate the centromeric regions. (*F*) Metaphase I. (*G*) Anaphase I. (*H*) Telophase I. (*I*) Prophase II. The arrow points to the organelle band between the two groups of chromosomes. (*J*) Metaphase II. (*K*) Anaphase II. (*L*) Telophase II. (*M*) Four newly formed nuclei. The images were obtained from DAPI-stained chromosome spreads following the fixation of floral buds. These images were provided by W. Li.

gene, a homolog of the yeast *REC8* gene (160, 221, 237), result in defects in chromosome condensation and pairing (13, 18), suggesting that the cohesion complex also plays a role in condensation, which may indirectly affect pairing. The localization pattern of this protein indicates that it plays an important role in maintaining chromosome arm cohesion and centromere cohesion during late stages of meiosis I (30). Other putative cohesion subunits have also been identified, but their roles in meiosis remain to be determined.

In addition to *SYN1*, genetic studies revealed that the *Arabidopsis SWI1/DYAD* gene is also essential for sister chromatid cohesion in male and female meiosis (4, 142). In *swi1/dyad* mutant male meiosis, sister chromatid cohesion is lost and 20 separated chromatids are seen at late prophase I to metaphase I instead of the normal 5 bivalents. This result indicates that sister chromatid cohesion is required to maintain association between homologs. Female meiosis in the *swi1* mutant undergoes a mitosis-like division, presumably because the defect in cohesion causes the centromeres to behave like mitotic centromeres.

Homeologs: corresponding chromosomes from different genomes in an allopolyploid, which is derived from hybridization of two related species

Chromosome Pairing, Synapsis, and Recombination

Chromosome pairing, synapsis, and recombination ensure the appropriate recognition and association of homologs and the proper segregation of genetic information into haploid meiotic products. The term "pairing" has been used by different investigators to describe various interactions or associations between homologs. Here it refers to the transient interaction at localized regions of homologs that occurs prior to synapsis (265). Pairing uses DNA homology search to identify homologs. Recent evidence supports the idea that the telomeres are important in initiating chromosome pairing (50, 192). In premeiotic interphase, telomeres are relatively dispersed in a number of organisms, including yeast, mice, wheat, and maize. Telomeres then move and attach to the nuclear envelope in a process called bouquet formation, resulting in the clustering of telomeres. Bouquet formation functions in initiating homologous chromosome pairing (17, 51, 193). Genetic studies in maize have identified mutants defective in telomere clustering and pairing, although the molecular identity of corresponding genes remains to be determined (16, 74). In *Arabidopsis*, the telomeres cluster to the nucleolus through interphase and undergo pairing prior to the chromosome synapsis (10). The paired telomeres become widely dispersed during leptotene as they lose the association with the nucleolus. These results suggest that telomere clustering during interphase might be important for homologous chromosome pairing in *Arabidopsis* (10).

In addition to telomeres, centromeres may also contribute to homolog pairing in some species, including polyploid species such as wheat (133). The wheat *Pairing homoeologous* (*Ph1*) locus is required for correct homolog pairing and recombination. In *ph1* mutant wheat meiotic cells, homeologous chromosomes (corresponding chromosomes from different parental genomes of an allopolyploid) may also pair (70, 134, 135), indicating that the normal *Ph1* function is to suppress pairing between homeologs. *Ph1* negatively regulates centrosome-microtubule interaction from analysis of wheat monosomic lines; it is possible that Ph1 regulates premeiotic distribution of homologous and homeologous centromeres (232). However, more recent studies showed that *Ph1* functions in correcting nonhomologous association among chromosomes in wheat, instead of predetermining chromosome pairing by premeiotic centromere association (56, 134). Therefore, the exact function of Ph1 requires further investigation.

Following pairing, homologs form the SC along the length of the chromosome. In light microscopy, synapsed chromosomes can be visualized as thick lines (**Figure 6*C***). SC is a tripartite proteinaceous structure between the homologs, as observed by transmission electron microscopy in a number of animals and plants (201). In early prophase I, each homolog condenses and forms a filamentous protein structure called the axial element. In *Arabidopsis*, full-length axial elements form prior to the formation of the SC and can be observed in using scanning electron microscopy (**Figure 7**); in contrast, in some organisms, the formation of axial element along part of the chromosome is followed by SC formation in that part before the axial element is formed along the entire chromosome (265). Following synapsis, the axial elements become the lateral elements of the SC (**Figure 7**). The central element forms in between and parallel to the lateral elements and is connected by the transverse elements to the lateral elements (201, 265).

In addition to pairing and synapsis, homologous recombination is another process that occurs during prophase I between homologs. Biochemical studies in yeast have led to a widely

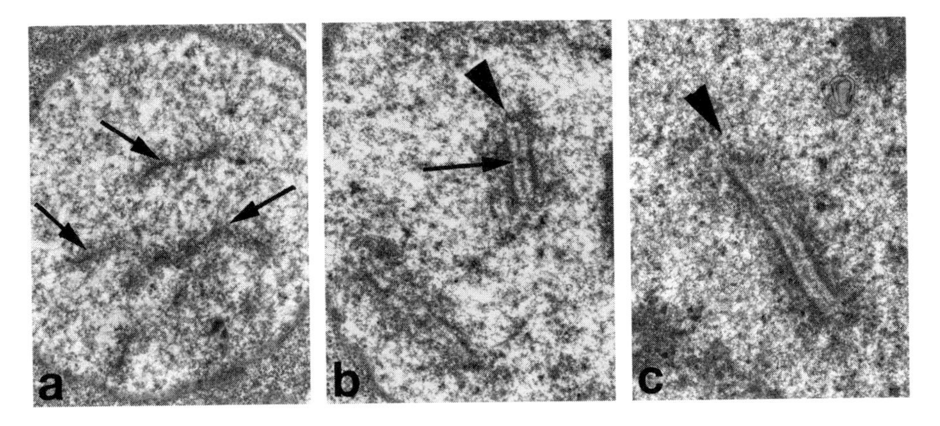

Figure 7

Transmission electron microscopy of *Arabidopsis* male meiotic prophase I. (*A*) A portion of a leptotene nucleus showing axial elements (*arrow*). (*B*) A portion of a zygotene nucleus shown in a short synaptonemal complex (SC) (*arrowhead*) with a recombination nodule (*arrow*). (*C*) A portion of a pachytene nucleus showing a long SC (arrowhead). These images were provided by L. Timofejeva.

accepted model for recombination, called a DSB repair model (**Figure 8**). Recombination is initiated by a DSB in one of two participating nonsister chromatids; the DSB then undergoes a 5′ resection to produce 3′ single-stranded DNAs, which invade the intact partner chromatid to form a D loop. DNA synthesis from the invading strands and ligation yield the double-Holliday junction. Resolution of the double-Holliday junction following cutting at alternative strands results in products with either crossovers or noncrossovers. Recently, an alternative model was proposed to explain results related to noncrossover and crossover events (19). The crossovers, along with the maintenance of sister chromatid cohesion, result in chiasmata. Homologous recombination is closely associated with the formation of recombination nodules, which are proteinaceous structures that are shown, using transmission electron microscopy, to be associated with progressing and completed SCs (50, 265).

In part because pairing, synapsis, and recombination are intimately associated events, mutants often have defects in more than one of these processes. Classically, mutants defective in synapsis are grouped as either asynaptic mutants, which fail to undergo synapsis, or desynaptic mutants, which can form SC but cannot maintain homolog association during late prophase I. In asynaptic and desynaptic mutants, univalents (single unattached chromosomes) rather than bivalents are observed at late prophase I (11, 16, 29, 33, 157, 182, 197). The *Arabidopsis asy1* mutant is defective in synapsis in male and female meiosis, resulting in the formation of only one to three bivalents per cell, instead of the normal five (33, 182). The *ASY1* gene is expressed at low levels in several tissues and encodes a protein whose N-terminal region is homologous to the yeast HOP1 protein (33, 80). The HOP1 protein plays a role in SC assembly and accumulates on axial elements (80). The *Arabidopsis asy1* mutant phenotypes are consistent with a defect in SC formation, suggesting that ASY1 and HOP1 may have a conserved function. Further support for the involvement of the ASY1 protein in SC formation comes from the observation that it is localized to chromatin regions that are associated with the axial/lateral elements (11). The rice *pair2* mutation affects a rice homolog of the *Arabidopsis ASY1* gene; the *pair2* mutant is defective in pairing at pachytene and has univalents at diakinesis (157). Another *Arabidopsis* gene, *AHP2*, was identified by screening for sterile mutants among a T-DNA population; the *ahp2* mutant is normal during vegetative

DSB: double-stranded break

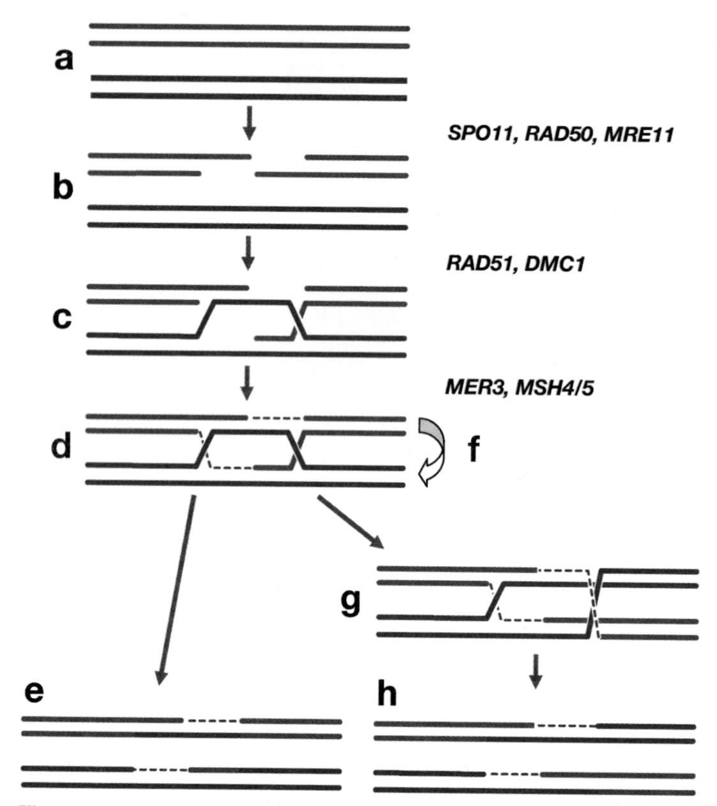

Figure 8

The double-stranded break (DSB) repair model of homologous recombination. (*A*) Double-stranded DNA molecules of two nonsister chromatids. (*B*) A DSB is generated and 5′ strands are resected to produce 3′ ssDNAs. (*C*) Strand invasion to produce a D-loop. (*D*) DNA synthesis followed by ligation results in a double-Holliday junction. (*E*) One way the double-Holliday junction is resolved to yield a noncrossover event. (*F,G*) Alternatively, the double-Holliday junction is resolved to produce a crossover event. On the right are a few of the yeast genes that are involved in some steps of this model; plant homologs of these genes are discussed in this review.

development but is male and female sterile (197). Further analysis suggests that the *ahp2* mutant is defective in pairing and bivalent formation. The *AHP2* gene encodes a protein with similarity to the human TBPIP (Tat-binding protein 1-interacting protein), *S. pombe* Meu13p, and the *S. cerevisiae* HOP2 proteins. The fact that *TBPIP* is expressed in the testis and the Meu13 and HOP2 proteins are involved in meiotic homolog pairing supports the idea that the AHP2 protein is also involved in pairing in *Arabidopsis*.

Reverse genetic approaches have also demonstrated the importance of *Arabidopsis* genes homologous to budding yeast genes required for meiotic recombination. In yeast, meiotic recombination is initiated by SPO11-dependent DSBs in one of the chromatids (102, 118). The SPO11 protein is related to the catalytic subunit of an archaebacterial type 2 topoisomerase and generates DSBs via a transesterase mechanism with the assistance of a set of other gene products (102). SPO11 homologs were identified in human, mouse, *Drosophila melanogaster*, *Caenorhabditis elegans*, and *Arabidopsis* (102, 201), suggesting that these organisms also may share the same mechanism for initiating meiotic recombination. A T-DNA insertion in the *AtSPO11-1* gene was isolated after screening T-DNA insertional lines; it has defects in homolog pairing, recombination, and bivalent formation (77). However, *atspo11-1* meiocytes can form occasional bivalents at diakinesis (77), suggesting that recombination may still occur at greatly reduced levels. Alternatively, bivalents might form in the absence of recombination in *Arabidopsis*.

Several yeast proteins are required for the SPO11-dependent meiotic recombination and participate in DSB repair (128). In particular, the RAD50 protein forms a complex with MRE11 and NBS1 and this complex is important for early steps of recombination immediately following the generation of DSBs by SPO11 (44). In *Arabidopsis*, a mutation in the *AtRAD50* gene, a homolog in the yeast *RAD50* gene, causes meiotic defects and hypersensitivity to DNA-damaging ionizing radiations and chemicals (69). Recently, a T-DNA insertion in the *Arabidopsis MRE11* homolog was shown to be defective in meiosis, resulting in chromosome fragmentation, a failure in normal pairing and bivalent formation, and abnormal chromosome segregation (174). Analysis of the *spo11-1 mre11* double mutant indicates that the chromosome fragmentation was due to a failure to repair SPO11-1-induced DSBs, suggesting that MRE11 is normally required for meiotic recombination. The *mre11* mutant plants are reduced in size and sometimes in height. They also have abnormal leaf shapes and other morphological defects. These mutant phenotypes and the observation that somatic genome stability is affected in the mutant indicate that the *MRE11* gene is also required for normal mitotic growth.

Other genes important for meiotic recombination include the yeast *RAD51* and *DMC1* genes that encode homologs of the RecA recombinase, which is a ssDNA-binding protein and DNA-dependent ATPase (20, 206). RecA is essential for homology recognition and repair of DSBs via homologous recombination. The yeast *RAD51* and *DMC1* genes are required for meiotic recombination and *RAD51* is also important for repairing DNA damage during the mitotic cell cycle (20, 206). Homologs of RAD51/DMC1 have been identified in *Arabidopsis*, maize, and other plants (6, 21, 53, 54, 67, 105, 190, 205). A T-DNA knockout mutant of *AtDMC1* was identified and the *atdmc1* mutant meiocytes have largely univalents instead of bivalents at late prophase I, indicating that *AtDMC1* is crucial for *Arabidopsis* meiosis (47).

Immunolocalization studies of the RAD51 protein in maize and *Arabidopsis* found that it is present in a large number of foci in the zygotene stage, and the number of foci is reduced dramatically in the pachytene stage (67, 143). The number of early sites and the observation that pairs of RAD51 sites are found when chromosomes are paired are consistent with the idea that RAD51 is important for homolog pairing; the number of late sites agrees with that of crossovers. Maize mutant cells defective in pairing have reduced RAD51 sites (66, 164). These results support the idea that RAD51 is important for homology search during pairing, in addition to its role in recombination. This hypothesis was recently supported by molecular genetic studies with a T-DNA insertional mutant in the *AtRAD51* gene (117). The *atrad51* mutant was completely male and female sterile and exhibited meiotic defects in pairing and synapsis, as suggested by fluorescence images of chromosome spreads and verified by transmission electron microscopy. *atrad51* mutant meiotic cells have severe chromosome fragmentation, suggesting a defect in DSB repair. Analysis of the *atspo11-1 atrad51* double mutant indicates that the fragmentation seen in the *atrad51* mutant requires the *SPO11-1* function (117), indicating that *AtRAD51* participates in the *SPO11-1*-dependent meiotic recombination and most likely participates in the formation of the double-Holliday junction of the DSB repair model. Therefore, *RAD51* function in meiosis seems to be conserved between yeast and plants. On the other hand, the *atrad51* mutant seems healthy under normal growth conditions and appears to undergo mitosis normally, unlike the mammalian *rad51* knockout, which is embryo lethal.

In addition to *RAD51* and *DMC1*, vertebrates possess additional *RAD51* homologs, including *XRCC2* and *XRCC3*, which are important for DSB repair in cell cultures (120). XRCC3 also associates with RAD51 in a yeast two-hybrid assay (195). To test the function of the *Arabidopsis XRCC3* homolog, *AtXRCC3*, a T-DNA insertional mutant was identified and shown to exhibit meiotic chromosome fragmentation and sterility (21), similar to those of the *atrad51* mutant. The *atxrcc3* mutant can develop normally under standard conditions, although in tissue culture the mutant cells are sensitive to DNA cross-linking agents but not to DSB-inducing chemicals. Another protein that participates in DSB repair is BRCA2; *brca2* knockout mice exhibit chromosome breaks and rearrangement and embryo lethality (256). BRCA2 is localized to human meiotic chromosomes, suggesting that it is involved in meiosis (37), but the embryo lethality of the mouse knockout has prevented a direct genetic test of this hypothesis. *Arabidopsis* has two highly similar (97% nucleotide identity) homologs of *BRCA2*; in addition, yeast two-hybrid assays with one of them showed that it interacts with AtRAD51 and AtDMC1 (207). When *BRCA2* expression in *Arabidopsis* meiocytes was reduced using a meiosis-specific RNAi construct, the cells formed univalents and had uneven chromosome segregation, similar to those of the *atdmc1* mutant cells.

RNAi: RNA interference

In mammals and birds, RAD51 and XRCC3 are essential for mitotic growth; in yeast and *Drosophila*, RAD51 is important for DSB repair during the mitotic cell cycle and mutant cells are hypersensitive to DSB-inducing agents, but not essential for growth under normal conditions. In *Arabidopsis*, *AtRAD51* and *AtXRCC3* are required for normal meiosis and fertility, but not for normal vegetative and flower development. It is possible that these genes and other RAD51 homologs have partially overlapping functions. For meiosis, high levels of AtRAD51 and AtXRCC3 activities are required because many DSBs are induced by the SPO11-1 protein and cofactors, whereas the normal mitotic S-phase only generates a very small number of DSBs that can be repaired in the single mutants when other homologs can partially substitute the functions. However, the existence of multiple *RAD51* homologs in fungal, plant, and animal kingdoms suggests that they may have evolved distinct functions. The fact that single mutants are embryo lethal in mammals and sterile in plants also supports the idea that they have different functions. Therefore, the meiotic functions of these genes might be conserved between yeast, plants, and probably animals, and their mitotic DSB repair function might not be universally critical. *RAD51* and *XRCC3* functions seem to be required for repairing even a single DSB in mammals. It is possible that in plants, DSBs do not cause programmed cell death; however, this is not supported by the phenotypes of *Arabidopsis rad50* and *mre11* mutants, which do have vegetative phenotypes. Alternatively, DSBs may be repaired by *RAD51*- and *XRCC3*-independent pathways. Further studies in animals and plants are needed to understand the possible evolutionary conservation and divergence of this family of important genes.

Forward genetic studies have also identified novel genes important for pairing, synapsis, and/or recombination. In maize, the *phs1* mutant exhibits synapsis between nonhomologous chromosomes, but not between homologs, indicating that the normal PHS1 function is required for synapsis between homologs (165). RAD51 localization is absent in *phs1* mutant chromosomes, indicating that there is a defect in the loading of recombination machinery. Therefore, the processes of pairing and recombination are decoupled from synapsis in the *phs1* mutant, in contrast to the highly coordinated situation in the wild-type cells. The *phs1* gene encodes a novel protein that has homologs in other plant genomes. Another example of novel meiotic genes is the rice *PAIR1* gene, identified by isolating sterile mutants following transposon mutagenesis (156). The *pair1* mutant had normal vegetative development but failed to produce microspores and megaspores. The mutant meiosis was defective in chromosome pairing and had only univalents. The *PAIR1* gene encoded a novel nuclear protein with a coiled-coil domain, suggesting that it may interact with other proteins. In *Arabidopsis*, a mutant called *solo dancers* (*sds*) was isolated because its greatly reduced fertility from a transposon insertional population (12). The *sds* mutant exhibited a severe defect in homolog pairing, recombination, and bivalent formation. These phenotypes are very similar to those of *atspo11-1* and *atdmc1* mutants, suggesting that it may be involved in the SPO11-dependent recombination pathway. However, the *sds* mutant does not show any chromosome fragmentation; therefore, unlike the *atrad51* or *atxrcc3* mutants, the *sds* mutant cells do not have any detectable defects in DSB repair. *SDS* is specifically expressed in male and female meiotic cells and encodes a putative novel cyclin (12, 234). Cyclins regulate multiple processes during the mitotic cell cycle by activating cyclin-dependent kinases, suggesting that SDS may coordinate pairing, synapsis, and recombination by regulating protein activities via phosphorylation.

Chromosome Separation and Segregation

The purpose of bivalent formation during prophase I is to ensure that homologs will segregate properly at anaphase I. Prior to homolog separation, the bivalents move to the metaphase plate. Subsequently, the cohesins along the chromosome arms must be removed to allow the separation of homologs. During anaphase I the separated homologs move toward the opposite poles of the spindle due to forces of the microtubules. Little is known about the control of homolog separation in plant meiosis. In *Arabidopsis*, the *ask1* mutant is abnormal in chromosome separation and segregation at anaphase I (249). Some chromosomes remain associated when they are pulled by the spindle and become stretched. Sometimes, chromosome fragments are observed, presumably due to the force of the spindle. Subsequent chromosome distribution is grossly uneven and the resulting microspores contain abnormal amounts of DNA. *ASK1* encodes a homolog of the SKP1 protein, which is an essential component of SCF complexes (for SKP1, cullin/CDC53, F-box protein) (249). The SCF complexes form the largest family of E3 ubiquitin-protein ligases, which control cell-cycle regulation, signal transduction, transcription, and other important biological processes through the selective proteolysis of regulatory proteins (261). It is possible that *ASK1* is involved in the removal of the cohesin complex along the arms prior to anaphase I.

Spindle is essential for normal chromosome segregation and starts to form at prometaphase when microtubules are quickly organized into a bipolar structure (65). An *Arabidopsis* mutant (*atk1*) with reduced male fertility was isolated from a population of transposon insertional lines and formed abnormal meiotic spindles and had uneven chromosome segregation (36). In the *atk1* mutant meiocytes, the spindle has unfocused poles, and microtubules are not well organized. The first detectable chromosomal phenotype is that bivalents seem not to congress at the metaphase I plate as tightly as those in the wild-type cells, suggesting that chromosome movement during early metaphase I is affected. In addition, homolog separation is not synchronous. The *ATK1* (previously named *KATA*) gene encodes a kinesin protein (36, 119, 146) and is homologous to the *Drosophila NCD* gene, which is also involved in meiotic spindle morphogenesis (58). Therefore, it is likely that the primary defect of the *atk1* mutant is abnormal spindle morphology and possibly other microtubule-based motor function. The abnormal *atk1* spindle is similar to the meiotic spindle of the maize *dv* (*divergent spindle*) mutant; however, defects in *dv* are caused by a failure in the breakdown of the nuclear envelop (217).

Meiotic Progression and Cytokinesis

Although most events during meiosis occur once, and the entire process is not repeated, or cyclic (unlike the mitotic cell cycle), it is necessary to coordinate multiple events such that they occur in the right order and for the proper duration. Therefore, the control of timing of meiotic events is as important as that in a mitotic cell cycle. Analyses of the *Arabidopsis MMD1/DUET* gene that encodes a protein with a C-terminal PHD-finger domain suggest that it may be a critical regulator of meiosis (175, 253). PHD fingers are cystein-containing domains that interact with proteins, DNA, and RNA; PHD-finger proteins are involved in a variety of cellular processes (3, 46, 104, 175, 253). Mutations in the *MMD1/DUET* gene were isolated separately by two labs, and found to be allelic; these mutations cause meiotic defects, a failure to produce normal pollen, and male sterility. During early meiosis, the *mmd1* (*male meiocyte death1*) mutant appears normal up to diakinesis (253). From diakinesis to telophase II, meiotic cells show signs of programmed cell death, including chromosome fragmentation and cytoplasmic shrinkage. All meiotic cells die before cytokinesis. The *mmd1* mutant was caused by a *Ds* transposon insertion just downstream of the ATG codon, suggesting that it is a null allele (253). The *duet* mutant carries an independent *Ds* insertion and exhibits a relatively mild phenotype of delayed meiotic cytokinesis and formation of largely dyads instead of tetrads (175). The abnormal microspores produced from the mutant meiocytes then undergo one or two mitotic divisions before degeneration. The *Ds* insertion in the duet allele is located more than 500 amino acid residues from the N terminus and upstream of the PHD finger; therefore, it may result in the production of a partially functional protein, providing a possible explanation for its mild phenotypes.

Because the *mmd1* meiotic cells undergo cell death without obvious chromosome abnormality in the normal course of chromosomal events, from condensation, to pairing, to synapsis, MMD1 may play a general regulatory role during meiotic progression.

Further support for the genetic control of timing of meiotic event comes from analysis of the *Arabidopsis tam* (*tardy asynchronous meiosis*) mutant, which exhibits defects in the timing of meiotic divisions and produces abnormal intermediates such as dyad meiotic products (127). An asynchrony in meiosis was most clearly observed in cells that are entering metaphase I or metaphase II, suggesting that the normal function of *TAM* may be to control entry into metaphase (127). The *tam* mutant is still fertile; an explanation for the scarcity of meiotic timing mutants might be that most screens for meiotic mutants rely on reduction of fertility. Mutations in genes that may function in controlling the progression of meiosis and the mitotic cell cycle might cause early defects not to be recognized as meiotic mutants. An *Arabidopsis* homolog of the yeast cell-cycle gene *CDC45* is expressed in early floral buds that include meiotic cells and RNAi transgenic plants with reduced levels of *CDC45* expression have fertility defects (219). Further analysis indicates that the RNAi lines produce polyads instead of tetrad from meiosis and subsequent nonviable pollen grains. Fluorescence microscopy reveals that meiotic chromosomes in the RNAi lines were first fragmented at late prophase I and more severe fragmentation occurred at metaphase I and anaphase I. Because CDC45 regulates mitotic DNA replication in yeast (266, 267), it is possible that the *Arabidopsis* homolog is involved in controlling the premeiotic DNA replication and abnormal DNA synthesis may cause chromosome fragmentation, which is revealed late in prophase I through anaphase I.

After completing meiotic nuclear division, four groups of chromosomes form four new nuclei. These newly formed nuclei are then enclosed by cell membranes and wall materials during cytokinesis to produce four haploid microspores. Several genes are genetically important for meiotic cytokinesis. Among these mutants are the *Arabidopsis stud* (*std*) and *tetraspore* (*tes*) mutants, which have similar phenotypes and are allelic (86, 215). In the mutant meiocytes, the nuclear meiotic events are normal, but cell plate formation is incomplete during cytokinesis, resulting in a giant microspore with four nuclei. During the subsequent pollen development, the four nuclei undergo mitotic divisions separately, producing four vegetative nuclei and up to eight sperm cells within a large abnormal "pollen grain." This pollen grain behaves as a single male gametophyte and produces only one pollen tube, resulting in greatly reduced male fertility (86). The *TES/STD* gene encodes a kinesin with an N-terminal motor domain (248), suggesting that it may affect the radial arrays of microtubules associated with male meiotic cytokinesis.

POLLEN DEVELOPMENT

Following meiosis, the development of pollen grains from the microspores involves cell divisions and cell differentiation to form a vegetative cell and two sperm cells (116, 138, 223). Early in pollen development, the microspore becomes vacuolized after release from the tetrad; the formation of a large centralized vacuole is accompanied by the migration of the microspore nucleus to one side of the cell. The first mitosis in pollen development then occurs and is asymmetric, producing a large vegetative cell and a small generative cell. The vegetative cell contains a dispersed nucleus and most of the cytoplasm from the microspore, whereas the generative cell has highly condensed chromatin and very little cytoplasm. Subsequently, the generative cell is completely engulfed by the cytoplasm of the vegetative cell. In *Arabidopsis*, maize, and other plants with trinucleate pollen, the second mitosis occurs before pollen maturation; it is this second mitosis that produce two sperm cells from the generative cell. In other plants, such as petunia and tomato, the mature pollen contains two cells and is referred to as binucleate; in such plants, the second mitosis occurs after pollen germination, resulting in two sperm cells in the pollen tube.

Sporophytic Functions for Pollen Development

Pollen development occurs within the anther walls and tissues of the anther wall are important for pollen development. In addition to providing the physical support for pollen development, it has long been thought that the anther wall supports pollen development by supplying signals and materials (71). Direct evidence for the importance of sporophytic tissues for pollen development came from molecular ablation studies and genetic analyses. In particular, the essential role of the tapetum layer was demonstrated using molecular cell ablation studies (72). A tapetum-specific promoter was fused to a gene encoding a RNase called barnase (132). In the transgenic plants harboring this construct, the tapetum layer was selectively destroyed and pollen development was severely defective, indicating that the tapetum is essential for normal pollen development.

Many postmeiotic sporophytic mutants affecting pollen development have been isolated as male-sterile mutants (35, 189). The *Arabidopsis MS1*, *AMS*, and *AtMYB103* genes mentioned earlier are expressed in the tapetum and behave genetically as sporophytic genes required for normal pollen development at the onset of pollen development (79, 90, 213, 245). Soon following the release of microspores from the tetrad, tapetal cells become morphologically abnormal and degenerate in mutants defective in any one of these genes; simultaneously, the microspores degenerate before further development. In particular, in the *ams* mutant, the microspore degenerates prior to the nuclear migration (213). All three genes encode putative transcriptional regulators, suggesting that they may control the expression of a number of genes critical to normal tapetum structure and/or function, which are essential for early stages of pollen development.

In the wild-type pollen, the pollen wall has two layers: the exine and the intine. The exine is primarily made up of sporopollenin consisting of simple aliphatic polymers containing aromatic or conjugated side chains and is important for cell-to-cell recognition during pollination and pollen germination.The surface of the exine exhibits a highly decorated pattern that is characteristic of a species. The intine is structurally simple and is comprised of cellulose, pectin, and proteins. Male-sterile mutations that disrupt tapetum function and affect exine and intine formation have been described (226). For example, the *Arabidopsis ms2* (*male sterile2*) mutant produces pollen grains that have a thin pollen wall with a reduced exine layer (2, 60). The *MS2* gene is expressed postmeiotically in the tapetum and encodes a protein with similarity to a jojoba protein that converts wax fatty acids to long-chain alcohol, suggesting that it may promote the synthesis of long-chain aliphatic molecules (2, 60). Also, in the *Arabidopsis nef1* (*no exine formation1*) mutant, sporopollenin is synthesized but not properly deposited onto the pollen membrane (8). Lipid content of the *nef1* tapetum plastids is reduced and leaf chloroplasts have abnormal structures. The predicted NEF1 protein has limited sequence similarity to membrane and transporter proteins, suggesting that NEF1 may facilitate the transport of molecules required for exine deposition. Another sporophytic *Arabidopsis* mutant, *ms33* (*male sterile33*), is defective in the formation of intine and the deposition of tryphine, which is a layer of proteinaceous and lipidic coating outside the exine (61). In addition, a T-DNA insertional mutant in the *AtGPAT1* gene encoding a membrane-bound glycerol-3-phosphate acyltransferase exhibits abnormal tapetum morphology and defects in the formation of the pollen wall (262). The *Arabidopsis dex1* (*defective in exine pattern formation*) mutant produces pollen grains with defective exine that has irregular sporopollenin patterns (166, 167). The *DEX1* gene encodes a novel protein that may bind to calcium and localize to the plasma membrane. Another gene required for normal exine formation is the *FLP1* gene, defined by the *flp1* (*faceless pollen-1*) mutation, that causes defective sporopollenin and abnormal tryphine (7). The predicted FLP1 protein shows similarity to the maize glossy1 and *Arabidopsis* CER1 proteins, which are important for cuticular wax deposition (1, 78).

MGU: Male Germ Unit

Gametophytic Genes Important for Pollen Development

In addition to genes expressed in the sporophytic anther tissues, thousands of genes are expressed in the developing pollen (82, 139). Because pollen is haploid, null mutations in a gene that is essential for pollen development cannot be transmitted to the next generation via pollen. Therefore, such mutations cannot be isolated as a homozygote after mutagenesis, but can be identified as heterozygotes that produce half abnormal pollen. These heterozygous plants are expected to be fertile because the number of pollen grains greatly exceeds that needed to fully fertilize the ovules. Therefore, gametophytic mutations affecting pollen development are usually isolated by looking for abnormal transmission of markers via pollen (83, 111).

A number of gametophytic pollen mutants have been described in *Arabidopsis*. The *sidecar pollen* (*scp*) mutation causes some of the developing pollen grains to undergo an extra mitosis that produces two vegetative cells prior to the asymmetric division, whereas other mutant pollen grains abort (41). Therefore, the *SCP* gene can regulate the pattern of cell division. Another *Arabidopsis* mutant, *gemini pollen1* (*gem1*), also exhibits defects in the divisional pattern. Among the mutant pollen grains, equal, unequal, and partial divisions can all be observed at the first mitotic division (162), suggesting that the mutation may affect the positioning of the nucleus. The *GEM1* gene encodes the MOR protein, which binds to microtubules (231), suggesting that the mutant may be defective in a microtubule-based function essential for the first pollen mitosis. In another mutant, *limpet pollen*, the generative cell remains outside the cytoplasm of the vegetative cell and against the pollen wall, suggesting that the migration of the generative cell is defective (83).

In the pollen grain, the vegetative nucleus and the two sperm cells form an integral structure called the MGU. Visual screens of pollen grains from mutagenized plants have yielded mutants that exhibit abnormal organization or position of the MGU (110). These mutants fall into two types: *gum* (*germ unit malformed*) and *mud* (*MGU displaced*) mutants. The *gum* mutants exhibit defects in the organization of the MGU such that the vegetative nucleus is localized near the pollen wall, away from the sperm cells. In *mud* mutant pollen grains, the entire MGU is mislocalized to near the pollen wall on one side of the cytoplasm. Further analysis of these mutants and the corresponding genes should provide new insights into the control of the integrity and positioning of the MGU. Recently, several *seth* mutants were isolated from *Ds* insertional populations using segregation distortion and exhibited various defects in pollen development (111). Sequence analysis of the DNA flanking the Ds elements suggests that the putative affected genes encode proteins that are potentially involved in diverse cellular processes (111).

Prior to anthesis, pollen needs to desiccate so that it can survive the much drier environment outside the anther locule. Following pollination, the pollen obtains water from the stigmatic cells and germinates. In the abovementioned *Arabidopsis ms33* mutant that is defective in pollen coat formation, pollen desiccation is also abnormal (61), suggesting that a normal pollen coat is required for desiccation. In tomato, the *LeProT1* gene encodes a homolog of *Arabidopsis* proline transporters and is specifically expressed in mature and germinating pollen. It was suggested that *LeProT1* facilitates the uptake of proline and compatible solutes for pollen development and desiccation (200). In the *Arabidopsis* gametophytic *rtg* (*raring-to-go*) mutant, pollen grains germinate precociously within the anther locule beginning at the binucleate stage (96). The growth of pollen tubes of the mutant pollen is further enhanced by high humidity. Several other mutants with abnormal pollen tube wall defects were also reported (96).

ANTHER DEHISCENCE

To facilitate pollination, mature pollen grains must be released from the locules of the anther; this is achieved through a process called anther dehiscence, which involves opening the anther wall (188). This requires the degeneration of specific anther tissues called septum and stomium (**Figure 2**). In addition, proper release of pollen requires expanding endothecial cells and thickening and strengthening their cell wall (189). Genetic studies have identified genes that are important for anther dehiscence (203). Some of the mutants completely fail to undergo anther dehiscence, including the *Arabidopsis ms35* and *nondehiscent1* mutants (52, 189). In the *ms35* mutant, the endothecium cell wall fails to thicken, and anther dehiscence does not occur after pollen maturation even though the stomium does degenerate, supporting the idea that the endothecium plays an important role in dehiscence. In anthers of the *nondehiscent1* mutant, abnormal degeneration of the endothecium and connective tissues occurs, whereas the epidermis remains intact, trapping the pollen inside. In both mutants, functional pollen grains are formed but not released. Further molecular characterization of these mutants is needed to understand the molecular nature of the affected genes.

More recently, a new *Arabidopsis* transposon insertional mutant was isolated for its sterility and found to have normal pollen, but its anthers failed to dehisce (218). Scanning electron microscopy revealed that the pollen of this mutant has normal morphology; the mutant pollen can pollinate the mutant pistil to produce normal seeds. Further analysis of anther development indicates that early development during phase 1 appears normal. During phase 2, the tapetum and the middle layer degenerate in a way similar to that in the wild type, but the endothecium did not expand. Additional mutant alleles were obtained following the excision of the transposon, and molecular analysis revealed that the mutations affect the *AtMYB26* gene in an R2R3 type of MYB DNA-binding domain. The expression of *AtMYB26* was detected using RT-PCR in floral buds, with the highest levels in medium buds at approximately the time of tapetum degeneration, but not in very early buds and vegetative organs. In a rice mutant called *anther indehiscence1* (*aid1*), about 55% of the flowers exhibited defective anther dehiscence, but produced functional pollen; about 25% of flowers had pollen development defects, and the other 20% of flowers were fertile (263). In the nondehiscent anthers, the degeneration of septum and stomium tissues was not observed. The *aid1* mutant was generated in a Ds-tagging population and molecular cloning indicates that the mutated gene encodes a protein with a single MYB domain, not closely related to AtMYB26. RT-PCR experiments indicate that the *AID1* gene is expressed in immature flowers, consistent with its function. The fact that both AtMYB26 and AID1 are putative transcription factors indicates that distinct transcriptional programs are needed in different tissues for proper anther dehiscence.

In addition to the nondehiscent mutants, several *delayed-dehiscence* mutants have been characterized. In these mutants, male fertility is reduced due to inefficient pollen release and pollination. Molecular cloning indicates several of these mutations are defective in genes for JA synthesis or signaling. JA is synthesized from linolenic acid, a triply unsaturated fatty acid (49), which has been proposed to be released from a phospholipid by a lipolytic reaction (89). The enzymes in the JA biosynthetic pathway include a lipoxygenase, allene oxide synthase, allene oxide cyclase, and 12-oxo-phytodienoic acid reductase. In *Arabidopsis*, the *fad3 fad7 fad8* triple mutant defective in three fatty acid desaturases was constructed that contains negligible levels of trienoic fatty acids. The triple mutant is normal in vegetative growth and development, but is male sterile due to a deficiency in the synthesis of precursors of the jasmonate pathway (137). The *Arabidopsis defective in anther dehiscence1* (*dad1*) mutation is in the gene that encodes a chloroplast phospholipase A1 that likely catalyzes the first step of JA synthesis (89). Furthermore, mutations in the allene oxide synthase also show delayed anther dehiscence and reduce fertility

JA: jasmonic acid

(161, 233). The *delayed dehiscence1/opr3* mutants exhibit a delay in the degeneration of stomium cells. Molecular cloning revealed that the *DELAYED DEHISCENCE1/OPR3* gene encodes a 12-oxo-phytodienoic acid reductase involved in the JA synthesis pathway (188, 220). Further evidence for the idea that JA is critical for anther dehiscence came from the observation that exogenous applications of JA can rescue these mutants and lead to seed production. In addition to the JA biosynthetic mutant, the *coi1* (*coronatine-insensitive1*) mutant defective in an F-box protein important for JA signaling is also unable to dehisce and cannot be rescued by JA applications. Anther dehiscence is associated with water loss from the anther wall tissues. This observation and the fact that mutants defective in JA synthesis have much less water loss in anther tissues support the idea that JA controls water transport in the anther and regulates dehiscence. The *coi1* mutant and some of the mutants defective in JA synthesis also show pollen defects, indicating that JA also plays a role in promoting normal pollen development (63, 137), although the precise function of JA is not well understood.

PERSPECTIVE

Successful male reproduction in flowering plants requires the specification of the stamen identity, proper anther cell division and differentiation, the regulation of male meiosis, sporophytic and gametophytic control of pollen development, and hormonal regulation of anther dehiscence. Molecular genetic studies have identified a large number of genes that regulate these processes, including genes that are conserved in a number of flowering plants, such as maize and *Arabidopsis*, as well as genes for meiosis that are similar to those in yeast and animals. The advances made in identifying genes using genetic and molecular methods have been extremely rapid and dramatic. At the same time, the availability of whole genome sequences and large-scale expression profiling capability will likely increase the pace in uncovering additional genes that are important for male reproduction. Traditional genetic approaches are limited by functional redundancy and possible early lethality of mutants in essential genes. The function of such redundant or essential genes in male reproduction can be investigated using dominant alleles, multiply mutated lines, and tissue- and stage-specific gene silencing. Further development of cell biological and biochemical tools will also be necessary to allow the understanding of molecular mechanisms and biochemical interactions of the identified gene products. The future for the analysis of plant male reproduction is exciting and challenging.

ACKNOWLEDGMENTS

I greatly appreciate the help from Y. Hu, W. Li, L. Timofejeva, and D. Zhao in providing images used in this review. I also thank X. Chen, C. Hendrix, W. Li, W. Ni, L. Timofejeva, and A. Wijeratne for helpful comments on this manuscript. I am grateful to the John Simon Guggenheim Memorial Foundation for their support on work during my sabbatical leave. The research in my laboratory is supported by grants from the USDA (2001-35,301-10,570 and 2003-35,301-13,313), NSF (IBN-00,77832, MCB-98,96340, MCB-00,92075, and DBI-01,15684), NIH (RO1 GM63871), and DOE (DE-FG02-02ER15332). Further support for our research was provided by the Biology Department and the Huck Institutes of the Life Sciences (including a LSC innovation seed grant) at the Pennsylvania State University.

LITERATURE CITED

1. Aarts MG, Keijzer CJ, Stiekema WJ, Pereira A. 1995. Molecular characterization of the *CER1* gene of Arabidopsis involved in epicuticular wax biosynthesis and pollen fertility. *Plant Cell* 7:2115–27

2. Aarts MGM, Hodge R, Kalantidis K, Florack D, Wilson ZA, et al. 1997. The *Arabidopsis* MALE STERILITY 2 protein shares similarity with reductases in elongation/condensation complexes. *Plant J.* 12:615–23
3. Aasland R, Gibson TJ, Stewart AF. 1995. The PHD finger: implications for chromatin-mediated transcriptional regulation. *Trends Biochem. Sci.* 20:56–59
4. Agashe B, Prasad CK, Siddiqi I. 2002. Identification and analysis of *DYAD*: a gene required for meiotic chromosome organisation and female meiotic progression in *Arabidopsis*. *Development* 129:3935–43
5. Ambrose BA, Lerner DR, Ciceri P, Padilla CM, Yanofsky MF, Schmidt RJ. 2000. Molecular and genetic analyses of the *SILKY1* gene reveal conservation in floral organ specification between eudicots and monocots. *Mol. Cell* 5:569–79
6. Anderson LK, Offenberg HH, Verkuijlen WMHC, Heyting C. 1997. RecA-like proteins are components of early meiotic nodules in lily. *Proc. Natl. Acad. Sci. USA* 94:6868–73
7. Ariizumi T, Hatakeyama K, Hinata K, Sato S, Kato T, et al. 2003. A novel male-sterile mutant of *Arabidopsis thaliana, faceless pollen-1*, produces pollen with a smooth surface and an acetolysis-sensitive exine. *Plant Mol. Biol.* 53:107–16
8. Ariizumi T, Hatakeyama K, Hinata K, Inatsugi R, Nishida I, et al. 2004. Disruption of the novel plant protein NEF1 affects lipid accumulation in the plastids of the tapetum and exine formation of pollen, resulting in male sterility in *Arabidopsis thaliana*. *Plant J.* 39:170–81
9. Armstrong SJ, Jones GH. 2001. Female meiosis in wild-type *Arabidopsis thialiana* and two meiotic mutants. *Sex. Plant Reprod.* 13:177–83
10. Armstrong SJ, Franklin FCH, Jones GH. 2001. Nucleolus-associated telomere clustering and pairing precede meiotic chromosome synapsis in *Arabidopsis thaliana*. *J. Cell Sci.* 114:4207–17
11. Armstrong SJ, Caryl AP, Jones GH, Franklin FC. 2002. ASY1, a protein required for meiotic chromosome synapsis, localizes to axis-associated chromatin in *Arabidopsis* and *Brassica*. *J. Cell Sci.* 115:3645–55
12. Azumi Y, Liu D, Zhao D, Li W, Wang G, et al. 2002. Homolog interaction during meiotic prophase I in *Arabidopsis* requires the *SOLO DANCERS* gene encoding a novel cyclin-like protein. *EMBO J.* 21:3081–95
13. Bai XF, Peirson BN, Dong FG, Xue C, Makaroff CA. 1999. Isolation and characterization of *SYN1*, a *RAD21*-like gene essential for meiosis in Arabidopsis. *Plant Cell* 11:417–30
14. Bao X, Franks RG, Levin JZ, Liu Z. 2004. Repression of *AGAMOUS* by BELLRINGER in floral and inflorescence meristems. *Plant Cell* 16:1478–89
15. Bartel DP. 2004. MicroRNAs: genomics, biogenesis, mechanism, and function. *Cell* 116:281–97
16. Bass HW, Bordoli SJ, Foss EM. 2003. The *desynaptic* (*dy*) and *desynaptic1* (*dys1*) mutations in maize (*Zea mays L.*) cause distinct telomere-misplacement phenotypes during meiotic prophase. *J. Exp. Bot.* 54:39–46
17. Bass HW, Marshall WF, Sedat JW, Agard DA, Cande WZ. 1997. Telomeres cluster de novo before the initiation of synapsis: a three-dimensional spatial analysis of telomere positions before and during meiotic prophase. *J. Cell Biol.* 137:5–18
18. Bhatt AM, Lister C, Page T, Fransz P, Findlay K, et al. 1999. The *DIF1* gene of *Arabidopsis* is required for meiotic chromosome segregation and belongs to the *REC8/RAD21* cohesin gene family. *Plant J.* 19:463–72
19. Bishop DK, Zickler D. 2004. Early decision; meiotic crossover interference prior to stable strand exchange and synapsis. *Cell* 117:9–15
20. Bishop DK, Park D, Xu L, Kleckner N. 1992. DMC1: a meiosis-specific yeast homolog of *Escherichia coli* recA required for recombination, synaptonemal complex formation, and cell cycle progression. *Cell* 69:439–56

21. Bleuyard JY, White CI. 2004. The Arabidopsis homologue of Xrcc3 plays an essential role in meiosis. *EMBO J.* 23:439–49
22. Bomblies K, Dagenais N, Weigel D. 1999. Redundant enhancers mediate transcriptional repression of *AGAMOUS* by *APETALA2*. *Dev. Biol.* 216:260–64
23. Bowman JL, Meyerowitz EM. 1991. Genetic control of pattern formation during flower development in Arabidopsis. *Symp. Soc. Exp. Biol.* 45:89–115
24. Bowman JL, Smyth DR, Meyerowitz EM. 1989. Genes directing flower development in Arabidopsis. *Plant Cell* 1:37–52
25. Bowman JL, Sakai H, Jack T, Weigel D, Mayer U, Meyerowitz EM. 1992. *SUPERMAN*, a regulator of floral homeotic genes in *Arabidopsis*. *Development* 114:599–615
26. Bradley D, Carpenter R, Sommer H, Hartley N, Coen ES. 1993. Complementary floral homeotic phenotypes result from opposite orientations of a transposon at the *plena* locus of Antirrhinum. *Cell* 72:85–95
27. Brunner AM, Rottmann WH, Sheppard LA, Krutovskii K, DiFazio SP, et al. 2000. Structure and expression of duplicate *AGAMOUS* orthologues in poplar. *Plant Mol. Biol.* 44:619–34
28. Busch MA, Bomblies K, Weigel D. 1999. Activation of a floral homeotic gene in *Arabidopsis*. *Science* 285:585–87
29. Cai X, Makaroff CA. 2001. The *dsy10* mutation of *Arabidopsis* results in desynapsis and a general breakdown in meiosis. *Sex. Plant Reprod.* 14:63–67
30. Cai X, Dong F, Edelmann RE, Makaroff CA. 2003. The *Arabidopsis* SYN1 cohesin protein is required for sister chromatid arm cohesion and homologous chromosome pairing. *J. Cell Sci.* 116:2999–3007
31. Canales C, Bhatt AM, Scott R, Dickinson H. 2002. EXS, a putative LRR receptor kinase, regulates male germline cell number and tapetal identity and promotes seed development in *Arabidopsis*. *Curr. Biol.* 12:1718–27
32. Caryl AP, Jones GH, Franklin FC. 2003. Dissecting plant meiosis using *Arabidopsis thaliana* mutants. *J. Exp. Bot.* 54:25–38
33. Caryl AP, Armstrong SJ, Jones GH, Franklin FCH. 2000. A homologue of the yeast *HOP1* gene is inactivated in the *Arabidopsis* meiotic mutant *asy1*. *Chromosoma* 109:62–71
34. Chaubal R, Zanella C, Trimnell MR, Fox TW, Albertsen MC, Bedinger P. 2000. Two male-sterile mutants of *Zea mays* (Poaceae) with an extra cell division in the anther wall. *Am. J. Bot.* 87:1193–201
35. Chaudhury AM. 1993. Nuclear genes controlling male fertility. *Plant Cell* 5:1277–83
36. Chen C, Marcus AI, Li WX, Hu Y, Calzada JP, et al. 2002. The *Arabidopsis ATK1* gene is required for spindle morphogenesis in male meiosis. *Development* 129:2401–9
37. Chen J, Silver DP, Walpita D, Cantor SB, Gazdar AF, et al. 1998. Stable interaction between the products of the BRCA1 and BRCA2 tumor suppressor genes in mitotic and meiotic cells. *Mol. Cell* 2:317–28
38. Chen X. 2004. A microRNA as a translational repressor of APETALA2 in *Arabidopsis* flower development. *Science* 303:2022–25
39. Chen X, Meyerowitz EM. 1999. *HUA1* and *HUA2* are two members of the floral homeotic *AGAMOUS* pathway. *Mol. Cell* 3:349–60
40. Chen X, Liu J, Cheng Y, Jia D. 2002. *HEN1* functions pleiotropically in *Arabidopsis* development and acts in C function in the flower. *Development* 129:1085–94
41. Chen YCS, McCormick S. 1996. *Sidecar pollen*, an *Arabidopsis thaliana* male gametophytic mutant with aberrant cell divisions during pollen development. *Development* 122:3243–53
42. Cheng Y, Kato N, Wang W, Li J, Chen X. 2003. Two RNA binding proteins, HEN4 and HUA1, act in the processing of AGAMOUS pre-mRNA in *Arabidopsis thaliana*. *Dev. Cell* 4:53–66

43. Coen ES, Meyerowitz EM. 1991. The war of the whorls: genetic interactions controlling flower development. *Nature* 353:31–37
44. Connelly JC, Leach DR. 2002. Tethering on the brink: the evolutionarily conserved Mre11-Rad50 complex. *Trends Biochem. Sci.* 27:410–18
45. Conner J, Liu Z. 2000. *LEUNIG*, a putative transcriptional corepressor that regulates *AGAMOUS* expression during flower development. *Proc. Natl. Acad. Sci. USA* 97:12902–7
46. Coscoy L, Ganem D. 2003. PHD domains and E3 ubiquitin ligases: viruses make the connection. *Trends Cell Biol.* 13:7–12
47. Couteau F, Belzile F, Horlow C, Grandjean O, Vezon D, Doutriaux MP. 1999. Random chromosome segregation without meiotic arrest in both male and female meiocytes of a *dmc1* mutant of Arabidopsis. *Plant Cell* 11:1623–34
48. Craig KL, Tyers M. 1999. The F-box: a new motif for ubiquitin dependent proteolysis in cell cycle regulation and signal transduction. *Prog. Biophys. Mol. Biol.* 72:299–328
49. Creelman RA, Mullet JE. 1997. Biosynthesis and action of jasmonates in plants. *Annu. Rev. Plant Physiol. Plant Mol. Biol.* 48:355–81
50. Dawe RK. 1998. Meiotic chromosome organization and segregation in plants. *Annu. Rev. Plant Physiol. Plant Mol. Biol.* 49:371–95
51. Dawe RK, Sedat JW, Agard DA, Cande WZ. 1994. Meiotic chromosome pairing in maize is associated with a novel chromatin organization. *Cell* 76:901–12
52. Dawson J, Sozen E, Vizir I, Van Waeyenberge S, Wilson ZA, Mulligan BJ. 1999. Characterization and genetic mapping of a mutation (*ms35*) which prevents anther dehiscence in *Arabidopsis thaliana* by affecting secondary wall thickening in the endothecium. *N. Phytol.* 144:213–22
53. Ding ZJ, Wang T, Chong K, Bai SN. 2001. Isolation and characterization of *OsDMC1*, the rice homologue of the yeast *DMC1* gene essential for meiosis. *Sex. Plant Reprod.* 13:285–88
53a. Ditta G, Pinyopich A, Robles P, Pelaz S, Yanofsky MF. 2004. The *SEP4* gene of *Arabidopsis thaliana* functions in floral organ and meristem identity. *Curr. Biol.* 14:1935–40
54. Doutriaux MP, Couteau F, Bergounioux C, White C. 1998. Isolation and characterisation of the RAD51 and DMC1 homologs from *Arabidopsis thaliana*. *Mol. Gen. Genet.* 257:283–91
55. Drews GN, Bowman JL, Meyerowitz EM. 1991. Negative regulation of the *Arabidopsis* homeotic gene *AGAMOUS* by the *APETALA2* product. *Cell* 65:991–1002
56. Dvorak J, Lukaszewski AJ. 2000. Centromere association is an unlikely mechanism by which the wheat *Ph1* locus regulates metaphase I chromosome pairing between homoeologous chromosomes. *Chromosoma* 109:410–14
57. Edlund AF, Swanson R, Preuss D. 2004. Pollen and stigma structure and function: the role of diversity in pollination. *Plant Cell* 16(Suppl.): S84–97
58. Endow SA, Chandra R, Komma DJ, Yamamoto AH, Salmon ED. 1994. Mutants of the *Drosophila* ncd microtubule motor protein cause centrosomal and spindle pole defects in mitosis. *J. Cell Sci.* 107:859–67
59. Fan H-Y, Hu Y, Tudor M, Ma H. 1997. Specific interactions between the K domains of AG and AGLs, members of the MADS domain family of DNA binding proteins. *Plant J.* 12:999–1010
60. Fei HM, Sawhney VK. 1999. *MS32*-regulated timing of callose degradation during microsporogenesis in *Arabidopsis* is associated with the accumulation of stacked rough ER in tapetal cells. *Sex. Plant Reprod.* 12:188–93
61. Fei HM, Sawhney VK. 2001. Ultrastructural characterization of *male sterile33* (*ms33*) mutant in *Arabidopsis* affected in pollen desiccation and maturation. *Can. J. Bot.* 79:118–29
62. Ferrandiz C, Pelaz S, Yanofsky MF. 1999. Control of carpel and fruit development in *Arabidopsis*. *Annu. Rev. Biochem.* 68:321–54

63. Feys B, Benedetti CE, Penfold CN, Turner JG. 1994. Arabidopsis mutants selected for resistance to the phytotoxin coronatine are male sterile, insensitive to methyl jasmonate, and resistant to a bacterial pathogen. *Plant Cell* 6:751–59
64. Flanagan CA, Ma H. 1994. Spatially and temporally regulated expression of the MADS-box gene *AGL2* in wild-type and mutant *Arabidopsis* flowers. *Plant Mol. Biol.* 26:581–95
65. Franklin AE, Cande WZ. 1999. Nuclear organization and chromosome segregation. *Plant Cell* 11:523–34
66. Franklin AE, Golubovskaya IN, Bass HW, Cande WZ. 2003. Improper chromosome synapsis is associated with elongated RAD51 structures in the maize *desynaptic2* mutant. *Chromosoma* 112:17–25
67. Franklin AE, McElver J, Sunjevaric I, Rothstein R, Bowen B, Cande WZ. 1999. Three-dimensional microscopy of the Rad51 recombination protein during meiotic prophase. *Plant Cell* 11:809–24
68. Franks RG, Wang C, Levin JZ, Liu Z. 2002. SEUSS, a member of a novel family of plant regulatory proteins, represses floral homeotic gene expression with LEUNIG. *Development* 129:253–63
69. Gallego ME, Jeanneau M, Granier F, Bouchez D, Bechtold N, White CI. 2001. Disruption of the *Arabidopsis RAD50* gene leads to plant sterility and MMS sensitivity. *Plant J.* 25:31–41
70. Gill KS, Gill BS. 1996. A PCR-based screening assay of *Ph1*, the chromosome pairing regulator gene of wheat. *Crop Sci.* 36:719–22
71. Goldberg RB, Beals TP, Sanders PM. 1993. Anther development: basic principles and practical applications. *Plant Cell* 5:1217–29
72. Goldberg RB, Sanders PM, Beals TP. 1995. A novel cell-ablation strategy for studying plant development. *Philos. Trans. R. Soc. London B. Biol. Sci.* 350:5–17
73. Golubovskaya IN. 1979. Genetic control of meiosis. *Int. Rev. Cytol.* 58:247–90
74. Golubovskaya IN, Harper LC, Pawlowski WP, Schichnes D, Cande WZ. 2002. The *PAM1* gene is required for meiotic bouquet formation and efficient homologous synapsis in maize (*Zea mays* L.). *Genetics* 162:1979–93
75. Goto K, Meyerowitz E. 1994. Function and regulation of the *Arabidopsis* floral homeotic gene *PISTILLATA*. *Genes Dev.* 8:1548–60
76. Goto K, Kyozuka J, Bowman JL. 2001. Turning floral organs into leaves, leaves into floral organs. *Curr. Opin. Genet. Dev.* 11:449–56
77. Grelon M, Vezon D, Gendrot G, Pelletier G. 2001. *AtSPO11-1* is necessary for efficient meiotic recombination in plants. *EMBO J.* 20:589–600
78. Hansen JD, Pyee J, Xia Y, Wen TJ, Robertson DS, et al. 1997. The *glossy1* locus of maize and an epidermis-specific cDNA from Kleinia odora define a class of receptor-like proteins required for the normal accumulation of cuticular waxes. *Plant Physiol.* 113:1091–100
79. Higginson T, Li SF, Parish RW. 2003. *AtMYB103* regulates tapetum and trichome development in *Arabidopsis thaliana*. *Plant J.* 35:177–92
80. Hollingsworth NM, Goetsch L, Byers B. 1990. The *HOP1* gene encodes a meiosis-specific component of yeast chromosomes. *Cell* 61:73–84
81. Honma T, Goto K. 2001. Complexes of MADS-box proteins are sufficient to convert leaves into floral organs. *Nature* 409:525–29
82. Honys D, Twell D. 2003. Comparative analysis of the Arabidopsis pollen transcriptome. *Plant Physiol.* 132:640–52
83. Howden R, Park SK, Moore JM, Orme J, Grossniklaus U, Twell D. 1998. Selection of T-DNA-tagged male and female gametophytic mutants by segregation distortion in *Arabidopsis*. *Genetics* 149:621–31

84. Huala E, Sussex IM. 1992. *LEAFY* interacts with floral homeotic genes to regulate Arabidopsis floral development. *Plant Cell* 4:901–13
85. Huang H, Tudor M, Su T, Zhang Y, Hu Y, Ma H. 1996. DNA binding properties of two Arabidopsis MADS domain proteins: binding consensus and dimer formation. *Plant Cell* 8:81–94
85a. Huang H, Tudor M, Weiss CA, Hu Y, Ma H. 1995. The *Arabidopsis* MADS-box gene *AGL3* is widely expressed and encodes a sequence-specific DNA-binding protein. *Plant Mol. Biol.* 28:549–67
86. Hulskamp M, Parekh NS, Grini P, Schneitz K, Zimmermann I, et al. 1997. The *STUD* gene is required for male-specific cytokinesis after telophase II of meiosis in *Arabidopsis thaliana*. *Dev. Biol.* 187:114–24
87. Irish VF. 1999. Petal and stamen development. *Curr. Top. Dev. Biol.* 41:133–61
88. Irish VF, Yamamoto YT. 1995. Conservation of floral homeotic gene function between Arabidopsis and Antirrhinum. *Plant Cell* 7:1635–44
89. Ishiguro S, Kawai-Oda A, Ueda J, Nishida I, Okada K. 2001. The *DEFECTIVE IN ANTHER DEHISCIENCE1* gene encodes a novel phospholipase A1 catalyzing the initial step of jasmonic acid biosynthesis, which synchronizes pollen maturation, anther dehiscence, and flower opening in Arabidopsis. *Plant Cell* 13:2191–209
90. Ito T, Shinozaki K. 2002. The *MALE STERILITY1* gene of *Arabidopsis*, encoding a nuclear protein with a PHD-finger motif, is expressed in tapetal cells and is required for pollen maturation. *Plant Cell Physiol.* 43:1285–92
91. Ito T, Wellmer F, Yu H, Das P, Ito N, et al. 2004. The homeotic protein AGAMOUS controls microsporogenesis by regulation of SPOROCYTELESS. *Nature* 430:356–60
92. Jack T. 2004. Molecular and genetic mechanisms of floral control. *Plant Cell* 16(Suppl.): S1–17
93. Jack T, Brockman LL, Meyerowitz EM. 1992. The homeotic gene *APETALA3* of *Arabidopsis thaliana* encodes a MADS box and is expressed in petals and stamens. *Cell* 68:683–97
94. Jack T, Fox GL, Meyerowitz EM. 1994. *Arabidopsis* homeotic gene *APETALA3* ectopic expression: transcriptional and posttranscriptional regulation determine floral organ identity. *Cell* 76:703–16
95. Jofuku KD, den Boer BG, Van Montagu M, Okamuro JK. 1994. Control of Arabidopsis flower and seed development by the homeotic gene *APETALA2*. *Plant Cell* 6:1211–25
96. Johnson SA, McCormick S. 2001. Pollen germinates precociously in the anthers of *raring-to-go*, an Arabidopsis gametophytic mutant. *Plant Physiol.* 126:685–95
97. Kachroo A, Nasrallah ME, Nasrallah JB. 2002. Self-incompatibility in the Brassicaceae: receptor-ligand signaling and cell-to-cell communication. *Plant Cell* 14(Suppl.): S227–38
98. Kamalay JC, Goldberg RB. 1980. Regulation of structural gene expression in tobacco. *Cell* 19:935–46
99. Kamalay JC, Goldberg RB. 1984. Organ-specific nuclear RNAs in tobacco. *Proc. Natl. Acad. Sci. USA* 81:2801–5
100. Kang HG, Jeon JS, Lee S, An G. 1998. Identification of class B and class C floral organ identity genes from rice plants. *Plant Mol. Biol.* 38:1021–29
101. Kater MM, Colombo L, Franken J, Busscher M, Masiero S, et al. 1998. Multiple *AGAMOUS* homologs from cucumber and petunia differ in their ability to induce reproductive organ fate. *Plant Cell* 10:171–82
102. Keeney S. 2001. Mechanism and control of meiotic recombination initiation. *Curr. Top. Dev. Biol.* 52:1–53

103. Kempin SA, Mandel MA, Yanofsky MF. 1993. Conversion of perianth into reproductive organs by ectopic expression of the tobacco floral homeotic gene *NAG1*. *Plant Physiol.* 103:1041–46
104. Kennison JA. 1995. The Polycomb and trithorax group proteins of Drosophila: trans-regulators of homeotic gene function. *Annu. Rev. Genet.* 29:289–303
105. Klimyuk VI, Jones JD. 1997. *AtDMC1*, the *Arabidopsis* homologue of the yeast *DMC1* gene characterization, transposon-induced allelic variation and meiosis-associated expression. *Plant J.* 11:1–14
106. Koltunow AM, Truong MT, Cox KH, Wallroth M, Goldberg RB. 1990. Different temporal and spatial gene expression patterns occur during anther development. *Plant Cell* 2:1201–24
107. Kramer EM, Irish VF. 1999. Evolution of genetic mechanisms controlling petal development. *Nature* 399:144–48
108. Kramer EM, Dorit RL, Irish VF. 1998. Molecular evolution of genes controlling petal and stamen development: duplication and divergence within the *APETALA3* and *PISTILLATA* MADS-box gene lineages. *Genetics* 149:765–83
109. Krizek BA, Meyerowitz EM. 1996. The *Arabidopsis* homeotic genes *APETALA3* and *PISTILLATA* are sufficient to provide the B class organ identity function. *Development* 122:11–22
110. Lalanne E, Twell D. 2002. Genetic control of male germ unit organization in Arabidopsis. *Plant Physiol.* 129:865–75
111. Lalanne E, Michaelidis C, Moore JM, Gagliano W, Johnson A, et al. 2004. Analysis of transposon insertion mutants highlights the diversity of mechanisms underlying male progamic development in *Arabidopsis*. *Genetics* 167:1975–86
112. Lee I, Wolfe DS, Nilsson O, Weigel D. 1997. A *LEAFY* co-regulator encoded by *UNUSUAL FLORAL ORGANS*. *Curr. Biol.* 7:95–104
113. Lenhard M, Bohnert A, Jürgens G, Laux T. 2001. Termination of stem cell maintenance in *Arabidopsis* floral meristems by interactions between *WUSCHEL* and *AGAMOUS*. *Cell* 105:805–14
114. Levin JZ, Meyerowitz EM. 1995. *UFO*: an Arabidopsis gene involved in both floral meristem and floral organ development. *Plant Cell* 7:529–48
115. Li J, Jia D, Chen X. 2001. *HUA1*, a regulator of stamen and carpel identities in Arabidopsis, codes for a nuclear RNA binding protein. *Plant Cell* 13:2269–81
116. Li W, Ma H. 2002. Gametophyte development. *Curr. Biol.* 12:R718–21
117. Li W, Chen C, Markmann-Mulisch U, Timofejeva L, Schmelzer E, et al. 2004. The *Arabidopsis AtRAD51* gene is dispensable for vegetative development but required for meiosis. *Proc. Natl. Acad. Sci. USA* 101:10596–601
118. Lichten M. 2001. Meiotic recombination: breaking the genome to save it. *Curr. Biol.* 11:R253–56
119. Liu B, Cyr RJ, Palevitz BA. 1996. A kinesin-like protein, KatAp, in the cells of Arabidopsis and other plants. *Plant Cell* 8:119–32
120. Liu N, Lamerdin JE, Tebbs RS, Schild D, Tucker JD, et al. 1998. XRCC2 and XRCC3, new human Rad51-family members, promote chromosome stability and protect against DNA cross-links and other damages. *Mol. Cell* 1:783–93
121. Liu Z, Meyerowitz EM. 1995. *LEUNIG* regulates *AGAMOUS* expression in *Arabidopsis* flowers. *Development* 121:975–91
122. Lohmann JU, Weigel D. 2002. Building beauty: the genetic control of floral patterning. *Dev. Cell* 2:135–42
123. Lohmann JU, Hong RL, Hobe M, Busch MA, Parcy F, et al. 2001. A molecular link between stem cell regulation and floral patterning in *Arabidopsis*. *Cell* 105:793–803

124. Ma H. 1994. The unfolding drama of flower development: recent results from genetic and molecular analyses. *Genes Dev.* 8:745–56
125. Ma H, dePamphilis C. 2000. The ABCs of floral evolution. *Cell* 101:5–8
126. Ma H, Yanofsky MF, Meyerowitz EM. 1991. *AGL1-AGL6*, an Arabidopsis gene family with similarity to floral homeotic and transcription factor genes. *Genes Dev.* 5:484–95
127. Magnard JL, Yang M, Chen YCS, Leary M, McCormick S. 2001. The Arabidopsis gene *Tardy Asynchronous Meiosis* is required for the normal pace and synchrony of cell division during male meiosis. *Plant Physiol.* 127:1157–66
128. Malkova A, Ross L, Dawson D, Hoekstra MF, Haber JE. 1996. Meiotic recombination initiated by a double-strand break in rad50 delta yeast cells otherwise unable to initiate meiotic recombination. *Genetics* 143:741–54
129. Mandel MA, Yanofsky MF. 1995. The Arabidopsis *AGL8* MADS box gene is expressed in inflorescence meristems and is negatively regulated by APETALA1. *Plant Cell* 7:1763–71
130. Mandel MA, Yanofsky MF. 1998. The *Arabidopsis AGL9* MADS box gene is expressed in young flower primordia. *Sex. Plant Reprod.* 11:22–28
131. Mandel MA, Bowman JL, Kempin SA, Ma H, Meyerowitz EM, Yanofsky MF. 1992. Manipulation of flower structure in transgenic tobacco. *Cell* 71:133–43
132. Mariani C, De Beuckeleer M, Truettner J, Leemans J, Goldberg RB. 1990. Induction of male sterility in plants by a chimaeric ribonuclease gene. *Nature* 347:737–41
133. Martinez-Perez E, Shaw PJ, Moore G. 2000. Polyploidy induces centromere association. *J. Cell Biol.* 148:233–38
134. Martinez-Perez E, Shaw P, Moore G. 2001. The *Ph1* locus is needed to ensure specific somatic and meiotic centromere association. *Nature* 411:204–7
135. Martinez-Perez E, Shaw P, Reader S, Aragon-Alcaide L, Miller T, Moore G. 1999. Homologous chromosome pairing in wheat. *J. Cell Sci.* 112:1761–69
136. Mayer KF, Schoof H, Haecker A, Lenhard M, Jürgens G, Laux T. 1998. Role of *WUSCHEL* in regulating stem cell fate in the *Arabidopsis* shoot meristem. *Cell* 95:805–15
137. McConn M, Browse J. 1996. The critical requirement for linolenic acid is pollen development, not photosynthesis, in an arabidopsis mutant. *Plant Cell* 8:403–16
138. McCormick S. 1993. Male gametophyte development. *Plant Cell* 5:1265–75
139. McCormick S. 2004. Control of male gametophyte development. *Plant Cell* 16(Suppl.): S142–53
140. Mena M, Mandel MA, Lerner DR, Yanofsky MF, Schmidt RJ. 1995. A characterization of the MADS-box gene family in maize. *Plant J.* 8:845–54
141. Mena M, Ambrose BA, Meeley RB, Briggs SP, Yanofsky MF, Schmidt RJ. 1996. Diversification of C-function activity in maize flower development. *Science* 274:1537–40
142. Mercier R, Vezon D, Bullier E, Motamayor JC, Sellier A, et al. 2001. SWITCH1 (SWI1): a novel protein required for the establishment of sister chromatid cohesion and for bivalent formation at meiosis. *Genes Dev.* 15:1859–71
143. Mercier R, Armstrong SJ, Horlow C, Jackson NP, Makaroff CA, et al. 2003. The meiotic protein SWI1 is required for axial element formation and recombination initiation in Arabidopsis. *Development* 130:3309–18
144. Meyerowitz EM, Bowman JL, Brockman LL, Drews GN, Jack T, et al. 1991. A genetic and molecular model for flower development in *Arabidopsis thaliana*. *Development* S1:157–67
145. Michaelis C, Ciosk R, Nasmyth K. 1997. Cohesins: chromosomal proteins that prevent premature separation of sister chromatids. *Cell* 91:35–45
146. Mitsui H, Yamaguchi-Shinozaki K, Shinozaki K, Nishikawa K, Takahashi H. 1993. Identification of a gene family (*kat*) encoding kinesin-like proteins in *Arabidopsis thaliana* and the characterization of secondary structure of KatA. *Mol. Gen. Genet.* 238:362–68

147. Mizukami Y, Ma H. 1992. Ectopic expression of the floral homeotic gene *AGAMOUS* in transgenic *Arabidopsis* plants alters floral organ identity. *Cell* 71:119–31
148. Mizukami Y, Huang H, Tudor M, Hu Y, Ma H. 1996. Functional domains of the floral regulator AGAMOUS: characterization of the DNA binding domain and analysis of dominant negative mutations. *Plant Cell* 8:831–45
149. Moon YH, Jung JY, Kang HG, An G. 1999. Identification of a rice *APETALA3* homologue by yeast two-hybrid screening. *Plant Mol. Biol.* 40:167–77
150. Mouradov A, Hamdorf B, Teasdale RD, Kim JT, Winter KU, Theissen G. 1999. A *DEF/GLO*-like MADS-box gene from a gymnosperm: *Pinus radiata* contains an ortholog of angiosperm B class floral homeotic genes. *Dev. Genet.* 25:245–52
151. Nasmyth K. 1999. Separating sister chromatids. *Trends Biochem. Sci.* 24:98–104
152. Ng M, Yanofsky MF. 2000. Three ways to learn the ABCs. *Curr. Opin. Plant Biol.* 3:47–52
153. Ng M, Yanofsky MF. 2001. Function and evolution of the plant MADS-box gene family. *Nat. Rev. Genet.* 2:186–95
154. Ni W, Xie D, Hobbie L, Feng B, Zhao D, et al. 2004. Regulation of flower development in Arabidopsis by SCF complexes. *Plant Physiol.* 134:1574–85
155. Nonomura K, Miyoshi K, Eiguchi M, Suzuki T, Miyao A, et al. 2003. The *MSP1* gene is necessary to restrict the number of cells entering into male and female sporogenesis and to initiate anther wall formation in rice. *Plant Cell* 15:1728–39
156. Nonomura K, Nakano M, Fukuda T, Eiguchi M, Miyao A, et al. 2004. The novel gene *HOMOLOGOUS PAIRING ABERRATION IN RICE MEIOSIS1* of rice encodes a putative coiled-coil protein required for homologous chromosome pairing in meiosis. *Plant Cell* 16:1008–20
157. Nonomura KI, Nakano M, Murata K, Miyoshi K, Eiguchi M, et al. 2004. An insertional mutation in the rice *PAIR2* gene, the ortholog of *Arabidopsis ASY1*, results in a defect in homologous chromosome pairing during meiosis. *Mol. Genet. Genomics* 271:121–29
158. Orr-Weaver TL. 1999. The ties that bind: localization of the sister-chromatid cohesin complex on yeast chromosomes. *Cell* 99:1–4
159. Parcy F, Nilsson O, Busch MA, Lee I, Weigel D. 1998. A genetic framework for floral patterning. *Nature* 395:561–66
160. Parisi S, McKay MJ, Molnar M, Thompson MA, vanderSpek PJ, et al. 1999. Rec8p, a meiotic recombination and sister chromatid cohesion phosphoprotein of the Rad21p family conserved from fission yeast to humans. *Mol. Cell. Biol.* 19:3515–28
161. Park JH, Halitschke R, Kim HB, Baldwin IT, Feldmann KA, Feyereisen R. 2002. A knockout mutation in allene oxide synthase results in male sterility and defective wound signal transduction in Arabidopsis due to a block in jasmonic acid biosynthesis. *Plant J.* 31:1–12
162. Park SK, Howden R, Twell D. 1998. The *Arabidopsis thaliana* gametophytic mutation *gemini pollen1* disrupts microspore polarity, division asymmetry and pollen cell fate. *Development* 125:3789–99
163. Park W, Li J, Song R, Messing J, Chen X. 2002. CARPEL FACTORY, a Dicer homolog, and HEN1, a novel protein, act in microRNA metabolism in *Arabidopsis thaliana*. *Curr. Biol.* 12:1484–95
164. Pawlowski WP, Golubovskaya IN, Cande WZ. 2003. Altered nuclear distribution of recombination protein RAD51 in maize mutants suggests the involvement of RAD51 in meiotic homology recognition. *Plant Cell* 15:1807–16
165. Pawlowski WP, Golubovskaya IN, Timofejeva L, Meeley RB, Sheridan WF, Cande WZ. 2004. Coordination of meiotic recombination, pairing, and synapsis by PHS1. *Science* 303:89–92

166. Paxson-Sowders DM, Owen HA, Makaroff CA. 1997. A comparative ultrastructural analysis of exine pattern development in wild-type *Arabidopsis* and a mutant defective in pattern formation. *Protoplasma* 198:53–65
167. Paxson-Sowders DM, Dodrill CH, Owen HA, Makaroff CA. 2001. DEX1, a novel plant protein, is required for exine pattern formation during pollen development in Arabidopsis. *Plant Physiol.* 127:1739–49
168. Peirson BN, Owen HA, Feldmann KA, Makaroff CA. 1996. Characterization of three male-sterile mutants of Arabidopsis thaliana exhibiting alterations in meiosis. *Sex. Plant Reprod.* 9:1–16
169. Pelaz S, Ditta GS, Baumann E, Wisman E, Yanofsky MF. 2000. B and C floral organ identity functions require *SEPALLATA* MADS-box genes. *Nature* 405:200–3
170. Pelaz S, Gustafson-Brown C, Kohalmi SE, Crosby WL, Yanofsky MF. 2001. APETALA1 and SEPALLATA3 interact to promote flower development. *Plant J.* 26:385–94
171. Pelaz S, Tapia-Lopez R, Alvarez-Buylla ER, Yanofsky MF. 2001. Conversion of leaves into petals in *Arabidopsis*. *Curr. Biol.* 11:182–84
172. Pnueli L, Hareven D, Rounsley SD, Yanofsky MF, Lifschitz E. 1994. Isolation of the tomato *AGAMOUS* gene *TAG1* and analysis of its homeotic role in transgenic plants. *Plant Cell* 6:163–73
173. Porat R, Lu P, O'Neill SD. 1998. *Arabidopsis* SKP1, a homologue of a cell cycle regulator gene, is predominantly expressed in meristematic cells. *Planta* 204:345–51
174. Puizina J, Siroky J, Mokros P, Schweizer D, Riha K. 2004. Mre11 deficiency in Arabidopsis is associated with chromosomal instability in somatic cells and Spo11-dependent genome fragmentation during meiosis. *Plant Cell* 16:1968–78
175. Reddy TV, Kaur J, Agashe B, Sundaresan V, Siddiqi I. 2003. The *DUET* gene is necessary for chromosome organization and progression during male meiosis in *Arabidopsis* and encodes a PHD finger protein. *Development* 130:5975–87
176. Reinhart BJ, Weinstein EG, Rhoades MW, Bartel B, Bartel DP. 2002. MicroRNAs in plants. *Genes Dev.* 16:1616–26
177. Riechmann JL, Meyerowitz EM. 1997. MADS domain proteins in plant development. *Biol. Chem.* 378:1079–101
178. Riechmann JL, Meyerowitz EM. 1998. The AP2/EREBP family of plant transcription factors. *Biol Chem* 379:633–46
179. Risseeuw EP, Daskalchuk TE, Banks TW, Liu E, Cotelesage J, et al. 2003. Protein interaction analysis of SCF ubiquitin E3 ligase subunits from *Arabidopsis*. *Plant J.* 34:753–67
180. Roeder GS. 1990. Chromosome synapsis and genetic recombination: their roles in meiotic chromosome segregation. *Trends Genet.* 6:385–89
181. Ross KJ, Fransz P, Jones GH. 1996. A light microscopic atlas of meiosis in *Arabidopsis thaliana*. *Chromosome Res.* 4:507–16
182. Ross KJ, Fransz P, Armstrong SJ, Vizir I, Mulligan B, et al. 1997. Cytological characterization of four meiotic mutants of *Arabidopsis* isolated from T-DNA-transformed lines. *Chromosome Res.* 5:551–59
183. Rounsley SD, Ditta GS, Yanofsky MF. 1995. Diverse roles for MADS box genes in Arabidopsis development. *Plant Cell* 7:1259–69
184. Rutledge R, Regan S, Nicolas O, Fobert P, Cote C, et al. 1998. Characterization of an *AGAMOUS* homologue from the conifer black spruce (*Picea mariana*) that produces floral homeotic conversions when expressed in Arabidopsis. *Plant J.* 15:625–34
185. Sakai H, Medrano LJ, Meyerowitz EM. 1995. Role of *SUPERMAN* in maintaining *Arabidopsis* floral whorl boundaries. *Nature* 378:199–203

186. Sakai H, Krizek BA, Jacobsen SE, Meyerowitz EM. 2000. Regulation of *SUP* expression identifies multiple regulators involved in arabidopsis floral meristem development. *Plant Cell* 12:1607–18
187. Samach A, Klenz JE, Kohalmi SE, Risseeuw E, Haughn GW, Crosby WL. 1999. The *UNUSUAL FLORAL ORGANS* gene of *Arabidopsis thaliana* is an F-box protein required for normal patterning and growth in the floral meristem. *Plant J.* 20:433–45
188. Sanders PM, Lee PY, Biesgen C, Boone JD, Beals TP, et al. 2000. The Arabidopsis *DELAYED DEHISCENCE1* gene encodes an enzyme in the jasmonic acid synthesis pathway. *Plant Cell* 12:1041–61
189. Sanders PM, Bui AQ, Weterings K, McIntire KN, Hsu YC, et al. 1999. Anther developmental defects in *Arabidopsis thaliana* male-sterile mutants. *Sex. Plant Reprod.* 11:297–322
190. Sato S, Hotta Y, Tabata S. 1995. Structural analysis of a recA-like gene in the genome of *Arabidopsis thaliana*. *DNA Res.* 2:89–93
191. Savidge B, Rounsley SD, Yanofsky MF. 1995. Temporal relationship between the transcription of two Arabidopsis MADS box genes and the floral organ identity genes. *Plant Cell* 7:721–33
192. Scherthan H. 2001. A bouquet makes ends meet. *Nat. Rev. Mol. Cell. Biol.* 2:621–27
193. Scherthan H, Weich S, Schwegler H, Heyting C, Haerle M, Cremer T. 1996. Centromere and telomere movements during early meiotic prophase of mouse and man are associated with the onset of chromosome pairing. *J. Cell Biol.* 134:1109–25
194. Schiefthaler U, Balasubramanian S, Sieber P, Chevalier D, Wisman E, Schneitz K. 1999. Molecular analysis of *NOZZLE*, a gene involved in pattern formation and early sporogenesis during sex organ development in *Arabidopsis thaliana*. *Proc. Natl. Acad. Sci. USA* 96:11664–69
195. Schild D, Lio YC, Collins DW, Tsomondo T, Chen DJ. 2000. Evidence for simultaneous protein interactions between human Rad51 paralogs. *J. Biol. Chem.* 275:16443–49
196. Schmidt RJ, Ambrose BA. 1998. The blooming of grass flower development. *Curr. Opin. Plant Biol.* 1:60–67
197. Schommer C, Beven A, Lawrenson T, Shaw P, Sablowski R. 2003. *AHP2* is required for bivalent formation and for segregation of homologous chromosomes in *Arabidopsis* meiosis. *Plant J.* 36:1–11
198. Schulman BA, Carrano AC, Jeffrey PD, Bowen Z, Kinnucan ER, et al. 2000. Insights into SCF ubiquitin ligases from the structure of the Skp1-Skp2 complex. *Nature* 408:381–86
199. Schultz EA, Pickett FB, Haughn GW. 1991. The *FLO10* gene product regulates the expression domain of homeotic genes *AP3* and *PI* in Arabidopsis flowers. *Plant Cell* 3:1221–37
200. Schwacke R, Grallath S, Breitkreuz KE, Stransky E, Stransky H, et al. 1999. LeProT1, a transporter for proline, glycine betaine, and gamma-amino butyric acid in tomato pollen. *Plant Cell* 11:377–91
201. Schwarzacher T. 2003. Meiosis, recombination and chromosomes: a review of gene isolation and fluorescent in situ hybridization data in plants. *J. Exp. Bot.* 54:11–23
202. Schwarz-Sommer Z, Huijser P, Nacken W, Saedler H, Sommer H. 1990. Genetic control of flower development: homeotic genes in *Antirrhinum majus*. *Science* 250:931–36
203. Scott RJ, Spielman M, Dickinson HG. 2004. Stamen structure and function. *Plant Cell* 16(Suppl.): S46–60
204. Shimamoto K, Kyozuka J. 2002. Rice as a model for comparative genomics of plants. *Annu. Rev. Plant Biol.* 53:399–419
205. Shimazu J, Matsukura C, Senda M, Ishikawa R, Akada S, et al. 2001. Characterization of a *DMC1* homologue, *RiLIM15*, in meiotic panicles, mitotic cultured cells and mature leaves of rice (*Oryza sativa* L.). *Theor. Appl. Genet.* 102:1159–63

206. Shinohara A, Ogawa H, Ogawa T. 1992. Rad51 protein involved in repair and recombination in S. cerevisiae is a RecA-like protein. *Cell* 69:457–70
207. Siaud N, Dray E, Gy I, Gerard E, Takvorian N, Doutriaux MP. 2004. Brca2 is involved in meiosis in *Arabidopsis thaliana* as suggested by its interaction with Dmc1. *EMBO J.* 23:1392–401
208. Sieburth LE, Meyerowitz EM. 1997. Molecular dissection of the AGAMOUS control region shows that cis elements for spatial regulation are located intragenically. *Plant Cell* 9:355–65
209. Skowyra D, Craig KL, Tyers M, Elledge SJ, Harper JW. 1997. F-box proteins are receptors that recruit phosphorylated substrates to the SCF ubiquitin-ligase complex. *Cell* 91:209–19
210. Smyth DR, Bowman JL, Meyerowitz EM. 1990. Early flower development in Arabidopsis. *Plant Cell* 2:755–67
211. Sommer H, Beltran J-P, Huijser P, Pape H, Lonnig W-E, et al. 1990. *Deficiens*, a homeotic gene involved in the control of flower morphogenesis in *Antirrhinum majus*: the protein shows homology to transcription factors. *EMBO J.* 9:605–13
212. Sorensen A, Guerineau F, Canales-Holzeis C, Dickinson HG, Scott RJ. 2002. A novel extinction screen in *Arabidopsis thaliana* identifies mutant plants defective in early microsporangial development. *Plant J.* 29:581–94
213. Sorensen AM, Krober S, Unte US, Huijser P, Dekker K, Saedler H. 2003. The *Arabidopsis ABORTED MICROSPORES* (*AMS*) gene encodes a MYC class transcription factor. *Plant J.* 33:413–23
214. Southerton SG, Marshall H, Mouradov A, Teasdale RD. 1998. Eucalypt MADS-box genes expressed in developing flowers. *Plant Physiol.* 118:365–72
215. Spielman M, Preuss D, Li FL, Browne WE, Scott RJ, Dickinson HG. 1997. *TETRASPORE* is required for male meiotic cytokinesis in *Arabidopsis thaliana*. *Development* 124:2645–57
216. Sridhar VV, Surendrarao A, Gonzalez D, Conlan RS, Liu Z. 2004. Transcriptional repression of target genes by LEUNIG and SEUSS, two interacting regulatory proteins for *Arabidopsis* flower development. *Proc. Natl. Acad. Sci. USA* 101:11494–99
217. Staiger CJ, Cande WZ. 1990. Microtubule distribution in *dv*, a maize meiotic mutant defective in the prophase to metaphase transition. *Dev. Biol.* 138:231–42
218. Steiner-Lange S, Unte US, Eckstein L, Yang C, Wilson ZA, et al. 2003. Disruption of *Arabidopsis thaliana MYB26* results in male sterility due to non-dehiscent anthers. *Plant J.* 34:519–28
219. Stevens R, Grelon M, Vezon D, Oh J, Meyer P, et al. 2004. A CDC45 homolog in Arabidopsis is essential for meiosis, as shown by RNA interference-induced gene silencing. *Plant Cell* 16:99–113
220. Stintzi A, Browse J. 2000. The *Arabidopsis* male-sterile mutant, *opr3*, lacks the 12-oxophytodienoic acid reductase required for jasmonate synthesis. *Proc. Natl. Acad. Sci. USA* 97:10625–30
221. StoopMyer C, Amon N. 1999. Meiosis: Rec8 is the reason for cohesion. *Nat. Cell Biol.* 1:E125–27
222. Sundstrom J, Carlsbecker A, Svensson ME, Svenson M, Johanson U, et al. 1999. MADS-box genes active in developing pollen cones of Norway spruce (*Picea abies*) are homologous to the B-class floral homeotic genes in angiosperms. *Dev. Genet.* 25:253–66
223. Tanaka I. 1997. Differentiation of generative and vegetative cells in angiosperm pollen. *Sex. Plant Reprod.* 10:1–7

224. Tandre K, Albert VA, Sundas A, Engstrom P. 1995. Conifer homologues to genes that control floral development in angiosperms. *Plant Mol. Biol.* 27:69–78
225. Tandre K, Svenson M, Svensson ME, Engstrom P. 1998. Conservation of gene structure and activity in the regulation of reproductive organ development of conifers and angiosperms. *Plant J.* 15:615–23
226. Taylor PE, Glover JA, Lavithis M, Craig S, Singh MB, et al. 1998. Genetic control of male fertility in *Arabidopsis thaliana*: structural analyses of postmeiotic developmental mutants. *Planta* 205:492–505
227. Theissen G, Saedler H. 1999. The golden decade of molecular floral development (1990–1999): a cheerful obituary. *Dev. Genet.* 25:181–93
228. Theissen G, Saedler H. 2001. Plant biology. Floral quartets. *Nature* 409:469–71
229. Trobner W, Ramirez L, Motte P, Hue I, Huijser P, et al. 1992. *GLOBOSA*: a homeotic gene which interacts with *DEFICIENS* in the control of *Antirrhinum* floral organogenesis. *EMBO J.* 11:4693–704
230. Tsuchimoto S, van der Krol AR, Chua NH. 1993. Ectopic expression of *pMADS3* in transgenic petunia phenocopies the petunia blind mutant. *Plant Cell* 5:843–53
231. Twell D, Park SK, Hawkins TJ, Schubert D, Schmidt R, et al. 2002. MOR1/GEM1 has an essential role in the plant-specific cytokinetic phragmoplast. *Nat. Cell Biol.* 4:711–14
232. Vega JM, Feldman M. 1998. Effect of the pairing gene *Ph1* on centromere misdivision in common wheat. *Genetics* 148:1285–94
233. von Malek B, van der Graaff E, Schneitz K, Keller B. 2002. The Arabidopsis male-sterile mutant *dde2-2* is defective in the *ALLENE OXIDE SYNTHASE* gene encoding one of the key enzymes of the jasmonic acid biosynthesis pathway. *Planta* 216:187–92
234. Wang G, Kong H, Sun Y, Zhang X, Zhang W, et al. 2004. Genome-wide analysis of the cyclin family in Arabidopsis and comparative phylogenetic analysis of plant cyclin-like proteins. *Plant Physiol.* 135:1084–99
235. Wang W, Chen X. 2004. HUA ENHANCER3 reveals a role for a cyclin-dependent protein kinase in the specification of floral organ identity in Arabidopsis. *Development* 131:3147–56
236. Wang Y, Wang X, Skirpan AL, Kao TH. 2003. S-RNase-mediated self-incompatibility. *J. Exp. Bot.* 54:115–22
237. Watanabe Y, Nurse P. 1999. Cohesin Rec8 is required for reductional chromosome segregation at meiosis. *Nature* 400:461–64
238. Weigel D. 1998. From floral induction to floral shape. *Curr. Opin. Plant Biol.* 1:55–59
239. Weigel D, Meyerowitz EM. 1994. The ABCs of floral homeotic genes. *Cell* 78:203–9
240. Weigel D, Jürgens G. 2002. Stem cells that make stems. *Nature* 415:751–54
241. Weigel D, Meyerowitz EM. 1993. Genetic hierarchy controlling flower development. In *Molecular Basis of Morphogenesis*, ed. M Bernfield, pp. 91–105. New York: Wiley-Liss
242. Weigel D, Alvarez J, Smyth DR, Yanofsky MF, Meyerowitz EM. 1992. *LEAFY* controls floral meristem identity in *Arabidopsis*. *Cell* 69:843–59
243. Western TL, Cheng Y, Liu J, Chen X. 2002. HUA ENHANCER2, a putative DExH-box RNA helicase, maintains homeotic B and C gene expression in *Arabidopsis*. *Development* 129:1569–81
244. Wilkinson MD, Haughn GW. 1995. *UNUSUAL FLORAL ORGANS* controls meristem identity and floral organ primordia fate in Arabidopsis. *Plant Cell* 7:1485–99
245. Wilson ZA, Morroll SM, Dawson J, Swarup R, Tighe PJ. 2001. The *Arabidopsis MALE STERILITY1 (MS1)* gene is a transcriptional regulator of male gametogenesis, with homology to the PHD-finger family of transcription factors. *Plant J.* 28:27–39

246. Winter KU, Becker A, Munster T, Kim JT, Saedler H, Theissen G. 1999. MADS-box genes reveal that gnetophytes are more closely related to conifers than to flowering plants. *Proc. Natl. Acad. Sci. USA* 96:7342–47
247. Wolfe KH, Gouy M, Yang YW, Sharp PM, Li WH. 1989. Date of the monocot-dicot divergence estimated from chloroplast DNA sequence data. *Proc. Natl. Acad. Sci. USA* 86:6201–5
248. Yang CY, Spielman M, Coles JP, Li Y, Ghelani S, et al. 2003. *TETRASPORE* encodes a kinesin required for male meiotic cytokinesis in *Arabidopsis*. *Plant J.* 34:229–40
249. Yang M, Hu Y, Lodhi M, McCombie WR, Ma H. 1999. The *Arabidopsis SKP1-LIKE1* gene is essential for male meiosis and may control homologue separation. *Proc. Natl. Acad. Sci. USA* 96:11416–21
250. Yang SL, Xie LF, Mao HZ, Puah CS, Yang WC, et al. 2003. *TAPETUM DETERMINANT1* is required for cell specialization in the Arabidopsis anther. *Plant Cell* 15:2792–804
251. Yang WC, Sundaresan V. 2000. Genetics of gametophyte biogenesis in *Arabidopsis*. *Curr. Opin. Plant Biol.* 3:53–57
252. Yang WC, Ye D, Xu J, Sundaresan V. 1999. The *SPOROCYTELESS* gene of *Arabidopsis* is required for initiation of sporogenesis and encodes a novel nuclear protein. *Genes Dev.* 13:2108–17
253. Yang X, Makaroff CA, Ma H. 2003. The Arabidopsis *MALE MEIOCYTE DEATH1* gene encodes a PHD-finger protein that is required for male meiosis. *Plant Cell* 15:1281–95
254. Yanofsky MF. 1995. Floral meristems to floral organs: genes controlling early events in *Arabidopsis* flower development. *Annu. Rev. Plant Physiol. Plant Mol. Biol.* 46:167–88
255. Yanofsky MF, Ma H, Bowman JL, Drews GN, Feldmann KA, Meyerowitz EM. 1990. The protein encoded by the *Arabidopsis* homeotic gene *agamous* resembles transcription factors. *Nature* 346:35–39
256. Yu VP, Koehler M, Steinlein C, Schmid M, Hanakahi LA, et al. 2000. Gross chromosomal rearrangements and genetic exchange between nonhomologous chromosomes following BRCA2 inactivation. *Genes Dev.* 14:1400–6
257. Zhao D, Yang M, Solava J, Ma H. 1999. The *ASK1* gene regulates development and interacts with the *UFO* gene to control floral organ identity in *Arabidopsis*. *Dev. Genet.* 25:209–23
258. Zhao D, Yu Q, Chen M, Ma H. 2001. The *ASK1* gene regulates B function gene expression in cooperation with *UFO* and *LEAFY* in *Arabidopsis*. *Development* 128:2735–46
259. Zhao DZ, Wang GF, Speal B, Ma H. 2002. The *EXCESS MICROSPOROCYTES1* gene encodes a putative leucine-rich repeat receptor protein kinase that controls somatic and reproductive cell fates in the *Arabidopsis* anther. *Genes Dev.* 16:2021–31
260. Zhao DZ, Yu QL, Chen CB, Ma H. 2001. Genetic control of reproductive meristems. In *Meristematic Tissues in Plant Growth and Development*, ed. MT McManus, BE Veit. pp. 89–141. Shefield: Academic
261. Zheng N, Schulman BA, Song L, Miller JJ, Jeffrey PD, et al. 2002. Structure of the Cul1-Rbx1-Skp1-F boxSkp2 SCF ubiquitin ligase complex. *Nature* 416:703–9
262. Zheng Z, Xia Q, Dauk M, Shen W, Selvaraj G, Zou J. 2003. Arabidopsis *AtGPAT1*, a member of the membrane-bound glycerol-3-phosphate acyltransferase gene family, is essential for tapetum differentiation and male fertility. *Plant Cell* 15:1872–87
263. Zhu QH, Ramm K, Shivakkumar R, Dennis ES, Upadhyaya NM. 2004. The *ANTHER INDEHISCENCE1* gene encoding a single MYB domain protein is involved in anther development in rice. *Plant Physiol.* 135:1514–25
264. Zickler D, Kleckner N. 1998. The leptotene-zygotene transition of meiosis. *Annu. Rev. Genet.* 32:619–97
265. Zickler D, Kleckner N. 1999. Meiotic chromosomes: integrating structure and function. *Annu. Rev. Genet.* 33:603–754

266. Zou L, Stillman B. 1998. Formation of a preinitiation complex by S-phase cyclin CDK-dependent loading of Cdc45p onto chromatin. *Science* 280:593–96
267. Zou L, Stillman B. 2000. Assembly of a complex containing Cdc45p, replication protein A, and Mcm2p at replication origins controlled by S-phase cyclin-dependent kinases and Cdc7p-Dbf4p kinase. *Mol. Cell. Biol.* 20:3086–96

Plant-Specific Calmodulin-Binding Proteins

Nicolas Bouché,[1] Ayelet Yellin,[2] Wayne A. Snedden,[3] and Hillel Fromm[2]

[1]Institut National de la Recherche Agronomique, Institut Jean-Pierre Bourgin, Laboratoire de Biologie Cellulaire, 78026 Versailles, France; email: bouche@versailles.inra.fr

[2]Department of Plant Sciences, Tel Aviv University, Tel Aviv, 69978 Israel; email: yellinay@post.tau.ac.il, hillelf@post.tau.ac.il

[3]Department of Biology, Queen's University, Kingston, Ontario, K7L 3N6, Canada; email: sneddenw@biology.queensu.ca

Annu. Rev. Plant Biol. 2005. 56:435–66

doi: 10.1146/annurev.arplant.56.032604.144224

First published online as a Review in Advance on January 18, 2005

1543-5008/05/0602-0435$20.00

Key Words

calcium, signal transduction, environmental stress, *Arabidopsis*

Abstract

Calmodulin (CaM) is the most prominent Ca^{2+} transducer in eukaryotic cells, regulating the activity of numerous proteins with diverse cellular functions. Many features of CaM and its downstream targets are similar in plants and other eukaryotes. However, plants possess a unique set of CaM-related proteins, and several unique CaM target proteins. This review discusses recent progress in identifying plant-specific CaM-binding proteins and their roles in response to biotic and abiotic stresses and development. The review also addresses aspects emerging from recent structural studies of CaM interactions with target proteins relevant to plants.

Contents

INTRODUCTION: THE LANGUAGE OF CALCIUM SIGNALING

Ca^{2+} as a Signal Carrier in Living Organisms

All organisms continually monitor their environment and respond to changes with adaptive mechanisms that are initiated at the molecular and biochemical levels and require the coordination of cellular events to ensure that a response is appropriate for a given stimulus. Therefore, complex intra- and intercellular communication networks have evolved to convey information about a perceived stimulus to the cellular machinery responsible for mediating the responses. Organisms use various small organic and inorganic molecules, termed second messengers (e.g., Ca^{2+}, cyclic nucleotides, phospholipids, sugars, amino acids), to encode information and deliver it to downstream effectors, which decode signals and initiate cellular responses including changes in enzyme activity, gene expression, transport across membranes, and cytoskeletal rearrangement. Ca^{2+} is one of the most prominent second messengers in eukaryotes, and its roles as a signal carrier are the subject of intensive investigations in both animals and plants. The reader is also referred to more specific reviews on different aspects of Ca^{2+} signaling in plants (73, 76, 145, 174).

In recent years, there has been a major effort directed toward elucidating the information carried in the Ca^{2+} signals evoked by exogenous stimuli. Spatial and temporal propagation of the Ca^{2+} signals, the amplitude of the signal, which is typically proportional to the strength of the stimulus, and the frequency of oscillations are all elements of information carried by the Ca^{2+} signal that must be decoded by the cellular machinery. Spatial distribution of the Ca^{2+} signal in the cell is controlled by a complex network of Ca^{2+}-permeable channels, Ca^{2+}-antiporters, and Ca^{2+}-pumps, which operate at the plasma membrane or in membranes of intracellular Ca^{2+} stores including the ER, mitochondria, chloroplast, and the vacuole, and are regulated by different second messengers (reviewed in 76). Of particular interest is the relationship between cytosolic and nuclear Ca^{2+}. Ca^{2+} signals in the cytosol and in the nucleus have distinct functions in both animals (13, 18, 71) and plants (169). Recent investigations (13, 132) suggest that although nuclear Ca^{2+} signals in many cases reflect the patterns of cytosolic Ca^{2+}, the nucleus can also generate stimulus-induced signals independently. There is evidence that the nucleus contains intrinsic Ca^{2+} signaling machinery that can release Ca^{2+} locally in discrete nuclear regions. Echevarría et al. (54) identified a nucleoplasmic reticulum in epithelial cells, composing a branching intranuclear network continuous with the nuclear envelope and the endoplasmic reticulum. These structures function as a nuclear Ca^{2+}-storing network that can give rise to localized Ca^{2+} gradients and thus Ca^{2+}-dependent events can be regulated differentially in the nucleus, just as they are in the cytosol. It is likely that plants possess similar intranuclear Ca^{2+} stores.

One important question concerning the role of Ca^{2+} as a second messenger is how cells control stimulus-response specificity. There is evidence that the frequency of Ca^{2+} oscillations carries information that is decoded by cellular targets in both plants and animals. In plants this was elegantly demonstrated by studying the *Arabidopsis det3* mutant, which is unable to close its stomata in response to exogenous Ca^{2+}. However, stomatal closure could be restored in *det3* by subjecting it to artificial Ca^{2+} oscillations (4). Changes in the frequency or duration of the Ca^{2+} oscillations influenced stomatal aperture. These studies indicate a specific mechanism to translate Ca^{2+} signals into a cellular response (3, 5). The nature of these decoders is intriguing. One can speculate that decoding Ca^{2+} signals requires proteins that can respond to Ca^{2+} oscillations by fine conformational changes. De Koninck & Schulman (44) found that the CaM-dependent protein Kinase-II is sensitive to the frequency of Ca^{2+} oscillations in vitro in a way that is reflected in the autonomous kinase activity, suggesting that CaMKII is a decoder of the frequency of Ca^{2+} oscillations. Other studies showed that oscillations increase both the efficacy and information content of Ca^{2+} signals that lead to gene expression and cell differentiation (50). Certain transcription factors are activated only by high-frequency Ca^{2+} oscillations, whereas others may be activated by infrequent oscillations (50) or both high-frequency and low-frequency oscillations. A recent review addressed the role of plant protein kinases in decoding Ca^{2+} signals (73). Given the importance of Ca^{2+}-modulated proteins in decoding Ca^{2+} signals, it is important to consider the repertoire of plant Ca^{2+}-modulated proteins.

Ca^{2+}-Modulated Proteins in Plants

Most proteins that function as transducers of Ca^{2+} signals contain a common structural motif, the "EF hand" (125), which is a helix-loop-helix structure that binds a single Ca^{2+} ion. These motifs typically, but not exclusively, occur in closely linked pairs, interacting through antiparallel β-sheets (125). This arrangement is the basis for cooperativity in Ca^{2+} binding. The superfamily of EF-hand proteins is divided into several classes based on differences in number and organization of EF-hand pairs, amino acid sequences within or outside the motifs, affinity to Ca^{2+}, and/or selectivity and affinity to target proteins (39, 125). Nakayama et al. (124) discussed the evolution of EF-hand proteins and divided them into 66 subfamilies. Day et al. (42) reported a comprehensive bioinformatic search for EF-hand-containing proteins in *Arabidopsis*. Other reviews discuss specific families of EF-hand proteins in plants including the CaM superfamily (119) and protein kinases (72, 73). Note that there are other protein motifs that bind Ca^{2+}. One is the 70-amino acid annexin fold, which is present in members of the membrane-associated annexin subfamily of Ca^{2+} sensors. Another Ca^{2+}-binding motif is the C2 domain of about 130–145 amino acids, present in membrane-associated proteins, of which over 140 are found in the *Arabidopsis* genome. Reddy & Reddy (145) discussed the

ER: endoplasmic reticulum

CaM: calmodulin

Calmodulin: CaM is a ubiquitous Ca^{2+} sensor protein (16 to 18 kD) with no catalytic activity that can, upon binding Ca^{2+}, activate target proteins involved in various cellular processes. The CaM prototype is comprised of two globular domains connected with a long flexible helix. Each globular domain contains a pair of intimately linked EF hands. One EF hand motif is composed of a specialized helix-loop-helix structure that binds one molecule of Ca^{2+}.

Calmodulin-like protein: CaM-like proteins (CMLs) are composed mostly of EF-hand Ca^{2+}-binding motifs, with no other identifiable functional domains, and are at least 16% identical to CaM, as defined by MacCormack & Braam (119). Plants possess many more CaMs and CMLs than any other known organism. For example, *Arabidopsis* possesses six *CAM* loci, encoding three CaM isoforms, and ~50 *CML* genes.

occurrence of these Ca^{2+}-binding motifs in plant proteins in more detail.

In plants, several classes of EF-hand-containing Ca^{2+}-modulated proteins were identified. The first of these is the family of CaMs, which have different isoforms and CaM-related proteins. According to the definitions of MacCormack & Braam (119), for CaM isoforms and CaM-like proteins (CMLs), *Arabidopsis* possesses 6 CaM loci, defined as *CAM*, which encode 3 isoforms, and ~50 additional genes, defined as *CAM-like* (*CML*), which are genes encoding proteins composed mostly of EF-hand Ca^{2+}-binding motifs, with no other identifiable functional domains, and at least 16% are identical to CaM. However, plants also possess species-specific CaM-related proteins that are not found in *Arabidopsis*. For example, petunia CaM PhCaM53 is highly similar to the rice CaM OSCaM61 because it has a stretch of basic amino acids as an extension at the C terminus. This extension undergoes prenylation in response to specific changes in carbon metabolism. PhCaM53 is targeted to the nucleus in the dark, or to the plasma membrane in the light. However, if a carbon source is provided exogenously in the dark, PhCaM53 will target to the plasma membrane (115, 147). There is no homolog of PhCaM53 in *Arabidopsis*, although there are other CMLs in *Arabidopsis* with different types of C-terminal extensions that might function similarly to PhCaM53. Another example of a species-specific CaM-related protein is the rgs-CaM, which was identified as a regulator of gene silencing in tobacco (6). The protein has 3 EF-hands and a 50-amino acid long N-terminal extension. The *Arabidopsis* genome does not contain a gene encoding a protein with a similar N-terminal extension. In cultivated hexaploid wheat, which has the largest known family of CaM genes in a single plant (188), one CaM isoform, designated TaCaM-III, lacks the first EF-hand and instead contains a hydrophobic domain with a tryptophan residue, which is typically absent from CaM (188). One of the fascinating aspects of Ca^{2+} signaling in plants is the occurrence of this large family of CaM-related proteins with species-specific isoforms, which is a sharp contrast to the situation in animals with only one CaM isoform encoded by three genes. It is likely that N- or C-terminal extensions of CaMs and CMLs underlie specific physiological roles. In addition to the functional roles of CaM extensions, many plant CaM isoforms differ from others by a few specific amino acids substitutions. The functional physiological relevance of these amino acid substitutions in providing stimulus-response specificity is beginning to be elucidated in vivo (75). In vitro studies support the idea that CaM isoforms differ in their ability to activate target enzymes (34, 35, 95, 106–108).

Another family of EF-hand-containing Ca^{2+}-modulated proteins in plants is the calcineurin B-like proteins (CBLs). A member of this family is SOS3, a regulator of salt tolerance. It activates the SOS2 protein kinase (69), which is a regulator of SOS1 and AtNHX1, the plasma membrane and tonoplast Na^+/H^+ antiporter, respectively. In *Arabidopsis* CBLs are encoded by 10 genes (115) that activate protein kinases related to the Suc-Non-Fermenting (SNF) protein kinase from yeast. In plants these protein kinases are referred to as CBL-Interacting Protein Kinases (CIPKs), of which *Arabidopsis* possesses 25 genes (115). A third family of EF-hand-containing Ca^{2+}-sensor proteins in plants are the SUB and SUL proteins involved in photomorphogenic responses (68). The SUB1 protein possesses a DNA-binding domain, which suggests that it may combine the function of Ca^{2+} sensing and transcriptional regulation, similar to the mammalian DREAM protein (27).

Other prominent Ca^{2+}-modulated proteins in plants are the Ca^{2+}-regulated protein kinases, which possess a catalytic kinase domain and a regulatory domain with EF-hand or visinin-like Ca^{2+}-binding motifs (72, 73). The Ca^{2+}-dependent protein kinases (CDPKs; reviewed in 32), the Ca^{2+}/CaM-regulated kinases, and the chimeric Ca^{2+} and Ca^{2+}/CaM regulated kinases (CCaMKs), are among this family of Ca^{2+}-regulated protein kinases. The CDPK family alone consists of 34 genes in *Arabidopsis* (72, 73).

A bioinformatic approach (42) revealed numerous new potential plant Ca^{2+}-binding proteins containing EF-hand domains, some of which lack functional domains other than the Ca^{2+}-binding sites (e.g., recoverin-like proteins), whereas some include diverse functional groups. Day et al. (42) estimated that, overall, *Arabidopsis* contains ~250 Ca^{2+}-responsive EF-hand-containing proteins, and these likely modulate the activity of an even larger repertoire of downstream targets. Among the identified Ca^{2+}-responsive proteins, some are predicted to be targeted to the chloroplast and mitochondria, others to the nucleus or cytosol. The estimated predicted number of EF-hand-containing proteins in *Arabidopsis* is the largest among all organisms with complete genomes sequenced (42). This may reflect the need of plants, as sessile organisms, to modulate their entire metabolism, growth, and development in response to changes in their environment.

CALMODULIN-TARGET INTERACTIONS

Structural Analysis and Functional Implications

CaM is one of the best characterized Ca^{2+}-responsive proteins. It has no catalytic activity of its own but, upon binding Ca^{2+}, activates numerous target proteins involved in a variety of cellular processes. CaM is an acidic EF-hand protein present in all eukaryotes. The CaM prototype is composed of 148 amino acids arranged in two globular domains connected with a long flexible helix (**Figure 1*a***). Each globular domain contains a pair of intimately linked EF hands. The majority of known target sites for CaM are composed of a stretch of 12–30 contiguous amino acids with positively charged amphiphilic characteristics, variability in primary sequence, and a propensity to form an α-helix upon binding to CaM. Early X-ray diffraction and NMR studies of CaM provided a model for the structural basis of CaM-target interactions (39). Binding Ca^{2+} to CaM (K_d in the range of 10^{-7} to 10^{-6} M) (**Figure 1*b***) exposes two hydrophobic surfaces surrounded by negative charges, one in each globular domain. Ca^{2+}/CaM may then bind to its targets mainly by hydrophobic interactions with long hydrophobic side chains in the target sites. Electrostatic interactions contribute to the stability of the CaM-target complex (CaM targets are depicted in red in **Figure 1**). The first 3D structures of CaM target peptide complexes to be resolved suggested that the two globular domains of CaM wrap around the target, forming an almost globular structure (e.g., CaMKII; **Figure 1*c***). However, as more 3D structures of CaM/target complexes have been resolved, different types of unexpected interactions have been revealed (77, 171, 181). In terms of CaM:target stochiometry, recent data revealed ratios of 1:1 (**Figure 1*c*,*d***; CaMKII and a Ca^{2+}-pump, respectively), 2:2 (**Figure 1*e***; a potassium channel), and 1:2 [**Figure 1*f***; petunia glutamate decarboxylase (GAD)]. Second, although some targets interact with both the C- and N-terminal lobes of CaM (**Figure 1*c*, *d*,*e*,*f***), others interact with only one lobe (**Figure 1*d***). Also note that Ca^{2+}-independent interactions of CaM, or CaM complexes with just two bound Ca^{2+} ions (e.g., **Figure 1*e***), have been resolved (reviewed in 77). The latter is consistent with genetic studies demonstrating that a yeast mutant expressing a CaM unable to bind any Ca^{2+} can still rescue an otherwise lethal mutant that completely lacks CaM (58). Recent structural findings also support earlier genetic studies of yeast CaM mutants bearing one or more substitutions of phenylalanine to alanine (129). This report showed that mutations affecting specific cellular functions such as CaM localization and nuclear division were all located in the N-terminal lobe of CaM, whereas mutations implicated in actin organization and bud emergence were located in the C-terminal half (129).

NMR: nuclear magnetic resonance

The recent data from 3D structures revealed several types of CaM-binding motifs (77) and provided the ability to cautiously predict, based on amino acid sequence, protein interactions with CaM (**http://calcium.uhnres.utoronto.ca**). However, there are clearly still unidentified

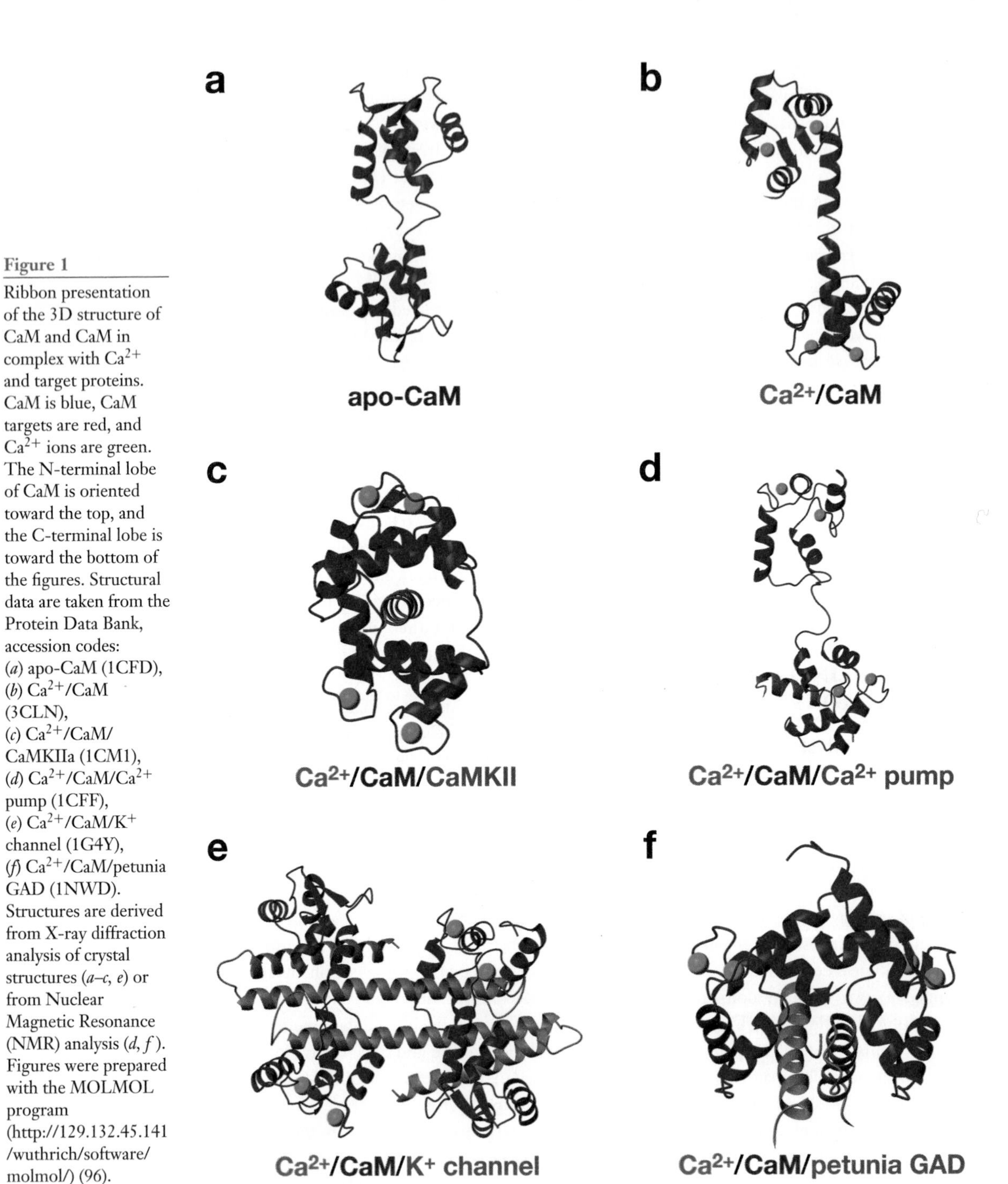

Figure 1

Ribbon presentation of the 3D structure of CaM and CaM in complex with Ca^{2+} and target proteins. CaM is blue, CaM targets are red, and Ca^{2+} ions are green. The N-terminal lobe of CaM is oriented toward the top, and the C-terminal lobe is toward the bottom of the figures. Structural data are taken from the Protein Data Bank, accession codes: (*a*) apo-CaM (1CFD), (*b*) Ca^{2+}/CaM (3CLN), (*c*) Ca^{2+}/CaM/CaMKIIa (1CM1), (*d*) Ca^{2+}/CaM/Ca^{2+} pump (1CFF), (*e*) Ca^{2+}/CaM/K^{+} channel (1G4Y), (*f*) Ca^{2+}/CaM/petunia GAD (1NWD). Structures are derived from X-ray diffraction analysis of crystal structures (*a–c*, *e*) or from Nuclear Magnetic Resonance (NMR) analysis (*d*, *f*). Figures were prepared with the MOLMOL program (http://129.132.45.141/wuthrich/software/molmol/) (96).

CaM-binding motifs, and some proteins known to bind CaM do not possess typical motifs. Recently, the first 3D structure of a CaM-binding domain (CaMBD) of a plant protein [glutamate decarboxylase (GAD)] associated with CaM was resolved (189). The 3D structure of the petunia CaM-binding peptide associated with CaM (**Figure 1*f***) shows the ability of CaM to interact with two GAD CaMBDs, suggesting a role for CaM in GAD dimerization and possibly in the formation of a high molecular weight oligomeric complex. These large GAD complexes (~500 kDa) are found in plants (16) and are stabilized by Ca^{2+} and destabilized in the presence of Ca^{2+}-chelators (16). Coimmunoprecipitation experiments show that CaM is associated with the native GAD complex in vivo (16). It remains unknown whether CaM's role in dimerization of GAD subunits and in GAD activation are separable functions. Possibly, activation of GAD requires the formation of a large oligomeric complex that depends on Ca^{2+}/CaM.

Recently, the crystal structure of a kinesin-like CaM-binding protein (KCBP) from potato was resolved (172). Together with biochemical data, the crystal structure of KCBP suggests that Ca^{2+}/CaM inhibits the binding of KCBP to microtubules by blocking their binding sites on KCBP.

The structures presented in **Figure 1**, together with biochemical data from other studies, reveal a number of protein activation/inhibition mechanisms by CaM, as depicted in **Figure 2**. Relieving autoinhibition has been demonstrated with several CaM-target interactions, including plant and animal Ca^{2+}-ATPases, CaM-dependent kinases, and plant GAD (156). Active-site remodeling is a new type of mechanism revealed by analyzing the endema factor protein from *Bacillus anthracis*, the causative agent of anthrax, in which interaction with CaM bound to two Ca^{2+} ions reorganizes the protein domains to construct the binding site (reviewed in 77). A similar mechanism may occur with other targets with different numbers of bound Ca^{2+} ions. Dimerization at a 2:2 ratio of CaM:target was revealed by studying the 3D structure of CaM bound to the small conductance Ca^{2+}-activated potassium (SK) channel. In this case, Ca^{2+} ions are bound only to the N-terminal EF-hands while the C-terminal EF motifs mediate tethering to the channel. Target activation may occur through dimerization and stabilization of a multimeric complex at a 1:2 ratio, as found for the plant GAD CaMBD with CaM (189). There is also evidence for target inactivation by occupying a ligand-binding site. This possible mechanism emerges from studies of plant CNGCs. Inactivation of CNGCs by Ca^{2+}/CaM occurs in plants and in other organisms. However, in plant CNGCs the binding sites for CaM and Cyclic Nucleotide Monophosphates (cNMPs) overlap (9), unlike in the mammalian proteins where their respective binding sites are well separated. In the plant CNGCs (of which 20 genes exist in the *Arabidopsis* genome) (118, 162), one of three conserved α-helices that constitute a conserved cNMP-binding domain retained the ability to bind CaM (9). Functional analysis of a plant CNGC in a heterologous system showed that Ca^{2+}/CaM inhibits cNMP-mediated channel activation (80). This functional model might also occur with KCBP, where Ca^{2+}/CaM occupies a domain that would otherwise bind microtubules (172).

Finally, note that because each plant possesses a large number of CaM and CML proteins (e.g., more than 50 in *Arabidopsis*), it is not always clear which of these proteins interacts in vivo with the identified CaMBPs. Interaction of different CMLs with the same CaMBPs may occur in vivo under different physiological situations. These interactions will likely result in different conformational changes of the CaMBPs, different sensitivities to Ca^{2+}, and different downstream cellular effects. A major challenge in the coming years will be to resolve these numerous types of interactions and the corresponding physiological responses.

CaMBD: calmodulin-binding domain

Calmodulin-binding domain: Target sites for CaM are defined as CaMBDs and are typically composed of a stretch of 12–30 contiguous amino acids with positively charged amphiphilic characteristics. Although their primary sequence varies, they usually form an α-helix upon binding to CaM. Recent structural studies revealed new and unexpected types of CaM-protein interactions and CaMBDs.

CNGC: cyclic-nucleotide gated channel

The Repertoire of CaM Target Proteins in Plants

Molecular approaches for screening cDNA expression libraries using labeled recombinant CaM as a probe (56, 110, 113) proved a powerful tool in identifying the cellular targets of CaM in plants as well as in other organisms. Screening cDNA libraries derived from various plant species, organs, and cell types, and from plants exposed to a variety of stimuli, revealed many clones encoding CaM-binding proteins

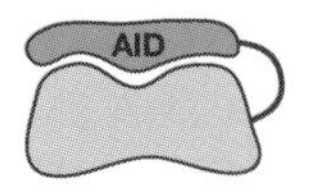

Inactive

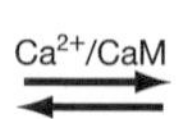

Active

Relief of auto-inhibition

(CaM:target ratio = 1:1)

b

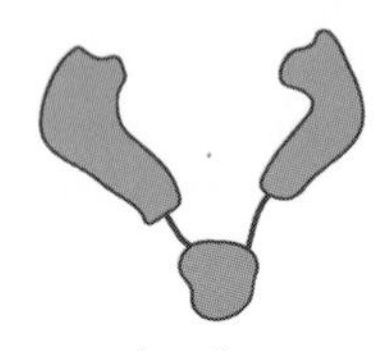

Inactive

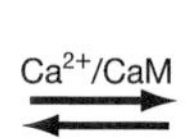

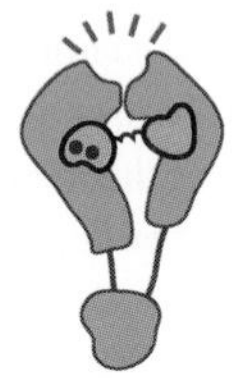

Active

Active site remodeling

(1:1)

c

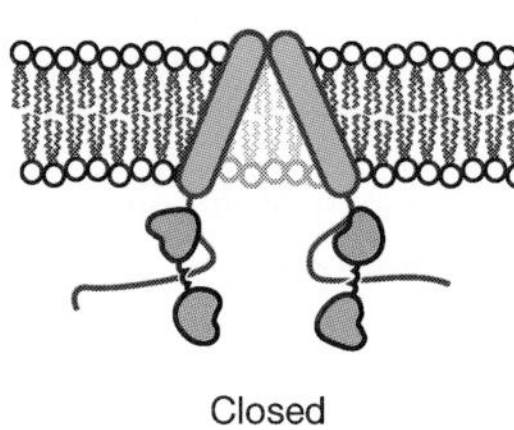

Closed

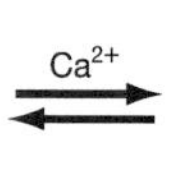

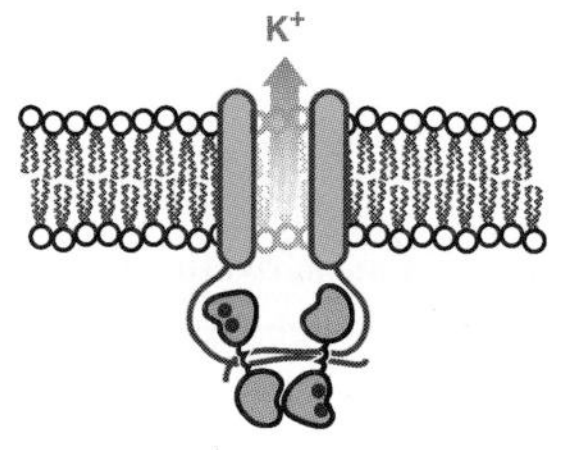

Open

Dimerization

(2:2)

d

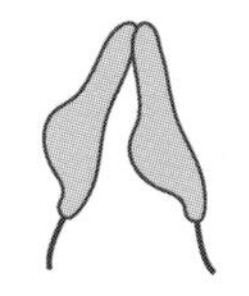

Inactive small complex

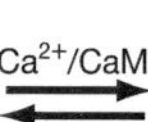

Active large complex

Stabilization of a multimeric complex

(1:2)

e

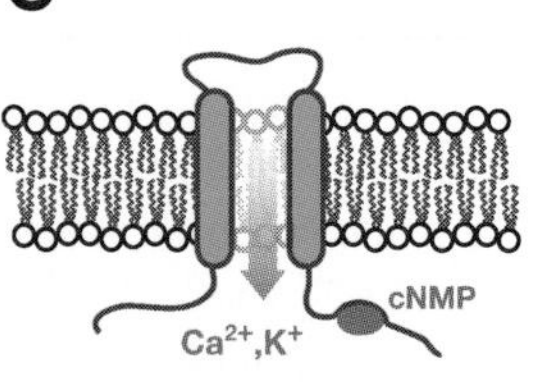

Open

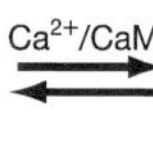

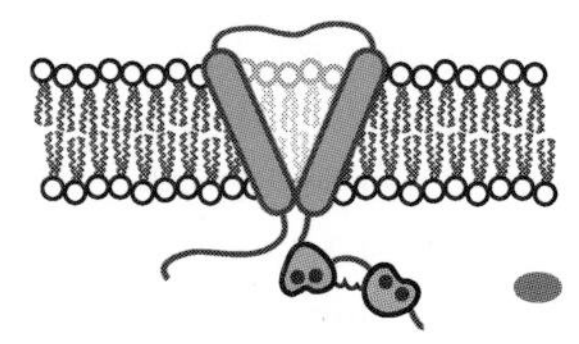

Closed

Inactivation by occupying a ligand-binding site

(1:1)

(CaMBP). Using bioinformatics tools has recently enabled the identification of several new genes encoding potential CaMBPs in the *Arabidopsis* genome (142). The reader is also referred to recent reviews on the roles of CaM as a Ca^{2+} signal transducer in plants (142, 155, 187, 195).

Table 1 (Follow the Supplemental Material link from the Annual Reviews home page at **http://www.annualreviews.org**) is an updated version of the list of CaMBPs described by Snedden & Fromm (155) summarizing the repertoire of plant proteins for which some evidence suggests regulation by (or association with) CaM. The table reflects the functional diversity of plant CaMBPs and reveals an involvement in regulation of metabolism, cytoskeleton, ion transport, protein folding, transcription, protein phosphorylation and dephosphorylation, phospholipid metabolism, and unknown functions. Considering that some of the proteins listed are encoded by relatively large gene families (e.g., 20 *CNGC* genes, 12 *ACA* genes in *Arabidopsis*), and given the potential for alternative splicing, the total number of different CaMBPs in a plant could easily reach several hundred.

Of the proteins listed in **Table 1** more than 20 are plant-specific CaMBPs with no obvious homologs in other organisms (proteins marked as *). This group includes, for example, the chimeric Ca^{2+}-resposive CaM-binding protein kinases (CCaMKs), a PP7 ser/thr phosphatase, the pollen-specific protein NPG1, and others. About 20 proteins have homologs in other organisms but are likely regulated by CaM only in plants (proteins marked as **). This group includes, for example, glutamate decarboxylase (although it may be CaM regulated in yeast), apyrase, and NAD kinase. The remainder are proteins whose homologs are regulated by CaM in other organisms as well as in plants (proteins marked as ***). Examples of this group include CaM-regulated plasma membrane Ca^{2+}-ATPases, protein kinases, and kinesin. In this review, we define plant-specific CaMBPs as those lacking a highly homologous protein in other organisms (*) or proteins whose probable homologs in other organisms do not bind CaM (**). The occurrence of many plant-specific CaMBPs likely reflects the extensive use of the CaM messenger system in plants, particularly in relation to responses to biotic and abiotic stress, and in development, as described later.

CALMODULIN AND PLANT RESPONSE TO ABIOTIC STRESS

Signaling pathways involved in the response to environmental stresses form interconnected networks in which Ca^{2+} plays a major role (reviewed in 94). Many abiotic stimuli can induce a transient cytosolic Ca^{2+} increase (reviewed in 76, 127, 174). Gene expression of Ca^{2+} sensors such as the CaMs and CMLs is often induced in response to various abiotic stresses (reviewed in 155), as has been observed for other Ca^{2+} sensors, such as CDPKs (116), calcineurin

Figure 2

Schematic presentation of the mechanisms of CaM activation or inhibition of target proteins. (*a*) Relieving autoinhibition: Ca^{2+}/CaM binding to the target displaces the autoinhibitory domain (AID) from the active site, thus allowing enzyme activation. (*b*) Active-site remodeling: Ca^{2+}/CaM binding confers a conformational change that stabilizes the active site and allows enzyme activation. (*c*) Target dimerization: Two CaM molecules interact with two K^+ channel domains. Upon Ca^{2+} binding the two channel domains dimerize, thus opening the channel for ion transport. (*d*) One Ca^{2+}/CaM binds two CaM-binding domains (CaMBDs) of petunia glutamate decarboxylase (GAD). One CaMBD interacts with the N-terminal lobe of CaM, the other with the C-terminal lobe. (*e*) Ca^{2+}/CaM binds to a site that coincides with the cyclic nucleotide monophosphate (cNMP)-binding site of plant cyclic-nucleotide gated channels (CNGCs). Thus, Ca^{2+}/CaM may displace bound cNMP and/or prevent binding of cNMP, causing the closure of the channel. Green, target proteins; pink, N-terminal lobe of CaM; orange, C-terminal lobe of CaM; blue, Ca^{2+} ions; blue arrows point to the direction of ion movement through channels; purple, cNMP (cAMP or cGMP).

B-like proteins (33), or annexins (104). For instance, three proteins encoded by genes induced by touch stimulation (*TCH* genes; i.e., touch-induced) encode CaM or CML proteins (26). Out of the conserved *CaMs* and the ~50 *CML* genes present in *Arabidopsis*, several seem well represented by ESTs in libraries generated from plants subjected to stress or hormone treatment (119). Collectively, these findings support the idea that CaM and CML proteins are involved in response to environmental stimuli likely via activation of specific CaMBPs. Evidence to support this hypothesis is discussed below.

Response to Osmotic Stress and Salinity

At the cellular level, one of the primary responses to environmental stresses such as an hyperosmotic stress (i.e., water deficit or high salinity) or a hypo-osmotic stress (i.e., an increase in environmental water potential) is a transient elevation of Ca^{2+} concentrations (reviewed in 174). Changes in Ca^{2+} levels can be monitored using transgenic plants expressing Ca^{2+} reporters such as aequorin. In these plants, repeated exposures to NaCl treatments provoke prolonged alterations of both cytoplasmic and apoplastic Ca^{2+} concentrations (57). This elevation of Ca^{2+} concentration is also observed in the cytoplasm when plants are exposed to hypo-osmotic stress (57). A stress-induced Ca^{2+} signal is likely detected by Ca^{2+} sensors such as annexins that are Ca^{2+}-dependent membrane-binding proteins (104) or proteins of the salt overly sensitive (SOS) signaling pathway in which the Ca^{2+}-binding protein SOS3 can activate the SOS2 protein kinase (reviewed in 64, 193, 194). CaM also contributes to the sensing of Ca^{2+} signals caused by osmotic or salt stresses. The transcription of *CaM* genes is induced when cultured cells from tomato (49) or young mung bean *Vigna radiate* (20) are exposed to salinity. Transgenic tobacco plants expressing a heterologous bovine CaM germinate faster on medium containing high levels of NaCl compared to the control (130). Thus, although CaM is probably a regulatory component during responses to osmotic and salinity stress, only a few CaMBPs involved in these pathways have been identified.

Some CaMBPs are induced by salt or osmotic stress treatment, based on gene expression data. In tobacco, a Ca^{2+}/CaM-binding kinase (NtCBK2) (82) is induced in high-salt conditions and GA treatment whereas other hormones (i.e., auxin, ABA), or exposure to cold or heat did not affect the expression of the *NtCBK2* (81). *In situ* hybridizations revealed that *NtCBK2* is expressed in tobacco anther, pistil, and embryo (81). In *Physcomitrella patens*, screening a cDNA library with radiolabeled CaM revealed a novel class of CaMBP (MCamb) sharing homologies with the mammalian ATP-sensitive potassium channels of the Kir family (161). The CaMBDs of both MCamb1 and MCamb2 were mapped and expression analysis revealed that MCamb1 is induced by mannitol and NaCl (161). There are other examples of CaMBPs whose transcription is induced by salt or osmotic stresses (**Table 1**), but the specificity of this activation is unclear and not exclusive to salinity or osmotic stress. For example, some members of the CAMTA (*CaM*-binding transcription activator) family are induced by salt but also by ethylene, methyl jasmonate, abscisic acid, H_2O_2, salicylic acid, UV, heat, and cold (185). As such, these proteins are probably involved in multiple pathways associated with environmental stress response.

The role of two CaMBPs has been specifically addressed in relation to salt or osmotic stresses. In *Arabidopsis*, *ACA4* encodes a vacuolar Ca^{2+}-ATPase containing a CaMBD in the N terminal part of the protein (59), unlike the animal counterpart where the CaMBD is located in the C terminal. The role of ACA4 was addressed in a yeast strain deficient for the transport of Ca^{2+} (i.e., lacking both endogenous Ca^{2+}-ATPases and the calcineurin regulatory subunit B). Only a truncated form of ACA4, lacking the N-terminal CaMBD, could restore the growth of the yeast mutant on medium containing low Ca^{2+}. ACA4 is an autoinhibited ATPase and, like the CaMBD, the AID is located in the N terminus. CaM binding likely relieves autoinhibition. Thus, the truncated form of ACA4 can restore the growth of yeast cells because the corresponding protein is constitutively active. Quantitative RT-PCR experiments revealed that the *ACA4* mRNA accumulates in a dose-dependent manner in *Arabidopsis* seedlings treated with increasing amounts of NaCl, raising the possibility that ACA4 could play a role in salt stress tolerance. Previous studies had shown that mRNA from Ca^{2+}-ATPases accumulates upon NaCl treatment in tomato (175) and soybean (37). To further address this question, the growth of the yeast mutant expressing the full-length or the truncated form of ACA4 was tested under different stress conditions. Both conferred protection against osmotic stress such as high levels of NaCl, KCl, and mannitol. The most adverse effect occurred in the deregulated form of ACA4 (i.e., lacking the N terminus). Thus, it is likely that during conditions of osmotic stress, ACA4 is stimulated by Ca^{2+}/CaM, leading to the rapid replenishment of Ca^{2+} stores such as the vacuole.

Recently, a new CaMBP (AtCaMBP25) involved in osmotic stress regulation was isolated from *Arabidopsis* (133). AtCaMBP25 is a nuclear-targeted protein that shows differential affinity to *Arabidopsis* CaM and CMLs. Northern blot analysis revealed that *AtCaMBP25* expression is induced by mannitol treatment, high salinity, cold, and seedling dehydration. The germination of transgenic seedlings overexpressing *AtCaMBP25* was inhibited on medium containing mannitol or salt, whereas antisense lines showed increased tolerance to these stresses. Similarly, root elongation of seedlings transferred from standard conditions to medium containing mannitol or NaCl was retarded for plants expressing the sense transgene and improved for antisense lines. However, the response to drought and freezing was not altered in these transgenic plants, implying that AtCaMBP25 plays a specific role in osmotic stress regulation. Overall, these data suggest that AtCAMBP25 is a negative regulator of osmotic stress responses.

Response to Cold and Heat Stresses

Ca^{2+} is also involved in plant responses to cold treatment. Tobacco seedlings expressing aequorin reveal that cold and wind can initiate specific Ca^{2+} signals that are spatially distinct (169). Genes encoding several CaMs and CMLs are induced by cold treatment and likely participate in translating these signals (26, 169). In addition, Ca^{2+} transporters have been implicated in cold response. Cold acclimation enhances the activity of a plasma membrane Ca^{2+}-ATPase in winter rye leaves (136), and in *Arabidopsis* a Ca^{2+}/H^{+} transporter (CAX1) controlled the expression of cold-induced genes (28). Compared to wild-type plants, *cax1* mutants showed enhanced freezing tolerance after cold acclimation. This directly correlated with the induction of *CBF/DREB1* genes that encode a family of transcription activators that mediate cold acclimation in *Arabidopsis*. CaM may function as a negative regulator of cold-induced gene expression given that cold-treated *Arabidopsis* plants overexpressing *CaM3* show decreased levels of *COR* (cold on regulated) transcripts (164). As noted above, some CaMBPs respond to multiple environmental stimuli, including cold treatment. Examples include proteins homologous to mammalian K^{+} channels (161) and the CAMTA transcription factors (185).

AID: autoinhibitory domain

Cells exposed to elevated temperatures respond by activating specific signals that mediate the heat shock response and mobilize heat shock proteins (HSPs). Ca^{2+} and CaM are important components in this process. In wheat, both the levels of *CaM* mRNA and CaM protein increased after heat shock (112). In addition, the expression of two *HSP* genes and the accumulation of the corresponding proteins were upregulated by the addition of Ca^{2+} and downregulated by a chelator of Ca^{2+}, Ca^{2+} channel blockers, or CaM antagonists (112). Therefore, changes in Ca^{2+} signals can mimic the effect of heat shock and induce the expression of *HSP* genes. Several CaMBPs function during the heat shock response. Two related high molecular weight FK506-binding proteins (FKBPs) (reviewed in 74) from wheat (FKBP73 and FKBP77) function in protein folding associated with their peptidyl prolyl *cis*-trans-isomerase activity. These FKBPs possess a predicted CaMBD (**Table 1**) and their homologs in *Arabidopsis* bind CaM (142). FKBP77 expression in wheat is induced by heat shock (97), and both wheat FKBPs form complexes with HSP90 (140). Recently, another FKBP from *Arabidopsis* (TWD1; i.e., AtFKBP42) bound both HSP90 and CaM (86), a property that is shared with human FKBPs (152). FKBP73 can also function in vitro as a molecular chaperone, a property also observed for some mammalian FKBPs (98). Transgenic wheat overexpressing FKBP77 present morphological abnormalities associated with higher levels of *HSP90* mRNA (99). In addition, cytosolic HSP70 from maize interacts with CaM, a phenomenon that inhibits its intrinsic ATPase activity (159). CaM also binds to a chloroplast chaperonin (182). Other CaMBPs of unknown function, whose expression is induced or suppressed by heat shock, have also been reported, namely TCB48 (114) or TCB60 (41, 113) in tobacco and its *Arabidopsis* homologs (142). Taken together, the data support a role for CaMBPs in modulating plant adaptation to temperature stress. However, more genetic and biochemical studies are needed to assign specific roles to the various CaMBPs and elucidate which pathways they participate in.

CaMBPs and Oxidative Stress

Plants exposed to environmental stresses often generate reactive oxygen intermediates (ROIs) whose levels must be tightly controlled to avoid cellular damage (reviewed in 7). There seems to be a relationship between the levels of ROIs and Ca^{2+} in the cell. In tobacco cell cultures expressing aequorin, a treatment with H_2O_2 triggers a biphasic Ca^{2+} elevation (103). Similarly, in *Arabidopsis* seedlings, Rentel & Knight (146) observed a biphasic Ca^{2+} signature composed of two independent peaks: an early one specific to cotyledons and a second one found in roots. More importantly, the magnitude of the Ca^{2+} signal correlates with the induction level of genes responding to oxidative stress and ROIs (146). It is thus likely that redox changes in the cell generate a Ca^{2+} signal that can modulate cellular responses to oxidative stress by activating specific genes, including certain CaMBPs. In addition, oxidative damage induced by heat stress in *Arabidopsis* seedlings is exacerbated by pretreating plants with CaM inhibitors (102), suggesting an important role for CaM in oxidative stress recovery. Some targets of CaM that might participate in oxidative stress response are described below.

In plants, at least one class of ROI-scavenging proteins seems to be regulated by CaM. Catalases (i.e., hydroperoxidases) are protective enzymes that degrade hydrogen peroxide (H_2O_2) to water and oxygen. By screening a cDNA expression library for CaMBPs, Yang & Poovaiah (186) isolated an *Arabidopsis* catalase isoform (AtCat3). The CaM-binding region of AtCat3 was mapped and the corresponding synthetic peptide binds a recombinant CaM in a Ca^{2+}-dependent manner (186). Additionally, a catalase from tobacco leaves could be purified using a CaM-sepharose column (186). The activity of the enzyme was stimulated about twofold by Ca^{2+}/CaM, but not by Ca^{2+} or CaM alone. The regulation of catalase by CaM is probably specific to plants because CaM had no effect on catalases from sources such as *Aspergillus niger*, human erythrocytes, and bovine liver (186). In addition, CaM was detected with specific antibodies in peroxisomes isolated from etiolated pumpkin cotyledons (186). Peroxisomes are specialized organelles involved in the catabolic oxidation of various biomolecules and the resultant H_2O_2 is consumed by peroxysomal catalases. Another class of ROI-scavenging enzymes, the superoxide dismutases (SODs), was suggested to be CaMBPs. A decade ago, SOD from maize germs was shown to reversibly bind CaM immobilized on a column in a Ca^{2+}-dependent manner (65). However, these findings were never confirmed in CaM-binding assays with recombinant SODs or SODs purified from other sources.

H_2O_2 has a dual role as both a strong oxidant that can damage cellular components and as a second messenger. Therefore, H_2O_2 levels must be tightly controlled. H_2O_2 as a signal mediates responses to biotic and abiotic stresses and can activate various pathways involved in adaptation to environmental stress (126). In plants, the transcription of genes encoding CaM-binding transcription factors is stimulated by H_2O_2. AtBTs (*Arabidopsis thaliana* BTB and TAZ domain proteins) represent a new family of proteins sharing homologies with transcription factors (51). All five AtBT proteins bind CaM and the transcription of some isoforms is modified within an hour in leaves exposed to H_2O_2 or salicylic acid. AtBT1 is localized to the nucleus and interacts with members of the fsh/Ring3 class of transcription regulators (51). Database searches reveal no homologies between *AtBT* and gene sequences of nonplant organisms. Similarly, some members of the CAMTA family of transcription factors are also induced rapidly by H_2O_2 (185). Plant CAMTAs resemble a group of putative transcription activators identified in the human genome (101). CAMTAs are specific to multicellular organisms and possess a conserved domain organization with a novel DNA-binding region termed CG-1, a TIG-like DNA-binding domain, and ankyrin repeats. Both human and *Arabidopsis* members of this family can activate transcription in yeast (24). Human CAMTA1 functions as a suppressor of neuroblastoma (87) and is transcriptionally regulated in a cell-cycle-dependent manner (123). However, CaM binding to human CAMTAs has yet to be demonstrated. The plant CAMTAs bind CaM in a Ca^{2+}-dependent manner (24, 183) and bind a DNA motif containing a CGCG element (185). No function has been attributed to BTs or CAMTAs; therefore, the link between these transcription activators and the perception of ROI contents is presently unclear. AtBTs and CAMTA transcription factors are induced not only by H_2O_2 but by many other treatments, including salicylic acid for AtBT (51) or UV-B, salt, ethylene, and wounding for AtCAMTAs (183). It is thus likely that these proteins are not directly involved in sensing ROIs, as are some transcription factors described in yeast (46) and bacteria (91), but rather are acting downstream of the ROI signal. Finding their gene targets should help to reveal their function in ROI and other signaling pathways.

Finally, CaM may participate indirectly in regulating ROI content through the CaM-regulated GABA-shunt metabolic pathway (see below). In *Arabidopsis*, mutants disrupted in the last enzyme (SSADH) of the pathway are more sensitive to environmental stress because they are unable to scavenge H_2O_2 (21). The reaction catalyzed by SSADH can provide both succinate and NADH to the respiratory chain. It was, therefore, hypothesized that the degradation of GABA could limit the accumulation of ROIs under oxidative stress conditions that inhibit certain enzymes of the TCA cycle.

GABA: γ-aminobutyric acid

Activation of Glutamate Decarboxylase by Environmental Stresses and Production of GABA

In plants, one of the best characterized Ca^{2+}/CaM regulated pathways involved in abiotic stress response is the GABA shunt (reviewed in 23). The first of the three enzymes of the pathway is GAD, which catalyzes the conversion of glutamate to GABA. Cloning the petunia *GAD* gene and characterizing the encoded enzyme as a Ca^{2+}/CaMBP (8, 15) revealed that GAD activity is upregulated by Ca^{2+}/CaM in many plant species, including *Vicia faba* (111), soybean (153), tobacco (16, 190, 191), petunia (156), rice (12), asparagus (36), and *Arabidopsis* (167, 196). These observations provide a model to explain the rapid stimulation of GAD activity in response to biotic and abiotic stresses, which elicit changes in cytosolic Ca^{2+} concentrations (reviewed in 151, 154). Thus, the production of GABA via the stimulation of GAD by CaM is directly associated with cytosolic Ca^{2+} fluxes.

The interaction between GAD and CaM seems to be plant specific because none of the mammalian GADs identified possess a Ca^{2+}/CaMBD. Recently, a rice GAD was described (OsGAD2) that lacks a CaMBD (1). The physiological relevance of GAD activation by CaM was first addressed in tobacco by using transgenic plants ectopically expressing either GAD or a truncated GAD that could not bind CaM (16). Disruption of the CaMBD of GAD resulted in constitutive GAD activity (i.e., Ca^{2+}-independent), abnormal steady-state levels of glutamate (low) and GABA (high) and aberrant plant development (16). Consequently, CaM tightly controls the activity of GAD in vivo and this regulation is necessary for normal plant development. Another approach to clarify the role of CaM-regulated GABA production in plants is by using *Arabidopsis* T-DNA insertion mutants for the *GAD* genes. For example, disrupting the root-specific *GAD1* gene revealed that it plays a major role in GABA synthesis under normal growth conditions and in response to heat stress (22). However, the roles of GAD and GABA in response to environmental stresses remain to be clarified.

Tolerance to Xenobiotic Compounds

CaMBPs expressed ectopically in transgenic plants can change the sensitivity of the plant to toxic compounds. For example, CNGCs may serve as entry pathways for certain heavy metals. CNGCs contain six putative transmembrane domains, a pore region, a cyclic-nucleotide-binding domain, and a CaMBD (reviewed in 162). These proteins are similar to the mammalian cyclic nucleotide gated nonselective cation channels, which are activated by cyclic nucleotides and involved in Ca^{2+} signal transduction (reviewed in 88). Tobacco transgenic plants overexpressing a member of the CNGC family (i.e., NtCBP4) exhibited hypersensitivity to Pb^{2+}. These transgenic lines were indistinguishable from WT plants under normal growth conditions but accumulated Pb^{2+} (11). Furthermore, transgenic lines expressing a truncated form of NtCBP4, lacking the CaMBD and part of the cyclic nucleotide-binding domain, displayed improved tolerance to Pb^{2+} associated with a lower Pb^{2+} content in the plant tissues as compared to the WT (160). A similar phenotype was observed in *Arabidopsis* mutant plants with a T-DNA inserted in the *AtCNGC1* gene, an ortholog of *NtCBP4* (160). Thus, NtCBP4 and AtCNGC1 are probably Ca^{2+}-permeable channels providing a route for Pb^{2+} entry across the plasma membrane. Nevertheless, transgenic plants overexpressing modified versions of CNGC proteins provide new tools to manipulate plant tolerance to heavy metals (10).

Apyrases are plasma membrane-associated enzymes with their hydrolytic activity directed to the extracellular matrix where they hydrolyze most nucleoside tri- and diphosphates. Functions of ecto-apyrases are not well defined in plants but are related to the hydrolysis of extracellular ATP and ADP. Although apyrases from animals have not been described as CaMBPs, the activity of an endogenous pea apyrase (psNTP9) is stimulated by Ca^{2+}/CaM (31), and the corresponding recombinant protein binds CaM in a Ca^{2+}-dependent manner (79). In *Arabidopsis*, only one of the two related apyrase genes cloned (i.e., *AtAPY1*) encodes a CaMBP (157). Transgenic *Arabidopsis* plants overexpressing the pea apyrase are more resistant to toxic concentrations of cycloheximide (163), plant growth regulators such as cytokinin (163), and herbicides of different chemical classes (176). WT plants grown in the presence of apyrase inhibitors become more sensitive to the herbicides (176). Thus, apyrase is involved in a multidrug resistance mechanism. Because the pea apyrase used in these studies is CaM regulated, it raises the interesting possibility that CaM may regulate extracellular apyrase activity. In addition, AtMRP1, a vacuolar multidrug resistance-associated (MRP/ABCC)-like ABC transporter, was recently shown to possess a functional CaMBD (60) that can also interact with the FK506-binding protein TWD1. CaM had no effect on the binding of TWD1 to the transporter in vitro, thus the mode of action of CaM versus TWD1 on AtMRP1 remains to be elucidated. Interestingly, TWD1 itself has a putative CaMBD (86) and interacts with two other multidrug resistance-like transporters (AtPGP1 and AtPGP19). *Arabidopsis twd1* mutants present a pleiotropic phenotype characterized by reduction of cell elongation and disorientated growth of all plant organs (61). It is thus likely that this protein is involved in regulating multidrug resistance-associated transporters although the biochemical role of CaM in this process is unclear.

CALMODULIN AND PLANT RESPONSE TO BIOTIC STRESS

Ca^{2+} likely functions as a second messenger during plant responses to biotic stimuli such as pathogen attacks. This section discusses recent findings that support a role for Ca^{2+} sensors such as CaM, CLMs, and their downstream targets in regulating the signaling events that follow biotic stimulus perception. For related discussions on CDPKs or reviews on plant host-pathogen interactions, the reader is referred to a number of recent articles (116, 128, 177).

When a plant cell detects a pathogen invader, a cascade of defensive events is triggered. Typical rapid responses to pathogen attack include ion fluxes, production of ROIs, cell wall fortification, synthesis of defense compounds, and changes in gene expression (reviewed in 40, 128). Ca^{2+} influx into the cytosol is among the earliest of these responses and likely serves as a messenger to regulate specific downstream defense pathways. Numerous studies have shown that influx across the plasma membrane from extracellular stores is a key source of Ca^{2+} in response to elicitors or pathogens (19, 62, 66, 109, 178, 197), but the participation of internal Ca^{2+} stores is also likely important (92, 103, 105). Note, however, that only a few of these studies (45, 66, 178) have been conducted in planta using living pathogens and although much has been learned using microbial elicitor preparations (or analogs) and cell cultures, we are a long way from understanding the dynamics of Ca^{2+} signaling during host-pathogen interaction. Nevertheless, one of the most intriguing findings that has emerged from these studies is that the cytosolic Ca^{2+} signal, which is evoked by pathogen attack, differs considerably from the spiking and oscillation patterns often seen during response to abiotic stimuli such as cold, drought, or salinity (reviewed in 76, 127, 174). In general, most studies suggest that the elevation in cytosolic Ca^{2+} that rapidly follows pathogen (or elicitor) perception is sustained for a prolonged period although it may be biphasic in character (19, 45, 66, 178). Importantly, the sustained level of cytosolic Ca^{2+} was not observed during compatible host-pathogen interactions, suggesting that early recognition of the pathogen is a prerequisite for this particular Ca^{2+} signature. How such an apparently simple signature contributes, if at all, to the regulation of the myriad downstream events such as the oxidative burst, the HR, the synthesis of defense compounds, and the induction of defense-related genes, remains largely unknown. And although Ca^{2+} has been implicated in many of these processes, typically through the use of pharmacological agents such as Ca^{2+} ionophores or chelators, the evidence for a regulatory role for Ca^{2+} is largely correlative. However, in addition to findings from Ca^{2+} imaging research, there is a growing body of data from gene expression analyses, biochemical studies, and reverse genetics to strongly suggest that Ca^{2+} and Ca^{2+} sensors and their downstream targets are important components in plant defense signal transduction. The following discussion focuses on recent work on CaMs, CMLs, and CaM targets during pathogen response. For the CML family, attention focuses on *Arabidopsis* CMLs [as annotated by McCormack & Braam (119)] and their putative homologs in other plants. Most examples are drawn from studies using microbial pathogens or elicitors, but a few examples involving herbivory response are also presented.

HR: hypersensitive response

CaMs and CMLs in Pathogen Response

A number of studies have shown the involvement of CaMs and CMLs during pathogen response. Two soybean CaMs, SCaM4 and SCaM5, which possess about 78% identity to a conserved CaM such as *Arabidopsis* CaM2, are rapidly induced by fungal elicitor preparations and this induction can be mimicked

SAR: systemic acquired resistance

PTGS: post-transcriptional gene silencing

using Ca^{2+} ionophores or prevented by Ca^{2+} chelators (75). Elegant functional analyses confirmed the involvement of these proteins in pathogen response by showing that overexpression in tobacco resulted in enhanced levels of SAR-associated transcripts, spontaneous formation of necrotic lesions, and increased broad-spectrum pathogen resistance (75). Several *CaMs* were upregulated in tobacco carrying the *N*-resistance gene by infection with tobacco mosaic virus (TMV) (180). Among the TMV-induced CaMs was a divergent isoform, *NtCaM13*, which is a putative ortholog of the pathogen-inducible soybean CaMs, *SCaM4* and *SCaM5*, described above. Although post-translational levels of several conserved CaMs were induced by wounding treatments, NtCaM13 was repressed, suggesting stimulus-specific responses. Another pathogen-inducible CML is Hra32 from *Phaseolus vulgaris* (84). Hra32 is about 53% identical to an *Arabidopsis* NaCl-inducible protein, AtCP1 (84), and about 23% identical to the conserved *Arabidopsis* CaM2. When plants were inoculated with an incompatible bacterial pathogen, *Hra32* transcript levels were temporally correlated with the onset of HR symptoms. Infection with a compatible pathogen resulted in a different temporal expression pattern of *Hra32* that was correlated with cell death associated with disease, whereas general stimuli such as wounding evoked transient *Hra32* expression (83). CaM transcript levels were induced in tomato in response to wounding or the wound signal molecule systemin and were constitutively high in transgenic plants overexpressing systemin (17). Expression-profiling studies on *Arabidopsis* identified a CML (At3g51920, herein referred to as CML9) as one of the earliest genes induced during a bacterial pathogen challenge (45). CML9 was also induced by other stimuli but in a temporally distinct manner. These authors speculated that the most rapidly induced genes may be part of an encoding signature of expression for infection that helps to establish the pathogenic context of the infection. The cell somehow decodes this information and responds accordingly.

Another pathogen-induced CML was identified in a high-throughput expression study using tobacco (carrying the tomato *Cf-9* resistance gene) that was challenged with fungal elicitor preparations. Among the genes rapidly induced by elicitors from an incompatible *Cladosporium fulvum* race was a *CML* designated *ACRE-31* (53). The ACRE-31 protein shows about 30% identity to *Arabidopsis*-conserved CaM2 and about 60% identity to both *Arabidopsis* CML42 and CML43. A similar transcript-profiling study, using the tomato and *Pseudomonas syringae* pv tomato host-pathogen system, identified a probable ortholog of *ACRE-31*, *APR134*, as an early upregulated gene during pathogen response (122). Functional analysis of *APR134* in tomato and *CML43* in *Arabidopsis* confirmed a role for these genes during the plant immune response. Suppression of *APR134* expression using virus-induced gene silencing (VIGS) results in loss of resistance to incompatible *P. syringae* pv tomato, whereas overexpression of *CML43* in *Arabidopsis* accelerates the HR (W.A. Snedden, unpublished data).

Another interesting study that shed light on the roles of CaMs during pathogen response describes a novel CML from tobacco that functions as a negative regulator of PTGS (6). PTGS is part of the defense strategy plants use against viral pathogens. rgs-CaM (regulator of gene silencing CaM) is about 26% identical to conserved *Arabidopsis* CaM2 at the amino acid level and about 50% identical to *Arabidopsis* CML37. Interestingly, the unusual N-terminal extensions of rgs-CaM and CML37 are quite different from one another. Tobacco rgs-CaM was isolated by a two-hybrid screen using the helper component protease (HC-Pro) from a plant potyvirus as bait. In tobacco, expression of HC-Pro from a transgene, or viral infection, induces the expression of rgs-CaM and suppresses PTGS, suggesting that rgs-CaM has a negative regulatory role in PTGS and is subverted during viral attack (6).

Collectively, the studies described above implicate CaM and related CMLs in plant responses to pathogens. Examining accessible databases from global expression studies suggests that many unstudied CMLs are induced during plant-pathogen interaction (W.A. Snedden, unpublished data), and these data should help direct future research efforts. The seemingly paradoxical concept of inducing the expression of a sensor for a signaling molecule, Ca^{2+}, whose presence is rapidly evoked and transitory, has not gone unnoticed. It has been suggested that increased levels of Ca^{2+} sensors during stress response prime cells for subsequent or long-term stimuli or create a stoichiometric level of the sensors to regulate downstream targets (195). Trewavas (166) suggested that the changes occurring in the levels of Ca^{2+} sensors upon the perception of stress stimuli is comparable to the process of learning in other organisms. A recent study in animal cells suggests that effectors compete for CaM in vivo and this competition is coordinated to simultaneously regulate the activities of various targets (165). Thus, given that CaMs and, presumably, CMLs are essentially noncatalytic regulators, identifying and characterizing their targets has become an important challenge. Our understanding of the role of CaM targets in pathogen response has made considerable progress in recent years.

CaM Targets and Pathogen Response

NAD kinase (NADK), which catalyzes the phosphorylation of NAD to NADP, was among the first CaM targets implicated in signaling during pathogen response in plants (70). NADKs are likely found in all organisms and were recently cloned from a number of prokaryotes and eukaryotes (see 168 and references therein). It is unclear whether other eukaryotes possess CaM-binding NADKs. Although the roles of NADKs remain obscure, their product, NADP, may indirectly contribute to the NADPH pool utilized by the important defense enzyme NADPH oxidase (135). A role in defense signaling for NADK was demonstrated (70). Transgenic tobacco plants overexpressing a mutant CaM, unable to be methylated at Lys-115 and that hyperactivates NADK, showed enhanced levels of NADPH and an accelerated production of ROIs in response to various elicitors or an incompatible bacterial pathogen (70). Recently, cDNAs were isolated from *Arabidopsis* encoding two isoforms of NADK, one of which is a CaM-binding isoform that is unique to plants (168). This work should help facilitate future transgenic studies aimed at assessing the roles of NADKs in pathogen response. Plant CaM-regulated catalases (186) and NADKs may function in concert to modulate ROI levels during Ca^{2+}-mediated responses to external stimuli.

GAD is a well studied CaM-activated enzyme whose product, GABA, is rapidly synthesized in plants in response to various abiotic stimuli (discussed above). Because GABA is an inhibitory neurotransmitter in insects it may also serve as a chemical deterrent against herbivory. Crawling or feeding by herbivorous insects stimulated rapid GABA production in tobacco and soybean (25) and transgenic plants that hyperaccumulate GABA received less herbivory than control transgenics or wild-type plants (117).

Further support for a role of CaM targets in pathogen response comes from the recent identification of MLO as a CaMBP (89, 90). The *Mlo* gene encodes a seven-transmembrane receptor-like protein that is unique to plants. Although related to mammalian G-protein-coupled receptors, MLO functions independently of heterotrimeric G-proteins (90). The *mlo* mutation is particularly interesting because it confers broad spectrum resistance in barley against powdery mildew disease. This suggests that MLO negatively regulates this defense pathway in wild-type plants. Expression of the rice homolog of *Mlo*, *OsMlo*, is significantly induced by a fungal pathogen or various defense signaling compounds and an influx of Ca^{2+} is necessary but not sufficient for this induction (89). Importantly, MLO mutations that impair its ability to bind CaM compromise its defense suppression activity in vivo (90). Future studies should focus on elucidating the biochemical activity of MLO and assessing whether other members of the MLO family are also CaM targets.

In *Arabidopsis*, the expression of various *CAMTA* genes is differentially induced by external stimuli, including signaling molecules such as ethylene, salicylic acid, and hydrogen peroxide, known to play a role during pathogen response (139, 185). A functional role for CAMTAs in pathogen response has not yet been established, but putative target genes, whose promoters contain CAMTA-binding *cis*-elements, include Ca^{2+}-binding proteins and other targets known to participate in stimulus-response pathways (185).

An important recent development in the role of CaM targets during pathogen response was the isolation of an *Arabidopsis* nitric oxide synthase (AtNOS1), which is a novel plant CaMBP (67). Although AtNOS1 is a CaM target that catalyzes the synthesis of NO using L-Arg as a substrate, it does not share significant sequence identity with mammalian NOS counterparts and thus represents a new class of NOS enzymes. NO is an important signaling molecule in animals and recently received similar attention in plant systems. AtNOS1 is involved in hormonal signaling associated with growth and development (67), and a role during pathogen response is emerging. Previous studies demonstrated a role for NO in the HR and the induction of defense genes during pathogen response (47, 48, 52), but the source of the NO remained obscure, especially given the absence of an obvious NOS ortholog in the *Arabidopsis* genome. Thus, isolating a plant NOS was a major breakthrough. NO production induces Ca^{2+} influx from intracellular stores during response to a fungal elicitor, suggesting a positive feedback of sorts for this CaM-dependent enzyme (100). Key evidence that AtNOS1 is an important component in plant defense was recently provided by a study that examined the effects of bacterial lipopolysaccharides on NO production and plant immunity (191a). This work revealed that the strong, rapid burst of NO observed during exposure of *Arabidopsis* to bacterial lipopolysaccharides was generated by AtNOS1 and was responsible for regulating downstream genes involved in pathogen response. Furthermore, mutant plants lacking AtNOS1 were more susceptible to pathogenic bacteria (191a). Overall, the data suggest that AtNOS1 plays an important role in plant immunity. However, there is still much to learn about the role of NO in different host-pathogen systems. Note that in addition to AtNOS1-mediated NO synthesis, there are other enzymatic and nonenzymatic mechanisms of generating NO in plants (99a). Consequently, it will require a multidisciplinary approach to elucidate the contributions of the various sources to NO-based defense signaling in plants.

Several other CaMBPs were recently identified where a role in pathogen response was suggested. *PICBP* encodes a plant-specific CaMBP of unknown biochemical function that is induced by incompatible bacterial strains as well as by a number of defense-related signaling molecules (141). It is noteworthy that PICBP is unique in possessing four Ca^{2+}-dependent CaMBDs. Because PICBP is encoded by a single gene in *Arabidopsis*, reverse genetic studies should help reveal its cellular role in defense (141). In a similar study, Ali et al. (2) identified additional CaM targets whose transcript expression patterns were responsive to pathogen infection or related stimuli. Several genes encoding isoforms of bean PvCBP-60 were strongly induced by incompatible bacteria (2). The CMP-60 family (also known as TCBP60s, see **Table 1**) is made up of plant-specific CaMBPs with no homology to any known protein. Although their biochemical function remains unknown, they are consistently among the most abundant proteins isolated in screening methods aimed at identifying novel CaM targets.

A CaM-binding cyclic-nucleotide gated channel, DND1 (AtCNGC2), was also recently shown to participate in defense signaling (38). Plants carrying a mutation in this gene fail to produce the HR when challenged with avirulent *P. syringae* but display gene-for-gene resistance and constitutive SAR (38). The underlying mechanism of the *dnd1* phenotype and whether DND1 participates directly or indirectly in pathogen response remains unclear but it seems to be positioned downstream of salicylic acid in the signal cascade. Complementation experiments using wild-type *AtCNGC2* restored the normal phenotype of *dnd1* mutants and it would be interesting to extend these studies using *AtCNGC2*-carrying mutations in the CaMBD (9) to assess the functional importance of this domain. Gene expression studies provide additional support for a role of specific CNGCs in defense response (2). A second member of this family, AtCNGC4, also participates in pathogen-response signaling (14). Plants carrying mutations in either *HLM1* (which encodes AtCNGC4) or *DND1* are phenotypically similar but there are some key differences in their responses to pathogen infection and in the expression patterns of these genes in wild-type plants, suggesting that although both function during pathogen response, they do not play redundant roles (14). Another CaM-regulated transporter, *SCA1*, which encodes a Ca^{2+}-ATPase, is induced in response to a fungal elicitor (37). Given the diversity of CaM-regulated Ca^{2+} pumps and channels in plants (see **Table 1**), it will be a challenge to determine which members contribute to the changes in Ca^{2+} dynamics observed in response to biotic stress.

Another novel CaMBP recently implicated in pathogen response is the MAPK phosphatase, NtMKP1 (179). MAPKs were previously shown to participate in defense signaling and it is likely that their activity is regulated by specific phosphatases (177). Consistent with this suggestion, transgenic plants overexpressing NtMKP1 showed reduced activity of several defense-related MAPKs in response to wounding. Although NtMKP1 showed differential binding to several CaMs, the role of CaM binding to NtMKP1 remains under investigation. It will be interesting to see if any of the other recently described CaM-binding phosphatases and kinases (see **Table 1**) have targets involved in pathogen response. Recently, *DMI3*, an important gene required for symbiotic nodule development in legumes, was shown to encode a kinase of the CCaMK class (120). This study reflects an importance of CaM targets in other aspects of plant-microbe communication beyond that of classic host-pathogen interaction.

It is likely that rapid advances in high-throughput transcript expression analyses will yield even more clues about which CaM targets participate in host-pathogen interaction. In addition, as proteomic methods evolve, it may soon be possible to screen an entire plant proteome for CaM targets, as was done with yeast (192). It will be particularly important to identify the targets of the unique CMLs to help understand why plants have evolved such a sophisticated array of these Ca^{2+} sensors. In general, an emphasis on functional analysis using the combined tools of genetics, biochemistry, and molecular and cell biology is needed to unravel the complexity of Ca^{2+}-mediated pathogen response.

CALMODULIN AND PLANT DEVELOPMENT

Much of our knowledge regarding a role for CaM in plant development has come from genetic studies. In particular, research on mutants with aberrant pollen tube growth or trichome morphogenesis has demonstrated that CaMBPs are key components in these developmental processes. These and other findings linking CaMBPs and plant development are discussed below.

CaMBPs Responding to Hormonal Treatment

CaM interacts with at least one group of proteins directly involved in response to hormonal changes: the small auxin up RNA (SAUR) proteins encoded by short unstable transcripts that accumulate rapidly and specifically after auxin treatment. By screening a cDNA library with a radiolabeled CaM, Yang & Poovaiah (184) isolated a cDNA encoding a SAUR protein (ZmSAUR1). CaM-binding SAURs are widely distributed in the plant kingdom, although their physiological role is unclear. *Northern* analysis confirmed that *ZmSAUR1* is an early auxin response gene, induced within 10 minutes after auxin treatment. Subsequently, another member of the SAUR family (ZmSAUR2) was isolated and shown to bind CaM (93). In addition, other genes encoding CaMBPs, such as members of the CAMTA transcription factor family, are induced by ethylene (183).

CaM and CaMBPs Involved in the Development of Polarized Cells

CaM and specific CaMBPs are involved in the development of at least two types of polarized cells specific to plants: pollen tubes and trichomes. Ca^{2+} fluxes are required for the normal growth and elongation of pollen tubes; furthermore, the Ca^{2+} gradient oscillations in pollen are very well documented (reviewed in 78). Unlike Ca^{2+}, CaM and CaM mRNA distribute uniformly throughout the pollen cell, as demonstrated with fluorescently labeled CaM or CaM RNA microinjected in living tubes and monitored by confocal microscopy (121). However, using a fluorescent analogue of CaM revealed that the activity of CaM is higher in the apical tip of the tube and oscillates, following the distribution and oscillation pattern of cytosolic Ca^{2+} (137).

There are many potential CaM targets that might be required for pollen tube growth, several of which have been identified by reverse genetic approaches in *Arabidopsis*. One member of the family of autoinhibited Ca^{2+}-ATPases (ACA9) is of particular importance for normal pollen tube growth and fertilization (150). Three alleles of the *aca9* mutation present the same phenotype: a reduced set of seeds due to the poor growth of *aca9* pollen tubes and a low frequency of fertilized ovules. When pollen tubes reach an ovule, more than 50% fail to develop an embryo. ACA9 is specifically expressed in the male gametophyte (150). One hypothesis to explain the phenotype of *aca9* mutants is that ACA9 is required to maintain the oscillations of Ca^{2+} observed in the tips of pollen tubes. Other CaMBPs specifically expressed in pollen play an essential role in pollen germination. The maize pollen calmodulin-binding protein (MPCBP) and NPG1 (no pollen germination1) in *Arabidopsis* are members of a plant-specific family of CaMBPs containing tetratricopeptide repeats (TPR), a type of protein-protein interaction domain (63, 149). Although the exact function of these proteins is unknown, the *npg1* mutant allele is not transmitted through the male gametophyte because *npg1* pollen germination is arrested (63). Another gene necessary for pollen tube growth encodes a transaminase (GABA-T) that degrades GABA. Although not a CaMBP, GABA-T is one component of the GABA-shunt pathway that, in plants, is CaM regulated (reviewed in 23). The *pollen-pistil interaction2* (*pop2*) mutant is deficient in a GABA-T, which ultimately leads to growth inhibition and misguidance of *pop2* pollen tubes in *pop2* pistils (131). In WT plants, the level of GABA increases along the path through which pollen tubes travel, whereas in *pop2* plants this gradient of GABA is disturbed (131). Thus, the growth of pollen tubes depends on GABA production and degradation, a process at least partially controlled by GAD, a CaMBP. The disruption of apyrases (see above) also inhibits pollen germination in *Arabidopsis* (158). T-DNA knockouts were isolated for each of the two *Arabidopsis* apyrases but they lacked a visible morphological change. However, pollen from double-knockout mutants failed to germinate. The double knockout can be complemented by either one of the two apyrases: AtAPY1, which is a CaMBP, and AtAPY2, which is not. Therefore, although apyrases are clearly needed for the development of pollen tubes, the role of CaM in the regulation of AtAPY1 activity remains unclear. In addition, CaM interacted with the cytosolic kinase domain of the S locus receptor kinase (SRK) involved in recognizing self-pollen during the self-incompatibility response (170). A CaMBD was identified in SRK, but the role of CaM in self-pollen recognition remains to be clarified.

CaM binds a kinesin protein that interacts with microtubules to play a major role in trichome morphogenesis. In *Arabidopsis*, *ZWICHEL* was identified by genetic screens for altered trichome morphology. *zwi* mutants are affected in trichome stalk expansion and branching, but not in other aspects of plant development. The ZWICHEL protein is a kinesin-like CaMBP (AtKCBP). Ca^{2+}/CaM inhibits the motor activity of AtKCBP and its interaction with microtubules (reviewed in 138). KCBP also plays a major role in cotton fiber elongation (85, 134). Kinesins from other organisms such as yeast, *Caenorhabditis*, or human do not contain a CaMBD. However, sea urchin does possess a KCBP with a CaMBD at the C terminus (148). The mechanism by which CaM inhibits plant KCBPs was addressed in a recent study where the CaMBD of a plant KCBP was fused to various kinesins from *Drosophila* that do not normally bind CaM (144). CaM was able to regulate, in a Ca^{2+}-dependent manner, both the binding to microtubules of these chimeric kinesins and their microtubule-stimulated ATPase activity. Finally, the crystal structure of a KCBP from potato was

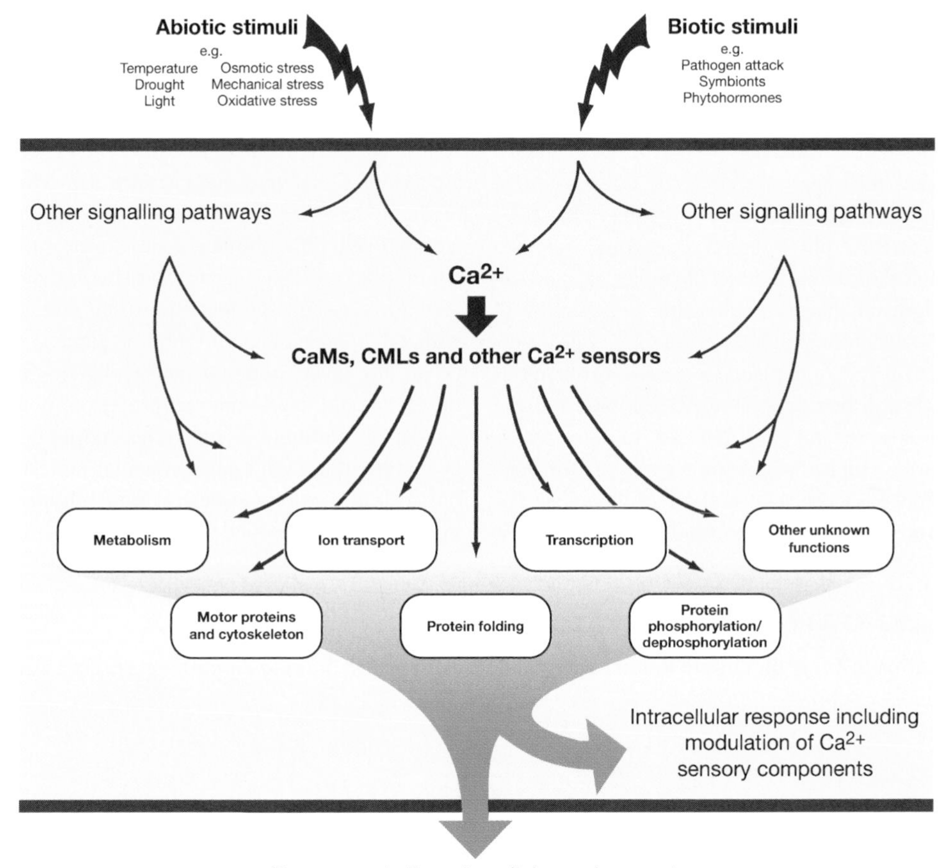

Figure 3

Schematic presentation of stimulus-response signaling mediated by Ca^{2+}/CaM. Various biotic and abiotic stimuli evoke a transient change in cytosolic and/or organelle Ca^{2+} levels, which are transduced by CaMs, CaM-like proteins (CMLs), and other Ca^{2+} sensors. CaMs and CMLs interact with downstream effectors that modulate numerous biochemical and cellular functions (see **Table 1** in the Supplemental Material link in the online version of this chapter or at http://www.annualreviews.org/) for details of the proteins involved in each depicted biochemical cellular activity). Exogenous stimuli also elicit signaling pathways independent of Ca^{2+}. These may interact with the Ca^{2+} signaling pathways at various points of the transduction cascades. Physiological response to external stimuli is comprised of the combined molecular and biochemical changes in both the intracellular and extracellular regions and includes changes in the Ca^{2+} sensing machinery itself.

resolved, suggesting that Ca^{2+}/CaM blocks the microtubule-binding sites on KCBP (172). KCBP is probably only one partner of a multiple protein complex because it also interacts with a protein kinase (43) and two proteins involved in trichome cell morphogenesis: ANGUSTIFOLIA (55) and KIC (143). KIC possesses a Ca^{2+}-binding EF-hand motif and both CaM and KIC inhibit KCBP activity in a Ca^{2+}-dependent manner, supporting a key role for Ca^{2+} in trichome morphogenesis.

CONCLUDING REMARKS AND FUTURE DIRECTIONS

The past decade has been fruitful in identifying the repertoire of plant CaMs, CMLs, and CaMBPs. However, only in very few cases have their physiological roles been revealed and, thus, more functional studies are needed. **Figure 3** depicts a schematic presentation of the main components of the Ca^{2+}/CaM messenger system in plants. Future research in this field will be driven by the developing tools of reverse and forward genetics, bioinformatics, various high-throughput techniques for gene expression analyses, and emerging technologies in proteomics to study protein expression and interactions. There should also be emphasis on real-time in vivo cellular studies of spatial and temporal protein dynamics and protein-protein interaction using microscopy techniques with fluorescently tagged CaMs, CMLs, and CaMBPs. Given that stimulus-response specificity is likely governed in part by fine-tuning the localization and expression of proteins that constitute the CaM messenger system, the challenge ahead is to pinpoint the specificities within the system and the interconnections with other signaling pathways. The dynamics and regulators of intracellular Ca^{2+} signals should also be further studied. Mathematical modeling of Ca^{2+} dynamics and response circuitry will help integrate the various aspects of signal transduction into a mechanistic model of how information is processed within a cell. The ultimate goal is to take the combined knowledge from related areas and, through a systems-biology approach, emerge with an understanding of how a plant perceives any given stimulus and reprograms its metabolic and developmental profiles to cope or adapt accordingly. As researchers target that goal, agricultural and environmental biotechnology will continue to benefit from advances in basic signal transduction research.

ACKNOWLEDGMENTS

We thank Zohar Bloom and Refael Ackermann for assistance in drawing the structural CaM-target models.

LITERATURE CITED

1. Akama K, Akihiro T, Kitagawa M, Takaiwa F. 2001. Rice (*Oryza sativa*) contains a novel isoform of glutamate decarboxylase that lacks an authentic calmodulin-binding domain at the C-terminus. *Biochim. Biophys. Acta* 1522:143–50
2. Ali GS, Reddy VS, Lindgren PB, Jakobek JL, Reddy AS. 2003. Differential expression of genes encoding calmodulin-binding proteins in response to bacterial pathogens and inducers of defense responses. *Plant Mol. Biol.* 51:803–15
3. Allen GJ, Chu SP, Harrington CL, Schumacher K, Hoffmann T, et al. 2001. A defined range of guard cell calcium oscillation parameters encodes stomatal movements. *Nature* 411:1053–57
4. Allen GJ, Chu SP, Schumacher K, Shimazaki CT, Vafeados D, et al. 2000. Alteration of stimulus-specific guard cell calcium oscillations and stomatal closing in *Arabidopsis det3* mutant. *Science* 289:2338–42

5. **Allen GJ, Schroeder JI. 2001. Combining genetics and cell biology to crack the code of plant cell calcium signaling. *Sciences STKE* 13:1–7**
6. Anandalakshmi R, Marathe R, Ge X, Herr JMJ, Mau C, et al. 2000. A calmodulin-related protein that suppresses posttranscriptional gene silencing in plants. *Science* 290:142–44
7. Apel K, Hirt H. 2004. Reactive oxygen species: metabolism, oxidative stress, and signal transduction. *Annu. Rev. Plant Physiol. Plant Mol. Biol.* 55:373–99
8. Arazi T, Baum G, Snedden WA, Shelp BJ, Fromm H. 1995. Molecular and biochemical analysis of calmodulin interactions with the calmodulin-binding domain of plant glutamate decarboxylase. *Plant Physiol.* 108:551–61
9. Arazi T, Kaplan B, Fromm H. 2000. A high affinity calmodulin-binding site in a tobacco plasma-membrane channel protein coincides with a characteristic element of cyclic nucleotide-binding domains. *Plant Mol. Biol.* 42:591–601
10. Arazi T, Kaplan B, Sunkar R, Dolev D, Fromm H. 2000. Cyclic nucleotide- and calcium/calmodulin-regulated channels in plants: targets for manipulating heavy metal tolerance, and possible physiological roles. *Biochem. Soc. Trans.* 28:471–75
11. Arazi T, Sunkar R, Kaplan B, Fromm H. 1999. A tobacco plasma membrane calmodulin-binding transporter confers Ni^{2+} tolerance and Pb^{2+} hypersensitivity in transgenic plants. *Plant J.* 20:171–82
12. Aurisano N, Bertani A, Reggiani R. 1995. Involvement of calcium and calmodulin in protein and amino-acid metabolism in rice roots under anoxia. *Plant Cell Physiol.* 36:1525–29
13. Badminton MN, Kenda JM, Rembold CM, Campbell AK. 1998. Current evidence suggests independent regulation of nuclear calcium. *Cell Calcium* 23:79–86
14. Balagué C, Lin B, Alcon C, Flottes G, Malmstrom S, et al. 2003. HLM1, an essential signaling component in the hypersensitive response, is a member of the cyclic nucleotide-gated channel ion channel family. *Plant Cell* 15:365–79
15. Baum G, Chen Y, Arazi T, Takatsuji H, Fromm H. 1993. A plant glutamate decarboxylase containing a calmodulin binding domain. Cloning, sequence, and functional analysis. *J. Biol. Chem.* 268:19610–17
16. Baum G, Lev-Yadun S, Fridmann Y, Arazi T, Katsnelson H, et al. 1996. Calmodulin binding to glutamate decarboxylase is required for regulation of glutamate and GABA metabolism and normal development in plants. *EMBO J.* 15:2988–96
17. Bergey DR, Ryan CA. 1999. Wound- and systemin-inducible calmodulin gene expression in tomato leaves. *Plant Mol. Biol.* 40:815–23
18. Berridge MJ, Lipp P, Bootman MD. 2000. Signal transduction: the calcium entry pas de deux. *Science* 287:1604–5
19. Blume B, Nurnberger T, Nass N, Scheel D. 2000. Receptor-mediated increase in cytoplasmic free calcium required for activation of pathogen defense in parsley. *Plant Cell* 12:1425–40
20. Botella JR, Arteca RN. 1994. Differential expression of two calmodulin genes in response to physical and chemical stimuli. *Plant Mol. Biol.* 24:757–66
21. Bouché N, Fait A, Bouchez D, Møller SG, Fromm H. 2003. Mitochondrial succinic-semialdehyde dehydrogenase of the γ-aminobutyrate shunt is required to restrict levels of reactive oxygen intermediates in plants. *Proc. Natl. Acad. Sci. USA* 100:6843–48
22. Bouché N, Fait A, Zik M, Fromm H. 2004. The root-specific glutamate decarboxylase (GAD1) is essential for sustaining GABA levels in *Arabidopsis*. *Plant Mol. Biol.* 55:315–25
23. Bouché N, Fromm H. 2004. GABA in plants: just a metabolite? *Trends Plant Sci.* 9:110–15

The paper reports for the first time that the period and duration of in vivo Ca^{2+} oscillations in plants (guard cells) translates into discrete physiological responses reflected in the aperture of the stomata.

24. Bouché N, Scharlat A, Snedden W, Bouchez D, Fromm H. 2002. A novel family of calmodulin-binding transcription activators in multicellular organisms. *J. Biol. Chem.* 277:21851–61
25. Bown AW, Hall DE, MacGregor KB. 2002. Insect footsteps on leaves stimulate the accumulation of 4-aminobutyrate and can be visualized through increased chlorophyll fluorescence and superoxide production. *Plant Physiol.* 129:1430–34
26. Braam J, Davis RW. 1990. Rain-, wind-, and touch-induced expression of calmodulin and calmodulin-related genes in *Arabidopsis*. *Cell* 60:357–64
27. Carrion AM, Link WA, Ledo F, Mellstrom B, Naranjo JR. 1999. DREAM is a Ca^{2+}-regulated transcriptional repressor. *Nature* 398:80–84
28. Catala R, Santos E, Alonso JM, Ecker JR, Martinez-Zapater JM, Salinas J. 2003. Mutations in the Ca^{2+}/H^+ transporter CAX1 increase CBF/DREB1 expression and the cold-acclimation response in *Arabidopsis*. *Plant Cell* 15:2940–51
29. Deleted in proof
30. Deleted in proof
31. Chen YR, Roux SJ. 1986. Characterization of nucleoside triphosphatase activity in isolated pea nuclei and its photoreversible regulation by light. *Plant Physiol.* 81:609–13
32. Cheng SH, Willmann MR, Chen HC, Sheen J. 2002. Calcium signaling through protein kinases. The *Arabidopsis* calcium-dependent protein kinase gene family. *Plant Physiol.* 129:469–85
33. Cheong YH, Kim KN, Pandey GK, Gupta R, Grant JJ, Luan S. 2003. CBL1, a calcium sensor that differentially regulates salt, drought, and cold responses in *Arabidopsis*. *Plant Cell* 15:1833–45
34. Cho MJ, Vaghy PL, Kondo R, Lee SH, Davis JP, et al. 1998. Reciprocal regulation of mammalian nitric oxide synthase and calcineurin by plant calmodulin isoforms. *Biochemistry* 37:15593–97
35. Choi JY, Lee SH, Park CY, Heo WD, Kim JC, et al. 2002. Identification of calmodulin isoform-specific binding peptides from a phage-displayed random 22-mer peptide library. *J. Biol. Chem.* 277:21630–38
36. Cholewa E, Cholewinski AJ, Sheen J, Snedden WA, Bown AW. 1997. Cold shock-stimulated γ-aminobutyrate synthesis is mediated by an increase in cytosolic Ca^{2+} not by an increase cytosolic H^+. *Can. J. Bot.* 75:375–82
37. Chung WS, Lee SH, Kim JC, Heo WD, Kim MC, et al. 2000. Identification of a calmodulin-regulated soybean Ca^{2+}-ATPase (SCA1) that is located in the plasma membrane. *Plant Cell* 12:1393–408
38. **Clough SJ, Fengler KA, Yu I-C, Lippok B, Smith RKJ, Bent AF. 2000. The *Arabidopsis dnd1* "defense, no death" gene encodes a mutated cyclic nucleotide-gated ion channel. *Proc. Natl. Acad. Sci. USA* 97:9323–28**

The paper describes a genetic study revealing that the "defense no death 1" (*DND1*) gene encodes the CNGC2 CaM-binding cyclic nucleotide gated channel.

39. Crivici A, Ikura M. 1995. Molecular and structural basis of target recognition by calmodulin. *Annu. Rev. Biophys. Biomol. Struct.* 24:85–116
40. Dangl JL, Jones JD. 2001. Plant pathogens and integrated defence responses to infection. *Nature* 411:826–33
41. Dash S, Niemaczura W, Harrington HM. 1997. Characterization of the basic amphiphilic α-helix calmodulin-binding domain of a 61.5 kDa tobacco calmodulin-binding protein. *Biochemistry* 36:2025–29
42. Day I, Reddy VS, Shad Ali G, Reddy ASN. 2002. Analysis of EF-hand-containing proteins in *Arabidopsis*. *Genome Biol.* 3:6.1–24
43. Day IS, Miller C, Golovkin M, Reddy AS. 2000. Interaction of a kinesin-like calmodulin-binding protein with a protein kinase. *J. Biol. Chem.* 275:13737–45

44. De Koninck P, Schulman H. 1998. Sensitivity of CaM Kinase II to the frequency of Ca^{2+} oscillations. *Science* 279:227–30
45. de Torres M, Sanchez P, Fernandez-Delmond I, Grant M. 2003. Expression profiling of the host response to bacterial infection: the transition from basal to induced defence responses in RPM1-mediated resistance. *Plant J.* 33:665–76
46. Delaunay A, Pflieger D, Barrault MB, Vinh J, Toledano MB. 2002. A thiol peroxidase is an H_2O_2 receptor and redox-transducer in gene activation. *Cell* 111:471–81
47. Delledonne M, Xia Y, Dixon RA, Lamb C. 1998. Nitric oxide functions as a signal in plant disease resistance. *Nature* 394:585–88
48. Delledonne M, Zeier J, Marocco A, Lamb C. 2001. Signal interactions between nitric oxide and reactive oxygen intermediates in the plant hypersensitive disease resistance response. *Proc. Natl. Acad. Sci. USA* 98:13454–59
49. Delumeau O, Morere-Le Paven M-C, Montrichard F, Laval-Martin DL. 2000. Effects of short-term NaCl stress on calmodulin transcript levels and calmodulin-dependent NAD kinase activity in two species of tomato. *Plant Cell Environ.* 23:329–36
50. Dolmetsch RE, Xu K, Lewis RS. 1998. Calcium oscillations increase the efficiency and specificity of gene expression. *Nature* 392:933–41
51. Du L, Poovaiah BW. 2004. A novel family of Ca^{2+}/calmodulin-binding proteins involved in transcriptional regulation: interaction with fsh/Ring3 class transcription activators. *Plant Mol. Biol.* 54:549–69
52. Durner J, Wendehenne D, Klessig DF. 1998. Defense gene induction in tobacco by nitric oxide, cyclic GMP, and cyclic ADP-ribose. *Proc. Natl. Acad. Sci. USA* 95:10328–33
53. Durrant WE, Rowland O, Piedras P, Hammond-Kosack KE, Jones JD. 2000. cDNA-AFLP reveals a striking overlap in race-specific resistance and wound response gene expression profiles. *Plant Cell* 12:963–77
54. Echevarria W, Leite MF, Guerra MT, Zipfel WR, Nathanson MH. 2003. Regulation of calcium signals in the nucleus by a nucleoplasmic reticulum. *Nat. Cell Biol.* 5:440–46
55. Folkers U, Kirik V, Schobinger U, Falk S, Krishnakumar S, et al. 2002. The cell morphogenesis gene *ANGUSTIFOLIA* encodes a CtBP/BARS-like protein and is involved in the control of the microtubule cytoskeleton. *EMBO J.* 21:1280–88
56. Fromm H, Chua N-H. 1992. Cloning of plant cDNAs encoding calmodulin-binding proteins using ^{35}S labeled recombinant calmodulin as a probe. *Plant Mol. Biol. Rep.* 10:199–206
57. Gao D, Knight MR, Trewavas AJ, Sattelmacher B, Plieth C. 2004. Self-reporting *Arabidopsis* expressing pH and $[Ca^{2+}]$ indicators unveil ion dynamics in the cytoplasm and in the apoplast under abiotic stress. *Plant Physiol.* 134:898–908
58. Geiser JR, van Tuinen D, Brockerhoff SE, Neff MM, Davis TN. 1991. Can calmodulin function without binding calcium? *Cell* 65:949–59
59. Geisler M, Frangne N, Gomes E, Martinoia E, Palmgren MG. 2000. The *ACA4* gene of *Arabidopsis* encodes a vacuolar membrane calcium pump that improves salt tolerance in yeast. *Plant Physiol.* 124:1814–27
60. Geisler M, Girin M, Brandt S, Vincenzetti V, Plaza S, et al. 2004. *Arabidopsis* immunophilin-like TWD1 functionally interacts with vacuolar ABC transporters. *Mol. Biol. Cell* 15:3393–405
61. Geisler M, Kolukisaoglu HU, Bouchard R, Billion K, Berger J, et al. 2003. TWISTED DWARF1, a unique plasma membrane-anchored immunophilin-like protein, interacts with *Arabidopsis* multidrug resistance-like transporters AtPGP1 and AtPGP19. *Mol. Biol. Cell.* 14:4238–49
62. Gelli A, Higgins VJ, Blumwald E. 1997. Activation of plant plasma membrane Ca^{2+}-permeable channels by race-specific fungal elicitors. *Plant Physiol.* 113:269–79

The authors describe the pollen-specific protein NPG1 (no pollen germination1) of *Arabidopsis*, which can bind CaM in a Ca^{2+}-dependent manner. The *npg1* mutation is not transmitted through the male gametophyte because NPG1 is an essential protein for pollen germination. The exact function of NPG1 remains undetermined.

This paper presents the first convincing case for functional specificity among CLMs during pathogen defense signaling.

63. **Golovkin M, Reddy ASN. 2003. A calmodulin-binding protein from *Arabidopsis* has an essential role in pollen germination. *Proc. Natl. Acad. Sci. USA* 100:10558–63**
64. Gong D, Guo Y, Schumaker KS, Zhu JK. 2004. The SOS3 family of calcium sensors and SOS2 family of protein kinases in *Arabidopsis*. *Plant Physiol.* 134:919–26
65. Gong M, Li Z-G. 1995. Involvement of calcium and calmodulin in the acquisition of heat shock induced thermotolerance in maize. *Phytochemistry* 40:1335–39
66. Grant M, Brown I, Adams S, Knight M, Ainslie A, Mansfield J. 2000. The RPM1 plant disease resistance gene facilitates a rapid and sustained increase in cytosolic calcium that is necessary for the oxidative burst and hypersensitive cell death. *Plant J.* 23:441–50
67. Guo F-Q, Okamoto M, Crawford NM. 2003. Identification of a plant nitric oxide synthase gene involved in hormonal signaling. *Science* 302:100–3
68. Guo H, Mockler T, Duong H, Lin C. 2001. SUB1, an *Arabidopsis* Ca^{2+}-binding protein involved in cryptochrome and phytochrome coaction. *Science* 291:487–90
69. Halfter U, Ishitani M, Zhu J-K. 2000. The *Arabidopsis* SOS2 protein kinase physically interacts with and is activated by the calcium-binding protein SOS3. *Proc. Natl. Acad. Sci. USA* 97:3735–40
70. Harding SA, Oh SH, Roberts DM. 1997. Transgenic tobacco expressing a foreign calmodulin gene shows an enhanced production of active oxygen species. *EMBO J.* 16:1137–44
71. Hardingham GE, Bading H. 1998. Nuclear calcium: a key regulator of gene expression. *BioMetals* 11:345–58
72. Harmon AC. 2003. Calcium-regulated protein kinases of plants. *Gravit. Space Biol. Bull.* 16:83–90
73. Harper JF, Breton G, Harmon AC. 2004. Decoding Ca^{2+} signals through plant protein kinases. *Annu. Rev. Plant Biol.* 55:263–88
74. Harrar Y, Bellini C, Faure JD. 2001. FKBPs: at the crossroads of folding and transduction. *Trends Plant Sci.* 6:426–31
75. **Heo WD, Lee SH, Kim MC, Kim JC, Chung WS, et al. 1999. Involvement of specific calmodulin isoforms in salicylic acid-independent activation of plant disease resistance responses. *Proc. Natl. Acad. Sci. USA* 96:766–71**
76. Hetherington AM, Brownlee C. 2004. The generation of Ca^{2+} signal in plants. *Annu. Rev. Plant Biol.* 55:401–27
77. Hoeflich KP, Ikura M. 2004. Calmodulin in action: diversity in target recognition and activation mechanisms. *Cell* 108:739–42
78. Holdaway-Clarke TL, Hepler PK. 2003. Control of pollen tube growth: role of ion gradients and fluxes. *New Phytol.* 159:539–63
79. Hsieh H-L, Song CJ, Roux SJ. 2000. Regulation of a recombinant pea nuclear apyrase by calmodulin and casein kinase II. *Biochim. Biophys. Acta* 1494:248–55
80. Hua BG, Mercier RW, Zielinski RE, Berkowitz GA. 2003. Functional interaction of calmodulin with a plant cyclic nucleotide gated cation channel. *Plant Physiol. Biochem.* 41:945–54
81. Hua W, Li R-J, Wang L, Lu Y-T. 2004. A tobacco calmodulin-binding protein kinase (NtCBK2) induced by high-salt/GA treatment and its expression during floral development and embryogenesis. *Plant Science* 166:1253–59
82. Hua W, Liang S, Lu Y. 2003. A tobacco (*Nicotiana tabaccum*) calmodulin-binding protein kinase, NtCBK2, is regulated differentially by calmodulin isoforms. *Biochem J.* 376:291–302
83. Jakobek JL, Smith-Becker JA, Lindgren PB. 1999. A bean cDNA expressed during a hypersensitive reaction encodes a putative calcium-binding protein. *Mol. Plant-Microbe Int.* 12:712–19

84. Jang HJ, Pih KT, Kang SG, Lim JH, Jin JB, et al. 1998. Molecular cloning of a novel Ca^{2+}-binding protein that is induced by NaCl stress. *Plant Mol. Biol.* 37:839–47
85. Ji SJ, Lu YC, Feng JX, Wei G, Li J, et al. 2003. Isolation and analyses of genes preferentially expressed during early cotton fiber development by subtractive PCR and cDNA array. *Nucleic Acids Res.* 31:2534–43
86. Kamphausen T, Fanghanel J, Neumann D, Schulz B, Rahfeld JU. 2002. Characterization of *Arabidopsis thaliana* AtFKBP42 that is membrane-bound and interacts with Hsp90. *Plant J.* 32:263–76
87. Katoh M. 2003. Identification and characterization of *FLJ10737* and *CAMTA1* genes on the commonly deleted region of neuroblastoma at human chromosome 1p36.31-p36.23. *Int. J. Oncol.* 23:1219–24
88. Kaupp UB, Seifert R. 2002. Cyclic nucleotide-gated ion channels. *Physiol. Rev.* 82:769–824
89. Kim MC, Lee SH, Kim JK, Chun HJ, Choi MS, et al. 2002. Mlo, a modulator of plant defense and cell death, is a novel calmodulin-binding protein. Isolation and characterization of rice Mlo homologue. *J. Biol. Chem.* 277:19304–14
90. **Kim MC, Panstruga R, Elliott C, Muller J, Devoto A, et al. 2002. Calmodulin interacts with MLO protein to regulate defence against mildew in barley. *Nature* 416:447–51**

MLO is a plasma membrane–localized seven-transmembrane-like receptor unique to plants and plays a role in defense against the powdery mildew fungus. The paper reports that mutations that abolish CaM binding halve the ability of MLO to negatively regulate defense against the fungus.

91. Kim SO, Merchant K, Nudelman R, Beyer WF Jr, Keng T, et al. 2002. OxyR: a molecular code for redox-related signaling. *Cell* 109:383–96
92. Klusener B, Young JJ, Murata Y, Allen GJ, Mori IC, et al. 2002. Convergence of calcium signaling pathways of pathogenic elicitors and abscisic acid in *Arabidopsis* guard cells. *Plant Physiol.* 130:2152–63
93. Knauss S, Rohrmeier T, Lehle L. 2003. The auxin-induced maize gene *ZmSAUR2* encodes a short-lived nuclear protein expressed in elongating tissues. *J. Biol. Chem.* 278:23936–43
94. Knight H, Knight MR. 2001. Abiotic stress signalling pathways: specificity and cross-talk. *Trends Plant Sci.* 6:262–67
95. Kondo R, Tikunova SB, Cho MJ, Johnson JD. 1999. A point mutation in a plant calmodulin is responsible for its inhibition of nitric-oxide synthase. *J. Biol. Chem.* 274:36213–18
96. Koradi R, Billeter M, Wüthrich K. 1996. MOLMOL: a program for display and analysis of macromolecular structures. *J. Mol. Graphics* 14:51–55
97. Kurek I, Aviezer K, Erel N, Herman E, Breiman A. 1999. The wheat peptidyl prolyl cis-trans-isomerase FKBP77 is heat induced and developmentally regulated. *Plant Physiol.* 119:693–704
98. Kurek I, Pirkl F, Fischer E, Buchner J, Breiman A. 2002. Wheat FKBP73 functions in vitro as a molecular chaperone independently of its peptidyl prolyl cis-trans isomerase activity. *Planta* 215:119–26
99. Kurek I, Stoger E, Dulberger R, Christou P, Breiman A. 2002. Overexpression of the wheat FK506-binding protein 73 (FKBP73) and the heat-induced wheat FKBP77 in transgenic wheat reveals different functions of the two isoforms. *Transgenic Res.* 11:373–79
99a. Lamattina L, Garcia-Mata C, Graziano M, Pagnussat G. 2003. Nitric oxide: the versatility of an extensive signal molecule. *Annu. Rev. Plant Biol.* 54:109–36
100. Lamotte O, Gould K, Lecourieux D, Sequeira-Legrand A, Lebrun-Garcia A, et al. 2004. Analysis of nitric oxide signaling functions in tobacco cells challenged by the elicitor cryptogein. *Plant Physiol.* 135:516–29
101. Lander ES, Linton LM, Birren B, Nusbaum C, Zody MC, et al. International Human Genome Sequencing Consortium. 2001. Initial sequencing and analysis of the human genome. *Nature* 409:860–921

102. Larkindale J, Knight MR. 2002. Protection against heat stress-induced oxidative damage in *Arabidopsis* involves calcium, abscisic acid, ethylene, and salicylic acid. *Plant Physiol.* 128:682–95
103. Lecourieux D, Mazars C, Pauly N, Ranjeva R, Pugin A. 2002. Analysis and effects of cytosolic free calcium increases in response to elicitors in *Nicotiana plumbaginifolia* cells. *Plant Cell* 14:2627–41
104. Lee EJ, Yang EJ, Lee JE, Park AR, Song WH, Park OK. 2004. Proteomic identification of annexins, calcium-dependent membrane binding proteins that mediate osmotic stress and abscisic acid signal transduction in *Arabidopsis*. *Plant Cell* 16:1378–91
105. Lee J, Klessig DF, Nurnberger T. 2001. A harpin binding site in tobacco plasma membranes mediates activation of the pathogenesis-related gene *HIN1* independent of extracellular calcium but dependent on mitogen-activated protein kinase activity. *Plant Cell* 13:1079–93
106. Lee SH, Johnson JD, Walsh MP, Van Lierop JE, Sutherland C, et al. 2000. Differential regulation of Ca^{2+}/calmodulin-dependent enzymes by plant calmodulin isoforms and free Ca^{2+} concentration. *Biochem. J.* 350:299–306
107. Lee SH, Kim MC, Heo WD, Kim JC, Chung WS, et al. 1999. Competitive binding of calmodulin isoforms to calmodulin-binding proteins: implication for the function of calmodulin isoforms in plants. *Biochim. Biophys. Acta* 1433:56–67
108. Lee SH, Seo HY, Kim JC, Heo WD, Chung WS, et al. 1997. Differential activation of NAD kinase by plant calmodulin isoforms. The critical role of domain I. *J. Biol. Chem.* 272:9252–59
109. Levine A, Pennell RI, Alvarez ME, Palmer R, Lamb C. 1996. Calcium-mediated apoptosis in a plant hypersensitive disease resistance response. *Curr. Biol.* 6:427–37
110. Liao B, Zielinski RE. 1995. Production of recombinant plant calmodulin and its use to detect calmodulin-binding proteins. *Methods Cell Biol.* 49:487–500
111. Ling V, Snedden WA, Shelp BJ, Assmann SM. 1994. Analysis of a soluble calmodulin binding protein from fava bean roots: identification of glutamate decarboxylase as a calmodulin-activated enzyme. *Plant Cell* 6:1135–43
112. Liu HT, Li B, Shang ZL, Li XZ, Mu RL, et al. 2003. Calmodulin is involved in heat shock signal transduction in wheat. *Plant Physiol.* 132:1186–95
113. Lu Y-T, Harrington HM. 1994. Isolation of tobacco cDNA clones encoding calmodulin-binding proteins and characterization of a known calmodulin-binding domain. *Plant Physiol. Biochem.* 32:413–22
114. Lu YT, Dharmasiri MA, Harrington HM. 1995. Characterization of a cDNA encoding a novel heat-shock protein that binds to calmodulin. *Plant Physiol.* 108:1197–202
115. Luan S, Kudla J, Rodriguez-Concepcion M, Yalovsky S, Gruissem W. 2002. Calmodulins and calcineurin B-like proteins: calcium sensors for specific signal response coupling in plants. *Plant Cell* 14:S389–400
116. Ludwig AA, Romeis T, Jones JD. 2004. CDPK-mediated signalling pathways: specificity and cross-talk. *J. Exp. Bot.* 55:181–88
117. MacGregor KB, Shelp BJ, Peiris S, Bown AW. 2003. Overexpression of glutamate decarboxylase in transgenic tobacco plants deters feeding by phytophagous insect larvae. *J. Chem. Ecol.* 29:2177–82
118. Maser P, Thomine S, Schroeder JI, Ward JM, Hirschi K, et al. 2001. Phylogenetic relationships within cation transporter families of *Arabidopsis*. *Plant Physiol.* 126:1646–67
119. McCormack E, Braam J. 2003. Calmodulins and related potential calcium sensors in *Arabidopsis*. *New Phytol.* 159:585–98

120. Mitra RM, Gleason CA, Edwards A, Hadfield J, Downie JA, et al. 2004. A Ca^{2+}/calmodulin-dependent protein kinase required for symbiotic nodule development: gene identification by transcript-based cloning. *Proc. Natl. Acad. Sci. USA* 101:4701–5
121. Moutinho A, Love J, Anthony J, Trewavas J, Malhó R. 1998. Distribution of calmodulin protein and mRNA in growing pollen tubes. *Sex Plant Reprod.* 11:131–39
122. Mysore KS, Crasta OR, Tuori RP, Folkerts O, Swirsky PB, Martin GB. 2002. Comprehensive transcript profiling of Pto- and Prf-mediated host defense responses to infection by *Pseudomonas syringae* pv. tomato. *Plant J.* 32:299–315
123. Nakatani K, Nishioka J, Itakura T, Nakanishi Y, Horinouchi J, et al. 2004. Cell cycle-dependent transcriptional regulation of calmodulin-binding transcription activator 1 in neuroblastoma cells. *Int. J. Oncol.* 24:1407–12
124. Nakayama S, Kawasaki H, Krestinger R. 2000. Evolution of EF-hand proteins. In *Calcium Homeostasis*, ed. E Carafoli, J Krebs. pp. 29–58. New York: Springer
125. Natalie C, Strynadka J, James MNG. 1989. Crystal structure of the helix-loop-helix calcium-binding proteins. *Annu. Rev. Biochem.* 58:951–58
126. Neill S, Desikan R, Hancock J. 2002. Hydrogen peroxide signalling. *Curr. Opin. Plant Biol.* 5:388–95
127. Ng CK-Y, Mc Ainsh MR. 2003. Encoding specificity in plant calcium signalling: hot-spotting the ups and downs and waves. *Ann. Bot.* 92:477–85
128. Nimchuk Z, Eulgem T, Holt BF 3rd, Dangl JL. 2003. Recognition and response in the plant immune system. *Annu. Rev. Genet.* 37:579–609
129. Ohya Y, Botstein D. 1994. Diverse essential functions revealed by complementing yeast calmodulin mutants. *Science* 263:963–66
130. Olsson P, Yilmaz JL, Sommarin M, Persson S, Bulow L. 2004. Expression of bovine calmodulin in tobacco plants confers faster germination on saline media. *Plant Sci.* 166:1595–604
131. Palanivelu R, Brass L, Edlund AF, Preuss D. 2003. Pollen tube growth and guidance is regulated by POP2, an *Arabidopsis* gene that controls GABA levels. *Cell* 114:47–59
132. Pauly N, Knight MR, Thuleau P, van der Luit AH, Moreau M, et al. 2000. Control of free calcium in plant cell nuclei. *Nature* 405:754–55
133. **Perruc E, Charpenteau M, Ramirez BC, Jauneau A, Galaud JP, et al. 2004. A novel calmodulin-binding protein functions as a negative regulator of osmotic stress tolerance in *Arabidopsis thaliana* seedlings. *Plant J.* 38:410–20**

The paper describes a CaMBP (designated AtCAMBP25) unique to plants isolated by screening an expression cDNA library with radiolabeled CaM. CaMBP25 has an unknown activity. The protein is localized in the nucleus and plays a role in the response to ionic and nonionic osmotic stress. It is also interesting that CaMBP25 binds differentially only to some CaM isoforms.

134. Preuss ML, Delmer DP, Liu B. 2003. The cotton kinesin-like calmodulin-binding protein associates with cortical microtubules in cotton fibers. *Plant Physiol.* 132:154–60
135. Pugin A, Frachisse JM, Tavernier E, Bligny R, Gout E, et al. 1997. Early events induced by the elicitor cryptogein in tobacco cells: involvement of a plasma membrane NADPH oxidase and activation of glycolysis and the pentose phosphate pathway. *Plant Cell* 9:2077–91
136. Puhakainen T, Pihakaski-Maunsbach K, Widell S, Sommarin M. 1999. Cold acclimation enhances the activity of plasma membrane Ca^{2+}-ATPase in winter rye leaves. *Plant Physiol. Biochem.* 37:231–39
137. Rato C, Monteiro D, Hepler PK, Malho R. 2004. Calmodulin activity and cAMP signalling modulate growth and apical secretion in pollen tubes. *Plant J.* 38:887–97
138. Reddy AS, Day IS. 2000. The role of the cytoskeleton and a molecular motor in trichome morphogenesis. *Trends Plant Sci.* 5:503–5
139. Reddy AS, Reddy VS, Golovkin M. 2000. A calmodulin binding protein from *Arabidopsis* is induced by ethylene and contains a DNA-binding motif. *Biochem. Biophys. Res. Commun.* 279:762–69

140. Reddy RK, Kurek I, Silverstein AM, Chinkers M, Breiman A, Krishna P. 1998. High-molecular-weight FK506-binding proteins are components of heat-shock protein 90 heterocomplexes in wheat germ lysate. *Plant Physiol.* 118:1395–401
141. Reddy VS, Ali GS, Reddy AS. 2003. Characterization of a pathogen-induced calmodulin-binding protein: mapping of four Ca^{2+}-dependent calmodulin-binding domains. *Plant Mol. Biol.* 52:143–59
142. Reddy VS, Ali GS, Reddy ASN. 2002. Genes encoding calmodulin-binding proteins in the *Arabidopsis* genome. *J. Biol. Chem.* 277:9840–52
143. Reddy VS, Day IS, Thomas T, Reddy AS. 2004. KIC, a novel Ca^{2+} binding protein with one EF-hand motif, interacts with a microtubule motor protein and regulates trichome morphogenesis. *Plant Cell* 16:185–200
144. Reddy VS, Reddy AS. 2002. The calmodulin-binding domain from a plant kinesin functions as a modular domain in conferring Ca^{2+}-calmodulin regulation to animal plus- and minus-end kinesins. *J. Biol. Chem.* 277:48058–65
145. Reddy VS, Reddy ASN. 2004. Proteomics of calcium-signaling components in plants. *Phytochemistry* 65:1745–76
146. Rentel MR, Knight MR. 2004. Oxidative stress-induced calcium signaling in *Arabidopsis thaliana*. *Plant Physiol.* 135:1471–79
147. Rodriguez-Concepcion M, Yalovsky S, Zik M, Fromm H, Gruissem W. 1999. The prenylation status of a novel plant calmodulin directs plasma membrane or nuclear localization of the protein. *EMBO J.* 18:1996–2007
148. Rogers GC, Hart CL, Wedaman KP, Scholey JM. 1999. Identification of kinesin-C, a calmodulin-binding carboxy-terminal kinesin in animal (*Strongylocentrotus purpuratus*) cells. *J. Mol. Biol.* 294:1–8
149. Safadi F, Reddy VS, Reddy ASN. 2000. A pollen-specific novel calmodulin-binding protein with tetratricopeptide repeats. *J. Biol. Chem.* 275:35457–70
150. Schiott M, Romanowsky SM, Baekgaard L, Jakobsen MK, Palmgren MG, Harper JF. 2004. A plant plasma membrane Ca^{2+} pump is required for normal pollen tube growth and fertilization. *Proc. Natl. Acad. Sci. USA* 101:9502–7
151. Shelp BJ, Bown AW, McLean MD. 1999. Metabolism and functions of gamma-aminobutyric acid. *Trends Plant Sci.* 4:446–52
152. Silverstein AM, Galigniana MD, Kanelakis KC, Radanyi C, Renoir JM, Pratt WB. 1999. Different regions of the immunophilin FKBP52 determine its association with the glucocorticoid receptor, hsp90, and cytoplasmic dynein. *J. Biol. Chem.* 274:36980–86
153. Snedden WA, Arazi T, Fromm H, Shelp BJ. 1995. Calcium/calmodulin activation of soybean glutamate decarboxylase. *Plant Physiol.* 108:543–49
154. Snedden WA, Fromm H. 1999. Regulation of the γ-aminobutyrate-synthesizing enzyme, glutamate decarboxylase, by calcium-calmodulin: a mechanism for rapid activation in response to stress. In *Plant Responses to Environmental Stresses: From Phytohormones to Genome Reorganization*, ed. HR Lerner, pp. 549–74 . New York: Marcel Dekker
155. Snedden WA, Fromm H. 2001. Calmodulin as a versatile calcium signal transducer in plants. *New Phytol.* 151:35–66
156. Snedden WA, Koutsia N, Baum G, Fromm H. 1996. Activation of a recombinant petunia glutamate decarboxylase by calcium/calmodulin or by a monoclonal antibody which recognizes the calmodulin binding domain. *J. Biol. Chem.* 271:4148–53
157. Steinebrunner I, Jeter C, Song C, Roux SJ. 2000. Molecular and biochemical comparison of two different apyrases from *Arabidopsis thaliana*. *Plant Physiol. Biochem.* 38:913–22
158. Steinebrunner I, Wu J, Sun Y, Corbett A, Roux SJ. 2003. Disruption of apyrases inhibits pollen germination in *Arabidopsis*. *Plant Physiol.* 131:1638–47

This article addresses a combination of an experimental approach and bioinformatic studies to identify genes encoding CaMBPs in the *Arabidopsis* genome.

This study addresses the mechanism by which CaM inhibits plant KCBPs.

This paper describes the phenotype of the *aca9 Arabidopsis* T-DNA mutants.

159. Sun XT, Li B, Zhou GM, Tang WQ, Bai J, et al. 2000. Binding of the maize cytosolic Hsp70 to calmodulin, and identification of calmodulin-binding site in Hsp70. *Plant Physiol.* 41:804–10
160. Sunkar R, Kaplan B, Bouche N, Arazi T, Dolev D, et al. 2000. Expression of a truncated tobacco *NtCBP4* channel in transgenic plants, and disruption of the homologous *Arabidopsis CNGC1* gene confer Pb^{2+} tolerance. *Plant J.* 24:533–42
161. Takezawa D, Minami A. 2004. Calmodulin-binding proteins in bryophytes: identification of abscisic acid-, cold-, and osmotic stress-induced genes encoding novel membrane-bound transporter-like proteins. *Biochem. Biophys. Res. Commun.* 317:428–36
162. Talke IN, Blaudez D, Maathuis FJ, Sanders D. 2003. CNGCs: prime targets of plant cyclic nucleotide signalling? *Trends Plant Sci.* 8:286–93
163. Thomas C, Rajagopal A, Windsor B, Dudler R, Lloyd A, Roux SJ. 2000. A role for ectophosphatase in xenobiotic resistance. *Plant Cell* 12:519–33
164. Townley HE, Knight MR. 2002. Calmodulin as a potential negative regulator of *Arabidopsis COR* gene expression. *Plant Physiol.* 128:1169–72
165. Tran QK, Black DJ, Persechini A. 2003. Intracellular coupling via limiting calmodulin. *J. Biol. Chem.* 278:24247–50
166. Trewavas A. 1999. How plants learn. *Proc. Natl. Acad. Sci. USA* 96:4216–18
167. Turano FJ, Fang TK. 1998. Characterization of two glutamate decarboxylase cDNA clones from *Arabidopsis*. *Plant Physiol.* 117:1411–21
168. Turner WL, Waller JC, Vanderbeld B, Snedden WA. 2004. Cloning and characterization of two NAD kinases from *Arabidopsis*: identification of a calmodulin binding isoform. *Plant Physiol.* 135:1243–55
169. van der Luit AH, Olivari C, Haley A, Knight MR, Trewavas AJ. 1999. Distinct calcium signaling pathways regulate calmodulin gene expression in tobacco. *Plant Physiol.* 121:705–14
170. Vanoosthuyse V, Tichtinsky G, Dumas C, Gaude T, Cock JM. 2003. Interaction of calmodulin, a sorting nexin and kinase-associated protein phosphatase with the *Brassica oleracea* S locus receptor kinase. *Plant Physiol.* 133:919–29
171. Vetter SW, Leclerc E. 2003. Novel aspects of calmodulin target recognition and activation. *Eur. J. Biochem.* 270:404–14
172. **Vinogradova MV, Reddy VS, Reddy ASN, Sablin EP, Fletterick RJ. 2004. Crystal structure of kinesin regulated by Ca^{2+}-calmodulin. *J. Biol. Chem.* 279:23504–9**

The paper describes the first crystal structure of a plant CaMBP (the kinesin-like protein from potato) and helps to explain the mechanism by which CaM inhibits the motor activity of plant kinesin-like proteins.

173. Deleted in proof
174. White PJ, Broadley MR. 2003. Calcium in plants. *Ann. Bot.* 92:487–511
175. Wimmers LE, Ewing NN, Bennett AB. 1992. Higher plant Ca^{2+}-ATPase: primary structure and regulation of mRNA abundance by salt. *Proc. Natl. Acad. Sci. USA* 89:9205–9
176. Windsor B, Roux SJ, Lloyd A. 2003. Multiherbicide tolerance conferred by AtPgp1 and apyrase overexpression in *Arabidopsis thaliana*. *Nat. Biotechnol.* 21:428–33
177. Xing T, Ouellet T, Miki BL. 2002. Towards genomic and proteomic studies of protein phosphorylation in plant-pathogen interactions. *Trends Plant Sci.* 7:224–30
178. Xu H, Heath MC. 1998. Role of calcium in signal transduction during the hypersensitive response caused by basidiospore-derived infection of the cowpea rust fungus. *Plant Cell* 10:585–98
179. Yamakawa H, Katou S, Seo S, Mitsuhara I, Kamada H, Ohashi Y. 2004. Plant MAPK phosphatase interacts with calmodulins. *J. Biol. Chem.* 279:928–36
180. Yamakawa H, Mitsuhara I, Ito N, Seo S, Kamada H, Ohashi Y. 2001. Transcriptionally and post-transcriptionally regulated response of 13 calmodulin genes to tobacco mosaic virus-induced cell death and wounding in tobacco plant. *Eur. J. Biochem.* 268:3916–29

181. Yamniuk AP, Vogel HJ. 2004. Calmodulin's flexibility allows for promiscuity in its interactions with target proteins and peptides. *Mol. Biotechnol.* 27:33–57

182. Yang T, Poovaiah BW. 2000. *Arabidopsis* chloroplast chaperonin 10 is a calmodulin-binding protein. *Biochem. Biophys. Res. Commun.* 275:601–7

183. Yang T, Poovaiah BW. 2000. An early ethylene up-regulated gene encoding a calmodulin-binding protein involved in plant senescence and death. *J. Biol. Chem.* 275:38467–73

184. Yang T, Poovaiah BW. 2000. Molecular and biochemical evidence for the involvement of calcium/calmodulin in auxin action. *J. Biol. Chem.* 275:3137–43

185. Yang T, Poovaiah BW. 2002. A calmodulin-binding/CGCG box DNA-binding protein family involved in multiple signaling pathways in plants. *J. Biol. Chem.* 277:45049–58

186. Yang T, Poovaiah BW. 2002. Hydrogen peroxide homeostasis: activation of plant catalase by calcium/calmodulin. *Proc. Natl. Acad. Sci. USA* 99:4097–102

187. Yang T, Poovaiah BW. 2003. Calcium/calmodulin-mediated signal network in plants. *Trends Plant Sci.* 8:505–12

188. Yang T, Segal G, Abbo S, Feldman M, Fromm H. 1996. Characterization of the calmodulin gene family in wheat: structure, chromosomal location, and evolutionary aspects. *Mol. Gen. Genet.* 252:684–94

189. Yap KL, Yuan T, Mal TK, Vogel HJ, Ikura M. 2003. Structural basis for simultaneous binding of two carboxy-terminal peptides of plant glutamate decarboxylase to calmodulin. *J. Mol. Biol.* 328:193–204

The paper describes the first 3D structure of a plant CaMBD associated with CaM. NMR analysis revealed that the structure shows a novel type of CaM-target interaction, with a stochiometry of 1:2, suggesting a role for CaM in glutamate decarboxylase subunit dimerization, and possibly in the formation of a larger oligomeric complex.

190. Yevtushenko DP, McLean MD, Peiris S, Van Cauwenberghe OR, Shelp BJ. 2003. Calcium/calmodulin activation of two divergent glutamate decarboxylases from tobacco. *J. Exp. Bot.* 54:2001–2

191. Yun SJ, Oh SH. 1998. Cloning and characterization of a tobacco cDNA encoding calcium/calmodulin-dependent glutamate decarboxylase. *Mol. Cells* 8:125–29

191a. Zeidler D, Zahringer U, Gerber I, Dubery I, Hartung T, et al. 2004. Innate immunity in *Arabidopsis thaliana*: Lipopolysaccharides activate nitric oxide synthase (NOS) and induce defense genes. *Proc. Natl. Acad. Sci. USA* 101:15811–16

192. Zhu H, Bilgin M, Bangham R, Hall D, Casamayor A, et al. 2001. Global analysis of protein activities using proteome chips. *Science* 293:2101–5

193. Zhu JK. 2002. Salt and drought stress signal transduction in plants. *Annu. Rev. Plant Biol.* 53:247–73

194. Zhu JK. 2003. Regulation of ion homeostasis under salt stress. *Curr. Opin. Plant Biol.* 6:441–45

195. Zielinski RE. 1998. Calmodulin and calmodulin-binding proteins in plants. *Annu. Rev. Plant Physiol. Plant Mol. Biol.* 49:697–725

196. Zik M, Arazi T, Snedden WA, Fromm H. 1998. Two isoforms of glutamate decarboxylase in *Arabidopsis* are regulated by calcium/calmodulin and differ in organ distribution. *Plant Mol. Biol.* 37:967–75

197. Zimmermann S, Nurnberger T, Frachisse JM, Wirtz W, Guern J, et al. 1997. Receptor-mediated activation of a plant Ca^{2+}-permeable ion channel involved in pathogen defense. *Proc. Natl. Acad. Sci. USA* 94:2751–55

Self-Incompatibility in Plants

Seiji Takayama and Akira Isogai

Laboratory of Intercellular Communications, Graduate School of Biological Sciences, Nara Institute of Science and Technology, 8916-5 Takayama, Ikoma, Nara 630-0101, Japan; email: takayama@bs.naist.jp, isogai@bs.naist.jp

Annu. Rev. Plant Biol.
2005. 56:467–89

doi: 10.1146/
annurev.arplant.56.032604.144249

First published online as a
Review in Advance on
January 18, 2005

1543-5008/05/0602-
0467$20.00

Key Words

self-/nonself-recognition, signal transduction, receptor kinase, F-box protein, Ca^{2+} signaling

Abstract

Sexual reproduction in many flowering plants involves self-incompatibility (SI), which is one of the most important systems to prevent inbreeding. In many species, the self-/nonself-recognition of SI is controlled by a single polymorphic locus, the *S*-locus. Molecular dissection of the *S*-locus revealed that SI represents not one system, but a collection of divergent mechanisms. Here, we discuss recent advances in the understanding of three distinct SI mechanisms, each controlled by two separate determinant genes at the *S*-locus. In the Brassicaceae, the determinant genes encode a pollen ligand and its stigmatic receptor kinase; their interaction induces incompatible signaling(s) within the stigma papilla cells. In the Solanaceae-type SI, the determinants are a ribonuclease and an F-box protein, suggesting the involvement of RNA and protein degradation in the system. In the Papaveraceae, the only identified female determinant induces a Ca^{2+}-dependent signaling network that ultimately results in the death of incompatible pollen.

Contents

INTRODUCTION

SI: self-incompatibility

GSI: gametophytic self-incompatibility

SSI: sporophytic self-incompatibility

Self-incompatibility (SI) is one of the most important systems used by many flowering plants to prevent self-fertilization and thereby generate and maintain genetic diversity within a species (6). The SI response is comprised of a self- and nonself-recognition process between pollen and pistil that is followed by selective inhibition of the self-pollen (tube) development. Classic genetic studies have established that the self-/nonself-recognition in most species is controlled by a single multiallelic locus, the *S*-locus, and that pollen inhibition occurs when the same "*S*-allele" specificity is expressed by both pollen and pistil.

After 20 years of intense molecular studies focused on the entities of the *S*-locus, the molecules involved in the SI recognition were finally identified in certain plant species. These identified determinant genes have diverse structures, suggesting that the SI does not represent one mechanism but encompasses a collection of divergent systems. The only unifying scheme that has emerged from these studies is that the *S*-locus consists of at least two linked transcriptional units arranged in pairs, with one functioning as the female determinant and the other as the male (**Figure 1**). This multigene complex at the *S*-locus is inherited as one segregating unit, and therefore the variants of the gene complex are called "*S*-haplotypes." Self-/nonself-recognition operates at the level of protein-protein interaction of the two determinants and the SI response occurs when both determinants are issued from the same *S*-haplotypes.

Both the female and male determinants were first identified in the Brassicaceae (**Figure 1**). Recent studies also identified the male determinant in the Solanaceae, Rosaceae, and Scrophulariaceae, all of which share the same female determinant molecule. Additionally, the female determinant was identified from studies on the Papaveraceae. In spite of this unifying scheme of a multiallelic two-gene recognition system, the identified determinants bear no similarity to one another, suggesting that SI evolved independently and probably multiple times in different lineages of the angiosperms. There are numerous excellent reviews of SI (6, 10, 12, 20, 27, 32, 47, 68, 80, 93). The present review highlights recent works to update the reader on our current understanding of the molecular mechanisms of SI in these plant species.

BRASSICACEAE-TYPE SELF-INCOMPATIBILITY

Classic genetic studies in the early 1950s unraveled two distinct forms of SI, the gametophytic (GSI) and the sporophytic (SSI), which were distinguished by the genetic behavior of the pollen's SI phenotype (6, 47, 68). The pollen SI phenotype in GSI is determined by its own haploid genome, whereas in SSI the pollen SI phenotype is determined by the diploid genome

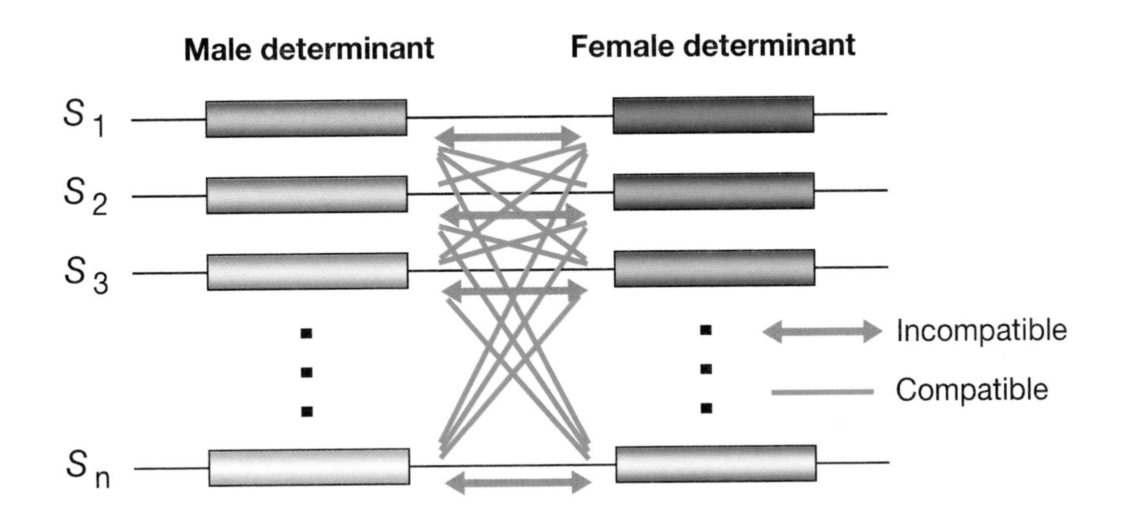

Family	Type of SI	Male determinant	Female determinant
Brassicaceae	SSI	SP11/SCR	SRK
Solanaceae, Rosaceae, Scrophulariaceae	GSI	SLF/SFB	S-RNase
Papaveraceae	GSI	(Unknown)	S-protein

Figure 1

A schematic drawing of the *S*-locus and a list of the identified female and male determinant genes. The *S*-locus contains at least two genes, one encoding the male determinant that is carried by the pollen grain, and the other encoding the female determinant that is expressed in the pistil. Both the male and female determinants are polymorphic and inherited as one segregating unit. The variants of this gene complex are called *S*-haplotypes. The recognition of self/nonself operates at the level of the protein-protein interactions between the two determinants and an incompatible response occurs when both determinants are issued from the same *S*-haplotype. Thus far, both determinants have been identified in the Brassicaceae and Solanaceae. Only the female determinant has been identified in the Papaveraceae.

of its parent (sporophyte). According to this classification, the SI in the Brassicaceae belongs to SSI and, so far, is the only SSI system in which the mechanism has been characterized at the molecular level (20, 27, 80, 93). More than 30 and 50 *S*-haplotypes have been identified in *B. rapa* (syn. *campestris*) and in *B. oleracea*, respectively (54, 56). In the self-incompatible plants of this family, pollen tubes do not develop properly on the stigma that express the same *S*-haplotypes as the pollen's parent. Self-pollen rejection results in abrogated pollen hydration, or a rapid arrest of the pollen tube growth at the stigma surface.

Female Determinant

Searches for the female determinant began with the immunological identification of an *S*-haplotype-specific antigen in the stigma, followed by the biochemical identification of stigma glycoproteins called *S*-locus glycoproteins (SLGs) that cosegregate with *S*-haplotypes. SLGs are 50–60-kDa secreted glycoproteins with several *N*-linked oligosaccharides and twelve conserved cysteine residues (52, 81). The identification of SLGs led to the isolation of the second *S*-locus gene, the *S*-locus receptor kinase (SRK) gene (72). SRK consists of an SLG-like extracellular domain (*S*-domain), a transmembrane domain, and an intracellular serine/threonine kinase domain.

SLG and SRK exhibit a number of characteristics that would be expected for the female determinant of SI. First, they are predominantly produced in the stigma papilla cells, which come into direct contact with pollen. Second, their expression occurs just prior to flower opening and coincides with the timing of SI acquisition by the stigma. Third, they exhibit allelic sequence diversity among all of the *S*-haplotypes examined.

A gain-of-function experiment clarified the involvement of SLGs and SRKs in the SI response (79). Transgenic *B. rapa* expressing SRK_{28} (*SRK* of the S_{28}-haplotype) acquired S_{28}-haplotype specificity in the stigma and rejected the S_{28} pollen. In contrast, transgenic plants expressing SLG_{28} did not display S_{28}-haplotype specificity. When both SLG_{28} and SRK_{28} were introduced, however, the transformants exhibited stronger incompatibility against S_{28} pollen and produced fewer seeds. These results demonstrate that *SRK* alone determines the *S*-haplotype specificity of the stigma, and that *SLG* enhances the activity of *SRK*. In another gain-of-function experiment performed in *B. napus*, the role of SRK as the female determinant was also confirmed (69). However, no enhancing role for SLG was detected in the experiment. Thus, the requirement for SLG in the SI response may be variable among the different *S*-haplotypes. In support of this view, several *S*-homozygous lines of *Brassica*, and other genera of the Brassicaceae, *Arabidopsis lyrata*, lack *SLG* expression, even though they still exhibit a strong SI phenotype (35, 77, 78).

Male Determinant

The first important clue for identifying the male determinant was obtained from a pollination bioassay, which demonstrated that the biological activity responsible for SI resides in a small protein fraction (<10 kDa) of the pollen coat (73). Although isolating the active component was unsuccessful, two genetic approaches, the cloning and sequencing of the *S*-locus region and the polymorphic gene search using fluorescent differential display, succeeded in identifying the male determinant genes, which were named *SP11* (*S-locus protein 11*) or *SCR* (*S-locus cysteine-rich*) (62, 76, 82).

SP11/SCR encodes the secreted forms of small, basic, cysteine-rich proteins. *SP11/SCR* exhibits an extensive *S*-haplotype-associated polymorphism in which the alleles share a relatively conserved signal sequence but have mature proteins that are highly variable (19.5% to 94% amino acid identity), suggesting a strong positive selection for diversification. Only a few residues are highly, but not absolutely, conserved among most *S*-haplotypes. Namely, these include eight cysteine residues (hereafter designated as C1 through C8), a glycine residue between C1 and C2, and an aromatic residue between C3 and C4 (63, 82, 92).

The identity of SP11/SCR as the male *S*-determinant was definitively established by gain-of-function experiments and by direct activity testing using a pollination bioassay (62, 65, 82, 83). In all of the gain-of-function experiments, pollen from the transformants with the *SP11/SCR* transgene acquired the *S*-haplotype specificity of the transgene. In the pollination bioassay, the pretreatment of the stigma with bacterially expressed or chemically synthesized "self" SP11/SCR inhibited the hydration and penetration of "cross" pollen. These results clearly suggest that SP11/SCR is the sole male

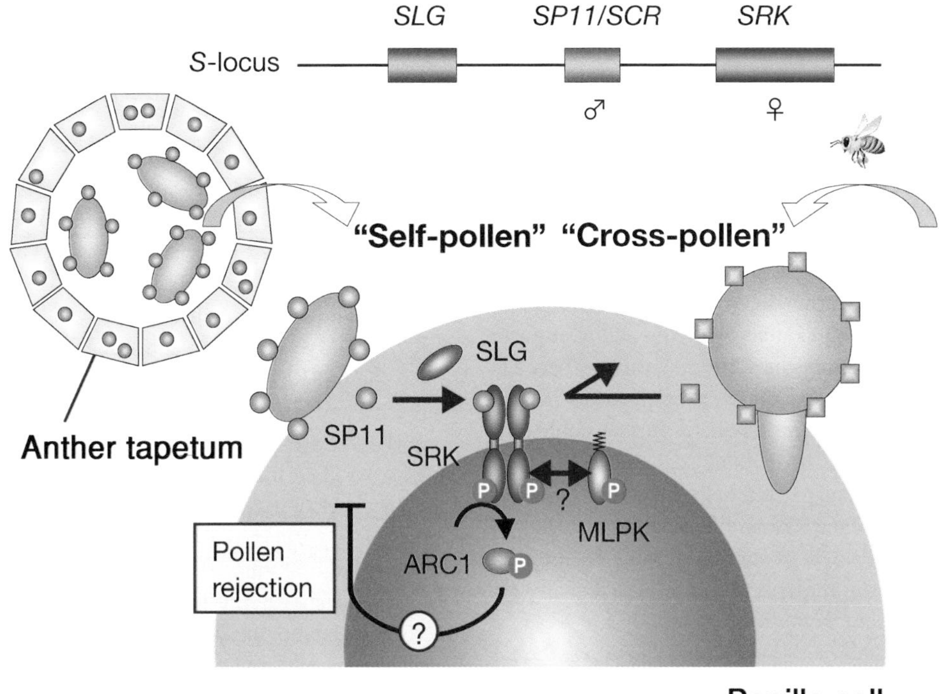

Figure 2

Molecular model of the self-incompatibility (SI) response in the Brassicaceae. The *S*-locus consists of three genes, *SRK*, *SP11*, and *SLG*. The SRK receptor kinase is the female determinant and spans the plasma membrane of the stigma papilla cell. SP11 is the male determinant and is predominantly expressed in the anther tapetum and accumulates in the pollen coat during pollen maturation. Upon pollination, SP11 penetrates the papilla cell wall and binds SRK in an *S*-haplotype-specific manner. This binding induces the autophosphorylation of SRK, triggering a signaling cascade that results in the rejection of self-pollen. SLG is not essential for the self-/nonself-recognition but localizes in the papilla cell wall and enhances the SI reaction in some *S*-haplotypes. The signaling cascade downstream of SRK has not yet been characterized, but the essential positive effectors include MLPK and ARC1. MLPK localizes papilla cell membrane and may form a signaling complex with SRK. ARC1, an E3 ubiquitin ligase, binds to the kinase domain of SRK in a phosphorylation-dependent manner and may target unknown substrates for ubiquitination. The proteasomal degradation of these substrates could result in pollen rejection.

determinant that directly induces incompatible reactions in the stigma papilla cells.

In situ hybridization analyses demonstrate that *SP11/SCR* is expressed sporophytically in the anther tapetum cells and gametophytically in microspores (82) (**Figure 2**). In some *S*-haplotypes, the *SP11/SCR* expression is detected only in the anther tapetal cell layer (63). Because the tapetal cell layer is a diploid tissue that nourishes developing pollen grains and provides the components of pollen coating, the expression pattern of *SP11/SCR* easily explains the sporophytic nature of *Brassica* SI. In fact, immunohistochemical studies suggest that the SP11/SCR protein is secreted in a cluster from the tapetal cells into the anther locule and translocated to the pollen surface (25).

Recently, NMR analysis revealed the tertiary solution structure of the SP11 protein of the S_8-haplotype (S_8-SP11) (48), which specifically induces the SI response on S_8 stigma at a dose of as little as 50 fmol per stigma (83). S_8-SP11 folds into an α/β sandwich structure made up of a twisted three-stranded β-sheet layer backed by another layer formed by an α-helix with flanking loops. Four disulfide bonds (C1–C8, C2–C5, C3–C6, and C4–C7) stabilize the structure. Although S_8-SP11 adopts a fold similar to plant defensins, the edges of the loop region are extensively stabilized by disulfide bonds and hydrophobic packing. This feature seems to be unique to SP11/SCR. The conserved aromatic (tyrosine) residue makes hydrophobic contact with the sulfur atom of C7, stabilizing the L1 loop structure between the α helix and the $\beta 2$ strand. Another conserved glycine residue helps form a stable type II β-turn at the flanking segment, connecting the $\beta 1$ strand to the α helix. Structure-based sequence alignment and homology modeling of allelic SP11/SCR suggest that the L1 loop region forms the hypervariable (HV) domain that bulges out from the protein body and serves as an *S*-haplotype-specific site. However, the L1 loop region contributes only the protein surface area of 470 $Å^2$, which is too small to confer the high affinity observed with the stigma receptor (see below), suggesting that an additional interface(s) is required (48). In support of this was an experiment that used site-directed (alanine-scanning) mutagenesis, and the results suggest that both the C3-C4 region (corresponding to L1 loop region) and the C5-C6 region (corresponding to L2 loop region) contribute to the SRK binding (4). Similarly, mutations of each of the conserved cysteines or a tyrosine abolished the activity of SP11/SCR, whereas the replacement of a conserved glycine residue with valine was tolerated. A domain-swapping experiment was also conducted, but the results were more perplexing. For example, the SCR_6 (SCR from the S_6-haplotype) protein in which the C5-C6 region is derived from SCR_{13} acquired the S_{13}-haplotype specificity, whereas the SCR_{13} protein in which the C5-C6 region is derived from SCR_6 retained the S_{13}-haplotype specificity. Although further studies are required, these results support the fact that the arrangement of specificity determinants might vary significantly between SCR variants.

Interaction between Male and Female Determinants

Two different biochemical approaches demonstrated *S*-haplotype-specific interactions between the male (SP11/SCR) and the female (SRK) determinants. In one experiment, the interactions between tagged versions of recombinant SRK and SP11/SCR were analyzed. The extracellular domain of SRK_6-FLAG ($eSRK_6$-FLAG) was expressed in tobacco leaves, and the SCR (SCR-Myc-His_6) was expressed in bacteria. The $eSRK_6$-FLAG was shown by both pull-down and enzyme-linked immunosorvent assay (ELISA) to interact more strongly with SCR_6-Myc-His_6 than with SCR_{13}-Myc-His_6 (28). In another experiment, ^{125}I-labeled S_8-SP11 (^{125}I-S_8-SP11) was used to monitor the interaction with the stigmatic receptor (83). The ^{125}I-S_8-SP11 specifically bound the stigmatic microsomal membranes of the S_8 homozygote. Scatchard analysis indicated the presence of both a high-affinity (K_d = 0.7 nM, B_{max} = 180 fmol/mg protein) and a low-affinity (K_d = 250 nM, B_{max} = 3 pmol/mg protein) binding site. Cross-linking experiments revealed that the high-affinity binding site consisted of proteins of 110 and 60 kDa. Immunoprecipitation experiments suggested that the 110-kDa protein is SRK_8. The 60-kDa protein is assumed to be SLG_8 or a truncated form of SRK (designated eSRKs), but currently its identity is not clear. Furthermore, in an in vitro phosphorylation assay, autophosphorylation of SRK_8 on the stigma plasma membrane was induced by S_8-SP11, but not by S_9-SP11, at a kinetically relevant concentration. This result clearly indicates that SP11/SCR alone can activate SRK in an *S*-haplotype-specific manner (83).

Signal Transduction in the Incompatible Stigmatic Cell

Now that the molecular basis of self-pollen recognition, which is the first step of the self-incompatibility response, has been established, the present focus of research has turned to characterizing the downstream signaling pathway(s) (**Figure 2**). The question that needs to be answered is how is the SRK activation transduced into the inhibition of self-related pollen? Presently, two molecules have been identified as positive signaling mediators of this pathway.

One is the arm repeat–containing protein, ARC1 (Armadillo-repeat-containing 1), a stigma protein first identified in a yeast two-hybrid screen as a protein interacting with the cytoplasmic domain of SRK (17, 51). This interaction, which requires the *C*-terminal arm repeats region of ARC1 and an active SRK kinase domain, results in the phosphorylation of ARC1 in vitro. Suppression of ARC1 expression by antisense cDNA causes a partial loss of the SI response, suggesting that ARC1 functions as a positive effector of SI signaling (75). However, the incomplete loss of the SI response implies that another branch of the signaling pathway exists, although it could also be attributed to residual *ARC1* expression in the transgenic stigmas. Scrutinizing the ARC1 sequence has suggested that it contains a U-box motif, a modified RING-finger; furthermore, recent analyses demonstrated that ARC1 has U-box-dependent E3 ubiquitin ligase activity (74). When expressed in cultured tobacco cells, ARC1 was distributed throughout the cytosol, but localized to the proteasome/COP9 signalosome in the presence of an active SRK kinase domain. In the pistil, levels of ubiquitinated protein increased after incompatible pollinations, but the increase was not apparent in ARC1 antisense-suppressed pistils. Furthermore, proteasome inhibition disrupts the SI response. Therefore, it was proposed that ARC1 is activated by SRK to promote the ubiquitination and proteasomal degradation of stigmatic proteins that support pollen germination and/or pollen tube growth. Other scenarios are possible, however, because ubiquitination has functions unrelated to protein degradation, such as subcellular targeting of proteins. Identifying ARC1 substrates will therefore be an essential next step in further dissecting the process of pollen rejection.

Another molecule is the *M* locus protein kinase (MLPK), which was recently identified after re-examining the *modifier* (*m*) gene, a recessive mutant gene responsible for the self-compatibility of *B. rapa* var Yellow Sarson (49), that was once thought to encode an aquaporin-like protein, MIP-MOD (24), although this turned out not to be the case (14). MLPK is a protein kinase belonging to the subfamily of the receptor-like cytoplasmic kinase (RLCK), which has a common monophyletic origin with receptor-like kinases but has no apparent signal sequence or transmembrane domain (66). MLPK from Yellow Sarson has a missense mutation in the conserved kinase subdomain VIa, resulting in the loss of kinase activity. The *mm* plants exhibit a completely self-compatible phenotype, and the transient expression of MLPK can restore the ability of *mm* papilla cells to reject self-pollen. These results suggest that MLPK is a positive mediator of SI signaling, and that MLPK localizes upstream in the signaling pathway, assuming that the pathway from SRK divides into multiple routes. In addition, MLPK has a typical plant *N*-myristoylation motif [MGXXXS/T(R)] at the *N* terminus, and is present in the plasma membrane of the stigma, suggesting that MLPK acts in the vicinity of SRK. If ARC1 is also the primary component of SRK signaling, then the SRK signal transferred to MLPK must return to SRK because ARC1 is a direct downstream effector of SRK. Taking the various factors together, MLPK may form a signaling complex with SRK that mediates the rejection response (49). Very little is known about the function of RLCKs, although the genome of *Arabidopsis thaliana* contains as many as 150 RLCKs (66). MLPK is the first example showing that a RLCK member mediates the signaling of receptor-like kinases. Further studies will address the precise relationship between SRK and MLPK in SI signaling.

In addition to these positive mediators of SI signaling, several identified components are expected to negatively regulate the pathway. Two thioredoxin-h proteins, THL1 and THL2, are interacting stigma proteins that were identified in a yeast two-hybrid screen. THL1 and THL2 interact with a conserved cysteine at the transmembrane domain of SRK in a phosphorylation-independent manner (1, 51). In vitro phosphorylation experiments demonstrate that THL1 inhibits the autophosphorylation activity of SRK in the absence of an "activating" component of the pollen coat (presumably SP11/SCR) (3). These results suggest that THL1 and THL2 may function as negative regulators, preventing the constitutive activation of the SI pathway. However, the relationship between THL1/2 and SRK activation requires further study because a different experimental system has not revealed any involvement of thioredoxins in the SP11/SCR-induced SRK activation process (83). Another negative-regulator candidate is a protein phosphatase, KAPP (kinase-associated protein phosphatase), which interacts with the kinase domains of many receptor-like kinases. In *Arabidopsis*, a series of transformation experiments suggest that KAPP plays a general role in the downregulation of various receptor kinases. Recently, a KAPP homolog was isolated from a stigma cDNA library of *B. oleracea* (90). This KAPP homolog interacts with, and is phosphorylated by, the kinase domain of SRK in vitro. In addition, KAPP dephosphorylated SRK as a substrate, suggesting its negative role in SRK downregulation. Recently, a yeast two-hybrid screen identified two kinds of additional interacting proteins, calmodulins 1 and 2, and a sorting nexin, SNX1. The calmodulins interact with an amphiphilic helix in the SRK subdomain VIa. In animal cells, both calmodulin and sorting nexin are implicated in the downregulation of receptor kinase activity. As with KAPP, the calmodulins and SNX1 interact in vitro with diverse members of plant receptor kinases, suggesting their general role in the receptor-kinase-mediated signaling pathways.

Another important finding was obtained from the comparative analysis of the self-compatible *A. thaliana* and its close relative the self-incompatible *A. lyrata* (35). Despite the fact that the *S*-locus region of *A. lyrata* contained functional *SRK* and *SP11/SCR* genes, the relevant genomic region of *A. thaliana* contained truncated and nonfunctional *SRK* and *SP11/SCR*, suggesting that the self-compatibility of *A. thaliana* might be due to the inactivation of the *S*-determinant genes. To support this, the transfer of *SRK* and *SCR* genes of *A. lyrata* are sufficient to impart a self-incompatible phenotype in *A. thaliana* (53). This transformation clearly demonstrates that the entire signaling cascade leading to inhibition of self-related pollen is retained in *A. thaliana*. Thus, the self-incompatible line of *A. thaliana* is an ideal tool for future genetic and molecular dissection of the SRK-mediated signal transduction cascade.

SOLANACEAE-TYPE SELF-INCOMPATIBILITY

The Solanaceae, Rosaceae, and Scrophulariaceae families all share a female *S*-determinant, an S-RNase (12, 32). The S-RNase was first identified in the Solanaceae so we refer to this S-RNase-mediated type of SI as Solanaceae-type SI. The Solanaceae-type SI is under gametophytic control (GSI) and the rejection of self-pollen occurs during pollen tube growth in the style. Recently, the genomic sequences around the S-RNase genes were thoroughly analyzed in these taxa, with the net result of finally identifying the elusive male *S*-determinant. The molecular nature of the identified male *S*-determinant suggests a new model of how these determinants are involved in the specific rejection of self-pollen.

Female Determinant

The female determinants were first identified in the self-incompatible *Nicotiana alata* as the style glycoproteins of ~30 kDa that cosegregate with the *S*-haplotype in genetic crosses. This enabled the identification and cloning of many related proteins from members of the Solanaceae and other families. Sequence data revealed that the style proteins contain a region homologous to the catalytic domain of the fungal T2-type ribonucleases. Further studies confirmed that these proteins possess ribonuclease activity and thus are referred to as S-RNases (45). S-RNases are expressed exclusively in the pistil, with the protein localized mostly in the upper segment of the style where inhibition of the self-pollen tubes occurs. The function of the S-RNases in SI was directly confirmed by gain- and loss-of-function experiments (38, 50). These experiments demonstrated that the S-RNase is the sole female determinant responsible for the *S*-haplotype specificity of the pistil. S-RNases are glycoproteins with one or more *N*-linked glycan chains. An engineered S_3-RNase of *Petunia inflata* that was engineered so that only the *N*-glycosylation site was knocked out, retained its activity to reject S_3 pollen, suggesting that the *S*-haplotype specificity determinant of S-RNases resides in the protein backbone and not in the glycan chains.

The S-RNases are highly divergent, with amino acid sequence identity ranging from 38% to 98% in the Solanaceae species (47). Despite this high sequence diversity, the S-RNases contain a number of conserved regions. S-RNases from the Solanaceae contain five highly conserved regions, designated C1 through C5, and those from the Rosaceae and the Scrophulariaceae have similar structural features except that the C4 region is absent. C2 and C3 are the regions similar to fungal RNase T2, and each contains a conserved catalytic histidine residue. There are two hypervariable regions in the S-RNases from the Solanaceae, termed HVa and HVb, whereas only one hypervariable region, corresponding to HVa of the solanaceous S_{f11}-RNases, has been detected in the rosaceous S-RNases. The crystal structures of a solanaceous S_{f11}-RNase from *Nicotiana alata* and a rosaceous S_3-RNase from *Pyrus pyrifolia* have been determined (23, 41). Both S-RNases have very similar structures consisting of eight helices and seven β-strands, and this topology is typical of the RNase T2 family of enzymes. The amino acid residues constituting the substrate-binding sites of these S-RNases can be geometrically superimposed over those of the RNase T2 enzymes. The most remarkable difference between the two S-RNases is in their HV regions. The HVa and HVb regions of solanaceous S_{f11}-RNase are composed of a long, positively charged loop followed by a part of an α-helix and a short, negatively charged α-helix, respectively. Because these HV regions are geometrically close to one another and exposed to the molecular surface, they both are expected to form a domain interacting with the male *S*-determinant. Domain-swapping experiments of S-RNases support this model (42, 43). S_{11}-RNase and S_{13}-RNase of *S. chacoense* differ only in 10 amino acids, 3 of which are located in the HVa and 1 in the HVb region. When the amino acids of HVa and HVb of S_{11}-RNase were changed to those of S_{13}-RNase, transgenic plants expressing this hybrid S-RNase rejected S_{13} pollen but not S_{11} pollen. These results clearly suggest that the HVa and HVb regions play a key role in determining the *S*-haplotype specificity, despite the fact that the involvement of other regions cannot be ruled out. In contrast to the solanaceous S-RNases, the HV region of the rosaceous S_3-RNase comprises a positively charged long loop followed by a short α-helix, but the α-helix corresponding to HVb does not exist (41). In the rosaceous S-RNase, the HVa region alone may form the interacting domain corresponding to the HVa and HVb regions of the solanaceous S-RNases.

S-haplotype-specific pollen rejection requires high levels of S-RNase expression. The concentration of S-RNase in the extracellular matrix is estimated at 10–50 mg/ml, and only the transformants with an equivalent amount of S-RNase expression are able to acquire new *S*-haplotype specificities. The ribonuclease activity of S-RNases is essential for pollen rejection (18). Furthermore, a radioactive tracer experiment showed that pollen RNA is degraded specifically after incompatible pollination (44). Thus, S-RNases function as highly specific cytotoxins that inhibit the growth of incompatible pollen (**Figure 3**).

Although S-RNase is the sole female factor determining the *S*-haplotype specificity of the pistil, a requirement of other stylar factors for the full function of S-RNase has been suggested (5). One such factor is HT-B, a

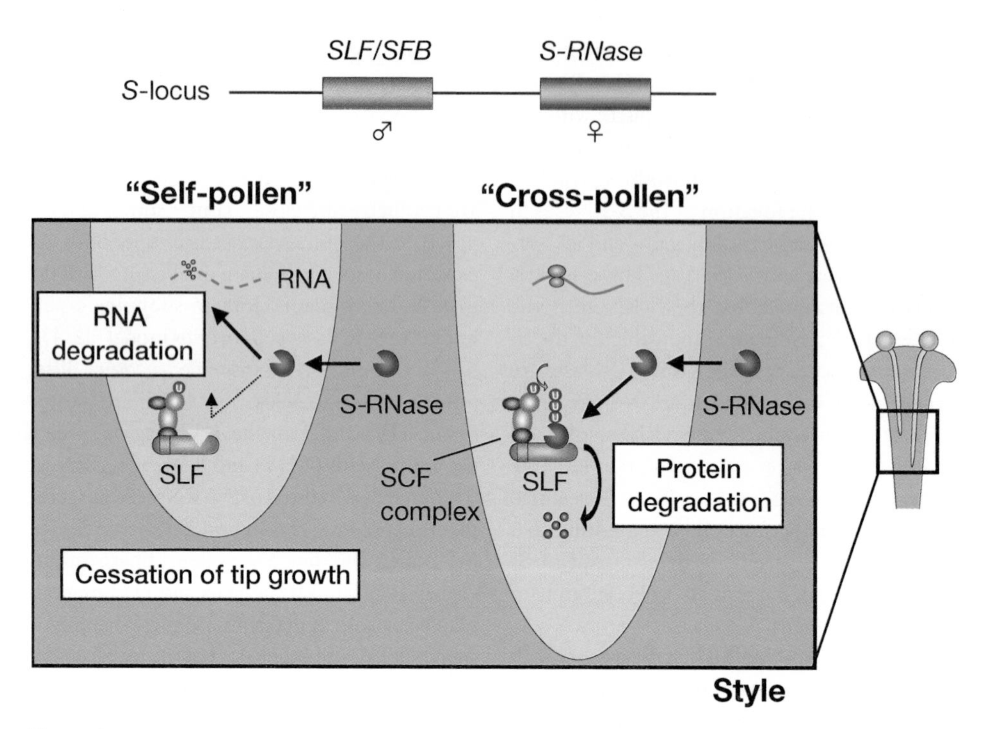

Figure 3

Molecular model of the self-incompatibility response in the Solanaceae, Rosaceae, and Scrophulariaceae. The *S*-locus consists of two genes, *S-RNase* and *SLF/SFB*. S-RNase is the female determinant and is secreted in large amounts into the extracellular matrix of the style. In a pollinated style, S-RNase is incorporated into the pollen tubes and functions as a cytotoxin that degrades pollen RNA. Although the S-RNase enters the pollen tubes regardless of their *S*-haplotypes, RNA degradation occurs only in self-pollen tubes. SLF/SFB is the male determinant and is a member of the F-box family of proteins, which generally function as a component of an E3-ubiquitin ligase complex. Thus, SLF/SFB is expected to be involved in ubiquitin-mediated protein degradation of nonself-S-RNases.

small asparagine-rich protein that was originally identified during a differential screen performed to identify stylar genes expressed in the self-incompatible *Nicotiana alata* but not in the self-compatible *Nicotiana plumbaginifolia* (46). Homologs of HT-B were also identified in two other genera of the Solanaceae, *Lycopersicon* and *Solanum* (33, 34, 55). In a comparative analysis of self-incompatible and self-compatible taxa of *Lycopersicon*, the expression of *HT-B* gene was not detected in all self-compatible taxa (33, 34). A requirement for HT-B protein was demonstrated by an RNAi suppression experiment in self-incompatible *Solanum chacoense*. Two *HT-B*-suppressed transformants expressed S-RNase normally but did not show *S*-haplotype-specific pollen rejection. These results suggest that the HT-B protein is implicated in the SI response, although its exact function remains unclear (55).

Male Determinant

The molecular nature of the male determinant and the molecular mechanisms of how S-RNases degrade pollen RNA in an *S*-haplotype-specific manner were long-standing mysteries of S-RNase-mediated SI. One plausible model was the "inhibitor model," in which the pollen *S*-determinant was postulated to be an inhibitor that could inhibit all S-RNases with the exception of the cognate S-RNase (31,

45). The immunocytochemical observations that all S-RNases could enter the pollen tube regardless of *S*-haplotype supported this model (40). Some refined "inhibitor models" were also proposed (39, 40), but such an S-RNase inhibitor has yet to be identified.

The male determinant was recently identified through genomic analyses of the *S*-locus region. Genomic analyses were first conducted on solanaceous species such as *Petunia inflata*, *Petunia hybrida*, *Lycopersicon peruvianum*, and *Nicotiana alata*. However, the *S*-locus of these species is located in the subcentromeric region and surrounded by abundant repetitive sequences that have hampered chromosomal walking (8, 91). The first clue for the male determinant was obtained from sequence analysis of the *S*-locus region of *Antirrhinum hispanicum*, a member of the Scrophulariaceae. The region of the S_2-haplotype contained a novel F-box protein gene, $AhSLF\text{-}S_2$ (*A. hispanicum* *S*-locus F-box of S_2-haplotype), which is specifically expressed in anther and pollen grains of S_2-haplotype (36). However, no gene allelic to $AhSLF\text{-}S_2$ has been identified in other *S*-haplotypes, and a gene with an extremely high sequence similarity (97.9% amino acid sequence identity) has been found in other lines with different *S*-haplotypes. It was thus unclear whether $AhSLF\text{-}S_2$ encoded the pollen *S*-determinant. Genomic analysis of the *S*-locus of *Prunus mume*, a member of the Rosaceae, reveals that the ~60-kb genomic region around the *S-RNase* gene contains as many as four F-box genes (7). Among them, only one F-box gene, termed *PmSLF*, fulfills the conditions of a pollen *S*-determinant gene: (*a*) it is located within the highly divergent genomic region of the *S*-locus, (*b*) it exhibits *S*-haplotype-specific diversity (78% to 81% amino acid identity), and (*c*) it is specifically expressed in pollen (7). Around the same time, polymorphic F-box genes were also found in the *S*-locus region of *Prunus dulcis*, *Prunus avium*, and *Prunus cerasus*, and were independently named SFB (*S*-haplotype-specific F-box) (86, 95). SLF/SFB from *Prunus* species fulfilled all conditions required of the pollen *S*-determinant. Aligning deduced amino acid sequences of SLF/SFBs of these *Prunus* species revealed the presence of two hypervariable regions, HVa and HVb, at the *C* terminus (32, 86). Two self-compatible haplotypes of *P. avium* and *P. mume* encoded partial loss-of-function mutations in SLF/SFB, which lack both HVa and HVb regions (87). This fact provides additional evidence that the SLF/SFB is the pollen *S*-determinant.

The conclusive evidence that SLF/SFB encodes the pollen *S*-determinant was finally obtained from transformation experiments in *Petunia inflata* (67). A thorough search of the pollen *S*-determinant in a huge *S*-locus region (328-kb BAC contig of S_2-haplotype) identified a polymorphic F-box gene, named *PiSLF*, ~161-kb downstream of the *S-RNase* gene. Although the genomic region outside this contig contained two more polymorphic F-box genes that were genetically linked to the *S*-locus, the *PiSLF* exhibited the highest sequence diversity. To ascertain whether *PiSLF* encodes the pollen *S*-determinant, a well-documented phenomenon termed "competitive interaction" was utilized. Competitive interaction is often observed in tetraploid plants. Among the diploid pollen grains produced, those carrying two different *S*-haplotypes (heteroallelic pollen), but not two of the same *S*-haplotypes (homoallelic pollen), fail to function in SI, although the molecular mechanism of the breakdown is unknown (6, 9). Consistent with this phenomenon, the transformation of S_1S_1, S_1S_2, and S_2S_3 plants with the S_2-allele of *PiSLF* ($PiSLF_2$) caused breakdown of their pollen function in SI. Furthermore, genotypic analyses of the progeny from self-pollinations of $S_1S_2/PiSLF_2$ and $S_2S_3/PiSLF_2$ revealed that only S_1 and S_3 pollen carrying the $PiSLF_2$ transgene (corresponding to heteroallelic pollen), but not S_2 pollen carrying $PiSLF_2$ (corresponding to homoallelic pollen), became self-compatible. These results conclusively demonstrate that SLF/SFB is the long-sought pollen *S*-determinant.

Mechanisms of *S*-haplotype-Specific Pollen Inhibition

In spite of the fact that both female and male determinants have been identified, the molecular mechanisms regulating how these molecules interact and specifically inhibit self-pollen growth remain unclear. The fact that RNase activity is required for the function of S-RNases, and that S-RNases are taken up by both self- and nonself-pollen tubes, suggests that S-RNases function inside pollen tubes as specific cytotoxins degrading the RNA of self-pollen (**Figure 3**). On the other hand, SLF/SFB contains a motif, called the F-box, which is best known for mediating interactions with other proteins that make up an enzyme complex referred to as the E3 ubiquitin ligase complex (15). E3 ubiquitin ligases act in conjunction with the E2 enzymes to ubiquitinate target proteins, which in many cases are degraded by the 26S proteasome. Recent biochemical studies suggest the involvement of AhSLF-S_2 in this protein degradation pathway, although it remains to be clarified whether *AhSLF-S_2* from *Antirrhinum* is an ortholog of *PiSLF* (57, 67). AhSLF-S_2 interacts with ASK1- and CULLIN1-like proteins, which are the expected components of the SCF complex. AhSLF-S_2 interacts with both self- and nonself-S-RNases, but appears to mediate degradation of only nonself-S-RNases. Although such interaction and degradation have not been reported for *Prunus* and *Petunia* SLF/SFBs, if this is the case then SLF/SFBs should interact with all S-RNases but ubiquitinate only nonself-S-RNases.

To explain the molecular mechanisms for this specificity, some hypothetical models that are compatible with the "inhibitor models" have been presented (7, 57, 67, 87). One model postulates that SLF/SFBs contain two separate interaction domains, like the classical "inhibitor model." One domain would bind to the hypervariable domain of its cognate S-RNase in an *S*-haplotype-specific way, and the other domain would bind to a domain common to all S-RNases. The *S*-haplotype-specific interaction is expected to somehow stabilize, or at least not alter, the S-RNase activity, and the general interaction would lead to the polyubiquitination and degradation of S-RNases. Another mechanism postulates the involvement of another molecule, such as a general inhibitor in a modified "inhibitor model." To support this, a pollen-expressed RING-finger protein, PhSBP1 (*P. hybrida* S-RNase-binding protein), interacts specifically with S-RNases in an *S*-haplotype-nonspecific manner (70). Because many RING-finger domain proteins also function as E3 ubiquitin ligases, PhSBP1 is postulated to be involved in the general degradation of S-RNases. In such cases, SLF/SFB is expected to bind to its cognate S-RNase as a pseudosubstrate and protect it from ubiquitination and subsequent degradation (12).

However, none of these models can explain the phenomenon of "competitive interaction." These models assume that the *S*-haplotype-specific binding between S-RNase and its cognate SLF/SFB is thermodynamically favored over general binding between S-RNase and nonself-SLF/SFB (or PhSBP1), and that the *S*-haplotype-specific binding somehow precludes S-RNases from ubiquitination, permitting RNase activity. In "competitive interaction," two SLF/SFBs in the heteroallelic pollen should each preferentially bind to their respective S-RNases in an *S*-haplotype-specific manner, leaving the RNases active. Therefore, in contrast to experimental observations, these models predict incompatibility for heteroallelic pollen. To explain the "competitive interaction," a refined version of the "inhibitor model" has been proposed in which the male *S*-determinants are predicted to form a multimer prior to interacting with the *S*-haplotype-specific binding site of S-RNases (39). However, SLF/SFBs are not likely to form a multimer during the interaction process. Thus, although both the female and male determinants have been identified, exactly how *S*-haplotype-specific pollen inhibition is achieved remains a mystery.

PAPAVERACEAE-TYPE SELF-INCOMPATIBILITY

SI in the field poppy, *Papaver rhoeas*, is also under gametophytic control (GSI) in that the *S*-phenotype of pollen is determined by its haploid *S*-genotype. However, the identified S-protein (female determinant) and the mechanisms involved in pollen inhibition differ dramatically from those in the Solanaceae. The overall number of *S*-haplotypes in *P. rhoeas* is estimated at around 66 (37). Although the exact nature of the male determinant is not known, a reliable in vitro bioassay system was developed in which pollen germination and pollen tube growth can be inhibited by the recombinant S-protein in an *S*-haplotype-specific manner, thus allowing the biochemical events that take place in the pollen following self-recognition to be studied in detail.

Female Determinant

The *bioassay* system that reproduces the SI reaction of *P. rhoeas* in vitro allowed the identification of biologically active stigmatic *S*-determinants (11). The S-proteins are small, secreted proteins (~15 kDa), some of which are modified by *N*-glycosylation (12, 47). Thus far, five allelic stigmatic *Papaver* S-protein genes have been cloned. The S-proteins are highly polymorphic and share between 51.3% and 63.7% amino acid sequence identity. Nevertheless, they have four conserved cysteine residues and a predicted conserved secondary structure that is comprised of six β-strands and two α-helices connected by seven hydrophilic surface loops. In contrast to the *S*–determinants in the Brassicaceae and Solanaceae, amino acid sequence variation is not found in hypervariable blocks, but rather throughout the S-proteins.

Because some recombinant S-proteins produced in *Escherichia coli* inhibit pollen germination in an *S*-haplotype-specific manner, the S-proteins must be the sole female determinant in *Papaver* SI. This also suggests that the glycan chains are not required for the *S*-determinant function. Further studies, using site-directed mutagenesis, revealed that some residues located in predicted surface loop 6 are crucial for pollen recognition. Mutations of the only amino acid residue (Asp-79 in loop 6 of S_1) that is variable across five available *S* alleles, and of adjacent highly conserved amino acids (Asp-77 and Asp-78), resulted in complete loss of ability of S_1-protein to inhibit S_1 pollen (30).

Male Determinant

When S-proteins were challenged on pollen tubes, the fastest change occurred in the shank region approximately 50 μm behind the pollen tube tip, where the concentration of intracellular free calcium ($[Ca^{2+}]_i$) increased within only a few seconds (**Figure 4**). Because this rapid alteration in $[Ca^{2+}]_i$ is induced by Ca^{2+} influx (13), the pollen *S*-determinant is expected to be a membrane-located receptor that is somehow associated with channels that conduct Ca^{2+}. Biochemical studies suggest the presence of a candidate receptor, which was named SBP (S-protein binding protein). SBP is a pollen-specific integral membrane proteoglycan of 70–120 kDa that binds specifically to stigmatic S-proteins but apparently does so in a non-*S*-haplotype-specific manner. Therefore, it has been proposed that SBP acts as an accessory receptor rather than as the pollen *S*-determinant (19). Biochemical analysis using site-directed mutagenesis reveals that all S-protein mutants that exhibit a reduced ability to inhibit incompatible pollen also exhibit reduced SBP binding activity, suggesting a direct involvement of SBP in the SI reaction (26).

Signaling Cascade in the Incompatible Pollen

The *S*-haplotype-specific interaction between stigmatic S-protein and its putative pollen receptor somehow induces a rapid increase of $[Ca^{2+}]_i$ in the pollen grains (tubes). The $[Ca^{2+}]_i$ increase occurs in the shank region of the pollen tube within a few seconds after the challenge of an incompatible S-protein and continues for several minutes (13) (**Figure 4**). The

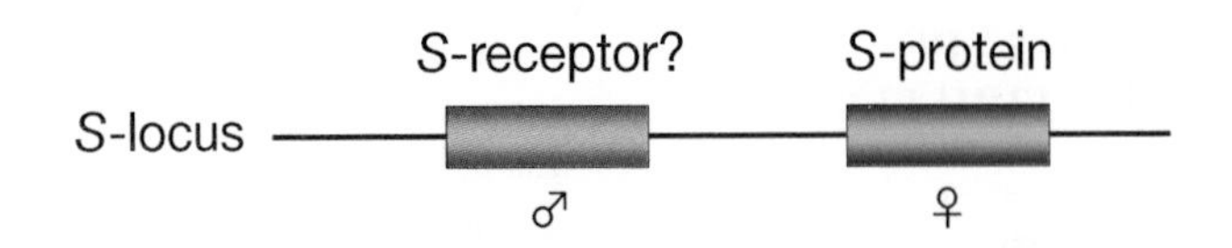

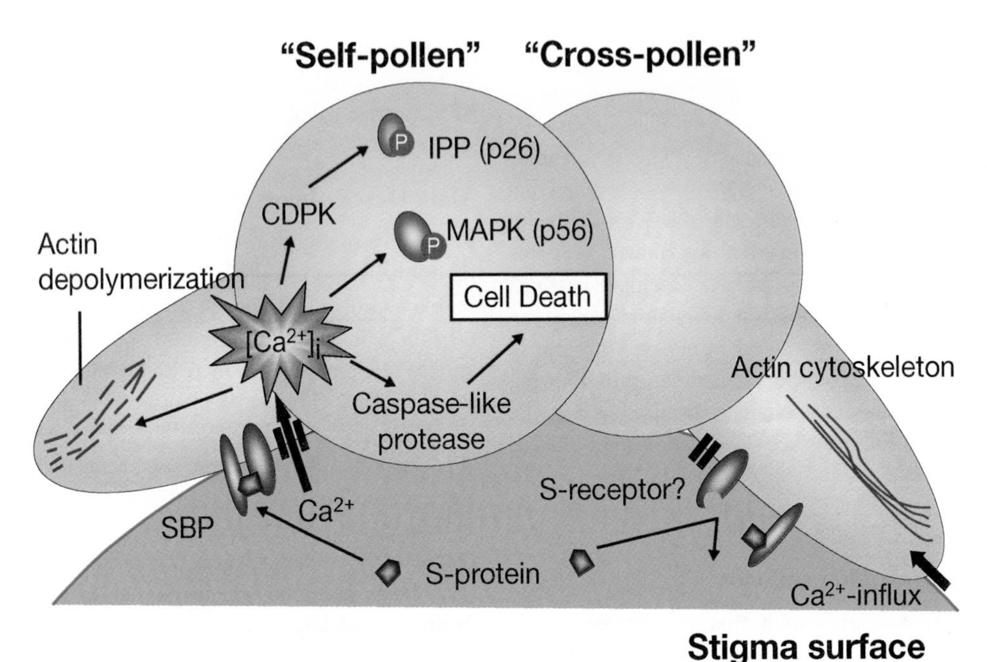

Figure 4

Molecular model of the self-incompatibility response in the Papaveraceae. Only the female determinant gene has been identified, which encodes a secreted stigma protein named S-protein. S-protein interacts with the assumed *S*-haplotype-specific pollen receptor (the putative male determinant) and induces Ca^{2+} influx in the shank of the pollen tube. SBP is an integral proteoglycan of the pollen plasma membranes and is expected to function as an accessory receptor. Ca^{2+}-influx stimulates increases in $[Ca^{2+}]_i$, with some contribution from the intracellular stores as well as from extracellular sources. These increases in $[Ca^{2+}]_i$ trigger the downstream signaling cascades that result in rapid growth inhibition and ultimately the death of incompatible pollen tubes.

increase of $[Ca^{2+}]_i$ in the subapical region accompanies the rapid loss of the oscillating high $[Ca^{2+}]_i$ gradient in the apical region of the pollen tubes. Most biochemical and physiological studies suggest that $[Ca^{2+}]_i$ acts as a second messenger that triggers multiple intracellular signaling cascades, resulting in rapid inhibition of pollen tube growth and ultimately the death of the incompatible pollen. Thus, recent studies have focused on identifying the signaling components downstream of the initial Ca^{2+} signals and their targets in pollen. These studies are expected to give insight into how the growth of the pollen tube is regulated, not only in the SI response, but also in a more general context.

The oscillating apical high $[Ca^{2+}]_i$ is typical of all tip-growing cells such as pollen tubes and neurons, although their biological significance is unclear. Thus, the loss of the oscillating $[Ca^{2+}]_i$ gradient will likely play a part in the initial inhibition of pollen tube growth.

One of the most rapid and dramatic physiological changes observed during the SI response is the dynamic rearrangement of the actin cytoskeleton in the pollen tube. Within 1–2 min postchallenge by an incompatible

S-protein, many F-actin bundles are lost and later punctate foci of actin accumulate in the cortex (16). Quantitative analysis of F-actin demonstrates that the SI-induced actin alterations are due to the depolymerization of F-actin, which starts within 1 min after SI challenge, with the reduction of F-actin reaching ~74% after 60 min (71). Similar depolymerization of actin was achieved in pollen by treatments that increase $[Ca^{2+}]_i$ artificially, suggesting that the actin depolymerization is not just a consequence of pollen tube growth arrest, but is part of the Ca^{2+}-mediated SI signaling cascade. Furthermore, recent biochemical studies have suggested that two kinds of actin-binding proteins, profilin and gelsolin, are involved in this Ca^{2+}-induced actin depolymerization process (22).

Another early target of the SI response is p26, a cytosolic 26-kDa pollen protein, the phosphorylation of which is induced within 90 sec of an incompatible S-protein challenge, with a further increase occurring during the next 400 sec. Because this timing coincides with the increase of $[Ca^{2+}]_i$ in SI-induced pollen and the protein kinase activity responsible for p26 phosphorylation is dependent on Ca^{2+} and calmodulin, the phosphorylation of p26 is also expected to be apart of the Ca^{2+}-mediated SI signaling cascade (12). Sequence analyses reveal that p26 shares approximately 80% amino acid sequence identity of plant-soluble inorganic pyrophosphatases, and its activity was confirmed by biochemical assays on the recombinant p26 protein. Furthermore, under conditions of raised $[Ca^{2+}]_i$, when p26 is phosphorylated, its pyrophosphatase activity is reduced (12, 59), indicating that p26 activity will likely be affected by the SI response. Soluble inorganic pyrophosphatases drive cellular biosynthetic reactions generating ATP and biopolymers, such as long-chain carbohydrates and proteins. Thus, it has been proposed that the inactivation of p26 during the SI response leads to inhibition of the pollen tube growth by depleting biopolymers required for tip growth. This proposal, however, needs to be experimentally tested.

Another target of SI signaling that displays a slightly delayed response is the putative mitogen-activated protein kinase (MAPK), p56 (60). In-gel kinase assays identified p56 as a 56-kDa protein kinase in pollen that is specifically activated after an SI challenge. The activation of p56 was detected 5 min after incompatible S-protein treatment, with peak activity at ~10 min. Pretreatment of growing pollen with the calcium channel blocker, lanthanum, inhibited the activation of p56, suggesting that p56 activation is also downstream of the Ca^{2+}-mediated SI signaling cascade. Although p56 has not yet been cloned, several pieces of biochemical evidence suggest that p56 is a MAPK. For example, p56 reacts with the TEY antibody that specifically binds activated MAPKs, and the kinase activity of p56 is abolished by apigenin, a specific inhibitor of MAPKs. Because the arrest of pollen tube growth precedes p56 activation, p56 is unlikely to play a role in the early inhibition events. One speculative hypothesis suggests that p56 might activate a PCD signaling cascade (see below) because some data have emerged suggesting a role for MAPK activation in the induction of PCD in plants.

PCD: programmed cell death

Recently, compelling evidence has suggested that SI challenge ultimately triggers PCD in incompatible pollen tubes (84). Nuclear DNA fragmentation, which is a hallmark of PCD, was observed in incompatible pollen tubes. Fragmentation was first detected 4 h after an incompatible S-protein challenge and increased to approximately 80% of affected pollen tubes within 14 h postchallenge. Recently, this DNA fragmentation was shown to be inhibited by pretreatment with the tetrapeptide DEVD, an inhibitor of caspase-3. Although no caspase homologue has been found in plant genomes, the result suggests the involvement of caspase-like activity in this signaling cascade. Furthermore, DEVD treatment revealed the biphasic nature of SI signaling. The SI-induced arrest of pollen tube growth is very rapid and occurs within 5 min of SI induction. The arrest was also observed in DEVD-pretreated pollen tubes. However, growth restarted in the tubes 15–45 min after the arrest, suggesting that by inhibiting the caspase-like activity, the initial growth arrest can be recommenced. Thus, there seems to be a biphasic SI response: rapid inhibition of pollen tube growth, followed by caspase-like protease activation, which presumably makes inhibition irreversible. Cytochrome c leakage from the mitochondria, a classic marker for PCD in many organisms, was also stimulated by SI induction in incompatible pollen tubes. Cytochrome c leakage was detected as early as 10 min after SI induction and increased until 120 min. To obtain further evidence of the involvement of caspase-like activity in this SI signaling, the enzymatic activity that cleaves poly(ADP-ribose) polymerase (PARP) was analyzed. The PARP-cleavage activity was detected in incompatible pollen tubes 2 h after SI induction and increased over time. Finally, each of these PCD hallmarks, i.e., nuclear DNA fragmentation, cytochrome c leakage, and PARP-cleavage activity, can be sequentially induced by artificially increasing $[Ca^{2+}]_i$ in the pollen tubes, suggesting that these are all a part of the series of reactions that make up the Ca^{2+}-mediated SI signaling cascade.

CONCLUSIONS AND FUTURE PERSPECTIVES

It is clear that the Brassicaceae, Solanaceae, and Papaveraceae have developed completely different self-/nonself-recognition systems. Nonetheless, the *S*-loci controlling these recognition reactions have structural commonalities, i.e., they contain at least two polymorphic determinant genes surrounded by highly divergent intergenic sequences (7, 10, 64, 94). Significant sequence heterogeneity of the *S*-locus explains how intergenic recombination is suppressed between two determinant genes, which results in the breakdown of SI. A more difficult issue to resolve is how new *S*-haplotypes evolve. In these two-gene systems, the female and male determinants must coevolve to maintain their interaction. Phylogenetic analyses of SRK and SP11/SCR in *Brassica* produced almost identical tree topologies for these two genes (61, 63, 92). These analyses provide evidence for the coevolution of the female and male determinant genes and suggest that new *S*-haplotypes arise from pre-existing *S*-haplotypes. Several interesting schemes outlining how such a process might occur have been proposed (4, 10, 42, 47, 88), although all schemes are hypothetical. Detailed comparative analyses characterizing more *S*-haplotypes, and additional structure-function relationship analyses of mutated determinants, should provide clues to this difficult issue.

Similar multiallelic two-gene recognition systems are also evident in the mating-type loci of fungi, which prevent self-matings (2, 29). The mating-type loci can be diallelic or multiallelic, similar to that seen for the *S*-loci. For example, in the phytopathogenic basidiomycete *Ustilago maydis*, the fusion of haploid cells and the maintenance of dikaryotic filamentous growth are controlled by two unlinked mating-type loci, termed *a* and *b*. Multiallelic genes at these loci (*a* locus exists in diallelic forms and *b* locus exists in at least 18 allelic forms) specify a large number of different mating types, and mating can only be completed between individuals that differ at both loci. The *a* locus contains two linked genes for lipopeptide ligands and their receptors, and is therefore analogous to the *S*-locus of the Brassicaceae. The *b* locus contains two genes for homeodomain proteins (bE and bW), and bE and bW from different alleles can form an active bE/bW heterodimer that is able to maintain filamentous growth. In multiallelic two-gene recognition systems, two contrary means can exist to lead to the same end of self-rejection, namely, "self-recognition" versus "nonself-recognition," or in other words, "opposition" versus "complementation" (**Figure 1**). In the mating-type recognition of fungi, the latter means are used, i.e., when two genes from different alleles are present in the mating cells, their interaction leads to compatible reactions. However, in the plant SI systems, at least in the Brassicaceae and the Papaveraceae types, the former means are used, i.e., when two genes from the same haplotypes meet, their interaction leads to incompatible reactions to reject self-pollen.

Classic genetic analyses reveal the presence of two SI types, GSI and SSI. Molecular analyses show that GSI contains at least two mechanisms, the Solanaceae type and the Papaveraceae type. For SSI, although the molecular mechanism has only been elucidated in the Brassicaceae, the presence of different mechanisms has been suggested. In the Convolvulaceae, the entire *S*-locus region has been cloned and sequenced in *Ipomoea trifida*, which has an SSI system (85). However, no homologous gene for the *Brassica SRK* or *SP11/SCR* has been found in the *S*-locus region. In the Asteraceae, a candidate gene for the female determinant, which was named SSP (stigma *S*-associated protein), has been identified, although it bears no resemblance to either SRK or SLG (21). Current evidence supports the view that SI evolved independently and probably multiple times in different lineages, with its recognition genes recruited in the family lineage by duplication and modification of preexisting genes that perform other functions (probably other cell-cell recognition or communication functions) in the plant. This view is also supported by the observation that the determinant genes identified thus far typically belong to large gene families that include members expressed in nonreproductive tissues (15, 58, 66, 89).

One great advantage of the studies on the self-/nonself-recognition mechanisms in SI is that both the female and male determinants involved in recognition are encoded in pairs in a single locus. Although not easy tasks due to the complex structures of the *S*-loci, several studies are attempting to identify novel *S*-determinant genes. These studies will not only lead to a greater understanding of how flowering plants discriminate self/nonself during fertilization, but should also shed light on the processes used by plants for cell-cell communication.

SUMMARY POINTS

1. In many species, the specificity of the SI response is determined by the haplotypes of the *S*-locus, which contains at least two separate multiallelic genes, the female and the male determinant genes.
2. SI does not represent one system, but rather a collection of divergent mechanisms, suggesting that SI evolved independently in several lineages.
3. In the Brassicaceae, the determinant genes encode a pollen ligand and its stigmatic receptor kinase, and their interaction induces incompatible signaling(s) within the stigma papilla cells.
4. In the Solanaceae, Rosaceae, and Scrophulariaceae, the determinants are a ribonuclease and an F-box protein, suggesting the involvement of RNA degradation and protein degradation within the system.
5. In the Papaveraceae, the only identified female determinant induces a Ca^{2+}-dependent signaling network that ultimately results in the death of incompatible pollen.

ACKNOWLEDGMENTS

We first and foremost would like to thank all of the researchers whose dedicated studies have greatly added to our current knowledge in this field and apologize for being unable to individually cite their work due to limited space. We would also like to thank former and current members of the Laboratory of Intercellular Communications at Nara Institute of Science and Technology for their various contributions, which enabled us to write this review. We are grateful to Emeritus Professor Kokichi Hinata at Tohoku University and Dr. Masao Watanabe at Iwate University for their helpful discussion. Research in the author's lab is supported by grants from the Ministry of Education, Culture, Sports, Science, and Technology of Japan, and from the Japan Society for the Promotion of Science.

LITERATURE CITED

1. Bower MS, Matias DD, Fernandes-Carvalho E, Mazzurco M, Gu T, et al. 1996. Two members of the thioredoxin-h family interact with the kinase domain of a *Brassica S* locus receptor kinase. *Plant Cell* 8:1641–50
2. Brown AJ, Casselton LA. 2001. Mating in mushrooms: increasing the chances but prolonging the affair. *Trends Genet.* 17:393–400
3. Cabrillac D, Cock JM, Dumas C, Gaude T. 2001. The *S*-locus receptor kinase is inhibited by thioredoxins and activated by pollen coat proteins. *Nature* 410:220–23
4. Chookajorn T, Kachroo A, Ripoll DR, Clark AG, Nasrallah JB. 2004. Specificity determinants and diversification of the *Brassica* self-incompatibility pollen ligand. *Proc. Natl. Acad. Sci. USA* 101:911–17
5. Cruz-Garcia F, Hancock CN, McClure B. 2003. S-RNase complexes and pollen rejection. *J. Exp. Bot.* 54:123–30
6. de Nettancourt D. 2001. *Incompatibility and Incongruity in Wild and Cultivated Plants*. Berlin/Heidelberg/New York: Springer-Verlag. 322 pp. 2nd ed.
7. Entani T, Iwano M, Shiba H, Che F-S, Isogai A, Takayama S. 2003. Comparative analysis of the self-incompatibility (*S*-) locus region of *Prunus mume*: identification of a pollen-expressed F-box gene with allelic diversity. *Genes Cells* 8:203–13
8. Entani T, Iwano M, Shiba H, Takayama S, Fukui K, Isogai A. 1999. Centromeric localization of an *S-RNase* gene in *Petunia hybrida* Vilm. *Theor. Appl. Genet.* 99:391–97
9. Entani T, Takayama S, Iwano M, Shiba H, Che F-S, Isogai A. 1999. Relationship between polyploidy and pollen self-incompatibility phenotype in *Petunia hybrida* Vilm. *Biosci. Biotechnol. Biochem.* 63:1882–88
10. Fobis-Loisy I, Miege C, Gaude T. 2004. Molecular evolution of the *S* locus controlling mating in the Brassicaceae. *Plant Biol.* 6:109–18
11. Foote HC, Ride JP, Franklin-Tong VE, Walker EA, Lawrence MJ, Franklin FC. 1994. Cloning and expression of a distinctive class of self-incompatibility (*S*) gene from *Papaver rhoeas* L. *Proc. Natl. Acad. Sci. USA* 91:2265–69
12. Franklin-Tong VE, Franklin FC. 2003. Gametophytic self-incompatibility inhibits pollen tube growth using different mechanisms. *Trends Plant Sci.* 8:598–605
13. Franklin-Tong VE, Holdaway-Clarke TL, Straatman KR, Kunkel JG, Hepler PK. 2002. Involvement of extracellular calcium influx in the self-incompatibility response of *Papaver rhoeas*. *Plant J.* 29:333–45
14. Fukai E, Nishio T, Nasrallah ME. 2001. Molecular genetic analysis of the candidate gene for *MOD*, a locus required for self-incompatibility in *Brassica rapa*. *Mol. Genet. Genomics* 265:519–25
15. Gagne JM, Downes BP, Shiu S-H, Durski A, Vierstra RD. 2002. The F-box subunit of the SCF E3 complex is encoded by a diverse superfamily of genes in *Arabidopsis*. *Proc. Natl. Acad. Sci. USA* 99:11519–24
16. Geitmann A, Snowman BN, Emons AMC, Franklin-Tong VE. 2000. Alterations in the actin cytoskeleton of pollen tubes are induced by the self-incompatibility reaction in *Papaver rhoeas*. *Plant Cell* 12:1239–52

17. Gu T, Mazzurco M, Sulaman W, Matias DD, Goring DR. 1998. Binding of an arm repeat protein to the kinase domain of the *S*-locus receptor kinase. *Proc. Natl. Acad. Sci. USA* 95:382–87
18. Huang S, Lee H-S, Karunanandaa B, Kao T-h. 1994. Ribonuclease activity of *Petunia inflata* S proteins is essential for rejection of self-pollen. *Plant Cell* 6:1021–28
19. Hearn MJ, Franklin FCH, Ride JP. 1996. Identification of membrane glycoprotein in pollen of *Papaver rhoeas* which binds stigmatic self-incompatibility (*S*-) proteins. *Plant J.* 9:467–75
20. Hiscock SJ, McInnis SM. 2003. Pollen recognition and rejection during the sporophytic self-incompatibility response: *Brassica* and beyond. *Trends Plant Sci.* 12:606–13
21. Hiscock SJ, McInnis SM, Tabah DA, Henderson CA, Brennan AC. 2003. Sporophytic self-incompatibility in *Senecio squalidus* L. (Asteraceae)—the search for *S*. *J. Exp. Bot.* 54:169–74
22. Huang S, Blanchoin L, Chaudhry F, Franklin-Tong VE, Staiger CJ. 2004. A gelsolin-like protein from *Papaver rhoeas* pollen (PrABP80) stimulates calcium-regulated severing and depolymerization of actin filaments. *J. Biol. Chem.* 279:23364–75
23. Ida K, Norioka S, Yamamoto M, Kumasaka T, Yamashita E, et al. 2001. The 1.55 Å resolution structure of *Nicotiana alata* S_{F11}-RNase associated with gametophytic self-incompatibility. *J. Mol. Biol.* 314:103–12
24. Ikeda S, Nasrallah JB, Dixit R, Preiss S, Nasrallah ME. 1997. An aquaporin-like gene required for the *Brassica* self-incompatibility response. *Science* 276:1564–66
25. Iwano M, Shiba H, Funato M, Shimosato H, Takayama S, Isogai A. 2003. Immunohistochemical studies on translocation of pollen *S*-haplotype determinant in self-incompatibility of *Brassica rapa*. *Plant Cell Physiol.* 44:428–36
26. Jordan ND, Kakeda K, Conner A, Ride JP, Franklin-Tong VE, Franklin FC. 1999. S-protein mutants indicate a functional role for SBP in the self-incompatibility reaction of *Papaver rhoeas*. *Plant J.* 20:119–25
27. Kachroo A, Nasrallah ME, Nasrallah JB. 2002. Self-incompatibility in the Brassicaceae: receptor-ligand signaling and cell-to-cell communication. *Plant Cell* 14 (Suppl.):S227–S38
28. Kachroo A, Schopfer CR, Nasrallah ME, Nasrallah JB. 2001. Allele-specific receptor-ligand interactions in *Brassica* self-incompatibility. *Science* 293:1824–26
29. Kahmann R, Bölker M. 1996. Self/nonself recognition in fungi: old mysteries and simple solutions. *Cell* 85:145–48
30. Kakeda K, Jordan ND, Conner A, Ride JP, Franklin-Tong VE, Franklin FCH. 1998. Identification of residues in a hydrophilic loop of the *Papaver rhoeas* S protein that play a crucial role in recognition of incompatible pollen. *Plant Cell* 10:1723–31
31. Kao T-h, McCubbin AG. 1996. How flowering plants discriminate between self and non-self pollen to prevent inbreeding. *Proc. Natl. Acad. Sci. USA* 93:12059–65
32. Kao T-h, Tsukamoto T. 2004. The molecular and genetic bases of S-RNase-based self-incompatibility. *Plant Cell* 16 (Suppl.):S72–S83
33. Kondo K, Yamamoto M, Itahashi R, Sato T, Egashira H, et al. 2002. Insights into the evolution of self-compatibility in *Lycopersicon* from a study of stylar factors. *Plant J.* 30:143–53
34. Kondo K, Yamamoto M, Matton DP, Sato T, Hirai M, et al. 2002. Cultivated tomato has defects in both *S-RNase* and *HT* genes required for stylar function of self-incompatibility. *Plant J.* 29:627–36
35. Kusaba M, Dwyer K, Hendershot J, Vrebalov J, Nasrallah JB, Nasrallah M. 2001. Self-incompatibility in the genus *Arabidopsis*: characterization of the *S* locus in the outcrossing *A. lyrata* and its autogamous relative *A. thaliana*. *Plant Cell* 13:627–43

36. Lai Z, Ma W, Han B, Liang L, Zhang Y, et al. 2002. An F-box gene linked to the self-incompatibility (*S*) locus of *Antirrhinum* is expressed specifically in pollen and tapetum. *Plant Mol. Biol.* 50:29–42
37. Lane MD, Lawrence MJ. 1993. The population genetics of the self-incompatibility polymorphism in *Papaver rhoeas*. *Heredity* 71:596–602
38. Lee H-S, Huang S, Kao T-h. 1994. S proteins control rejection of incompatible pollen in *Petunia inflata*. *Nature* 367:560–63
39. Luu D-T, Qin X, Laublin G, Yang Q, Morse D, Cappadocia M. 2001. Rejection of S-heteroallelic pollen by a dual-specific S-RNase in *Solanum chacoense* predicts a multimeric SI pollen component. *Genetics* 159:329–35
40. Luu D-T, Qin X, Morse D, Cappadocia M. 2000. S-RNase uptake by compatible pollen tubes in gametophytic self-incompatibility. *Nature* 407:649–51
41. Matsuura T, Sakai H, Unno M, Ida K, Sato M, et al. 2001. Crystal structure at 1.5-Å resolution of *Pyrus pyrifolia* pistil ribonuclease responsible for gametophytic self-incompatibility. *J. Biol. Chem.* 276:45261–69
42. Matton DP, Luu DT, Xike Q, Laublin G, O'Brien M, et al. 1999. Production of an S RNase with dual specificity suggests a novel hypothesis for the generation of new *S* alleles. *Plant Cell* 11:2087–97
43. Matton DP, Maes O, Laublin G, Xike Q, Bertrand C, et al. 1997. Hypervariable domains of self-incompatibility RNases mediate allele-specific pollen recognition. *Plant Cell* 9:1757–66
44. McClure BA, Gray JE, Anderson MA, Clarke AE. 1990. Self-incompatibility in *Nicotiana alata* involves degradation of pollen rRNA. *Nature* 347:757–60
45. McClure BA, Haring V, Ebert PR, Anderson MA, Simpson RJ, et al. 1989. Style self-incompatibility gene products of *Nicotiana alata* are ribonucleases. *Nature* 342:955–57
46. McClure BA, Mou B, Canevascini S, Bernatzky R. 1999. A small asparagine-rich protein required for *S*-allele-specific pollen rejection in *Nicotiana*. *Proc. Natl. Acad. Sci. USA* 96:13548–53
47. McCubbin AG, Kao T-h. 2000. Molecular recognition and response in pollen and pistil interactions. *Annu. Rev. Cell Dev. Biol.* 16:333–64
48. Mishima M, Takayama S, Sasaki K-i, Jee J-g, Kojima C, et al. 2003. Structure of the male determinant factor for *Brassica* self-incompatibility. *J. Biol. Chem.* 278:36389–95
49. **Murase K, Shiba H, Iwano M, Che F-S, Watanabe M, et al. 2004. A membrane-anchored protein kinase involved in *Brassica* self-incompatibility signaling. *Science* 303:1516–19**

This paper describes the identification and characterization of MLPK as a positive mediator of the SI signaling in *Brassica*.

50. Murfett J, Atherton TL, Mou B, Gasser CS, McClure BA. 1994. S-RNase expressed in transgenic *Nicotiana* causes *S*-allele-specific pollen rejection. *Nature* 367:563–66
51. Muzzurco M, Sulaman W, Elina H, Cock JM, Goring DR. 2001. Further analysis of the interactions between the *Brassica S* receptor kinase and three interacting proteins (ARC1, THL1 and THL2) in the yeast two-hybrid system. *Plant Mol. Biol.* 45:365–76
52. Nasrallah JB, Kao T-h, Chen C-H, Goldberg ML, Nasrallah ME. 1987. Amino-acid sequence of glycoproteins encoded by three alleles of the *S* locus of *Brassica oleracea*. *Nature* 326:617–19
53. Nasrallah ME, Liu P, Nasrallah JB. 2002. Generation of self-incompatible *Arabidopsis thaliana* by transfer of two *S* locus genes from *A. lyrata*. *Science* 297:247–49
54. Nou IS, Watanabe M, Isogai A, Hinata K. 1993. Comparison of *S*-alleles and *S*-glycoproteins between two wild populations of *Brassica campestris* in Turkey and Japan. *Sex. Plant Reprod.* 6:79–86

55. O'Brien M, Kapfer C, Major G, Laurin M, Bertrand C, et al. 2002. Molecular analysis of the stylar-expressed *Solanum chacoense* small asparagine-rich protein family related to the HT modifier of gametophytic self-incompatibility in *Nicotiana*. *Plant J.* 32:985–96
56. Ockendon DJ. 2000. The *S*-allele collection of *Brassica oleracea*. *Acta Hort.* 539:25–30
57. Qiao H, Wang H, Zhao L, Zhou J, Huang J, et al. 2004. The F-box protein AhSLF-S_2 physically interacts with S-RNases that may be inhibited by the ubiquitin/26S proteasome pathway of protein degradation during compatible pollination in Antirrhinum. *Plant Cell* 16:582–95
58. Ride JP, Davies EM, Franklin FCH, Marshall DF. 1999. Analysis of *Arabidopsis* genome sequence reveals a large new gene family in plants. *Plant Mol. Biol.* 39:927–32
59. Rudd JJ, Franklin-Tong VE. 2003. Signals and targets of the self-incompatibility response in pollen of *Papaver rhoeas*. *J. Exp. Bot.* 54:141–48
60. Rudd JJ, Osman K, Franklin FCH, Franklin-Tong VE. 2003. Activation of a putative MAP kinase in pollen is stimulated by the self-incompatibility (SI) response. *FEBS Lett.* 547:223–27
61. Sato K, Nishio T, Kimura R, Kusaba M, Suzuki T, et al. 2002. Coevolution of the *S*-locus genes *SRK*, *SLG*, and *SP11/SCR* in *Brassica oleracea* and *B. rapa*. *Genetics* 162:931–40
62. Schopfer CR, Nasrallah ME, Nasrallah JB. 1999. The male determinant of self-incompatibility in *Brassica*. *Science* 286:1697–700
63. Shiba H, Iwano M, Entani T, Ishimoto K, Shimosato H, et al. 2002. The dominance of alleles controlling self-incompatibility in *Brassica* pollen is regulated at the RNA level. *Plant Cell* 14:491–504
64. Shiba H, Kenmochi M, Sugihara M, Iwano M, Kawasaki S, et al. 2003. Genomic organization of the *S*-locus region of *Brassica*. *Biosci. Biotechnol. Biochem.* 67:622–26
65. Shiba H, Takayama S, Iwano M, Shimosato H, Funato M, et al. 2001. A pollen coat protein, SP11/SCR, determines the pollen *S*-specificity in the self-incompatibility of *Brassica* species. *Plant Physiol.* 125:2095–103
66. Shiu SH, Bleecker AB. 2001. Receptor-like kinase from *Arabidopsis* form a monophyletic gene family related to animal receptor kinases. *Proc. Natl. Acad. Sci. USA* 98:10763–68
67. **Sijacic P, Wang X, Skirpan AL,Wang Y, Dowd PE, et al. 2004. Identification of the pollen determinant of S-RNase-mediated self-incompatibility. *Nature* 429:302–5**
68. Silva NF, Goring DR. 2001. Mechanisms of self-incompatibility in flowering plants. *Cell. Mol. Life Sci.* 58:1988–2007
69. Silva NF, Stone SL, Christie LN, Sulaman W, Nazarian KAP, et al. 2001. Expression of the *S* receptor kinase in self-compatible *Brassica napus* cv. Westar leads to the allele-specific rejection of self-incompatible *Brassica napus* pollen. *Mol. Genet. Genomics* 265:552–59
70. Sims TL, Ordanic M. 2001. Identification of a S-ribonuclease-binding protein in *Petunia hybrida*. *Plant Mol. Biol.* 47:771–83
71. Snowman BN, Kovar DR, Shevchenko G, Franklin-Tong VE, Staiger CJ. 2002. Signal-mediated depolymerization of actin in pollen during the self-incompatibility response. *Plant Cell* 14:2613–26
72. Stein JC, Howlett B, Boyes DC, Nasrallah ME, Nasrallah JB. 1991. Molecular cloning of a putative receptor protein kinase gene encoded at the self-incompatibility locus of *Brassica oleracea*. *Proc. Natl. Acad. Sci. USA* 88:8816–20
73. Stephenson AG, Doughty J, Dixon S, Elleman C, Hiscock S, Dickinson HG. 1997. The male determinant of self-incompatibility in *Brassica oleracea* is located in the pollen coating. *Plant J.* 12:1351–59

This paper describes the identification of *SLF* gene in *Petunia inflata* (Solanaceae) and provides clear evidence that it encodes the male determinant of SI through the transformation experiments.

74. Stone SL, Anderson EM, Mullen RT, Goring DR. 2003. ARC1 is an E3 ubiquitin ligase and promotes the ubiquitination of proteins during the rejection of self-incompatible *Brassica* pollen. *Plant Cell* 15:885–98
75. Stone SL, Arnoldo M, Goring DR. 1999. A breakdown of *Brassica* self-incompatibility in ARC1 antisense transgenic plants. *Science* 286:1729–31
76. Suzuki G, Kai N, Hirose T, Fukui K, Nishio T, et al. 1999. Genomic organization of the *S* locus: identification and characterization of genes in *SLG/SRK* region of S_9 haplotype of *Brassica campestris* (syn. *rapa*). *Genetics* 153:391–400
77. Suzuki G, Kakizaki T, Takada Y, Shiba H, Takayama S, et al. 2003. The *S* haplotypes lacking *SLG* in the genome of *Brassica rapa*. *Plant Cell Rep.* 21:911–15
78. Suzuki T, Kusaba M, Matsushita M, Okazaki K, Nishio T. 2000. Characterization of *Brassica* *S*-haplotypes lacking *S*-locus glycoprotein. *FEBS Lett.* 482:102–8
79. Takasaki T, Hatakeyama K, Suzuki G, Watanabe M, Isogai A, Hinata K. 2000. The *S* receptor kinase determines self-incompatibility in *Brassica* stigma. *Nature* 403:913–16
80. Takayama S, Isogai A. 2003. Molecular mechanism of self-recognition in *Brassica* self-incompatibility. *J. Exp. Bot.* 54:149–56
81. Takayama S, Isogai A, Tsukamoto C, Ueda Y, Hinata K, et al. 1987. Sequences of *S*-glycoproteins, products of *Brassica campestris* self-incompatibility locus. *Nature* 326:102–5
82. Takayama S, Shiba H, Iwano M, Shimosato H, Che F-S, et al. 2000. The pollen determinant of self-incompatibility in *Brassica campestris*. *Proc. Natl. Acad. Sci. USA* 97:1920–25
83. Takayama S, Shimosato H, Shiba H, Funato M, Che F-S, et al. 2001. Direct ligand-receptor complex interaction controls *Brassica* self-incompatibility. *Nature* 413:534–38
84. **Thomas SG, Franklin-Tong VE. 2004. Self-incompatibility triggers programmed cell death in *Papaver* pollen. *Nature* 429:305–9**

This paper demonstrates the involvement of programmed cell death in the Ca^{2+}-mediated SI signaling of *Papaver rhoeas*.

85. Tomita RN, Suzuki G, Yoshida K, Yano Y, Tsuchiya T, et al. 2004. Molecular characterization of a 313-kb genomic region containing the self-incompatibility locus of *Ipomoea trifida*, a diploid relative of sweet potato. *Breed. Sci.* 54:165–75
86. Ushijima K, Sassa H, Dandekar AM, Gradziel TM, Tao R, Hirano H. 2003. Structural and transcriptional analysis of the self-incompatibility locus of almond: identification of a pollen-expressed F-box gene with haplotype-specific polymorphism. *Plant Cell* 15:771–81
87. Ushijima K, Yamane H, Watari A, Kakehi E, Ikeda K, et al. 2004. The *S* haplotype-specific F-box protein gene, *SFB*, is defective in self-compatible haplotypes of *Prunus avium* and *P. mume*. *Plant J.* 39:573–86
88. Uyenoyama MK, Zhang Y, Newbigin E. 2001. On the origin of self-incompatibility haplotypes: transition through self-compatible intermediates. *Genetics* 157:1805–17
89. Vanoosthuyse V, Miege C, Dumas C, Cock JM. 2001. Two large *Arabidopsis thaliana* gene families are homologous to the *Brassica* gene superfamily that encodes pollen coat proteins and the male component of the self-incompatibility response. *Plant Mol. Biol.* 16:17–34
90. Vanoosthuyse V, Tichtinsky G, Dumas C, Gaude T, Cock JM. 2003. Interaction of calmodulin, a sorting nexin and kinase-associated protein phosphatase with the *Brassica oleracea S* locus receptor kinase. *Plant Physiol.* 133:919–29
91. Wang Y, Wang X, McCubbin AG, Kao T-h. 2003. Genetic mapping and molecular characterization of the self-incompatibility (*S*) locus in *Petunia inflata*. *Plant Mol. Biol.* 53:565–80
92. Watanabe M, Ito A, Takada Y, Ninomiya C, Kakizaki T, et al. 2000. Highly divergent sequences of the pollen self-incompatibility (*S*) gene in class-I *S* haplotypes of *Brassica campestris* (syn. *rapa*) L. *FEBS Lett.* 473:139–44

93. Watanabe M, Takayama S, Isogai A, Hinata K. 2003. Recent progresses on self-incompatibility research in *Brassica* species. *Breed. Sci.* 53:199–208
94. Wheeler MJ, Armstrong SA, Franklin-Tong VE, Franklin FCH. 2003. Genomic organization of the *Papaver rhoeas* self-incompatibility S_1 locus. *J. Exp. Bot.* 54:131–39
95. Yamane H, Ikeda K, Ushijima K, Sassa H, Tao R. 2003. A pollen-expressed gene for a novel protein with an F-box motif that is very tightly linked to a gene for S-RNase in two species of cherry, *Prunus cerasus* and *P. avium*. *Plant Cell Physiol.* 44:764–69

Remembering Winter: Toward a Molecular Understanding of Vernalization

Sibum Sung and Richard M. Amasino

Department of Biochemistry, University of Wisconsin, Madison, Wisconsin 53706;
email: sbsung@biochem.wisc.edu, amasino@biochem.wisc.edu

Annu. Rev. Plant Biol.
2005. 56:491–508

doi: 10.1146/
annurev.arplant.56.032604.144307

1543-5008/05/0602-
0491$20.00

Key Words

cold acclimation, flowering, histone modifications, epigenetics, Polycomb-mediated gene repression

Abstract

Exposure to the prolonged cold of winter is an important environmental cue that favors flowering in the spring in many types of plants. The process by which exposure to cold promotes flowering is known as vernalization. In *Arabidopsis* and certain cereals, the block to flowering in plants that have not been vernalized is due to the expression of flowering repressors. The promotion of flowering is due to the cold-mediated suppression of these repressors. Recent work has demonstrated that covalent modifications of histones in the chromatin of target loci are part of the molecular mechanism by which certain repressors are silenced during vernalization.

Contents

INTRODUCTION

"I could while away the hours
Conferrin' with the flowers,
Consultin' with the rain.
And my head I'd be scratchin'
While my thoughts were busy hatchin'
If I only had a brain."

—Written by E.Y. Harburg and sung by the scarecrow in the movie version of "The Wizard of Oz"

Like the scarecrow, plants do not have a brain, but plants have nevertheless evolved the ability to "remember" a past exposure to winter. The flowering of many plants is either dependent on or promoted by prior exposure to the prolonged cold of a winter season. The process by which exposure to cold promotes flowering is known as vernalization. Many crop species need to undergo vernalization prior to flowering. In certain crops, for example beet and cabbage, the vernalization requirement prevents flowering in the first growing season and ensures that the desirable vegetative parts of the plants proliferate. The vernalization requirement of winter cereals permits them to be planted in the fall season to take maximum advantage of the favorable growing conditions in the spring. Not surprisingly, the first reports of the effect of cold exposure on flowering are from agricultural sources. For example, in 1619 it was noted in an agricultural bulletin that winter barley planted in the fall could sometimes be destroyed by a harsh winter; therefore, as a backup in case a damaged field needed to be replanted in the spring, imbibed barley seed could be placed outdoors in the late winter to acquire the ability to flower (as cited in 43). The term vernalization was coined by the infamous Russian agriculturalist Trofim Lysenko, who referred to the process as "jarovization." Jarovization was later translated from Russian into vernalization; vernal is derived from *vernum*, the Latin word for spring (as cited in 15, 53). There are many excellent reviews that provide a thorough coverage of the history of research on the process of vernalization (e.g., 9, 15, 53).

PHYSIOLOGY AND GENETICS OF VERNALIZATION

Physiology of Vernalization

Plants that need to experience winter to flower the next spring are either perennials, biennials, or winter annuals. Biennials and winter annuals senesce after flowering. The biennial designation is sometimes used to refer to plants with an obligate vernalization requirement, whereas winter annuals are sometimes defined as plants that exhibit a quantitative response to cold exposure (9, 15, 53). (Summer annuals do not need vernalization to flower, senesce after flowering, and complete their life cycle in one growing season.)

In many species, vernalization is not sufficient to induce flowering. Rather, vernalization renders plants competent to flower. This is illustrated by the classic experiments of Lang & Melchers with a biennial type of henbane (*Hyoscyamus niger*) (reviewed in 53). Biennial henbane has an absolute requirement for vernalization followed by a requirement for inductive long-day photoperiods to flower. Vernalized henbane plants kept in a noninductive short-day photoperiod grow vegetatively, but do not flower. The vernalized henbane plants readily flower, however, when shifted to inductive photoperiods. Therefore, during the period of vegetative growth in noninductive photoperiods, the plants "remember" that they had been exposed to cold. The duration of this "memory" varies from days to over a year depending on species. There are vernalizataion-requiring plants that do not exhibit memory (in these plants, the floral transition must occur during prolonged exposure to cold) (9). A useful definition of vernalization that covers the broad range of responses was provided by Chouard (15) in the last review on this topic in this series: "the acquisition or acceleration of the ability to flower by a chilling treatment."

Grafting and localized cooling studies demonstrate that the shoot apex is the site of cold perception during vernalization (53, 109). In most species, vernalization is a localized response; i.e., it is not graft transmissible. Yet there are examples in which the flowering of a nonvernalized shoot is promoted by grafting to a vernalized plant of the same genotype, or to a related variety of plant that does not require vernalization [e.g., pea (88) and henbane (53)]. In these graft-transmissible situations, it is possible that vernalization directly leads to the production of a flowering hormone, which is called "vernalin" (53). Alternatively, the graft donors may be producing a graft-transmissible substance that can bypass the need for vernalization.

Another classic study that demonstrated both the site of vernalization and the memory effect involved the in vitro regeneration of plants from various tissues of *Lunaria biennis* that had been exposed to a vernalizing cold treatment (108, 109). Plants regenerated from mature leaves of vernalized plants acted as if they had not been vernalized; only tissues that contained dividing cells (including root meristems and young developing leaves) regenerated into vernalized plants. Thus, only dividing cells (or perhaps cells in which DNA replication is occurring) are capable of becoming vernalized, and these cells maintain the vernalized state through cell divisions in tissue culture during the regeneration process. This type of experiment has also been done in *Arabidopsis* (11). The mitotically stable cellular memory of the vernalized state in these experiments, and in those with henbane described above, illustrate the epigenetic nature of vernalization in many species, namely the acquisition of a state of competence to flower that is mitotically stable in the absence of the inducing signal. This memory is lost as vernalized cells pass through meiosis and into the next generation. The loss of memory ensures that the vernalization requirement is re-established in each generation in monocarpic species.

Low Temperature Response—Cold Acclimation versus Vernalization

Temperature variation is a major seasonal change in temperate climates. Except for summer-annual plants, which finish their life cycle before winter arrives, it is critical for plants to have a mechanism to survive freezing temperatures during winter. Freezing tolerance is induced by exposure to low but nonfreezing temperature. Plants usually do not exhibit freezing tolerance during the warm growing season, but as the seasons change, plants sense the falling temperatures that signal the approaching winter and develop freezing tolerance. The process of preparing to withstand freezing temperatures is known as cold acclimation (reviewed in 102, 103).

Plants typically become cold-acclimated within a short period of time. Usually, one or two days of low but nonfreezing temperature is sufficient to cause cold acclimation in most plant species that are capable of cold acclimation (29, 32, 33, 62, 72, 102). This relatively quick establishment of freezing tolerance is necessary to deal with the sudden temperature

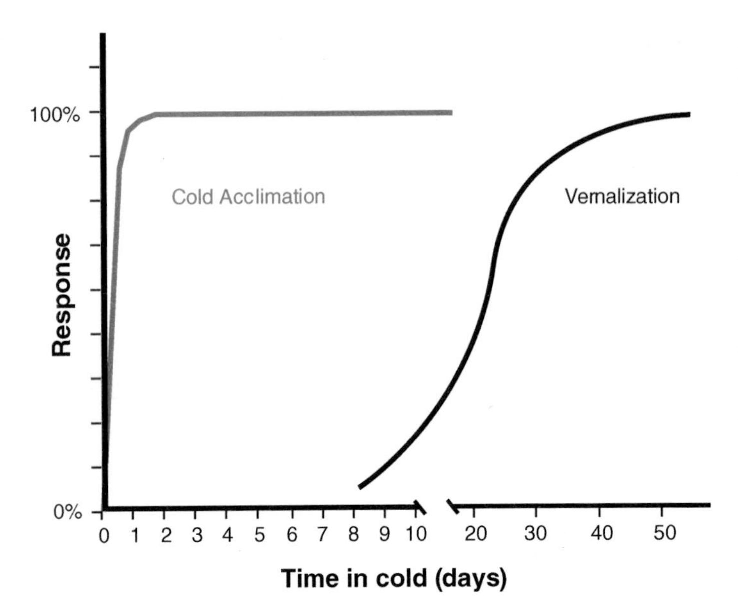

Figure 1

Typical time course of the cold acclimation and vernalization responses in *Arabidopsis thaliana*. The acquisition of freezing tolerance occurs within days, whereas vernalization requires several weeks of cold exposure.

CBF: C-repeat-/DRE-binding factor

fluctuations that are often characteristic of seasonal change. During cold acclimation, a suite of genes encoding a range of proteins are rapidly induced by cold (3, 29, 32, 33, 49, 62, 72). Comparing and analyzing promoters of these cold-induced genes have led to the discovery of a family of transcriptional activators, CBFs, that are necessary for their induction (26, 39). The CBFs are also cold induced and at least one CBF is induced by another cold-activated Myc-related transcriptional activator, ICE1 (14). Thus, cold acclimation results, at least in part, from a cascade of cold-induced transcription factors which, in turn, induce downstream genes encoding proteins involved in cold protection. As discussed above, this response is rapid: The activation of ICE1 and induction of CBFs occur within minutes of cold exposure and reach a peak after several hours (14, 25, 39, 66).

Although both cold acclimation and vernalization are responses to sensing low temperature, the duration of cold exposure that is required to initiate these responses is distinct (**Figure 1**). As discussed above, a rapid induction of cold-protective proteins is essential for surviving the colder temperatures that the plant may soon encounter. One advantage of a vernalization requirement is to prevent flowering in the fall season, i.e., to ensure that flowering does not occur until spring. Temperatures often fluctuate in the fall and it is critical that a short cold spell followed by warm weather is not sufficient for vernalization; otherwise, flowering could be initiated before the onset of winter, which would be disastrous for most plants. Thus, plants typically achieve a vernalized state only after exposure to a period of cold of sufficient duration to indicate that winter has passed. The flowering of many vernalization-requiring plants is also promoted by long days. This provides an extra level of insurance that flowering does not occur in the fall when the day lengths become shorter.

Temperatures below ∼−2°C are typically not effective for vernalization, and thus there may be periods of extreme cold in certain climates that are not "counted" in the measurement of the cold duration (53). Different species are adapted to a variety of winter climates; therefore, it is not surprising that the range of effective cold temperatures varies among species. For example, 2–4°C is optimal in *Arabidopsis*, whereas 8–17°C is optimal in onion (*Allium cepa*) (53). There is also variation in the duration of cold required for effective vernalization. In the two accessions of *Arabidopsis* tested, a strong vernalization response requires 40 days of continuous exposure to 4°C (57), whereas in a variety of radish (*Raphanus sativus*), 8 days of continuous exposure to 6°C is sufficient (20). It is important to note that, in the laboratory, the duration of cold required to achieve the vernalized state is often measured as the time of continuous exposure to the optimum cold temperature, which does not mimic conditions in the field.

Another process that, in many plants, requires exposure to prolonged cold during winter is the release of bud dormancy in the spring. Vernalization and the cold-mediated release of bud dormancy were previously compared (15) and will not be further considered in this review because there have not been any molecular developments in the study of bud dormancy that relate to recent findings in vernalization. An interesting area of future research will be to determine if there are parallels, at a molecular level, between vernalization and the release of bud dormancy.

Genetics of the Vernalization Requirement

In many species, there are both annual and biennial varieties, and crosses between annual and biennial varieties can establish the genetic basis for the vernalization requirement in a given species. In the few species that have been studied, the biennial or winter-annual habit is governed by a relatively small number of loci, either dominant or recessive depending on the species (**Table 1**). With henbane, a single dominant locus is responsible for the biennial habit (54), whereas a single recessive locus is responsible for the biennial habit in sugar beet (2). The vernalization requirement of many cereals, including wheat and barley, is controlled by one dominant and one recessive locus (18, 55, 104).

There are also summer-annual and winter-annual varieties (accessions) of *Arabidopsis thaliana*. The summer-annual types are typically used as models to study various aspects of plant biology because the ability to flower rapidly without vernalization provides for a conveniently short life cycle. The winter-annual types exhibit a facultative vernalization response. Napp-Zinn (76) first showed that in certain crosses of summer-annual and winter-annual accessions, a single dominant locus, which he named *FRIGIDA* (*FRI*), plays a major role in conferring the winter-annual habit in *Arabidopsis* (although other loci contribute). Subsequent genetics studies comparing various winter-annual and summer-annual accessions of *Arabidopsis* have revealed that variation at one or both of two loci, *FLOWERING LOCUS C* (*FLC*) and *FRI*, can account for a large portion of the winter-annual habit in *Arabidopsis* (12, 16, 46, 58, 59). *FLC* and *FRI* synergistically delay flowering in winter-annual accessions of *Arabidopsis*, and a loss-of-function mutation in either gene results in the loss of the late-flowering phenotype (41, 68).

Cloning *FLC* (68, 96) provided the first insight into the molecular nature of vernalization in *Arabidopsis*. *FLC* is a repressor of flowering that encodes a MADS-box transcriptional regulator. The presence of a dominant allele of *FRI* elevates *FLC* expression to a level that inhibits flowering (68, 96). Vernalization overcomes the effect of *FRI* by repressing *FLC* expression, and this repression is stably maintained after plants are returned to warm growth conditions (68, 96). Thus, the epigenetic repression of *FLC* is part of the vernalization-mediated "memory of winter" in *Arabidopsis*.

FLC is expressed predominantly in mitotically active regions (31, 67, 101), such as shoot and root apicies, which, as discussed above, are the sites of cold perception and the tissues that achieve the vernalized state. Although most of the flowering promotion by vernalization in *Arabidopsis* is due to *FLC* repression, there is clearly a component that is *FLC* independent (69). As discussed below, much of this *FLC*-independent component involves repressing other MADS-box genes related to *FLC*.

Table 1 Loci that confers vernalization requirement

Species	Loci
Arabidopsis thaliana[a]	FRI* FLC*
Brassicca rapa[b]	VFR1 (BrFLC1*) VFR2 (BrFLC5*)
Brassica napus[b]	VFN1 VFN2
Triticum aestivum[c] (Wheat)	VRN-Am1 (VRN1*) VRN-Am2 (VRN2*)
Hordeum vulgare[d] (Barley)	VRN-H1 VRN-H2 (VRN2*)
Hyocyamus niger[e] (Henbane)	Single dominant locus
Beta vulagris[f] (Beet)	b (recessive)
Pisum sativum[g] (Pea)	Late flowering Vegetative 2

[a](12, 16, 41, 46, 58, 59, 68, 96); [b](81, 93); [c](104, 111, 112); [d](55, 111); [e](54); f (2); [g](74).
*Genes have been cloned.

Summer-annual accessions of *Arabidopsis* often contain loss-of-function mutations in *FRI* (41). Therefore, some summer-annual types of *Arabidopsis* are derived from winter annuals by loss of *FRI*. Lesions in *FRI* have arisen independently several times (e.g., 23). Recently it was shown that certain summer-annual types contain an active *FRI* allele but also contain an allele of *FLC* that is not upregulated by *FRI* (23, 70). Thus, there are at least two routes by which winter-annual types of *Arabidopsis* have become summer annuals. Presumably these summer-annual-causing genetic changes resulted in an adaptation to a particular niche.

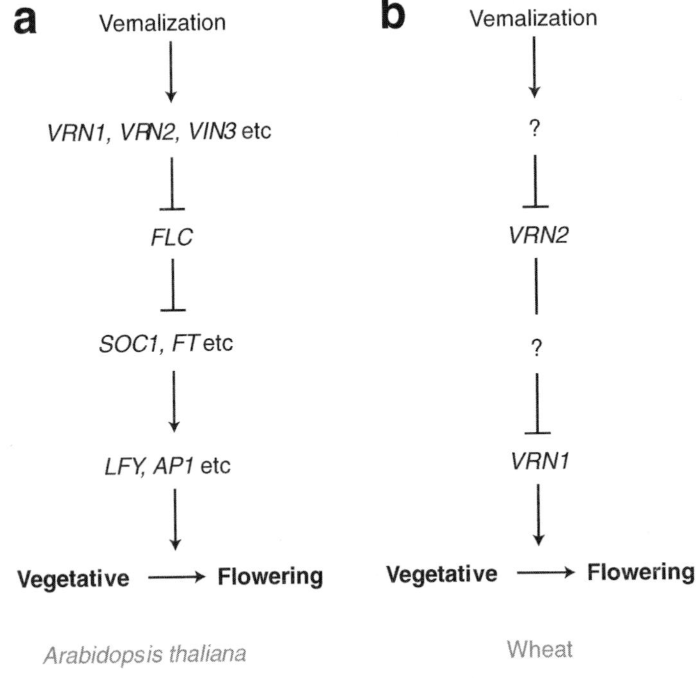

Figure 2

The status of genes currently known to be involved in the pathway from vernalization to flowering in (*A*) *Arabidopsis thaliana* and (*B*) wheat. Vernalization results in the epigenetic repression of *FLC* in *Arabidopsis* and *VRN2* in wheat. *FLC* and *VRN2* are both repressors of flowering. In *Arabidopsis*, *FLC* repression permits the activation of floral integrators such as *FT* and *SOC1* (these genes are called floral integrators because their expression is also affected by other flowering cues such as photoperiod) and, in turn, *FT* and *SOC1* activation leads to the activation of floral meristem identity genes, such as *LFY* and *AP1*. In wheat, the repression of *VRN2* results in *VRN1* activation and the promotion of flowering. It is not known whether *VRN1* is directly repressed by *VRN2*.

The inability of one allele of *FLC* to be upregulated by *FRI* is due to the presence of a transposable element in an *FLC* intron (70). The presence of the transposable element is associated with an island of heterochromatin (64) that may attenuate *FLC* expression by a mechanism similar to that which occurs during vernalization (see below). The idea that transposable element insertions can result in the "transposition of heterochromatin" was first noted by McClintock (65).

The facultative vernalization requirement of *Arabidopsis* can be changed to an obligate requirement simply by increasing the number of copies of the native *FLC* locus in the genome (67). This shows that there need not be a fundamental difference in the mechanism of an obligate versus a facultative requirement; rather, different levels of a repressor can be the cause.

Although the involvement of *FLC* and *FRI* in establishing the vernalization requirement seems to be conserved in other crucifers (81, 93), cereals have different genes that confer winter-annual behavior. Genetic analyses of the vernalization requirement in spring and winter varieties of wheat and barley have revealed two genes that act antagonistically. Dominant alleles of *VRN1* from wheat and barley cause a spring growth habit, whereas dominant alleles of *VRN2* from wheat and barley are necessary for a winter growth habit (55, 104, 111, 112). *VRN1* encodes an AP1-type MADS box gene that promotes flowering (112). *VRN2* encodes a ZCCT family zinc-finger protein that is downregulated by cold exposure (111). *VRN2* is a repressor that prevents flowering by repressing *VRN1* expression. Prolonged cold exposure promotes flowering by shutting off *VRN2* and relieving the repression of *VRN1*. Thus, *VRN2* may play a role analogous to that of *FLC*: Both are repressors that target genes required for flowering, and both repressors are turned off by cold exposure (**Figure 2**).

Dominant alleles of *VRN1* in wheat confer a spring habit because these alleles are not repressed by *VRN2* (112). One dominant *VRN1* allele contains a deletion that may remove a *cis* element required for *VRN2*-mediated repression (112). It is not known whether VRN2 directly interacts with *VRN1* regulatory elements.

One definition of the vernalization pathway is the system that measures the duration of cold exposure and, when a sufficient duration of cold exposure has been perceived, initiates changes in gene expression. By this definition, the vernalization pathway of both cereals and *Arabidopsis* acts to downregulate flowering repressors, but cereals and *Arabidopsis* appear to have evolved different targets of the vernalization pathway (**Figure 2**). Whether the actual cold-sensing pathways are conserved between cereals and *Arabidopsis* remains to be determined. It is possible that different groups of flowering plants evolved in a warm climate in which a vernalization response was not needed. If this were the case, then the vernalization response evolved independently when different groups of plants radiated into climates in which a vernalization response was advantageous. Yet these plants had related genomic "parts catalogs" from which to assemble a cold measurement system, and it will be interesting to learn more about how this system works in a range of plants.

MOLECULAR BASIS OF VERNALIZATION

As discussed above, the identification of *FLC* in *Arabidopsis* and *VRN2* in wheat provides clues about how vernalization affects meristem competence. Expression of repressors like *FLC* and *VRN2* in the shoot apex suppresses flowering. Vernalization provides competence to flower by repressing these floral repressors.

FLC repression in *Arabidopsis* and *VRN2* repression in cereals is stably maintained through mitotic cell divisions and serves as part of the "memory of winter" in vernalization (67, 111). Some insight into the basis of this epigenetic repression has come from genetic studies of the vernalization response in *Arabidopsis*.

Unlike biennial plants, which have an obligate vernalization requirement, vernalization promotes flowering in a facultative manner in winter-annual *Arabidopsis* accessions. The facultative nature of the promotion of flowering by vernalization in winter-annual accessions makes *Arabidopsis* an attractive system in which to study vernalization because plants containing mutations in this process will eventually flower and thus homozygous mutants are amenable to genetic studies. Genetic screens designed to find mutations that impair the vernalization response in *Arabidopsis* have led to the identification of three genes: *VRN2* (*VERNALIZATION 2*), *VRN1* (*VERNALIZATION 1*), and *VIN3* (*VERNALIZATION INSENSITIVE 3*). *VRN2* encodes a homolog of the *Drosophila* Supressor of Zeste 12 (Su(z)12), which is a Polycomb group (PcG) transcriptional repressor present in the Enhancer of Zeste (E(z)) complex (13, 17, 24, 51, 73). *VRN1* encodes a plant-specific DNA-binding protein (60). *VIN3* encodes a PHD finger containing protein (101). PHD-finger proteins are often found in protein complexes involved in chromatin remodeling (1, 28). The nature of the proteins encoded by *VRN2* and *VIN3* provide a clue that chromatin remodeling is involved in "memory of winter" (6, 101).

PHD: plant homeodomain

The Histone Code and Vernalization

The establishment and maintenance of cell identity involves pathways that activate and silence specific sets of genes in a tissue- and development-specific manner. Maintaining stable states of gene expression is a form of cellular memory. In the early stages of animal development, the genes of the PcG and trithorax group (trxG) are part of a widely conserved cell memory system that maintains cell identity patterns via cell type–specific transcription patterns (27, 79, 82). PcG and trxG control, respectively, the repressed and active transcriptional states of many developmentally and cell cycle–regulated genes in animals. Although *FLC* is repressed in response to an environmental cue rather than a developmental program, there are several similarities between the vernalization-mediated repression of *FLC* and the PcG-mediated repression of animal genes: (*a*) both are stable through mitotic cell divisions, (*b*) both are relieved during meiosis so that the repression system is reset for the next generation, (*c*) both repress genes in euchromatin, and (*d*) both require Su(z)12.

A breakthrough in determining how cellular memory is achieved was the discovery that Enhancer of zeste (E(z)) of the PcG complex is a histone methyltransferase (13, 17, 51, 73). Histone methyltransferases are one of many examples of enzymes that catalyze covalent modifications of the amino-terminal "tails" of histones (47). There are a range of modifications (methylations, acetylations, phosphorylations, ubiquitinations, etc.) that occur on specific amino acid residues of histone tails that can influence whether a locus assumes a transcriptionally active (euchromatin) or silent (heterochromatin) state (8, 19, 36, 40, 94, 105). The histone modifications are recognized by a variety of chromatin-associated proteins that govern transitions between active or silent chromatin (61, 107). For example, methylation of lysine 27 (K27) on histone H3 by E(z) leads to gene repression via the binding of the Polycomb Repression Complex 1 (PRC1), which contains the Polycomb protein (PC). The spectrum of histone modifications at a given locus is often called the "histone code" because the combination of modifications can create a unique state of gene activity.

During vernalization, *FLC* chromatin undergoes a series of histone modifications. Deacetylation of lysine 9 (K9) and lysine 14 (K14) on histone H3 (101) is followed by dimethylation of K27 and K9 on histone H3 (6, 101). In the *vrn1* mutant, the repressed state of *FLC* is not maintained, although dimethylation of K27 still occurs, suggesting that dimethylation of K27 is not sufficient to cause a stable repressed state of *FLC*. Vernalization-mediated dimethylation of K9 on histone H3 does not occur in *vrn1*, *vrn2*, or *vin3* mutants (6, 101), indicating that dimethylation of K9 on histone H3 plays a critical role in the maintenance of stable *FLC* repression. Thus, a series of histone modifications is associated with encoding the "memory of winter" during vernalization.

Establishment and Maintenance of Heterochromatin on *FLC*

Although there are broad similarities between vernalization-mediated target gene repression in *Arabidopsis* and PcG-mediated gene repression in animals, some aspects of these repression systems are different. Methylation of K27 on histone H3 is likely a specific mark for PcG-mediated repression in animals, as this chromatin modification is highly enriched at PcG target sites and poorly represented in regions of constitutive heterochromatin (17). Constitutive heterochromatin refers to regions of the genome that are always maintained as heterochromatin such as centromeres and telomeres. In PcG-mediated repression, regions that were euchromatin are converted to a heterochromatin-like state (79). The mitotic stability of this heterochromatin-like state is postulated to be mediated by the binding of methylated K27 by PC (21, 22, 98). The repression of *FLC* during vernalization is another example of converting euchromatin into heterochromatin. However, dimethylation of K9 may be more important than dimethylation of K27 in the maintenance of stable *FLC* repression (6, 101).

Another difference is that homologs of PC have not been found in plants (34), raising the issue of the components required for mitotic stability of vernalization-repressed genes. PC shares a region of homology called the "chromodomain" with heterochromatin protein 1 (HP1) (10). HP1 exists both in animals and plants. The chromodomains of HP1 and PC are responsible for the recognition and binding to modified histone tails (5, 21, 38, 52, 107). In animals, HP1 recognizes and binds to methylated K9 on histone H3 and is involved mainly in maintaining constitutive heterochromatin (5, 38, 52, 92). The lack of PC in plants and the importance of dimethyl K9 raises the possibility that HP1 might be involved in the maintenance of the vernalization-mediated repressed state of *FLC* chromatin (100). Consistent with this hypothesis, vernalization-mediated *FLC* repression is not stable in *Arabidopsis hp1* mutants (S. Sung & R.M. Amasino, unpublished). Thus, it is likely that HP1 is involved in maintaining *FLC* repression after vernalization. Su(z)12, which is a mammalian homolog of VRN2, can interact with HP1 protein in vitro, further supporting the possible role of the plant homolog of HP1 in vernalization (110).

Repression of *FLC* is quantitative (97), i.e., *FLC* can be partially repressed by shorter periods of cold exposure than that which is optimal for a maximum vernalization response. Dosage-dependent repression is characteristic of PcG repression (84). It is possible that suboptimal exposure to cold creates a situation during vernalization similar to the dosage effect in PcG repression: Partial suppression is presumably due to less than saturating levels of a component of repression. The molecular details of how partial repression of *FLC* can occur during a suboptimal exposure to cold remain to be determined, but a candidate for a component that might be limiting after shorter than optimal periods of cold exposure is VIN3 (101).

Activation of *FLC* Transcription

There are two genetic situations in which *FLC* is actively transcribed to levels that significantly delay flowering unless the plants are vernalized. One, discussed above, is due to the presence of active alleles of *FRI*. The other is due to mutations in autonomous-pathway genes (69). The autonomous pathway is defined by a group of genes involved in *FLC* repression. In the absence of *FRI*, the autonomous pathway keeps *FLC* levels low and is thus responsible for the rapid flowering, summer-annual-type habit in many accessions of *Arabidopsis*. In autonomous-pathway mutants, *FLC* repression is abrogated and such mutants exhibit the vernalization-responsive late flowering characteristic of winter annuals. The vernalization-responsiveness of autonomous-pathway mutants indicates that the vernalization-mediated suppression of *FLC* operates independently of the autonomous pathway. Recent studies indicate that the autonomous pathway may operate at several different levels to regulate *FLC* expression (31, 45), i.e., it is not a linear pathway but a collection of genes involved in different levels of *FLC* repression. Two of the autonomous-pathway genes appear to be involved in deacetylation of *FLC* chromatin (4, 31).

When *FLC* is actively transcribed due either to the presence of *FRI* or to the absence of an autonomous-pathway gene, trimethylation of K4 on histone H3 is increased in *FLC* chromatin at the 5′ region around the transcription start site (30). This situation is similar to the maintenance of gene activation by TrxG proteins in animals. The histone methyltransferases TRX and ASH1, two SET-domain-containing trxG gene products, methylate histone H3 at K4 (90). Methylation at histone H3 K4 is generally considered a mark of active genes (90). Furthermore, K4 trivalent methylation is necessary for trxG-dependent transcriptional activation, and it appears to be incompatible with the binding of PC and HP1 repressor proteins (7). Mutations in an ASH1 homolog (EFS) (99) in *Arabidopsis* abolish trimethylation of K4 on histone H3 in *FLC* chromatin, and *efs* mutants fail to activate transcription of *FLC* either in the presence of *FRI* or in the absence of autonomous-pathway genes (S. Michaels, Y. He & R. Amasino, unpublished).

Arabidopsis homologs of members of a complex known in yeast as the PAF1 complex (RNA *p*olymerase II *a*ssociated *f*actor 1) are also involved in activating *FLC*. In yeast, the PAF1 complex interacts with SET1, the yeast ASH1/EFS homolog, resulting in recruitment of SET1 to transcription start sites and an increase in histone trimethylation in these regions (50, 77). Lesions in components of the PAF1 complex in *Arabidopsis* result in the loss of trimethylation of K4 on histone H3 and a loss of the ability to activate *FLC* transcription (30).

Recently it was shown in yeast that trimethylation of K4 on histone H3 could be recognized and bound by ISW1p (91). ISW1p contains

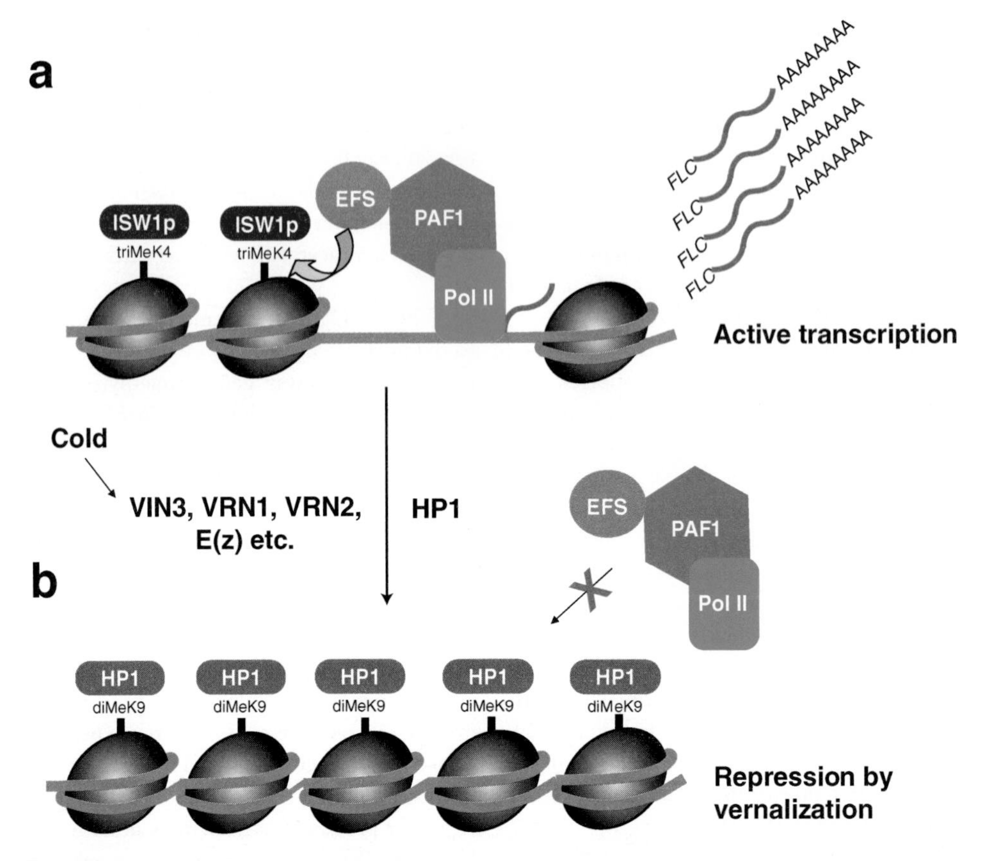

Figure 3

Maintenance of active and repressed states of *FLC* transcription by chromatin modification. (*A*) Activation of *FLC* transcription by either the presence of *FRI* or the loss of autonomous-pathway genes requires a suite of proteins that are involved in chromatin remodeling. (*B*) Vernalization results in heterochromatin formation at the *FLC* locus. Heterochromatin formation requires the induction of VIN3 by a prolonged period of cold exposure, and involves a series of histone modifications, such as deacetylation and methylation. VRN2 and VRN1 are also involved in the methylation of K9 of histone H3. Enhancer of Zeste (E(z)) is histone methyltransferase that is responsible for methylation on histone H3 in PcG-mediated repression; there are three homologs of E(z) in *Arabidopsis* that might play a role in this process. HP1 may recognize and bind to dimethyl K9 on histone H3, to participate in maintaining the heterochromatin-like state of *FLC* through mitotic cell divisions.

an ATPase domain and is a member of a class of chromatin-remodeling proteins that use the energy of ATP hydrolysis to position nucleosomes into an arrangement conducive to transcription (91). Mutations in a relative of ISW1p (*PIE1*) in *Arabidopsis* also result in failure to activate *FLC* transcription (78). It is possible that *PIE1* serves the same role in *Arabidopsis* as ISW1p does in yeast (**Figure 3**). However, PIE1 is also related to yeast SWR1, an ATP-dependent chromatin-remodeling protein involved in Histone H2A variant replacement (71a, 44a). It is possible that this type of activity is also required for *FLC* expression, and that a different ISW1p relative is also involved in *FLC* activation.

In both *FRI*-containing lines and in autonomous-pathway mutants, vernalization can overcome the activation of *FLC*. Interestingly, trimethylation of K4 on histone H3 in *FLC* chromatin is reduced by vernalization (Y. He & R. Amasino, unpublished). Because expression at the mRNA level of components that are required for histone H3 K4 trimethylation of *FLC* chromatin is not repressed by vernalization, one simple model for the vernalization-mediated decrease of *FLC* H3 K4 trimethylation is that vernalization-mediated HP1 association with *FLC* chromatin (and thus the formation of heterochromatin) renders *FLC* chromatin inaccessible to the components that are required for *FLC* activation (**Figure 3**). This is analogous to the template exclusion of the ATP-dependent chromatin-remodeling protein hSWI/SNF by the interaction of components of PRC1 with target genes (22).

FLC-Independent Vernalization Response in *Arabidopsis*

Although a major target for vernalization in *Arabidopsis* is the repression of *FLC*, the observation that, in noninductive photoperiods, vernalization causes plants to flower earlier than nonvernalized *flc* null mutants, demonstrates that additional targets are regulated by vernalization (69). Some, perhaps all, of these additional targets are relatives of *FLC*. There are five *FLC*-related genes in *Arabidopsis*, and mutations in two of the *FLC*-related genes (*FLM/MAF1* and *MAF2*) cause earlier flowering (mutants in others have not been identified) and all five *FLC*-related genes act as floral repressors when they are ectopically expressed (85, 86, 95). Vernalization represses *FLM/MAF1* and *MAF4* (85, 86). Thus, a clad of MADS-box genes are targets for vernalization.

Measurement of Duration of Cold

Essentially nothing is known about the mechanism by which plants measure the duration of cold in processes such as vernalization or the release of bud dormancy. Unlike cold acclimation, which can be established by a very short period of cold exposure, vernalization and the release of bud dormancy typically require an extended period of cold exposure.

Although cold acclimation and vernalization differ in many aspects, it is possible that some early events in cold signaling could be shared by both processes. However, to date, the only possible link between cold acclimation and vernalization comes from the study of *HIGH EXPRESSION OF OSMOTICALLY RESPONSIVE GENES 1* (*HOS1*). A mutation in *HOS1* causes elevated *CBF* expression as well as early flowering and reduced *FLC* expression (56). The *hos1* phenotype indicates that *HOS1* is a negative regulator of cold signaling. However, in a *hos1* mutant, vernalization promotes flowering by further reducing *FLC* expression (37, 56). Thus, the loss of *HOS1* either does not entirely mimic the vernalization response or the *hos1* lesion might affect *FLC* expression independently of the vernalization pathway. *HOS1* encodes a RING-finger protein, and such proteins are usually associated with the ubiquitin protein-degradation pathway. Lesions in *ESD4*, which encode small ubiquitin-related modifier (SUMO), a specific nuclear protease, also cause early flowering partly due to the reduction in *FLC* expression in a vernalization-independent manner (75, 87). Whether lesions in *SUMO* affect *CBF* expression has not been reported.

Part of the cold-acclimation process involves modifying membrane lipid composition to adjust membrane fluidity, rearrangements of microfilaments, Ca^{2+} fluxes, and changes of protein phosphorylation patterns (35, 42, 44, 48, 71, 72, 80, 83, 89, 106). These cold-induced biochemical changes occur rapidly, as expected for any process related to cold acclimation. Therefore, the timing of these changes is not likely directly involved in the system that measures the duration of cold during the vernalization response. Furthermore, constitutive expression of a *CBF1*, which causes elevated expression of downstream cold-acclimation genes and provides a degree of cold resistance without cold exposure, has no effect on the vernalization response (63). However, it is possible that some aspect of cold acclimation provides a component of the cold measurement system. For example, the altered lipid composition of a cold-acclimated membrane might provide a cold-specific substrate for a measurement system.

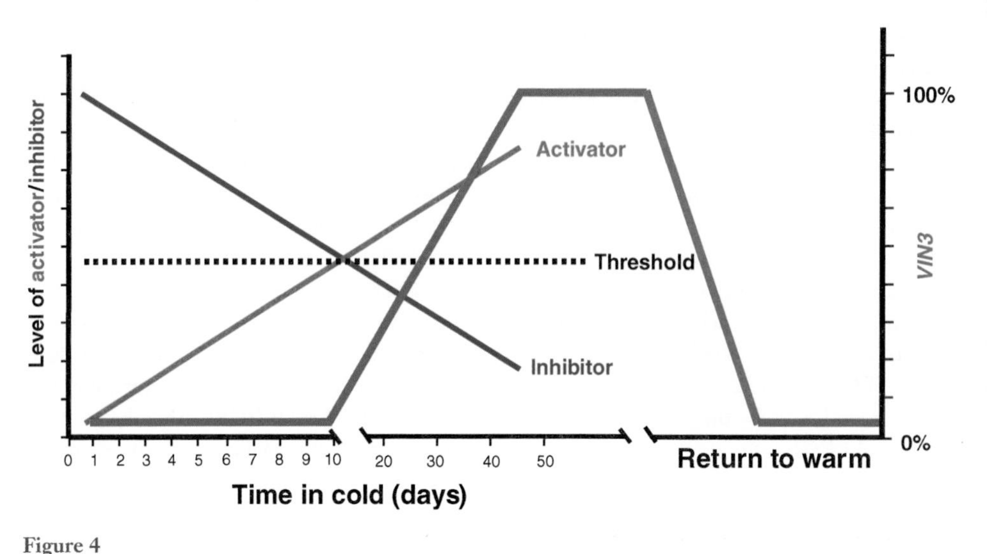

Figure 4

A general and hypothetical model for measuring the duration of cold. During exposure to low temperatures, the level of an activator (*red*) and/or inhibitor (*blue*) might slowly reach a certain threshold level (*dotted line*). When the threshold is reached in *Arabidopsis*, the induction of *VIN3* (*green*) leads to epigenetic changes in gene expression.

How might a cold measurement system operate? Nothing is known about the cold-sensing mechanism, and thus we can only speculate. General models have been proposed in which there is a slow rise or decline in the levels of a compound or compounds during prolonged cold exposure, and, when the levels cross a threshold, the processes that provide competence to flower are initiated (**Figure 4**). The time required to reach the threshold provides a system to measure the duration of cold. What might cause such a slow rise or decline to a threshold? One possibility is differential cold sensitivity of competing enzyme activities (e.g., different Q_{10} values). For example, a kinase that does not lose activity as the temperature is lowered as rapidly as does a phosphatase that acts on the same substrate would lead to the accumulation of a phosphorylated product in the cold. The substrate could be a small molecule or a protein. Another possibility is that the process of cold acclimation provides a greater amount of a specific substrate, for example a membrane lipid, which leads to an increase in the levels of a signaling molecule derived from that substrate. The system may have multiple cold-sensing components such that a cascade of signaling events is required to measure a complete winter duration of cold.

To date, the most direct connection between the system that measures the duration of cold exposure and the acquisition of competence to flower is the induction of *VIN3* expression. *VIN3* is induced only during cold exposures of sufficient duration to affect flowering (101). Thus, the induction of *VIN3* is an output of the cold-sensing system (**Figure 4**). As discussed above, *VIN3* expression initiates a series of histone modifications of *FLC* chromatin that are required for repressing *FLC* (101). However, constitutive expression of *VIN3* in the absence of cold exposure is not sufficient to establish stable repression of *FLC* (101). Thus, cold exposure must do more than induce *VIN3* to achieve the vernalized state. Chromatin-remodeling activity is commonly mediated by protein complexes and some of the other components may be cold-induced (although the *Arabidopsis VRN1* and *VRN2* genes are not cold-induced). Another possibility is that cold-induced biochemical changes must accompany *VIN3* induction. VIN3 contains a

PHD-finger motif, and this motif can bind phosphoinositides (28). It is conceivable that differences in the spectrum of phosphoinositides that arise during a vernalizing cold exposure are required to establish *FLC* repression.

CONCLUSIONS AND FUTURE DIRECTIONS

We now understand, in outline, how vernalization affects the competence of the meristem to flower in *Arabidopsis*. Competence involves the epigenetic repression of genes encoding MADS-box flowering repressors, and this epigenetic repression is associated with a series of histone modifications. A similar system appears to operate in cereals with a different type of protein acting as the vernalization-repressed repressor. Many unanswered questions remain. For example, what is the nature of the chromatin-remodeling complexes that accomplish repression? How are these complexes targeted to *FLC* and other flowering repressors? How widespread is this type of system in plants that have a vernalization response? Do all flowering plants use either MADS-box genes or ZCCT family zinc-finger proteins to establish a vernalization requirement, or will other types of transcriptional regulators be involved in this process in different plant lineages? Furthermore, nothing is known regarding the mechanism by which the duration of cold is measured during vernalization, and this is a key area for future research efforts to address.

ACKNOWLEDGMENTS

R.A. is grateful to the National Science Foundation, the U.S. Department of Agriculture National Research Initiative Competitive Grants Program, and the College of Agricultural and Life Sciences and the Graduate School of the University of Wisconsin for their generous support of our flowering research.

LITERATURE CITED

1. Aasland R, Gibson TJ, Stewart AF. 1995. The PHD finger: implications for chromatin-mediated transcriptional regulation. *Trends Biochem. Sci.* 20:56–59
2. Abegg FA. 1936. A genetic factor for the annual habit in beets and linkage relationship. *J. Argicultural Res.* 53:493–511
3. Anderson BE, Ward JM, Schroeder JI. 1994. Evidence for an extracellular reception site for abscisic acid in commelina guard cells. *Plant Physiol.* 104:1177–83
4. Ausin I, Alonso-Blanco C, Jarillo JA, Ruiz-Garcia L, Martinez-Zapater JM. 2004. Regulation of flowering time by FVE, a retinoblastoma-associated protein. *Nat. Genet.* 36:162–66
5. Bannister AJ, Zegerman P, Partridge JF, Miska EA, Thomas JO, et al. 2001. Selective recognition of methylated lysine 9 on histone H3 by the HP1 chromo domain. *Nature* 410:120–24
6. Bastow R, Mylne JS, Lister C, Lippman Z, Martienssen RA, Dean C. 2004. Vernalization requires epigenetic silencing of FLC by histone methylation. *Nature* 427:164–67
7. Beisel C, Imhof A, Greene J, Kremmer E, Sauer F. 2002. Histone methylation by the Drosophila epigenetic transcriptional regulator Ash1. *Nature* 419:857–62
8. Berger SL. 2002. Histone modifications in transcriptional regulation. *Curr. Opin. Genet. Dev.* 12:142–48
9. Bernier G, Kinet J-M, Sachs RM. 1981. *The Physiology of Flowering*. Boca Raton, FL: CRC Press
10. Breiling A, Bonte E, Ferrari S, Becker PB, Paro R. 1999. The Drosophila polycomb protein interacts with nucleosomal core particles in vitro via its repression domain. *Mol. Cell. Biol.* 19:8451–60

11. Burn JE, Bagnall DJ, Metzger JD, Dennis ES, Peacock WJ. 1993. DNA methylation, vernalization, and the initiation of flowering. *Proc. Natl. Acad. Sci. USA* 90:287–91
12. Burn JE, Smyth DR, Peacock WJ, Dennis ES. 1993. Genes conferring late flowering in *Arabidopsis thaliana*. *Genetica* 90:147–55
13. Cao R, Wang L, Wang H, Xia L, Erdjument-Bromage H, et al. 2002. Role of histone H3 lysine 27 methylation in Polycomb-group silencing. *Science* 298:1039–43
14. Chinnusamy V, Ohta M, Kanrar S, Lee BH, Hong X, et al. 2003. ICE1: a regulator of cold-induced transcriptome and freezing tolerance in Arabidopsis. *Genes Dev.* 17:1043–54
15. Chouard P. 1960. Vernalization and its relations to dormancy. *Annu. Rev. Plant Physiol.* 11:191–238
16. Clarke JH, Dean C. 1994. Mapping *FRI*, a locus controlling flowering time and vernalization response in *Arabidopsis thaliana*. *Mol. Gen. Genet.* 242:81–89
17. Czermin B, Melfi R, McCabe D, Seitz V, Imhof A, Pirrotta V. 2002. Drosophila enhancer of Zeste/ESC complexes have a histone H3 methyltransferase activity that marks chromosomal Polycomb sites. *Cell* 111:185–96
18. Dubcovsky J, Luo MC, Zhong GY, Bransteitter R, Desai A, et al. 1996. Genetic map of diploid wheat, Triticum monococcum L., and its comparison with maps of Hordeum vulgare L. *Genetics* 143:983–99
19. Dutnall RN. 2003. Cracking the histone code: one, two, three methyls, you're out! *Mol. Cell.* 12:3–4
20. Erwin JE, Warner RM, Smith AG. 2002. Vernalization, photoperiod and GA3 interact to affect flowering of Japanese radish (Raphanus sativus Chinese Radish Jumbo Scarlet). *Physiol. Plant* 115:298–302
21. Fischle W, Wang Y, Jacobs SA, Kim Y, Allis CD, Khorasanizadeh S. 2003. Molecular basis for the discrimination of repressive methyl-lysine marks in histone H3 by Polycomb and HP1 chromodomains. *Genes Dev.* 17:1870–81
22. Francis NJ, Saurin AJ, Shao Z, Kingston RE. 2001. Reconstitution of a functional core polycomb repressive complex. *Mol. Cell.* 8:545–56
23. Gazzani S, Gendall AR, Lister C, Dean C. 2003. Analysis of the molecular basis of flowering time variation in Arabidopsis accessions. *Plant Physiol.* 132:1107–14
24. Gendall AR, Levy YY, Wilson A, Dean C. 2001. The VERNALIZATION 2 gene mediates the epigenetic regulation of vernalization in Arabidopsis. *Cell* 107:525–35
25. Gilmour S, Hajela RK, Thomashow MF. 1988. Cold acclimation in *Arabidopsis thaliana*. *Plant Physiol.* 87:745–50
26. Gilmour SJ, Zarka DG, Stockinger EJ, Salazar MP, Houghton JM, Thomashow MF. 1998. Low temperature regulation of the Arabidopsis CBF family of AP2 transcriptional activators as an early step in cold-induced COR gene expression. *Plant J.* 16:433–42
27. Gould A. 1997. Functions of mammalian Polycomb group and trithorax group related genes. *Curr. Opin. Genet. Dev.* 7:488–94
28. Gozani O, Karuman P, Jones DR, Ivanov D, Cha J, et al. 2003. The PHD finger of the chromatin-associated protein ING2 functions as a nuclear phosphoinositide receptor. *Cell* 114:99–111
29. Guy CL, Niemi KJ, Brambl R. 1985. Altered gene expression during cold acclimation of spinach. *Proc. Natl. Acad. Sci. USA* 82:3673–77
30. He Y, Doyle MD, Amasino RM. 2004. PAF1 complex-mediated histone methylation of *FLOWERING LOCUS C* chromatin is required for the vernalization-responsive, winter-annual habit in *Arabidopsis*. *Genes Dev.* 18:2774–84
31. He Y, Michaels SD, Amasino RM. 2003. Regulation of flowering time by histone acetylation in Arabidopsis. *Science* 302:1751–54

32. Hong BM, Barg R, Ho THD. 1992. Developmental and organ-specific expression of an ABA-induced and stress-induced protein in barley. *Plant Mol. Biol.* 18:663–74
33. Houde M, Danyluk J, Laliberte JF, Rassart E, Dhindsa RS, Sarhan F. 1992. Cloning, characterization, and expression of a carrier DNA encoding a 50-kilodalton protein specifically induced by cold acclimation in wheat. *Plant Physiol.* 99:1381–87
34. Hsieh TF, Hakim O, Ohad N, Fischer RL. 2003. From flour to flower: how Polycomb group proteins influence multiple aspects of plant development. *Trends Plant Sci.* 8:439–45
35. Ichimura K, Mizoguchi T, Yoshida R, Yuasa T, Shinozaki K. 2000. Various abiotic stresses rapidly activate Arabidopsis MAP kinases ATMPK4 and ATMPK6. *Plant J.* 24:655–65
36. Iizuka M, Smith MM. 2003. Functional consequences of histone modifications. *Curr. Opin. Genet. Dev.* 13:154–60
37. Ishitani M, Xiong L, Lee H, Stevenson B, Zhu JK. 1998. HOS1, a genetic locus involved in cold-responsive gene expression in arabidopsis. *Plant Cell* 10:1151–61
38. Jacobs SA, Khorasanizadeh S. 2002. Structure of HP1 chromodomain bound to a lysine 9-methylated histone H3 tail. *Science* 295:2080–83
39. Jaglo-Ottosen KR, Gilmour SJ, Zarka DG, Schabenberger O, Thomashow MF. 1998. Arabidopsis CBF1 overexpression induces COR genes and enhances freezing tolerance. *Science* 280:104–6
40. Jenuwein T, Allis CD. 2001. Translating the histone code. *Science* 293:1074–80
41. Johanson U, West J, Lister C, Michaels S, Amasino R, Dean C. 2000. Molecular analysis of *FRIGIDA*, a major determinant of natural variation in *Arabidopsis* flowering time. *Science* 290:344–47
42. Jonak C, Kiegerl S, Ligterink W, Barker PJ, Huskisson NS, Hirt H. 1996. Stress signaling in plants: a mitogen-activated protein kinase pathway is activated by cold and drought. *Proc. Natl. Acad. Sci. USA* 93:11274–79
43. Kim YJ. 1998. Studies on vernlization treatment of winter barley in the 17th century that are described in Ko's "Monthly Farming Guide." *Proc. Korean Soc. Cereal Res.* 5:117–25
44. Knight MR, Campbell AK, Smith SM, Trewavas AJ. 1991. Transgenic plant aequorin reports the effects of touch and cold-shock and elicitors on cytoplasmic calcium. *Nature* 352:524–26
44a. Kobor MS, Venkatasubrahmanyam S, Meneghini MD, Gin JW, Jennings JL, et al. 2004. A protein complex containing the conserved Swi2/Snf2-related ATPase Swr1p deposits histone variant H2A.Z into euchromatin. *PLoS Biol.* 2:587–99
45. Koornneef M, Alonso-Blanco C, Peeters AJ, Soppe W. 1998. Genetic control of flowering time in Arabidopsis. *Ann. Rev. Plant Physiol. Plant Mol. Biol.* 49:345–70
46. Koornneef M, Blankestijn-de Vries H, Hanhart C, Soppe W, Peeters T. 1994. The phenotype of some late-flowering mutants is enhanced by a locus on chromosome 5 that is not effective in the Landsberg *erecta* wild-type. *Plant J.* 6:911–19
47. Kouzarides T. 2002. Histone methylation in transcriptional control. *Curr. Opin. Genet. Dev.* 12:198–209
48. Kovtun Y, Chiu WL, Tena G, Sheen J. 2000. Functional analysis of oxidative stress-activated mitogen-activated protein kinase cascade in plants. *Proc. Natl. Acad. Sci. USA* 97:2940–45
49. Krishna P, Sacco M, Cherutti JF, Hill S. 1995. Cold-induced accumulation of hsp90 transcripts in Brassica napus. *Plant Physiol.* 107:915–23
50. Krogan NJ, Dover J, Wood A, Schneider J, Heidt J, et al. 2003. The Paf1 complex is required for histone H3 methylation by COMPASS and Dot1p: linking transcriptional elongation to histone methylation. *Mol. Cell.* 11:721–29
51. Kuzmichev A, Nishioka K, Erdjument-Bromage H, Tempst P, Reinberg D. 2002. Histone methyltransferase activity associated with a human multiprotein complex containing the enhancer of Zeste protein. *Genes Dev.* 16:2893–905

52. Lachner M, O'Carroll D, Rea S, Mechtler K, Jenuwein T. 2001. Methylation of histone H3 lysine 9 creates a binding site for HP1 proteins. *Nature* 410:116–20
53. Lang A. 1965. Physiology of flower initiation. In *Encyclopedia of Plant Physiology*, ed. W Ruhland, pp. 1371–536. Berlin: Springer-Verlag
54. Lang A. 1986. Hyoscyamus niger. In *CRC Handbook of Flowering*, ed. AH Halevy, pp. 144–86. Boca Raton, FL: CRC Press
55. Laurie DA. 1997. Comparative genetics of flowering time. *Plant Mol. Biol.* 35:167–77
56. Lee H, Xiong L, Gong Z, Ishitani M, Stevenson B, Zhu JK. 2001. The Arabidopsis HOS1 gene negatively regulates cold signal transduction and encodes a RING finger protein that displays cold-regulated nucleo–cytoplasmic partitioning. *Genes Dev.* 15:912–24
57. Lee I, Amasino RM. 1995. Effect of vernalization, photoperiod and light quality on the flowering phenotype of Arabidopsis plants containing the *FRIGIDA* gene. *Plant Physiol.* 108:157–62
58. Lee I, Bleecker A, Amasino R. 1993. Analysis of naturally occurring late flowering in *Arabidopsis thaliana*. *Mol. Gen. Genet.* 237:171–76
59. Lee I, Michaels SD, Masshardt AS, Amasino RM. 1994. The late-flowering phenotype of *FRIGIDA* and *LUMINIDEPENDENS* is suppressed in the Landsberg *erecta* strain of Arabidopsis. *Plant J.* 6:903–9
60. Levy YY, Mesnage S, Mylne JS, Gendall AR, Dean C. 2002. Multiple roles of Arabidopsis VRN1 in vernalization and flowering time control. *Science* 297:243–46
61. Li Y, Kirschmann DA, Wallrath LL. 2002. Does heterochromatin protein 1 always follow code? *Proc. Natl. Acad. Sci. USA* 99(Suppl.)4:16462–69
62. Lin CT, Thomashow MF. 1992. A cold-regulated Arabidopsis gene encodes a polypeptide having potent cryoprotective activity. *Biochem. Biophys. Res. Commun.* 183:1103–8
63. Liu J, Gilmour SJ, Thomashow MF, Van Nocker S. 2002. Cold signalling associated with vernalization in Arabidopsis thaliana does not involve CBF1 or abscisic acid. *Physiol. Plant* 114:125–34
64. Liu J, He Y, Amasino RM, Chen X. 2004. siRNA targeting an intronic transposon in the regulation of natural flowering behavior in *Arabidopsis*. *Genes Dev.* 18:2873–78
65. McClintock B. 1950. The origin and behavior of mutable loci in Maize. *Proc. Natl. Acad. Sci. USA* 36:344–55
66. Medina J, Bargues M, Terol J, Perez-Alonso M, Salinas J. 1999. The Arabidopsis CBF gene family is composed of three genes encoding AP2 domain-containing proteins whose expression is regulated by low temperature but not by abscisic acid or dehydration. *Plant Physiol.* 119:463–70
67. Michaels S, Amasino R. 2000. Memories of winter: vernalization and the competence to flower. *Plant Cell Environ.* 23:1145–54
68. Michaels SD, Amasino RM. 1999. FLOWERING LOCUS C encodes a novel MADS domain protein that acts as a repressor of flowering. *Plant Cell* 11:949–56
69. Michaels SD, Amasino RM. 2001. Loss of FLOWERING LOCUS C activity eliminates the late-flowering phenotype of FRIGIDA and autonomous pathway mutations but not responsiveness to vernalization. *Plant Cell* 13:935–41
70. Michaels SD, He Y, Scortecci KC, Amasino RM. 2003. Attenuation of FLOWERING LOCUS C activity as a mechanism for the evolution of summer-annual flowering behavior in Arabidopsis. *Proc. Natl. Acad. Sci. USA* 100:10102–7
71. Mizoguchi T, Irie K, Hirayama T, Hayashida N, Yamaguchi-Shinozaki K, et al. 1996. A gene encoding a mitogen-activated protein kinase kinase kinase is induced simultaneously with genes for a mitogen-activated protein kinase and an S6 ribosomal protein kinase by touch, cold, and water stress in Arabidopsis thaliana. *Proc. Natl. Acad. Sci. USA* 93:765–69

71a. Mizuguchi G, Shen X, Landry J, Wu WH, Sen S, Wu C. 2004. ATP-driven exchange of histone H2AZ variant catalyzed by SWR1 chromatin remodeling complex. *Science* 303:343–48
72. Monroy AF, Castonguay Y, Laberge S, Sarhan F, Vezina LP, Dhindsa RS. 1993. A new cold-induced alfalfa gene is associated with enhanced hardening at subzero temperature. *Plant Physiol.* 102:873–79
73. Muller J, Hart CM, Francis NJ, Vargas ML, Sengupta A, et al. 2002. Histone methyltransferase activity of a Drosophila Polycomb group repressor complex. *Cell* 111:197–208
74. Murfet IC. 1989. Flowering genes in *Pisum*. In *Plant Reproduction: From Floral Induction to Pollination*, ed. E Lord, G Bernier, pp. 10–18. Rockville, MD: Am. Soc. Plant Physiol.
75. Murtas G, Reeves PH, Fu YF, Bancroft I, Dean C, Coupland G. 2003. A nuclear protease required for flowering-time regulation in Arabidopsis reduces the abundance of SMALL UBIQUITIN-RELATED MODIFIER conjugates. *Plant Cell* 15:2308–19
76. Napp-Zinn K. 1987. Vernalization: environmental and genetic regulation. In *Manipulation of Flowering*, ed. JG Atherton, pp. 123–32. London: Butterworths
77. Ng HH, Robert F, Young RA, Struhl K. 2003. Targeted recruitment of Set1 histone methylase by elongating Pol II provides a localized mark and memory of recent transcriptional activity. *Mol. Cell.* 11:709–19
78. Noh YS, Amasino RM. 2003. PIE1, an ISWI family gene, is required for FLC activation and floral repression in Arabidopsis. *Plant Cell* 15:1671–82
79. Orlando V. 2003. Polycomb, epigenomes, and control of cell identity. *Cell* 112:599–606
80. Orvar BL, Sangwan V, Omann F, Dhindsa RS. 2000. Early steps in cold sensing by plant cells: the role of actin cytoskeleton and membrane fluidity. *Plant J.* 23:785–94
81. Osborn TC, Kole C, Parkin IAP, Sharpe AG, Kuiper M, et al. 1997. Comparison of flowering time genes in *Brassica rapa, B. napus* and *Arabidopsis thaliana. Genet. Soc. Am.* 146:1123–29
82. Pirrotta V. 1997. PcG complexes and chromatin silencing. *Curr. Opin. Genet. Dev.* 7:249–58
83. Plieth C, Hansen UP, Knight H, Knight MR. 1999. Temperature sensing by plants: the primary characteristics of signal perception and calcium response. *Plant J.* 18:491–97
84. Poux S, Horard B, Sigrist CJ, Pirrotta V. 2002. The Drosophila trithorax protein is a coactivator required to prevent re-establishment of polycomb silencing. *Development* 129:2483–93
85. Ratcliffe OJ, Kumimoto RW, Wong BJ, Riechmann JL. 2003. Analysis of the Arabidopsis MADS AFFECTING FLOWERING gene family: MAF2 prevents vernalization by short periods of cold. *Plant Cell* 15:1159–69
86. Ratcliffe OJ, Nadzan GC, Reuber TL, Riechmann JL. 2001. Regulation of flowering in Arabidopsis by an FLC homologue. *Plant Physiol.* 126:122–32
87. Reeves PH, Murtas G, Dash S, Coupland G. 2002. Early in short days 4, a mutation in Arabidopsis that causes early flowering and reduces the mRNA abundance of the floral repressor FLC. *Development* 129:5349–61
88. Reid JB, Murfet IC. 1975. Flowering in Pisum: the sites and possible mechanisms of the vernalization response. *J. Exp. Bot.* 26:860–67
89. Sangwan V, Orvar BL, Beyerly J, Hirt H, Dhindsa RS. 2002. Opposite changes in membrane fluidity mimic cold and heat stress activation of distinct plant MAP kinase pathways. *Plant J.* 31:629–38
90. Santos-Rosa H, Schneider R, Bannister AJ, Sherriff J, Bernstein BE, et al. 2002. Active genes are tri-methylated at K4 of histone H3. *Nature* 419:407–11
91. Santos-Rosa H, Schneider R, Bernstein BE, Karabetsou N, Morillon A, et al. 2003. Methylation of histone H3 K4 mediates association of the Isw1p ATPase with chromatin. *Mol. Cell.* 12:1325–32

92. Schotta G, Ebert A, Krauss V, Fischer A, Hoffmann J, et al. 2002. Central role of Drosophila SU(VAR)3–9 in histone H3-K9 methylation and heterochromatic gene silencing. *EMBO J.* 21:1121–31
93. Schranz ME, Quijada P, Sung SB, Lukens L, Amasino R, Osborn TC. 2002. Characterization and effects of the replicated flowering time gene FLC in Brassica rapa. *Genetics* 162:1457–68
94. Schreiber SL, Bernstein BE. 2002. Signaling network model of chromatin. *Cell* 111:771–78
95. Scortecci KC, Michaels SD, Amasino RM. 2001. Identification of a MADS-box gene, FLOWERING LOCUS M, that represses flowering. *Plant J.* 26:229–36
96. Sheldon CC, Burn JE, Perez PP, Metzger J, Edwards JA, et al. 1999. The FLF MADS box gene: a repressor of flowering in arabidopsis regulated by vernalization and methylation. *Plant Cell* 11:445–58
97. Sheldon CC, Rouse DT, Finnegan EJ, Peacock WJ, Dennis ES. 2000. The molecular basis of vernalization: the central role of FLOWERING LOCUS C (FLC). *Proc. Natl. Acad. Sci. USA* 97:3753–58
98. Simon JA, Tamkun JW. 2002. Programming off and on states in chromatin: mechanisms of Polycomb and trithorax group complexes. *Curr. Opin. Genet. Dev.* 12:210–18
99. Soppe WJ, Bentsink L, Koornneef M. 1999. The early-flowering mutant efs is involved in the autonomous promotion pathway of Arabidopsis thaliana. *Development* 126:4763–70
100. Sung S, Amasino RM. 2004. Vernalization and epigenetics: how plants remember winter. *Curr. Opin. Plant Biol.* 7:4–10
101. Sung S, Amasino RM. 2004. Vernalization in Arabidopsis thaliana is mediated by the PHD finger protein VIN3. *Nature* 427:159–64
102. Thomashow MF. 1999. PLANT COLD ACCLIMATION: freezing tolerance genes and regulatory mechanisms. *Annu. Rev. Plant Physiol. Plant Mol. Biol.* 50:571–99
103. Thomashow MF. 2001. So what's new in the field of plant cold acclimation? Lots! *Plant Physiol.* 125:89–93
104. Tranquilli G, Dubcovsky J. 2000. Epistatic interaction between vernalization genes Vrn-Am1 and Vrn-Am2 in diploid wheat. *J. Hered.* 91:304–6
105. Turner BM. 2002. Cellular memory and the histone code. *Cell* 111:285–91
106. Vigh L, Los DA, Horvath I, Murata N. 1993. The primary signal in the biological perception of temperature: Pd-catalyzed hydrogenation of membrane lipids stimulated the expression of the desA gene in Synechocystis PCC6803. *Proc. Natl. Acad. Sci. USA* 90:9090–94
107. Wang L, Brown JL, Cao R, Zhang Y, Kassis JA, Jones RS. 2004. Hierarchical recruitment of polycomb group silencing complexes. *Mol. Cell.* 14:637–46
108. Wellensiek SJ. 1962. Dividing cells as the locus for vernalization. *Nature* 195:307–8
109. Wellensiek SJ. 1964. Dividing cells as the prerequisite for vernalization. *Plant Physiol.* 39:832–35
110. Yamamoto K, Sonoda M, Inokuchi J, Shirasawa S, Sasazuki T. 2004. Polycomb group suppressor of zeste 12 links heterochromatin protein 1alpha and enhancer of zeste 2. *J. Biol. Chem.* 279:401–6
111. Yan L, Loukoianov A, Blechl A, Tranquilli G, Ramakrishna W, et al. 2004. The wheat VRN2 gene is a flowering repressor down-regulated by vernalization. *Science* 303:1640–44
112. Yan L, Loukoianov A, Tranquilli G, Helguera M, Fahima T, Dubcovsky J. 2003. Positional cloning of the wheat vernalization gene VRN1. *Proc. Natl. Acad. Sci. USA* 100:6263–68

New Insights to the Function of Phytopathogenic Bacterial Type III Effectors in Plants

Mary Beth Mudgett

Department of Biological Sciences, Stanford University, Stanford, California 94305-5020; email: mudgett@stanford.edu

Annu. Rev. Plant Biol. 2005. 56:509–31

doi: 10.1146/annurev.arplant.56.032604.144218

First published online as a Review in Advance on January 18, 2005

1543-5008/05/0602-0509$20.00

Key Words

avirulence proteins, TTSS, pathogenesis

Abstract

Phytopathogenic bacteria use the type III secretion system (TTSS) to inject effector proteins into plant cells. This system is essential for bacteria to multiply in plant tissue and to promote the development of disease symptoms. Until recently, little was known about the function of TTSS effectors in bacterial-plant interactions. New studies dissecting the molecular and biochemical action of TTSS effectors show that these proteins contribute to bacterial pathogenicity by interfering with plant defense signal transduction. These investigations provide us with a fresh view of how bacteria manipulate plant physiology to colonize their hosts.

Contents

INTRODUCTION

TTSS: type III secretion system

NBS: nucleotide binding site

LRR: leucine rich repeat

HR: hypersensitive response

SAR: systemic acquired resistance

Many bacterial pathogens of plants and animals use a conserved TTSS to infect their eukaryotic hosts during pathogenesis (17). For phytopathogenic bacteria in the genera *Pseudomonas*, *Xanthomonas*, *Ralstonia*, *Erwinia*, and *Pantoea*, the TTSS is encoded by the hypersensitive response and pathogenicity (*hrp*) gene cluster. These pathogens use the TTSS within plant tissue at early stages of infection to establish cell-to-cell contact. Once contact is made, the TTSS facilitates the direct secretion and translocation of a diverse group of bacterial proteins, known as TTSS effectors, into host plant cells. Mutant bacteria lacking the TTSS cannot multiply within plant tissue, preventing their colonization and subsequent spread. Thus, the TTSS is a critical virulence determinant for many phytopathogenic bacteria.

Inside the host cell, TTSS effectors are expected to modulate plant physiology in susceptible hosts to sustain the growth of the pathogen outside the cell. Yet virtually nothing is known about the role of TTSS effectors in acquiring or scavenging nutrients from infected plant tissue (**Figure 1*a***). In contrast, much is known about how plants have evolved resistances to combat individual TTSS effectors. The plant innate immune pathway monitors the entry of TTSS effectors into plant cells. Disease resistance proteins of the NBS-LRR class directly or indirectly detect the action of some TTSS effectors [originally referred to as avirulence (Avr) proteins] (74). This leads to the activation of several defense signal transduction pathways that ultimately limit bacterial growth (**Figure 1*b***). Plant responses associated with the recognition of TTSS Avr effectors include the activation of localized cell death classically referred to as the HR (32), the release of reactive oxygen intermediates (57) and nitric oxide (101), the fortification of plant cell walls (9, 12), the production of numerous antimicrobial secondary metabolites and defense-related proteins (24), and the activation of SAR (25).

The precise functions of TTSS effectors have remained a mystery for many years. The ultimate challenge has been to explain how individual effectors contribute to the pathogenic lifestyle of bacteria in susceptible and resistant hosts. There have been significant breakthroughs recently that have increased our understanding of TTSS effector function. These advances have stemmed from seminal investigations revealing that phytopathogenic bacteria deliberately suppress basal defenses (42) and strategically interfere with the activation of disease resistance protein-mediated pathways (41, 84, 86, 98). Moreover, they have come from the development of several model pathosystems (*e.g.*, *Pseudomonas-Arabidopsis*, *Pseudomonas*-tomato, *Xanthomonas*-pepper, and *Ralstonia-Arabidopsis*), which have been greatly impacted by the recent and near completion of several sequenced genomes.

In this review, I highlight the most exciting advances in elucidating the molecular and biochemical basis of TTSS effector action inside plant cells. We are far from determining all of the plant processes manipulated by TTSS

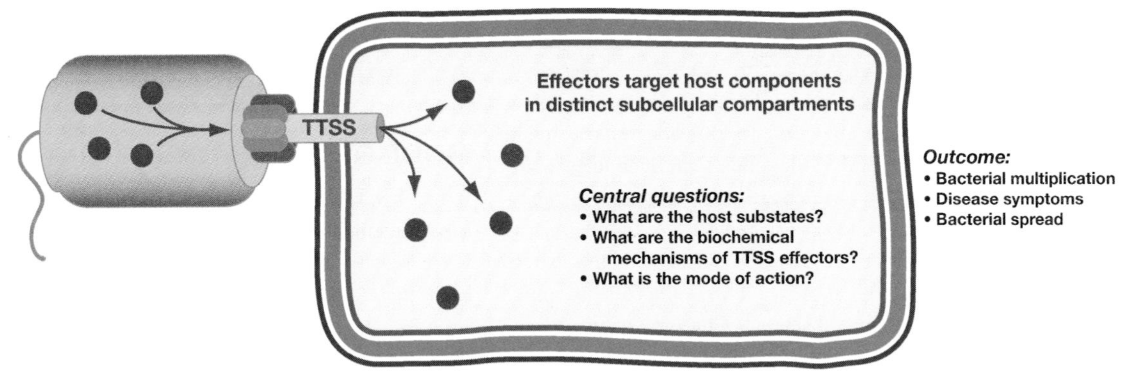

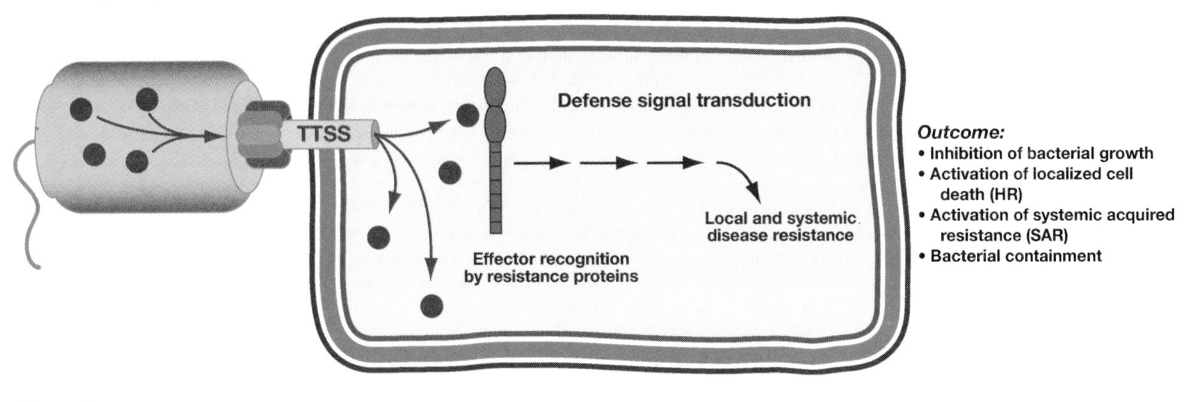

Figure 1

Phytopathogenic bacteria inject TTSS effector proteins into plant cells, resulting in (*a*) bacterial multiplication and disease symptoms in susceptible plants and (*b*) the activation of local and systemic disease resistance in resistant plants.

effectors. However, the investigations described herein provide essential clues that will enable new investigations to further dissect the intricate interactions that have evolved between pathogen TTSS effectors and the host plant cellular machinery.

IDENTIFICATION OF TTSS EFFECTORS

Several approaches have been employed to identify TTSS effectors. The first approach was to purify proteins secreted into culture fluids in a TTSS-dependent manner. This led to the identification of several proteins including harpin (38) and subunits of the TTSS Hrp pilus (106). This approach was difficult because low concentrations of the effectors are secreted by the pathogen in the absence of its host. It was also difficult to ascertain if the secreted proteins were functioning at the surface of the plant cell wall or inside the plant cell.

At the same time, many groups were busy isolating new bacterial Avr proteins and their respective plant resistance proteins. The pivotal observation that some *avr* genes and *hrp* genes

IVET: in vitro expression technology

are coregulated (31, 40, 81) suggested that Avr proteins were TTSS substrates. This was finally confirmed by the development of TTSS translocation assays (14, 71) and in planta transient expression assays (90, 97). These assays were used to prove that Avr proteins, as well as candidate effectors, are translocated into plant cells in a TTSS-dependent manner and are sufficient for the activation of resistance protein-mediated defense responses.

Several researchers suspected that there might be more TTSS effectors in phytopathogenic bacteria. The limitation in finding an Avr effector was discovering rare host plants, mutants, or natural variants, which are resistant to a normally virulent pathogen. Thus, several genome-wide screens were conducted to identify bacterial genes that are coregulated with TTSSs, with the hope of finding new TTSS effectors. Genetic screens using IVET (10) and reporter-based transposons (29, 65) were performed in *P. syringae* while a cDNA-AFLP (76) analysis was done in *X. campestris* pathovar (pv.) *vesicatoria*. The availability of genome sequences for *P. syringae* pv. *tomato* (13), *X. campestris* pv. *campestris* (19), *X. axonopodis* pv. *citri* (19), and *R. solanacearum* (89) enabled researchers to employ computer-based algorithms to identify TTSS substrates. Some of these genomes were specifically mined for orthologs of known TTSS effectors, proteins containing *hrp*-dependent promoter elements implicating coregulation with the TTSS regulon (18, 29, 110), and/or features of N-terminal export-associated signal peptides defined by known TTSS effectors (80). Furthermore, an elegant functional screen was employed in *P. syringae* pv. *maculicola* to isolate TTSS-dependent translocated proteins (34). A similar screen was done in *X. campestris* pv. *vesicatoria* (87), expanding our knowledge of TTSS effectors in this pathogen.

Using all of these approaches, hundreds of TTSS-associated genes have been identified. This indicates that any particular phytopathogen may deploy at least 40–50 TTSS effectors during infection. Consistent with this prediction, more than 40 TTSS effectors have been experimentally confirmed in *P. syringae* pv. *tomato*. This will likely be true for the other phytopathogens as well. How, then, is this large army of proteins altering plant physiology during bacterial pathogenesis?

MODE OF EFFECTOR ACTION

It has been known for many years that phytopathogenic bacteria can suppress plant defense responses. For example, *P. syringae* pv. *phaseolicola* strains that cause halo blight in bean suppress the expression of defense genes and the biosynthesis of phytoalexins, inducible antimicrobial compounds (42). This shows that successful growth of bacteria in a plant host is directly correlated with their ability to interfere with plant defense responses. Bacterial interference with host immunity was subsequently linked, in part, to the action of specific TTSS effectors within infected plant cells. In *Arabidopsis*, *P. syringae* pv. *tomato* AvrRpt2 directly interferes with the activation of RPM1-dependent disease resistance triggered by two other effectors, AvrB and AvrRpm1 (84, 86). In specific cultivars of bean, VirPphA suppresses host defenses (41) and AvrPphC masks the ability of AvrPphF to trigger an HR (98). Together, these insightful studies predict that, in addition to other potential virulence roles, a primary role of some TTSS effectors is to directly suppress plant defenses from within the host cell.

The investigations described below support this hypothesis by providing critical molecular evidence for TTSS effector action in planta. The evidence has emerged from the application of a number of approaches, including novel genetic screens, physical interaction studies, protein structure analysis, effector mutagenesis, and plant gene expression profiling.

Suppressors of Programmed Cell Death

The recent study of *P. syringae* pv. *tomato* AvrPtoB (homolog of VirPphA) shows that this effector directly interferes with the Pto disease resistance pathway in *N. benthamiana* by

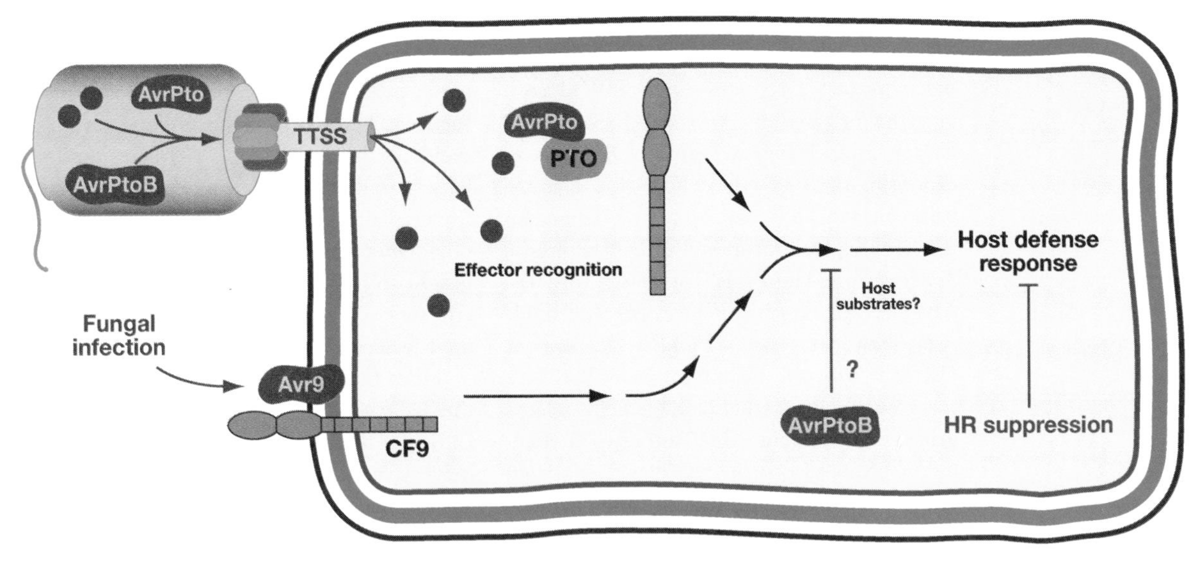

Figure 2

Pseudomonas syringae pv. *tomato* AvrPtoB functions downstream of disease resistance proteins to suppress the HR in plant cells.

operating as a general PCD suppressor (1). Both AvrPtoB and AvrPto interact with the tomato Pto serine/threonine kinase (51). Pto-mediated recognition of AvrPtoB and AvrPto in tomato activates a Prf-dependent signal transduction pathway that leads to disease resistance (79). Coexpression of AvrPto and Pto in *N. benthamiana* activates HR, indicating that the Pto-mediated defense pathway is conserved in this host. Surprisingly, coexpression of AvrPtoB and Pto fails to activate HR in *N. benthamiana*. This suggests that AvrPtoB might act as a specific suppressor of the Pto defense pathway in *N. benthamiana* but not in tomato. In fact, expression of AvrPtoB in *N. benthamiana* is sufficient to block the HR triggered by a constitutively activated mutant Pto kinase (1). Deletion analysis of AvrPtoB shows that the N- and C-terminal domains are sufficient for Pto recognition and anti-PCD activity, respectively. AvrPtoB suppression activity in *N. benthamiana* is not restricted to the Pto defense pathway. AvrPtoB also inhibits HR triggered by the tomato Cf-9 resistance protein in *N. benthamiana* in response to the *Cladosporium fulvum* peptide elicitor Avr9 (1). Taken together, these data show that AvrPtoB functions downstream of disease resistance proteins to suppress the HR (**Figure 2**).

PCD: programmed cell death

Given that PCD is associated with many cellular processes in eukaryotes (33, 56), Abramovitch and colleagues also explored the possibility that AvrPtoB may be a general cell death inhibitor. They found that AvrPtoB protects plants from HR-like PCD induced by Bax, a proapoptotic protein in the Bcl-2 family that initiates PCD in animal cells (1). Furthermore, AvrPtoB protects yeast cells from stress-induced PCD mediated by hydrogen peroxide, menadione, and heat shock, but curiously not Bax (1). It is not clear how AvrPtoB mediates such cross-kingdom PCD protection because we currently know very little about the machinery that regulates PCD in plants (56). However, AvrPtoB inhibition of PCD in both plants and yeast suggests that different eukaryotic PCD pathways may be regulated by conserved mechanisms.

AvrPtoB and four additional *P. syringae* pv. *tomato* effectors were subsequently shown to suppress HR responses triggered by another effector, HopPsyA (43). This was demonstrated by infecting tobacco plants with a

SA: salicylic acid
JA: jasmonic acid
ET: ethylene
COR: coronatine

nonpathogenic *P. fluorescens* strain carrying pHIR11 and a broad host range vector expressing individual TTSS effectors. pHIR11 contains the *P. syringae* pv. *syringae* TTSS *hrp* gene cluster and the HopPsyA effector. *P. fluorescens* expressing pHIR11 genes elicited a HR in tobacco in a TTSS- and HopPsyA-dependent manner. Expression of AvrPtoB, $AvrPphE_{Pto}$, $AvrPpiB1_{Pto}$, HopPtoE, or HopPtoF (an AvrPphF homolog) in *P. fluorescens* (pHIR11) completely suppressed HopPsyA-dependent HR and, with the exception of $AvrPpiB1_{Pto}$, the expression of the pathogenicity-related gene *PR1*. Transient expression of each effector in planta also protected plant cells from HopPsyA-induced HR, demonstrating that HR suppression occurs within the plant cell.

Suppressor activity for some effectors was not restricted to HopPsyA-induced PCD. $AvrPphE_{Pto}$, HopPtoE, HopPtoF, and HopPtoG inhibited PCD induced by Bax in plants and yeast. Unlike AvrPtoB, these effectors could not protect yeast from hydrogen peroxide stress. The fact that *P. syringae* uses a number of effectors to suppress PCD in plants triggered by different internal and external cues suggests that this pathogen may have evolved distinct suppressor activities to inhibit PCD at different stages of plant infection. Is this phenomenon unique to *P. syringae*? Or is the suppression of PCD a central requirement for bacterial pathogenesis? Identifying the mechanism(s) of PCD suppression is the next challenge.

Activators of the JA Pathway

A growing body of literature indicates that pathogens modulate the SA, JA, and ET defense pathways in different ways to colonize their hosts (54). This is confounded by the fact that these pathways influence each other by a complex series of positive and/or negative interactions. For instance, plant SA and JA pathways are mutually antagonistic (54). Bacterial pathogens, in particular *P. syringae*, have exploited this fact to overcome SA-dependent defenses. SA-dependent signaling pathways play a critical role in establishing local and systemic resistance to some phytopathogenic bacteria (25). However, activating JA-dependent signaling, a pathway that is typically induced in response to herbivores and wounding, negatively regulates the expression of SA-dependent defenses, rendering plants more susceptible to bacterial attack (26, 30, 107). The recent study of mutants in the SA and JA pathways provides insight to how bacterial pathogens use TTSS effectors, as well as the phytotoxin COR, to coopt SA-dependent immune responses.

Hints that *Pseudomonas* strains may activate the JA pathway came from studying COR. COR is a nonhost-specific, chlorosis-inducing polyketide sharing structural similarity with JA and methyl-JA (8). The structure of COR suggested that it might activate the JA pathway and thereby inhibit or delay SA-mediated defenses (85). Consistent with this hypothesis, tomato and *Arabidopsis* plants with an intact JA signal transduction pathway are sensitive to the effects of COR, whereas JA-insensitive tomato (*jai1*) and *Arabidopsis* (*coi1*) mutants do not respond to COR and exhibit enhanced resistance to *P. syringae* pv. *tomato* (52, 107). JAI1 and COI1 encode homologous F-box proteins that are expected to be part of a functional E3 ubiquitin ligase. This E3 ligase is predicted to regulate the expression of JA-responsive genes by ubiquitinating repressors of the JA pathway, an event that would promote their degradation by the proteosome (23). Treatment of wild-type tomato plants with exogenous methyl-JA is sufficient to complement a virulence defect of a *P. syringae* pv. *tomato* strain defective in COR production (107). This shows that a functional tomato JA-dependent signaling pathway is necessary and sufficient for susceptibility to wild-type *P. syringae* pv. *tomato*. It also strongly supports the hypothesis that COR contributes to *Pseudomonas* pathogenicity by mimicking endogenous jasmonate (**Figure 3**).

Several lines of evidence suggested that COR is not the only *Pseudomonas* factor manipulating the JA pathway. First, COR-deficient mutants produce small necrotic lesions in tomato even though they do not grow as efficiently as wild-type *P. syringae* pv. *tomato*.

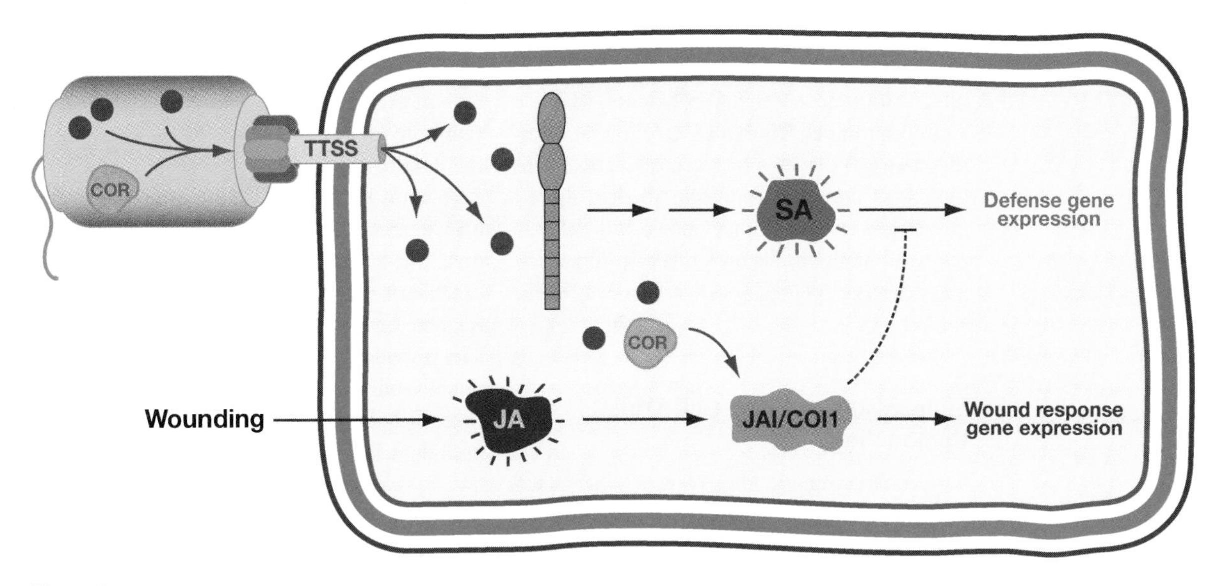

Figure 3

Pseudomonas syringae pv. *tomato* uses the phytotoxin coronatine (COR) and TTSS effectors to activate the jasmonic acid (JA) pathway and thereby inhibit or delay salicylic acid (SA)-mediated defense responses in plant cells.

However, the same COR deficient mutants do not produce visible lesions in *jai1* tomato plants, even though their growth in leaves is comparable to that of wild-type bacteria (107). This indicates that bacterial growth alone is not sufficient for disease development and that unknown bacterial factors and a competent JA signaling pathway are required for plant susceptibility. Second, both COR and a functional TTSS are required for *P. syringae* pv. *tomato* to induce JA response genes and repress *PR* genes in susceptible tomato plants (107). This shows that the TTSS and/or its effectors modify JA and SA signaling networks (**Figure 3**).

The development of a transgenic *Arabidopsis* reporter line that specifically monitors both COR and the TTSS during *Pseudomonas*-plant interactions led to the identification of five TTSS effectors that augment the JA pathway (37). It was discovered that COR and the TTSS are required for *P. syringae* pv. *tomato* to induce the early expression of *RAP2.6*, an *Arabidopsis* ET response factor gene, in a COI1-dependent manner. This indicated that *RAP2.6* expression could serve as a reporter to assess the influence of individual TTSS effectors on JA signaling during infection (37). Compared to wild-type *P. syringae* pv. *tomato* strains, mutants lacking *hopPtoK* or *avrPphE*$_{Pto}$ poorly induced the expression of a *RAP2.6* promoter-luciferase reporter gene in transgenic *Arabidopsis*, while strains expressing an additional effector (AvrB, AvrRpt2, or AvrPphB) accelerated the timing and magnitude of the reporter expression (37). None of the five effectors was alone sufficient to activate *RAP2.6* expression, indicating that a combination of pathogen factors is required to adequately manipulate the JA pathway.

PAMP: pathogen-associated molecular pattern

Finally, it is intriguing that virulent *P. syringae* strains actively suppress *Arabidopsis NHO1*, a gene required for general resistance to nonhost and avirulent isolates of *P. syringae* (66). Suppression of *NHO1* was abolished in *coi1* mutants, demonstrating that JA signaling also mediates this interaction (48). Constitutive overexpression of *NHO1* relieved plants from pathogen suppression, resulting in enhanced plant disease resistance. It is currently unknown which pathogen factors suppress *NHO1* expression, although PAMPs (*e.g.*, flagellin, DNA,

LPS: lipopolysaccharides

or LPSs) are suspected (66). It is also not clear how *NHO1* contributes to plant immunity. *NHO1* encodes a glycerol kinase, which is a rate-limiting enzyme in glycerol metabolism that converts glycerol to glycerol 3-phosphate (66). It is tempting to speculate that altered glycerol metabolism may affect the cellular pools of lipids, molecules recently linked to many facets of plant disease resistance (28, 45, 47, 69).

Suppressors of Plant Cell Wall Remodeling

Early microscopic studies revealed that plant hosts infected with the TTSS mutant but not wild-type phytopathogenic bacteria can remodel their cell walls, resulting in the formation of thick protrusions known as papillae (9, 12). Hauck and colleagues (36) elegantly determined the TTSS factors that prevent plant cells from mounting such cell wall–based defenses. Using cDNA microarray analysis, they set out to define a host gene expression profile that reflects the virulent functions associated with the *P. syringae* pv. *tomato* TTSS. They identified a large set of *Arabidopsis* genes that were repressed in a SA-independent and TTSS-dependent manner (36). These genes encode or are predicted to encode secreted cell wall and defense proteins, including extensions and germin-like proteins, two known components of papillae. This is the first molecular evidence showing that the TTSS is globally suppressing extracellular plant processes in addition to papillae formation (36). The nature of the *Pseudomonas* component that triggers plant cell wall remodeling is not clear, although LPS produced by *X. campestris* pv. *vesicatoria* induces a similar response in pepper (50).

Hauck and colleagues showed that the *P. syringae* pv. *tomato* effector AvrPto is one of the pathogen factors capable of suppressing papillae formation (36). This was demonstrated using a transgenic *Arabidopsis* line expressing AvrPto. Expression of AvrPto in planta prevented the host from appropriately responding to a *P. syringae* pv. *tomato* TTSS secretion mutant. These plants failed to deposit callose, a component of papillae. As a result, the TTSS mutant grew but failed to induce water-soaking symptoms or necrosis in leaves (36). In addition, the AvrPto-expressing plants exhibited a gene expression profile remarkably similar to that of plants infected with the pathogen (36). Surprisingly, an *avrPto* bacterial mutant was also capable of modulating the same set of TTSS-regulated genes in *Arabidopsis*. Together, these data provide direct evidence that AvrPto functionally mimics TTSS-dependent phenotypes elicited by the pathogen in the natural context of a plant infection. It also reinforces the fact that additional TTSS effectors are required to suppress plant cell wall defenses and to promote disease symptoms.

Two additional *P. syringae* pv. *tomato* effectors were characterized as suppressors of plant cell wall–based defenses. DebRoy and colleagues (20) showed that both HopPtoM and AvrE suppress callose deposition in *Pseudomonas*-infected *Arabidopsis* plants. These proteins along with HopPtoA1 and HrpW are encoded by the conserved effector locus (CEL), a gene cluster that is widely conserved among diverse *P. syringae* pvs. (3). CEL deletion mutants (ΔCEL) are severely impaired in virulence in tomato (3). It is now clear that this loss in virulence is due primarily to the action of HopPtoM and AvrE. Importantly, HopPtoM and AvrE can each suppress SA-dependent cell wall defenses (20). This is distinct from the SA-independent suppression of cell wall defenses mediated by AvrPto described above. *Arabidopsis* microarray analysis showed that the CEL locus does not suppress the classic SA-responsive defense genes activated by disease resistance proteins (20). This indicates that HopPtoM and AvrE are specific suppressors of basal SA-dependent plant defense responses. Importantly, this mechanism of suppression promotes host cell death, an event that is critical for the development of disease necrosis. Analysis of the AvrE ortholog, DspA/E, in the CEL of *E. amylovora* confirms that DspA/E also acts as a suppressor of cell wall–based defenses in *Erwinia*-apple interactions (20). This suggests that deployment of HopPtoM and AvrE-like suppressors may be a conserved virulence strategy used by many phytopathogens to overcome basal plant defenses.

Activators of Plant Transcription

Members of the AvrBs3 effector family are highly related and are found in many *Xanthomonas* spp. (60). In several susceptible *Xanthomonas*-plant interactions, AvrBs3-like effectors are actively involved in promoting disease symptoms. For example, AvrBs3 from *X. campestris* pv. *vesicatoria* induces hypertrophy of pepper mesophyll cells (70), PthA from *X. citri* elicits canker lesions in citrus (93), AvrXa7 from *X. oryzae* contributes to lesion length in rice (7), and Avrb6 from *X. campestris* pv. *malvacearum* increases watersoaking in cotton leaves (105). These distinct phenotypes are mediated by strikingly similar polypeptides containing highly conserved N- and C-terminal domains and a variable central domain comprised of tandem repeats. All members possess three NLSs and an AAD in their C terminus, suggesting that they function in the plant nucleus to modulate plant transcription during infection. Consistent with this hypothesis, AvrBs3, Avrb6, AvrXa7, and PthA possess functional NLSs (99, 103, 104) and the AAD of AvrXa10 activates transcription in yeast and *Arabidopsis* (109). Interestingly, AvrBs3 uses one of its NLSs to interact with importin α, a component of the nuclear import machinery (95, 96). This suggests that AvrBs3, and possibly other members of this family, may hijack a ride into the plant nucleus via importin α. The set of plant genes directly regulated by each AvrBs3 family member has yet to be defined. However, AvrXa7 preferentially binds to dA/dT-rich double-stranded DNA sequences (103) and AvrBs3 induces the expression of auxin-induced expansin-like genes in pepper (70). Taken together, these data strongly suggest the AvrBs3 effector family functions in the plant nucleus to alter transcription. Determining if the effectors directly or indirectly interact with the plant transcription machinery is still a major question.

NLS: nuclear localization signal

AAD: acidic transcriptional activation domain

Suppressor of Resistance Protein Activation

The *P. syringae* pv. *tomato* AvrRpt2 effector plays many elusive roles in plant cells. However, it is clear that one of its roles in establishing *Pseudomonas-Arabidopsis* interactions is to directly suppress the activation of the RPM1-dependent disease resistance pathway. A wealth of new genetic and biochemical evidence suggests that AvrRpt2 interferes with RPM1 activation by functioning as a cysteine protease (5, 6, 67). Details supporting the biochemical basis of AvrRpt2's function are described in detail below.

BIOCHEMICAL ACTIVITY AND HOST TARGETS

Critical insight to TTSS effector enzymatic function recently emerged from the investigation of five effector families: XopD, YopJ, YopT, AvrRpt2, and HopPtoD2. By employing protein structure prediction programs, researchers identified unique protein folds and conserved catalytic residues buried within these proteins. In doing so, they elucidated that some TTSS effectors encode enzymes with distinct catalytic activities. These discoveries allowed researchers to test the mechanism of effector action and to dissect how individual effectors activate resistance protein-mediated defense signal transduction.

XopD and YopJ SUMO Protease Families

The prevalence of YopJ-like TTSS effectors in pathogenic bacteria prompted Orth and colleagues to investigate whether these molecules use a conserved mechanism to alter host physiology during infection. Using CPH models and Threader to predict protein structure, they found that YopJ from *Y. pseudotuberculosis* and AvrBsT from *X. campestris* pv. *vesicatoria* resemble cysteine proteases in the C55 protease family (78). Consistent with this protease assignment, the predicted catalytic core of YopJ is required for the inhibition of MAPK and NF-κB signaling in animals. Similarly, the predicted core of AvrBsT is required for activating HR in resistant plants (78). This supported the hypothesis that these effectors encode cysteine proteases, but the nature of the host substrates was still a mystery. Upon further inspection, Orth and colleagues showed that YopJ-like effectors share limited structural similarity with the yeast ubiquitin-like protease ULP1, a cysteine protease in the C48 family (61). This was a pivotal observation because it revealed that TTSS effectors might interfere with the SUMO conjugation pathway in both animal and plant cells.

SUMO is a small ubiquitin-like modifier that is posttranslationally linked to proteins by a conjugation system analogous to the ubiquitin conjugation system (46) (**Figure 4**). ULPs function as SUMO proteases that reversibly control the attachment of SUMO to protein substrates. First, as peptidases, ULPs cleave SUMO after the C-terminal sequence (-GG-XXX) generating the mature form (-GG). Then the terminal glycine forms an isopeptide bond with the epsilon amino group of a lysine residue in a protein substrate. Second, as isopeptidases, ULPs cleave the isopeptide bond, releasing SUMO from SUMO-protein conjugates. YopJ (78), AvrBsT (A. Hotson & M.B. Mudgett, unpublished), and *X. campestris* pv. *vesicatoria* AvrXv4 (88) reduce the accumulation of SUMO-protein conjugates in host cells, suggesting that each protein possesses isopeptidase activity in vivo. However, enzymatic activity has not yet been detected for purified recombinant protein.

Direct biochemical evidence demonstrating that TTSS effectors act as SUMO proteases came from the study of the *X. campestris* pv. *vesicatoria* effector XopD (39). Unlike YopJ-like effectors, XopD shares significant structural homology with ULPs, classifying it as a cysteine protease in the C48 family. Biochemical analysis of XopD showed that the protein encodes a constitutively active cysteine protease with SUMO peptidase and isopeptidase activity. Furthermore, XopD cleaves tomato SUMO and not mammalian

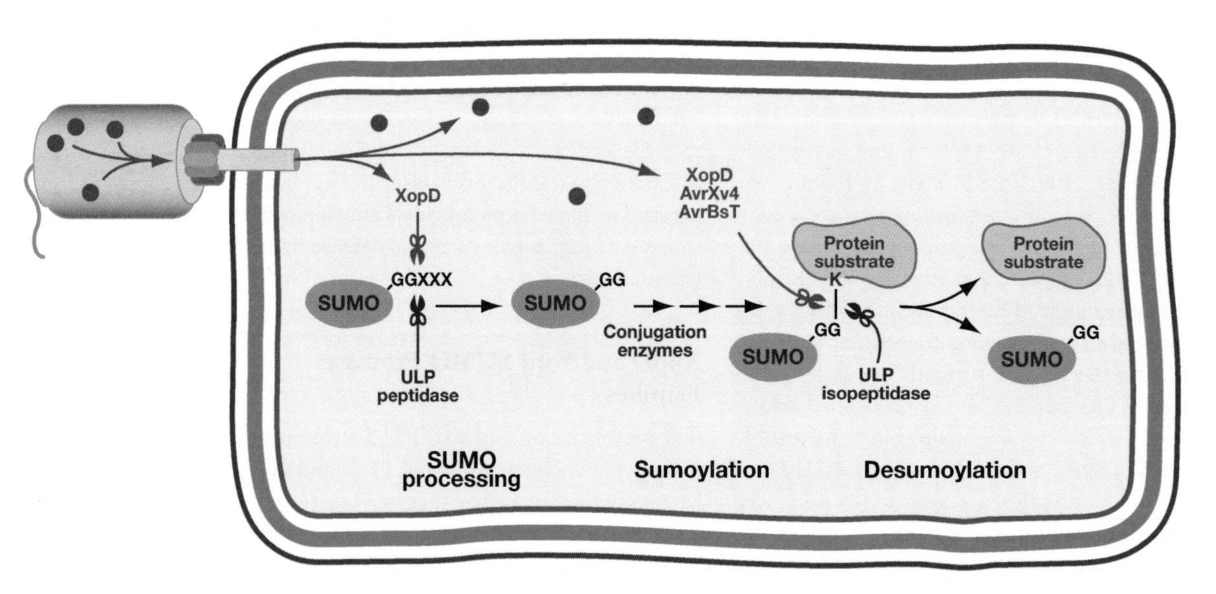

Figure 4

Xanthomonas campestris pv. *vesicatoria* TTSS effectors XopD, AvrXv4, and AvrBsT encode cysteine proteases that mimic plant ULP proteases to interfere with the SUMO protein conjugation pathway.

SUMO, demonstrating that XopD's substrate specificity is plant-specific. The fact that XopD, AvrBsT, and AvrXv4 primarily disrupt protein sumoylation in planta suggests that the primary role of these TTSS effectors is to interfere with plant pathways regulated by SUMO (**Figure 4**).

Which SUMO-protein conjugates are targeted by pathogen SUMO proteases? More importantly, how does interference with SUMO posttranslational modification contribute to bacterial pathogenicity? We await the answers to these questions. A recent report shows that a diverse array of sumoylated proteins exists in plants and that some of these are isoform-specific (55). In *Arabidopsis*, the SUMO pathway consists of at least eight SUMO isoforms and twelve SUMO proteases. These data suggest that numerous protein signal transduction pathways in plants may be controlled by SUMO, in an isoform- and protease-dependent manner. So far in plants, SUMO has been linked to pathogen attack (35), abiotic stress (55), abscisic acid signaling (64), and flowering time (73). In other eukaryotes, SUMO controls many diverse cellular processes including nuclear transport, enzyme activity, transcription, and the cell cycle (46, 100). Thus, pathogens could potentially target numerous SUMO-protein conjugates, possibly disrupting many cellular events. Elucidating effector substrate specificity will be a significant biochemical challenge.

The plant subcellular localization of pathogen SUMO proteases might provide hints to possible host targets. For instance, XopD (39) and PopP2, a YopJ-like effector from *R. solanacearum* (21), are targeted to the plant nucleus whereas AvrXv4 appears to be restricted to the plant cytoplasm (88). Physiological partitioning of ULPs has been reported in yeast. ULP1 is confined to the periphery of the nuclear membrane while ULP2 is dispersed throughout the nucleus (62). The noncatalytic N-terminal domain of ULP1 controls its localization, restricting it to a subset of sumolyated proteins not accessible to ULP2 (63). Interestingly, XopD also contains an N-terminal regulatory domain that controls its nuclear targeting (39) and isopeptidase substrate specificity (Hotson and Mudgett, unpublished). Thus, it appears that phytopathogens may target distinct effectors to different cellular compartments to mimic the plethora of endogenous SUMO proteases found in plants. This is consistent with the fact that *Xanthomonas* (4, 16, 75, 77, 102) and *Ralstonia* (21, 58, 59) inject multiple SUMO protease-like effectors into plant cells during infection.

Some plants have evolved resistance proteins to recognize YopJ-like effectors. The *Arabidopsis* RRS1 protein provides resistance to strains of *R. solanacearum* expressing PopP2 (22). RRS1 physically interacts with PopP2 in yeast (21), suggesting that the molecular recognition of PopP2 *in planta* is direct. RRS1 is an atypical resistance protein that codes for a TIR-NBS-LRR protein with a WRKY domain (22). WRKY domains are primarily found in plant transcription factors. Thus, it is intriguing that cytoplasmic RRS1 moves into the plant nucleus in the presence of PopP2 (21). This implies that RRS1 trafficking to the nucleus might activate the transcription of defense genes in response to PopP2. Given the importance of SUMO in transcriptional regulation (46, 100), it is imperative to molecularly link RRS1 function with PopP2 virulence activity *in planta*. Positional cloning projects are currently underway to identify the molecular basis of AvrXv4 and AvrBsT resistance in tomato (4) and *Arabidopsis* (Wilson and Mudgett, unpublished), respectively. Recognition of these effectors will likely be indirect since protease catalytic core mutants do not elicit a HR in resistant hosts (78, 88).

YopT Cysteine Protease Family

To explore the function of the *Yersinia* TTSS effector YopT, Shao and colleagues used multiple PSI-Blast iterations to identify homologous proteins in the available genome databases (92). They discovered a large effector family with at least 19 members conserved among pathogenic bacteria that infect animal or plant cells, including two plant symbionts. The predicted secondary structure of each member identified structural conservation surrounding three invariant residues (Cys/His/Asp) that are found in the catalytic core of cysteine proteases in the C58 family. The protease assignment for this family was confirmed upon the release of the crystal structure of AvrPphB from *P. syringae* pv. *phaseolicola* (108). As predicted, AvrPphB's core structure closely resembles that of papain-like cysteine proteases. More importantly, the solved structure defined substrate-binding residues in AvrPphB that are divergent in the other YopT family members. This predicts that each member will display different substrate specificities, dictated by the proteins encountered during host invasion. So far this appears to be true seeing that AvrPphB and YopT target distinct proteins in plant and animal cells, respectively (91, 92).

The genetic and molecular dissection of the AvrPphB-dependent *RPS5* disease resistance pathway in *Arabidopsis* supported a protease role for AvrPphB and provided clues to candidate plant substrates. First, the fact that AvrPphB is proteolyzed in the pathogen (83) suggested that AvrPphB might encode a protease that cleaves itself as well as host substrates. Second, the cleavage of AvrPphB and AvrPphB-dependent activation of the RPS5 disease resistance pathway requires a functional AvrPphB protease catalytic core (92). This clearly indicated that proteolysis of AvrPphB is autocatalytic and that AvrPphB-dependent cleavage of a plant substrate is required for activation of RPS5 signal transduction. Third, RPS5-mediated resistance requires the PBS1 protein kinase possessing autophosphorylating activity (94), suggesting that PBS1 itself is the target of AvrPphB.

Based on these observations, Shao and colleagues exquisitely showed that PBS1 is an AvrPphB substrate (91). *In vitro*, constitutively active AvrPphB cleaves itself and recombinant PBS1 after the sequence 'Gly-Asp-Lys', a motif that defines enzyme specificity. Furthermore, AvrPphB physically interacts with PBS1 *in planta* cleaving it into two polypeptides. Curiously, PBS1 kinase activity is not required for its cleavage by AvrPphB; however, PBS1 cleavage and PBS1 kinase activity are both required for the activation of RPS5. Together, these studies suggest a novel mechanism for the activation of RPS5-mediated disease resistance. The model proposes that a phosphorylated PBS1 cleavage product, possibly in a complex with AvrPphB, binds to and activates RPS5. Such a protein complex is expected to reside at the host plasma membrane considering that upon entry into the plant cell AvrPphB is immediately fatty acylated and trafficked to the plasma membrane (**Figure 5**).

At present, it is unclear how AvrPphB protease activity contributes to bacterial pathogenesis. Elucidating the function of PBS1 in plant cells and/or isolating additional AvrPphB substrates should reconcile this matter. In comparison, YopT protease activity explains in part its cytotoxic role during *Yersinia* pathogenesis. The YopT protease cleaves the carboxy termini of lipid-modified Rho GTPases releasing the prenyl moieties that anchor them to the host plasma membrane (92). This irreversible inactivation of Rho GTPases leads to the disruption of the actin cytoskeleton and impairs mammalian cells from properly executing phagocytosis and bacteria internalization. Although plants lack Rho GTPases, they have maintained a subfamily of Rho-related GTPases (ROPs), some of which are linked to plant defense responses (2). It is thus tempting to speculate that, in addition to PBS1, AvrPphB may also target ROPs during bacterial-plant interactions.

AvrRpt2 Cysteine Protease

The AvrRpt2 protein is a unique effector found only in *P. syringae* pv. *tomato*. AvrRpt2 is translocated into plant cells where it is amino-terminally processed into two polypeptides (72). The C-terminal AvrRpt2 polypeptide is sufficient to activate defense signal transduction in plants containing the RPS2 resistance protein. In addition, AvrRpt2 surprisingly triggers an RPS2-independent disappearance of RIN4 (6, 67), a novel protein required for RPM1-mediated disease resistance (68). This

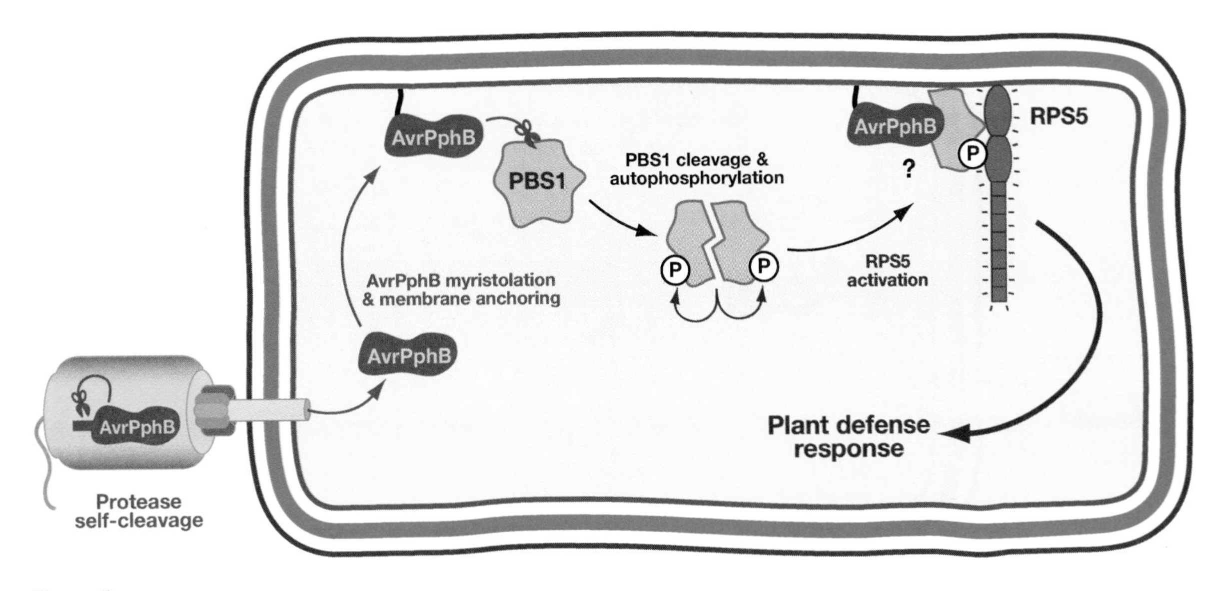

Figure 5

Model of AvrPphB cysteine protease action in resistant RPS5 *Arabidopsis* plants infected with *Pseudomonas syringae* pv. *phaseolicola*.

prompted Axtell and colleagues to suspect that AvrRpt2 encodes a RIN4 protease. Using the 3D-PSSM structure prediction program, they established that the AvrRpt2 protein contains a catalytic core with structural folds similar to staphopain, a cysteine protease in the C47 family (5). Consistent with this prediction, AvrRpt2 catalytic core mutants prevent the cleavage of AvrRpt2, the degradation of RIN4, and the activation of RPS2 (5). This strongly suggests that AvrRpt2 cleavage *in planta* is autocatalytic and that RIN4 is an AvrRpt2 substrate. Moreover, this supports other experimental evidence that the activation of RPS2 is coupled to the disappearance of RIN4 (6, 67) (**Figure 6**). Despite much effort, there is no direct evidence that AvrRpt2 possesses protease activity or that RIN4 is its substrate. This may be due to the fact that AvrRpt2 cleavage, an event that likely activates this protease, requires a plant factor (44). Identification of this elusive factor is expected to reveal the mechanisms for AvrRpt2 activation and substrate cleavage.

The AvrRpt2-dependent elimination of RIN4 and interference with RPM1 activation suggests that one of AvrRpt2's virulence functions is to suppress the early steps in defense signal transduction. In the absence of AvrRpt2, two other effectors, AvrRpm1 and AvrB, trigger the RIN4-RPM1 defense cascade during *Pseudomonas* infection. AvrRpm1 and AvrB physically interact with RIN4 *in planta*, an event that leads to RIN4 phosphorylation and the activation of RPM1 at the plasma membrane (68). In the presence of AvrRpt2, RIN4 is degraded (6, 67); therefore RPM1 remains inactive because it cannot associate with RIN4 at the plasma membrane (68). Thus, AvrRpt2 appears to subvert the host's recognition of AvrRpm1 and AvrB, allowing the invading pathogen to go unnoticed in an *rps2* genetic background.

AvrRpt2 protease activity explains how AvrRpt2 interferes with RPM1 signal transduction. Still, it does not explain how AvrRpt2 independently contributes to *Pseudomonas* pathogenicity (15). AvrRpt2 increases *P. syringae* pv. *tomato* pathogenicity in *Arabidopsis* plants lacking *rps2* and *rin4* (111). This indicates that AvrRpt2 likely has plant targets in addition to RIN4. The identification and characterization of AvrRpt2 host susceptibility factors should resolve the role of AvrRpt2

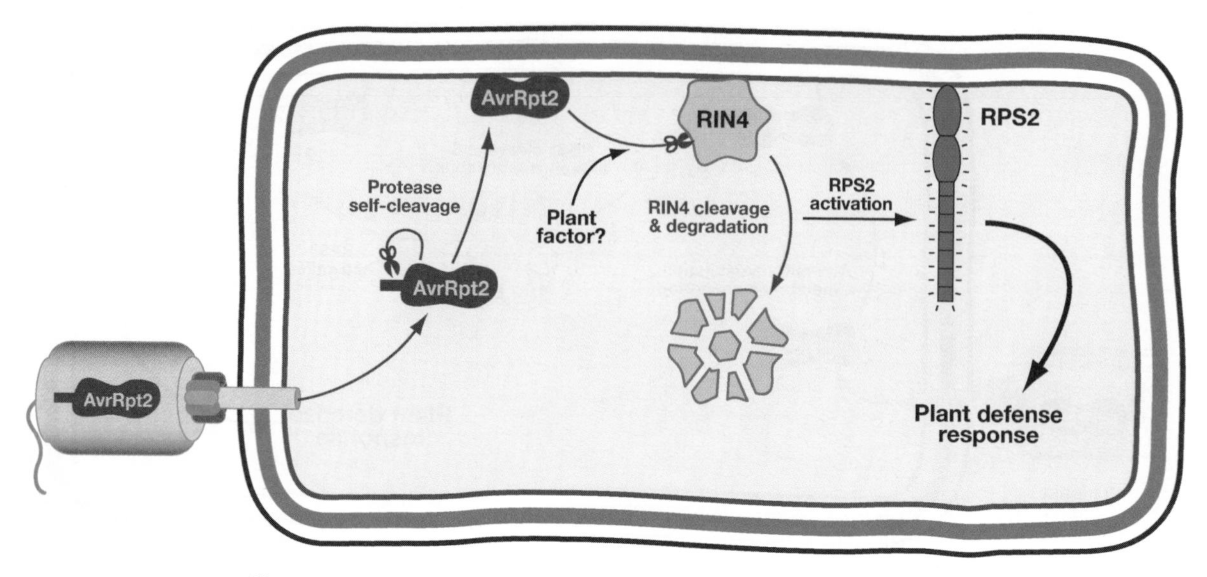

Figure 6

Model of AvrRpt2 cysteine protease action in resistant RPS2 *Arabidopsis* plants infected with *Pseudomonas syringae* pv. *tomato*.

in *Pseudomonas*-plant interactions. The exciting finding that AvrRpt2 modulates free auxin levels within bacterially-infected plants gives us a new clue as to how AvrRpt2 may promote pathogen growth and disease formation (53).

HopPtoD2 Protein Phosphatase

By employing BLAST and 3D-PSSM analysis, two groups established that the *P. syringae* pv. *tomato* HopPtoD2 encodes a novel, modular TTSS effector (11, 27). The N terminus of HopPtoD2 is homologous to the N terminus of AvrPphD, an effector from *P. syringae* pv. *phaseolicola*. The C terminus of HopPtoD2 exhibits a predicted protein fold conserved in many protein tyrosine phosphatases (PTPs). This suggests that HopPtoD2 encodes a chimeric AvrPphD-PTP effector with two potential functions. Biochemical analysis of purified, recombinant HopPtoD2 confirms that the PTP domain encodes a tyrosine phosphatase. HopPtoD2 hydrolyzes the artificial PTP substrate para- nitrophenyl phosphate and two phosphotyrosine-containing peptides derived from the insulin and EGF receptors (11, 27). No serine phosphatase activity was detected. Thus, HopPtoD2 is a tyrosine-specific protein phosphatase.

The precise role of HopPtoD2 in planta is not known; however, it clearly plays an important role in *Pseudomonas*-host interactions. In susceptible interactions, *P. syringae* pv. *tomato hopPtoD2* mutants exhibit reduced growth in *Arabidopsis* and tomato (11, 27). Optimal pathogen growth requires HopPtoD2 phosphatase activity (27), indicating that this effector is required for pathogenicity. In otherwise resistant interactions, HopPtoD2 phosphatase activity in planta decreases hydrogen peroxide production and *PR1* expression (11) and blocks the induction of the HR in plants infected with avirulent *P. syringae* strains lacking HopPtoD2 (11, 27). HopPtoD2 phosphatase activity also suppresses the HR induced in *N. benthamiana* by ectopic expression of NtMEK2^{DD}, a constitutively active mitogen-activated protein kinase (MAPK)/extracellular-signal-regulated kinase (ERK) involved in plant defense signaling (27). Together, these studies indicate that HopPtoD2 prevents plants from mounting a defense response against other injected

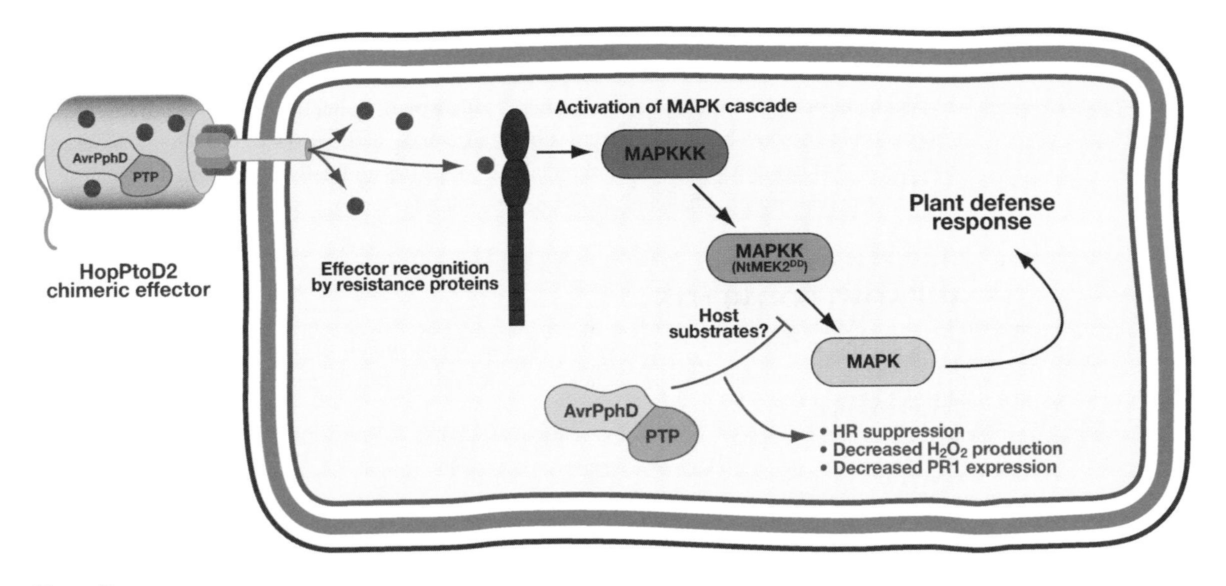

Figure 7

Model of HopPtoD2 tyrosine-specific protein phosphatase action in suppressing HR responses elicited in resistant *Nicotiana benthamiana* plants infected with avirulent *Pseudomonas syringae* strains.

TTSS effectors and suggests that HopPtoD2 dephosphophorylates a substrate downstream of NtMEK2 in the MAPK pathway (**Figure 7**). Further dissection of MAPK cascades controlling PCD associated with plant immunity and plant susceptibility (82) should identify key targets for HopPtoD2 and possibly other TTSS effectors.

Note that PTP sequences appear to be limited to a few *P. syringae* spp. whereas AvrPphD sequences are widely distributed in *P. syringae* spp. and distantly related phytopathogenic bacteria (11, 27). This suggests that sequences encoding PTP domains may have been recently acquired by *P. syringae* genomes, directly or indirectly creating the chimeric HopPtoD2 effector. Similar protein phosphatases are present in the animal pathogens *Yersinia* (YopH) and *S. typhimurium* (SptP). Interestingly, SptP is a chimeric effector containing GTPase-activating protein activity and PTP activity in its N- and C-terminal domains, respectively (49). Further study of HopPtoD2 should provide insight to its evolution and the possibility that it may encode one polypeptide with two virulence activities.

CONCLUDING REMARKS

These investigations have greatly deepened our understanding of how phytopathogenic bacteria use TTSS effectors to colonize plant tissue. We now have molecular and biochemical evidence that several TTSS effectors contribute to pathogenesis by targeting distinct plant processes that are required for disease resistance. Specifically, TTSS effectors interfere with resistance protein activation, repress the SA pathway by activating the JA pathway, modulate plant transcription, and suppress cell wall–based defenses as well as the execution of PCD. In addition, we have new insight to some of the biochemical signatures used by TTSS effectors to manipulate these pathways. TTSS effectors interfere with the regulation of host proteins by mimicking the activity of plant proteases and phosphatases. Modification of host targets may be reversible considering the biochemical mechanism of the XopD/YopJ-like SUMO proteases and the HopPtoD2 phosphatase, or irreversible, considering the destruction imposed by the YopT and AvrRpt2 cysteine proteases.

Despite this progress, a clear link has not been yet established between the mode of effector action and its biochemical properties. Several effectors manipulate plant defenses, but the biochemical mechanisms are not known, whereas others obviously encode enzymes but their host substrates and mode of action remain to be determined. Future challenges will be to decipher the mechanism by which TTSS effectors, independently and/or collectively, manipulate plant signal transduction pathways during bacterial pathogenesis. Identifying additional substrates and pathways targeted by pathogens will be essential.

ACKNOWLEDGMENTS

Work in the author's lab is supported by the Department of Energy (Grant DE-FG02-03ER15443). I thank members of my laboratory for thoughtful discussions and critical reading of this manuscript.

SUMMARY POINTS

1. Molecular evidence shows that individual TTSS effectors suppress plant innate immunity by interfering with resistance protein activation, MAPK signaling, and the execution of PCD.
2. TTSS effectors cooperate to suppress SA-dependent and SA-independent plant cell wall–based defenses.
3. TTSS effectors work with the phytotoxin COR to activate the JA pathway in order to down-regulate SA-mediated defenses.
4. Four TTSS effector families encode cysteine proteases, indicating that proteolysis of host proteins is a conserved mechanism used by phytopathogenic bacteria to manipulate plants.
5. The biochemical activity of TTSS effectors shows that pathogens can reversibly and irreversibly manipulate plant signal transduction.

FUTURE ISSUES TO BE RESOLVED

1. What is the biochemical function and mode of action for each TTSS effector?
2. What is the timing and order of effector injection into plant cells? Does it matter?
3. How do so many TTSS effectors contribute to the coevolution of bacterial-plant interactions?

LITERATURE CITED

AvrPtoB acts as a general PCD suppressor.

1. **Abramovitch RB, Kim YJ, Chen S, Dickman MB, Martin GB. 2003. *Pseudomonas* type III effector AvrPtoB induces plant disease susceptibility by inhibition of host programmed cell death. *EMBO J.* 22:60–69**
2. Agrawal GK, Iwahashi H, Rakwal R. 2003. Small GTPase 'Rop': molecular switch for plant defense responses. *FEBS Lett.* 546:173–80
3. Alfano JR, Charkowski AO, Deng WL, Badel JL, Petnicki-Ocwieja T, et al. 2000. The *Pseudomonas syringae hrp* pathogenicity island has a tripartite mosaic structure composed of a

cluster of type III secretion genes bounded by exchangeable effector and conserved effector loci that contribute to parasitic fitness and pathogenicity in plants. *Proc. Natl. Acad. Sci. USA* 97:4856–61

4. Astua-Monge G, Minsavage GV, Stall RE, Vallejos CE, Davis MJ, Jones JB. 2000. Xv4-AvrXv4: a new gene-for-gene interaction identified between *Xanthomonas campestris* pv. vesicatoria race T3 and the wild tomato relative *Lycopersicon pennellii*. *Mol. Plant Microbe. Interact.* 13:1346–55
5. **Axtell MJ, Chisholm ST, Dahlbeck D, Staskawicz BJ. 2003. Genetic and molecular evidence that the *Pseudomonas syringae* type III effector protein AvrRpt2 is a cysteine protease. *Mol. Microbiol.* 49:1537–46**
6. Axtell MJ, Staskawicz BJ. 2003. Initiation of RPS2-specified disease resistance in *Arabidopsis* is coupled to the AvrRpt2-directed elimination of RIN4. *Cell* 112:369–77
7. Bai J, Choi SH, Ponciano G, Leung H, Leach JE. 2000. *Xanthomonas oryzae* pv. *oryzae* avirulence genes contribute differently and specifically to pathogen aggressiveness. *Mol. Plant Microbe. Interact.* 13:1322–29
8. Bender CL, Alarcon-Chaidez F, Gross DC. 1999. *Pseudomonas syringae* phytotoxins: mode of action, regulation, and biosynthesis by peptide and polyketide synthetases. *Microbiol. Mol. Biol. Rev.* 63:266–92
9. Bestwick CS, Bennett MH, Mansfield JW. 1995. Hrp mutant of *Pseudomonas syringae* pv. *phaseolicola* induces cell wall alterations but not membrane damage leading to the hypersensitive reaction in lettuce. *Plant. Physiol.* 108:503–16
10. Boch J, Joardar V, Gao L, Robertson TL, Lim M, Kunkel BN. 2002. Identification of *Pseudomonas syringae* pv. *tomato* genes induced during infection of *Arabidopsis thaliana*. *Mol. Microbiol.* 44:73–88
11. **Bretz JR, Mock NM, Charity JC, Zeyad S, Baker CJ, Hutcheson SW. 2003. A translocated protein tyrosine phosphatase of *Pseudomonas syringae* pv. *tomato* DC3000 modulates plant defence response to infection. *Mol. Microbiol.* 49:389–400**
12. Brown I, Mansfield J, Bonas U. 1995. Hrp genes in *Xanthomonas campestris* pv. *vesicatoria* determine ability to suppress papilla deposition in pepper mesophyll cells. *Mol. Plant Microbe. Interact.* 8:825–36
13. Buell CR, Joardar V, Lindeberg M, Selengut J, Paulsen IT, et al. 2003. The complete genome sequence of the *Arabidopsis* and tomato pathogen *Pseudomonas syringae* pv. *tomato* DC3000. *Proc. Natl. Acad. Sci. USA* 100:10181–86
14. Casper-Lindley C, Dahlbeck D, Clark ET, Staskawicz BJ. 2002. Direct biochemical evidence for type III secretion-dependent translocation of the AvrBs2 effector protein into plant cells. *Proc. Natl. Acad. Sci. USA* 99:8336–41
15. Chen Z, Kloek AP, Boch J, Katagiri F, Kunkel BN. 2000. The *Pseudomonas syringae avrRpt2* gene product promotes pathogen virulence from inside plant cells. *Mol. Plant Microbe. Interact.* 13:1312–21
16. Ciesiolka LD, Hwin T, Gearlds JD, Minsavage GV, Saenz R, et al. 1999. Regulation of expression of avirulence gene *avrRxv* and identification of a family of host interaction factors by sequence analysis of *avrBsT*. *Mol. Plant Microbe. Interact.* 12:35–44
17. Cornelis GR, Van Gijsegem F. 2000. Assembly and function of type III secretory systems. *Annu. Rev. Microbiol.* 54:735–74
18. Cunnac S, Occhialini A, Barberis P, Boucher C, Genin S. 2004. Inventory and functional analysis of the large Hrp regulon in *Ralstonia solanacearum*: identification of novel effector proteins translocated to plant host cells through the type III secretion system. *Mol. Microbiol.* 53:115–28

AvrRpt2 has properties of a cysteine protease suggesting it cleaves *Arabidopsis* RIN4.

HopPtoD2 encodes a tyrosine phosphatase whose activity is required for the suppression of HR and PR expression.

19. da Silva AC, Ferro JA, Reinach FC, Farah CS, Furlan LR, et al. 2002. Comparison of the genomes of two *Xanthomonas* pathogens with differing host specificities. *Nature* 417:459–63

20. DebRoy S, Thilmony R, Kwack YB, Nomura K, He SY. 2004. A family of conserved bacterial effectors inhibits salicylic acid-mediated basal immunity and promotes disease necrosis in plants. *Proc. Natl. Acad. Sci. USA* 101:9927–32

HopPtoM, AvrE, and DspA/E are SA-independent suppressors of cell wall–based defenses.

21. Deslandes L, Olivier J, Peeters N, Feng DX, Khounlotham M, et al. 2003. Physical interaction between RRS1-R, a protein conferring resistance to bacterial wilt, and PopP2, a type III effector targeted to the plant nucleus. *Proc. Natl. Acad. Sci. USA* 100:8024–29

22. Deslandes L, Olivier J, Theulieres F, Hirsch J, Feng DX, et al. 2002. Resistance to *Ralstonia solanacearum* in *Arabidopsis thaliana* is conferred by the recessive RRS1-R gene, a member of a novel family of resistance genes. *Proc. Natl. Acad. Sci. USA* 99:2404–9

23. Devoto A, Nieto-Rostro M, Xie D, Ellis C, Harmston R, et al. 2002. COI1 links jasmonate signalling and fertility to the SCF ubiquitin-ligase complex in *Arabidopsis*. *Plant J.* 32:457–66

24. Dixon RA, Harrison MJ. 1990. Activation, structure, and organization of genes involved in microbial defense in plants. *Adv. Genet.* 28:165–234

25. Durrant WE, Dong X. 2004. Systemic acquired resistance. *Annu. Rev. Phytopathol.* 42:185–209

26. Ellis C, Karafyllidis I, Turner JG. 2002. Constitutive activation of jasmonate signaling in an *Arabidopsis* mutant correlates with enhanced resistance to *Erysiphe cichoracearum*, *Pseudomonas syringae*, and *Myzus persicae*. *Mol. Plant Microbe. Interact.* 15:1025–30

27. Espinosa A, Guo M, Tam VC, Fu ZQ, Alfano JR. 2003. The *Pseudomonas syringae* type III-secreted protein HopPtoD2 possesses protein tyrosine phosphatase activity and suppresses programmed cell death in plants. *Mol. Microbiol.* 49:377–87

HopPtoD2 encodes a tyrosine phosphatase whose activity is required for the suppression of HR.

28. Falk A, Feys BJ, Frost LN, Jones JDG, Daniels MJ, Parker JE. 1999. EDS1, an essential component of R gene-mediated disease resistance in *Arabidopsis* has homology to eukaryotic lipases. *Proc. Natl. Acad. Sci. USA* 96:3292–97

29. Fouts DE, Abramovitch RB, Alfano JR, Baldo AM, Buell CR, et al. 2002. Genome wide identification of *Pseudomonas syringae* pv. *tomato* DC3000 promoters controlled by the HrpL alternative sigma factor. *Proc. Natl. Acad. Sci. USA* 99:2275–80

30. Glazebrook J, Chen W, Estes B, Chang HS, Nawrath C, et al. 2003. Topology of the network integrating salicylate and jasmonate signal transduction derived from global expression phenotyping. *Plant J.* 34:217–28

31. Gopalan S, Bauer DW, Alfano JR, Loniello AO, He SY, Collmer A. 1996. Expression of the *Pseudomonas syringae* avirulence protein AvrB in plant cells alleviates its dependence on the hypersensitive response and pathogenicity (Hrp) secretion system in eliciting genotype-specific hypersensitive cell death. *Plant Cell* 8:1095–105

32. Greenberg JT, Yao N. 2004. The role and regulation of programmed cell death in plant-pathogen interactions. *Cell Microbiol.* 6:201–11

33. Greenberg JT, Guo A, Klessig DF, Ausubel FM. 1994. Programmed cell death in plants: a pathogen-triggered response activated coordinately with multiple defense functions. *Cell* 77:551–63

34. Guttman DS, Vinatzer BA, Sarkar SF, Ranall MV, Kettler G, Greenberg JT. 2002. A functional screen for the type III (Hrp) secretome of the plant pathogen *Pseudomonas syringae*. *Science* 295:1722–26

35. Hanania U, Furman-Matarasso N, Ron M, Avni A. 1999. Isolation of a novel SUMO protein from tomato that suppresses EIX-induced cell death. *Plant J.* 19:533–41

36. **Hauck P, Thilmony R, He SY. 2003. A *Pseudomonas syringae* type III effector suppresses cell wall-based extracellular defense in susceptible *Arabidopsis* plants. *Proc. Natl. Acad. Sci. USA* 100:8577–82**

AvrPto is a SA-dependent suppressor of cell wall–based defenses.

37. **He P, Chintamanani S, Chen Z, Zhu L, Kunkel BN, et al. 2004. Activation of a COI1-dependent pathway in *Arabidopsis* by *Pseudomonas syringae* type III effectors and coronatine. *Plant J.* 37:589–602**

AvrB, AvrRpt2, AvrPphB, HopPtoK, and AvrPphE$_{Pto}$ augment the JA pathway.

38. He SY, Huang HC, Collmer A. 1993. *Pseudomonas syringae* pv. *syringae harpinPss*: a protein that is secreted via the Hrp pathway and elicits the hypersensitive response in plants. *Cell* 73:1255–66
39. **Hotson A, Chosed R, Shu H, Orth K, Mudgett MB. 2003. *Xanthomonas* type III effector XopD targets SUMO-conjugated proteins in planta. *Mol. Microbiol.* 50:377–89**

XopD is a constitutively active cysteine protease that hydrolyzes plant-specific SUMO-protein conjugates.

40. Huynh TV, Dahlbeck D, Staskawicz BJ. 1989. Bacterial blight of soybean: regulation of a pathogen gene determining host cultivar specificity. *Science* 245:1374–77
41. Jackson RW, Athanassopoulos E, Tsiamis G, Mansfield JW, Sesma A, et al. 1999. Identification of a pathogenicity island, which contains genes for virulence and avirulence, on a large native plasmid in the bean pathogen *Pseudomonas syringae* pathovar *phaseolicola*. *Proc. Natl. Acad. Sci. USA* 96:10875–80
42. Jakobek JL, Smith JA, Lindgren PB. 1993. Suppression of bean defense responses by *Pseudomonas syringae*. *Plant Cell* 5:57–63
43. **Jamir Y, Guo M, Oh HS, Petnicki-Ocwieja T, Chen S, et al. 2004. Identification of *Pseudomonas syringae* type III effectors that can suppress programmed cell death in plants and yeast. *Plant J.* 37:554–65**

AvrPtoB, AvrPphE$_{Pto}$, AvrPpiB1$_{Pto}$, HopPtoE, HopPtoF, and HopPtoG suppress cell death.

44. Jin P, Wood MD, Wu Y, Xie Z, Katagiri F. 2003. Cleavage of the *Pseudomonas syringae* type III effector AvrRpt2 requires a host factor(s) common among eukaryotes and is important for AvrRpt2 localization in the host cell. *Plant. Physiol.* 133:1072–82
45. Jirage D, Tootle TL, Reuber TL, Frost LN, Feys BJ, et al. 1999. *Arabidopsis thaliana* PAD4 encodes a lipase-like gene that is important for salicylic acid signaling. *Proc. Natl. Acad. Sci. USA* 96:13583–88
46. Johnson ES. 2004. Protein modification by sumo. *Annu. Rev. Biochem.* 73:355–82
47. Kachroo P, Shanklin J, Shah J, Whittle EJ, Klessig DF. 2001. A fatty acid desaturase modulates the activation of defense signaling pathways in plants. *Proc. Natl. Acad. Sci. USA* 98:9448–53
48. Kang L, Li J, Zhao T, Xiao F, Tang X, et al. 2003. Interplay of the *Arabidopsis* nonhost resistance gene NHO1 with bacterial virulence. *Proc. Natl. Acad. Sci. USA* 100:3519–24
49. Kaniga K, Uralil J, Bliska JB, Galan JE. 1996. A secreted protein tyrosine phosphatase with modular effector domains in the bacterial pathogen *Salmonella typhimurium*. *Mol. Microbiol.* 21:633–41
50. Keshavarzi M, Soylu S, Brown I, Bonas U, Nicole M, et al. 2004. Basal defenses induced in pepper by lipopolysaccharides are suppressed by *Xanthomonas campestris* pv. *vesicatoria*. *Mol. Plant Microbe. Interact.* 17:805–15
51. Kim YJ, Lin NC, Martin GB. 2002. Two distinct *Pseudomonas* effector proteins interact with the Pto kinase and activate plant immunity. *Cell* 109:589–98
52. Kloek AP, Verbsky ML, Sharma SB, Schoelz JE, Vogel J, et al. 2001. Resistance to *Pseudomonas syringae* conferred by an *Arabidopsis thaliana* coronatine-insensitive (coi1) mutation occurs through two distinct mechanisms. *Plant J.* 26:509–22
53. Kunkel BN, Agnew J, Collins JJ, Cohen J, Chen Z. 2004. Molecular genetic analysis of AvrRpt2 activity in promoting virulence of *Pseudomonas syringae*. In *Genomic and Genetic*

Analysis of Plant Parasitism and Defense, ed. S Tsuyumu, J Leach, T Shiraishi, T Wolpert. APS Press. In press

54. Kunkel BN, Brooks DM. 2002. Cross talk between signaling pathways in pathogen defense. *Curr. Opin. Plant Biol.* 5:325–31
55. Kurepa J, Walker JM, Smalle J, Gosink MM, Davis SJ, et al. 2003. The small ubiquitin-like modifier (SUMO) protein modification system in *Arabidopsis*. Accumulation of SUMO1 and -2 conjugates is increased by stress. *J. Biol. Chem.* 278:6862–72
56. Lam E. 2004. Controlled cell death, plant survival and development. *Nat. Rev. Mol. Cell. Biol.* 5:305–15
57. Lamb C, Dixon RA. 1997. The oxidative burst in plant disease resistance. *Annu. Rev. Plant. Physiol. Plant Mol. Biol.* 48:251–75
58. Lavie M, Seunes B, Prior P, Boucher C. 2004. Distribution and sequence analysis of a family of type III-dependent effectors correlate with the phylogeny of *Ralstonia solanacearum* strains. *Mol. Plant Microbe. Interact.* 17:931–40
59. Lavie M, Shillington E, Eguiluz C, Grimsley N, Boucher C. 2002. PopP1, a new member of the YopJ/AvrRxv family of type III effector proteins, acts as a host-specificity factor and modulates aggressiveness of *Ralstonia solanacearum*. *Mol. Plant Microbe. Interact.* 15:1058–68
60. Leach JE, Vera Cruz CM, Bai J, Leung H. 2001. Pathogen fitness penalty as a predictor of durability of disease resistance genes. *Annu. Rev. Phytopathol.* 39:187–224
61. Li SJ, Hochstrasser M. 1999. A new protease required for cell-cycle progression in yeast. *Nature* 398:246–51
62. Li SJ, Hochstrasser M. 2000. The yeast ULP2 (SMT4) gene encodes a novel protease specific for the ubiquitin-like Smt3 protein. *Mol. Cell. Biol.* 20:2367–77
63. Li SJ, Hochstrasser M. 2003. The Ulp1 SUMO isopeptidase: distinct domains required for viability, nuclear envelope localization, and substrate specificity. *J. Cell. Biol.* 160:1069–81
64. Lois LM, Lima CD, Chua NH. 2003. Small ubiquitin-like modifier modulates abscisic acid signaling in *Arabidopsis*. *Plant Cell* 15:1347–59
65. Losada L, Sussan T, Pak K, Zeyad S, Rozenbaum I, Hutcheson SW. 2004. Identification of a novel *Pseudomonas syringae* Psy61 effector with virulence and avirulence functions by a HrpL-dependent promoter-trap assay. *Mol. Plant Microbe. Interact.* 17:254–62
66. Lu M, Tang X, Zhou JM. 2001. *Arabidopsis* NHO1 is required for general resistance against *Pseudomonas* bacteria. *Plant Cell* 13:437–47
67. Mackey D, Belkhadir Y, Alonso JM, Ecker JR, Dangl JL. 2003. *Arabidopsis* RIN4 is a target of the type III virulence effector AvrRpt2 and modulates RPS2-mediated resistance. *Cell* 112:379–89
68. Mackey D, Holt BF, Wiig A, Dangl JL. 2002. RIN4 interacts with *Pseudomonas syringae* type III effector molecules and is required for RPM1-mediated resistance in *Arabidopsis*. *Cell* 108:743–54
69. Maldonado AM, Doerner P, Dixon RA, Lamb CJ, Cameron RK. 2002. A putative lipid transfer protein involved in systemic resistance signalling in *Arabidopsis*. *Nature* 419:399–403
70. Marois E, Van den Ackerveken G, Bonas U. 2002. The *Xanthomonas* type III effector protein AvrBs3 modulates plant gene expression and induces cell hypertrophy in the susceptible host. *Mol. Plant Microbe. Interact.* 15:637–46
71. Mudgett MB, Chesnokova O, Dahlbeck D, Clark ET, Rossier O, et al. 2000. Molecular signals required for type III secretion and translocation of the *Xanthomonas campestris* AvrBs2 protein to pepper plants. *Proc. Natl. Acad. Sci. USA* 97:13324–29

72. Mudgett MB, Staskawicz BJ. 1999. Characterization of the *Pseudomonas syringae* pv. *tomato* AvrRpt2 protein: demonstration of secretion and processing during bacterial pathogenesis. *Mol. Microbiol.* 32:927–41
73. Murtas G, Reeves PH, Fu YF, Bancroft I, Dean C, Coupland G. 2003. A nuclear protease required for flowering-time regulation in *Arabidopsis* reduces the abundance of small ubiquitin-related modifier conjugates. *Plant Cell* 15:2308–19
74. Nimchuk Z, Eulgem T, Holt BF 3rd, Dangl JL. 2003. Recognition and response in the plant immune system. *Annu. Rev. Genet.* 37:579–609
75. Noel L, Thieme F, Gabler J, Buttner D, Bonas U. 2003. XopC and XopJ, two novel type III effector proteins from *Xanthomonas campestris* pv. *vesicatoria*. *J. Bacteriol.* 185:7092–102
76. Noel L, Thieme F, Nennstiel D, Bonas U. 2001. cDNA-AFLP analysis unravels a genome-wide hrpG-regulon in the plant pathogen *Xanthomonas campestris* pv. *vesicatoria*. *Mol. Microbiol.* 41:1271–81
77. Noel L, Thieme F, Nennstiel D, Bonas U. 2002. Two novel type III-secreted proteins of *Xanthomonas campestris* pv. *vesicatoria* are encoded within the *hrp* pathogenicity island. *J. Bacteriol.* 184:1340–48
78. Orth K, Xu ZH, Mudgett MB, Bao ZQ, Palmer LE, et al. 2000. Disruption of signaling by *Yersinia* effector YopJ, a ubiquitin-like protein protease. *Science* 290:1594–97
79. Pedley KF, Martin GB. 2003. Molecular basis of Pto-mediated resistance to bacterial speck disease in tomato. *Annu. Rev. Phytopathol.* 41:215–43
80. Petnicki-Ocwieja T, Schneider DJ, Tam VC, Chancey ST, Shan L, et al. 2002. Genomewide identification of proteins secreted by the Hrp type III protein secretion system of *Pseudomonas syringae* pv. *tomato* DC3000. *Proc. Natl. Acad. Sci. USA* 99:7652–57
81. Pirhonen MU, Lidell MC, Rowley DL, Lee SW, Jin S, et al. 1996. Phenotypic expression of *Pseudomonas syringae* avr genes in *E. coli* is linked to the activities of the hrp-encoded secretion system. *Mol. Plant Microbe. Interact.* 9:252–60
82. Pozo Od O, Pedley KF, Martin GB. 2004. MAPKKKalpha is a positive regulator of cell death associated with both plant immunity and disease. *EMBO J.* 23:3072–82
83. Puri N, Jenner C, Bennett M, Stewart R, Mansfield J, et al. 1997. Expression of *avrPphB*, an avirulence gene from *Pseudomonas syringae* pv. *phaseolicola*, and the delivery of signals causing the hypersensitive reaction in bean. *Mol. Plant Microbe. Interact.* 10:247–56
84. Reuber TL, Ausubel FM. 1996. Isolation of *Arabidopsis* genes that differentiate between resistance responses mediated by the *RPS2* and *RPM1* disease resistance genes. *Plant Cell* 8:241–49
85. Reymond P, Farmer EE. 1998. Jasmonate and salicylate as global signals for defense gene expression. *Curr. Opin. Plant Biol.* 1:404–11
86. Ritter C, Dangl JL. 1996. Interference between two specific pathogen recognition events mediated by distinct plant disease resistance genes. *Plant Cell* 8:251–57
87. Roden J, Belt B, Ross J, Tachibana T, Vargas J, Mudgett MB. 2004. A genetic screen to isolate type III effectors translocated into pepper cells during *Xanthomonas* infection. *Proc. Natl. Acad. Sci. USA* 101:16624–29
88. Roden J, Eardley L, Hotson A, Cao Y, Mudgett MB. 2004. Characterization of the *Xanthomonas* AvrXv4 effector, a SUMO protease translocated into plant cells. *Mol. Plant Microbe. Interact.* 17:633–43
89. Salanoubat M, Genin S, Artiguenave F, Gouzy J, Mangenot S, et al. 2002. Genome sequence of the plant pathogen *Ralstonia solanacearum*. *Nature* 415:497–502

90. Scofield SR, Tobias CM, Rathjen JP, Chang JH, Lavelle DT, et al. 1996. Molecular basis of gene-for-gene specificity in bacterial speck disease of tomato. *Science* 274:2063–65

91. Shao F, Golstein C, Ade J, Stoutemyer M, Dixon JE, Innes RW. 2003. Cleavage of *Arabidopsis* PBS1 by a bacterial type III effector. *Science* 301:1230–33

AvrPphB cleaves *Arabidopsis* PBS1 kinase during infection.

92. Shao F, Merritt PM, Bao Z, Innes RW, Dixon JE. 2002. A *Yersinia* effector and a *Pseudomonas* avirulence protein define a family of cysteine proteases functioning in bacterial pathogenesis. *Cell* 109:575–88

93. Swarup S, De Feyter R, Brlansky RH, Gabriel DW. 1991. A pathogenicity locus from *Xanthomonas citri* enables strains from several pathovars of *Xanthomonas campestris* to elicit cankerlike lesions on citrus. *Phytopathology* 81:802–9

94. Swiderski MR, Innes RW. 2001. The *Arabidopsis* PBS1 resistance gene encodes a member of a novel protein kinase subfamily. *Plant J.* 26:101–12

95. Szurek B, Marois E, Bonas U, Van den Ackerveken G. 2001. Eukaryotic features of the *Xanthomonas* type III effector AvrBs3: protein domains involved in transcriptional activation and the interaction with nuclear import receptors from pepper. *Plant J.* 26:523–34

96. Szurek B, Rossier O, Hause G, Bonas U. 2002. Type III-dependent translocation of the *Xanthomonas* AvrBs3 protein into the plant cell. *Mol. Microbiol.* 46:13–23

97. Tang X, Frederick RD, Zhou J, Halterman DA, Jia Y, Martin GB. 1996. Initiation of plant disease resistance by physical interaction of AvrPto and Pto kinase. *Science* 274:2060–63

98. Tsiamis G, Mansfield JW, Hockenhull R, Jackson RW, Sesma A, et al. 2000. Cultivar-specific avirulence and virulence functions assigned to *avrPphF* in *Pseudomonas syringae* pv. *phaseolicola*, the cause of bean halo-blight disease. *EMBO J.* 19:3204–14

99. Van den Ackerveken G, Marois E, Bonas U. 1996. Recognition of the bacterial avirulence protein AvrBs3 occurs inside the host plant cell. *Cell* 87:1307–16

100. Verger A, Perdomo J, Crossley M. 2003. Modification with SUMO. A role in transcriptional regulation. *EMBO Rep* 4:137–42

101. Wendehenne D, Durner J, Klessig DF. 2004. Nitric oxide: a new player in plant signalling and defence responses. *Curr. Opin. Plant Biol.* 7:449–55

102. Whalen MC, Stall RE, Staskawicz BJ. 1988. Characterization of a gene from a tomato pathogen determining hypersensitive resistance in non-host species and genetic analysis of this resistance in bean. *Proc. Natl. Acad. Sci. USA* 85:6743–47

103. Yang B, Zhu W, Johnson LB, White FF. 2000. The virulence factor AvrXa7 of *Xanthomonas oryzae* pv. *oryzae* is a type III secretion pathway-dependent nuclear-localized double-stranded DNA-binding protein. *Proc. Natl. Acad. Sci. USA* 97:9807–12

104. Yang Y, Gabriel DW. 1995. *Xanthomonas* avirulence/pathogenicity gene family encodes functional plant nuclear targeting signals. *Mol. Plant Microbe. Interact.* 8:627–31

105. Yang Y, Yuan Q, Gabriel DW. 1996. Watersoaking function(s) of XcmH1005 are redundantly encoded by members of the *Xanthomonas avr/pth* gene family. *Mol. Plant Microbe. Interact.* 9:105–13

106. Yuan J, He SY. 1996. The *Pseudomonas syringae* Hrp regulation and secretion system controls the production and secretion of multiple extracellular proteins. *J. Bacteriol.* 178:6399–402

107. Zhao Y, Thilmony R, Bender CL, Schaller A, He SY, Howe GA. 2003. Virulence systems of *Pseudomonas syringae* pv. *tomato* promote bacterial speck disease in tomato by targeting the jasmonate signaling pathway. *Plant J.* 36:485–99

Evidence linking COR and TTSS effectors to activation of JA pathway and repression of SA pathway.

108. Zhu M, Shao F, Innes RW, Dixon JE, Xu Z. 2004. The crystal structure of *Pseudomonas* avirulence protein AvrPphB: a papain-like fold with a distinct substrate-binding site. *Proc. Natl. Acad. Sci. USA* 101:302–7

109. Zhu W, Yang B, Chittoor JM, Johnson LB, White FF. 1998. AvrXa10 contains an acidic transcriptional activation domain in the functionally conserved C terminus. *Mol. Plant Microbe. Interact.* 11:824–32

Evidence AvrXa10, an AvrBs3 family member, activates plant transcription.

110. Zwiesler-Vollick J, Plovanich-Jones AE, Nomura K, Bandyopadhyay S, Joardar V, et al. 2002. Identification of novel hrp-regulated genes through functional genomic analysis of the *Pseudomonas syringae* pv. *tomato* DC3000 genome. *Mol. Microbiol.* 45:1207–18

111. Lim MTS, Kunkel BN. 2004. Mutations in the *Pseudomonas syringae avrRpt2* gene that dissociate its virulence and avirulence activities lead to decreased efficiency in the AvrRpt2-induced disappearance of RIN4. *Mol. Plant Microbe Interact.* 17:313–21

Subject Index

B

C

D

E

F

N

O

Q

R

S

T

U

V

W

X

Y

Z

Cumulative Indexes

Contributing Authors, Volumes 46–56

D

E

F

G

H

I

X

Y

Z

Chapter Titles, Volumes 46–56

Genetics and Molecular Biology

Cell Differentiation

Tissue, Organ, and Whole Plant Events

Acclimation and Adaptation

Methods